NMR SPECTROSCOPY EXPLAINED

THE WILEY BICENTENNIAL–KNOWLEDGE FOR GENERATIONS

Each generation has its unique needs and aspirations. When Charles Wiley first opened his small printing shop in lower Manhattan in 1807, it was a generation of boundless potential searching for an identity. And we were there, helping to define a new American literary tradition. Over half a century later, in the midst of the Second Industrial Revolution, it was a generation focused on building the future. Once again, we were there, supplying the critical scientific, technical, and engineering knowledge that helped frame the world. Throughout the 20th Century, and into the new millennium, nations began to reach out beyond their own borders and a new international community was born. Wiley was there, expanding its operations around the world to enable a global exchange of ideas, opinions, and know-how.

For 200 years, Wiley has been an integral part of each generation's journey, enabling the flow of information and understanding necessary to meet their needs and fulfill their aspirations. Today, bold new technologies are changing the way we live and learn. Wiley will be there, providing you the must-have knowledge you need to imagine new worlds, new possibilities, and new opportunities.

Generations come and go, but you can always count on Wiley to provide you the knowledge you need, when and where you need it!

WILLIAM J. PESCE
PRESIDENT AND CHIEF EXECUTIVE OFFICER

PETER BOOTH WILEY
CHAIRMAN OF THE BOARD

NMR SPECTROSCOPY EXPLAINED

Simplified Theory, Applications and Examples for Organic Chemistry and Structural Biology

Neil E. Jacobsen, Ph.D.
University of Arizona

A JOHN WILEY & SONS, INC., PUBLICATION

Copyright © 2007 by John Wiley & Sons, Inc. All rights reserved

Published by John Wiley & Sons, Inc., Hoboken, New Jersey
Published simultaneously in Canada

No part of this publication may be reproduced, stored in a retrieval system, or transmitted in any form or by any means, electronic, mechanical, photocopying, recording, scanning, or otherwise, except as permitted under Section 107 or 108 of the 1976 United States Copyright Act, without either the prior written permission of the Publisher, or authorization through payment of the appropriate per-copy fee to the Copyright Clearance Center, Inc., 222 Rosewood Drive, Danvers, MA 01923, (978) 750-8400, fax (978) 750-4470, or on the web at www.copyright.com. Requests to the Publisher for permission should be addressed to the Permissions Department, John Wiley & Sons, Inc., 111 River Street, Hoboken, NJ 07030, (201) 748-6011, fax (201) 748-6008, or online at http://www.wiley.com/go/permission.

Limit of Liability/Disclaimer of Warranty: While the publisher and author have used their best efforts in preparing this book, they make no representations or warranties with respect to the accuracy or completeness of the contents of this book and specifically disclaim any implied warranties of merchantability or fitness for a particular purpose. No warranty may be created or extended by sales representatives or written sales materials. The advice and strategies contained herein may not be suitable for your situation. You should consult with a professional where appropriate. Neither the publisher nor author shall be liable for any loss of profit or any other commercial damages, including but not limited to special, incidental, consequential, or other damages.

For general information on our other products and services or for technical support, please contact our Customer Care Department within the United States at (800) 762-2974, outside the United States at (317) 572-3993 or fax (317) 572-4002.

Wiley also publishes its books in a variety of electronic formats. Some content that appears in print may not be available in electronic formats. For more information about Wiley products, visit our web site at http://www.wiley.com.

Wiley Bicentennial Logo: Richard J. Pacifico

Library of Congress Cataloging-in-Publication Data:

Jacobsen, Neil E.
  NMR Spectroscopy Explained : Simplified Theory, Applications and Examples for Organic Chemistry and Structural Biology / Neil E. Jacobsen, Ph.D.
    p. cm.
  ISBN 978-0-471-73096-5 (cloth)
 1. Nuclear magnetic resonance spectroscopy. 2. Chemistry, Organic. 3. Molecular biology. I. Title.
    QD96.N8J33 2007
543'.66- -dc22

2007006911

Printed in the United States of America
10 9 8 7 6 5 4 3 2

CONTENTS

Preface xi
Acknowledgments xv

1 Fundamentals of NMR Spectroscopy in Liquids 1

    1.1 Introduction to NMR Spectroscopy, 1
    1.2 Examples: NMR Spectroscopy of Oligosaccharides and Terpenoids, 12
    1.3 Typical Values of Chemical Shifts and Coupling Constants, 27
    1.4 Fundamental Concepts of NMR Spectroscopy, 30

2 Interpretation of Proton (^{1}H) NMR Spectra 39

    2.1 Assignment, 39
    2.2 Effect of B_o Field Strength on the Spectrum, 40
    2.3 First-Order Splitting Patterns, 45
    2.4 The Use of ^{1}H–^{1}H Coupling Constants to Determine Stereochemistry and Conformation, 52
    2.5 Symmetry and Chirality in NMR, 54
    2.6 The Origin of the Chemical Shift, 56
    2.7 J Coupling to Other NMR-Active Nuclei, 61
    2.8 Non-First-Order Splitting Patterns: Strong Coupling, 63
    2.9 Magnetic Equivalence, 71

3 NMR Hardware and Software 74

    3.1 Sample Preparation, 75
    3.2 Sample Insertion, 77
    3.3 The Deuterium Lock Feedback Loop, 78

- 3.4 The Shim System, 81
- 3.5 Tuning and Matching the Probe, 88
- 3.6 NMR Data Acquisition and Acquisition Parameters, 90
- 3.7 Noise and Dynamic Range, 108
- 3.8 Special Topic: Oversampling and Digital Filtering, 110
- 3.9 NMR Data Processing—Overview, 118
- 3.10 The Fourier Transform, 119
- 3.11 Data Manipulation Before the Fourier Transform, 122
- 3.12 Data Manipulation After the Fourier Transform, 126

4 Carbon-13 (^{13}C) NMR Spectroscopy — 135

- 4.1 Sensitivity of ^{13}C, 135
- 4.2 Splitting of ^{13}C Signals, 135
- 4.3 Decoupling, 138
- 4.4 Heteronuclear Decoupling: ^{1}H Decoupled ^{13}C Spectra, 139
- 4.5 Decoupling Hardware, 145
- 4.6 Decoupling Software: Parameters, 149
- 4.7 The Nuclear Overhauser Effect (NOE), 150
- 4.8 Heteronuclear Decoupler Modes, 152

5 NMR Relaxation—Inversion-Recovery and the Nuclear Overhauser Effect (NOE) — 155

- 5.1 The Vector Model, 155
- 5.2 One Spin in a Magnetic Field, 155
- 5.3 A Large Population of Identical Spins: Net Magnetization, 157
- 5.4 Coherence: Net Magnetization in the x–y Plane, 161
- 5.5 Relaxation, 162
- 5.6 Summary of the Vector Model, 168
- 5.7 Molecular Tumbling and NMR Relaxation, 170
- 5.8 Inversion-Recovery: Measurement of T_1 Values, 176
- 5.9 Continuous-Wave Low-Power Irradiation of One Resonance, 181
- 5.10 Homonuclear Decoupling, 182
- 5.11 Presaturation of Solvent Resonance, 185
- 5.12 The Homonuclear Nuclear Overhauser Effect (NOE), 187
- 5.13 Summary of the Nuclear Overhauser Effect, 198

6 The Spin Echo and the Attached Proton Test (APT) — 200

- 6.1 The Rotating Frame of Reference, 201
- 6.2 The Radio Frequency (RF) Pulse, 203
- 6.3 The Effect of RF Pulses, 206
- 6.4 Quadrature Detection, Phase Cycling, and the Receiver Phase, 209

6.5 Chemical Shift Evolution, 212
6.6 Scalar (*J*) Coupling Evolution, 213
6.7 Examples of *J*-coupling and Chemical Shift Evolution, 216
6.8 The Attached Proton Test (APT), 220
6.9 The Spin Echo, 226
6.10 The Heteronuclear Spin Echo: Controlling *J*-Coupling Evolution and Chemical Shift Evolution, 232

7 Coherence Transfer: INEPT and DEPT 238

7.1 Net Magnetization, 238
7.2 Magnetization Transfer, 241
7.3 The Product Operator Formalism: Introduction, 242
7.4 Single Spin Product Operators: Chemical Shift Evolution, 244
7.5 Two-Spin Operators: *J*-coupling Evolution and Antiphase Coherence, 247
7.6 The Effect of RF Pulses on Product Operators, 251
7.7 INEPT and the Transfer of Magnetization from ^{1}H to ^{13}C, 253
7.8 Selective Population Transfer (SPT) as a Way of Understanding INEPT Coherence Transfer, 257
7.9 Phase Cycling in INEPT, 263
7.10 Intermediate States in Coherence Transfer, 265
7.11 Zero- and Double-Quantum Operators, 267
7.12 Summary of Two-Spin Operators, 269
7.13 Refocused INEPT: Adding Spectral Editing, 270
7.14 DEPT: Distortionless Enhancement by Polarization Transfer, 276
7.15 Product Operator Analysis of the DEPT Experiment, 283

8 Shaped Pulses, Pulsed Field Gradients, and Spin Locks: Selective 1D NOE and 1D TOCSY 289

8.1 Introducing Three New Pulse Sequence Tools, 289
8.2 The Effect of Off-Resonance Pulses on Net Magnetization, 291
8.3 The Excitation Profile for Rectangular Pulses, 297
8.4 Selective Pulses and Shaped Pulses, 299
8.5 Pulsed Field Gradients, 301
8.6 Combining Shaped Pulses and Pulsed Field Gradients: "Excitation Sculpting", 308
8.7 Coherence Order: Using Gradients to Select a Coherence Pathway, 316
8.8 Practical Aspects of Pulsed Field Gradients and Shaped Pulses, 319
8.9 1D Transient NOE using DPFGSE, 321
8.10 The Spin Lock, 333
8.11 Selective 1D ROESY and 1D TOCSY, 338
8.12 Selective 1D TOCSY using DPFGSE, 343
8.13 RF Power Levels for Shaped Pulses and Spin Locks, 348

9 Two-Dimensional NMR Spectroscopy: HETCOR, COSY, and TOCSY — 353

- 9.1 Introduction to Two-Dimensional NMR, 353
- 9.2 HETCOR: A 2D Experiment Created from the 1D INEPT Experiment, 354
- 9.3 A General Overview of 2D NMR Experiments, 364
- 9.4 2D Correlation Spectroscopy (COSY), 370
- 9.5 Understanding COSY with Product Operators, 386
- 9.6 2D TOCSY (Total Correlation Spectroscopy), 393
- 9.7 Data Sampling in t_1 and the 2D Spectral Window, 398

10 Advanced NMR Theory: NOESY and DQF-COSY — 408

- 10.1 Spin Kinetics: Derivation of the Rate Equation for Cross-Relaxation, 409
- 10.2 Dynamic Processes and Chemical Exchange in NMR, 414
- 10.3 2D NOESY and 2D ROESY, 425
- 10.4 Expanding Our View of Coherence: Quantum Mechanics and Spherical Operators, 439
- 10.5 Double-Quantum Filtered COSY (DQF-COSY), 447
- 10.6 Coherence Pathway Selection in NMR Experiments, 450
- 10.7 The Density Matrix Representation of Spin States, 469
- 10.8 The Hamiltonian Matrix: Strong Coupling and Ideal Isotropic (TOCSY) Mixing, 478

11 Inverse Heteronuclear 2D Experiments: HSQC, HMQC, and HMBC — 489

- 11.1 Inverse Experiments: ^{1}H Observe with ^{13}C Decoupling, 490
- 11.2 General Appearance of Inverse 2D Spectra, 498
- 11.3 Examples of One-Bond Inverse Correlation (HMQC and HSQC) Without ^{13}C Decoupling, 501
- 11.4 Examples of Edited, ^{13}C-Decoupled HSQC Spectra, 504
- 11.5 Examples of HMBC Spectra, 509
- 11.6 Structure Determination Using HSQC and HMBC, 517
- 11.7 Understanding the HSQC Pulse Sequence, 522
- 11.8 Understanding the HMQC Pulse Sequence, 533
- 11.9 Understanding the Heteronuclear Multiple-Bond Correlation (HMBC) Pulse Sequence, 535
- 11.10 Structure Determination by NMR—An Example, 538

12 Biological NMR Spectroscopy — 551

- 12.1 Applications of NMR in Biology, 551
- 12.2 Size Limitations in Solution-State NMR, 553
- 12.3 Hardware Requirements for Biological NMR, 558
- 12.4 Sample Preparation and Water Suppression, 564
- 12.5 ^{1}H Chemical Shifts of Peptides and Proteins, 570

12.6	NOE Interactions Between One Residue and the Next Residue in the Sequence, 577	
12.7	Sequence-Specific Assignment Using Homonuclear 2D Spectra, 580	
12.8	Medium and Long-Range NOE Correlations, 586	
12.9	Calculation of 3D Structure Using NMR Restraints, 590	
12.10	^{15}N-Labeling and 3D NMR, 596	
12.11	Three-Dimensional NMR Pulse Sequences: 3D HSQC–TOCSY and 3D TOCSY–HSQC, 601	
12.12	Triple-Resonance NMR on Doubly-Labeled (^{15}N, ^{13}C) Proteins, 610	
12.13	New Techniques for Protein NMR: Residual Dipolar Couplings and Transverse Relaxation Optimized Spectroscopy (TROSY), 621	

Appendix A: A Pictorial Key to NMR Spin States   **627**
Appendix B: A Survey of Two-Dimensional NMR Experiments   **634**
Index   **643**

PREFACE

Nuclear magnetic resonance (NMR) is a technique for determining the structure of organic molecules and biomolecules in solution. The covalent structure (what atoms are bonded to what), the stereochemistry (relative orientation of groups in space), and the conformation (preferred bond rotations or folding in three dimensions) are available by techniques that measure direct distances (between hydrogens) and bond dihedral angles. Specific NMR signals can be identified and assigned to each hydrogen (and/or carbon, nitrogen) in the molecule.

You may have seen or been inside an MRI (magnetic resonance imaging) instrument, a medical tool that creates detailed images (or "slices") of the patient without ionizing radiation. The NMR spectroscopy magnet is just a scaled-down version of this huge clinical magnet, rotated by 90° so that the "bore" (the hole that the patient gets into) is vertical and typically only 5 cm (2 in.) in diameter. Another technique, solid-state NMR, deals with solid (powdered) samples and gives information similar to solution NMR. This book is limited to solution-state NMR and will not cover the fields of NMR imaging and solid-state NMR, even though the theoretical tools developed here can be applied to these fields.

NMR takes advantage of the magnetic properties of the nucleus to sense the proximity of electronegative atoms, double bonds, and other magnetic nuclei nearby in the molecular structure. About one half of a micromole of a pure molecule in 0.5 mL of solvent is required for this nondestructive test. Precise structural information down to each atom and bond in the molecule can be obtained, information rivaled only by X-ray crystallography. Because the measurement can be made in aqueous solution, we can also study the effects of temperature, pH, and interactions with ligands and other biomolecules. Uniform labeling (^{13}C, ^{15}N) permits the study of large biomolecules, such as proteins and nucleic acids, up to 30 kD and beyond.

Compared to other analytical techniques, NMR is quite insensitive. For molecules of the size of most drugs and natural products (100–600 Da), about a milligram of pure material is required, compared to less than 1 μg for mass spectrometry. The intensity of NMR signals is directly proportional to concentration, so NMR "sees all" and "tells all," even giving multiple signals for stereoisomers or slowly interconverting conformations. This complexity is very rich in information, but it makes mixtures very difficult to analyze. Finally, the NMR instrument is quite expensive (from US $200,000 to more than $5 million depending on the magnet strength) and can only analyze one sample at a time, with some experiments requiring a few minutes and the most complex ones requiring up to 4 days to acquire the data. But used in concert with complementary analytical techniques, such as light spectroscopy and mass spectrometry, NMR is the most powerful tool by far for the determination of organic structure. Only X-ray crystallography can give a comparable kind of detailed information on the precise location of atoms and bonds within a molecule.

The kind of information NMR gives is always "local": The world is viewed from the point of view of one atom in a molecule, and it is a very myopic view indeed: This atom can "see" only about 5 Å or three bonds away (a typical C–H bond is about 1 Å or 0.1 nm long). But the point of view can be moved around so that we "see" the world from each atom in the molecule in turn, as if we could carry a weak flashlight around in a dark room and try to put together a picture of the whole room. The information obtained is always coded and requires a complex (but very satisfying) puzzle-solving exercise to decode it and produce a three-dimensional model of a molecule. In this sense, NMR does not produce a direct "picture" of the molecule like an electron microscope or an electron density map obtained from X-ray crystallography. The NMR data are a set of relationships among the atoms of the molecule, relationships of proximity either directly through space or along the bonding network of the molecule. With a knowledge of these relationships, we can construct an unambiguous model of the molecular structure. To an organic chemist trained in the interpretation of NMR data, this process of inference can be so rapid and unconscious that the researcher really "sees" the molecule in the NMR spectrum. For a biochemist or molecular biologist, the data are much more complex and the structural information emerges slowly through a process of computer-aided data analysis.

The goal of this book is to develop in the reader a real understanding of NMR and how it works. Many people who use NMR have no idea what the instrument does or how the experiments manipulate the nuclei of the molecule to reveal structural information. Because NMR is a technique involving the physics of magnetism and superconductivity, radio frequency electronics, digital data processing, and quantum mechanics of nuclear spins, many researchers are understandably intimidated and wish only to know "which button to push." Although a simple list of instructions and an understanding of data interpretation are enough for many people, this book attempts to go deeper without getting buried in technical details and physical and mathematical formalism. It is my belief that with a relatively simple set of theoretical tools, learned by hands-on problem solving and experience, the organic chemist or biologist can master all of the modern NMR techniques with a solid understanding of how they work and what needs to be adjusted or optimized to get the most out of these techniques.

In this book we will start with a very primitive model of the NMR experiment, and explain the simplest NMR techniques using this model. As the techniques become more complex and powerful, we will need to expand this model one step at a time, each time avoiding formal physics and quantum mechanics as much as possible and instead relying on analogy and common sense. Necessarily, as the model becomes more sophisticated, the comfortable physical analogies become fewer, and we have to rely more on symbols and math. With lots of examples and frequent reminders of what the practical result (NMR spectrum) would be at each stage of the process, these symbols become familiar and useful tools. To understand NMR one only needs to look at the interaction of at most two nearby nuclei in a molecule, so the theory will not be developed beyond this simplest of relationships. By the end of this book, you should be able to read the literature of new NMR experiments and be able to understand even the most complex biological NMR techniques. My goal is to make this rich literature accessible to the "masses" of researchers who are not experts in physics or physical chemistry. My hope is that this understanding, like all deep understanding of science, will be satisfying and rewarding and, in a research environment, empowering.

ACKNOWLEDGMENTS

I would like to thank my NMR mentors: Paul A. Bartlett (University of California, Berkeley), who taught me the beauty of natural product structure elucidation by NMR and chemical methods through group meeting problem sessions and graduate courses; Krish Krishnamurthy, who introduced me to NMR maintenance and convinced me of the power and usefulness of product operator formalism during a postdoc in John Casida's lab at Berkeley; Rachel E. Klevit (University of Washington), who gave me a great opportunity to get started in protein NMR; and Wayne J. Fairbrother (Genentech, Inc.), who taught me the highest standards of excellence and thoroughness in structural biology by NMR. Through this book I hope to pass on some of the knowledge that was so generously given to me.

This book grew out of my course in NMR spectroscopy that began in 1987 as an undergraduate course at The Evergreen State College, Olympia, WA, and continued in 1997 as a graduate course at the University of Arizona. Those who helped me along the way include Professors Michael Barfield, F. Ann Walker, and Michael Brown, as well as my teaching assistants, especially Igor Filippov, Jinfa Ying, and Liliya Yatsunyk.

1

FUNDAMENTALS OF NMR SPECTROSCOPY IN LIQUIDS

1.1 INTRODUCTION TO NMR SPECTROSCOPY

NMR is a spectroscopic technique that relies on the magnetic properties of the atomic nucleus. When placed in a strong magnetic field, certain nuclei resonate at a characteristic frequency in the radio frequency range of the electromagnetic spectrum. Slight variations in this resonant frequency give us detailed information about the molecular structure in which the atom resides.

1.1.1 The Classical Model

Many atoms (e.g., ^{1}H, ^{13}C, ^{15}N, ^{31}P) behave as if the positively charged nucleus was spinning on an axis (Fig. 1.1). The spinning charge, like an electric current, creates a tiny magnetic field. When placed in a strong external magnetic field, the magnetic nucleus tries to align with it like a compass needle in the earth's magnetic field. Because the nucleus is spinning and has angular momentum, the torque exerted by the external field results in a circular motion called precession, just like a spinning top in the earth's gravitational field. The rate of this precession is proportional to the external magnetic field strength and to the strength of the nuclear magnet:

$$\nu_o = \gamma B_o / 2\pi$$

where ν_o is the precession rate (the "Larmor frequency") in hertz, γ is the strength of the nuclear magnet (the "magnetogyric ratio"), and B_o is the strength of the external magnetic field. This resonant frequency is in the radio frequency range for strong magnetic fields

---

NMR Spectroscopy Explained: Simplified Theory, Applications and Examples for Organic Chemistry and Structural Biology, by Neil E. Jacobsen
Copyright © 2007 John Wiley & Sons, Inc.

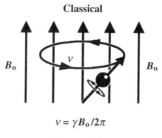

Figure 1.1

and can be measured by applying a radio frequency signal to the sample and varying the frequency until absorbance of energy is detected.

1.1.2 The Quantum Model

This classical view of magnetic resonance, in which the nucleus is treated as a macroscopic object like a billiard ball, is insufficient to explain all aspects of the NMR phenomenon. We must also consider the quantum mechanical picture of the nucleus in a magnetic field. For the most useful nuclei, which are called "spin ½" nuclei, there are two quantum states that can be visualized as having the spin axis pointing "up" or "down" (Fig. 1.2). In the absence of an external magnetic field, these two states have the same energy and at thermal equilibrium exactly one half of a large population of nuclei will be in the "up" state and one half will be in the "down" state. In a magnetic field, however, the "up" state, which is aligned with the magnetic field, is lower in energy than the "down" state, which is opposed to the magnetic field. Because this is a quantum phenomenon, there are no possible states in between. This energy separation or "gap" between the two quantum states is proportional to the strength of the external magnetic field, and increases as the field strength is increased. In a large population of nuclei in thermal equilibrium, slightly more than half will reside in the "up" (lower energy) state and slightly less than half will reside in the "down" (higher energy) state. As in all forms of spectroscopy, it is possible for a nucleus in the lower energy state to absorb a photon of electromagnetic energy and be promoted to the higher energy state. The energy of the photon must exactly match the energy "gap" (ΔE) between

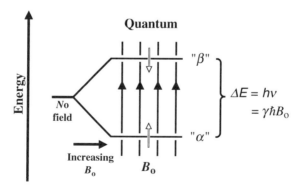

Figure 1.2

the two states, and this energy corresponds to a specific frequency of electromagnetic radiation:

$$\Delta E = h\nu_o = h\gamma B_o/2\pi$$

where h is Planck's constant. The resonant frequency, ν_o, is in the radio frequency range, identical to the precession frequency (the Larmor frequency) predicted by the classical model.

1.1.3 Useful Nuclei for NMR

The resonant frequencies of some important nuclei are shown below for the magnetic field strength of a typical NMR spectrometer (Varian Gemini-200):

Nucleus	Abundance (%)	Sensitivity	Frequency (MHz)
^{1}H	100	1.0	200
^{13}C	1.1	0.016	50
^{15}N	0.37	0.001	20
^{19}F	100	0.83	188
^{31}P	100	0.066	81
^{57}Fe	2.2	3.4×10^{-5}	6.5

The spectrometer is a radio receiver, and we change the frequency to "tune in" each nucleus at its characteristic frequency, just like the stations on your car radio. Because the resonant frequency is proportional to the external magnetic field strength, all of the resonant frequencies above would be increased by the same factor with a stronger magnetic field. The relative sensitivity is a direct result of the strength of the nuclear magnet, and the effective sensitivity is further reduced for those nuclei that occur at low natural abundance. For example, ^{13}C at natural abundance is 5700 times less sensitive ($1/(0.011 \times 0.016)$) than ^{1}H when both factors are taken into consideration.

1.1.4 The Chemical Shift

The resonant frequency is not only a characteristic of the type of nucleus but also varies slightly depending on the position of that atom within a molecule (the "chemical environment"). This occurs because the bonding electrons create their own small magnetic field that modifies the external magnetic field in the vicinity of the nucleus. This subtle variation, on the order of one part in a million, is called the chemical shift and provides detailed information about the structure of molecules. Different atoms within a molecule can be identified by their chemical shift, based on molecular symmetry and the predictable effects of nearby electronegative atoms and unsaturated groups.

The chemical shift is measured in parts per million (ppm) and is designated by the Greek letter delta (δ). The resonant frequency for a particular nucleus at a specific position within a molecule is then equal to the fundamental resonant frequency of that isotope (e.g., 50.000 MHz for ^{13}C) times a factor that is slightly greater than 1.0 due to the chemical shift:

$$\text{Resonant frequency} = \nu(1.0 + \delta \times 10^{-6})$$

4 FUNDAMENTALS OF NMR SPECTROSCOPY IN LIQUIDS

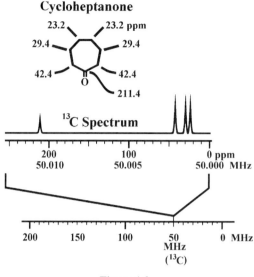

Figure 1.3

For example, a ^{13}C nucleus at the C-4 position of cycloheptanone (δ23.3 ppm) resonates at a frequency of

$$50.000 \text{ MHz} (1.0 + 23.2 \times 10^{-6}) = 50.000(1.0000232) = 50,001,160 \text{ Hz}$$

A graph of the resonant frequencies over a very narrow range of frequencies centered on the fundamental resonant frequency of the nucleus of interest (e.g., ^{13}C at 50.000 MHz) is called a *spectrum*, and each peak in the spectrum represents a unique chemical environment within the molecule being studied. For example, cycloheptanone has four peaks due to the four unique carbon positions in the molecule (Fig. 1.3). Note that symmetry in a molecule can make the number of unique positions less than the total number of carbons.

1.1.5 Spin–Spin Splitting

Another valuable piece of information about molecular structure is obtained from the phenomenon of spin–spin splitting. Consider two protons (^{1}H$_a$C–C^{1}H$_b$) with different chemical shifts on two adjacent carbon atoms in an organic molecule. The magnetic nucleus of H$_b$ can be either aligned with ("up") or against ("down") the magnetic field of the spectrometer (Fig. 1.4). From the point of view of H$_a$, the H$_b$ nucleus magnetic field perturbs the external magnetic field, adding a slight amount to it or subtracting a slight amount from it, depending on the orientation of the H$_b$ nucleus ("up" or "down"). Because the resonant frequency is always proportional to the magnetic field *experienced* by the nucleus, this changes the H$_a$ frequency so that it now resonates at one of two frequencies very close together. Because roughly 50% of the H$_b$ nuclei are in the "up" state and roughly 50% are in the "down" state, the H$_a$ resonance is "split" by H$_b$ into a pair of resonance peaks of equal intensity (a "doublet") with a separation of J Hz, where J is called the coupling constant. The relationship is mutual so that H$_b$ experiences the same splitting effect (separation of J Hz) from H$_a$. This effect is transmitted through bonds and operates only when the two nuclei are very close (three bonds or less) in the bonding network. If there is more than one "neighbor"

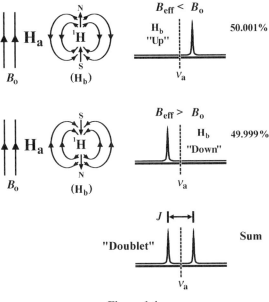

Figure 1.4

proton, more complicated splitting occurs so that the number of peaks is equal to one more than the number of neighboring protons doing the splitting. For example, if there are two neighboring protons ($H_aC-CH^b_2$), there are four possibilities for the H_b protons, just like the possible outcomes of flipping two coins: both "up," the first "up" and the second "down," the first "down" and the second "up," and both "down." If one is "up" and one "down" the effects cancel each other and the H_a proton absorbs at its normal chemical shift position (ν_a). If both H_b spins are "up," the H_a resonance is shifted to the right by J Hz. If both are "down," the H_a resonance occurs J Hz to the left of ν_a. Because there are two ways it can happen, the central resonance at ν_a is twice as intense as the outer resonances, giving a "triplet" pattern with intensity ratio 1 : 2 : 1 (Fig. 1.5). Similar arguments for larger numbers of neighboring spins lead to the general case of n neighboring spins, which split the H_a resonance peak into $n + 1$ peaks with an intensity ratio determined by *Pascal's triangle*. This triangle of numbers is created by adding each adjacent pair of numbers to get the value below it in the triangle:

```
singlet                  1                    (no neighbors)
doublet                 1 1                   (one neighbor)
triplet                1 2 1                  (two neighbors)
quartet               1 3 3 1                 (three neighbors)
quintet              1 4 6 4 1                (four neighbors)
sextet             1 5 10 10 5 1              (five neighbors)
septet            1 6 15 20 15 6 1            (six neighbors)
```

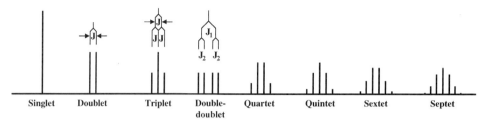

Figure 1.5

The strength of the spin–spin splitting interaction, measured by the peak separation ("*J* value") in units of hertz, depends in a predictable way on the dihedral angle defined by H_a–C–C–H_b, so that information can be obtained about the stereochemistry and conformation of molecules in solution. Because of this dependence on the geometry of the interceding bonds, it is possible to have couplings for two neighbors with different values of the coupling constant, *J*. This gives rise to a splitting pattern with four peaks of equal intensity: a double doublet (Fig. 1.5).

1.1.6 The NOE

A third type of information available from NMR comes from the nuclear Overhauser enhancement or NOE. This is a direct through-space interaction of two nuclei. Irradiation of one nucleus with a weak radio frequency signal at its resonant frequency will equalize the populations in its two energy levels. This perturbation of population levels disturbs the populations of nearby nuclei so as to enhance the intensity of absorbance at the resonant frequency of the nearby nuclei. This effect depends only on the distance between the two nuclei, even if they are far apart in the bonding network, and varies in intensity as the inverse sixth power of the distance. Generally the NOE can only be detected between protons (^{1}H nuclei) that are separated by 5 Å or less in distance. These measured distances are used to determine accurate three-dimensional structures of proteins and nucleic acids.

1.1.7 Pulsed Fourier Transform (FT) NMR

Early NMR spectrometers recorded a spectrum by slowly changing the frequency of a radio frequency signal fed into a coil near the sample. During this gradual "sweep" of frequencies the absorption of energy by the sample was recorded by a pen in a chart recorder. When the frequency passed through a resonant frequency for a particular nucleus in the sample, the pen went up and recorded a "peak" in the spectrum. This type of spectrometer, now obsolete, is called "continuous wave" or CW. Modern NMR spectrometers operate in the "pulsed Fourier-transform" (FT) mode, permitting the entire spectrum to be recorded in 2–3 s rather than the slow (5 min) frequency sweep. The collection of nuclei (sample) is given a strong radio frequency pulse that aligns the nuclei so that they precess in unison, each pointing in the same direction at the same time. The individual magnetic fields of the nuclei add together to give a measurable rotating magnetic field that induces an electrical voltage in a coil placed next to the sample. Over a period of a second or two the individual nuclei get out of synch and the macroscopic signal dies down. This "echo" of the pulse, observed in the coil, is called the free induction decay (FID), and it contains all of the resonant frequencies of the sample nuclei combined in one cacophonous reply. These data are digitised, and a

INTRODUCTION TO NMR SPECTROSCOPY   7

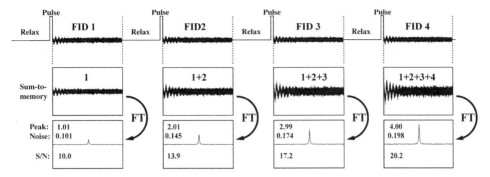

Figure 1.6

computer performs a Fast Fourier Transform to convert it from an FID signal as a function of time (time domain) to a plot of intensity as a function of frequency (frequency domain). The "spectrum" has one peak for each resonant frequency in the sample. The real advantage of the pulsed-FT method is that, because the data is recorded so rapidly, the process of pulse excitation and recording the FID can be repeated many times, each time adding the FID data to a sum stored in the computer (Fig. 1.6). The signal intensity increases in direct proportion to the number of repeats or "transients" (1.01, 2.01, 2.99, 4.00), but the random noise tends to cancel because it can be either negative or positive, resulting in a noise level proportional to the square root of the number of transients (0.101, 0.145, 0.174, 0.198). Thus the signal-to-noise ratio increases with the square root of the number of transients (10.0, 13.9, 17.2, 20.2). This signal-averaging process results in a vastly improved sensitivity compared to the old frequency sweep method.

The pulsed Fourier transform process is analogous to playing a chord on the piano and recording the signal from the decaying sound coming out of a microphone (Fig. 1.7). The chord consists of three separate notes: the "C" note is the lowest frequency, the "G" note is the highest frequency, and the "E" note is in the middle. Each of these pure frequencies gives a decaying pure sine wave in the microphone, and the combined signal of three frequencies is a complex decaying signal. This time domain signal ("FID") contains all three of the frequencies of the piano chord. Fourier transform will then convert the data to a "spectrum"—a graph of signal intensity as a function of frequency, revealing the three frequencies of the chord as well as their relative intensities. The Fourier transform allows us to record all of the signals simultaneously and then "sort out" the individual frequencies later.

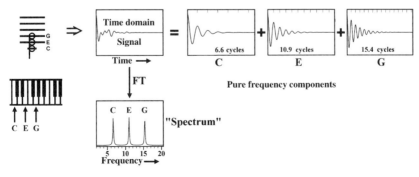

Figure 1.7

8 FUNDAMENTALS OF NMR SPECTROSCOPY IN LIQUIDS

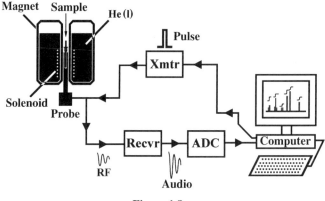

Figure 1.8

1.1.8 NMR Hardware

An NMR spectrometer consists of a superconducting magnet, a probe, a radio transmitter, a radio receiver, an analog-to-digital converter (ADC), and a computer (Fig. 1.8). The magnet consists of a closed loop ("solenoid") of superconducting Nb/Ti alloy wire immersed in a bath of liquid helium (bp 4 K). A large current flows effortlessly around the loop, creating a strong continuous magnetic field with no external power supply. The helium can ("dewar") is insulated with a vacuum jacket and further cooled by an outer dewar of liquid nitrogen (bp 77 K). The probe is basically a coil of wire positioned around the sample that alternately transmits and receives radio frequency signals. The computer directs the transmitter to send a high-power and very short duration pulse of radio frequency to the probe coil. Immediately after the pulse, the weak signal (FID) received by the probe coil is amplified, converted to an audio frequency signal, and sampled at regular intervals of time by the ADC to produce a digital FID signal, which is really just a list of numbers. The computer determines the timing and intensity of pulses output by the transmitter and receives and processes the digital information supplied by the ADC. After the computer performs the Fourier transform, the resulting spectrum can be displayed on the computer monitor and plotted on paper with a digital plotter. The cost of an NMR instrument is on the order of $120,000–$5,000,000, depending on the strength of the magnetic field (200–900 MHz proton frequency).

1.1.9 Overview of ^{1}H and ^{13}C Chemical Shifts

A general understanding of the trends of chemical shifts is essential for the interpretation of NMR spectra. The chemical shifts of ^{1}H and ^{13}C signals are affected by the proximity of electronegative atoms (O, N, Cl, etc.) in the bonding network and by the proximity to unsaturated groups (C=C, C=O, aromatic) directly through space. Electronegative groups shift resonances to the left (higher resonant frequency or "downfield"), whereas unsaturated groups shift to the left (downfield) when the affected nucleus is in the plane of the unsaturation, but have the opposite effect (shift to the right or "upfield") in regions above and below this plane. Although the range of chemical shifts in parts per million is much larger for ^{13}C than for ^{1}H (0–220 ppm *vs.* 0–13 ppm), there is a rough correlation between the shift of a proton and the shift of the carbon it is attached to (Fig. 1.9). For a "hydrocarbon" environment with no electronegative atoms or unsaturated groups nearby, the shift is

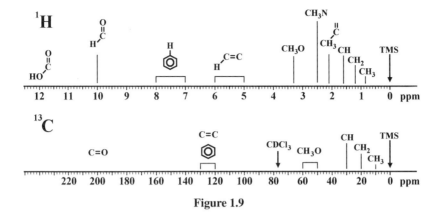

Figure 1.9

near the upfield (right) edge of the range, with a small downfield shift for each substitution: CH > CH$_2$ > CH$_3$ (^{1}H: 1.6, 1.2, 0.8; ^{13}C: 30, 20, 10 ppm). Oxygen has a stronger downfield-shifting effect than nitrogen due to its greater electronegativity: 3–4 ppm (^{1}H) and 50–85 (^{13}C) for CH–O. As with the hydrocarbon environment, the same downfield shifts are seen for increasing substitution: C$_q$–O (quaternary) > CH–O > CH$_2$O > CH$_3$O (^{13}C around 85, 75, 65, and 55 ppm, respectively). Proximity to an unsaturated group usually is downfield shifting because the affected atom is normally in the plane of the unsaturation: CH$_3$ attached to C=O moves downfield to 30 (^{13}C) and 2.1 ppm (^{1}H), whereas in HC=C (closer to the unsaturation) ^{13}C moves to 120–130 ppm and ^{1}H to 5–6 ppm. The combination of unsaturation *and* electronegativity is seen in H–C=O: 190 ppm ^{13}C and 10 ppm ^{1}H. There are some departures from this correlation of ^{1}H and ^{13}C shifts. Aromatic protons typically fall in the 7–8 ppm range rather than the 5–6 ppm range for olefinic (HC=C for an isolated C=C bond) protons, whereas ^{13}C shifts are about the same for aromatic or olefinic carbons. Because carbon has more than one bond, it is sensitive to distortion of its bond angles by the steric environment around it, with steric crowding usually leading to downfield shifts. Hydrogen has no such effect because it has only one bond, but it is more sensitive than carbon to the through-space effect of unsaturations. For example, converting an alcohol (CH–OH) to an ester (CH–OC(O)R) shifts the ^{1}H of the CH group downfield by 0.5 to 1 ppm, but has little effect on the ^{13}C shift.

1.1.10 Equivalence in NMR

Nuclei can be equivalent (have the same chemical shift) by symmetry within a molecule (e.g., the two methyl carbons in acetone, CH$_3$COCH$_3$), or by rapid rotation around single bonds (e.g., the three methyl protons in acetic acid, CH$_3$CO$_2$H). The intensity (integrated peak area or *integral*) of ^{1}H signals is directly proportional to the number of equivalent nuclei represented by that peak. For example, a CH$_3$ peak in a molecule would have three times the integrated peak area of a CH peak in the same molecule.

1.1.11 Proton Spectrum Example

The first step in learning to interpret NMR spectra is to learn how to predict them from a known chemical structure. An example of a ^{1}H (proton) NMR spectrum is shown for

10   FUNDAMENTALS OF NMR SPECTROSCOPY IN LIQUIDS

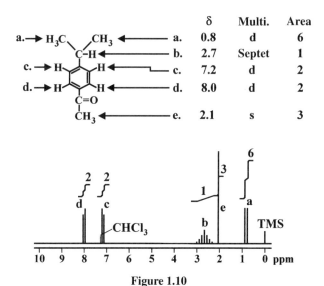

Figure 1.10

4-isopropylacetophenone (Fig. 1.10). The two isopropyl methyl groups are equivalent by symmetry, and each methyl group has three protons made equivalent by rapid rotation about the C–C bond. This makes all six H_a protons equivalent. Because they are far from any electronegative atom, these protons have a chemical shift typical of an isolated CH_3 group: 0.8 ppm (see Fig. 1.9). The absorbance is split into two peaks (a doublet) by the single neighboring H_b proton. The six H_a protons do not split each other because they are equivalent. The integrated area of the doublet is 6.0 because there are six H_a protons in the molecule. The H_b proton is split by all six of the H_a protons, so its absorbance shows up as a septet (seven peaks with intensity ratio 1:6:15:20:15:6:1). Its integrated area is 1.0, and its chemical shift is downfield of an isolated CH_2 (1.2 ppm) because of its proximity to the unsaturated aromatic ring (close to the plane of the aromatic ring so the effect is a downfield shift). The H_e methyl group protons are all equivalent due to rapid rotation of the CH_3 group, and their chemical shift is typical for a methyl group adjacent to the unsaturated C=O group (2.1 ppm). There are no neighboring protons (the H_d proton is five bonds away from it, and the maximum distance for splitting is three bonds) so the absorbance appears as a single peak ("singlet") with an integrated area of 3.0. The H_c and H_d protons on the aromatic ring appear at a chemical shift typical for protons bound directly to an aromatic ring, with the H_d protons shifted further downfield by proximity to the unsaturated C=O group. Each pair of aromatic protons is equivalent due to the symmetry of the aromatic ring. The H_c absorbance is split into a doublet by the neighboring H_d proton (note that from the point of view of either of the H_c protons, only one of the H_d protons is close enough to cause splitting), and the H_d absorbance is split in the same way. Note that the J value (separation of split peaks) is the same for the H_c and H_d doublets, but slightly different for the H_a–H_b splitting. In this way we know, for example, that H_a is not split by either H_c or H_d.

1.1.12 Carbon Spectrum Example

The ^{13}C spectrum of the same compound is diagramed in Figure 1.11. Several differences can be seen in comparison with the ^1H spectrum. First, there is no spin–spin splitting due

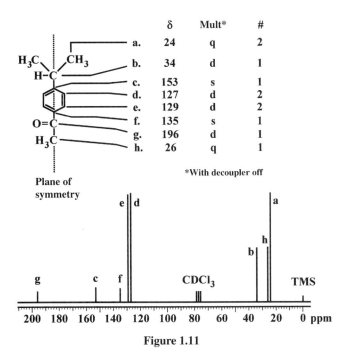

Figure 1.11

to adjacent carbons. This is because of the low natural abundance of ^{13}C, which is only 1.1%. Thus the probability of a ^{13}C occurring next to another ^{13}C is very low, and splitting is not observed because ^{12}C has no magnetic properties. Second, there is no spin–spin splitting due to the protons attached to each carbon. This is prevented intentionally by a process called *decoupling*, in which all the protons in the molecule are simultaneously irradiated with continuous low-power radio frequency energy at the proton resonance frequency. This causes each proton to flip rapidly between the upper and lower (disaligned and aligned) energy states, so that the ^{13}C nucleus sees only the average of the two states and appears as a singlet, regardless of the number of attached protons. The lack of any spin–spin splitting in decoupled ^{13}C spectra means that each carbon always appears as a singlet. The multiplicity (s, d, t, q) indicated for each carbon in the diagram is observed only with the decoupler turned off and is not shown in the spectrum. Third, the peaks are not integrated because the peak area does not indicate the number of carbon atoms accurately. This is because ^{13}C nuclei *relax* more slowly than protons, so that unless a very long relaxation delay between repetitive pulses is used, the population difference between the two energy states of ^{13}C is not reestablished before the next pulse arrives. Quaternary carbons, which have no attached protons, relax particularly slowly and thus show up with very low intensity.

The molecular symmetry, indicated by a dotted line (Fig. 1.11) where the mirror plane intersects the plane of the paper, makes the two isopropyl methyl carbons C_a equivalent. Their chemical shift is a bit downfield of an isolated methyl group due to the steric crowding of the isopropyl group. Unlike protons, ^{13}C nuclei are sensitive to the degree of substitution or branching in the immediate vicinity, generally being shifted downfield by increased branching. C_b is shifted further downfield because of direct substitution (it is attached to three other carbons) and proximity to the aromatic ring. C_h is in a relatively uncrowded

12 FUNDAMENTALS OF NMR SPECTROSCOPY IN LIQUIDS

environment, but is shifted downfield by proximity to the unsaturated and electronegative carbonyl group. With the decoupler turned off, CH_3 carbons appear as quartets because of the three neighboring protons. The aromatic CH carbons C_d and C_e are in nearly identical environments typical of aromatic carbons, and each resonance peak represents two carbons due to molecular symmetry. With the decoupler turned off, these peaks turn into doublets due to the presence of a single attached proton. The two quaternary aromatic carbons C_c and C_f are shifted further downfield by greater direct substitution (they are attached to three other carbons) and by steric crowding (greater remote substitution) in the case of C_c and proximity to a carbonyl group in the case of C_f. The chemical shift of the carbonyl carbon C_g is typical for a ketone. All three of the quaternary carbons C_c, C_f, and C_g have low peak intensities due to slow relaxation (reestablishment of population difference) in the absence of directly attached protons.

1.2 EXAMPLES: NMR SPECTROSCOPY OF OLIGOSACCHARIDES AND TERPENOIDS

A few real-world examples will illustrate the use of ^1H and ^{13}C chemical shifts and J couplings, as well as introduce some advanced methods we will use later. Two typical classes of complex organic molecules will be introduced here to familiarize the reader with the elements of structural organic chemistry that are important in NMR and how they translate into NMR spectra. Terpenoids are typical of natural products; they are relatively nonpolar (water insoluble) molecules with a considerable amount of "hydrocarbon" part and only a few functional groups—olefin, alcohol, ketone—in a rigid structure. Oligosaccharides are polar (water soluble) molecules in which every carbon is functionalized with oxygen—alcohol, ketone, or aldehyde oxidation states—and relatively rigid rings are connected with flexible linkages. In both cases, rigid cyclohexane-chair ring structures are ideal for NMR because they allow us to use J-coupling values to determine stereochemical relationships of protons (*cis* and *trans*). The molecules introduced here will be used throughout the book to illustrate the results of the NMR experiments.

1.2.1 Oligosaccharides

A typical monosaccharide (single carbohydrate building block) is a five or six carbon molecule with one of the carbons in the aldehyde or ketone oxidation state (the "anomeric" carbon) and the rest in the alcohol oxidation state (CH(OH) or CH_2OH). Thus the anomeric carbon is unique within the molecule because it has two bonds to oxygen whereas all of the other carbons have only one bond to oxygen. Normally the open-chain monosaccharide will form a five- or six-membered ring as a result of the addition of one of the alcohol groups (usually the second to last in the chain) to the ketone or aldehyde, changing the C=O double bond to an OH group.

The six-membered ring of glucose prefers the chair conformation shown in Figure 1.12, with nearly all of the OH groups arranged in the equatorial positions (sticking out and roughly in the plane of the ring) with the less bulky H atoms in the axial positions (pointing up or down, above or below the plane of the ring). This limits the dihedral angles between neighboring protons (vicinal or three-bond relationships) to three categories: axial–axial (*trans*): 180° dihedral angle, large J coupling (∼10 Hz); axial-equatorial (*cis*): 60° dihedral angle, small J (∼4 Hz); and equatorial–equatorial (*trans*): 60° dihedral angle, small J (∼4 Hz).

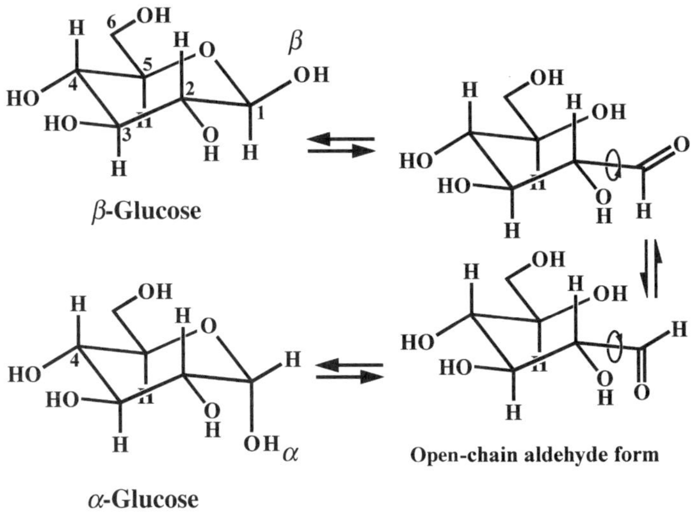

Figure 1.12

The third category is rare in carbohydrates because the bulky OH groups prefer the equatorial position, pushing the H into the axial position.

In this ring form, the anomeric carbon (C1) of an aldehyde sugar (aldose) has one bond to the oxygen of the ring and another to an OH group external to the ring. Also external to the ring is the CH$_2$OH group of the last carbon in the chain. The anomeric OH group can either be *cis* or *trans* to the external CH$_2$OH group, depending on which side of the aldehyde or ketone group the OH group is added to. If it is *cis*, we call this isomer the β-anomer, and if it is *trans* we call it the α-anomer. When a crystalline monosaccharide is dissolved in water, these two ring forms rapidly form an equilibrium mixture of α and β anomers with very little of the open-ring aldehyde existing in solution (Fig. 1.12).

It is possible to link a monosaccharide to an alcohol at the anomeric carbon, so that instead of an OH group the anomeric carbon is connected to an OR group (e.g., OCH$_3$) that is external to the ring. This is called a "glycoside," and the anomeric carbon is now a full acetal or ketal. The ring can no longer freely open into the open-chain aldehyde or ketone, so there is no equilibration of α and β forms. Thus a β-glycoside (OR group *cis* to the CH$_2$OH group) will remain locked in the β form when dissolved in water. If the alcohol used to form the glycoside is the alcohol of another monosaccharide, we have formed a disaccharide with the two monosaccharides connected by a glycosidic linkage (Figure 1.13). Usually the alcohol comes from one of the alcohol carbons of the second sugar, but it is also possible to form a glycosidic linkage to the *anomeric* carbon of the second sugar. In this case we have a linkage C–O–C from one anomeric carbon to another, and both monosaccharides are "locked" with no possibility of opening to the aldehyde or ketone form.

1.2.2 NMR of Carbohydrates: Chemical Shifts

NMR chemical shifts give us information about the proximity of electronegative atoms (e.g., oxygen) and unsaturated groups (double bonds and aromatic rings). In this discussion we will ignore the protons attached directly to oxygen (OH) because they provide little

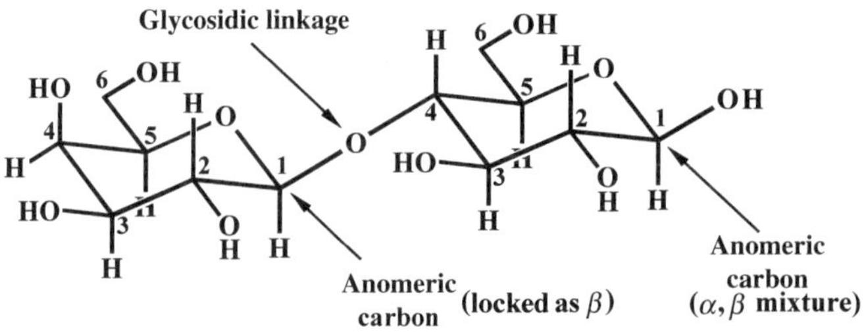

Figure 1.13

chemical information in NMR and are exchanged for deuterium by the solvent if we use deuterated water (D_2O). In the case of carbohydrates, nearly all of the protons attached to carbon are in a similar environment: one oxygen attached to the carbon (C<u>H</u>OH or C<u>H</u>$_2$OH). These protons all have similar chemical shifts, in the range of 3.3–4.1 ppm, so there is often a great deal of overlap of these signals in the ^{1}H NMR of carbohydrates, even at the highest magnetic fields achievable. For this reason carbohydrate NMR (and NMR of nucleic acids RNA and DNA, which have a sugar-phosphate backbone) has been limited to relatively small molecules because the complexity of overlapping signals is limiting. The anomeric proton, however, is in a unique position because the carbon it is attached to has *two* bonds to oxygen. This additional inductive pull of electron density away from the hydrogen atom leads to a further downfield shift of the NMR signal, so that anomeric protons resonate in a distinct region at 5–6 ppm. A similar effect is seen for anomeric carbons, which have ^{13}C chemical shifts in the range of 90–110 ppm, whereas their neighbors with only *one* bond to oxygen resonate in the normal alcohol region of 60–80 ppm. Because each monosaccharide unit in a complex carbohydrate has only one anomeric carbon, we can count up the number of monosaccharide building blocks by simply counting the number of NMR signals in this anomeric region. Thus the analysis of carbohydrate NMR spectra is greatly simplified if we focus on the anomeric region of the ^{1}H or ^{13}C spectrum. The "alcohol" (nonanomeric) carbons of a sugar (H–C–O or H$_2$C–O) are sensitive to steric crowding, so that the CH$_2$OH carbons appear at higher field (60–70 ppm) than the more crowded CHOH carbons (70–80 ppm). This steric effect is also seen at the alcohol side of a glycosidic linkage (–O–CH–O–C<u>H</u>–C): this carbon is shifted downfield by as much as 10 ppm from the rest of the "alcohol" carbons (HO–CH–C) that are not involved in glycosidic linkages.

1.2.3 ^{1}H NMR: Coupling Constants

In the proton NMR spectrum, each signal is "split" into a multiple peak pattern by the influence of its "neighbors," the protons attached to the next carbon in the chain. These protons are three bonds away from the proton being considered and are sometimes called "vicinal" protons. For example, the anomeric proton in a cyclic aldose has only one neighbor: the proton on the next carbon in the chain (carbon 2). Note that because of rapid exchange processes or deuterium replacement in D_2O, we seldom see splitting by the OH protons. Because it has only one neighbor, the anomeric proton will always appear as a *doublet* in the NMR spectrum. Also, because of its unique chemical shift position (5–6 ppm) and relatively rare occurrence (only one anomeric position per monosaccharide unit), the anomeric proton signal is usually not overlapped so we can see its splitting pattern clearly. The distance

(J, in frequency units of Hz) between the two component peaks of the doublet is a measure of the intensity of the splitting (or J coupling) interaction. For vicinal ("next-door neighbor") protons the value of J depends on the dihedral angle of the C–C bond between them. This angle is fixed in six-membered ring (pyranose) sugars because the ring adopts a stable *chair* conformation. For many common sugars (glucose, galactose, etc.) all or nearly all of the bulky groups on the ring (OH or CH_2OH) can be oriented in the less crowded equatorial position in one of the two chair forms. Thus the sugar ring is effectively "locked" in this one chair form and we can talk about each proton on the ring as being in an axial or equatorial orientation. This is important for NMR because two neighboring (vicinal) protons that are both in axial positions ("*trans*-diaxial" relationship) have a dihedral angle at the maximum value of 180°, and this leads to the maximum value of the coupling constant J (about 10 Hz separation of the two peaks of the doublet). This does not make intuitive sense because in this arrangement the two protons are as far apart as possible; however, it is the parallel alignment of the two C–H bonds that leads to the strong coupling because the J-coupling (splitting) interaction is transmitted through bonds and not through space. Two vicinal protons in a locked chair with an axial–equatorial or an equatorial–equatorial relationship will have a much smaller coupling constant (much narrower pair of peaks in the doublet) in the range of 4 Hz. Thus we can use NMR coupling constants to determine the stereochemistry of sugars.

Here is how we can use this in the analysis of carbohydrate ^1H NMR spectra: most naturally occurring sugars have an equatorial OH at the 2 position (numbering starts with the anomeric position as number 1), so the proton at carbon 2 is axial in a six-membered ring sugar. In addition, the CH_2OH group is also equatorial in most pyranose sugars. So if the anomeric proton is axial, we should see it in the ^1H NMR spectrum as a doublet with a large coupling (10 Hz), because the H_1–H_2 relationship is axial–axial. If the anomeric proton is axial, then the anomeric OH or OR substituent is equatorial and the sugar is in the β configuration (anomeric OH or OR *cis* to the CH_2OH group at C_5). If we see an anomeric proton with a small (4 Hz) coupling, then the anomeric proton is equatorial, the OH or OR group is axial, and we have an α sugar (anomeric OH or OR *trans* to the CH_2OH group at C_5). This reasoning works *only* if we are dealing with an aldopyranose (six-membered ring sugar based on an open-chain aldehyde) with an equatorial OH at C_2; fortunately, nature seems to favor this situation.

1.2.4 Reducing Sugars

If the anomeric carbon of a sugar in the ring form bears an OH substituent instead of OC (glycosidic linkage), it will have the possibility of opening to the open-chain aldehyde or ketone form and reclosing in either the α or the β configuration. This is called a "reducing sugar" because the open-chain aldehyde form is accessible and can be oxidized to the carboxylic acid. The two isomers (α and β) are in equilibrium and we usually see about a 2:1 ratio of β to α forms. The equilibration is slow on the NMR timescale (milliseconds) and so we see two distinct NMR peaks for the two isomers. The anomeric proton for the major β form will be a doublet with a large coupling constant (10 Hz) and for the minor form a doublet at a different chemical shift with a small coupling constant (4 Hz). The ratio of integrals for these two peaks will be about 2:1 (0.67:0.33 for normalized integrals). This pattern is a dead giveaway that you have a free (reducing) aldopyranose sugar. This monosaccharide could still be linked to other sugars by formation of a glycosidic linkage with one of the nonanomeric OH groups.

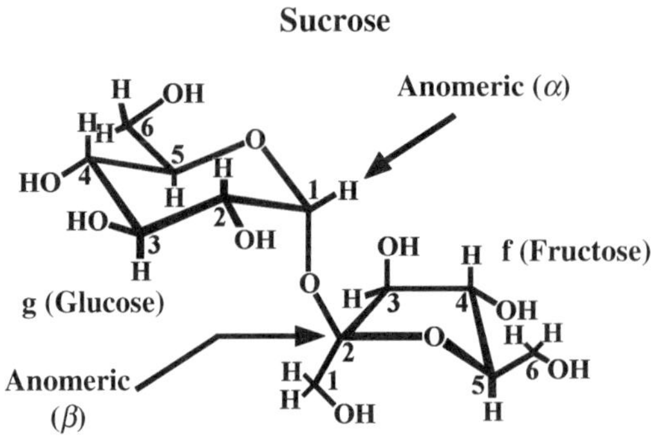

Sucrose

Figure 1.14

1.2.5 Keto Sugars

A ketose or keto sugar is a sugar based on a ketone rather than aldehyde functional group for its anomeric carbon. In this case the anomeric carbon is not C_1 and there is no proton attached to the anomeric carbon (i.e., it is a *quaternary* carbon). The most common naturally occurring ketose is fructose, a 6 carbon sugar with the anomeric (ketone) carbon at position 2 in the chain. It forms a five-membered ring hemiketal (furanose) with the C_1 and C_6 CH_2OH groups external to the ring. For a keto sugar you will not see an anomeric proton signal in the ^1H NMR because the anomeric carbon has no hydrogen bonded to it. The only evidence will be the quaternary carbon in the ^{13}C spectrum that appears at the typical chemical shift (90–110 ppm) for an anomeric carbon (two bonds to oxygen). Furanose (five-membered ring) sugars pose another problem for NMR analysis: five-membered rings are generally flexible and do not adopt a stable chair-type conformation. For this reason we cannot speak of "axial" and "equatorial" protons or substituents in a furanose, so that stereochemical analysis by ^1H NMR is very difficult.

1.2.6 Sucrose

A classic example of a keto sugar occurs in sucrose, a disaccharide formed from glucose in a six-membered ring linked to fructose in a five-membered ring, with the glycosidic linkage between the anomeric carbon of glucose (α configuration) and the anomeric carbon of fructose (β configuration) (Fig. 1.14). In the ^1H spectrum of sucrose (Fig. 1.15) we see the "alcohol" CH protons in the chemical shift range 3.4–4.2 ppm and the glucose anomeric proton at about 5.4 ppm. Fructose has no anomeric proton signal because the anomeric carbon is quaternary (keto sugar). The g1 (glucose position 1) proton signal occurs as a doublet (coupled only to g2) with a small coupling constant (3.8 Hz) indicating that it is in the equatorial position (equatorial–axial coupling). This confirms that the glucose configuration is α because the glycosidic oxygen is pointing "down," opposite to the g6 CH_2OH group. There is a double doublet at 3.5 ppm that can be broken down into two couplings: a doublet coupling of 10.0 Hz is further split by another doublet coupling of 3.8 Hz. The 3.8 Hz coupling matches the H-g1 doublet (also 3.8 Hz), so we can assign this peak to H-g2. Because the other coupling (to H-g2's other neighbor H-g3) is large, we know that H-g3 is axial and we confirm that H-g2 is also axial, further confirming that H-g1 is equatorial. There are three triplets with large coupling constants (3.4, 3.7, and

EXAMPLES: NMR SPECTROSCOPY OF OLIGOSACCHARIDES AND TERPENOIDS 17

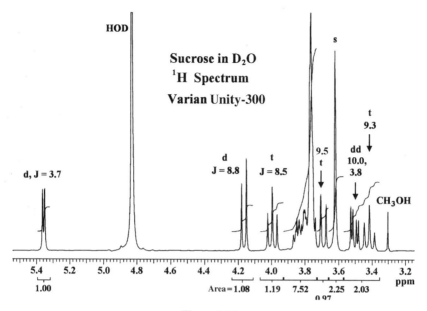

Figure 1.15

4.0 ppm), and it is likely that they represent axial protons in a cyclohexane chair structure with an axial proton on each side. Because all of the OH groups and the CH_2OH group are in equatorial positions in the glucose portion, the nonanomeric H's are in axial positions, and we expect triplets with large couplings (~10 Hz) for H–g3 and H–g4 because both are in axial positions with one neighbor on each side in an axial position. These two large (axial–axial) couplings, if identical, would lead to a triplet pattern. Because we see three such triplets in the ^1H spectrum, each one with normalized integral area 1, one of them must belong to the fructose part. Only H–f4 can be a triplet because it is the only fructose position with a single neighbor on each side. The doublet at 4.2 ppm ($J = 8.8$) can be assigned to H–f3 because it is next to the quaternary (anomeric) carbon C–f2 and therefore has only one coupling partner: H-f4. Note that this is the only doublet besides H-g1, which can be assigned because of its chemical shift in the anomeric region. Of the three resolved triplets, careful examination of the coupling constants reveals that one has a slightly smaller J value (8.5 Hz) that closely matches the H–f3 doublet splitting. Thus we can assign this triplet at 4.0 ppm to H–f4. A sharp singlet at 3.6 ppm (integral area 2) corresponds to the only CH_2 group (H-f1) that is isolated from coupling by the quaternary carbon (C–f2). Because this is a chiral molecule, the two protons of CH_2–f1 *could* have different chemical shifts, leading to a pair of doublets, but in this case they coincidentally have the same chemical shift and give a singlet. Two protons of the same carbon atom (CH_2) are called "geminal" (twins), and if they have the same chemical shift in a chiral molecule they are called "degenerate." The overlapped group of signals between 3.75 and 3.9 ppm integrates to six protons and must contain the glucose CH_2OH (H–g6), the other fructose CH_2OH (H–f6), and the more complex H–g5 and H–f5 signals (each with one coupling partner at position 4 and two at position 6). Thus the only ambiguity remains the two resolved (not overlapped) triplet signals at 3.4 and 3.7 ppm that correspond to H–g3 and H–g4. To solve this puzzle, we will

18 FUNDAMENTALS OF NMR SPECTROSCOPY IN LIQUIDS

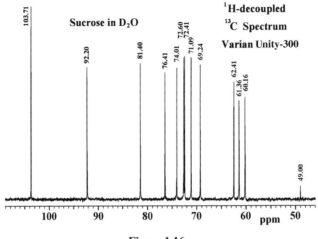

Figure 1.16

need more information from more advanced NMR experiments such as two-dimensional NMR.

The ^{13}C spectrum of sucrose is shown in Figure 1.16. Because it is proton decoupled, we see only one peak for each unique carbon in the molecule: 12 peaks for the $C_{12}H_{22}O_{12}$ molecule of sucrose. We see two peaks in the anomeric (90–110 ppm) region, and we can assign the more substituted C–f2 (two bonds to carbon) to the more downfield of the two at 103.7 ppm. The less substituted C–g1 (one bond to carbon) appears at 92.2 ppm, about 10 ppm upfield of C–f2. This is a rule of thumb: about 10 ppm downfield shift each time an H is replaced with a C in the four bonds to a carbon atom. We see a tight group of three peaks at 60–63 ppm; these are the three CH_2OH groups C–g6, C–f1 and C–f6. The remaining peaks are more spread out over the range 69–82 ppm; these are the nonanomeric "alcohol" or H–C–O carbons that constitute the majority of sugar positions. Again we see the roughly 10 ppm downfield shift due to substitution of an H with a C on the carbon atom of interest: CH_2OH to C–CH–OH. How can we be sure that the CH_2 and CH carbons are so neatly divided into chemical shift regions? More advanced one-dimensional ^{13}C experiments called APT and DEPT allow us to determine the precise number of hydrogens attached to each carbon in the spectrum. To specifically assign the carbons within these three categories will require two-dimensional experiments.

1.2.6.1 Two-Dimensional Experiments A full NMR analysis of a carbohydrate, in which each ^{1}H and ^{13}C peak in the spectrum is assigned to a particular position in the molecule, requires the use of two-dimensional (2D) NMR. In a 2D spectrum, there are two chemical shift scales (horizontal and vertical) and a "spot" appears in the graph at the intersection of two chemical shifts when two nuclei (^{1}H or ^{13}C) in the molecule are close to each other in the structure. For example, one type of 2D spectrum called an HSQC spectrum presents the ^{1}H chemical shift scale on the horizontal (*x*) axis and the ^{13}C chemical shift scale on the vertical (*y*) axis. If proton H_a is directly bonded to carbon C_a, there will be a spot at the intersection of the ^{1}H chemical shift of H_a (horizontal axis) and the ^{13}C chemical shift of C_a (vertical axis). Because the peaks are spread out into two dimensions, the chances of overlap of peaks are much less and we can count up the number of anomeric and

EXAMPLES: NMR SPECTROSCOPY OF OLIGOSACCHARIDES AND TERPENOIDS   19

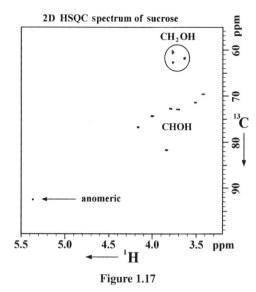

Figure 1.17

nonanomeric peaks very quickly. The HSQC spectrum of sucrose is shown in Figure 1.17. There are 11 "spots" representing the 11 carbons that have at least one hydrogen attached. Quaternary carbons do not show up in the spectrum because the H has to be directly bonded to the C to generate a "spot." Note that the crosspeaks ("spots") fall roughly on a diagonal line extending from the lower left to the upper right. This is because there is a rough correlation between ^{1}H chemical shifts and ^{13}C chemical shifts: the same things that lead to downfield or upfield shifts of protons also affect the carbon they are attached to in the same way. We can also see that the small "triangle" of CH$_2$OH peaks at the top is shifted "up" from the other nonaromatic peaks, due to the reduced steric crowding of the less-substituted CH$_2$ (methylene) carbon compared to CH (methine) carbons. The ^{1}H chemical shifts fall in the range of 3.5–4.2 ppm regardless of the degree of substitution.

A variation of this experiment, called HMBC (MB stands for multiple bond), shows spots only when the carbon and the proton are separated by two or three bonds in the structure. For example, for a monosaccharide we would see a spot at the chemical shift of the anomeric proton (H-1, horizontal axis) and the chemical shift of the C-3 carbon (vertical axis). Working together with data from the HSQC and HMBC 2D spectra, we can "walk" through the bonding structure of a carbohydrate, even "jumping" across the glycosidic linkages and establishing the points of connection of each monosaccharide unit.

Figure 1.18 shows a portion of the ^{1}H spectrum of the trisaccharide D-raffinose in D$_2$O. From just this portion we can conclude that, most probably, one of the sugars is a keto sugar and the other two are aldoses locked in the α configuration. The presence of two anomeric protons, each with a small doublet coupling (3.6 Hz) indicates that two of the sugars have the anomeric proton in the equatorial orientation. This assumes that we have the common pyranose arrangement with H-2 axial and the CH$_2$OH group equatorial. The exact 1:1 ratio of integrals and the absence of major and minor (β and α) anomeric peaks prove that these anomeric centers are locked in a glycosidic linkage. The absence of a third proton in the anomeric region means that the third sugar is most likely a keto sugar, with a quaternary anomeric carbon.

20 FUNDAMENTALS OF NMR SPECTROSCOPY IN LIQUIDS

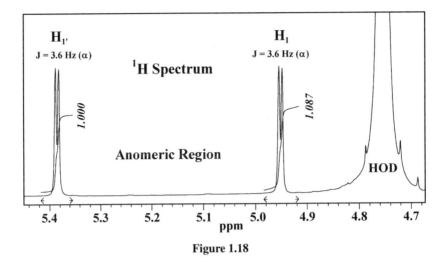

Figure 1.18

1.2.7 Terpenoids

A vast variety of plant and animal natural products are based on a repeating 5-carbon unit called isoprene: C–C(–C)–C–C. The end of the chain nearest the branch can be called the "head" and the other end is the "tail." Two isoprene units connected together make up a "monoterpene" or 10 carbon natural product (e.g., menthol, Fig. 1.19). Six isoprene units make a "triterpene" with 30 carbons. Cholesterol loses three of these in the biosynthetic process to give a 27 carbon "steroid" with four rings (Fig. 1.20). The *trans* ring junctures and the planar olefin "lock" the cyclohexane chairs into a single rigid conformation with well-defined axial and equatorial positions, just as we saw for the glucose ring in sucrose. Another triterpene skeleton that retains all 30 carbons is shown in Figure 1.21; the D and E rings are also locked in cyclohexane chair conformations.

1.2.8 Menthol

Menthol (Fig. 1.19) is a monoterpene natural product obtained from peppermint oil. Typical of terpenoids, menthol is only slightly soluble in water and is soluble in most organic solvents. The *trans* arrangement of the methyl and isopropyl substituents on the cyclohexane

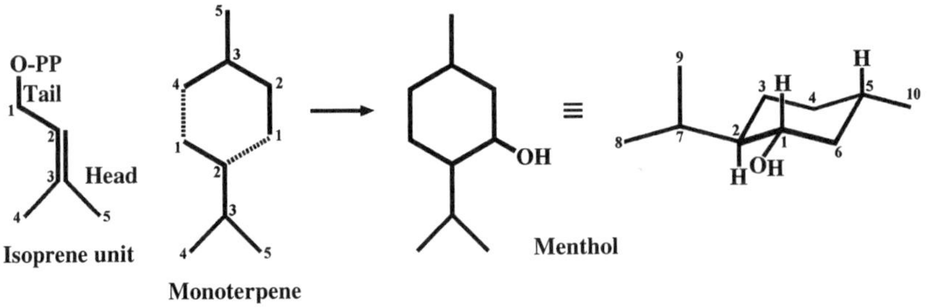

Figure 1.19

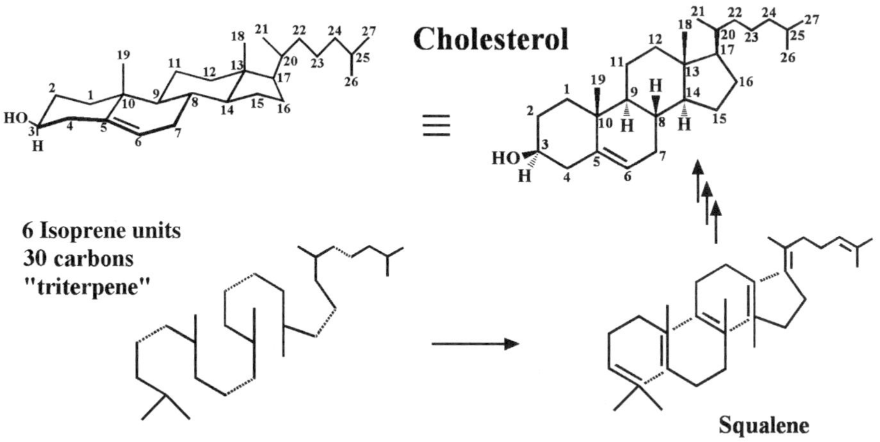

EXAMPLES: NMR SPECTROSCOPY OF OLIGOSACCHARIDES AND TERPENOIDS   21

Figure 1.20

ring lock the ring in a single chair conformation with all of the substituents in the equatorial position.

The 250 MHz ^{1}H spectrum of menthol is shown in Figure 1.22. We see that even at 250 MHz a number of single proton signals are resolved (i.e., not overlapped with any other signals): "h," "l," "m," and "n." Integral values (normalized to one for the smallest resolved peaks) add up to 19.88 or 20 protons, consistent with the molecular formula $C_{10}H_{20}O$. The tall, sharp peaks at the right-hand side ("a," "b," and "c") represent the methyl groups, which usually give the most intense peaks because there are three equivalent protons. The most downfield signal ("n") corresponds to the proton closest to the single functional group, the H–C–OH proton. The OH proton chemical shift depends on concentration because of hydrogen bonding with the OH oxygen of other menthol molecules in solution—looking at different samples it can be identified as the singlet peak at 1.55 ppm. It is a singlet because J-coupling interactions are averaged to zero by exchange: a particular OH proton on one menthol molecule jumps to another menthol molecule rapidly so it is constantly exposed to different H–C–OH protons at position 1, some in the α state and some in the β state, so it sees only a blur and appears as a singlet instead of a doublet. The H-1 proton at 3.37 ppm appears at a chemical shift typical for "alcohol" protons: protons attached to an sp^3 hybridized (i.e., tetrahedral) carbon with a single bond to oxygen (3–4 ppm). Its coupling

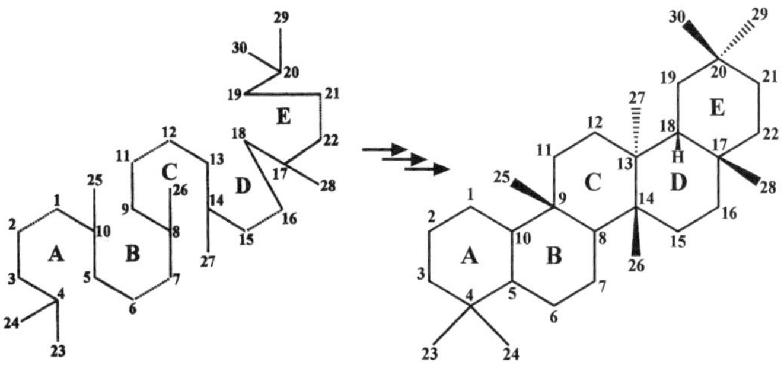

Figure 1.21

22   FUNDAMENTALS OF NMR SPECTROSCOPY IN LIQUIDS

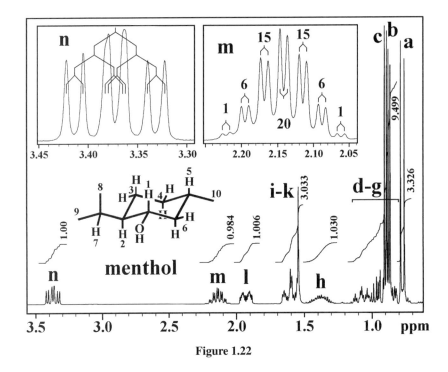

Figure 1.22

pattern (inset, Fig. 1.22) shows two nearly equal large couplings ($J = 9.9$ and 10.9 Hz) due to the axial–axial relationships to H–2 and H–6_{ax}. Because these two couplings are not equal, the double-triplet (1:1:2:2:1:1) pattern is distorted, widening the two center peaks and making them shorter (less than twice the height of the four outer peaks). This is an example of an unresolved splitting: we should be seeing eight peaks, but we see only six because the separation of the third and fourth peaks (and of the fifth and sixth) is comparable to the peak width. This separation is about 1.0 Hz (10.9–9.9) and the peak width (measured at half-height) of the outer peaks is 1.3 Hz. Later on we will see how resolution enhancement can be used to make the peaks sharper and at least begin to see the separation of this multiplet into eight peaks. The third coupling of the double-doublet-doublet (ddd) is 4.3 Hz, due to the interaction with H-6_{eq}. This coupling is axial–equatorial (*gauche* relationship), so it is smaller, in the middle range of observed couplings.

The peak at 2.14 ppm is a double septet, with an intensity ratio 1:1:6:6:15:15:20:20:15:15:6:6:1:1 and J couplings of 7.0 Hz for the septet and 2.6 Hz for the doublet. The only proton with six coupling partners is the CH proton of the isopropyl group, H-7. A J coupling near 7.0 Hz is typical of a vicinal coupling with free rotation (of the methyl group) averaging the dihedral angle effects. The additional coupling of 2.6 Hz is due to its interaction with H-2. The outer peaks of the septet are only one twentieth of the intensity of the center peaks, so unless you have very good signal-to-noise you might miss these peaks and mistake it for a quintet. The intensity ratio for this "quintet" is 1:2.5:3.3:2.5:1, instead of the expected 1:4:6:4:1. The remaining resolved single-proton peaks ("l" and "h") cannot be assigned without advanced experiments. The strong, sharp peaks at the right-hand side of the spectrum correspond to the methyl groups. All three methyl groups are attached to CH carbons ("methine" carbons) so they will appear as doublets. One doublet ("a") is separate

EXAMPLES: NMR SPECTROSCOPY OF OLIGOSACCHARIDES AND TERPENOIDS 23

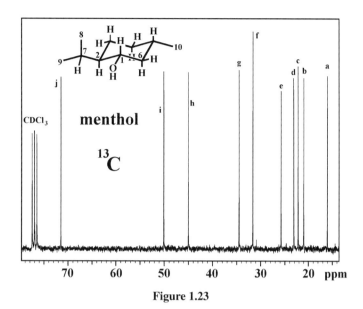

Figure 1.23

from the other two ("b" and "c"), but we cannot make the assumption that it represents the "lone" methyl group H-10. Because this is a chiral molecule, the isopropyl group can have distinct environments and widely different chemical shifts for the two methyls. The J couplings for these three doublets are all around 7.0 Hz due to free rotation of the C–C bond, although one is slightly lower (6.6 Hz) and this corresponds to the CH–C$\underline{H}_3$ group, C-10. Chemical shifts for the methyl groups are a bit less than 1 ppm, typical for methyl groups in a saturated hydrocarbon environment, far from any functional group. The same is true for the four proton signals buried in the overlapped region between 0.75 and 1.15 ppm: they are shifted downfield of the methyl groups slightly because of the higher degree of substitution (CH$_2$ and CH), but they are not close to any functional group.

The ^{1}H-decoupled ^{13}C spectrum of menthol (Fig. 1.23) has ten peaks in addition to the three solvent peaks. All we can say about it is that the most downfield peak ("j") corresponds to the carbon with the alcohol oxygen: C-1. We can see a bit of a gap between this peak and the rest of the peaks, and we expect singly oxygenated sp^3 carbons in the range 50–90 ppm, with methine carbon (CHOH) typically in the range 70–80 ppm. Every time we replace an H with C we add about 10 ppm to the chemical shift, so compared to CH$_3$O (50–60 ppm) we can add about 20 ppm to get the range of CHOH. The rest of the carbons can only be assigned if we can assign the attached protons and then correlate the ^{13}C shifts with the ^{1}H shifts by a 2D spectrum such as HSQC.

1.2.9 Cholesterol

Cholesterol (Fig. 1.20) is a steroid, the same rigid five-ring backbone used for the mammalian sex hormones. There are only two functional groups: an olefin (C-5, C-6) and an alcohol (C-3). The bulk of the molecule can be described as saturated hydrocarbon. There are five methyl groups: two are attached to quaternary carbons so they should appear as singlets; and three are attached to CH carbons so they should appear as doublets. Most of the protons in the A, B, and C rings can be described as "axial" or "equatorial" due to the rigid,

24 FUNDAMENTALS OF NMR SPECTROSCOPY IN LIQUIDS

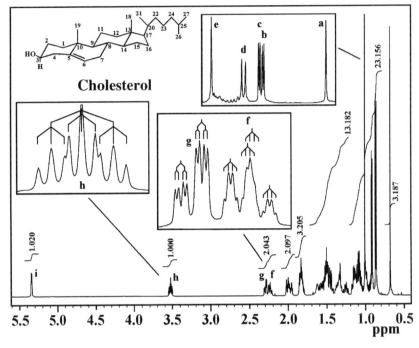

Figure 1.24

locked cyclohexane ring structure. The 600 MHz ^{1}H spectrum is shown in Figure 1.24. The total integration adds up to 48.89 protons, a bit high for the molecular formula $C_{27}H_{46}O$ but consistent with the fact that the resolved peaks in the upfield part of the spectrum integrate several percent above the expected integer values. The olefin functional group (C-5 and C-6) has a single proton, H-6, which we expect in the region 5–6 ppm. Thus the one-proton signal at 5.35 ppm (peak "i") can be assigned to H-6 (only the resolved peaks are identified with letters). The other functional group is an alcohol, and we expect the H–C–OH proton at 3–4 ppm; we can assign the one-proton signal at 3.52 ppm (peak "h") to H-3. The splitting pattern of H-3 can be described as a triplet of triplets, with a small triplet coupling of 4.6 Hz and a large triplet coupling of 11.2 Hz (Fig. 1.24, proton h inset). Because the OH group is equatorial, H-3 is axial and is split by its two equatorial neighbors, H-2$_{eq}$ and H-4$_{eq}$. Because both the relationships are axial–equatorial (*gauche*), the couplings are identical and in the medium range (4.6 Hz). H-3 is also split by its two axial neighbors, H-2$_{ax}$ and H-4$_{ax}$. Each of these relationships is axial–axial (*anti*), so the couplings are identical and large (11.2 Hz). Taken together, we get a large triplet (1:2:1 intensity ratio, $J = 11.2$), with each of the three arms split into a smaller triplet (1:2:1 ratio, $J = 4.6$). These coupling relationships are shown in the partial structure in Figure 1.25.

Moving from left to right, the next resolved peak is a two-proton multiplet at 2.20–2.32 ppm (peaks "g" and "f"). The most likely assignment for these peaks would be H-4$_{ax}$ and H-4$_{eq}$, since C-4 lies between the two functional groups and we expect the minor downfield-shifting effects of both groups to add together, pulling the H-4 resonances out of the "pack" of saturated hydrocarbon peaks (0.6–1.7 ppm). We cannot be absolutely sure of this assignment until we see two-dimensional data, but this is a reasonable guess. Looking at the fine structure of these two peaks (inset, Fig. 1.24) and ignoring the smaller couplings, we

EXAMPLES: NMR SPECTROSCOPY OF OLIGOSACCHARIDES AND TERPENOIDS 25

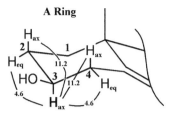

Figure 1.25

see a triplet on the right (peak "g") and a doublet on the left (peak "f"). These two peaks are "leaning" toward each other, with the outer peaks reduced in intensity and the inner peaks increased relative to a "standard" doublet (1:1) or triplet (1:2:1). This distortion of peak intensities is a common feature when the chemical shift difference (in Hz) is relatively small compared to the J coupling between the two protons. In this case, the chemical shift difference is 0.053 ppm or 32 Hz and the large geminal ($^2J_{HH}$) coupling is 13.0 Hz, leading to a large distortion of peak intensities. The basic doublet and triplet patterns are further split by smaller couplings: each side of the doublet is split into a double doublet ($J = 5.0$ and 2.1 Hz) and each of the three peaks of the triplet on the right is split into a quartet ($J = 2.8$ Hz). Ignoring the "small" couplings, we can ask how many large couplings each proton experiences and in this way count the number of geminal and axial–axial relationships. The "doublet" peak ("g") has only the geminal ($^2J_{HH}$) coupling, which is always large for saturated (sp^3 hybridized) carbons. So it must be the equatorial proton, H-4$_{eq}$. The "triplet" peak ("f") has the geminal coupling and one axial-axial coupling, so it must be the axial proton, H-4$_{ax}$, which has an axial–axial coupling to H-3. The smaller couplings can be explained as follows: H-4$_{eq}$ has one equatorial–axial coupling (5.0 Hz) to H-3$_{ax}$ and one "W" coupling ($^4J_{HH}$) to H-2$_{ax}$ (2.1 Hz). A "W" coupling occurs in a series of saturated carbons when the H–C–C–C–H network is rigidly aligned in a plane in the form of a "W." H-4$_{ax}$ has small long-range couplings to H-6, H-7$_{ax}$, and H-7$_{eq}$, all around 2.8 Hz. These long-range couplings will be discussed later, but you can think of the C=C double bond as a kind of "conductor" for J couplings that allows these small interactions to occur over four or five bonds as long as the double bond is in the path: H–C–C=C–H ("allylic coupling") and H–C–C=C–C–H ("bis-allylic" coupling). In each case, if you remove the C=C from the path, you have a close bonding relationship of two bonds ("geminal") or three bonds ("vicinal").

The five methyl groups of cholesterol give rise to tall, sharp peaks in the upfield region of the ^{1}H spectrum (inset, Fig. 1.24, peaks a–e). We can see two singlet methyl signals ("a" and "e") that correspond to the "angular" methyls attached to the quaternary carbons at the A-B and C-D ring junctures (C-18 and C-19). Later on we will use an NOE experiment to assign these two peaks specifically, taking advantage of the proximity of CH$_3$-19 to the H-4$_{ax}$ proton. There are also three doublet methyl signals ("b," "c," and "d") that correspond to the three methyl groups in the side chain attached to CH carbons: C-21, C-26, and C-27. Specific assignments for these signals will require two-dimensional experiments such as HSQC and HMBC.

The 125 MHz ^{1}H-decoupled ^{13}C spectrum of cholesterol is shown in Figure 1.26. Because the ^{13}C nuclear magnet is only about one fourth as strong as the ^{1}H nuclear magnet, the ^{13}C resonant frequency is always about one fourth of the ^{1}H frequency in the same magnetic field. Thus on a "500 MHz" NMR spectrometer (i.e., an 11.74 T B_o field in which ^{1}H resonates at 500 MHz) the ^{13}C frequency is about 125 MHz. The CDCl$_3$ peaks (a 1:1:1

26 FUNDAMENTALS OF NMR SPECTROSCOPY IN LIQUIDS

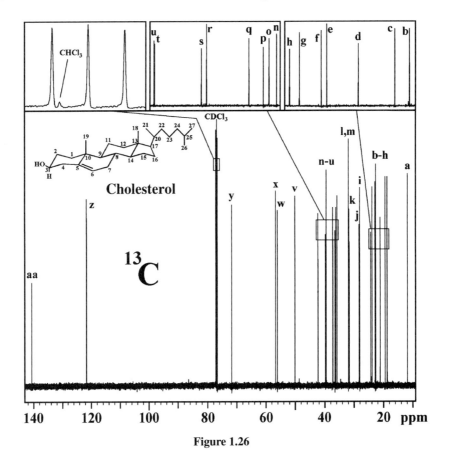

Figure 1.26

triplet at 77.0 ppm) appear at the center of the spectrum. Note that there is a small peak due to $CHCl_3$ at 77.21 ppm (upper left inset, Fig. 1.26). This may be residual $CHCl_3$ in the $CDCl_3$ (0.2%) or $CHCl_3$ residue in the solid cholesterol sample. Such a small amount of $CHCl_3$ is visible in the spectrum due to the effects of relaxation and decoupling. Because the ^1H nuclear magnet is about seven times stronger than the ^2H nuclear magnet, the ^{13}C in $CHCl_3$ relaxes faster than the ^{13}C in $CDCl_3$ and thus gives a stronger NMR peak. In addition, due to ^1H decoupling there is only one peak for $CHCl_3$, and this makes for a taller peak than these for $CDCl_3$, whose ^{13}C intensity is divided into three peaks. Note also that there is a deuterium isotope effect on the ^{13}C chemical shift: $CHCl_3$ appears 0.21 ppm downfield of $CDCl_3$.

In addition to these solvent peaks, we can count 26 peaks in the spectrum. Because there are 27 carbons in the cholesterol molecule (three are lost in the biosythesis from a triterpene precursor), there must be one peak that accounts for two carbons. The tallest peak (labeled "l, m") in fact corresponds to two different carbons with nearly identical chemical shifts. The most downfield peaks ("aa" and "z") are in the olefin/aromatic region of the ^{13}C spectrum (120–140 ppm), so they must correspond to C-5 and C-6. Peak "aa" is less intense ("shorter") than all of the other peaks because of slow relaxation: it must be a quaternary carbon. We will see that the proximity of protons is the primary means of relaxation of ^{13}C nuclei, so carbons lacking a proton relax much more slowly and give less intense peaks, especially if the relaxation delay is short (in this case the recycle delay was only 1.74 s (1.04 s acquisition time and 0.7 s relaxation delay). So we can assign peak "aa" (140.75) to C-5 and peak "z" (121.69) to C-6. Note also that the more substituted carbon, C-5 (three bonds to carbon) is shifted downfield relative to C-6 (two bonds to carbon) due to the steric crowding effect.

Peak "y" (71.78 ppm) is in the "alcohol" region (C–O) in the range expected for methine carbon (CH–O), so it can be assigned to C-3. The next three carbons (peaks "x," "w," and "v," 50–57 ppm) could be methoxy (CH_3O) groups, but because we have accounted for all the functional groups of cholesterol they must be either close to these functional groups (inductive effect) or shifted downfield due to steric crowding. The inductive effect (electron withdrawing and donating groups) is most important for ^1H chemical shifts, so let us consider the steric effects. The most sterically crowded carbons in the cholesterol structure are the methine (CH) groups next to an sp^3-hybridized quaternary carbon: C-9, C-14, and C-17. These three carbons account for this group of downfield-shifted peaks. The rest of the ^{13}C peaks (a–u) lie in the region of saturated hydrocarbon (sp^3 carbon with no functional groups) and cannot be assigned without more advanced experiments such as DEPT and 2D HSQC/HMBC.

1.3 TYPICAL VALUES OF CHEMICAL SHIFTS AND COUPLING CONSTANTS

1.3.1 Typical Values of ^1H Chemical Shifts

The chemical shift scale can be roughly divided into regions that correspond to specific chemical environments (olefinic, aromatic, etc.). Knowing these regions gives you a useful first guess as to the interpretation of a resonance, but you must keep in mind that more than one functional group might contribute in an additive fashion to the chemical shift. For example, we can estimate the chemical shift of a CH_2 group situated between an olefin and a carbonyl group (C=C–CH_2–C=O) as follows: A CH_3 group next to an olefin or carbonyl resonates at 2.1 ppm (see below under "b"). This represents a downfield shift of 1.25 ppm from a "hydrocarbon" CH_3 group (0.85 ppm, under "a" below). Thus we can estimate the shift for this CH_2 as follows:

    1.2 CH_2 in hydrocarbon environment
+2.5   effect of neighboring C=C or C=O (+1.25 ppm) times 2
    3.7   total: predicted chemical shift of C=C–CH_2–C=O

If we saw a resonance at 3.7 ppm, our first guess would be a proton on a singly oxygenated carbon, -CH_2–O- (part "d" below), but it is dangerous to get "locked into" that idea because the possibility exists of smaller effects adding together, as shown in the example above.

(a) *"Hydrocarbon"*: attached to an sp^3-hybridized carbon and many bonds away from any unsaturation or electronegative atom. The same differences between methyl, methylene, and methine are observed in all other environments.
  1. CH_3 0.85,
  2. CH_2 1.2,
  3. CH 1.8.
(b) α to a carbonyl, olefin, or aromatic group: H–C–C=O or H–C–C=C: 2.1.
(c) *Next to a nitrogen*: H–C–N (attached to an sp^3-hybridized carbon with one single bond to nitrogen): 2.6.

(d) *Next to an oxygen*: H–C–O (attached to an sp^3-hybridized carbon with one single bond to oxygen):
   1. alcohol or ether (H–C–OH or H–C–O–C): 3.3
   2. ester (H–C–O–CO–R): 3.8.
(e) *"Olefinic"*: H–C=C: 5–6 ppm (where C=C is not part of an aromatic ring). Resonance effects can shift out of this range: up to 1 ppm upfield for electron-donating groups (e.g., H–C=C–O–) and 1 ppm downfield for electron-withdrawing groups (e.g., H–C=C–C=O). This is a result of increased or decreased electron density at the carbon bearing the proton in resonance structures such as H–C$^-$–C=O$^+$– (electron donation: vinyl ether) and H–C$^+$–C=C–O$^-$ (electron withdrawal: α,β-unsaturated ketone).
(f) *"Anomeric"*: H–C(–O)–O (attached to an sp^3-hybridized carbon that has two single bonds to oxygen): 5–6 ppm.
(g) *"Aromatic"*: attached to carbon of a benzene, furan, pyrrole, pyridine, indole, naphthalene, and so on, ring: generic 7–8 ppm. The effect of substituents due to resonance effects (strongest at *ortho* position):
   1. electron-rich carbon (e.g., *ortho* or *para* to O or N of phenol, aniline, phenolic ether, or in an electron-rich heteroaromatic: pyrrole, furan): 6–7 ppm;
   2. electron-poor carbon (e.g., *ortho* or *para* to C=O or NO$_2$, or in the two or four position of pyridine): 8–9 ppm.
(h) *Aldehyde*: H–C=O: 10 ppm.
(i) Carboxylic acid: HO–C(=O) or *phenolic*: HO–C(aromatic): 12–14 ppm.

Note that there are other types of protons not listed here that can fall into the same chemical shift ranges listed above. The above categories are simply the most common ones. Also, through-space ("anisotropic") effects of unsaturated groups (C=C, C=O, and aromatic rings) can change chemical shifts from the above categories in ways that depend on conformation.

1.3.2 Typical Values of ^{1}H–^{1}H Coupling Constants (J)

A superscipt preceding the letter J refers to the number of bonds between the two nuclei: ^3J means three-bonds or vicinal (H–C–C–H) and ^2J means two bonds or geminal (H–C–H). Sometimes a subscript is used to clarify which types of nuclei are coupled: J_{HH} means proton-to-proton coupling.

(a) $^3J_{HH}$ (vicinal):
   1. In freely rotating alkyl groups (e.g., CH$_3$–CH$_2$–): 7.0 Hz
   2. In benzene rings: $^3J_{HH} = 7.5$, $^4J_{HH} = 1.5$, $^5J_{HH} = 0.7$ Hz
   3. In a pyridine ring: $J_{2,3} = 5.5$, $J_{3,4} = 7.6$, $J_{3,5} = 1.6$, $J_{2,5} = 0.9$, $J_{2,6} = 0.4$ Hz
   4. In a furan (pyrrole) ring: $J_{2,3} = 1.8$ (2.6), $J_{3,4} = 3.4$ (3.5), $J_{2,4} = 0.9$ (1.3), $J_{2,5} = 1.5$ (2.1) Hz
   5. In a chair cyclohexane ring: $J_{1,2} = 12$ (ax-ax), 3 Hz (eq–ax or eq–eq)
   6. In a chair six-membered ring sugar, $J_{1,2}$(eq–ax) $= 4$, $J_{1,2}$(ax–ax) $= 9$ Hz.
   7. In an isolated olefin C$_1$H-C$_2$H=C$_3$H-C$_4$H: $J_{2,3} = 8$–12(cis), 14–17(trans), $J_{1,2} = 7$ Hz
   8. In a cyclopropane, 7–13(cis), 4–9 Hz (trans)

(b) $^2J_{HH}$ (geminal):
   1. In a terminal olefin C=CH$_2$, $J_{1,1} = 0$–2 Hz
   2. On a saturated (sp^3) carbon: 12–15 Hz (12.5 in a cyclohexane chair)
(c) Long-range ($^4J_{HH}$ and $^5J_{HH}$):
   1. Isolated olefin C$_1$H–C$_2$H=C$_3$H–C$_4$H: $J_{1,3}$ and $J_{2,4}$ ("allylic") 0–3; $J_{1,4}$ ("bis-allylic") 1–2 Hz
   2. "W" coupling (saturated chain in rigid planar W conformation):

$$\begin{array}{c} \diagup C \diagdown \diagup C \diagdown \\ H \quad\quad C \quad\quad H \end{array} \text{ or } \begin{array}{c} \diagup C \diagdown \diagup CH_3 \\ H \quad\quad C \end{array}$$

   1–4 Hz (2.5 in a cyclohexane chair: $J_{1,3}$ eq–eq)

1.3.3 Typical Values of ^{13}C Chemical Shifts

^{13}C chemical shifts are more sensitive to steric crowding effects and less sensitive to through-space effects of double bonds than ^{1}H chemical shifts. Increasing the substitution of a carbon (CH$_3$ to CH$_2$ to CH to C) leads to downfield shifts of about 10 ppm in each step.

(a) *Carbonyl* (C=O) shifts are far downfield (155–210 ppm) and the peaks are generally weak due to slow relaxation of quaternary carbons (except aldehydes, which are not quaternary). Ketones and aldehydes: 200–210 (isolated), 190–200 (α,β unsaturated), Carboxylic acids, esters, amides: 170–180, Urethanes (NC(O)O): 150–160.

(b) *Aromatic* carbons are typically 120–130 ppm for unsubstituted positions (i.e., CH) and 136–150 at the position of alkyl substitution (weak quaternary peak). Strong electron-withdrawing groups (O, N, NO$_2$, F) can shift the substituted (*ipso*) carbon to 150–160. Substituents that can donate to the ring by resonance (O, N) shift the *ortho* carbons and, to a lesser extent, the *para* carbon upfield to 110–120. Likewise, substituents that are electron-withdrawing by resonance (CO, CN) shift the *ortho* and *para* carbons downfield to 130–140. *Meta* carbons are unaffected because resonance structures cannot place + or − charges at these positions. Nitro (NO$_2$) is unusual in that it shifts the *ortho* carbon upfield about 5 ppm and the *para* carbon downfield about 6 ppm. At the point of attachment of the substituent ("*ipso*" carbon) the range is 130–140 for "neutral" substituents and farther downfield (150–160) for electron-withdrawing substituents (e.g., O).

(c) *Nitrile* (CN with triple bond): 110–120.

(d) *Olefinic* carbons (isolated C=C) fall in the same range as aromatic CH: 120–130. Substitution pulls this value downfield: a quaternary olefinic carbon resonates in the range of 140 ppm. They can also be shifted by resonance effects when electron withdrawing or donating groups are attached, just like in aromatic systems. For example, a quaternary β carbon of an α, β-unsaturated ketone resonates in the 170–180 ppm range, making it easy to confuse with an ester carbonyl carbon. This is due to the resonance structure: $-C_\beta^+-C_\alpha=C-O^-$.

(e) *Anomeric* carbons of sugars (O–C–O) and in acetals and ketals: 90–110.

(f) *Singly oxygenated* carbons (C–O single bond): 50–85. CH$_2$OH carbon is in the upfield range (60–70) and quaternary carbons in the downfield range (75–85). A methoxy group (CH$_3$O) is even farther upfield: 50–60.

(g) Carbons with a single bond to *nitrogen*: 50–70.

(h) *Saturated* carbons with no nearby electronegative atoms or double bonds: 10–50, with CH_3 on the upfield side and quaternary carbons on the downfield side. Strained rings (cyclobutane, cyclopropane) show significant upfield shifts.

1.4 FUNDAMENTAL CONCEPTS OF NMR SPECTROSCOPY

1.4.1 Spin

The atomic nucleus can be viewed as a positively charged sphere that is spinning on its axis. This spin is an inherent property of the nucleus, and because charge is being moved it creates a small magnetic field aligned with the axis of spinning. Thus we can consider the nucleus as a tiny, permanent bar magnet. Because different isotopes of a given atom (e.g., ^{12}C, ^{13}C, ^{14}C) have different numbers of neutrons in the nucleus, they have different magnetic properties. For this reason we only talk about specific isotopes in NMR: ^1H, ^{19}F, ^{11}B, and so on, and our attention is focused on the nucleus of these isotopes.

The nucleus of each isotope has the following intrinsic properties:

1. Magnetogyric ratio, γ. This is essentially the strength of the nuclear magnet. Different nuclei have different magnet strengths; for example, the ^{13}C nuclear magnet is only one-fourth as strong as the ^1H nuclear magnet, and the ^{15}N nuclear magnet has only one-tenth of the strength of the ^1H magnet. The γ is the same for every nucleus of a given type (e.g., ^{19}F), regardless of its position within a molecule.

2. "Spin." This determines the number of quantum states available for the nucleus.

spin-0	no magnetic properties	
spin-½	2 states:	1/2, −1/2
spin-1	3 states:	1, 0, −1
spin-³⁄₂	4 states:	3/2, 1/2, −1/2, 3/2
etc.		

For example, a spin-½ nucleus can be viewed as having two quantum states: one with the spin axis at a 45° angle to the external magnetic field and one with the spin axis at a 135° angle to the external field. A spin-1 nucleus can be viewed as having three possible states: 45°, 90°, and 135°. In this book we will be concerned primarily with spin-½ nuclei.

Here are some examples showing the composition of the nucleus (p = protons, n = neutrons):

Spin-0	spin-½	spin > 1/2
^{12}C (6p + 6n)	^1H (1p + 0n)	^2H (1p + 1n)
^{16}O (8p + 8n)	^3H (1p + 2n)	^{14}N (7p + 7n)
^{18}O (8p + 10n)	^{13}C (6p + 7n)	^{17}O (8p + 9n)
	^{15}N (7p + 8n)	
	^{19}F (9p + 10n)	
	^{29}Si (14p + 15n)	
	^{31}P (15p + 16n)	

Note that there is a pattern: Nuclei with an even number of protons and neutrons (even–even) have spin zero; "odd–even" and "even–odd" nuclei tend to be spin-½; and "odd–odd" nuclei tend to have a spin greater than 1/2. This is just a rule of thumb (e.g., ^{17}O violates the "rule"). Nuclei with spin greater than 1/2 are more difficult to observe than spin-½ nuclei because they have a "nuclear quadrupole moment" that makes their NMR peaks very broad. For this reason, most NMR work is focused on the spin-½ nuclei. Because NMR is usually done in deuterated solvents (D$_2$O, CD$_3$OD, etc.), we will have to occasionally consider the effects of a spin-1 (three quantum states) nucleus.

1.4.2 Precession

When we place a spin-½ nucleus in a strong external magnetic field, the nucleus wants to align itself with the magnetic field, just like a compass needle moves to align with the earth's magnetic field. But because the nucleus is spinning (i.e., it has an intrinsic property of angular momentum), it cannot simply change its angle with the magnetic field from 45° to 0°. The torque it experiences from the external magnetic field instead causes the spin axis to "wobble" or precess around the magnetic field direction. This is analogous to a spinning top or gyroscope, which responds to the torque produced by the earth's gravitational field by describing a circle with its spin axis. The precession rate of the nucleus in a magnetic field is the *resonant frequency* referred to in the name "nuclear magnetic resonance." The precession rate is in the range of radio frequency, tens or hundreds of megahertz, or millions of rotations per second. In this classical model the torque exerted on the nucleus is proportional to both the laboratory magnetic field strength, B_o, and to the strength of the nuclear magnet, γ. The rate of precession is proportional to the torque, so we have:

$$\omega_o = 2\pi \nu_o = \gamma B_o$$

It cannot be emphasized too much that the resonant frequency in NMR is proportional to the magnetogyric ratio, γ, and to the laboratory magnetic field strength, B_o. This relationship forms the basis of nearly every phenomenon observed in NMR. There are two ways to measure the precession rate: the angular velocity, ω_o, in units of radians per second and the frequency, ν_o, in units of cycles per second or hertz. In this book we will use frequencies in hertz. This frequency is sometimes called the Larmor frequency, and the zero subscript refers to this fundamental frequency, which results from the laboratory magnetic field interacting with the nucleus' magnetic field.

As an example, consider a proton (^{1}H nucleus) in a 7.05 T laboratory magnetic field:

$$\gamma_H = 2.675 \times 10^8 \text{ T}^{-1} \text{ rad s}^{-1}$$

$$B_o = 7.05 \text{ T}$$

$$\nu_o = \gamma B_o/2\pi = 3.001 \times 10^8 \text{ Hz} = 300.1 \text{ MHz}$$

Such a magnet would be called a "300 MHz" magnet because the ^{1}H nucleus precesses at a rate of 300 MHz in this magnet. NMR magnets are almost never described in tesla but rather by their ^{1}H resonance frequency. This can be confusing because if you are observing ^{13}C nuclei on a 500-MHz NMR instrument, you are operating at a resonant frequency of 125 MHz, not 500 MHz. Because the resonant frequency for a given magnet (NMR magnets have a fixed magnetic field strength) is proportional to the magnetogyric ratio, γ, of the nucleus being observed, the NMR frequencies for different nuclei will always be in the same ratio: the ratio determined by their relative γ values.

	$B_o = 7.05$ T	$B_o = 11.74$ T	γ/γ_H(%)	γ_H/γ	Abund. (%)
^{1}H	300.0 MHz	500.0 MHz	100.0	1.000	99.98
^{2}H	46.05	76.75	15.35	6.515	0.015
^{13}C	75.43	125.72	25.14	3.977	1.11
^{15}N	30.40	50.66	10.13	9.870	0.37
^{19}F	282.23	470.39	94.08	1.063	100
^{31}P	121.44	202.40	40.48	2.470	100

Proton (^{1}H) is the king of the nuclei (radioactive tritium, ^{3}H, is actually 6.7% stronger) and all other nuclei can be viewed in terms of their magnet strength (γ) relative to proton. ^{19}F is a bit weaker than proton (94%) and ^{31}P is about 40% of the proton frequency. Proton is about four times stronger than ^{13}C, seven times stronger than ^{2}H, and 10 times stronger than ^{15}N. Of all these spin-½ nuclei, three have very low natural abundance: 1.11% for ^{13}C, 0.37% for ^{15}N, and 0.015% for ^{2}H. This makes them difficult to observe because the signal strength in NMR is proportional to the number of NMR-active nuclei in the sample: for ^{13}C, only one in every 100 carbon atoms is participating in the NMR experiment. However, we will see that there are advantages to having a "dilute" nucleus—one that is "sprinkled" lightly over the collection of molecules in the sample. We can improve on nature by isotopically labeling or enriching the sample either by synthesis from labeled starting materials or by biosynthesis on labeled growth media. Many compounds can be purchased with nearly 100% abundance of ^{13}C, ^{2}H, or ^{15}N either at one site in the molecule or at all sites ("uniformly labeled"). This can be very costly, but the benefits often justify the cost. For example, with uniform ^{13}C labelling, the ^{13}C signal can be increased by a factor of 100, reducing the experiment time by a factor of 10,000. It should be noted that all three of these isotopes are stable, that is, they are not radioactive.

1.4.3 Chemical Shift

Because at any given field strength each nucleus has a characteristic resonant frequency, we can "tune" the radio dial to any nucleus we are interested in observing. We can think of the various NMR-active nuclei in the sample as "radio stations" that we can tune into very accurately, just as stations come into tune in a very narrow range of frequencies on an FM radio. Having chosen a "station" to listen to, what can we learn by observing a particular type of nucleus? The resonant frequency is always, always, always proportional to the magnetic field:

$$\nu_o = \gamma B_o / 2\pi$$

but the exact magnetic field *experienced by the nucleus* may be slightly different than the external magnetic field. The nucleus is located at the center of a cloud of electrons, and we know that electrons are easily pulled away or pushed toward an atom, changing the electron density around that nucleus. Furthermore, electron clouds can begin to circulate under the influence of the laboratory field, creating their own magnetic fields, which subtract from or add to the external field. So the nucleus "feels" a slightly different field, depending on its position within a molecule (its "chemical environment"):

$$B_{\text{eff}} = B_o(1 - \sigma) \quad \nu_o = \gamma B_{\text{eff}} = \gamma B_o(1 - \sigma)$$

Where σ is a "shielding constant" in units of parts per million, which reflects the extent to which the electron cloud around the nucleus "shields" it from the external magnetic field. These differences, which we call "chemical shifts" are really tiny: for a ^{1}H nucleus the "spread" of resonant frequencies around the fundamental frequency is only about 10 ppm. That means that on a 500 MHz NMR instrument, the protons in a molecule might have a range of resonant frequencies between 499.9975 and 500.0025 MHz (0.0025 MHz is 5 ppm of 500 MHz), depending on their location within the molecule. Thus we tune in to a "station" (499.9975–500.0025 MHz) and study the tiny variations (chemical shifts) in resonant frequency to learn something about the chemical structure of the molecule. In this way, physics (and radio electronics) comes to the aid of chemistry in helping us determine a molecule's structure. An NMR spectrum is just a graph of intensity versus frequency for the narrow range of frequencies corresponding to the particular nucleus we are interested in. Each "peak" in this graph corresponds to a particular environment within the molecule, such as a particular hydrogen atom position in an organic structure. When each position in a molecule has a different chemical shift, we can "talk" to these atoms individually in NMR experiments, looking around at the local environment from the point of view of one atom in the structure at a time.

1.4.4 The Energy Diagram

If we consider the energy of a nucleus as it interacts with the external magnetic field, we see that there are two energy levels for a spin-½ nucleus. The "aligned" state (or α state) has the nuclear magnet aligned with the laboratory field, giving it a lower energy (more stable) state (Fig. 1.2). The "disaligned" state (or β state) is aligned opposite to the external field, resulting in a higher energy. The energy "gap" between these two levels is:

$$\Delta E = h\nu_o = h\gamma B_o/2\pi$$

where h is Planck's constant and ν_o is the Larmor ("resonant") frequency. This relationship between the energy gap between two quantum states and the frequency of electromagnetic radiation ("photons"), which can excite a particle from the lower energy level to the higher one, is fundamental to all forms of spectroscopy. The Larmor frequency, ν_o, is the same as the rate of precession of the spinning nucleus in the classical model (Fig. 1.1). Note that the size of the energy gap is proportional to the strength of the nuclear magnet (γ) and also to the strength of the laboratory magnetic field (B_o). Much effort and expense is put into getting the largest possible energy gap, as we design and build bigger and stronger superconducting magnets for NMR. We will see that a larger energy gap results in a more sensitive NMR experiment and better separation of the resonant frequencies of like nuclei in different chemical environments.

1.4.5 Populations

In an NMR sample there are a very large number of identical spins, a number approaching Avogadro's number. Even though there may be different types of spins (^{1}H, ^{13}C, ^{15}N, etc.)

within a molecule and different environments (H_1, H_2, H_3, etc.) within a molecule for each type of spin, we can view each molecule in a sample of a pure compound as identical and experiencing the same magnetic field. This is because the magnetic field has a very high degree of spatial homogeneity (on the order of parts per billion variation in B_o) and each molecule is tumbling very rapidly and has no preferred orientation in the magnetic field. Let us focus on one type of nucleus (^1H) and one position within the molecule (H_2). If there are N molecules in the sample (e.g., for a 1 mM sample, $N = 3 \times 10^{17}$), then we can talk about the N ^1H nuclei at position H_2 in the molecule: each one will be either aligned with the B_o field (lower energy or α state) or disaligned with the B_o field (higher energy or β state). At thermal equilibrium, there will be a tendency for the spins to prefer the lower energy state, but because the energy difference ($\Delta E = h\gamma B_o/2\pi$, where h is Planck's constant) is small compared to the average energy available at room temperature (kT), the populations are very nearly equal in the α and β states. The population of the more stable α state is $N/2 + \delta$, and the population of the less stable β state is $N/2 - \delta$, where δ is a very small number roughly equal to $N\Delta E/4kT$.

For example, at 7.05 T magnetic field (a 300 MHz NMR instrument) and 25 °C, the population difference for protons is 0.00064% of the number of nuclei N. This equilibrium population difference is a constant throughout the NMR experiment and, as we perturb the equilibrium, the spins will always try to return to this equilibrium population distribution. Because the measureable signal from a nucleus in the β state is exactly cancelled by the signal from a nucleus in the α state, it is this population difference that is the only material we have to work with and to detect in the NMR experiment. Because the difference is so small, the sensitivity of NMR is in many orders of magnitude lower than all other analytical techniques; so low, in fact, that NMR is not considered a branch of "analytical chemistry" but rather a tool used by organic chemists and biologists.

1.4.6 Net Magnetization at Equilibrium

At thermal equilibrium, the Boltzmann distribution determines the populations in various energy levels. For any two quantum states, the ratio of populations between the higher energy state and the lower energy state at equilibrium will always be:

$$P_\beta/P_\alpha = e^{-\Delta E/kT}$$

where k is the microscopic gas constant, T is the absolute temperature in kelvin (K), and ΔE is the difference in energy between the two states—the "energy gap." We can think of kT as the average amount of total energy that a molecule has—analogous to the amount of money the average person is carrying in his or her pocket. ΔE is analogous to the price difference between a hamburger and a cheeseburger. If the amount of money the average person has (kT) is very small and the price difference (ΔE) is large, then nearly everyone will take the hamburger. But if the average person is carrying around a lot of money and the price difference is very small, there will be only a very slight preference for the hamburger. Just how big is kT compared to the energy difference in NMR? At 25 °C (298 K), kT is equal to 2478 J/mol. For a proton (^1H) in a 7.05 T magnetic field ($\nu_o = 300$ MHz), the energy gap is:

$$\Delta E = h\nu_o = \hbar(\gamma B_o) = 0.0315 \, \text{J/mol}$$

So the energy gap is very, very small compared to the average energy that a molecule has at room temperature. Another way of saying this is that $\Delta E/kT$ is a number much, much less than 1. The exponential function can be simplified by approximation if the argument is a very small number compared to 1:

$$e^{-x} \sim 1 - x, \quad \text{if } x \ll 1$$

We can now simplify the Boltzmann equation:

$$P_\beta/P_\alpha = e^{-\Delta E/kT} \sim 1 - \Delta E/kT$$

The population difference, $P_\alpha - P_\beta$, is the most interesting thing for us because the magnetism of every "up" nuclear magnet cancels the magnetism of every "down" nuclear magnet, and it is only the difference in population that results in a "net magnetization" of the sample.

$$P_\beta/P_\alpha = 1 - \Delta E/kT; \quad 1 - P_\beta/P_\alpha = \Delta E/kT; \quad P_\alpha/P_\alpha - P_\beta/P_\alpha = \Delta E/kT$$

$$(P_\alpha - P_\beta)/P_\alpha = \Delta E/kT; \quad P_\alpha - P_\beta = P_\alpha \Delta E/kT = N\Delta E/2kT$$

The last equality is obtained by substituting $N/2$ for P_α because both P_α and P_β are very close to half the total number of spins in the sample. Finally, substituting $\hbar\gamma B_o$ for ΔE we obtain:

$$P_\alpha - P_\beta = N\hbar\gamma B_o/2kT$$

Thus the population difference is proportional to the total number of spins in the sample and to the strength of the nuclear magnet (γ) and inversely proportional to the absolute temperature (T). If we add together all of the nuclear magnets, each spin in the β state cancels one in the α state and we end up with only $P_\alpha - P_\beta$ spins in the α state, aligned with the magnetic field. These add together to give a *net magnetization*, which is equal to the net number of spins pointing "up" times the magnet strength of each individual spin, γ. The magnitude of this net magnetization is called M_o,

$$M_o = \gamma(P_\alpha - P_\beta) = N\hbar\gamma^2 B_o/2kT$$

The net magnetization of the sample at equilibrium is proportional to the amount of sample (N), the square of the nuclear magnet strength (γ^2), and the field strength (B_o), and inversely proportional to the absolute temperature (T).

1.4.7 Absorption of Radio Frequency Energy

In order to measure the resonant frequency of each nucleus within a molecule, we need to have some way of getting the nuclei to absorb or emit RF energy. If we subject the sample to an oscillating magnetic field provided by a coil (the equivalent of a radio transmitter's antenna), a spin in the lower energy state can be "bumped" into the higher energy state if the radio frequency is exactly equal to the Larmor frequency, ν_o. Formally, one spin jumps up to the higher energy level and one "photon" of electromagnetic radiation (energy $h\nu_o$) is absorbed. Unfortunately, there is another process that is equally likely, called "stimulated emission," in which one photon ($h\nu_o$) is absorbed by a spin in the upper (β) energy state, kicking it down to the lower state with the emission of two photons. So as long as our RF energy is applied at the resonant frequency, spins are jumping up (absorption of one photon) and down (emission of one photon) constantly. The rate of these processes is proportional to the population of spins in each of the two states: absorption occurs at a rate proportional to the number of spins in the sample that are in the lower energy state, and emission occurs at a rate proportional to the number of spins in the upper energy state.

In order to understand the net behavior of this system, we have to think about the populations (number of spins in the sample) in each of the two states. At thermal equilibrium, there will be a slight preference for the lower energy state according to the Boltzmann distribution. For now we will only think about this preference qualitatively; it turns out to be very small indeed at room temperature—a population difference of about 1 in 10^6 spins. But as long as there are more spins in the lower energy state, we will see a net absorption of RF energy when we turn on an RF energy source at the Larmor frequency. As there is a net migration of spins from the lower energy state to the upper energy state (absorption exceeds emission), we will quickly see the two populations become equal:

$$N/2 - \delta \quad\quad N/2 \quad (\beta)$$
$$\rightarrow$$
$$N/2 + \delta \quad\quad N/2 \quad (\alpha)$$

where N is the total number of identical spins in the sample and δ is a very small fraction of this number. With the equal populations, the rate of absorption equals the rate of emission and we no longer have any net absorption of RF energy. This condition is called *saturation*. If there were no other way for the spins to drop down to the lower energy state, this would be the end of the NMR experiment: a quick burst of absorption and then nothing. But there is a pathway to reestablish the Boltzmann distribution: spins can drop down from the higher energy state to the lower energy state with the energy appearing as thermal energy (molecular motion) instead of in the form of a photon. This process is called *relaxation* and is an extremely important phenomenon that will be discussed in detail. If our source of RF energy is weak enough, we can reach a steady state in which the absorption of RF energy is exactly equal to the rate of relaxation. The amount of energy absorbed is very small, and the heating of the sample resulting from relaxation is not even noticeable.

1.4.8 A Continuous Wave Spectrometer

So now we have a way to construct a simple NMR spectrometer: We have a weak source of RF energy (a transmitter) and we gradually decrease the frequency, with the magnetic field strength (B_o) remaining constant. A detector in the transmitter circuit monitors the amount of RF energy absorbed, and this signal is applied to a pen, which moves up and down. The pen moves from left to right across the paper as the frequency is gradually decreased, and when we reach the Larmor frequency (ν_o), there is a net absorption of energy and the pen moves up. As we pass through the Larmor frequency, the resonance condition is no longer met and absorption stops, so the pen moves back down. The spins never reach the saturated state because the RF energy level is very low, and after passing through the resonance condition they quickly reestablish the equilibrium energy difference through the process of relaxation. The result is an NMR spectrum: a graph of absorption of RF energy (vertical axis) versus frequency (horizontal axis). The range of frequencies "scanned" by the spectrometer is very narrow—for example, from 500.0025 MHz down to 499.9975 MHz, and the position of the absorption peak on the spectrum (its "chemical shift") tells us something about the chemical environment of the spin within the molecule. This technique is called "continuous wave" (CW) NMR because the radio frequency energy is applied continuously as the frequency is gradually varied. The first commercial NMR spectrometers (e.g., the Varian T-60 operating at 60 MHz) were all continuous wave. In the earliest CW instruments, the radio frequency was held constant and the field (B_o) was gradually changed ("swept"). This gave the same result because the absorption of RF energy led to a peak when the field reached a value that satisfied the resonance condition ($\nu_o = \gamma B_o / 2\pi$). The left-hand side of the spectrum was called "low field" and the right-hand side was called "high field." The chemical shift scale was in ppm units of τ ($\tau = 10 - \delta$), which increased from left to right. To this day we use the terms "downfield" and "upfield" to refer to the left-hand and right-hand side of the spectrum, respectively, and the frequency scale runs from right-hand to left-hand side, contrary to all other graphical scales. This is because a higher frequency in the frequency-swept spectrum corresponds to a lower field ("downfield") in the old field-swept instruments.

1.4.9 Pulsed Fourier transform NMR

All modern spectrometers now use a "pulsed Fourier transform" method, which is much faster and allows repeating the experiment many times and summing the resulting data to increase sensitivity. A very brief pulse of high-power radio frequency energy is used to excite all of the nuclei in the sample of a given type (e.g., ^{1}H). Immediately after the pulse is over, the nuclei are organized in such a way that their precessing magnets sum together to form a net magnetization of the sample, which rotates at the Larmor frequency. The coil that was used to transmit RF is now used as a receiver, and a signal is observed at the precise Larmor frequency, ν_o. This signal, which oscillates in time at the Larmor frequency, is recorded by a computer and a mathematical calculation called the Fourier transform converts it to a spectrum, a graph of intensity versus frequency. Essentially the Fourier transform measures the frequency of oscillation of the signal. If there are a number of slightly different Larmor frequencies, corresponding to different positions within a molecule, their signals add together to give the recorded signal, and the Fourier transform can sort out all the signals into a spectrum with many peaks at different frequencies. The whole experiment (pulse followed by recording the "echo" signal) takes only a few seconds and can be repeated as many times as desired, summing the data to get a stronger signal.

1.4.10 Sensitivity of the NMR Experiment

Although techniques like mass spectrometry require only nanograms (10^{-9} gram) of sample, NMR requires milligrams (10^{-3} gram) of a typical organic molecule. This insensitivity stems primarily from the fact that only the *difference* in population at thermal equilibrium is active in the experiment. That means that only approximately one spin in 10^6 is actually detected. We saw this in the CW experiment, where absorption of RF energy is almost completely cancelled by stimulated emission. Another important aspect is the relative sensitivity of different nuclei: because of the inherent differences between different nuclei in the strength of the nuclear magnet (γ), the signal strength received can be very much weaker than a proton signal. There are three ways in which γ affects the sensitivity of the experiment ("the three gammas"):

1. The *population difference* at thermal equilibrium is proportional to the energy gap, which is in turn proportional to γB_o, and inversely proportional to absolute temperature. This population difference is the only thing we can observe by NMR.
2. As the nuclear magnet precesses, it induces a signal in the receiver coil. The amplitude of this signal is proportional to the *strength of the rotating magnet*, which is the magnetogyric ratio γ.
3. The *rate at which the nuclear magnet precesses* (ν_o) is also proportional to γB_o. As with any electrical generator, if you turn the crank faster you get a higher voltage out of the generator.

Factors 1 and 2 taken together give the net magnetization at equilibrium, M_o, so we can also think of a large magnet of strength M_o rotating in the *x-y* plane when we consider the final factor, the rate of rotation (3). Either way we can say that the amplitude of the NMR signal (sensitivity) for a spin-½ nucleus is proportional to:

$$[N \times \gamma B_o/T] \times [\gamma] \times [\gamma B_o] = N\gamma^3 B_o^2/T$$

where N is the number of identical spins in the sample. This tells us that sensitivity depends on the third power of γ as well as the square of B_o. So it is worth a lot of money to build larger and more powerful magnets, and we will pay a big price in sensitivity to study nuclei with relatively small γ. Consider some of the most useful nuclei for organic chemistry and biological research:

	$B_o = 7.05$ T	$B_o = 11.74$ T	$\gamma/\gamma_H(\%)$	γ^3/γ_H^3	Abundance (%)
^{1}H	300.0 MHz	500.0 MHz	1.000	1.000	99.98
^{13}C	75.43	125.72	0.2514	0.0159	1.11
^{15}N	30.40	50.66	0.1013	0.00104	0.37

Using our rule of thumb that ^{13}C has a γ value four times smaller than ^{1}H and that ^{15}N has a γ value 10 times smaller than ^{1}H, we can see that the FID signal will be $4^3 = 64$ times less with ^{13}C and $10^3 = 1000$ times less with ^{15}N when compared to ^{1}H with the same number of identical nuclei (N) in the sample. But even at the same sample concentration we do not have the same number of nuclei because for ^{13}C only about one in 100 carbon atoms is ^{13}C and for ^{15}N only about one in 300 nitrogen atoms is ^{15}N. Accounting for this smaller value of N, the signal strength (sensitivity) is 5670 times less than ^{1}H for ^{13}C and 260,000 times less than ^{1}H for ^{15}N at natural abundance. For this reason, commercial continuous wave NMR spectrometers could only detect ^{1}H. With pulsed Fourier transform NMR it became possible to detect ^{13}C with long experiments (1 h or more) and concentrated samples (30 mg or more of a typical organic molecule). Detection of ^{15}N is still very difficult without isotopic labeling of ^{15}N in the sample. Biological NMR experiments (proteins, nucleic acids, etc.) now typically involve preparation of uniformly ^{13}C and ^{15}N labeled samples by biosynthesis (e.g., protein expression in *E. coli*) on labeled media (e.g., ^{15}NH$_4$Cl and U-^{13}C-glucose). We will see that NMR tricks can also allow us to avoid the disadvantage of the first γ (e.g., the DEPT experiment, Chapter 7), or even to avoid the disadvantage of all three gammas (e.g., ^{1}H-detected two-dimensional experiments, Chapter 11). Without isotopic labeling, however, there is no trick that can overcome the disadvantage of low isotopic abundance.

2

INTERPRETATION OF PROTON (^{1}H) NMR SPECTRA

2.1 ASSIGNMENT

There are two things we can do with a proton spectrum: try to figure out the structure of an unknown compound or try to assign the peaks to the hydrogen positions of a known compound. The latter process is called *assignment*: pairing each resonance in a ^{1}H spectrum with a hydrogen or a group of equivalent hydrogens in the chemical structure. A "resonance" is a single chemical shift position in the spectrum; it can be a single peak (a singlet) or it may be "split" by J coupling into a complex pattern of peaks—a triplet or a double septet, for example. Sometimes we refer to a "resonance" as a peak, but this can be confusing because it may consist of many peaks of a multiplet pattern. The best way to learn to interpret NMR spectra is to assign the peaks in a spectrum of a known compound. This is much easier than dealing with unknowns and teaches the same principles that will be necessary to analyze unknown spectra. The vast majority of examples in this book will be discussed as assignment problems rather than unknown problems.

We will see that with complex molecules chemical shift is not enough to arrive at a unique assignment; normally, there will be several or many ^{1}H resonances with similar chemical shifts, and we can only put these resonances into categories (e.g., ^{1}H α to a carbonyl group or olefinic proton) rather than unique assignments. To uniquely assign we will need to *correlate* protons to other protons or other spins (e.g., ^{13}C) within the molecule, either by through-bond relationships (i.e., J couplings) or by through-space relationships (i.e., NOEs). *Chemical shift correlation* is a process of pairing a proton with another spin that is nearby in the bonding network (number of bonds) or by direct distance through space (Å), usually by a two-dimensional (2D) experiment. In establishing these relationships, we only "know" a spin by its precise chemical shift. That is why we call the process of correlating two spins *chemical shift* correlation. So the chemical shift of a proton is

---

NMR Spectroscopy Explained: Simplified Theory, Applications and Examples for Organic Chemistry and Structural Biology, by Neil E. Jacobsen
Copyright © 2007 John Wiley & Sons, Inc.

not only an imprecise description of its chemical environment but also a precise "address" or "label" by which we can "talk to" that proton and ask questions about its immediate environment in the molecule, in terms of nearby spins. In this sense we can think of each proton as a probe or flashlight, which we can shine on the immediate environment of the molecule to see what is around it. The flashlight is very weak, however, and can only "see" up to about 5 Å in space and about three bonds through the bonding network, to identify its neighbors. The neighbors are, of course, only identified by their chemical shifts.

2.2 EFFECT OF B_0 FIELD STRENGTH ON THE SPECTRUM

A large part of the history of NMR instrument development concerns the effort to attain higher and higher magnetic fields by building stronger and stronger magnets. The first widely available commercial NMR instruments were 60 MHz continuous-wave (CW) instruments (e.g., Varian A-60 and T-60) that only did ^{1}H spectra. These magnets were simple electromagnets: copper wire wound on an iron core with a large current passed through the coil. A large amount of heat was generated by the current, so water was passed through the magnet to cool it. Newer instruments came out with 90 MHz (Varian EM-390) and 100 MHz (Varian XL-100) proton frequencies. These were also CW, although by the mid-1970s the strongest electromagnets were being used for the first pulsed Fourier-transform instruments. An early ^{13}C instrument (FT-80) used an 80-MHz electromagnet with RF pulse electronics and computer built by Nicolet. It operated at 20 MHz for ^{13}C, used 8-mm sample tubes, and had a whopping 8192 bytes of memory!

One hundred megahertz (2.35 T) was the "brick wall" for electromagnets, and it was necessary to develop an entirely new technology to go beyond that limit. Superconductivity is the phenomenon of zero resistance for electrical conductors at low temperature. Special alloys including niobium and titanium can be made into wires that when cooled to 4.2 K (the boiling point of liquid helium) can support large electrical currents without any resistance. This means that if a coil of this wire is immersed in liquid He and a current is passed through the coil, we can connect the end of the coil to the beginning and get the current to flow in a closed loop without any resistance. The large current will produce a very strong magnetic field, and because there is no resistance, there is no loss of energy to heat and the current will be stable. Superconducting magnets can run for decades without any significant loss of magnetic field strength as long as the superconducting coil is kept at liquid He temperature the whole time. The first superconducting NMR instrument I used (in the late 1970s) was a 180 MHz instrument built by Alex Pines at the University of California at Berkeley. The "pulse programmer" was set up using a teletype terminal; the Nicolet computer had to be "bootstrapped" by setting an array of switches (bits) to a specific binary number (address) and hitting the start button, and the audio filters had to be set to the spectral width by twirling dials. Shims were adjusted with a vast array of knobs, and data could be saved on a computer "disk" the size of a large dinner plate. Soon commercial magnets began to climb in field strength: 200, 250, 300, 400, and 500 MHz. Finally, the same technology was extended to 600 MHz, but this was the limit at 4.2 K, the boiling point of He at atmospheric pressure. By reducing the pressure in the helium can, the temperature was lowered and magnets reached 750, 800, and finally 900 MHz. A 900-MHz magnet looks like a space shuttle on its launch pad and requires a whole building devoted to one NMR instrument. Many groups are struggling to come up with the first 1-GHz

(1000 MHz) magnet, but this has become a very difficult goal to achieve. It would seem that the current superconductivity technology has been pushed to a limit, and there is a great need for a fundamental breakthrough. Ceramic superconductors have been developed that can achieve superconductivity at much higher temperatures (77 K and higher), but it has been difficult to form these materials into wires, and the current carrying capacity is very small.

NMR spectrometers cost roughly $1000 per MHz of field strength in the lower range, but above 600 MHz the cost rises exponentially to more than $5 million for a 900-MHz system. Why do people pay so much money to get higher magnetic fields? There is an obvious advantage in sensitivity because increasing the B_0 field increases the population difference between the α and β states proportionately. This increases the net magnetization of the sample at equilibrium and thus increases the FID signal received after the pulse. Another factor enters in during the recording of the FID: The Larmor frequency, ν_0, is proportional to B_0, so we have the nuclear magnets precessing at a higher speed when we increase the magnetic field. Just as turning the crank on a generator faster produces a higher voltage in the output, spinning the net magnetization faster generates a bigger FID. So we expect the sensitivity to be proportional to B_0^2, but in reality you cannot increase B_0 while keeping everything else constant, so it works out in a practical sense to about $B_0^{1.5}$. That means that a ^{13}C acquisition on a 200-MHz instrument would require 27 times as long as the same experiment on a 600-MHz instrument to achieve the same signal-to-noise ratio (600/200 to the 1.5 power, then squared because signal-to-noise ratio varies with the square root of the number of scans).

But it turns out that there is a much more important advantage to stronger magnets: resolution. What do we mean by resolution? In a technical sense, resolution is the width of an NMR line measured in hertz at one half the height of the peak. The peak width depends on the rate of decay of the FID, which is determined by the homogeneity of the magnetic field (shimming) and the inherent rate of decay of the net magnetization in the x–y plane (determined by a relaxation parameter of that proton called T_2). In this sense, resolution is the same at 1.41 T (60 MHz) as it is at 21.1 T (900 MHz): about 1 Hz for a "small" molecule in organic solvent. But there is a broader and more important meaning of "resolution" that has to do with the ability to separate one proton resonance (chemical shift with splitting pattern) from another without overlap. We say that two proton signals ("signal" is another word for resonance) are "resolved" if there is no overlap between the group of peaks associated with one chemical shift and the group of peaks associated with another. As molecules become larger, the corresponding ^1H spectra become more complex because a larger number of resonances (chemical shift positions) is spread out over the same range of chemical shifts: roughly 0–10 ppm. As this happens there is more and more chance for overlap because many chemical shifts fall very close to each other. The spread or footprint of a ^1H resonance is determined by the J couplings, which are measured in units of hertz (the total width of the multiplet pattern is roughly equal to the sum of all J couplings to that proton). The larger this footprint, the fewer the unique ^1H signals that can be squeezed into the fixed 0–10 ppm territory without overlap. This is where a fundamental difference between the J coupling and chemical shift becomes crucial to this discussion: J couplings represent interactions between a pair of nuclei and as such their strength is always measured in hertz and is independent of magnetic field. Chemical shifts (expressed as frequency in hertz) are proportional to magnetic field, which is why we normally use units of parts per million (millionths of the Larmor frequency), so we have the same ppm value regardless of hertz value. Normally, we look at a proton spectrum with a horizontal

42 INTERPRETATION OF PROTON (^{1}H) NMR SPECTRA

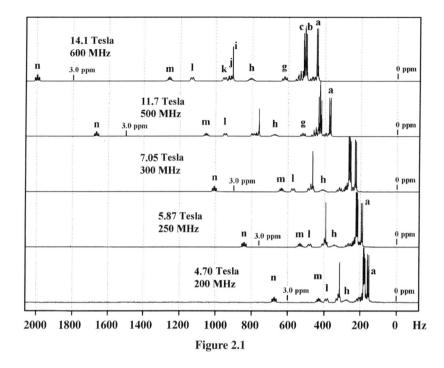

Figure 2.1

scale in parts per million, not hertz, so the positions of peaks will be the same at any field strength.

This is illustrated by the spectra of menthol at 200, 250, 300, 500, and 600 MHz. As we saw in Chapter 1, NMR signals are frequencies measured in hertz, based on the audio frequencies detected in the FID by the Fourier transform. If we plot the spectra on a frequency scale in hertz (Fig. 2.1), we see that the chemical shifts (positions of the proton resonances a–n) in hertz units are proportional to the magnetic field strength, as they should be:

$$\nu_o = \gamma B_o / 2\pi$$

The linewidths and J couplings are independent of field strength, so the appearance of each proton multiplet is the same in each spectrum, regardless of field strength (compare, for example, peak n). If we expand and align all of the peak n multiplets (Fig. 2.2), we can see that they are identical, with a linewidth of about 1.5 Hz and three coupling constants of 10.7, 10.0, and 4.3 Hz (the two broader peaks in the center are unresolved pairs of lines). But this method of displaying the spectrum is impractical because the chemical shift (in hertz) would be different on different NMR instruments, and it would be confusing to make comparisons unless everyone had the same field strength. This is why the ppm scale was developed: to make chemical shifts independent of field strength so that they could be reported on a universal scale. One part per million is one millionth of the fundamental frequency being used for the nucleus being observed. For example, on a Bruker DRX-600, the ^{1}H frequency is 600.13 MHz, so 1 ppm is 600 Hz (600×10^6 Hz $\times 10^{-6}$). In Figure 2.1 the frequency corresponding to 3.00 ppm is shown on each spectrum: $3 \times 200 = 600$ Hz on the 200-MHz instrument, $3 \times 250 = 750$ Hz on the 250, 900 Hz on the 300, 1500 Hz on the 500, and 1800 Hz on the 600. It is important to realize that the conversion from Hz to ppm depends on the

EFFECT OF B_0 FIELD STRENGTH ON THE SPECTRUM   43

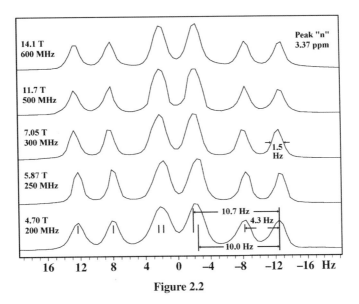

Figure 2.2

nucleus as well as the field strength: on a "600" the ^{13}C resonant frequency is 150 MHz (roughly one fourth of 600 because γ_C/γ_H is about 0.25), so 1 ppm is 150 Hz, not 600 Hz.

Figure 2.3 shows the spectra lined up on a ppm scale rather than a hertz scale (the vertical scale is increased and the tall methyl and OH peaks are clipped off). This is the universally accepted format for presenting NMR data. Although all the peaks (resonances) appear at the same place in the spectrum (same chemical shift in ppm), the multiplet patterns appear to "shrink" horizontally as we go to higher field strength because the J couplings in hertz get smaller and smaller on the ppm scale. For example, on a 200-MHz spectrometer, a typical

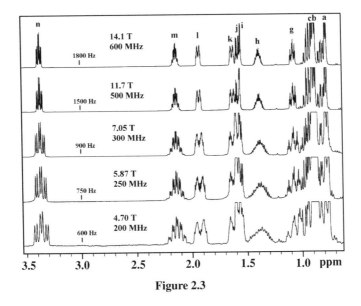

Figure 2.3

7.0-Hz coupling appears as a separation of 7/200 or 0.035 ppm. The same coupling on a 250, 300, 500, or 600 MHz spectrometer appears as a separation of 0.028, 0.023, 0.014, and 0.012 ppm, respectively. As the ^{1}H coupling patterns "shrink," the footprint of each resonance gets smaller and the chances of overlap get smaller. We say that peaks that are overlapped on the 200-MHz spectrometer are "resolved" (separated by a region of baseline with no intensity) on the 600-MHz spectrometer. This is clearly illustrated by peak g, which at low field is spread out and overlapped with some peaks on its upfield (right-hand) side. This is the more general and more important meaning of "resolution" and explains why people are willing to spend enormous sums of money to achieve even modest gains in magnetic field strength. With smaller footprints we can move to larger, more complex molecules to make use of that "empty space" between the peaks. There are, of course, other ways to avoid overlap—primarily by using 2D and even 3D and 4D experiments, but in each case it is always better to have higher field because the "footprint" size is reduced on the ppm scale. For peaks that are not overlapped at any field strength, there is no change in the appearance of the peak as we move from 200 to 600 MHz, as long as we expand the peak to the same range of chemical shifts in hertz (Fig. 2.2). In this case, the structure of the multiplet depends only on the linewidth (a function of shimming and T_2) and J-coupling values, all of which are independent of the field strength, B_0.

A closer look at the methyl region (Fig. 2.4) shows how we can easily mistake two resonances for one. The pattern observed for CH_3(b) and CH_3(c) at 600 MHz looks like a double doublet. At 500 MHz it looks like a triplet. The reason it changes its form with magnetic field is that it really represents two different resonances with their own chemical shift positions, that is, two doublets. The two vertical lines show the peak positions in ppm, which do not change with field strength. At 600 MHz the two doublets are separated, and

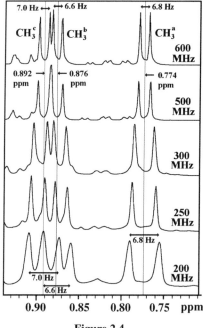

Figure 2.4

we can read the coupling constants from the left-hand side pair (7.0 Hz) and the right-hand side pair (6.6 Hz). At 500 MHz the doublet splittings are "wider" (on the ppm scale) so that the two inner peaks are now overlapped. At 300 MHz and below, the two doublets become intertwined, so that we have to measure the coupling constants from the first and third peaks (7.0 Hz) and the second and fourth peaks (6.6 Hz). If we only saw the spectrum at one field strength, we might be fooled into thinking it was a single resonance split into a triplet or a double doublet.

2.3 FIRST-ORDER SPLITTING PATTERNS

All of this assumes that the proton in question is only coupled to other protons that are far away in chemical shift, so that its coupling pattern is simple ("first order" or "weak coupling"). If it is coupled to nearby peaks, distortions of the peak intensities and more complex patterns can result, and this effect is strongest at lower field strengths ("second order" or "strong coupling"). To state this more precisely, the J coupling in hertz between two spins must be much less than the chemical shift difference *in hertz* to see a simple first-order pattern. We could write this as

$$\Delta v / J > 5 \quad \text{for first-order (weak) coupling}$$

where we arbitrarily divide it where J is one fifth of the chemical shift difference. Note that the chemical shift difference (Δv) has to be expressed in hertz in order to directly compare it to the J coupling. This means that the criterion depends on field strength: a pattern that is second order at low field can be resolved into a simple first-order pattern at high field—yet another reason to spend the big bucks. For example, if two protons are coupled with a 7.0-Hz coupling constant and have a chemical shift difference of 0.1 ppm, they would be in a "second-order" splitting pattern at 200 MHz ($\Delta v = 20$ Hz, $\Delta v/J = 2.86$) and a "first-order" pattern at 600 MHz ($\Delta v = 60$ Hz, $\Delta v/J = 8.57$). Of course, the transition from first order to second order is a gradual process, so the cutoff of a factor of 5 is arbitrary, but you can see that higher field means fewer problems with distorted and more complex splitting patterns.

First-order patterns are easy to analyze because each splitting by a proton divides the pattern into two equal patterns separated by the coupling constant, J. To predict the splitting pattern, you can draw a diagram starting with the chemical shift position of the resonance. Arrange the coupling constants in descending order and write next to each one its multiplicity (doublet, d; triplet, t; quartet, q). For example, if a proton H_a has a chemical shift of 3.56 ppm and has two "neighbors," H_b and H_c, with coupling constants $J_{ab} = 10.0$ Hz and $J_{ac} = 4.0$ Hz, we have

$$H_a: \delta\, 3.56\,\text{ppm}\,(d,\ 10.0\,\text{Hz};\ d,\ 4.0\,\text{Hz})$$

We first divide the resonance position (3.56 ppm) into two equal peaks (1:1 ratio) by moving left 5.0 Hz ($J/2$) and right 5.0 Hz ($J/2$) (Fig. 2.5). Then each of these peaks is divided again into two equal peaks (1:1 ratio) by the 4.0-Hz coupling: 2.0 Hz to the left and 2.0 Hz to the right. This results in a pattern we call a doublet of doublets or (more concisely) a double doublet (abbreviated "dd"). In the literature we would report the peak like this: $\delta 3.56$ (dd, 10.0, 4.0). To "deconstruct" (analyze) the pattern, we first note that all four peaks are of the same height, and because 4 is a power of 2 ($2^2 = 4$), we assume that there is no overlap of

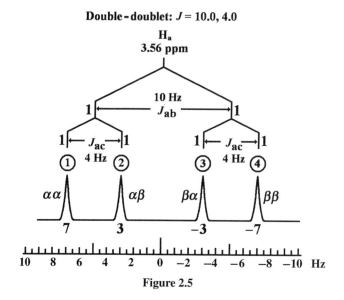

Figure 2.5

peaks. This implies that there are two coupling constants. The smaller one ($J_{ac} = 4.0$ Hz) can be measured as the separation of peaks 1 and 2 (left-hand side pair, numbering from left to right) or peaks 3 and 4 (right-hand side pair). The larger coupling ($J_{ab} = 10.0$ Hz) can be measured as the separation between peaks 1 and 3 or between peaks 2 and 4. The full "footprint" of the pattern (separation between peaks 1 and 4) is the sum of all coupling constants: $10 + 4 = 14$ Hz. The more the time you spend diagraming coupling patterns, the easier it will be to recognize and deconstruct these patterns in real NMR spectra. In a more theoretical sense, we can see that the effect on H_a of H_b being in the α state is a downfield shift of its resonant frequency by 5.0 Hz ($J_{ab}/2$). The effect on H_a of H_b being in the β state is an upfield shift by 5.0 Hz. The effect on H_a of H_c being in the α state is a downfield shift of 2.0 Hz ($J_{ac}/2$), and the effect of H_c being in the β state is a upfield shift of 2.0 Hz. Thus, each component of the multiplet pattern can be viewed as a particular spin state of H_b and H_c and its effect on the resonant frequency of H_a. Peak 1 (leftmost) is labeled $\alpha\alpha$ ($H_b = \alpha$, $H_c = \alpha$), peak 2 is labeled $\alpha\beta$ ($H_b = \alpha$, $H_c = \beta$), and so on, and we can calculate the position of each peak relative to the center by adding $J/2$ for α and subtracting $J/2$ for β. For example, peak 2 (+3.0 Hz) represents the H_a resonance where H_b is in the α state ($+10.0/2 = 5.0$ Hz) and H_c is in the β state ($-4.0/2 = -2.0$ Hz). Adding the two effects, we get $5.0 - 2.0 = 3.0$ Hz.

Note: In this book we will use the convention that the α state leads to a downfield shift (higher resonant frequency) for all coupled spins and the β state leads to an upfield shift. In fact, this may be reversed depending on the sign of the coupling constant J and the sign of the magnetogyric ratios, γ. J couplings can be either negative or positive, as can magnetogyric ratios. For simplicity, we will ignore this detail.

A triplet can be viewed as a special case of a double doublet where $J_{ab} = J_{ac}$. In this case the two inner peaks (peaks 2 and 3) have the same resonant frequency and combine

FIRST-ORDER SPLITTING PATTERNS 47

Double-triplet: $J = 10.0, 5.0$

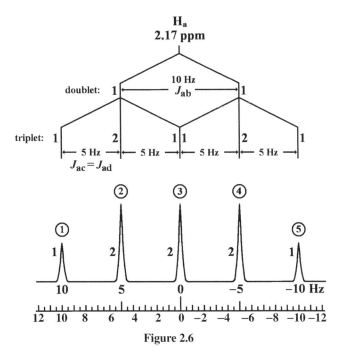

Figure 2.6

to form a single peak with twice the intensity of the outer peaks. This can occur by true equivalence (i.e., by molecular symmetry making the H_a–H_b relationship identical to the H_a–H_c relationship) or by coincidence: the two coupling constants may just happen to be nearly or exactly the same. In the case of the triplet (1:2:1 ratio), the central peaks represent both the $\alpha\beta$ and the $\beta\alpha$ states of the H_b/H_c system, and we have less than 2^n peaks (where n is the number of coupled spins affecting H_a) because of overlap.

More complex coupling patterns are dealt with in the same way. Generally, it is the overlap or near overlap of the 2^n original components that makes the pattern complex and more challenging to take apart. Figure 2.6 shows a double triplet with coupling constants of 10.0 Hz (doublet coupling) and 5.0 Hz (triplet coupling). This means that H_a (2.17 ppm) is coupled to H_b ($J_{ab} = 10.0$ Hz) and to two equivalent protons, H_c and H_d, each with a coupling to H_a of 5.0 Hz. The two triplet patterns meet in the center so that an intensity 1 peak of the left-hand triplet combines with an intensity 1 peak of the right-hand triplet to give a peak of intensity 2. Thus, the overall pattern is five equally spaced peaks with intensity ratio 1:2:2:2:1. This might be mistaken for a quintet with a single coupling of 5.0 Hz, but that would give an intensity pattern of 1:4:6:4:1, very different from the pattern we observe. To analyze this pattern, first note that there are five peaks in the multiplet, so we must have at least three couplings ($2^2 = 4$; $2^3 = 8$). If there are three couplings, then we have eight peaks that are reduced to five peaks by overlap. The intensities are clearly not all the same: the outer peaks are smaller and the inner peaks all look the same. So 1:2:2:2:1 would be a good estimate of relative intensities. These numbers add up to eight, confirming that there are three couplings. Measuring frequency differences from the

48 INTERPRETATION OF PROTON (^{1}H) NMR SPECTRA

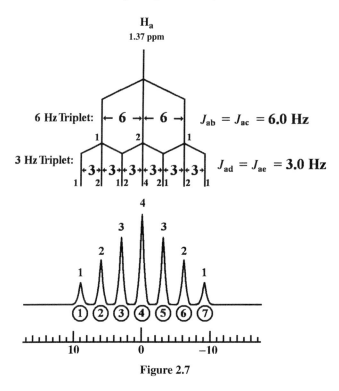

Figure 2.7

outermost peaks, we see that peak 2 is 5.0 Hz from peak 1, and peak 3 is 10.0 Hz from peak 1. This suggests that we have couplings of 5 and 10 Hz. The equal spacing and the 1:2 ratio of peaks 1 and 2 suggest a triplet, and two triplets that meet in the center explain the pattern. Diagraming the coupling pattern gives us the complete story: a double triplet with couplings of 10.0 and 5.0 Hz. We report this as δ 2.17 (dt, J = 10.0, 5.0). The simpler splitting comes first (d), and the coupling constants are listed in the same order as the coupling patterns, so we know that the 10.0-Hz coupling goes with the "d" and the 5.0-Hz coupling goes with the "t".

Figure 2.7 shows a triple–triplet pattern where one triplet coupling is exactly twice the other triplet coupling. As always, we start the diagram with the larger coupling and then split each of the peaks again with the smaller coupling. The three narrow triplet patterns grow out of the three peaks of the wider triplet, and we write the intensities according to the intensities of the "parent" peaks they grow out of: 1:2:1 for the outer triplets and 2:4:2 for the inner triplet derived from the intensity 2 peak of the wide triplet. The narrow triplets overlap in two places, and we add the intensities of the two peaks that combine in each case. The final intensity pattern is 1:2:3:4:3:2:1. This is distinct from a septet (splitting by six equivalent protons), which has intensity pattern 1:5:10:15:10:5:1. To analyze this pattern, we have to rule out the possibility that there are only three couplings ($2^3 = 8$), even though there are less than eight peaks. With three couplings, there would only be one overlap and the intensity ratio would have to be 1:1:1:2:1:1:1. With four couplings ($2^4 = 16$) we would have nine overlaps. In fact, the intensity ratio 1:2:3:4:3:2:1 adds up to 16, accounting for all

FIRST-ORDER SPLITTING PATTERNS

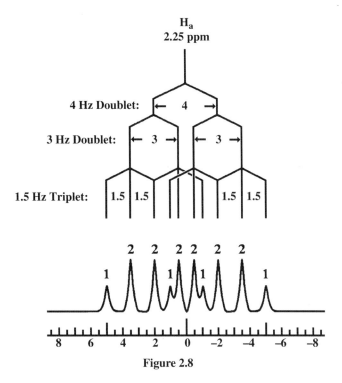

Figure 2.8

of the overlap. The "triangular" shape of the multiplet envelope suggests the 1:2:3:4:3:2:1 intensity ratio. Measuring in from the outer peaks, we see couplings of 3.0 Hz (peak 1 to peak 2) and 6.0 Hz (peak 1 to peak 3). One can visualize a triplet on each edge of the pattern (1:2:1) and a more intense triplet in the center (2:4:2), and the even spacing, leading to overlap, explains the anomolous intensity 3 peaks.

Figure 2.8 shows a double-double triplet (ddt) pattern. We see 10 peaks, not all evenly spaced. With four couplings ($2^4 = 16$) we would have six overlaps, and because we see four small peaks and six larger peaks, the intensity ratio might be 1:2:2:1:2:2:1:2:2:1, adding up to 16. Starting from the leftmost peak, we measure separations of 1.5, 3.0, and 4.0 Hz. The 1:2 ratio at the left edge suggests a triplet, so we know that the third peak (intensity 2) has another peak of intensity 1 from another triplet, suggested by the peak of intensity 1 (the fourth peak from the right-hand side) located 3.0 Hz to its right. It is useful when a component such as the 1:2:1 triplet is identified to mark the pattern with three lines on a piece of paper and then move the paper around on the multiplet to see if there are other identical patterns in the multiplet. In this case, we can locate four 1.5-Hz triplets, and their centers describe the double-doublet ($J = 4.0, 3.0$) that we see in the upper part of the diagram. Another way to look at this pattern is the superposition of two doublet-triplets (see Fig. 2.6, 1:2:2:2:1 ratio) with a separation of 4.0 Hz. Notice also what happens when the separation between two "lines" (components of the multiplet) comes close together: the smaller peak "climbs" up the larger one and becomes a "shoulder" on the side of the peak. We do not have "baseline separation" between the fourth and fifth peaks (1 and 0.5 Hz on the scale)

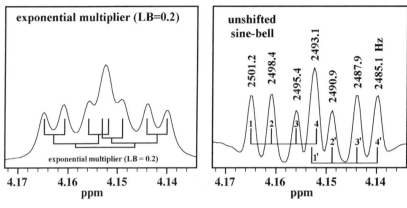

Figure 2.9

because their linewidths (about 0.5 Hz) are similar to their separations (0.5 Hz). Baseline separation means that the intensity level comes all the way to zero (to the noise baseline of the spectrum) between the peaks. When there is partial overlap, the separation of the two peaks in hertz will be a bit less than their actual frequency difference, and this can lead to errors in measuring coupling constants. If there are multiple places to measure a frequency difference, always avoid measuring it to a partially overlapped peak and choose two peaks that are baseline separated.

Some real-world examples will help to reinforce these concepts. Figure 2.9 shows a one-proton multiplet from the 600 MHz ^{1}H spectrum of a testosterone (steroid) metabolite. The chemical shift is measured at the precise center of the symmetrical pattern: 4.153 ppm, in the region of singly oxygenated CH groups. The spectrum on the right is processed with resolution enhancement: a sine-bell function starting at zero and ending at 180° of the sine function. The spectrum on the left is processed without resolution enhancement. The intensity ratio appears to be 1:1:1:2:1:1:1, which adds up to eight and suggests three couplings ($2^3 = 8$) with one overlap (seven peaks). If you mark the two leftmost peak positions on a piece of paper, you can see that the two rightmost peaks have the same spacing. Measuring from the third peak, we can make a mark on the right shoulder of the fourth (center) peak with the same spacing. These four lines (1–4 on the right-side spectrum) make up a double doublet. The coupling constants can be extracted from the line frequencies in hertz: $J_1 = 2501.2 - 2498.4 = 2.8$ Hz; $J_2 = 2501.2 - 2495.4 = 5.8$ Hz. This entire four-line pattern can then be transferred to the other side of the multiplet from your piece of paper, lining up the rightmost line with the rightmost peak (Fig. 2.9, right, 1'–4'). When this is done, the two lines in the middle do not perfectly overlap, resulting in a broader peak in the center with an intensity more like 1.5 rather than 2. It is still possible, however, to extract the third coupling constant by measuring the separation of the "second" lines (2 and 2') in the two four-peak (dd) patterns, or the separation of the two "third" lines (3 and 3'). The first (1 to 1') and fourth (4 to 4') peak separations require measurement to an overlapped pair of peaks (1' and 4) that are not perfectly aligned, so we avoid these measurements. Using the line frequencies, we have from the second peaks (lines 2 and 2'): $J_3 = 2498.4 - 2490.9 = 7.5$ Hz; or from the third peaks (lines 3 and 3'): $J_3 = 2495.4 - 2487.9 = 7.5$ Hz. The complete diagram is shown on the left-side spectrum: δ 4.153 (ddd, $J = 7.5, 5.8, 2.8$). Note that the method of measuring

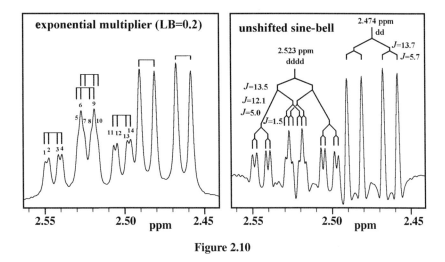

Figure 2.10

the separation from the outermost peak fails in this case: The first two couplings can be obtained in this way but the third, which would come from the separation of peaks 1 and 1'/4, would be incorrect because peak 1'/4 is not a perfectly aligned overlap of two peaks. It is always better to completely analyze and understand the coupling diagram and measure from the resolved, single line peaks as much as possible.

A more complex pattern with some overlap is shown in Figure 2.10, another steroid metabolite. First-order splitting patterns are always symmetric about the center, so this cannot be a single resonance. Integration shows two protons, and the centers of the two symmetric patterns can be measured as 2.523 ppm (left-hand side pattern with 14 lines) and 2.474 ppm (right-hand side pattern with four lines). These two peaks are barely resolved at 600 MHz; at lower field the multiplets would expand about the same two chemical shift positions, and it would be very difficult to analyze the two overlapping patterns. The right-hand side multiplet is easy to analyze: it is a double-doublet with couplings of 13.7 and 5.7 Hz. The left-hand side pattern is more challenging. It has 14 lines (or "multiplet components") that suggests four couplings ($2^4 = 16$) and two overlaps: 1:1:1:1:1:2:1:1:2:1:1:1:1:1. Peaks 1–4 can be diagrammed as a double-doublet because the 1–2 spacing is the same as the 3–4 spacing. Marking down this pattern, we can transfer it to peaks 11–14 exactly. The center pattern can be viewed as two of these double-doublet patterns offset by the 1–2 separation. This gives overlap at peaks 6 and 9, which are twice the intensity of the others. We can trace peaks 1–4, 5, 6, 8, and 9 onto a piece of paper and verify that the pattern lines up with peaks 6, 7, 9, 10, and 11–14. This confirms that we have a double-double-double-doublet (dddd), and we only have to measure the four coupling constants. The smallest coupling can be most easily measured as the difference between 1 and 2, 3 and 4, 11 and 12, or 13 and 14. In each case, however, the two peaks are far from baseline resolved, so the difference we measure, even in the resolution-enhanced spectrum, is an underestimate of the true coupling constant. The two peaks "ride up" on each other, and the measured separation is reduced by the overlap. A more accurate method is to make a computer simulation of two ideal shaped (Lorentzian) peaks added together, and vary the peak width and J coupling to get the best fit to the data in the lower trace (because of the resolution enhancement, the upper trace peaks do not have ideal peak shape). This analysis, using a non linear least-squares fit, gives a linewidth of

1.46 Hz and a J coupling of 1.47 Hz. This is slightly larger than the difference measured on the lower spectrum (1.37 Hz) but very close to the distance measured on the resolution-enhanced spectrum (1.51 Hz). Computer-generated peak lists can also be inaccurate due to the distortion of peak shapes by "grainy" digitization: It is better to use cursors positioned in the center of a peak by "eyeball" than to rely on an algorithm that simply looks for the highest intensity data point in the peak.

To measure the larger couplings, we measure from one peak of the double doublet to the corresponding peak of another double doublet. For example, we can measure the second-to-largest coupling between peaks 1 and 5, 2 and 6, 3 and 8, or 4 and 9. But it is best to avoid measuring to the "shoulder" peaks (5, 7, 8, and 10) because they are distorted toward the nearby taller (more intense) peak they are riding on. So we can measure this coupling best between peaks 2 and 6 or between 4 and 9. Peaks 6 and 9 are not distorted because they are pulled equally toward the two shoulders on either side. This separation gives us a coupling of 12.1 Hz, and measuring between the first and third group of four we get 13.5 Hz (distance between 1 and 6 or between 3 and 9). The peak can be reported as δ 2.523 (dddd, J = 13.5, 12.1, 5.0, 1.5). It is important to note that none of the couplings are between the two nearly overlapped resonances. If any coupling were between two peaks with such a small chemical shift difference ($\Delta \nu = (2.523 - 2.474) \times 600 = 29.4$ Hz), we would see some distortion of the peak intensities and possibly some additional, weak lines due to strong coupling (second-order pattern). In fact, the peaks can be assigned to H-6β (left-hand side peak) and H-1α (right-hand side peak), which are far apart in the steroid structure and have no mutual couplings. In this case we can often fully analyze two resonances even if they are overlapped, as long as we can recognize the coupling patterns and assign each individual line to one or the other resonance.

2.4 THE USE OF ^{1}H–^{1}H COUPLING CONSTANTS TO DETERMINE STEREOCHEMISTRY AND CONFORMATION

Clearly we can extract important information from coupling patterns about the number and equivalence groupings of other protons that are nearby (generally two or three bonds away) in the bonding network. But we can also get valuable information from the magnitude of the coupling constants, which tells us about the geometric relationship of the bonds connecting the two protons. Three-bond (vicinal) relationships are the most useful because the coupling constant is related in a predictable way to the dihedral angle between the bonds attached to the protons. For example, for two protons attached to neighboring saturated (sp^3 hybridized) carbons (H–C–C–H), rotation of the C–C bond leads to different relationships of the two C–H bonds: *anti* if they are opposite each other (180° dihedral angle) and *gauche* if they are next to each other in a staggered conformation (60° dihedral angle). If you look directly down the C–C bond, with one carbon right behind the other, the angle described by the two C–H bonds is the dihedral angle. Because J coupling is transmitted through bonds, and more specifically through electrons in bonding orbitals, the magnitude of the coupling constant depends on orbital overlap. The largest coupling constant actually corresponds to the *anti* conformation (180° dihedral angle), which is counterintuitive in terms of a through-space interaction but makes sense in terms of oribtal overlap. The minimum J coupling is observed when the two C–H bonds are exactly perpendicular (90° dihedral angle) because the orbital overlap is at a minimum for perpendicular molecular orbitals. This relationship between dihedral angle and coupling constant has been formalized into a mathematical relationship

THE USE OF ^{1}H–^{1}H COUPLING CONSTANTS   53

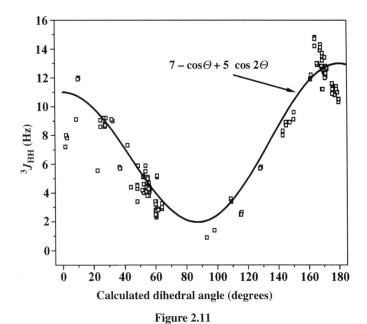

Figure 2.11

called the Karplus relation or the Karplus curve. This is really an empirical relationship that has been "parameterized" into many different equations for different specific situations. A general equation used for organic molecules with two saturated carbons (H–C–C–H) can be written as follows:

$$J = 7 - \cos\Phi + 5\cos(2\Phi)$$

where Φ is the dihedral angle. This equation is plotted in Figure 2.11, with experimental J values for nine different steroid metabolites plotted against calculated dihedral angles obtained from energy-minimized structures. The clustering of experimental points around 60° and 180° reflects the preference for staggered conformations (in this case cyclohexane chair conformations) with either *gauche* (60°) or *anti* (180°) relationships. The eclipsed conformation (0° dihedral angle), though rare, gives a second maximum in coupling constant that is a bit smaller than the maximum at 180°. The minimum J values are observed for dihedral angles near 90°, also a rare occurrence. More specific subsets of vicinal relationships can be fit more accurately to yield specific Karplus equations. For example, in NMR of peptides and proteins, the H–N–C$_\alpha$–H dihedral angle (related to the Φ angle that, along with the Ψ angle, defines the backbone conformation of the polypeptide) can be related quite accurately to the H$_N$–H$_\alpha$ J value by comparing dihedral angles measured in X-ray crystal structures of proteins to J values measured by NMR. There is another Karplus equation for three-bond couplings between ^{1}H and ^{13}C, relating to the dihedral angle H–C–C–C, so that long-range heteronuclear couplings, $^3J_{CH}$, can be used to obtain stereochemical and conformational information. J couplings are also sensitive to electronegative substituents, so we must be careful not to overinterpret the general Karpus relation in specific situations.

Steroids are rigid molecules, particularly in the locked six-membered A, B, and C rings. More flexible molecules give rise to conformational averaging, whereby NMR measurables such as J values are really weighted averages of the values expected for each of the multiple conformations available, weighted by the percent of time spent in each conformation. Usually, these conformational changes are rapid on the timescale of the NMR experiment ($1/J$), so we see only the average values. For example, a vicinal ^{1}H–^{1}H coupling across a C–C bond with free rotation will average to about 7 Hz, which is the average of the J values expected for the three staggered conformations: 4 Hz (60°), 4 Hz (−60°), and 13 Hz (180°). If the conformational change happens on a scale comparable to $1/J$, we will see broadening of the NMR lines due to the uncertainty in the J value.

If dihedral angles are measured from energy-minimized structures, it is important to consider whether the structure is rigid (a steep potential well) or flexible (a broad minimum) or if there are multiple steep minima (multiple interconverting conformations). Most energy minimization programs simply search from the starting conformation for an energy minimum and then stop—this may be a broad minimum signifying little about the actual structure, or it may be one of several minima. Solvents also affect conformation, and most structure calculations do not specifically include solvent.

2.5 SYMMETRY AND CHIRALITY IN NMR

Proton NMR spectra are considerably simplified by the equivalence of many protons: A group of protons may have exactly the same chemical shift ("chemical equivalence") and/or exactly the same J couplings ("magnetic equivalence"). This can happen in two ways: by molecular symmetry (mirror plane or rotation axis) or by rapid conformational change (bond rotation). The simplest example of bond rotation is a methyl (-CH$_3$) group. If it were stationary, we might expect different chemical shifts for each of the three protons and different J couplings due to differences in dihedral angle. In fact, all methyl groups rotate very rapidly about the bond connecting to the rest of the molecule, making all three protons equivalent. The same can be said of a *tert*-butyl group (-C(CH$_3$)$_3$), which has nine equivalent protons and three equivalent ^{13}C nuclei. Cyclohexane has only one proton chemical shift at room temperature because the chair conformation (with axial and equatorial protons) rapidly interchanges with the other chair conformation, exchanging the roles of axial and equatorial so rapidly that all protons experience a single, average chemical shift.

Another way to achieve equivalence is by symmetry. For flexible molecules, we always arrange the molecule in the most symmetric conformation to examine its symmetry properties. Diethyl ether (CH$_3$–CH$_2$–O–CH$_2$–CH$_3$) has only two proton chemical shifts (Fig. 2.12): The four CH$_2$ protons are equivalent and the six CH$_3$ protons are equivalent due to symmetry. Each of the methyl groups contains three equivalent protons due to rotation of the CH$_3$–CH$_2$ bond, and the mirror plane in the center (perpendicular to the plane of the paper) reflects the H$_f$ methyl group into the H$_e$ methyl group, making them equivalent. The mirror plane also converts H$_a$ (coming out of the paper) into H$_c$ (also coming out of the paper), making them equivalent, and H$_b$ (going into the paper) into H$_d$. Finally, there is another mirror plane in the plane of the paper that converts H$_a$ into H$_b$ and H$_c$ into H$_d$. Thus, all four methylene (CH$_2$) protons are equivalent. The spectrum consists of a triplet at about 1.2 ppm (area = 6) and a quartet at about 3.4 ppm (area = 4), with a coupling constant of about 7 Hz (free rotation). Note that protons within an equivalent group do not split each other—we will see why this is when we consider the effect of strong coupling (second-order splitting).

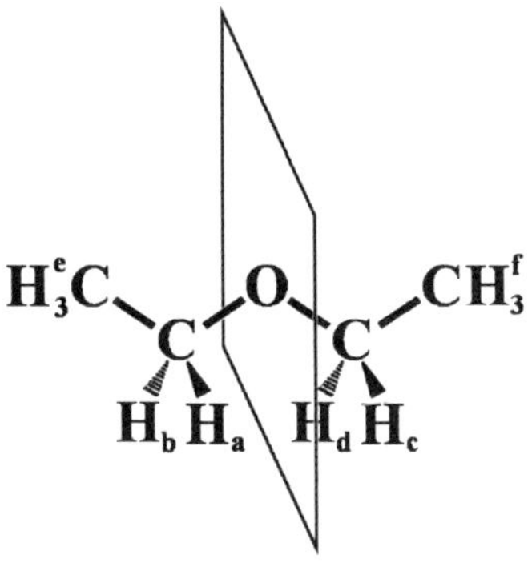

SYMMETRY AND CHIRALITY IN NMR   55

H_3^eC—C—O—C—CH_3^f
    $H_b H_a$ $H_d H_c$

Figure 2.12

In many cases you do not need to talk about formal symmetry elements to see the equivalence of protons. Just look at the world from the point of view of the proton: What kind of environment does it find itself in? If the world looks exactly the same (or is a mirror image) from the point of view of one proton as it does from another, they will be chemically equivalent, that is, they will have the same chemical shift. For example, in 4-bromotoluene (4-bromo-1-methylbenzene), the two protons adjacent to the bromine are chemically equivalent. Imagine sitting at a six-sided table, and you see a bromine seated at your right and a proton to its right, a proton seated to your left and a methyl group to its left, and another proton across from you. Now, if you sit in the other position next to the bromine, you have a bromine to your left and a proton to its left, a proton seated on your right with a methyl group to its right, and a proton across from you. Except for the mirror image relationships, which have no effect on chemical shift, you are in the same environment. If there is a difference, no matter how far away in the molecule, the two protons are not chemically equivalent. If the difference is far enough away, there may be little or no difference in chemical shift, but there is no chemical equivalence in the formal sense.

Chiral molecules do not have mirror planes, but they can have rotation axes as symmetry elements. The molecule shown in Figure 2.13 is chiral in its central oxygen-bridged tricyclic ring system as well as in the two R group substituents. But rotation about the C_2 axis by 180° yields the identical molecule, transforming H_a into H_b and H_c into H_d. Even the chiral R groups are interchanged by the rotation, so that there are exactly half the number of unique ^1H resonances as one would predict just by counting protons. Without mass spectrometric verification of the molecular weight, we might propose a "monomer" structure rather than the dimer structure shown. Integration of peak areas only gives us the relative number of protons represented by each peak, not the absolute number, so it is often difficult to confirm the presence of symmetrical dimers and higher multimers by NMR alone.

A methyl group always forms an equivalent group of three protons, but saturated methylene (X–CH_2–Y) groups are more complicated. In an achiral molecule, such as diethyl ether, they are always a chemically equivalent pair. But if the molecule has a chiral center, it cannot have a mirror plane, and in most cases the two protons of the CH_2 group will not be chemically equivalent. Thus, in most of the interesting molecules such as natural products and biological molecules, each CH_2 group will give rise to two proton resonances unless they have coincidentally the same chemical shift (a "degenerate" pair). The two

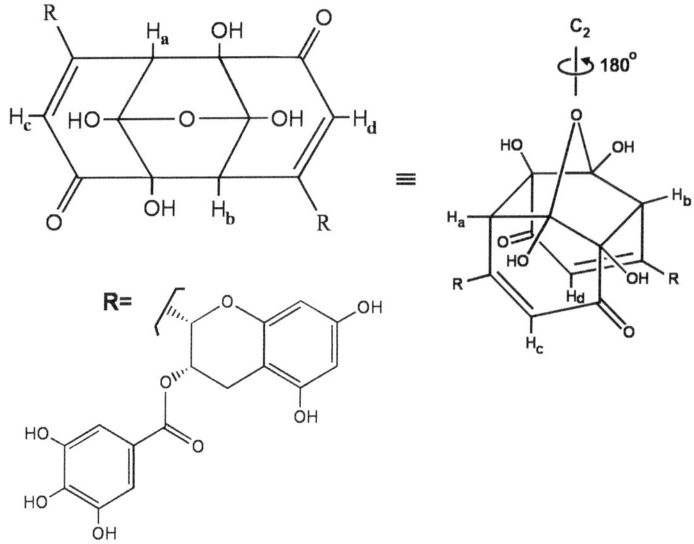

Figure 2.13

nonequivalent protons will show a large two-bond coupling (13–16 Hz) in addition to any other (vicinal and/or long-range) couplings. The nonequivalent protons of a methylene group are often referred to as "diastereotopic" protons. Formally, if you do not have a mirror plane in the X–C–Y plane or a C_2 axis bisecting the H–C–H angle, the two methylene protons are nonequivalent. The methylene protons of a terminal olefin ((R,R')C=CH$_2$) are a different case; they are equivalent only if the two R groups are identical (C_2 axis bisecting the H–C–H angle) or mirror images (mirror plane through C=C and bisecting the H–C–H angle). If they are nonequivalent, the two-bond coupling is small (0–2 Hz).

2.6 THE ORIGIN OF THE CHEMICAL SHIFT

Each type of nucleus (each specific isotope like ^{1}H) has a characteristic resonant frequency (precession frequency or Larmor frequency) in a given external magnetic field, B_o. The simple relationship $\nu_o = \gamma B_o/2\pi$ shows that the Larmor frequency depends only on the "magnet strength" of the nuclear magnet (the magnetogyric ratio γ) and the strength of the external magnetic field B_o. This is how we can "tune in" to a particular nucleus on the spectrometer, by setting the base frequency (e.g., 500 MHz for ^{1}H and 125 MHz for ^{13}C on an 11.7 T spectrometer). But if all protons in a molecule had exactly the same resonant frequency, the technique would be useless because we would see a single peak in the spectrum representing all of the protons. In fact, as we have seen, there are slight differences in resonant frequency depending on the chemical environment of the nucleus within a molecule. The relationship still holds that resonant frequency is exactly proportional to external field strength, but it is the local magnetic field strength at the position of the nucleus that is important: the effective field B_{eff},

$$\nu_o = \gamma B_{\text{eff}}/2\pi$$

This local magnetic field is slightly less than the applied magnetic field, B_o, due to the effect of the electron cloud (bonding and nonbonding electrons) surrounding the nucleus. This cloud of electrons "shields" the nucleus from the applied magnetic field by a tiny factor, on the order of parts per million, of the applied field:

$$B_{\text{eff}} = B_o(1 - \sigma)$$

The shielding factor, σ, is related to the chemical shift in parts per million:

$$\delta = 10^6 \times (\sigma_o - \sigma)$$

where σ_o is the shielding factor for a reference compound such as tetramethylsilane (TMS) that defines the zero of the chemical shift scale. Note that δ gets smaller as there is more shielding and larger as there is less shielding. We can view the right-hand side of the spectrum as relatively "shielded" (upfield, small δ) and the left-hand side as relatively "deshielded" (downfield, large δ).

The physical origin of this shielding by electrons is relatively easy to explain. The cloud of electrons surrounding a nucleus begins to circulate when the sample is placed in a magnetic field. This is a general phenomenon of physics: If you place a closed circle of wire in a magnetic field, a current will be induced to flow around the circle. This induced current will create a new magnetic field, just as any coil of wire with a current, and the direction of the induced current will always be such that the new magnetic field will oppose the original magnetic field that created it. This is called Lenz's law. The stronger the original magnetic field, the more the current will flow in the wire loop and the stronger will be the new, opposing magnetic field. Thus, the opposing field is proportional to the original field. Returning to the nucleus in a cloud of electrons, the electrons are mobile and thus form a kind of circle of wire around the nucleus (Fig. 2.14). When the sample is inserted in the magnetic field, the electrons begins to circulate around the nucleus (the induced current) and produce a magnetic field that opposes the B_o field at the center of the current (i.e., at the nucleus). This induced field (B_i) is proportional to the external field and subtracts from it, reducing the effective field felt by the nucleus:

$$B_{\text{eff}} = B_o - B_i = B_o - \sigma B_o = B_o(1 - \sigma)$$

$$\nu_o = \gamma B_{\text{eff}}/2\pi = \gamma B_o(1 - \sigma)/2\pi = \gamma B_o/2\pi - \gamma B_o \sigma/2\pi \, (\sigma \ll 1)$$

The change in resonant frequency (in hertz) is thus proportional to B_o for a given shielding constant σ, that is, for a given nucleus at a particular position in a molecule. This is exactly the effect we saw in Figure 2.1, as the field strength B_o is increased. To make chemical shifts the same regardless of magnet strength, we use the δ scale in parts per million, where the proportionality to B_o is already taken into account:

$$1 \, \text{ppm} = \gamma B_o/2\pi \times 10^{-6}$$

Shielding is just a combination of electron density in the vicinity of the nucleus and the ease of circulation of those electrons. For a proton, there is only one bond to the rest of the molecule, and the electron density around the proton is affected primarily by the electron-withdrawing effect of electronegative atoms that are nearby in the bonding network. For

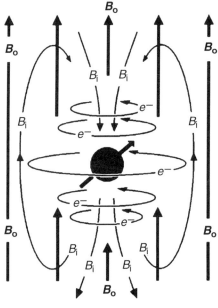

Figure 2.14

example, a proton bound to an oxygenated carbon (H–C–O) experiences a "deshielding" effect (downfield shift) because the electronegative oxygen pulls electron density toward it, making the carbon atom slightly positive and displacing electron density away from the proton and toward the carbon. This reduction of electron density reduces the circulating current around the proton, leading to a reduction in the opposing magnetic field created by that current. Thus, the proton is less shielded from (more "exposed" to) the applied magnetic field, B_o, and its resonant frequency increases (downfield shift). The magnitude of this change in electron density is miniscule, on the order of a few parts per million.

2.6.1 Through-space Effects

The deshielding effect of electronegative groups operates by displacing electrons in bonds, effectively decreasing the electron density immediately surrounding the proton. Another effect arises when induced magnetic fields are strong enough to extend through space from other atoms or molecular subunits to the point occupied by the proton. The classic example of this is the benzene ring, which has two mobile clouds of π electrons, one above and one below the plane of the ring (Fig. 2.15). In the external (B_o) magnetic field, the π electrons circulate according to Lenz's law, generating an induced magnetic field (B_i) that opposes B_o *at the center* of the circulating current. The induced field lines are circular, however, extending from the bottom upward around the outside of the benzene ring and then descending again into the center of the ring. At the position of the proton, at the outside of the benzene ring, the induced field is aligned with the B_o field so that it adds to the laboratory field. This strong deshielding effect shifts the benzene protons downfield to more than 7 ppm on the δ scale. These larger scale electron currents due to loosely bound π electrons in double bonds and conjugated systems generate stronger induced fields that can extend

THE ORIGIN OF THE CHEMICAL SHIFT

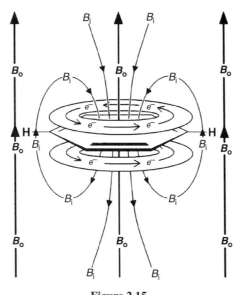

Figure 2.15

through space several Angstroms and influence chemical shifts. For the benzene ring, the effect is shielding (upfield shifting) in the region directly above and below the ring, while it is deshielding (downfield shifting) in the plane of the aromatic ring. One can imagine two cones extending from the center of the ring, one above the ring and another cone below the ring; protons inside one of the cones will be shielded (upfield shifted) and protons outside of the cones will be deshielded (downfield shifted). The effect diminishes as the proton moves away from the center of the ring. The same pair of cones can be visualized above and below the plane of a single unsaturation (C=C or C=O), leading to zones of "anisotropic" shielding and deshielding.

In rigid molecules, we sometimes see large chemical shift differences (up to 2 ppm) between the two protons of a CH_2 group if there is a double bond or aromatic ring nearby. The inductive effects of electronegative groups will be the same for these two protons because each one has the same through-bond relationship to the rest of the molecule, but the difference in chemical shift is largely due to their different positions in space relative to the plane of the unsaturated group (π bond). In globular proteins the monomer unit (-NH–CH(R)–CO-) of the biological polymer can have unique chemical shifts even when there are many of the same monomer unit (e.g., alanine: R = CH_3) in the polypeptide chain. For example, if there are six alanine residues (A4, A15, A26, A78, A92, and A126) in a protein, there might be six different chemical shifts for the six methyl doublets. This "dispersion" or spreading out of chemical shifts for identical monomer units is due in large part to the proximity and orientation of aromatic rings (side chains of nearby aromatic amino acids). Proteins with few aromatic amino acids usually show more overlap and are more difficult to work with in NMR structure determination. In NMR of natural products, the use of d_6-benzene (C_6D_6) as a solvent can sometimes "spread out" overlapped chemical shifts in the same way, due to the anisotropic (relative orientation dependent) effects of π-electron circulation in the solvent molecules on solute chemical shifts.

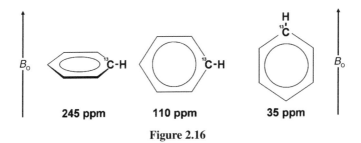

Figure 2.16

2.6.2 Chemical Shift Anisotropy

The amount of electron circulation (and thus the intensity of the induced field) is dependent on the orientation of the molecule with respect to the B_o field direction. In the above example of the benzene ring, we assumed that the plane of the aromatic ring is perpendicular to the B_o field vector. In fact, in solution the molecule is rapidly reorienting itself and samples all orientations equally over time (rapid isotropic tumbling). If we could lock the ring in place and measure chemical shifts, we would see three different chemical shifts for the three principle orthogonal orientations of the molecule. For example, the ^{13}C chemical shift of benzene is 245 ppm for the orientation with the ring plane perpendicular to B_o (Fig. 2.16) because this orientation gives the maximum electron circulation. For the orientation with the ring plane parallel to the B_o field and the ^{13}C–H vector perpendicular to B_o, the chemical shift is 110 ppm, and for the other parallel orientation with the ^{13}C–H vector parallel to B_o, the shift is only 35 ppm. In the two parallel cases, the electron circulation would have to cross the plane of the benzene ring, which is a node (zero electron density) in the π orbital. The observed chemical shift in solution (the "isotropic" chemical shift) is the average of these three fixed-orientation chemical shifts:

$$\delta_{\text{iso}} = (245 + 110 + 35)/3 = 130\,\text{ppm}$$

The amount of variation of chemical shift with the orientation is called the *chemical shift anisotropy*, or CSA. CSA is simply the difference between the smallest fixed-position chemical shift and the average of the other two fixed-position chemical shifts:

$$\text{CSA} = (245 + 110)/2 - 35 = 177.5 - 35 = 142.5\,\text{ppm}$$

By comparison, a saturated methine carbon (C–H) has a CSA of only 25 ppm because the mobility of electrons around the carbon nucleus is much less in an sp^3-hybridized carbon and depends much less on the orientation of the C–H bond with respect to B_o. In solution-state NMR we only see the isotropic chemical shift, δ_{iso}, and the fixed-position chemical shifts and the CSA value are obtained from solid-state NMR measurements. Although CSA does not affect chemical shifts in solution, it does contribute to NMR relaxation and can be exploited to sharpen peaks of large molecules such as proteins in solution. For large molecules, such as proteins, nucleic acids, and polymers, or in viscous solutions, molecular tumbling is slow and CSA broadens NMR lines due to incomplete averaging of the three principle chemical shift values on the NMR timescale. Like isotropic chemical shifts, CSA in parts per million is independent of magnetic field strength B_o but is proportional to B_o when expressed in hertz. Because linewidths are measured in hertz, the line-broadening effect of CSA becomes more significant as we increase the field strength.

2.7 *J* COUPLING TO OTHER NMR-ACTIVE NUCLEI

Proton signals will also be split by (*J* coupled to) other NMR-active nuclei that are nearby in the bonding network. Two commonly encountered spin-½ nuclei are ^{19}F and ^{31}P (both 100% abundance). Common NMR-active nuclei that must be introduced by isotopic labeling include deuterium (^{2}H, spin 1), and ^{13}C and ^{15}N (both spin ½).

Couplings to ^{31}P are similar to ^{1}H–^{1}H couplings in magnitude for two- and three-bond relationships. For example, dimethyl methylphosphonate ($CH_3P(O)(OCH_3)_2$) gives two doublets in the ^{1}H NMR spectrum: CH_3P at 1.481 ppm (integral 3H) and CH_3O at 3.741 ppm (integral 6H). The CH_3O doublet is 11.0 Hz wide ($^3J_{HP} = 11.0$) and the CH_3P doublet is 17.4 Hz wide ($^2J_{HP} = 17.4$). In the ^{13}C spectrum we would expect two peaks because there are only two distinct ^{13}C chemical shifts: CH_3P and CH_3O. But instead we see four peaks: a doublet at 52.19 ppm (CH_3O, $^2J_{CP} = 6$ Hz) and another doublet at 9.83 ppm (CH_3P, $^1J_{CP} = 144$ Hz). We do not expect to see splitting in ^{13}C spectra because they are ^{1}H decoupled, but we have to remember that ^{1}H is the only NMR-active nucleus that is decoupled: all other splittings will show up in routine ^{13}C spectra. In particular, the one-bond ^{13}C–^{31}P coupling (144 Hz) is large enough that, especially on lower field instruments, the doublet is easily mistaken for two different carbon chemical shifts. As with ^{1}H–^{1}H couplings, long-range (>3 bond) couplings are observed in conjugated systems. Triphenylphosphine oxide (($C_6H_5)_3P=O$) shows coupling from ^{31}P to the *ipso* carbon ($^1J_{CP} = 103$ Hz), to the *ortho* carbon ($^2J_{CP} = 10$ Hz), to the *meta* carbon ($^3J_{CP} = 12.5$ Hz), and to the *para* carbon ($^4J_{CP} = 3.5$ Hz).

Coupling to ^{19}F is also similar to ^{1}H–^{1}H couplings, but there are some unusually large couplings as well. Geminal ^{1}H–^{19}F couplings on saturated (sp^3-hybridized) carbons are around 50 Hz, and on unsaturated (sp^2-hybridized) carbons they can be around 80 Hz. Vicinal ^{1}H–^{19}F couplings in rigid saturated systems with an *anti* relationship (180° dihedral angle) are around 40 Hz, and in a fluoro-olefin the ^3J values are around 20 for *cis* and 50 for *trans*. In flexible saturated systems, vicinal couplings are similar to ^{1}H–^{1}H couplings. Long-range couplings can be significant: in aromatic rings the 4-bond or *meta* coupling (6–8 Hz) is similar to the vicinal or *ortho* coupling (8–10 Hz), and the 5-bond or *para* coupling is significant (~2 Hz). Coupling of ^{19}F is significant in ^{13}C spectra, again because only coupling to ^{1}H is removed by decoupling. In fluorobenzene, for example, the J_{CF} couplings are 245, 21, 8, and 3 Hz for ^1J, ^2J, ^3J, and ^4J, respectively. A CF_3 group will be split into a quartet (1:3:3:1) with very wide coupling. Trifluoroacetic acid, for example, gives two quartets in the ^{13}C spectrum: $^1J_{CF} = 282$ Hz (CF_3) and $^2J_{CF} = 44$ Hz (CO_2H). This splitting and the loss of the heteronuclear NOE can cause a fluorinated carbon to "disappear" into the noise. It is possible to decouple ^{19}F, but most spectrometers do not have the capability to decouple both ^{1}H and ^{19}F simultaneously.

Coupling to deuterium, ^{2}H, is observed for deuterated solvents and their residual peaks (from solvent molecules with one ^{2}H replaced by ^{1}H). ^{1}H–^{2}H coupling constants are proportional to the corresponding ^{1}H–^{1}H *J* value, reduced by a factor of about 7 ($\gamma_H/\gamma_D = 6.51$) due to the weaker nuclear magnet of deuterium. Because deuterium has a spin of 1, it has three spins states almost equally populated, and so it splits the ^{1}H signal into three equal peaks centered on the ^{1}H chemical shift position. Multiple ^{2}H splittings can be built up by diagramming just like with ^{1}H splittings, as long as each ^{2}H results in a split into three peaks of intensity ratio 1:1:1. In ^{13}C spectra we see the splitting by the directly bound ^{2}H nuclei, reduced by a factor of 6.5 from the corresponding ^{1}H–^{13}C couplings. For example, without ^{1}H decoupling CHCl$_3$ gives a doublet in the ^{13}C spectrum with $^1J_{CH}$ = 209 Hz, and CDCl$_3$ gives a 1:1:1 triplet in the ^{13}C spectrum with $^1J_{CD}$ = 32 Hz (209/6.5 = 32.15). Exchange of ^{2}H between the sample molecule and deuterated solvent can lead to loss of ^{1}H peak intensity, for example, at an activated position next to a ketone functional group in CD$_3$OD solvent. The carbon peak may disappear entirely from the ^{13}C spectrum due to splitting by ^{2}H into many smaller peaks that sink into the noise. In addition, carbons are much slower to relax when bound to ^{2}H (relative to ^{1}H), again due to the smaller nuclear magnet of deuterium, further reducing their intensity in the ^{13}C spectrum.

Coupling to ^{13}C is observed as weak satellites on either side of each ^{1}H resonance. Because the natural abundance of ^{13}C is only 1.1%, we have to consider two separate molecules and add together their contributions to the ^{1}H spectrum: one with our ^{1}H bound to ^{12}C (98.9% of the sample) and another with ^{1}H bound to ^{13}C (1.1% of the sample). In the ^{12}C case, we see the normal ^{1}H resonance with its splitting pattern from coupling to other protons (triplet, double-doublet, singlet, etc.). In the ^{13}C case, we see this same splitting pattern but divided into two identical patterns, one about 75 Hz downfield of the ^{1}H-(^{12}C) pattern and one about 75 Hz upfield of this pattern (Fig. 2.17). The exact distance is one half of $^1J_{CH}$, which is typically in the range of 150 Hz for a one-bond coupling between ^{13}C and ^{1}H. This doublet is 1.1% of the intensity of the central ^{1}H–^{12}C resonance, so each component of the doublet (each side) is 0.55% of the intensity of the central peak. For isotopically enriched ("labeled") compounds, the satellites are larger due to the greater abundance of ^{13}C. For example, a sample of ^{13}C-enriched methyl iodide (CH$_3$I) with 50% of ^{13}C will show a ^{12}CH$_3$I singlet (50% of total intensity) and a ^{13}CH$_3$I doublet centered on the same chemical shift (each side 25% of total intensity). This will look like a triplet with spacing of $^1J_{CH}$ and intensity ratio 1:2:1, but it is not a triplet and it is important to realize that this pattern is the superposition of two separate spectra from two different species (isotopomers) in solution.

Long-range (2 or 3 bond) coupling to ^{13}C is difficult to observe unless the sample is enriched. A sample of ethyl acetate (CH$_3$–C*O–OCH$_2$CH$_3$) with 100% ^{13}C label at the carbonyl carbon would give a ^{1}H spectrum with a doublet for the acetate methyl group ($^2J_{CH}$), a double-quartet for the methylene group ($^3J_{CH}$), and the normal triplet for the

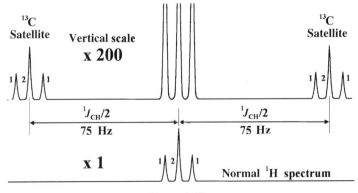

Figure 2.17

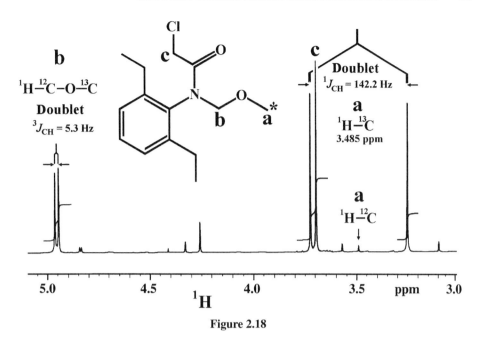

Figure 2.18

ethyl CH_3 group ($^4J_{CH} = 0$). These long-range couplings ($^2J_{CH}$ and $^3J_{CH}$) are similar in magnitude to ^1H–^1H couplings (0–10 Hz), whereas couplings over more than three bonds are extremely rare. If the enrichment was less than 100%, this spectrum would be superimposed on the normal spectrum of ethyl acetate (singlet, quartet, triplet) with intensities proportional to the quantity of each isotopic species (^{12}C or ^{13}C). Figure 2.18 shows the ^1H spectrum of alachlor herbicide with 99% ^{13}C at the methyl group: N–CH_2–O–$^{13}CH_3$. The methoxy group singlet is split into a doublet ($^1J_{CH}$ = 142 Hz) and the N–CH_2–O methylene singlet is also split into a doublet ($^3J_{CH}$ = 5.3 Hz). The 1% ^1H–^{12}C peak for the methoxy group can be seen at the center of the ^1H–^{13}C doublet. It is important to understand that we are observing ^1H here, not carbon nuclei. You cannot observe ^{12}C by NMR because it has spin 0 (no magnetic properties of the nucleus), but you can observe the protons attached to ^{12}C.

All heteronuclear couplings (couplings between two different kinds of nuclei—two different isotopes) are first order, without any distortion of peak intensities. The chemical shift difference between two different isotopes is usually on the order of megahertz to hundreds of megahertz, so even with large J couplings (hundreds of hertz), the shift difference is much, much larger than the J coupling, and there are no second-order (strong coupling) effects.

2.8 NON-FIRST-ORDER SPLITTING PATTERNS: STRONG COUPLING

The simple splitting patterns discussed above appear only when we have *weak coupling*: when the chemical shift difference between two nuclei (expressed in hertz) is much greater than the J coupling between them. In this case, the coupling pattern is symmetric with a maximum of 2^n peaks (for n coupled spin ½ nuclei), and the chemical shift is at the exact center of the pattern. When the chemical shift difference in hertz is on the same order

64 INTERPRETATION OF PROTON (^{1}H) NMR SPECTRA

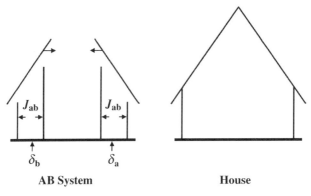

Figure 2.19

of magnitude as the *J* coupling, the quantum mechanical situation is more complex and we see distortions of peak intensity (nonsymmetric patterns), and in some cases new lines arise in the pattern. These second-order patterns are often observed as simple "leaning" of the classical (doublet, triplet, quartet, ...) patterns: the peaks on the "inside" (nearest to the chemical shift of the coupled nucleus) are taller (more intense) and the peaks on the "outside" (away from the chemical shift of the coupled nucleus) are shorter (weaker). For the simplest case of two protons strongly coupled to each other and no other protons, we call this an AB system. We use A and B because they are next to each other in the alphabet and indicate that the chemical shifts are very close together (Fig. 2.19). If we think of the sloping doublets as rooflines, we can use the analog to a cartoon house to remember the effect. The tall side of the leaning doublet always "points" toward the other spin that is splitting it. More complicated patterns can also "lean" when the *J*-coupled resonances get close to each other. In Figure 2.20 we see the simplest possible pattern from two adjacent methylene groups: X–CH$_2$–CH$_2$–Y. The two triplets "lean" toward each other so that the outer lines of the triplets are less than 1 in relative area and the inner lines are more than 1; the center lines still have relative area 2. Three methine (CH) groups in a row (Fig. 2.21) lead to doublets for the outer protons (H$_a$ and H$_c$) and a double doublet for the middle proton (H$_b$). The two doublets will "lean" toward their coupling partner in the center, and the H$_b$ pattern will lean both ways: the large coupling (J_{ab}) "leans" toward H$_a$ and the small coupling (J_{bc}) "leans" toward H$_c$. Another example is the case of three protons in a row on an aromatic ring, with

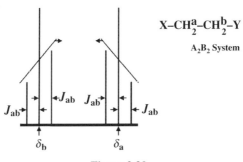

Figure 2.20

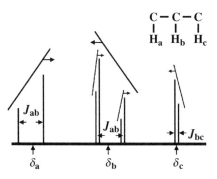

ABC System: $J_{ac} = 0$

Figure 2.21

the center proton being the most downfield (Fig. 2.22). If the center proton (H_b) has the same coupling constant to H_a and H_c (i.e., if $J_{ab} = J_{bc}$), it will appear as a triplet, leaning toward the chemical shift positions of H_a and H_c. H_a and H_c will appear as doublets, each leaning toward the H_b triplet. In reality, there is a smaller long-range coupling ($^4J_{HH}$) between H_a and H_c (the "*meta*" coupling), and we would see the H_a and H_c patterns further split by the smaller coupling. The distortion of this smaller coupling "points" to the right-hand side in the H_c pattern and to the left-hand side in the H_a pattern. It is important to keep in mind that these patterns can be considerably more complicated, but the "leaning" principle can often be recognized and used to interpret coupling patterns even when they are more complex.

Let's look at the simple AB system in more detail. Consider that the J coupling is held constant and the chemical shift difference $\Delta\nu$ is gradually reduced (Fig. 2.23). We

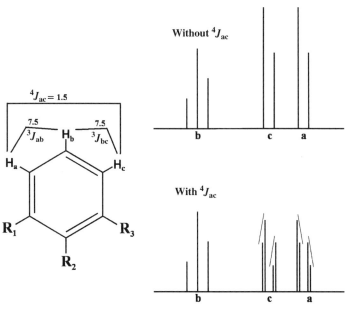

Figure 2.22

66 INTERPRETATION OF PROTON (^{1}H) NMR SPECTRA

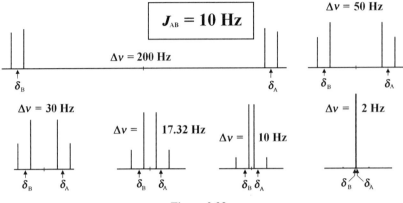

Figure 2.23

can do this by simply reducing the B_o field strength, so that the chemical shift difference (which is constant in units of parts per million) is reduced in units of hertz, while the J coupling is constant because it is independent of B_o. When the shift difference is very large ($\Delta \nu \gg J$), we see little or no "leaning" of the two doublets. As the shift difference gets smaller, the inner peaks grow and the outer peaks shrink. Another important point is that the chemical shift position is no longer halfway between the two lines of the doublet: it is now the weighted average of the two line positions, weighted by the peak intensities:

$$\delta = (\delta_1 I_1 + \delta_2 I_2)/(I_1 + I_2)$$

where δ_1 and δ_2 are the two line positions and I_1 and I_2 are the two line heights. You can get the exact line positions from a peak list and the line heights by measuring in millimeters with a ruler. Another way to say this is that as the two chemical shift positions, δ_a and δ_b, are brought closer together, the outer lines get shorter and farther away from the chemical shift positions and the inner lines get taller and closer to the chemical shift positions. The distance between the two lines of each doublet does not change, however. It is always equal to the coupling constant J_{ab}.

There is a point in this process where the distance between the inner lines is exactly J_{ab} so that the patterns look very much like a quartet, the 1:3:3:1 pattern resulting from a single resonance split by three equivalent protons (Fig. 2.23, $\Delta \nu = 17.32$ Hz). In this case $\Delta \nu / J = 1.732$ and the ratio of peaks in the AB system is exactly 1:3:3:1, because the outer lines have been reduced by 50% and the inner lines have been increased in intensity by 50%. Some people report this AB pattern as an "AB quartet," but this terminology is misleading and should not be used. The AB system always has two different chemical shifts coming from two distinct proton resonances, unlike a true quartet that comes from a single resonance. There is no way to distinguish a true quartet from an AB system with this coincidental spacing; you have to consider both possibilities and use the context of what you already know about the molecule to decide.

As we continue to reduce the B_o field, moving the δ_a position even closer to the δ_b position, the two inner lines become very close to each other, but they never meet or cross. The outer lines become so weak that they may appear as tiny bumps, or they may disappear into the noise, depending on the signal-to-noise ratio of the spectrum. At this point of near-equivalence of chemical shifts, the pattern looks very much like a doublet with a small coupling constant, especially if you do not notice the weak outer lines. This can be very confusing if you do not consider the possibility of a "tight AB" ($\Delta \nu < J$) system. The separation between the two inner peaks is not a J-coupling at all—it is a complex function of $\Delta \nu$ and J_{ab}. In the weak coupling limit, this distance is close to $\Delta \nu - J$, but in the strong coupling limit, it becomes close to $\Delta \nu$ because the inner lines are just inside the two chemical shift positions. If you cannot find the outer lines, there is really no way to determine δ_a, δ_b, and J_{ab}.

Finally, when the two chemical shifts are exactly equal ($\Delta \nu = 0$), the two inner lines become one and the two outer lines have zero intensity. The two protons H_a and H_b are chemically equivalent, that is, they have the same chemical shift, and we see no splitting at all. This explains, in a way, why equivalent protons do not split each other: they do split each other but the inner lines coincide and the outer lines have zero intensity. Theoretically, the pattern still has four lines, but we only observe one: a singlet. The same applies to any number of chemically equivalent protons, for example, a methyl group (CH_3). All three protons have the same chemical shift, and in the absence of any other coupling (e.g., CH_3O or CH_3C_q, where C_q is a quaternary carbon), we will see a singlet with area proportional to three.

The distance between the first and second line of the AB system, and between the third and fourth line, is always exactly equal to J_{ab}, even though the chemical shift positions are not exactly in the middle of these pairs. Here are some exact values for the intensities, the error in chemical shift (in hertz) if you just calculate the simple average of line positions for each doublet, and the hertz spacing of the two inner lines:

$\Delta\nu/J$	I(outer) (%)	I(inner) (%)	ν error/J	inner spacing/J
20	95	105	0.01	19.02
5	80	120	0.05	4.10
3	68	132	0.08	2.16
1.732	50	150	0.13	1.00
1	29	171	0.21	0.414
0.5	11	189	0.31	0.118
0.2	2	198	0.41	0.02
0	0	200	0.5	0

For a J coupling of 10 Hz, the $\Delta\nu/J = 1.732$ case above (1:3:3:1 ratio) would be observed when $\Delta \nu = 17.32$ Hz, which is 0.29 ppm on a 60 MHz instrument, 0.058 ppm on a 300 MHz instrument, and 0.029 ppm on a 600 MHz instrument. Clearly, a much closer similarity of chemical shifts in parts per million is required to see strong coupling on high-field instruments. This is one reason to pay the big money for higher field—to simplify spectra and see mostly first-order splitting patterns. The chemical shift error due to simply averaging the line positions of each doublet of the AB pattern would be 1.3 Hz, which on a 60 MHz instrument is 0.022 ppm (a significant error) and on a 600 MHz instrument it is only 0.002 ppm.

In general, if we define $\Delta\nu' = [J^2 + \Delta\nu^2]^{1/2}$, then relative to the center of the overall pattern the inner lines are $(\Delta\nu' - J)/2$ away and the outer lines are $(\Delta\nu' + J)/2$ away, with intensities of $1 + J/\Delta\nu'$ (inner) and $1 - J/\Delta\nu'$ (outer).

The AB system is a basic building block that can be expanded to a more complex system with additional weak couplings by simply building the splitting diagram (Fig. 2.24). Start with the distorted AB system and then diagram in the additional coupling by moving $J/2$

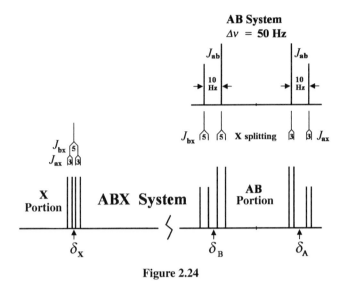

Figure 2.24

to one side and $J/2$ to the other side of each line in the AB system. This is called an ABX system: The X is chosen far away in the alphabet to indicate that its chemical shift is far away (weak coupling) from the H_a and H_b resonances, which are close together relative to J_{ab}. There are two additional coupling constants: J_{ax} and J_{bx}, and generally they will not be the same. Simply split the A half of the AB pattern with the J_{ax} coupling and split the B half of the pattern with the J_{bx} coupling. Because it is a weak coupling, the additional splittings result in precise 50:50 intensity ratios, but they retain the distorted intensities of the parent lines. This is an extremely common system; for example, the CH–CH$_2$ fragment in a chiral molecule will lead to an ABX system if it is isolated from any other couplings: CH_x–CH_aH_b. A classic example of this occurs in the amino acid unit of peptides and proteins: in D$_2$O the NH proton exchanges with D from solvent, so we have ND–CH–CH$_2$–X for many of the amino acids: Asp, Asn, Cys, Ser, His, Phe, Trp, and Tyr. These systems are also referred sometimes as AMX systems, implying that the A and M chemical shifts are farther apart than in an ABX system. Because amino acids are chiral, the two protons of the CH$_2$ group are nearly always nonequivalent and the CH (alpha proton) is usually considerably downfield of the pair due to the bond to electronegative nitrogen. If H_a and H_b are geminal protons on a saturated carbon, they will have a large coupling J_{ab} (2-bond coupling 16–18 Hz) and the vicinal couplings J_{ax} and J_{bx} will likely be smaller: for amino acids, 8–10 Hz for *anti*, 3–6 Hz for *gauche*, and 6–7.5 Hz for averaged conformations. The X position (H_x) should appear as a double doublet: δ_x split by J_{ax} and then by J_{bx}. If δ_a and d_b get very close ($\Delta\nu_{ab} < J_{ab}$), it is possible that new lines will appear in the H_x pattern, making it more complex than a simple double doublet.

Another example that illustrates both the consequences of asymmetry and the use of the AB pattern as a building block is shown in Figure 2.25. The dihydropyridine has a mirror plane perpendicular to the plane of the double bonds, passing through the nitrogen and the CH–CH$_3$ group. This makes the two ethyl groups chemically equivalent, and we can label the methylene (CH$_2$) protons coming out of the paper H_a and the ones pointing back into the paper H_b. But we cannot say that H_a is equivalent to H_b because there is no

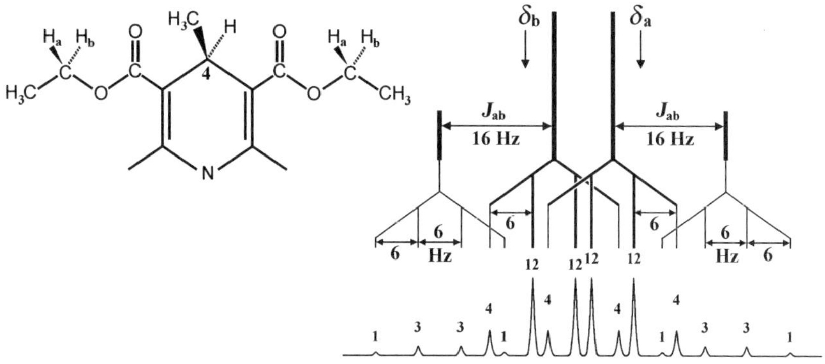

NON-FIRST-ORDER SPLITTING PATTERNS: STRONG COUPLING

Figure 2.25

plane of symmetry in the plane of the double bonds (the plane of the paper)—such a plane would reflect the H into the CH_3 at the 4-position of the dihydropyridine ring. This is an interesting case because the molecule as a whole is not chiral, as it possesses a mirror plane perpendicular to the plane of the double bonds. Because there is no mirror plane between the geminal pairs of the CH_2 groups, these pairs represent two equivalent AB systems, each with two distinct chemical shifts. This makes sense because H_a is on the side of the CH_3 group at C-4 and H_b is on the side of the H atom at C-4: two different chemical environments. We can predict the spectrum by first constructing an AB pattern ($J_{ab} = 16$ for geminal protons on an sp^3 carbon) from the two chemical shift values, δ_a and δ_b, and then building a 1:3:3:1 quartet ($J = 6$ Hz) from each line of the AB pattern, to generate the AB part of an ABX_3 system. In this case the outer lines are one fourth of the height of the inner lines, so the two outer quartets have intensities 1:3:3:1 whereas the two inner quartets have intensity ratio 4:12:12:4. This is a surprisingly complex pattern for a pair of equivalent ethyl esters, for which we expect a simple quartet (CH_2) and triplet (CH_3) pattern.

There are many computer programs available for calculating spectra from the chemical shifts and J coupling values. NMR is unique in that all line positions and intensities are easily calculated as long as the NMR parameters-chemical shifts and J values-are known. There are programs that will simulate the spectrum and compare it to a real spectrum, incrementing the NMR parameters in an iterative process until the greatest possible similarity is obtained between calculated and observed spectra. In this way all of the shifts and couplings can be determined even in systems that are too complex to analyze directly by diagraming and measuring peak separations.

2.8.1 Virtual Coupling

A common phenomenon occurs at lower magnetic fields when one nucleus (H_a) is coupled to another (H_m) that is far away from it in chemical shift but coupled to a third spin, H_n, that is very close to H_m in chemical shift:

$$H_a\text{------------}H_m\text{—}H_n$$

We say that H_m and H_n are *strongly coupled*, which will distort the multiplet patterns of these two spins. We expect a simple doublet for H_a (weak coupling to H_m only), but instead

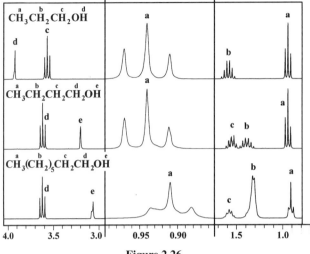

Figure 2.26

the H_a resonance is a more complex multiplet, as if it were coupled to H_n as well as to H_m. This is called "virtual coupling" because the H_a nucleus, which has no J coupling to H_n, appears to be coupled to it because of the strong coupling between H_m and H_n. It is like a very tight social group: If you get to know one of them, you end up knowing all of them. A general way of stating this is that *any nucleus that is coupled to one member of a strongly coupled group of nuclei will behave as if it is J coupled to all of the members of the group.* This phenomenon can lead to some very baffling results if you do not take it into account. Consider, for example, the straight-chain alcohols $CH_3-(CH_2)_n-CH_2OH$ (Fig. 2.26). For n-propanol ($n = 1$), the CH_3 resonance (0.94 ppm) is well separated from the neighboring CH_2 resonance (1.59 ppm), which in turn is well separated from the CH_2OH resonance (3.57 ppm). Both vicinal couplings ($J_{ab} \sim 7$ Hz, $J_{bc} \sim 7$ Hz) can be described as weak couplings because the chemical shift differences are fairly large ($v_b - v_a = (1.59 - 0.94) \times 250 = 163$ Hz; $v_c - v_b = (3.57 - 1.59) \times 250 = 495$ Hz). A nearly perfect triplet (H_a), sextet (H_b), and triplet (H_c) are observed with only slight "leaning." The OH resonance (d) does not show J coupling due to rapid exchange. For n-butanol ($n = 2$), the CH_2 group next to the CH_3 is farther from the electronegative oxygen, so it resonates farther upfield (1.38 ppm), closer to the "generic" hydrocarbon CH_2 chemical shift of around 1.3 ppm. Still, the chemical shift difference ($v_a - v_b = (1.38 - 0.94) \times 250 = 110$ Hz) is significantly larger than the J value, and we can describe the H_a–H_b coupling as "weak." The next CH_2 resonance (H_c), however, is still somewhat distant from the oxygen and resonates at 1.55 ppm. The H_b–H_c coupling (~ 7 Hz) is not ideally weak because the chemical shift separation is only $(1.55 - 1.38) \times 250 = 42.5$ Hz, so that $\Delta v/J = 6$. Some distortion of the H_b sextet and the H_c quintet can be seen in the spectrum (Fig. 2.26, right).

The spectrum of *n*-octanol ($n = 6$, Figure 2.26, bottom) is dramatically different. The first five CH_2 groups after the CH_3 group all resonate at nearly the same generic hydrocarbon shift, at 1.32 ppm (peak b). The CH_3 peak falls at 0.83 ppm, essentially the generic hydrocarbon value for a methyl group. Only the difference in substitution (CH_3 vs. CH_2) provides a chemical shift difference between the *a* and *b* resonances because the OH group is far away and has no effect. Still, this difference is fairly large relative to

J: $\Delta v = (1.32 - 0.83) \times 250 = 122.5$ Hz $= 17.5 J$. The CH$_3$ to CH$_2$ coupling is a weak coupling. But the CH$_3$ triplet is broadened and highly distorted, in contrast to the clean, sharp triplets observed for *n*-propanol and *n*-butanol. The reason is that the CH$_3$ resonance (C8) is coupled to a CH$_2$ group (C7) that is very strongly coupled to the next CH$_2$ resonance in the chain (C6), leading the CH$_3$ resonance to show "virtual coupling" to the protons on C6. In fact, the protons on C2–C6 (five CH$_2$ groups) all have nearly the same chemical shift (1.32 ppm) and each vicinal relationship has a J coupling of around 7 Hz. This makes the CH$_2$ resonance next to the CH$_3$ group part of an extremely strongly coupled family of 10 protons, and coupling to two of these (the CH$_2$ next to the CH$_3$) is like coupling a little bit to all of them. This explains the very broad components of the CH$_3$ triplet. This phenomenon is seen in all long, straight hydrocarbon chains: a very broad and distorted CH$_3$ resonance around 0.83 ppm and a very tall and broad peak for all of the "hydrocarbon" CH$_2$ protons around 1.3 ppm. Even at higher field (e.g., 600 MHz), the "pack" of CH$_2$ groups is very strongly coupled and the "ugly" CH$_3$ resonance is not improved.

2.9 MAGNETIC EQUIVALENCE

Two or more protons may have identical chemical shifts (chemical equivalence) but may not have the same coupling constant (J) to another proton in the molecule. In this case they are chemically equivalent but not magnetically equivalent. In this case we usually label protons with a "prime": H$_a$ and H$_{a'}$, to indicate their chemical equivalence but still distinguish them. For example, in a para-disubstituted benzene ring, X–*p*-C$_6$H$_4$–Y, we have chemical equivalence of the two protons *ortho* to group X (H$_a$ and H$_{a'}$) and of the two protons *ortho* to group Y (H$_b$ and H$_{b'}$) because of the symmetry of the benzene ring. There is an ortho coupling ($^3J_{HH}$) of 8–10 Hz between H$_a$ and H$_b$, and between H$_{a'}$ and H$_{b'}$ (Fig. 2.27), and if the two systems were completely isolated from each other and identical, we could analyze this as an AB system. In fact, many such compounds show fairly clean AB systems in the ^{1}H spectrum, but a closer look reveals that there are other lines, although the whole pattern is symmetrical. The reason for the additional lines is that there is a significant *meta* ($^4J_{HH}$) coupling between H$_a$ and H$_{a'}$ and between H$_b$ and H$_{b'}$. Even though these pairs are chemically equivalent and should not split each other, their coupling complicates the overall

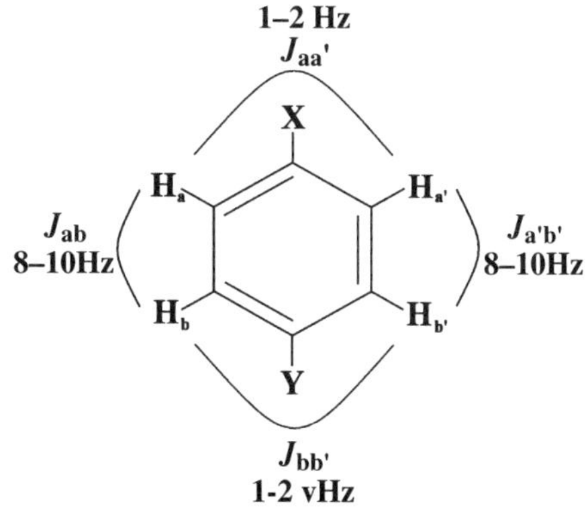

Figure 2.27

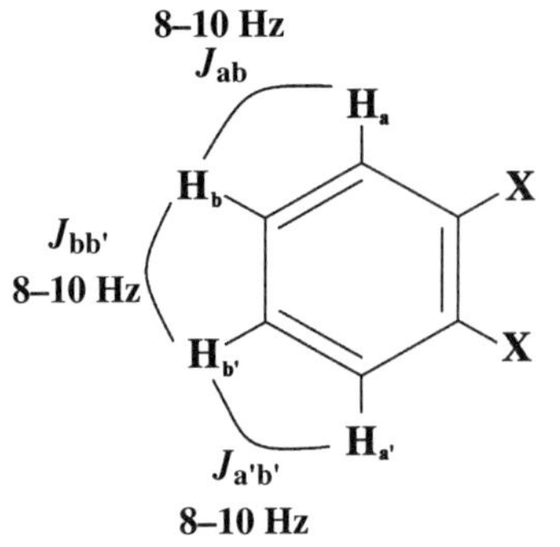

Figure 2.28

pattern. This is because H_a and $H_{a'}$ are chemically equivalent but are not magnetically equivalent since we can find a third proton, H_b, that has a different coupling to H_a (8–10 Hz—*ortho*) than it does to $H_{a'}$ (~0 Hz—*para*). Instead of calling this an AB system (or two identical AB systems), we have to call it an AA′BB′ system, which has a more complicated analysis. Computer programs can calculate the spectrum precisely, including the additional lines, given the values of δ_a, δ_b, J_{ab}, $J_{aa'}$ and $J_{bb'}$.

A more dramatic example occurs for an *ortho*-disubstituted benzene with two identical substituents: X–*o*-C_6H_4–X (Fig. 2.28). In this case we can label the four adjacent protons on the benzene ring as H_a, H_b, $H_{b'}$ and $H_{a'}$ in that order. The two systems are very tightly connected because H_b and $H_{b'}$ have a large coupling (*ortho* or $^3J_{HH}$ = 8–10 Hz), similar to J_{ab} and $J_{a'b'}$. This pattern is very distorted, and in many instances there is no recognizable underlying AB pattern.

Earlier on in this chapter, the X–CH_2–CH_2–Y system was mentioned and predicted to give a pair of leaning triplets (A_2B_2 pattern). In fact, it can be much more complex because this is actually an AA′BB′ system. This might seem surprising, because there is no chiral center, and the molecule can be drawn with a mirror plane interchanging H_a and $H_{a'}$ (Fig. 2.29). But again the criterion for magnetic equivalence of H_a and $H_{a'}$ is not met: The coupling from H_a to H_b is not in general the same as the coupling between

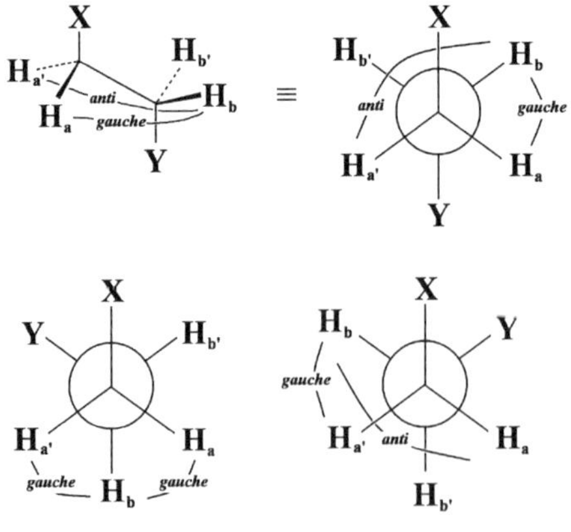

Figure 2.29

$H_{a'}$ and H_b. This is because the dihedral angle (in the conformation where X and Y are *anti*) between H_a and H_b is 60° (*gauche*), and between $H_{a'}$ and H_b it is 180° (*anti*). This would lead us to expect a smaller coupling constant (~5 Hz) for J_{ab} and a larger J value (8–10 Hz) for $J_{a'b}$. One might argue that we have picked an arbitrary conformation—what about the two *gauche* conformations? In one of these conformations, the criterion for magnetic equivalence is approximately satisfied (both dihedral angles are *gauche*). But if H_a and $H_{a'}$ are not magnetically equivalent in any one of the conformations, they cannot be considered magnetically equivalent, and we have to analyze the whole system as an AA'BB' system, not an A_2B_2 system. In fact, in many simple cases the X–CH$_2$–CH$_2$–Y system gives a very complex pair of resonances, symmetrical about the center $((\delta_a + \delta_b)/2)$ but not resembling in any way a pair of triplets leaning toward each other. In some cases, the outer lines of the individual resonances are more intense than the inner lines, giving them a very odd appearance because we always see overlap in the inner parts of a first-order multiplet, not in the outer lines.

3

NMR HARDWARE AND SOFTWARE

In this chapter, we will follow the process of acquiring and processing an NMR spectrum in chronological order: preparing the sample, inserting it in the spectrometer, locking, shimming, acquiring the free induction decay (FID) (with a detailed look at the hardware), and processing the data. The goal is to obtain a general understanding of how the spectrometer works and how the data are processed, independent of any specific spectrometer (Bruker, Varian, etc.) or NMR software package. As these steps are the same in all spectrometers and with all software, you can apply this knowledge to the specific instrumentation and programs available to you even though the terminology and specific software commands and parameters may be different.

At this point, we need to discuss briefly the instrument manufacturers ("vendors") and the models (generations) of NMR spectrometers in use today. There are three main vendors right now: Bruker (based in Germany and Switzerland), Varian (Palo Alto, CA), and JEOL (Japan). We will concentrate on Bruker and Varian due to lack of experience with JEOL. The earliest commercial instruments (continuous wave) are gone to the scrap heap. The first significant generation of Fourier-transform (FT) instruments was the Bruker WM and AM, and the Varian VXR and Gemini. These had built-in computers and two channels: a transmitter for ^1H, ^{13}C, and other nuclei, and a decoupler devoted to ^1H only. The Varian Unity came along with an "industry standard" UNIX computer (made by Sun Microsystems) connected to the spectrometer using an SCSI interface, and two equivalent "broadband" channels (covering ^1H and all other nuclei). Bruker introduced the AMX with a separate UNIX computer made by Silicon Graphics (SGI) interfaced to the spectrometer with an Ethernet link. The AMX had capabilities for shaped pulses and optional pulsed field gradients (Chapter 8), as well as three radio frequency (RF) channels. Varian introduced the Unity-Plus with these features as options. They were integrated into the design of the next generation: Bruker DRX and Varian Inova models, which also included oversampling and

---

NMR Spectroscopy Explained: Simplified Theory, Applications and Examples for Organic Chemistry and Structural Biology, by Neil E. Jacobsen
Copyright © 2007 John Wiley & Sons, Inc.

digital filtering (Section 3.8). These have a modular design that allows for any number of RF channels, so that the designations "transmitter" and "decoupler" are no longer relevant; for example, you might have ^{1}H, ^{13}C, ^{15}N, and ^{2}H channels all working together in one experiment.

In the discussion of parameter names and software commands, upper case (SW, TO, etc.) will be used in describing the oldest generation (Bruker AM and Varian Gemini) and lower case (*sw*, *tof*, etc.) will be used for the newer (UNIX based) models. Software is constantly being upgraded, so that the reader will have to refer to the vendor's manuals for precise information; the parameter and command names used in this book are only illustrative.

3.1 SAMPLE PREPARATION

3.1.1 Solvent

For liquid-state NMR, you will need to dissolve your sample in a solvent. The solvent molecules should have all hydrogen atoms replaced with deuterium atoms (^{2}H) for two reasons. First, if you are doing proton (^{1}H) NMR, you do not want the solvent resonance to dominate your spectrum. Solvent molecules typically outnumber solute molecules by 1000 to 1, so you would not really see your solute spectrum at all. Second, the spectrometer needs a deuterium (^{2}H) signal to "lock" the magnetic field strength and keep it from changing with time. Because the NMR experiment usually adds together a number of FIDs (scans), if the field changes during the experiment the frequency changes with it and the NMR peaks will not add together correctly. The deuterium NMR signal is used to monitor "drift" of the field and to correct it (more about this later). For ordinary lipophilic ("greasy") organic molecules, deuterochloroform ($CDCl_3$) is the ideal solvent. For hydrophilic molecules (e.g., salts) the ideal solvent is D_2O. For molecules that are in-between in polarity or have both polar and nonpolar parts (e.g., organic acids), there are a number of more expensive solvents to try. d_6-DMSO (CD_3-SO-CD_3) is a very good solvent, but it is difficult to recover your sample from the solvent afterward. Fully deuterated versions of acetone, methanol, acetonitrile, benzene, and THF are available at prices that increase in that order. Of course, test your sample compound for solubility with the cheap, nondeuterated solvents first before wasting the expensive stuff! Acetone, D_2O, methanol, DMSO, and acetonitrile all absorb H_2O from the atmosphere when open, giving an H_2O peak in the spectrum at a chemical shift that depends on the solvent.

3.1.2 Concentration and Volume

The optimal concentration depends on the nucleus: For routine ^{1}H NMR, 5–10 mg is typical for medium-sized (MW 50–400) organic molecules; for ^{13}C NMR, which is 5700 times less sensitive, about 30–40 mg of sample is best if your molecule is soluble enough to get this amount in less than 1 mL. Too high a concentration can cause problems: It may cause overloading of the receiver in ^{1}H spectra (this can be fixed, though); it will weaken the lock signal because there is less solvent present; and it can increase the viscosity of the solution, which will lead to broader peaks. The lower limits of concentration depend on field strength and the type of probe: For ^{1}H you can get a good spectrum of an organic molecule with as little as 0.1 mg (500 MHz or higher); for ^{13}C with a probe optimized for ^{13}C detection you can get away with as little as 2–3 mg. The sample volume should be about 0.65 mL, which gives a 4.0 cm depth in a standard 5-mm NMR tube. Smaller volumes will require that you position the tube on the spinner turbine very carefully to center the sample volume in the probe coils of the spectrometer. Small-volume samples will also require a lot more time to shim properly. Too large a volume wastes solvent, reduces concentration, and can sometimes cause problems with spinning the sample because the weight of the sample and turbine is larger. The sample tube should be of high quality to avoid wobbling and breakage in the probe. Cheap or damaged tubes can lead to a broken tube in the probe, which can cost thousands of dollars in repairs and possibly lead to weeks of downtime for the spectrometer. NMR tubes with cracked or broken tops should be discarded or "sawed off" in a glass shop—they are dangerous and can lead to very serious injuries to the hands when capping or removing caps. If there are any solid "specks," chunks, or crystals in the sample solution, it should be filtered through a plug of glass wool placed in a disposable pipette. These chunks can degrade the linewidth and quality of your spectrum, even though solid material is "invisible" in the liquid NMR experiment. Fine crystals or powders may have no effect if they are evenly distributed throughout the sample volume or if they collect in an area above or below the probe coil. Cloudiness indicates that the sample molecule is only marginally soluble ("unhappy") in the solvent, which leads to *aggregation* of solvent molecules into large (but still microscopic) globs that tumble slowly in the solvent, leading to broad NMR peaks. Adding a cosolvent may solve the problem; for example, CD_3OD in $CDCl_3$ for molecules too polar for $CDCl_3$, or CD_3CN in D_2O for molecules too nonpolar for D_2O. If you add a cosolvent, be sure to measure it accurately and keep track of the volume ratio of the solvents used. You will need to report this ratio in the literature because chemical shifts depend on the ratio of solvents used. Also, you will need to use tetramethylsilane (TMS) as a reference because residual solvent peaks no longer will have the standard chemical shifts in a mixed solvent. If TMS is not used, at least report the solvent peak and chemical shift used as a reference.

3.1.3 Chemical Shift Reference

A standard is usually added to provide a sharp peak of known chemical shift in the NMR spectrum, in a region of chemical shifts that does not interfere with the sample peaks. For organic solvents, TMS is ideal because its ^1H chemical shift is upfield of nearly all organic signals, it gives a strong, sharp singlet, and its volatility makes it easy to remove. TMS is not soluble in D_2O, so a related sodium salt (sodium 2,2',3,3'-d_4-3-trimethylsilylpropionate or TSP) is used as the standard. The solvent ^{13}C peak is usually used as the ^{13}C chemical shift reference because the TMS peak is usually too weak (see below). Since D_2O does not contain carbon, you will need to add a standard (such as methanol, acetonitrile, or dioxane) to the sample for a ^{13}C reference. A very common error is to add too much of a standard—this makes the standard peak dominate the spectrum and limits the sensitivity and dynamic range of the sample peaks as the receiver gain has to be reduced to accommodate the huge standard signal. The best way to add a standard is to "spike" a bottle (100 g) of deuterated solvent with a single drop of standard and mark the bottle accordingly. Fourier-transform NMR is very sensitive, and the standard peak can be very small and still be easily detected.

3.1.4 Sample Recovery

Unlike many other analytical techniques, NMR is a nondestructive test. You can recover your sample by removing the solution from the tube and evaporating the solvent. Samples in organic solvents such as $CDCl_3$ can be removed from the tube by simply inverting the NMR tube into a vial and touching the top of the tube on the bottom of the vial to start the flow of solvent; rinse the tube with organic solvent. D_2O has too much surface tension to be poured out of an NMR tube; Wilmad sells extra-long disposable glass pipettes that will reach all the way to the bottom of the NMR tube. With care it is possible to remove relatively volatile solvents *in* the NMR tube. The tube is clamped in a fume hood at a 45° angle and a long glass pipette is introduced so that the end is near the top of the solvent. A very gentle stream of dry air or nitrogen is introduced into the pipette and as the solvent evaporates the pipette is moved down to keep it near the solvent level. This is usually done when NMR data are needed in a different solvent for comparison with literature data—chemical shifts are solvent dependent and must be compared in the same solvent. Care of NMR tubes is important: They should be washed with solvents and water only, and dried with a stream of dry air or nitrogen. NMR tubes can be cleaned by repeatedly filling them with solvent ($CHCl_3$, acetone, methanol, or water) and emptying. Aldrich sells an NMR tube cleaner that uses pump or aspirator vacuum to pull solvent into the tube. Never use paramagnetic solutions (e.g., chromium-containing glass cleaners) to clean NMR tubes. Drying the tubes is very important—residues of nondeuterated solvent can ruin your spectrum or cause you to make erroneous assignments. Drying NMR tubes in a drying oven not only is ineffective at removing solvents but also warps the tubes. The best method is to invert an extra-long glass pipette and run a slow stream of clean, dry air or nitrogen through it, and place the inverted NMR tube on the pipette so that the pipette reaches all the way to the bottom of the tube. A few seconds or minutes of gas flow should flush out all of the solvent residues without any heating. One final caveat: storing NMR samples in D_2O in the freezer will crack or weaken the glass as the "water" expands.

3.2 SAMPLE INSERTION

At the top center of each NMR magnet is a round vertical hole, called the bore, which extends all the way to the bottom of the magnet (Fig. 3.1). At the bottom, the bore is filled with the probe, which is inserted from below, and the room temperature shim coils, which form a concentric cylinder around the probe. The probe has a small vertical hole just large enough to admit the sample tube, and inside there is a set of RF coils that surround the sample. These are aligned with the center of the superconducting magnet. Wires connect these coils to the probe head, at the bottom of the probe, where connectors lead RF power into and out of the probe from cables. The probe coil acts as a radio transmitter antenna during the exciting pulse and as a radio receiver antenna during acquisition of the FID. The room temperature shims are just coils of wires wound in various directions and spacings around the probe so that adjusting the currents in these coils adds or subtracts magnetic field strength to the space occupied by the sample to make up for any lack of homogeneity in the main (superconducting) magnetic field. The sample tube is held by a plastic spinner turbine or "spinner," which is ejected from the probe out through the top of the bore with a cushion of air pressure, and inserted by gradually reducing the air pressure. When the spinner turbine and sample are resting in the probe, a small current of air can be directed at a skewed angle toward the spinner turbine, causing the spinner to lift slightly and spin on the vertical axis. For one-dimensional (1D) NMR spectra, samples are usually spun at about 20 Hz (revolutions per second) in order to average out any lack of magnetic field homogeneity along the X and Y (horizontal) axes.

78 NMR HARDWARE AND SOFTWARE

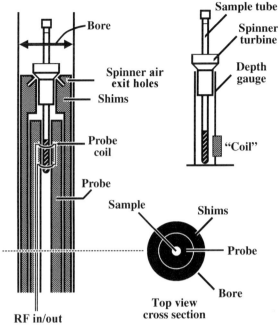

Figure 3.1

To insert the sample, first push the sample tube gently into the spinner turbine and adjust the vertical position of the tube using a gauge to assure that the actual sample solution will be centered in the probe inside the RF coils. Place the sample on the air cushion at the top of the magnet bore and deactivate the eject air, and the sample will gently descend into the bore until the spinner rests on the probe with the bottom of the NMR tube inserted into the probe.

3.3 THE DEUTERIUM LOCK FEEDBACK LOOP

3.3.1 The Lock Channel

Although the magnetic field of a superconducting magnet is very stable, there is a tendency for the field strength to change gradually or "drift" by very small (parts per billion) amounts. If this tendency were not corrected, it would be impossible to sum a number of FID acquisitions because each FID would have a slightly different frequency than the previous one. Drift is prevented by a separate channel in the probe and the spectrometer that detects deuterium (^{2}H). This can be done independently of proton or carbon acquisition because deuterium nuclei resonate at a very different frequency (e.g., 30.7 MHz compared to 200 MHz for ^{1}H on a 200 MHz instrument). The lock channel continuously detects the deuterium signal of the deuterated solvent and monitors its chemical shift position. You can think of this as a separate NMR spectrometer dedicated to ^{2}H detection, which runs continuously in the background. Because the resonance frequency of any nucleus is proportional to the magnetic field strength ($\nu_o = \gamma B_o/2\pi$), any drift in the magnetic field (decrease/increase)

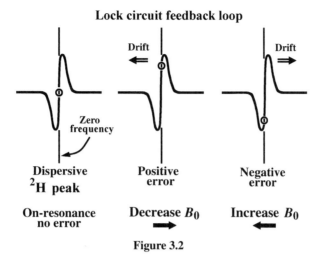

Figure 3.2

will cause a shift (upfield: lower frequency/downfield: higher frequency) of the deuterium frequency detected. This shift in frequency is connected to a feedback loop that adjusts the field strength (by changing the current through a room temperature coil in the shim cylinder) so that the deuterium frequency does not change. This mechanism is called the "lock" system, and it maintains a constant magnetic field strength throughout your NMR acquisition. Regardless of the lock display presented to the user, the lock circuitry sees a dispersive (up/down) deuterium signal centered on the zero frequency (null point) of the feedback circuit (Fig. 3.2). The magnetic field strength is manually adjusted (Z_o or *field* knob) to center the signal at the zero frequency. When the lock is turned on, the feedback loop is activated and control of the magnetic field strength (B_o) is given over to the control circuit. If the magnetic field increases slightly, the ^{2}H signal is shifted to the left (higher frequency) leading to a positive error signal. This signal decreases the current in the Z_o (field) coil in the shim stack, which decreases the magnetic field, correcting the drift. A slight decrease in magnetic field leads to the opposite error signal and a compensating increase in current sent to the shim coil. The system cannot achieve lock unless the null point is between the two maxima of the dispersive peak when the feedback loop is activated. The proper *lock phase* setting assures a symmetrical (equally up and down) dispersive signal in the feedback loop.

3.3.2 Locking

As soon as your sample drops into the probe, the ^{2}H signal will become visible on the screen. You may need to increase the lock power and gain, and adjust the field (Bruker: Field, Varian: Z_o) setting to see the lock signal. If the homogeneity of the magnetic field is very poor ("bad shims") you may not even see a signal! There is another major difference (Fig. 3.3) between Varian and Bruker in the way the lock signal is displayed: Varian shows a time-domain signal, which is a sine wave whose frequency is the audio frequency of the deuterium signal. Bruker shows a frequency-domain signal, which is "swept" by moving the deuterium excitation frequency back and forth repeatedly over a range of frequencies. When the excitation frequency matches the deuterium resonance frequency, you get a peak

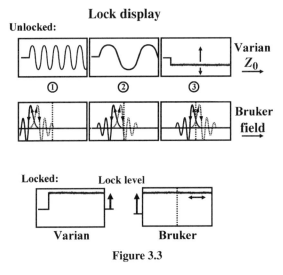

Figure 3.3

that dies away in wiggles ("ringing") as the excitation frequency moves away from resonance. The same peak and ringing is observed as the excitation sweeps back the other way across the resonance position. This display is in the "unlocked" state: the feedback loop is inactivated and the deuterium signal is simply observed on the screen. If the deuterium frequency is far from the locking position, you will not see any signal. For Varian this is because the time-domain frequency is very high and the signal is weak; for Bruker it is because the deuterium resonance position is outside the range of frequencies being swept. On the Varian spectrometers, you adjust the field strength (Z_o) until you begin to see a sine-wave signal ("wiggles") and continue to adjust until the frequency (number of cycles of sine wave displayed) decreases to zero and you have a horizontal line instead of a sine wave. On Bruker, you adjust the field strength (*field*) until the pattern of peaks and ringing is exactly centered on the screen. You are now ready to activate the lock feedback loop. Turning on the lock leads to a horizontal line that rises above the baseline. You can think of the height of this line as the peak height of the deuterium NMR peak of the solvent. Once locked, the deuterium frequency is no longer swept (Bruker) and the magnetic field strength (B_o) should be rock-stable over time. Mixed solvents (e.g., d_6-acetone/$CDCl_3$) or solvents with more than one deuterium resonance (e.g., CD_3OD) can lead to problems if you lock on the wrong ^2H signal, so be sure to verify which signal is centered in the ^2H spectral window before turning on the lock.

3.3.3 Lock Parameters

The field setting required to center the lock signal depends on the deuterium chemical shift, which is roughly proportional to the proton chemical shift. Thus, the deuterium resonance of $CDCl_3$ (^1H δ 7.24 ppm) is downfield of the deuterium resonance of d_6-acetone (^1H δ 2.04 ppm) but very similar to that of d_6-benzene (^1H δ 7.15 ppm). The field settings for various common deuterated solvents are often posted near the spectrometer or in a logbook to allow easy access to "ballpark" settings. The lock level (the height of the lock signal on the screen) represents the height of the deuterium peak. As shimming (i.e., homogeneity of the magnetic field) is improved, the deuterium resonance becomes sharper and the height of the deuterium "peak" increases, since the area (amount of deuterium in the sample) remains constant. Two other factors affect the lock level: The *lock power* affects how much ^{2}H signal is fed into the probe to excite the deuterium nuclei and the *lock gain* affects how much the detected signal is amplified in the receiver before being presented on the screen. Increasing either one will increase the lock signal level, but increasing the lock power will eventually "saturate" or overload the ^{2}H nuclei with RF energy, causing the lock level to "breathe" or oscillate slowly up and down. Since the lock level is used to monitor changes in field homogeneity while shimming, a randomly oscillating level will interfere with shimming. If this happens you need to turn down the lock power and then pump up the lock gain as necessary to get a good lock level. A "good" lock level is about 80% of the maximum, allowing room for improvement during the shimming process. The lock signal should have a little bit of noise, indicating that the lock power is not excessive, but it should not have so much noise that small changes in the level cannot be readily observed. It is important to realize that the lock level is arbitrary; you can increase it or decrease it at any time by adjusting the lock gain and the lock power. It is only the *changes* in lock level resulting from changes in the shim settings that are important. If shimming brings the lock level above 100%, just reduce the lock gain to bring it back to 80% and continue shimming.

3.4 THE SHIM SYSTEM

Shimming is the process of adjusting the magnetic field to achieve the best possible homogeneity. By homogeneous we mean that the magnetic field strength does not vary significantly from one part of the sample to another. Because the resonance frequency (chemical shift) is directly proportional to magnetic field strength, a variation of 1 ppm in field strength from one location within the sample volume to another would lead to a peak with a 1 ppm linewidth (or 200 Hz on a 200-MHz instrument!). Since linewidths of 1.0 Hz can be routinely obtained, the magnetic field when well shimmed does not vary more than 5 ppb within the sample volume. The dictionary meaning of a "shim" is a thin piece of wood or metal placed within a gap to make two parts fits snugly. The NMR shim plays the same role—it increases the magnetic field strength in certain volumes of space so that it is uniform throughout. The inhomogeneity of the field is a complex function in three dimensions; the goal of the shim system is to exactly match that function with a function of opposite sign so that all of the inhomogeneity cancels out (Fig. 3.4). To match that function, a number of simple 3D functions (Z, Z^2, Z^3, XY, XZ^2, etc.) are summed with different coefficients. For example, the Z (or $Z1$) shim coil creates a linear gradient of magnetic field along the vertical (Z) axis, which is the axis of the NMR sample tube. When the Z shim is set to zero, there is no effect; when it is set to a positive value, the field strength is slightly increased in the upper part of the sample and slightly decreased in the lower part of the sample, to an extent proportional to the vertical distance from the center of the sample. Increasing the Z shim setting increases the current to the Z shim coil and makes the gradient "steeper" so that the field varies more for a given vertical distance. Changing the Z shim setting to a negative value reverses the sense of the gradient so that the field is decreased in the upper part of the sample and increased in the lower part. Likewise, the other shim coils control other simple gradient functions, and the current put through each coil controls its contribution to the field correction (or, mathematically speaking, its coefficient). These shim currents are set by computer and can be saved and recalled as files. Your job as operator is to search the

82 NMR HARDWARE AND SOFTWARE

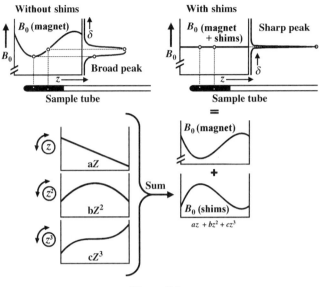

Figure 3.4

n-dimensional space (where n is the number of shims available) to find the global optimum of homogeneity. To get instant feedback on homogeneity, you have the lock level as a guide. As the field becomes more homogeneous, the ^{2}H peak of your deuterated solvent becomes sharper, and its peak height (which is equivalent to the lock level) becomes higher as the same peak area gets squeezed into a narrower and narrower peak. The newer and higher field instruments have a large number of shim functions (or shim "gradients") available, so that you may be trying to find an optimum in a 28-dimensional space! This can be a daunting task, and only those with experience and great patience attempt to adjust any but the simple low-order shims.

3.4.1 Shimming for Beginners

Routine users usually adjust only Z and Z^2, whereas experienced users might attempt to adjust $Z^1, Z^2, Z^3, Z^4, Z^5, X, Y, XZ, YZ, XZ^2, YZ^2, XY$, and X^2-Y^2. For beginners, the first thing to learn is to move the shim far enough past the optimum so that there is a significant and observable drop in lock level (e.g., 10%), and then move back to the setting that gives the highest lock level. The territory on both sides of the optimal setting needs to be explored, and the optimal setting is near the center of the two settings that degrade the lock level by 10% on either side. The other problem for beginners is understanding that you are not finished shimming until you cannot improve the lock level any more with the shims you are adjusting (e.g., Z^1 and Z^2). There is no universal "good" lock level at which you can stop shimming, only the "best" lock level for your sample achieved after going over the shims several times (e.g., Z^1, Z^2, Z^1, Z^2, etc.). Although you should not be shy to make large changes, if you change any shim too rapidly there may be a transient response that obscures the true effect of the shim. For example, increasing Z^2 might give an increase in lock level, but when you stop changing the shim this effect goes away. A rapid change in shim setting in the opposite direction will cause a transient decrease in lock level. In this situation you have to make changes slowly and be sure to wait for a moment after making a change to see if the lock level has really changed. As the shim settings are changed, the lock phase may be affected. Improperly set lock phase can lead to a situation where the optimal lock level does not correspond to the best peak shape in your spectrum. Since you need to use the lock level as a criterion for shimming, it is important to reoptimize the lock phase from time to time as you change the shims. This can be done simply by adjusting the lock phase for maximum lock level, just as you would do with a shim.

3.4.2 Shimming and Peak Shape

The effect of large errors in Z shim settings on peak shape is simulated in Figure 3.5 for a singlet NMR peak with a natural linewidth of 1.0 Hz. Note that a linear Z gradient (bad Z shim setting) simply "stretches" the peak horizontally. This peak is now a map of the sample molecules, with the molecules at the top of the sample having a resonant frequency at the left edge of the peak and the molecules at the bottom giving rise to the right edge of the peak (actually the limits of the peak are the top and bottom of the probe coil, since only that part of the sample within the coil is "seen"). This is an NMR imaging experiment (MRI) and illustrates the principle of making images by NMR using a linear gradient of magnetic field. For example, if there were bubbles in the sample these would show up as dips in the peak at the point corresponding to the position of the bubble along the vertical axis, since sample molecules would be missing at these points. Note that the higher order "odd" Z gradients (Z^3 and Z^5, Fig. 3.5, left) have symmetrical "pedestals" at the base of the peak. These pedestals are lower relative to the top of the peak as the shim order increases from Z^3 to Z^5. The "even" Z gradients (Z^2, Z^4, and Z^6, Fig. 3.5, right) have "porches" or "verandas" at the base of the peak on one side only. If the shim error is reversed (e.g., from Z^2 too high to Z^2 too low) the porch will move to the opposite side of the peak. Each instrument will usually have a different polarity of the even shim gradients, so you will

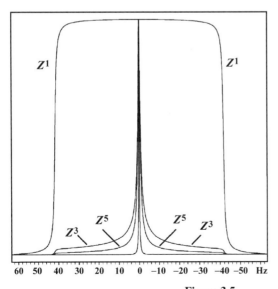
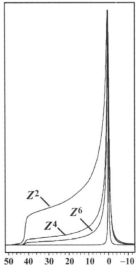

Figure 3.5

84 NMR HARDWARE AND SOFTWARE

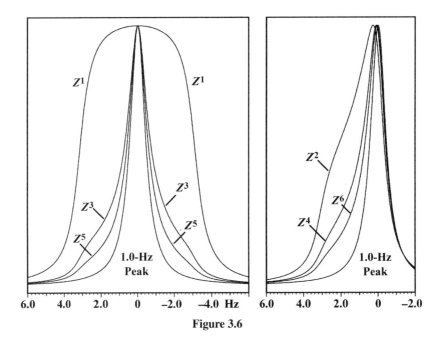

Figure 3.6

have to experiment to see which way to move the shim in order to move the porch into the peak. As with the odd Z gradients, as we move to higher order even Z gradients, the wide portion appears lower down at the base of the peak. Figure 3.6 shows the effect on lineshape of smaller errors in shim settings on an ideal peak with 1.0 Hz linewidth. A Z^1 error just broadens the peak evenly from top to bottom, while a Z^2 error leads to a bulge or shoulder on one side of the peak. Similar effects are seen with higher order Z shims, with odd shims leading to symmetrical bulges and even shims leading to bulges on one side. The bulges move down lower in the peaks as we go to higher order shims. All of this assumes that the shim coils (Z^1, Z^2, Z^3, etc.) actually deliver pure mathematical changes in magnetic field (z, z^2, z^3, etc.) as a function of position in the sample (z coordinate). In reality, there is a good deal of mixing of these pure functions, such that the Z^1 shim knob actually changes all the other shims a little bit as well. This means that in practice you may not see these ideal changes in lineshape, and it is more difficult to diagnose which shim needs to be adjusted just by looking at peak shape. Furthermore, when you make a change in a higher order shim, such as Z^4, you will need to readjust the lower order shims, especially Z^2, because the Z^4 shim "contains" a bit of Z^2. Likewise, changes in Z^3 and Z^5 will require readjustment of Z^1 and Z^2.

3.4.3 More Advanced Shimming

In general, high-order shim errors do not affect the linewidth at half-height, but they do affect the linewidth at the base of the peak and reduce the peak height by spreading the peak intensity into the pedestals and porches at the base. To assess the overall quality of shimming, it is best to measure the peak width in several places; typically for a singlet peak such as chloroform ($CHCl_3$), the width in hertz is measured at 50% of peak height, 0.55% of peak height, and 0.11% of peak height. The 0.55% is conveniently determined by the

THE SHIM SYSTEM 85

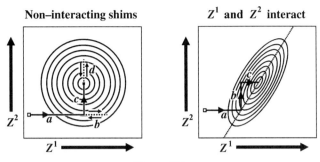

Figure 3.7

height of the doublet produced by the 1.1% ^{13}CHCl$_3$ present at natural abundance. These ^{13}C "satellite" peaks are found at 105 Hz ($^1J_{CH}/2$) downfield and upfield from the main ^{12}CHCl$_3$ peak. For example, the quality of shimming could be reported as 0.5/8/11 Hz at 50/0.55/0.11% of peak height. If linewidth is only measured at the half-height level, you will be ignoring the higher order shim errors.

Often the effect of two shims is interactive, such that changing one shim setting affects the optimal setting of the other. This is commonly observed for Z and Z^2 with short (low volume) samples. For example, you might visualize the effect of Z and Z^2 on the lock level as a two-dimensional plot, like a topographic map (Fig. 3.7). Climbing a simple round peak (Fig. 3.7, left) is easy: just optimize Z (the east-west direction) by moving to the right (segment a) as long as the lock level continues to increase. The lock level reaches a maximum and begins to go down again (segment b), so you turn around and move left to return to the maximum. It is important to really see the lock level go down significantly, so you know you have reached the maximum for the noisy lock level. Next optimize Z^2 (the north-south direction) by finding the direction (up or down on the map) that increases the lock level and then moving straight to the top of the peak (segment c). Again, to be sure you are at the maximum you need to go significantly beyond and return to the peak (segment d). But what if the surface is more like a ridge that runs from the southwest to the northeast (Fig. 3.7, right)? You might use Z^1 to climb to the top of the ridge (segment a), but Z^2 would not give any further improvement, even though the peak is a long way up the ridge. What you need to do is simultaneously adjust Z^1 and Z^2 so that you can move diagonally, like trying to draw a diagonal line with an Etch-a-Sketch. To do this on a spectrometer, note the lock level and then arbitrarily move Z^2 away from the maximum in one direction to degrade the lock level by a certain amount, like 10% (segment b). Then use Z^1 to optimize again (segment c). If the optimized lock level is better than where you started, you have made progress up the ridge. Now you only need to continue making small changes in Z^2 in the same direction away from the Z^2 optimum, followed by optimization of Z^1. If the optimized lock level is worse, try an arbitrary change in Z^2 in the opposite direction. The process is a zigzag approach to the peak, alternately going downhill and then uphill to the top of the ridge. Many shim pairs behave in this way, so you can see how shimming is an art requiring a lot of patience.

Another way to shim is to use the FID as a criterion for homogeneity instead of the lock level. The goal is to get a smooth exponential decay curve with the longest time constant (slowest decay) possible for the FID. Bruker uses the command *GS* to enter an interactive mode where the FID is acquired over and over again, displaying it each time

86   NMR HARDWARE AND SOFTWARE

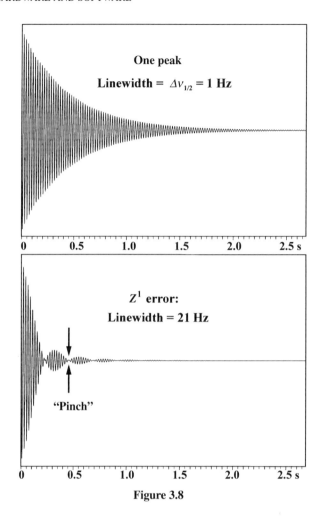

Figure 3.8

without summing in the sum to memory. On the Varian you can do the same thing by selecting *FID* instead of *SHIM* in the *acqi* window. A simple Z^1 shim error (Fig. 3.8, bottom) will give an FID that not only decays faster but also goes though a series of evenly spaced nulls ("pinches") in the FID as it decays. At each null the FID signal reverses phase. Mathematically this can be described as a "sinc" function ($y = \sin(x)/x$), which gives a rectangle in frequency domain when you do the Fourier transform. The rectangle is the peak shape you get from a linear Z gradient, the MRI experiment (Fig. 3.5). The sinc function and the rectangle can be viewed as a "Fourier pair" since Fourier transformation of one shape gives the other and vice versa. We will encounter this and other Fourier pairs later in the course of this book. As you improve the Z shim you will see the "pinches" move to the right, to longer times, corresponding to a narrower rectangle in frequency domain, and eventually off the "end" of the FID. After the last "pinch" is moved off the end, you should be able to maximize the signal at the end of the FID to get the slowest decay. The lock level may actually be going down as you do this! The reason is that either the lock phase is not adjusted right or there are other shims, especially higher order shims, which are interacting with Z^1. If Z^1 were the only misadjusted shim, the lock signal would go up as you improve the FID. You may be able to alternate between optimizing other shims (Z^2, Z^3, etc.) using the lock level and optimizing Z^1 using the FID. You may also find that the Z^2 shim moves the pinches, contrary to theory. Go with it! Shimming is an art, not a science. Sometimes if the FID has no "pinches" but the shape is not exponential, you can adjust higher order shims (Z^3, Z^4, Z^5, Z^6) to try to get an exponential FID shape. Then you can go back to the lower order shims to get a slower exponential decay. Shimming on the FID is easiest if you have a single peak dominating the spectrum so that the FID is a simple decay of one sine wave. For example, a small amount of $CHCl_3$ in $CDCl_3$ or d_6-acetone, or of H_2O in D_2O is good for this kind of shimming.

The "pure" Z shims are called "axial" shims. Other shims contain X or Y in their names; for example, X, Y, XZ, YZ^2, and so on. Spinning does not correct these "off-axial" errors; it simply moves the intensity to spinning sidebands, which are satellite peaks separated on either side of the main peak. If the sample is spinning at 20 Hz (20 revolutions per second), spinning sidebands will appear at 20, 40, 60 Hz, and so on, away from the main peak on either side. If there are large errors in the off-axial shims there will be large spinning sidebands. Another indication of poor off-axial shims is the increase in lock level observed when the spinning is turned on. If this is more than about 15% of the nonspinning lock level, the X and Y contributions to the shims are not optimal. The off-axial shims must be adjusted without spinning the sample. In general, low-order shims (Z, Z^2, X, and Y) should be adjusted with small changes ("fine" setting), and all higher order shims can be adjusted using large changes in the shim value ("coarse" setting). A typical approach for off-axis shims is to adjust them in the order X, Y, XZ, YZ, XZ^2, YZ^2, XY, and X^2–Y^2, and then reoptimize X and Y.

If shims are really bad, you may be able to recall a recent shim file from the disk. Ideally, a spectrometer should be shimmed regularly (daily or weekly) by an expert, and these "current" shims should be saved. Rather than waste a lot of time trying to get home from someone else's *n*-dimensional wanderings away from the optimum, you can just read the latest shim file and start from there. Of course, you will still need to adjust Z^1 and Z^2 because these change a great deal based on the sample volume, position, and polarity of the solvent.

3.4.4 Autoshimming

There are two automated methods for shimming. The first, simplex autoshimming, has been around as long as shims have been controlled by computers. A computer program simply does what you do—moves a shim by a certain amount and notes the effect on lock level. If there is an improvement, it moves again in the same direction. The whole tedious process can be written into a computer method and you can choose which shims you want to adjust in what order, including turning the spinner on and off. Because there is no human element, the process is slow but it is perfect to set up overnight and check in the morning. The second automated shimming method is gradient shimming, which is only available if you have pulsed field gradient capability and a gradient probe. This is an imaging (MRI) experiment that actually makes a physical map of the magnetic field strength as a function of position within the sample volume. Medical MRI uses the water in the human body to make images, relying on the fact that from an NMR standpoint the human body has just one peak: water (fat is a minor peak). Likewise, for gradient shimming you need a sample with one dominating peak. For ^{1}H imaging, that sample would be water, so we are limited to biological NMR samples that are typically dissolved in 90% H_2O/10% D_2O. A more recent development is the convenient use of deuterium gradient shimming, which makes an image of the sample using the single, very strong ^{2}H peak of the deuterated solvent. In either case, a one-dimensional map of field strength variations along the Z axis is created using a Z gradient, so the variation of field is known and what remains is to figure out how much change in each shim would cancel out that variation. Stored in the computer is a "shim map," which gives the exact effect at each point in the sample of a given change in each of the Z shims. For example, we expect that the Z^2 shim map is a parabolic function of position along the Z axis, but the shim map for Z^2 gives the exact values of this function at each point. The computer then calculates using these maps the exact combination of shim value changes that will best correct the measured inhomogeneities. These changes are applied and the process is repeated: map the inhomogeneities, calculate the new changes using the shim map, and apply the shim value changes. After several rounds of this process the homogeneity cannot be improved and we have the "best" shims for that sample. The whole process requires only a minute or two. For "triple axis" gradient probes, which have the capability of creating imaging gradients along the X and Y axes as well as the Z axis, gradient shimming can be used to optimize all of the shims, including the off-axis shims. One round of 3D gradient shimming might require about 5 min. Gradient shimming is rapidly eliminating the need for experienced shimming "experts"! This topic is covered in detail in Chapter 12, Section 12.3.

3.5 TUNING AND MATCHING THE PROBE

Tuning the probe assures that the resonant frequency of the probe coil is the same as the RF frequency you will be using and matching the probe matches the probe coil as a load to the impedance (internal electrical resistance) of the amplifiers. This gives maximum efficiency of transfer of RF power from the amplifiers to your sample nuclei and maximum sensitivity in detecting the FID. Each sample modifies the resonant frequency and matching of the probe, so these have to be reoptimized with each new sample. Tuning the probe is not necessary for routine ^{1}H spectra, but for advanced experiments it is important if you wish to use standard values for pulse widths without the need to calibrate for each sample.

3.5.1 Matching and Tuning

Every electrical device that supplies power (such as a battery) has an internal resistance associated with it. For example, if you short out the two terminals of a battery, you will get a very large current but not the infinite current that would result from zero resistance (Ohm's law: current = voltage/resistance). This is because the battery's voltage is applied across the total resistance of the circuit: the sum of the external load (zero resistance) and the internal load (the internal resistance of the battery). In the case of the short circuit, all of the power (power = current^2 × resistance) is delivered to the battery itself since there is no resistance in the "load," and the battery heats up. If, on the contrary, the resistance of the load is very high, there will be very little current flow and very little power will be transferred to the load. It turns out that the maximum transfer of power from the source (the battery) to the load (e.g., a light bulb) is obtained if the resistance of the load equals the internal resistance of the battery. In RF electronics we use the term *impedance* instead of resistance, but the principle is the same: *Matching* the impedance of the load to the impedance of the source will maximize the transfer of power to the load. The internal impedance of the RF amplifiers is 50 Ω, so we are always trying to match the probe coil to 50 Ω. The other factor to consider is the resonant frequency of the probe. The probe circuit may be very complex, but we can view it as an inductance (the coil) connected in parallel with a variable capacitor (the tuning capacitor). The resonant frequency of this tuned circuit is determined by the amount of inductance in the coil and the (variable) amount of capacitance in the capacitor. By rotating the tuning rod we change the capacitance and move the resonant frequency to higher or lower frequency.

If the probe is not properly tuned and matched, the pulse will "reflect" off of the probe and return to the amplifier rather than reaching the sample. The simplest way to tune a probe is to introduce a continuous RF source at very low power and measure the amount of reflected power using a bridge circuit that compares the probe to a 50 Ω resistor. The probe tuning and matching knobs are adjusted to minimize the reflected power reading on the bridge. This is a bit like shimming, except that we are trying for a minimum signal rather than a maximum signal. As with shimming, sometimes the tune and match interact, so it is necessary to "detune" one of the settings a bit and readjust the other to get a lower minimum. Varian still uses this method with a built-in tune display on the preamplifier. The cable from the probe has to be moved from the preamplifier connector to the "probe" connector of the tune interface. Bruker (AMX, DRX) uses a "wobble" tuning method that sweeps the tune frequency back and forth around the desired frequency and records the probe response as a graph on the computer screen. A "dip" in the curve occurs at the resonant frequency of the probe, and the tune knob can be adjusted to position the bottom of the dip (left and right) exactly at the desired frequency. The match knob makes the dip sharper and deeper, so this can be independently adjusted to bring the dip to its lowest value (best impedance match).

The probe tuning rods are long extensions of the variable capacitors located at the top of the probe, near the probe coil. The capacitors are delicate and there are two ends of the travel of the knob: If any force at all is applied at the end of the travel, the capacitor will break. This will usually require that the probe be sent back to the manufacturer for repair, a process requiring a week or two and costing many thousands of dollars. For this reason many NMR labs do not allow users to tune the probe!

3.5.2 Types of Probes

All multinuclear probes have more than one probe coil: one is dedicated to ^{1}H and the other is for one or more heteronuclei (e.g., ^{13}C). They are positioned concentrically about the NMR tube, with a cylindrical glass insert separating the tube from the inner coil and another larger insert separating the inner coil from the outer coil. The inner coil is much more sensitive for detection of the FID, so usually we detect using this coil. RF pulses can be applied on either coil, but pulses applied on the inner coil will require less power to excite the nuclei. The inner coil is also more sensitive to the electronic disturbance of the sample, so it is much more important to tune the inner coil when a new sample is introduced. Probes with the heteronucleus (e.g., ^{13}C) coil on the inside are called "direct" probes and those with the ^{1}H coil on the inside are called "inverse" probes. In many probes the coils are "double tuned" so that more than one nucleus can be detected. For example, ^{19}F and ^{1}H, ^{13}C and ^{31}P, or ^{13}C and ^{15}N can be paired together. Some double-tuned probes can have as many as eight tuning knobs at the probe head, and getting a good compromise between the two nuclei can be very complicated. "Broadband" probes try to cover a very wide range of frequency, nearly all of the NMR-detectable nuclei in a single probe coil. This often involves "tuning rods": long rods with a fixed capacitor at the end. The rod is inserted into the probe head from the bottom, and the capacitor screws into the circuit near the probe coil. By using a set of rods with different capacitor values, the entire range of NMR frequencies can be tuned with the tuning knob of the heteronuclear coil. This kind of probe is essential for working with "exotic" nuclei such as ^{57}Fe, ^{29}Si, and ^{77}Se. Usually a spectrometer will have a number of probes optimized for different purposes; for example, a direct ^{13}C probe (^{13}C inside, ^{1}H outside) for ^{13}C spectra, a direct broadband probe for "exotic" nuclei, an inverse ^{13}C probe (^{1}H inside, ^{13}C outside) for heteronuclear 2D experiments, and an "HCN" or "triple-resonance" probe (^{1}H inside, double-tuned ^{13}C/^{15}N outside) for biological work. Changing probes takes about 15 min, but it should only be done by expert users. For biological samples (usually in 90% H$_2$O/10% D$_2$O) you need a "water suppression probe" with shielded wires coming from the probe to avoid picking up the very strong H$_2$O signal on these wires.

For a heteronuclear experiment, such as a 1D ^{13}C spectrum with ^{1}H decoupling, you need to tune and match both the ^{1}H and the ^{13}C coil. First, set the spectrometer frequency to the ^{1}H frequency and tune and match the ^{1}H coil at the probe head. Then set the spectrometer frequency to the ^{13}C frequency and tune and match the ^{13}C coil. If you are using a direct probe (the most sensitive for ^{13}C detection) the ^{1}H tuning is less important because it varies only slightly from sample to sample (outer coil). If you try to get a ^{13}C spectrum with an inverse probe you will get poor sensitivity, but if you have lots of sample you may be able to overcome this. The baseline may not be flat since you are observing on a coil not designed for observing the FID. If you do not tune and match the ^{1}H coil, you may get no spectrum at all because ^{1}H decoupling will not work if the probe is very badly tuned for ^{1}H, and as the inner coil, the ^{1}H coil is most sensitive to sample differences.

3.6 NMR DATA ACQUISITION AND ACQUISITION PARAMETERS

The process of data acquisition results in an FID signal residing in the computer of the NMR instrument. In order to properly set up the acquisition parameters, it is helpful to understand a little about how this is accomplished. We will follow the sequence of events involved in the acquisition of the raw data for a simple 1D ^{1}H spectrum on a 200-MHz instrument through a simplified diagram of the spectrometer:

(a) Wait for a period of time called the relaxation delay for spins to reach thermal equilibrium;
(b) send a high-power short-duration RF (200 MHz) pulse to the probe coil;
(c) receive the resulting FID signal from the probe coil;
(d) amplify this weak RF ($\sim$200 MHz) signal;
(e) convert the RF (MHz) signal to a "stereo" audio (kHz) signal;
(f) sample the audio (analog) signal at regular intervals and convert it to a list of integers;
(g) add the digital FID "list" to a sum FID in memory;
(h) repeat steps (a)–(g) for as many "scans" or "transients" as desired.

Each of these steps will be discussed in more detail with the goal of understanding the basic parameters needed to set up an NMR experiment. The FID contains all of the frequencies of the sample protons, which represent a range of chemical shift values. The range of radio frequencies in the FID is extremely narrow: 199.999–200.001 MHz for a 10 ppm range of chemical shifts. We are only interested in this tiny slice of frequencies, so it is convenient to subtract out the fundamental frequency (200.000 MHz) and look only at the differences, ranging from −0.001 to +0.001 MHz, or from −1000 to +1000 Hz. These much lower frequencies are called "audio" frequencies, since they are in the range of sound waves that can be detected by the human ear. In fact, the audio signal of an NMR can be connected to a pair of speakers so that you can listen to the FID in stereo!

The audio signals must be converted into a list of numbers, which is the only language that a computer understands. This is done by sampling the voltage of each signal at regular intervals of time and converting each analog voltage level into an integer number. Thus, an FID becomes a long list of numbers, which is stored in the computer memory. As the same FID is acquired over and over again, repeating the sequence (relaxation delay—pulse-acquire FID), each new list of numbers is added to the list stored in memory. This process is called "sum to memory." As more and more "scans" or "transients" are acquired, the signal-to-noise ratio of this sum improves.

Now let's look in detail at each process, so that we can understand the NMR acquisition parameters needed to set up the experiment. After the pulse, we will follow through the hardware devices in a block diagram and try to understand a little about the NMR hardware. It turns out that processing of the NMR signal in the hardware is strictly analogous to the theoretical steps we will use in viewing the NMR experiment with the vector model, so it is essential to understand it in general.

3.6.1 Relaxation Delay: D1 (Varian) or RD (Bruker)

Usually the acquisition (pulse-FID) is repeated a number of times in order to sum the individual FIDs and increase the signal-to-noise ratio. In this case, a delay must be inserted before each pulse-FID sequence to allow the populations to return to a Boltzmann distribution ("relaxation"). Without this delay the nuclei will become saturated (equal populations in the two energy levels), and there will be little or no signal in each FID. Ideally, you should wait for about five times the characteristic relaxation time (T_1) before starting the next pulse, but in practice the relaxation delay is quite a bit less and you live with a certain reduction of signal. This is a compromise value because pulsing more often gives you more data per unit time. In this case, you rapidly reach a steady state where the nuclei are not completely relaxed but are at least at the same degree of relaxation each time a new pulse arrives. Relaxation is going on during the FID acquisition period as well, and sometimes with long FID acquisition times, there is no need for a specific relaxation delay. Another strategy for slow-relaxing nuclei (e.g., quaternary carbons in ^{13}C-NMR) is to reduce the amount of RF excitation (the pulse width) so that the perturbation from equilibrium resulting from each pulse is reduced.

3.6.2 The Pulse

The RF pulse is simply a high-power RF signal turned on for a very short period of time, on the order of microseconds (μs). The duration of the RF pulse in microseconds is called the pulse width (PW) (Fig. 3.9). As with all sine waves, the pulse has characteristics of frequency, amplitude, and phase. The *frequency* of the pulse is set at the center of the

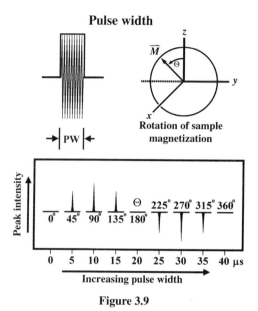

Figure 3.9

spectral window (range of resonant frequencies expected), but because of its short duration and high power it is capable of exciting all of the sample nuclei within the spectral window simultaneously. Using our example of a 200 MHz ^{1}H experiment, the pulse frequency would be 200.000 MHz, but all nuclei resonating in the range 199.999–200.001 MHz would be equally excited by it. The pulse shape is rectangular, meaning that the RF power turns on more or less instantaneously to full power, and then PW microseconds later turns off. A "90° pulse" is the pulse width required to exactly tip the sample magnetization from the z axis into the x-y plane, where it rotates at the resonant frequency in the x-y plane, leading to a maximum-intensity FID signal. The sample magnetization is just the vector sum of all the individual nuclear magnets. The *amplitude* of the pulse (height of the rectangular envelope) can be adjusted but is usually set near the maximum for simple 1D spectra. RF *power* is the square of the amplitude, and usually we talk about pulse power (in watts or decibels) rather than amplitude. The duration (pulse width) of the 90° pulse depends on the RF power and, to some extent, on the characteristics of the probe and the sample. With higher power (higher pulse amplitude) we need less time to rotate the magnetization by 90°, so the 90° pulse width is shorter. For some experiments, calibration of the 90° pulse width is essential for the experiment to work right. For simple one-pulse experiments, an approximate value is sufficient. In more sophisticated experiments that use more than one pulse separated by various time delays, the pulse duration parameters are *P1*, *P2*, *P3*, and so on for the various pulses in the sequence. Pulses are always entered in microseconds (μs) and should not be made very long (more than a few hundred microseconds) because at full power the amplifiers can "burn up" if left on continuously for too long. Finally, we have control over the pulse *phase* as well. Relative to an RF signal that starts at the zero degree point of the sine function, we can shift the phase by any amount we choose, although 90°, 180°, and 270° phase shifts are the most common. For example, a 90° phase shift means that the wave starts at the top of the cycle and goes down to zero and then to negative. We will see that setting the pulse phase is equivalent to placing the pulse vector on the x, y, $-x$, or $-y$ axis of the rotating

NMR DATA ACQUISITION AND ACQUISITION PARAMETERS   93

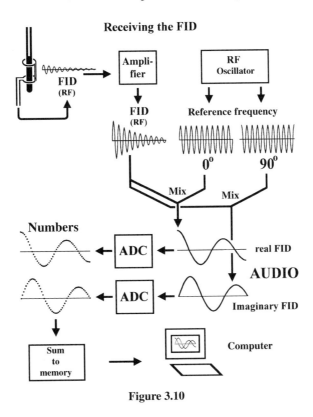

Figure 3.10

frame of reference (0°, 90°, 180°, or 270° phase shifts, respectively). This gives us much more flexibility in controlling the complex "dance" of sample magnetization in advanced experiments. In the actual pulse sequence code, which is written to tell the spectrometer the exact sequence of events in an NMR experiment, the phases are represented by 0, 1, 2, and 3 for the x, y, $-x$, and $-y$ axes, respectively, in the rotating frame of reference.

The remaining steps involved in receiving the FID signal are diagramed in Figure 3.10, showing the process of amplification, quadrature detection, digitisation, and summation to give the final FID in the computer.

3.6.3 Receiving the FID from the Sample

As the sample magnetization rotates in the x-y plane, the same probe coil that transmitted the high-power RF pulse to the sample experiences a very weak induced signal. This signal decays to nothing over a period of a second or two, and the full-time course of this induced signal is called the free induction decay or FID. Each type of nucleus in the molecule (e.g., the CH_3, CH_2, and OH protons in ethanol) has its own resonant frequency, so the FID consists of a superposition of a number of pure frequencies, corresponding to a number of peaks in the spectrum. All of the information of the NMR spectrum is contained in the FID, and a large part of the spectrometer is devoted to amplifying, recording, and analyzing this signal. In cryogenic probes (Chapter 12, Section 12.3), the probe coil is cooled to very low temperatures (e.g., 25 K) resulting in a 3–4 times reduction in thermal electronic noise and a concomitant 3–4 times increase in sensitivity (signal-to-noise ratio).

94 NMR HARDWARE AND SOFTWARE

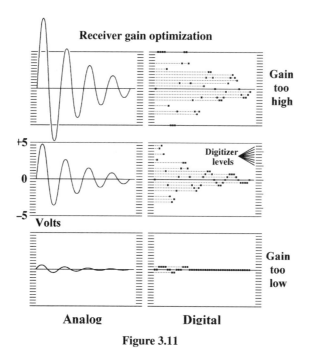

Figure 3.11

3.6.4 The Receiver

The receiver consists of preamplifier, detector, audio filters, analog-to-digital converter (ADC), and sum to memory. It amplifies the RF FID signal coming from the probe, converts it to an audio frequency signal by subtracting out the RF at the center of the spectral window, amplifies it some more, converts it to a list of numbers, and adds these numbers over a number of repeated scans (Fig. 3.10). The total amplification given to the FID in the receiver is called the receiver gain (Varian: *GAIN* or Bruker: *RG*). The intensity of the FID signal induced in the probe coil depends on the sample concentration, so the amount of gain or amplification in the receiver must be adjusted for each new sample. The audio signal coming into the digitization stage (ADC) should ideally be of the same magnitude for all samples, regardless of concentration. The ADC has a maximum range of integer values that it can give to the signal as it comes in, for example $-32,767$ to $+32,768$ (Fig. 3.11). If the signal is amplified too little before digitization, the numbers will get "grainy": they might range from -1 to $+1$ with only three possible values (Fig. 3.11, bottom). In this case, it would be very difficult to find a small peak in the spectrum in the presence of big ones (this is called a "dynamic range" limitation). If, on the contrary, the signal is amplified too much, it might exceed the digitizer limits and get truncated or "cut off" (Fig. 3.11, top). For example, a signal that would give a value of 52,314 would be read as 32,768 because the digitizer cannot respond to any larger value. This cutting off or "clipping" has very drastic effects on the spectrum: the baseline gets huge oscillations ("wiggles") that cannot be corrected in any way. So it is clear that the receiver gain has to be set correctly for each sample to get the best results. More concentrated samples (or samples with large solvent peaks) will require smaller receiver gain values, whereas dilute samples are best run with large gain. Both Varian and Bruker allow for an automatic receiver gain adjustment. On the Varian, simply set *GAIN* to "*N*" (not used) and start the acquisition; a number of trial FIDs will be recorded to determine the best gain value and then the acquisition will begin. On the Bruker, the command *RGA* (receiver gain adjust) will do the same thing but will not automatically start acquisition. To better understand how the NMR hardware works, we will look in detail at five essential stages of the receiver (Fig. 3.10).

3.6.4.1 The Preamplifier This is a physical box that sits on the floor next to the magnet or is part of the "magnet leg," which either supports the magnet (Varian Gemini-200) or stands alone next to the magnet (Varian Unity or Inova). Its job is to amplify the RF FID signal immediately, before any thermal electronic noise has accumulated from the connecting cables. That is why it is located very close to the magnet, so the cables from the probe are short. Once the FID has been amplified, any thermal noise that is added to it later will be less important relative to the NMR signal. In cryogenic probes (Chapter 12, Section 12.3) the preamplifier is actually moved *into* the probe head where it is cooled to the same low temperature (e.g., 25 K) as the probe coil. This further limits the introduction of thermal noise at the first stage of amplification. Any noise that comes into the preamplifier is there "forever" because amplification at that stage will amplify the noise just as much as the signal (the FID). Thus, it is essential to limit this noise as much as possible in the early stages.

The preamplifier also contains a "send-receive" switch that allows the high-power pulse going into the probe and the very low-power FID coming out to travel on the same cable connecting the preamp to the probe. This "switch" is actually a solid-state device with no moving parts.

3.6.4.2 The Detector This converts the RF FID signal into an audio frequency FID signal by "mixing" it with a reference RF signal that has a single pure frequency at the center of the spectral window (v_r, the reference frequency). Subtraction of frequencies is accomplished by an electronic process called "mixing" using an analog device called a "mixer," "phase-sensitive detector," or "modulator." It actually involves analog multiplication of the FID signal by a reference frequency signal (200.000 MHz in this example), with the resulting signal having frequency components representing both the sum and the difference of the FID frequency and the reference frequency.

$$(199.999 - 200.001 \text{ MHz range}) \times 200.000 \text{ MHz reference}$$
$$= \underbrace{-1000 \text{ to } +1000 \text{ Hz range}}_{\text{(difference)}} + \underbrace{399.999 - 400.001 \text{ MHz range}}_{\text{(sum)}}$$

The multiplication sign represents a multiplication of the two signal amplitudes at each instant in time to give a momentary product amplitude. The sum frequency is eliminated by an electronic filter, leaving only the desired (difference) audio signal. In case any of you are wondering about this bit of electronic magic, it can be explained by high-school trigonometry as follows:

$$\sin(\alpha t)\cos(\beta t) = 1/2\{[\sin(\alpha t)\cos(\beta t) + \sin(\beta t)\cos(\alpha t)]$$
$$+ [\sin(\alpha t)\cos(-\beta t) + \sin(-\beta t)\cos(\alpha t)]\}$$
$$= 1/2\{[\sin((\alpha + \beta)t)] + [\sin((\alpha - \beta)t)]\}$$

This just says that the product of two waves of different frequency α and β is the same as the sum of two waves of frequency $\alpha + \beta$ (sum) and $\alpha - \beta$ (difference).

In reality this is done twice, first to "mix down" to an intermediate frequency (IF) still in the megahertz range [e.g., 20.5 MHz (Varian) or 22 MHz (Bruker)] and then to mix to the final audio frequency by subtracting out the IF. The reason for this is that we want to detect a wide range of different NMR frequencies (different nuclei), but it is difficult to have all of the amplification stages "broadband" so that they can deal equally well with all frequencies. By mixing down to an IF that is the same regardless of which nucleus we are observing, the electronics can be optimized to that frequency alone (narrow band) with greater efficiency. For example, a 300-MHz spectrometer with an IF of 20.5 MHz will mix a ^{1}H FID (300 MHz) with a reference frequency (also called the "local oscillator" or "LO") of 320.5 to get an IF of 20.5 MHz. This signal is amplified and then split into two signals that are mixed with 20.5 MHz reference signals (0° and 90° phase) to give the real and imaginary audio FIDs. To observe ^{13}C, the LO is changed from 320.5 to 95.5 MHz and mixed with the 75 MHz ^{13}C FID to give the 20.5 MHz IF. As before, the IF is split and mixed with the 20.5-MHz reference signals to give the real and imaginary ^{13}C FIDs. Regardless of the NMR frequency being observed (300 or 75 MHz), the IF (20.5 MHz) is the same. Radio receivers use the same principle of mixing to a common IF regardless of which station you are tuned to, and then mixing again to the audio frequency that you hear. The newest spectrometers produced today have an ADC that is fast enough to *directly sample* the RF FID, eliminating the need for analog mixing of any kind! The detection step (conversion to audio) is done by digital multiplication of the sampled RF FID.

The resulting audio signal has frequencies that represent the difference between the actual resonant frequencies of the sample nuclei and the reference frequency ($v_o - v_r$). This means that the audio frequency at the center of the spectral window is zero (reference frequency minus reference frequency = 0); the downfield half of the spectrum represents positive audio frequencies (FID frequency > reference frequency) and the upfield half represents negative audio frequencies (FID frequency < reference frequency). It is important to recognize that this audio frequency scale has nothing to do with the chemical shift (ppm) scale; that scale is added by the software after we find a reference peak and assign it a value on the ppm scale. Subtracting out the reference frequency from the RF FID corresponds to rotating the coordinate system used to represent the sample magnetization about the z (vertical) axis at the reference frequency. In this *rotating frame of reference*, a nucleus that resonates at the reference frequency (v_r) would have its magnetization vector stand still in the x'–y' plane after the pulse, since it is rotating in the same direction and at the same speed as the x' and y' axes of the rotating frame of reference. Nuclei that resonate in the downfield half of the spectral window have their magnetization rotating counter-clockwise in the x'–y' plane after the pulse, and those that resonate in the upfield half give rise to magnetization that rotates clockwise in the rotating frame of reference (Fig. 3.12).

Placing the zero of our audio frequency scale in the center of the spectral window has many advantages, but it requires that we have a way to tell the difference between positive frequencies and negative frequencies. This is accomplished by using a technique called *quadrature detection*. The RF FID signal is split into two and mixed with two different reference RF signals, one of which is phase shifted by 90° (one fourth of a cycle) with respect to the other (Fig. 3.10). The different frequency is selected in both cases, resulting in two audio signals that are 90° out of phase with each other. These signals are traditionally called the "real" and the "imaginary" FIDs, but there is nothing more or less real about either one. The best way to think about these "stereo" signals is to imagine that there are two receiver coils in the spectrometer: one placed on the x' axis of the rotating frame (recording the

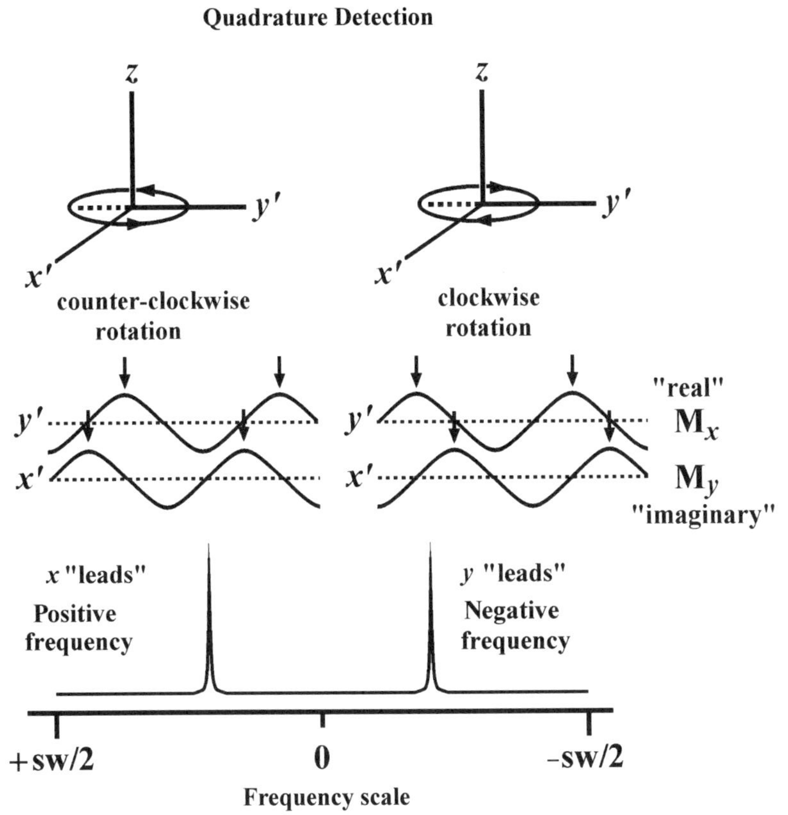

Quadrature Detection

Figure 3.12

x-component M_x of the sample magnetization) and one placed on the y'-axis of the rotating frame (recording the y-component M_y). If the magnetization vector resulting from a nucleus in the sample is rotating counter-clockwise in the rotating frame (i.e., faster than the axes are rotating), it will generate a maximum signal in the x-axis detector just before it reaches a maximum signal in the y-axis detector, so that the "real" (x-axis) signal will lead ahead of the "imaginary" (y-axis) signal by 90° (Fig. 3.12, left). We can thus determine that this frequency is a positive audio frequency, and place the peak in the spectrum in the downfield (left) half of the spectral window. If instead the magnetization vector is rotating clockwise in the rotating frame (i.e., slower than the x' and y' axes), it will generate a maximum signal in the y-axis detector just before it reaches a maximum signal in the x-axis detector, so the "imaginary" (y-axis) signal will lead ahead of the "real" (x-axis) signal by 90° (Fig. 3.12, right). In this case the frequency is a negative audio frequency, and the peak belongs in the upfield (right) half of the spectral window. Imagine a carrousel with one person riding on it near the edge. If you have two observers, one on the north side and one on the east side, and each observer calls out the direction as the rider goes by, you can tell which way the carrousel is rotating because you would hear "north east … …north east … …" for the clockwise direction and "east north … …east north … …" for the counter-clockwise direction. With only one observer you would hear, for example, "north … … …north … … …" and you would not be able to tell which direction the carrousel is rotating. The direction of rotation of the carrousel is analogous to the sign of the frequency of an NMR peak in your spectrum. If we only had one "detector," which is the equivalent of having only one FID channel, we could not distinguish between positive and negative audio frequencies. The two FIDs, real and imaginary, are processed by the computer using a complex (i.e., both real and imaginary) Fourier transform, which sorts out the positive and negative frequencies mathematically and spits out the correct NMR spectrum.

3.6.4.3 Audio Filters The audio stage amplifies the audio signal and noise and also tries to block by analog filtering any signals that are outside of the spectral window, that is, which have frequencies greater than the maximum frequency you have set up to observe (set by the ADC, see below). Generally you would not have any signals outside these limits, but you do have noise frequencies that extend in both directions from zero to positive and negative infinity. If these noise frequencies are not blocked, they will "fold in" to the desired range of frequencies and add to the noise that is mixed in with your desired signals. Without audio filters, the signal-to-noise ratio would be very near to zero. The audio filter response should ideally be flat throughout the desired range of frequencies and fall to zero very rapidly beyond the maximum frequency. This is not possible with analog filter devices (made up of electronic components such as capacitors, inductors, and resistors), so there is a certain amount of reduction in response ("droop") within the spectral window near the edges, and the response falls to zero gradually rather than suddenly for frequencies above the maximum (see Fig. 3.20). There are no parameters to adjust since the computer automatically adjusts the audio filters to a response that fits the width of the spectral window, as defined by the parameter *SW* (spectral width).

3.6.4.4 The ADC The computer cannot understand anything but numbers. The audio frequency FID is a continuous, smooth function of voltage (electrical intensity) versus time. The ADC or digitizer samples the FID voltage at regular intervals of time and assigns an integer value (positive or negative) to the intensity at each sample time (Fig. 3.13). These numbers go into a continuous list of numbers that constitute the digital FID. The spectrometer does not actually just acquire a single value for each time point—it is more like a stereo receiver. There are two channels in the receiver, one that effectively records signals along the x'-axis of the rotating frame and one that records signals along the y'-axis (Fig. 3.10). So the list of numbers is really twice as long because both FIDs are sampled by the ADC, and the numbers are loaded into the list in pairs: real (1), imaginary (1), real (2), imaginary (2), ..., and so on. Bruker and Varian originally had different ways of sampling, which led to some differences in processing and interpretation of data. Varian samples the two FIDs simultaneously at each time value, and Bruker alternates between real and imaginary samples in time; for example,

Time (μs)	Varian (simultaneous)		Bruker (alternate)	
	Real	Imaginary	Real	Imaginary
0	1. 23435	2. −2344	1. 13465	
80				2. 9354
160	3. 6509	4. 3496	3. −3546	
240				4. 31593
320	5. 5673	6. −234	5. 23486	
400				6. −14367

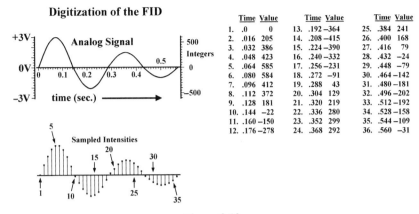

Figure 3.13

We will see that in 2D NMR, the sampling in the second dimension can also be done either way, except that this choice is up to the user and is not "hard wired." The alternate ("Bruker-like") sampling method is called "TPPI" (time proportional phase incrementation), and the simultaneous ("Varian-like") method is called "States" or "States-Haberkorn" (after the originators of the technique). The consequences for processing and interpretation of the data are the same in the second dimension of 2D spectra as they are in 1D NMR.

This leads to confusion over two parameters: the number of points collected (do you mean the total number of data points, or the number of real/imaginary pairs or "complex" pairs?) and the time spacing between data points. Both Bruker and Varian list the number of data points (Varian: *NP*, Bruker: *TD*) as the total number of points, counting both real and imaginary. Some independent NMR software packages (e.g., Felix) count points as "complex pairs": One "point" corresponds to one pair of numbers (real and imaginary). The time spacing between successive data points sampled in the FID is called the *dwell time* (Bruker: *DW*). In the above example, the dwell time is clearly equal to 80 μs between samples for the Bruker data, but in the Varian case we need to think of the dwell time as the average time per sample, which is still 80 μs because two samples are collected in a period of 160 μs. Varian does not have a parameter corresponding to dwell time, leaving the sampling process hidden from the user. The two types of data (alternate and simultaneous) must be processed by a different Fourier transform algorithm, but this is transparent as long as you process the data on the instrument that acquired it. If you transfer the data to another computer and use "third party" software (e.g., Felix, MestRec, Acorn-NMR, NMR-pipe, etc.) to process it, you need to choose the correct Fourier transform method for the type of FID data (alternating or simultaneous) being processed.

How rapidly do we need to sample the data? Clearly this is limited by how fast the hardware can convert analog to digital, but in most cases this limitation is not serious. It turns out that the rate of sampling is determined by the highest frequency signal you need to describe by the digitized data. In other words, what peak in your spectrum is farthest from the center frequency (the reference frequency)? For the sake of simplicity, consider a spectrometer from the middle ages that does not use quadrature detection, so that the audio frequency scale runs from zero on the right to the maximum detectable audio frequency (F_{max}) on the left. The highest frequency signal needs to be sampled at least twice during each cycle of its sine wave, meaning that the number of samples has to be twice the number of cycles in the highest frequency signal allowed.

$$\text{Number of samples in 1 s} = 1/\text{DW} = 2 \times \text{number of cycles in 1s}$$

$$1/\text{DW} = 2 \times F_{\max}$$

Once we have chosen a particular dwell time DW, the maximum frequency we can accurately determine (since the computer does not know anything about the what the signal does in between the samples) is $1/(2 \times \text{DW})$. What happens if the frequency of a signal exceeds $1/(2 \times \text{DW})$? The signal will not simply disappear; instead it is misinterpreted as a signal of lower frequency. For example, if the dwell time is 1/6 of a second, we will get six samples in 1 s from a signal of 1.5 Hz for a total of four samples per cycle, describing the sine wave quite accurately (Fig. 3.14, top). If we keep the sampling rate constant and increase the frequency to 3 Hz, we now have two samples per cycle, which is the minimum to describe its frequency: one sample at each trough and one sample at each crest of the wave (Fig. 3.14, middle). The frequency is now equal to $F_{\max} = 1/(2 \times \text{DW}) = 1/(2 \times (1/6)) = 3$ Hz, and the peak will appear in the spectrum at the left edge of the spectral window. If we now increase the frequency of the FID signal to 4.5 Hz, we have 1.33 samples per cycle, which is not sufficient to describe the sine wave and accurately determine its frequency (Fig. 3.14, bottom). Instead, a simpler interpretation would be to connect the dots to reveal a different sine wave of frequency 1.5 Hz, since we do not know what is going on between samples. Thus the peak would appear in the center of the spectrum, at $F_{\text{obs}} = F_{\max} - (F - F_{\max}) = 3 - (4.5 - 3) = 3 - 1.5 = 1.5$ Hz. This process is called "aliasing" or "folding" because the peak appears at the wrong position in the NMR spectrum. Anyone who has watched Western movies or television shows has seen the phenomenon of aliasing. A film (or videotape) of a moving stagecoach will often show the wheels slowing, coming to a

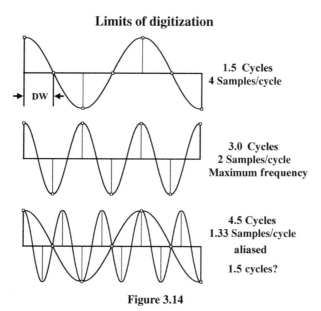

Figure 3.14

stop, or reversing direction even though the stagecoach is still obviously moving forward at full speed. The film is sampling the position of the spokes at a rate of 30 frames (samples) per second. If the wheels move fast enough, the motion of the spokes exceeds the sampling rate and we interpret the motion as being at a lower frequency than it really is. If this occurs in an NMR spectrum, we need to increase the sampling rate (decrease the dwell time DW) until we have two or more samples per cycle of the aliased frequency. Usually the aliased peak can be identified because it is lower in intensity and cannot be correctly phased.

The limits of frequency imposed by a fixed sampling rate lead directly to the concept of the "spectral window" (Fig. 3.15). In the case of quadrature detection, the center of the window is the zero point of audio frequency, which is determined by the reference frequency. The width of the spectral window is called the spectral width (SW), which is determined by the sampling rate and corresponds to F_{max} in the nonquadrature example. The extremes of the spectral window are +SW/2 at the left edge and −SW/2 at the right edge, and we can replace F_{max} with SW in the equation: SW = 1/(2 × DW). The spectral window can be moved to the left or right by adjusting the offset (Bruker: *O1*; Varian: *TO*), which changes the exact value of the reference frequency. The offset frequency (in hertz) is added to the fundamental resonance frequency for the nucleus of interest to obtain the reference frequency. For example, a 250 MHz instrument set up for proton acquisition might have a fundamental ^{1}H frequency of 250.13 MHz. Adding an offset (*O1*) of 10,000 Hz (0.01 MHz) would yield a reference frequency of 250.14 MHz. To move the spectral window downfield by 1 ppm (250 Hz), one would simply add 250 Hz to the offset value (*O1*), changing the value of this parameter from 10,000 to 10,250.

Why would you need to move the spectral window upfield or downfield? The lock system changes the magnetic field strength of the spectrometer (B_o) slightly to center the ^{2}H frequency of the solvent at the null point of the lock feedback circuit. Changing the field changes all of the resonant frequencies of the spectrum by the same amount, effectively moving the whole spectrum upfield or downfield by as much as 5 ppm when you

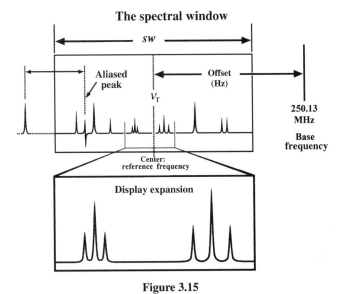

Figure 3.15

change from one deuterated solvent (e.g., CDCl$_3$) to another (e.g., d$_6$-acetone). If this is not corrected by changing the offset by an equal and opposite amount, the spectrum will move out of the spectral window and some peaks will be aliased. For routine work, this hassle has been removed in two ways. On the old Bruker (AM) instruments, you had to have a list of *O1* values for different solvents in order to keep the center of the spectral window at the same value (e.g., 5 ppm) for all solvents. On the Varian, the correction is made automatically by entering the lock solvent as the parameter "*SOLVNT*." This changes the fundamental resonance frequency so that the offset (*TO*) is always the same for a given ppm value at the center of the window. This can be frustrating if you neglect to change the *SOLVNT* parameter for solvents other than the default setting (e.g., CDCl$_3$). The newer Bruker instruments (DRX) use a parameter in the lock system called *lock shift*, which is the ppm value of the lock solvent (for example, 7.24 for CDCl$_3$), and this corrects the reference frequency internally. If you use the automatic lock and specify the lock solvent, this parameter is automatically set to the correct value. Sometimes the spectral window needs to be changed for unusual samples with chemical shifts outside the standard (for example, 11 ppm to -1 ppm for ^{1}H) spectral window. If you have a carboxylic acid with an OH resonance at 13 ppm, you would like to have a spectral window from -1 to 17 ppm. That means you need to increase the spectral width by 6 ppm (from 12 ppm to 18 ppm) and move the center of the spectral window downfield (to higher frequency) by 3 ppm (from 5 ppm to 8 ppm). On a 200 MHz instrument that would mean adding $6 \times 200 = 1200$ Hz to the spectral width (SW) parameter and adding $3 \times 200 = 600$ Hz to the transmitter offset (*TO* or *O1*) parameter. You will have to repeat the acquisition, of course, because these parameters have no effect, except at the time that the FID is acquired.

With quadrature detection, the range of audio frequencies detected runs from $+$SW/2 to $-$SW/2, with zero in the center. The same relationship exists between the maximum frequency detectable and the dwell time, except that we substitute SW for F_{max}:

$$1/(2 \times DW) = SW$$

The last equation tells us what value of the dwell time we have to use to establish a particular spectral width. In practice, the user enters a value for SW and the computer calculates DW and sets up the ADC to digitize at that rate. It is important to understand that with the simultaneous (Varian-type) acquisition mode, there is a wait of $2 \times$ DW between acquisition of successive *pairs* of data points. The average time to acquire a data point (DW) is the total time to acquire a data set divided by the number of data points acquired whether they are acquired simultaneously or alternately. The spectral window is fixed once the sampling rate and the reference frequency have been set up. The spectral window must not be confused with the "display window," which is simply an expansion of the acquired spectrum displayed on the computer screen or printed on a paper spectrum (Fig. 3.15, bottom). The display window can be changed at will but the spectral window is fixed once the acquisition is started.

Any peak outside the spectral window will be aliased ("folded") into the spectral window at a position the same distance from the edge of the window. Aliased peaks are usually reduced in intensity (by the audio filter) and impossible to correctly phase; increasing the spectral width will eliminate them and reveal the peak in its correct position. The manner of aliasing depends on the type of acquisition. With the "Bruker-type" acquisition (alternating acquisition of real and imaginary data samples), aliased peaks appear reflected at equal distance from the same edge of the spectral window ("folded"), as shown in Figure 3.15 (upper left). With the "Varian-type" acquisition (simultaneous acquisition of a real, imaginary pair of samples), aliased peaks appear at equal distance inside the opposite edge of the spectral window ("aliased"). The terms "folding" and "aliasing" are often used interchangeably, but it would be more accurate to use "folding" for the alternating mode (reflecting across the nearest edge of the spectral window) and "aliasing" for the simultaneous mode (appearing at the same distance inside the other edge of the spectral window as the frequency is from the nearest edge).

The same phenomenon applies to aliasing in the second dimension of a 2D spectrum: Alternating (TPPI) acquisition in the second dimension will lead to aliasing on the same side of the spectral window ("folding"); simultaneous (States) acquisition will lead to aliasing from the opposite edge of the spectral window ("aliasing").

An example of the real part of an actual audio frequency FID of a sample of chloroform and dichloromethane (recorded on a Varian Gemini-200) is shown in Figure 3.16. The full real FID (acquisition time (AT) = 2.9 s) is shown in the inset at the upper right: a decaying oscillating signal is clearly visible with a frequency of about 34 cycles per 2.9 s = 11.7 Hz. A horizontal expansion of the first 0.43 s (below the inset) makes it clear that there are two different frequencies, with their oscillating signals added together in the FID. The slower (lower frequency) oscillation completes one cycle in 87.5 ms, corresponding to a CH_2Cl_2 frequency of 11.5 Hz (1/0.0875 s), whereas the faster (higher frequency) signal completes a cycle in 2.67 ms, corresponding to a $CHCl_3$ frequency of 375 Hz (1/0.00267 s). Figure 3.17 shows a horizontal expansion of the first 36 ms of the real FID, capturing one-half cycle of the 11.5 Hz signal and 13.5 cycles of the 375 Hz signal. Now we can see the "grain" of the actual digital samples: one every 360 μs (0.36 ms) for a dwell time of 180 μs, since there are two data points, real and imaginary, for each point shown. The NMR software draws straight lines to connect the data points (□), but in fact, we know nothing about the signal in between these samples. One cycle of the highest frequency signal (375 Hz) corresponds to

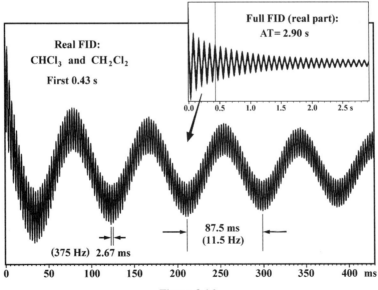

Figure 3.16

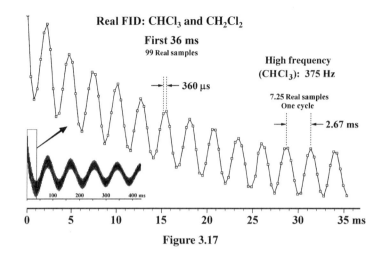

Figure 3.17

7.25 real samples, well over the minimum of two samples per cycle. The spectral width (SW) is $1/(2 \times 180\ \mu s)$ or 2777.8 Hz. The high-frequency signal (from $CHCl_3$) has a frequency of 375 Hz (one cycle = $1/375$ s = 2.67 ms), and the low-frequency signal (from CH_2Cl_2) has a frequency of -11.5 Hz. The sign of the frequency can be determined only by examining the relative phases of the real and imaginary parts of the FID (quadrature detection). When you set the spectral reference using a standard such as TMS, you establish a third frequency scale (in addition to the absolute RF scale and the audio frequency scale relative to the reference frequency), which is the chemical shift scale in parts per million. Because the data were acquired on a 200 MHz spectrometer, an audio frequency of 200 Hz is 1 ppm away from the center of the spectral window. In this case the center of the spectral window is 5.37 ppm, so that the $CHCl_3$ chemical shift is $5.37 + (375/200) = 7.24$ ppm and the CH_2Cl_2 chemical shift is $5.37 - (11.5/200) = 5.31$ ppm.

The 99 data points shown in Figure 3.17 are part of a total FID of 8000 complex pairs (total number of data points NP = 16,000). Since a single data point takes 180 μs (the dwell time) to acquire on average, 16,000 points require $16,000 \times 180\ \mu s = 2,880,000\ \mu s$ or 2.88 s to acquire. This is called the acquisition time (Bruker: *AQ*; Varian: *AT*), and it represents the time required to record the entire FID once. This is not the time required for the entire spectrum to be acquired, since it does not include the relaxation delay and the pulse width, and it does not take into account the number of times the whole sequence is repeated (i.e., the number of scans or transients). In general,

Acquisition time

= number of points (real and imaginary) × time required per data point

$$AT = NP \times DW$$

But the dwell time (DW) is determined by the spectral width: $DW = 1/(2 \times SW)$. Substitution of $1/(2 \times SW)$ for DW gives

$$AT = NP \times DW = NP \times (1/(2 \times SW))$$

Multiplying by (2 × SW) on both sides:

$$NP = 2 \times SW \times AT \text{ (Varian)}$$

$$TD = 2 \times SW \times AQ \text{ (Bruker)}$$

Number of data points = 2 × spectral width × acquisition time

This is the fundamental equation of NMR data acquisition (the mnemonic "swat" is useful). It tells us that the three parameters NP, SW, and AT (or TD, SW, and AQ in Bruker) are wedded by this equation such that changing any one of the three will require changing another to maintain the equality. For example, if we double the spectral width, either the number of points will double or the acquisition time will be cut by half. This is because the larger spectral width requires a faster sampling rate (half the dwell time) to assure that all of the frequencies in the spectral window are sampled at least twice in each cycle. With twice the sampling rate, you will either complete sampling the fixed number of points in half the time or keep the acquisition time constant and sample twice as many points. Bruker keeps the number of points constant and changes the acquisition time; Varian leaves the acquisition time unchanged and calculates a new value for the number of points. This can be frustrating because parameters you thought you had not changed are changing before your eyes!

The spectrum resulting from Fourier transformation of this FID is diagramed in Figure 3.18. The three frequency scales shown illustrate the progression in recording the FID from RF (actual frequency observed) to audio frequency (after subtracting out the reference RF signal, $v_r = 200.010$ MHz in this example) to a referenced chemical shift scale (after setting the spectral reference of TMS).

3.6.4.5 The Sum to Memory The sequence: (relaxation delay–pulse–acquisition of FID) is repeated a number of times (Fig. 3.19) with the acquired and digitized FID added each time to a "sum" FID stored in memory. The figure shows Bruker parameter names with Varian names in parentheses. The "recycle time" is the total time required to acquire one scan: relaxation delay + pulse width + acquisition time. The total experiment time is the

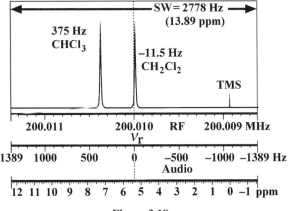

Figure 3.18

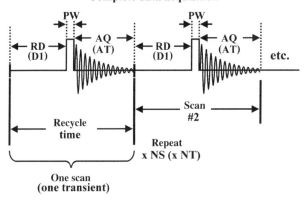

Figure 3.19

product of the recycle time, RD + PW + AQ (Varian: D1 + PW + AT), and the number of scans, NS (Varian: NT). Each individual FID contains the same signal (sum of decaying sine waves for all the sample nuclei), but the noise is different in each FID because it is random. The signal intensity increases directly with the number of repeats ("scans" or "transients"), but the noise increases with the square root of the number of repeats (Chapter 1, Fig. 1.6). This is like two people walking from the same starting point: one is sober and walks continuously in a straight line and the other is drunk and changes direction regularly in a random fashion. The distance from the start is directly proportional to the time for the sober one, but the drunk walk is less efficient, gradually drifting farther and farther from the start. The signal-to-noise ratio (S/N) is thus proportional to the number of scans divided by the square root of the number of scans:

$$S/N = \text{signal/noise} \, \alpha \, NS/\sqrt{NS} = \sqrt{NS}$$

This means that if you want to improve the S/N by a factor of 2, you will need to acquire four times as many scans. Since the total experiment time is proportional to the number of scans

$$\text{Time required} = NS\,(RD + PW + AQ)\,\text{(Bruker)}$$
$$= NT\,(D1 + PW + AT)\text{(Varian)}$$

you will need four times as much time on the spectrometer to get a factor of 2 improvement in S/N.

Because in each repeated acquisition the observed data is simply added into the accumulated sum in memory, the size of the data file is not changed by increasing the number of scans. The individual FIDs are lost as their data values are added to memory. Consider a simplified example in which four FIDs are summed in memory. Although these calculations are always done in the computer with binary numbers, we will use decimal numbers in this example for clarity. Assume that the digitizer has only one decimal digit (typically there are 16 binary digits) available and that the memory allotment for each data point is two decimal digits (typically there are 16, 24, or 32 binary digits). Thus, the FID data coming out of the digitizer can range from a value of -9 to a value of $+9$, and the sum-to-memory value at each time point can range from -99 to $+99$. Although this list may be very long (16384 or 32768 data points in all for a 1D spectrum), we will consider only the first six real data points (DW = 40 μs).

time (μs)	80	160	240	320	400	480	...
FID1	8	3	0	−2	−3	−1	...
memory	8	3	0	−2	−3	−1	...
FID2	7	2	0	−1	−3	0	...
memory	15	5	0	−3	−6	−1	...
FID3	9	3	1	−3	−2	−1	...
memory	24	8	1	−6	−8	−2	...
FID4	8	3	−1	−1	−4	0	...
memory	32	11	0	−7	−12	−2	...

In this case the receiver would overflow ("clip") with any FID value greater than +9 or less than −9. Notice that the FID values for different scans at any given time point are roughly the same, since only the noise is different. At each time point the FID value is added to the running total in memory; for example, the 160 μs time point of FID3 has a value of 3, which is added to the previous sum value of 5 to give the new sum value of 8. As more and more FIDs are acquired, the sum increases steadily and will overflow the number of digits allotted to it in memory after a certain number of scans (sum greater than 99 or less than −99). On Varian instruments this will stop acquisition, resulting in the error message "maximum number of transients accumulated." This will only occur on long (e.g., overnight) acquisitions and can be avoided by setting the variable DP (double precision) to *Y* (Yes). This doubles the number of digits used in memory (from 16 to 32 binary digits) and also doubles the size of the data file. On the Bruker a memory overflow (beyond the 24 or 32 binary digits reserved) results in the whole FID sum in memory being divided by 2; acquisition continues with the new FIDs being divided by 2 before being added in. In this way Bruker never has a problem with memory overflow, but accuracy is lost in the division process because 1 bit is discarded with each overflow.

The number of scans needed is primarily determined by the concentration of the sample and the desired signal-to-noise ratio. Another factor to consider is the *phase cycle*. Artifacts that are inherent in the electronics of the spectrometer can be canceled out by changing the phase of the RF pulse in a fixed pattern (e.g., 0°, 90°, 180°, and 270° in scans 1, 2, 3, and 4) and changing the phase of the receiver (by subtracting the signal instead of adding, or switching the real and imaginary parts) to follow this progression. The number of scans should be an integer multiple of the phase cycle length (a multiple of four for simple 1D acquisition) to assure optimal cancelation of artifacts. Some experiments, which subtract undesired signals from desired ones, will not work if the number of scans is set wrong. The phase cycle cancelation can also be screwed up if the first scan or two are acquired with the nuclei not in the "steady state" in terms of relaxation. Often the relaxation delay is not long enough for complete return of all spins to the equilibrium state, so the spins reach a steady state after a few scans where the degree of relaxation is always the same at the start of each scan. This steady state can be established by using dummy (or steady-state) scans. These are scans that include a relaxation delay, pulse, and acquisition just like a normal scan, but the data are not added into memory. The number of dummy (steady-state) scans is *DS* (Bruker) or *SS* (Varian).

3.6.4.6 The Computer On newer (Bruker AMX, Varian Unity and newer) NMR spectrometers there is an acquisition computer that runs the NMR console and a data processing computer (usually a UNIX system purchased off the shelf from Sun Microsystems or Silicon Graphics) that communicates with the NMR console through an Ethernet (Internet-like) or SCSI (device interface) communication cable. When the data acquisition is complete, the FID data in the sum to memory is transferred to the acquisition computer in the console, and this data is then sent to the "master" (or "host") computer that the user is running. All the data processing—display on the screen, weighting functions, Fourier transform, phase correction, baseline correction, peak and integral analysis, and plotting—is done on this computer using the vendor's own software package.

3.7 NOISE AND DYNAMIC RANGE

Two terms that are very important in assessing the quality of NMR data are signal to noise and dynamic range. The first has already been discussed, but it is useful to think about where noise comes from and what can be done to reduce it. The signal is proportional to the concentration of the sample and to the number of scans. Although signal is often measured as peak height, we must remember that it is the peak area that is proportional to concentration and number of scans; the peak height is very sensitive to shimming. As the shimming improves, the peak gets taller because the constant peak area is squeezed into a narrower and narrower peak. It is the peak height that determines whether a peak can be seen "over" the noise: If a peak is very broad, it is much easier to "lose" it in the noise than a sharp peak that "sticks up" above the level of the noise. If we compare two one-proton signals from the same molecule, the one with fewer splittings (J couplings) will be easier to see above the noise. For example, a fully resolved doublet of doublets (four peaks of equal height) is only one fourth of the height of a singlet with the same linewidth and area. Especially in low-sensitivity 2D experiments, the complex multiplet signals are often lost in the noise while the sharp singlet signals are easy to see. When measuring signal-to-noise ratio with a standard sample (0.1% Ethylbenzene in $CDCl_3$ is the ^1H standard), the result is not meaningful unless the shimming is superb. With high-order shim errors, all of those "porches" and "pedestals" that may not even show above the noise are robbing intensity from the peak and lowering its peak height.

3.7.1 Analog Noise

The source of noise is another consideration. As already discussed, noise that comes into the preamplifier is there forever and will be amplified along with the signal throughout the receiver. But with each successive stage of amplification (RF, IF, and audio) new thermal noise is introduced, and this noise only gets amplified by the successive stages of amplification that follow. So noise introduced late in the process contributes less to degrading the signal-to-noise ratio because the signal was amplified already quite a bit before the noise was introduced. A quantitative way of measuring the noise introduced in the receiver is the "noise figure." This is a measure of all noise added to the initial signal introduced to the input of the preamplifier. As technology improves, the noise figure goes down and NMR instruments become more sensitive because the signal-to-noise ratio improves. The noise figure can be measured by introducing a calibrated source of noise into the preamplifier and measuring the noise level in the fake "FID" recorded. A cheaper but less reliable method involves measuring noise from a 50-Ω resistor attached to the preamplifier input and comparing the "FID" noise with the resistor at room temperature and at 77 K (in a bath of liquid nitrogen). Since the increase in noise from the resistor in going from 77 K to room temperature is known, the increase in noise in the FID should reflect a constant (the noise introduced by the receiver) plus this variable of known ratio. If the noise figure is small, signal and noise get about the same amplification in the receiver, and signal-to-noise ratio will not change much as the receiver gain is increased, but if the noise figure is large there will be a significant increase in signal-to-noise ratio as the gain is increased, since noise that is introduced at later stages of amplification will not be boosted as much as the signal, which is amplified at all of the stages. This is one reason why you want to increase the receiver gain as much as possible to a level just below where you start overloading the digitizer ("clipping" the FID).

3.7.2 Digitizer Noise

Finally, noise can be introduced digitally by the sampling process of the ADC. Since the ADC must select an integer value for the intensity of the FID at each moment it samples, an input voltage right at the dividing point for choosing one integer or the next larger value leads to an uncertainty or "jitter" of one integer unit in the digital output. If the real thermal noise is less than one integer value in the ADC, the thermal noise is masked and we only see this digital noise. In this case, the thermal noise fluctuations are too small to "flip" the ADC to the next integer value, so the noise information is completely lost. Starting with a very low value of the receiver gain (RG or GAIN), increasing the gain leads to a steady increase in signal-to-noise ratio, since the noise is fixed at the value of the digital noise. At some point, however, the noise amplitude gets large enough that it is being accurately digitized by the ADC. Beyond this point any increase in receiver gain boosts both the signal and the noise, and the signal-to-noise ratio no longer increases as steeply. Using an ADC with more "bits" (a finer division of integer values with respect to the input voltages) or sampling many points and averaging them to get each single data point (oversampling) can reduce the step size of the ADC so that this maximum of signal to noise is achieved at lower gain values. This is important if there are large signals like solvents (e.g., H_2O in biological samples) that dominate the FID and limit the gain to a low value.

3.7.3 Dynamic Range

Dynamic range is the range of concentrations or signal intensities over which you can detect samples in a single measurement. If you are trying to find a "needle in a haystack"—for example, observing the ^1H spectrum of a 1 mM protein sample in 55 M H_2O—you need to have dynamic range. Signal to noise is an absolute limitation: It sets a minimum of signal height that can be observed for a "weak" signal, regardless of any strong signals in the sample. Dynamic range is a relative limitation: It determines how small a signal can be detected relative to the largest signal in the sample. The receiver gain is limited by the largest signal in the sample because the digitizer limits will be exceeded first by that large signal as you increase the receiver gain. Thus at this limit, the "top" of the digitizer (largest integer value it can assign to a signal) is set to the signal strength of the largest signal in the sample. We say that the digitizer is "dominated" by this large signal. The small signal that "rides" on top of this large signal FID has to be accurately described by the digitizer. If it is smaller than one integer step in the ADC output, the signal is lost. If it is only a few integer steps, it will be picked up but the peak in the spectrum will be "blocky"—described by large square integer steps in intensity instead of by smooth curves, similar to a bitmap drawing with very low resolution. The ratio between the largest integer and the smallest (one unit) is the dynamic range of the digitizer. A 12-bit ADC uses 12 bits to digitally measure the analog voltage at each sample point, so the dynamic range is 4096 to 1 (2^{12} to 1). More modern instruments use 16-bit digitizers, so they have a dynamic range of 65,536 to 1 (2^{16} to 1). This can be further increased by oversampling (acquiring many samples for each data point and averaging them), since this allows partial integer values when the average is computed. For example, if you digitize four equally spaced data points during an 80 μs dwell time and average the value to one data point, you can have a result that is, for example, 645.00, 645.25, 645.50, or 645.75. You now have four times as many intensity levels to choose from and you have increased your digitizer resolution by 2 bits. A modern NMR spectrometer can typically oversample by a factor of 32, leading to 5 additional bits or a total of 21 effective bits in the digitizer and a dynamic range of 2,097,152 to 1!

3.8 SPECIAL TOPIC: OVERSAMPLING AND DIGITAL FILTERING

The sampling rate is the rate at which the ADC samples the raw analog FID audio signal and converts the intensities (voltages) into numbers. The delay between samples is called the dwell time (DW) so that the sampling rate can also be expressed as 1/DW in units of hertz. Because we must have at least two samples per cycle of a sinusoidal signal to properly define its frequency without aliasing, the sampling rate is determined by the highest frequency we need to digitize. The user defines the spectral width (SW) and the spectrometer calculates the sampling rate:

$$\text{Rate} = 1/\text{DW} = 2 \times \text{SW} \quad (\text{DW} = 1/(2 \times \text{SW}))$$

For a typical spectral width of 6250 Hz (12.5 ppm for a ^{1}H spectrum on a 500 MHz spectrometer), the sampling rate is 12,500 samples per second. For a ^{13}C spectrum the spectral width is larger, so that a ^{13}C spectrum with a 250 ppm spectral width on a 600 MHz spectrometer would require a sampling rate of

$$\text{Rate} = 2 \times \text{SW} = 2 \times 250\,\text{ppm} \times 150\,\text{MHz} = 75{,}000\,\text{samples per second}$$

since the ^{13}C frequency on a 600 MHz spectrometer is 600 × (γ_C/γ_H) = 150 MHz. This may seem like an impressive feat, but it is well within the capabilities of even an older generation ADC. Even an inexpensive modern ADC can sample at 400,000 samples per second, so that the capacity is 5.33 times greater than that needed for the ^{13}C spectrum example and 32 times greater than that needed for the ^{1}H spectrum example. The question naturally arises: Is there anything useful we can do with the excess sampling capacity?

3.8.1 Oversampling

In the ^{1}H spectrum example, if we sample at 400,000 samples per second, we will have 32 samples for every data point that we actually need for the FID. The simplest thing to do with all of this extra data is to divide them into groups of 32 consecutive data samples and average each group to give a single data value. Is this any better than sampling 12,500 samples per second? Yes, because any time you repeat a measurement many times and average the results, you get a more accurate measurement. Furthermore, since each measurement is an integer value with a limited dynamic range (e.g., −32,767 to 32,768 for a 16-bit ADC) you would have a much finer range of possible intensities because each bit (0 or 1) can now have 32 possible values after averaging 32 measurements (0, 1/32, 2/32, ..., 31/32). This finer "graininess" of the intensity values might be useful if we are trying to find a very weak signal in the presence of a very strong one (needle in a haystack problem). Essentially we now have more significant figures (precision) in each measurement, increasing the *dynamic range* (ability to detect small signals in the presence of large ones).

This simple process of averaging each set of 32 raw measurements to get a single value is called *decimation*, and we would define the *decimation factor* as 32 in this case. Thus we see a part of the overall strategy: Oversampling produces many more data points than we need, and decimation averages them to give us the required sampling rate determined by the spectral width. But we can do much more than just increase the accuracy of our measurements: We can use digital methods to construct a *filter* that rejects signals outside of the spectral window without affecting the desired signals within the spectral window. With *digital filtering*, we can set a narrow spectral window that covers only part of the spectrum, and none of the other peaks in the spectrum will alias or "fold in" to the narrow window because they are removed by the digital filter. To understand how this works, we need to first understand the "old fashioned" analog filter used in an NMR spectrometer.

3.8.2 Analog Filtering

We have already seen that the digitization of the FID signal (sampling at regular intervals) sets a limit on the frequencies that can be detected without aliasing. Any frequency larger than SW/2 (with quadrature detection, the frequency is zero at the center of the spectral window, so the edges are at ±SW/2) will be aliased back into the spectral window. This applies to noise as well as to peaks, so that without some way of rejecting signals outside the spectral window, we would have a huge amount of noise aliased into the spectral window and the signal-to-noise ratio would be abysmally low. To avoid this, an analog audio filter is included before the ADC to remove any frequencies with absolute value greater than SW/2. Analog filters are constructed from capacitors, resistors, and inductors and have switches to match the "bandpass" (region of frequencies passed through) of the filter to the spectral width set by the user. The *frequency response curve* shows how effectively a filter blocks the signals outside the spectral window and to what extent it affects signals within the spectral window. The ideal filter response would be "flat" throughout the spectral window and would drop instantly to zero outside the spectral window (Fig. 3.20). Unfortunately, real audio filters tend to attenuate signals in the spectral window that are near the edge, and drop off only gradually outside the spectral window. Peaks that are outside the spectral window are aliased with some attenuation into the spectral window, along with the noise.

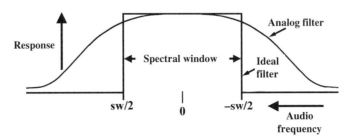

Figure 3.20

Thus we expect less signal and more noise near the edges of the spectral window, and this will degrade the sensitivity as well as the accuracy of integration. Peaks and noise that are very far outside the spectral window are effectively blocked by the analog filter. The goal of digital filtering is to achieve a nearly ideal filter response curve, matched to the spectral window, with only the computational tools of a computer chip rather than the cumbersome and imperfect electronic components of an analog circuit. The filter will be moved from its place in the analog stream before the ADC and placed in the digital stream after the ADC.

3.8.3 Decimation with Digital Filtering

Let us return to our example of a ^{1}H FID oversampled 32 times and decimated by a simple average of each set of 32 data values. It turns out that this method of decimation by simple averaging actually discriminates among frequencies in the FID, so that a high frequency that changes sign many times during the 32 samples will be nearly eliminated (positive swings cancel the negative swings) and a low frequency that is nearly constant for the 32 samples is unaffected. So this simple filter is a crude kind of *low-pass filter*: It cuts out the high frequencies and passes (leaves unchanged) the signals with low frequencies.

To understand how this works, consider a simpler example: a filter that averages groups of four data points to give a single filtered value. If the data is oversampled with a decimation factor of 4, the sampling rate (1/DW) is eight times the spectral width (8 × SW) instead of twice the spectral width (2 × SW). Consider the effect of this process on a pure sine wave FID with frequency 4 × SW, which is sampled at a rate of 8 × SW (delay between samples is 1/(8 × SW)). The raw data has sampled values of 1, −1, 1, −1, ... and has a frequency of 4 × SW since it goes through a complete sine wave cycle in two data points (period = 1/(4 × SW)). Such a signal would be reduced to zero by the filter, which averages each group of four data points. A zero response would also be obtained for a signal of frequency 2 × SW (0, 1, 0, −1, ...). A signal with frequency SW (data values 0, 0.707, 1, 0.707, 0, −0.707, −1, −0.707) would be retained with reduced intensity because decimation would give two data points: 0.604 and −0.604. This pair would repeat leading to a correct frequency measurement of SW. A signal of zero frequency (1, 1, 1, 1) would also be retained with unchanged intensity (1.0 for each averaged value). We can map out the frequency response curve for this digital filter as shown in Figure 3.21: the response is maximum at the left edge of the spectral window, drops to 71% at the right edge, then drops to zero and oscillates and decays for larger frequencies (for simplicity we are not considering real and imaginary data acquired with quadrature detection, so the edge of the spectral window is at frequency SW rather than SW/2). The response is a "sinc" function (sin(x)/x) that effectively passes low frequencies

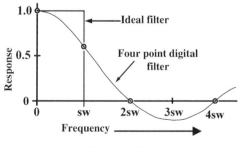

Figure 3.21

and discriminates against high frequencies, but it has many undesirable characteristics. The response is far from constant in the interval 0–SW, and there is significant nonzero response outside of the spectral window. Furthermore, the filtered data switches sign in some regions, indicating an alteration of the original phase. Clearly we need to design a better filter, and to do so we must examine various digital filters with different sizes and shapes.

3.8.4 Digital Filtering and the Convolution Theorem

Consider a filter that averages in groups of three data points *without any decimation*. Every group of three consecutive raw data points is summed together, and the sum is applied to the center value (second of three data points) in a new data set—the digitally filtered FID. We can think of the filter as a rectangle-shaped window, three data points wide, which moves through the FID data, stopping to add together three points and deposit the sum in the new data set and then moving one data point to the right-hand side and repeating the process. What effect will this have on a simple sine function FID? The math is shown in Figure 3.22(a). Each data value on the bottom is the sum of the three data values above it: one above and to the left-hand side, one directly above, and one above and to the right-hand side. Note that the filtered data has one extra data point at each end since the filter is three data points wide and begins to encounter data when the first raw data point is reached. After two anomalous points at the beginning (the "group delay" of the filter) the filtered data are identical to the raw data. Thus this frequency ($2 \times$ SW sampled at a rate of $8 \times$ SW) is passed without any change by the filter. With a raw FID of frequency $4 \times$ SW we see that the data are passed by the filter but with inverted phase (Fig. 3.22(b)). A frequency of zero (all data points equal to 1) is passed with high efficiency (all data points equal to three). Thus the frequency response for this digital filter is a sinc function with a maximum response at zero frequency, a smaller positive response at $2 \times$ SW, and a negative response at $4 \times$ SW.

A wider filter function gives a narrower frequency response. Consider a digital filter that is four points wide, with all four values given equal weight. A frequency of $4 \times$ SW is

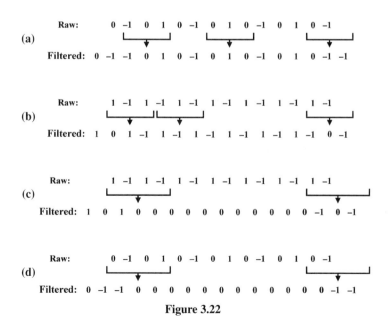

Figure 3.22

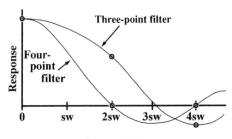

Figure 3.23

blocked by the filter after a brief transient response (Fig. 3.22(c)). Likewise a frequency of $2 \times$ SW is also blocked (Fig. 3.22(d)). As before, a zero frequency (all data points equal to 1) is passed with maximum response. Similar arguments show that odd multiples of SW (SW, $3 \times$ SW, etc.) are passed by this filter. The filter response curve for the four-point filter is a sinc function with null points at $2 \times$ SW and $4 \times$ SW, so it is narrower than the response curve for the three-point filter (Fig. 3.23). Note that the shape of the frequency response curve (frequency domain) is the Fourier transform of the shape of the filter function (time domain). A filter function that is rectangular in shape (three or four equally weighted points with all other points weighted zero) leads to a frequency response curve that is a sinc $(\sin(x)/x)$ function, and the sinc function is narrower in frequency as the filter function is made wider in time. This is just what we expect for the Fourier transform of a rectangular shape in time domain. This principle can be stated more generally if we consider that the filter function need not be rectangular, that is, the points in the filter do not have to be weighted equally. A general digital filter has N coefficients or weighting factors $c_1, c_2, c_3, \ldots, c_N$, and it is passed through the raw data stopping at each new position where the weighted average is calculated (Fig. 3.24, top), where $r_1, r_2, r_3, \ldots$ represent the raw (unfiltered) FID data and the filtered data value for point 7 is

$$d_7 = c_1 r_3 + c_2 r_4 + c_3 r_5 + c_4 r_6 + c_5 r_7$$

The filter is then moved to the next position and the weighted average is again calculated (Fig. 3.24, bottom). The value for point 8 is

$$d_8 = c_1 r_4 + c_2 r_5 + c_3 r_6 + c_4 r_7 + c_5 r_8$$

This process of moving the filter function through the raw data and calculating weighted averages is called *convolution*, and the digitally filtered data $d_1, d_2, d_3, \ldots$ are called the

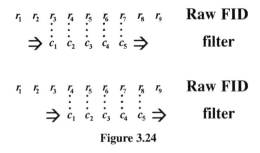

Figure 3.24

convolution of the raw data $(r_1, r_2, r_3, \ldots)$ with the filter function $(c_1, c_2, c_3, \ldots)$. In mathematical terms,

$$d(t) = c(t) \otimes r(t)$$

where $d(t)$ is the digitally filtered FID, $c(t)$ is the filter function, and $r(t)$ is the raw FID, all of them digital time-domain functions. The process of digital filtering is the same as the mathematical operation of convolution, represented by the symbol $\otimes$.

The *convolution theorem* states that the Fourier transform of the convolution (d) is simply the *product* of the Fourier transforms of the two functions (c and r) that are combined by convolution to make d. Thus convolution in time domain is equivalent to simple multiplication in frequency domain:

$$d(t) = c(t) \otimes r(t) \qquad D(f) = C(f) \times R(f)$$

where $D(f)$ is the spectrum obtained by Fourier transformation of the digitally filtered FID $d(t)$, $R(f)$ is the spectrum obtained by Fourier transform (FT) of the raw FID $r(t)$, and $C(f)$ is the frequency response curve obtained by FT of the filter function $c(t)$. To determine the effect of any digital filter on the spectrum, we simply look at the Fourier transform of the digital filter's shape (its coefficients).

Now that we understand the exact relationship between the shape of the weighting factors (coefficients) used in the digital filter and the frequency response curve it produces in the spectrum, we can begin to design a digital filter with ideal properties. The ideal frequency response curve is flat throughout the spectral window and falls off instantly to zero outside the window. In mathematical terms, this is a rectangular shape in frequency domain. The design question then boils down to this: What time-domain function (digital filter shape) will give, after FT, a rectangular function (frequency response curve)? The answer is simple: A sinc ($\sin(t)/t$) time-domain function gives a rectangular frequency-domain function upon FT. The narrower we make the sinc function in time domain, the wider will be the rectangular frequency response curve. So this is our goal: to construct a set of digital filter coefficients that correspond to a sinc function in time domain.

3.8.5 Optimizing the Digital Filter

The digital filter cannot be infinitely long, so we will have to cut off (truncate) the sinc function at some point. This will affect the frequency response curve, so it will not be a perfect rectangle, and some of the proprietary (trade secret) information guarded by NMR instrument makers (Bruker and Varian) has to do with the optimization of finite-sized filter functions to give optimal frequency response. We can start with a fairly simple filter: a 15-point sinc function with "sinc" coefficients (Fig. 3.25). Note that this is a symmetrical sinc shape with a maximum at the center (c_8) and two null points (c_2 and c_5, c_{11} and c_{14}) on each side. The effect of this filter was tested on a "fake" FID that gives, after FT, a spectrum with 41 equally spaced peaks of equal height and width. This raw FID was digitally filtered by sliding the 15-point "sinc" filter (Fig. 3.25) through it, calculating the sum of 15 products ($c_i \times r_j$) at each stop. The Fourier transform of the digitally filtered FID is shown in Fig. 3.26. Clearly the effect of truncating the sinc function (using only 15 points) is dramatic: the response sags in the center, and the cutoff is not very sharp at the edges of the spectral window. In addition, there is significant intensity outside the spectral window with alternating phase. But this frequency response curve is much better than the simple

15-Point "sinc" filter

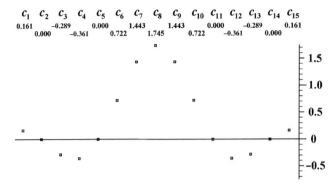

Figure 3.25

sinc functions obtained with "flat" (rectangular) digital filters (Fig. 3.23). The filters in use on modern spectrometers use many more coefficients and are optimized to compensate for truncation effects, so that rejection of signals outside the spectral window is excellent and the cutoff is very sharp.

3.8.6 Combining Decimation with Digital Filtering

Our discussion of digital filtering was inspired by a need to reduce the sampling rate from the maximum possible permitted by the ADC to the rate desired for the spectral window of interest ($2 \times$ SW). We found that a simple average is not a good way to decimate

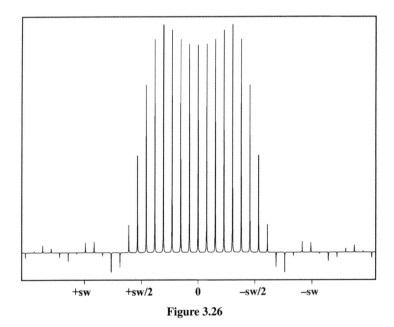

Figure 3.26

the oversampled data because it introduces a sinc-shaped frequency response curve into the spectrum. After a detailed examination of the effect of filter weighting on frequency response, we found that this seeming disadvantage can be used to construct a frequency filter that is far better than any analog audio filter. Thus we can get the advantages of oversampling (greater accuracy and dynamic range) as well as the advantages of digital filtering (very sharp or "brick wall" audio filters) by using a carefully planned shaped digital filter to average the oversampled data and reduce (decimate) it to the desired sampling rate. The only difference in our convolution process is that the filter function does not stop at every point in the raw FID; instead, it jumps ahead by many points each time. For example, if the raw FID is oversampled by a factor of 24 (sampling rate 48 × SW), the filter function will jump 24 points forward each time and calculate a weighted average. The filter function can contain many points: for example, 3000 points for a decimation factor of 24. The sinc-shaped "footprint" of the filter function moves forward through the raw FID jumping 24 points forward each time and calculating the weighted average over the whole filter function width (3000 points) at each stop. This weighted average becomes the data value for the digitally filtered FID at each stop, so that the new FID has only 1/24 the number of points as the raw FID (decimation factor = 24).

3.8.7 Practical Considerations and Applications

Digital filtering is more or less invisible to the routine user. You will notice that the filtered FID has a "dead time" at the beginning during which intensities are very low, and then the normal FID "blossoms out" after this *group delay*. The group delay is the time necessary for the digital filter function to "walk into" the raw FID and start generating significant intensity. For a sinc function, most of the intensity of the function is at the center, so the digitally filtered FID does not start to show intensity until this part of the filter function reaches the beginning of the raw FID. This may be as far as 64 points into the digitally filtered FID. The effect of this delay is the same as the effect of a delay in the start of acquisition after a pulse: It introduces a very large first-order (chemical shift dependent) phase error into the spectrum. This will appear as a lot of "squiggles" in the baseline of the spectrum in a shape similar to a sinc function centered at the center of the spectral window. First-order phase errors in the order of 30,000 are typical, so that it is nearly impossible to correct them manually. On the spectrometer, the NMR software calculates this phase correction from the decimation factor and automatically applies it, so the spectrum never shows any unusual phase errors. When using a "third-party" software package (e.g., Felix), the decimation factor must be supplied so that the software can calculate these phase corrections. The only other noticeable difference between digitally filtered data and analog filtered data is that the Bruker "brick wall" filter function produces a slight downturn in the baseline at the extreme edges of the spectral window. This "Bruker frown" is more preferable to the old "Bruker smile" baseline distortion because the baseline is extremely flat through nearly all the spectral window.

Because digital filtering can produce a "brick wall" frequency response, any peak that falls outside the spectral window is removed completely and will not alias. This can be a problem if you set the spectral window too narrow: You will never be aware of the peaks you miss. If you accidentally set the spectral window to include nothing but noise, you will get just that in the spectrum: nothing but noise! The good news is that if we are only interested in a small part of the 1D spectrum, we can "cut out" the rest of the spectrum using the digital filter. For example, in a 2D ^{15}N-^{1}H HSQC spectrum of a protein, we are only

118 NMR HARDWARE AND SOFTWARE

interested in the H_N (H of the peptide NHCO linkage) region of the spectrum (7–11 ppm). The rest of the spectrum, including the intense water resonance at 4.7 ppm, can be cut out by setting our spectral window to include only the H_N region. That does not allow you to turn up the receiver gain, however, since it is the raw, unfiltered analog FID that is being digitized by the ADC. In 2D NMR the digital filter only applies to the directly detected dimension (F_2). Any excitation that occurs outside the F_1 (vertical) spectral window *will* alias into the spectral window.

3.9 NMR DATA PROCESSING—OVERVIEW

When you have finished acquiring your NMR data, you will need to process the data into a spectrum and plot that spectrum on paper with a ppm axis, and possibly with integrals, peak lists, and other features. You may want to expand interesting or complex regions of the spectrum as insets or on a separate plot so that the fine structure of peaks (splitting patterns, J values, etc.) can be analyzed. Each NMR instrument has its own software for data processing, and it can be daunting to try to learn all of the different commands and operations. The actual data processing task, however, is the same in all cases and the learning curve will be more efficient if we first deal with these tasks in general without discussing individual NMR programs. Starting with an FID (raw time-domain data), we need to carry out the following operations:

(a) Multiply the FID by a multiplier or window function.
(b) Fourier transform the time domain data to obtain a frequency domain spectrum.
(c) Correct for phase errors by adjusting the phase.
(d) Find a reference standard peak and set its chemical shift to the reference value in parts per million.
(e) Expand the desired region of the full spectral window to be plotted.
(f) Plot the spectrum.

In addition, there are several optional operations we might want to perform:

(g) Add zeroes to the end of the FID to increase digital resolution ("zero fill").
(h) Flatten the spectrum baseline (average of noise regions where there are no peaks).
(i) Measure the area under individual peaks by integration.
(j) Plot the chemical shift values of peaks on the spectrum, or print a separate list.
(k) List the acquisition and processing parameters on the spectrum or in a printout.
(l) Expand and plot smaller regions of the spectral window.

To gain a better understanding of what is involved in these steps, we will start with a look at the raw time-domain data (FID).

3.9.1 What is NMR Data?

Raw time-domain NMR data (the FID) consists of a list of numbers, usually negative and positive integers, as a function of time in equal time increments. The list is usually quite long, with as many as 16,000 or 32,000 entries. There are two types of data values (reflecting the two channels of the NMR receiver): real and imaginary. Data are arranged in the order: real, imaginary, real, imaginary, ..., regardless of the acquisition mode (alternating or simultaneous) of the pairs. This is the "raw" data of the NMR experiment. The data are contained in a computer disk file (if you saved it!) as a binary file containing a header (with some information about the spectrometer settings—not used on Bruker AMX and DRX instruments) and a list of numbers without the time values. For newer instruments, the NMR data are saved as a directory that contains the binary FID file (*fid*), and a number of text files containing parameters (Bruker *acqu*, Varian *procpar*) and other information relevant to the experiment. On the older instruments (Bruker AM and Varian Gemini) you simply save the FID as a single binary file.

Some experiments involve more than one FID: For example, DEPT analysis (Chapter 7) performs a ^{13}C experiment four times with different parameter settings; 2D experiments involve collections of up to 750 similar FIDs. These can be combined in a single binary file. The FIDs are just listed one after the other in a single continuous list of data that Bruker calls a *serial* file. Varian treats these multiple FID files in the same way as single-FID files, so they can be used for "arrayed" experiments (a set of 1D experiments acquired by varying some parameter such as pulse width) or for 2D experiments. The file name is the same in either case that is, "*fid*." Bruker uses the filename "*fid*" only for single FIDs, and instead uses "*ser*" for the binary data of all serial files.

3.10 THE FOURIER TRANSFORM

The raw data or FID is a series of intensity values collected as a function of time: *time-domain* data. A single proton signal, for example, would give a simple sine wave in time with a particular frequency corresponding to the chemical shift of that proton. This signal dies out gradually as the protons recover from the pulse and relax. To convert this time-domain data into a spectrum, we perform a mathematical calculation called the Fourier transform (FT), which essentially looks at the sine wave and analyzes it to determine the frequency. This frequency then appears as a peak in the spectrum, which is a plot in *frequency domain* of the same data (Fig. 3.27). If there are many different types of protons with different chemical shifts, the FID will be a complex sum of a number of decaying sine waves with different frequencies and amplitudes. The FT extracts the information about each of the frequencies:

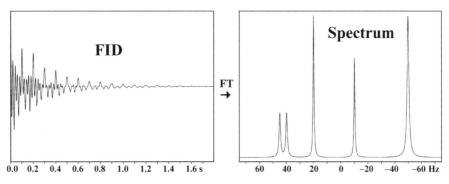

Figure 3.27

120 NMR HARDWARE AND SOFTWARE

their intensities, phases and even the rate at which they decay, which determines the linewidth of each peak in the spectrum (signals that decay quickly transform into broad peaks, whereas signals that last a long time transform into sharp peaks). The method of data collection ("Bruker" sequential vs. "Varian" simultaneous) will affect the type of Fourier transform calculation you must perform. This difference is invisible if you process your data on the instrument on which it was acquired, but if you transfer data to a separate workstation and use independent processing software, you need to tell the software which kind of data you have. For example, with the Felix software package you will have to specify Bruker Fourier transform (*bft*) or complex Fourier transform (*ft*) for Bruker or Varian data, respectively.

3.10.1 How the FT Works

It is actually very easy to visualize how the Fourier transform works. Consider an FID with a single frequency (one peak in the spectrum). The goal of the Fourier transform is to determine the value of that frequency. First, we pick a "guess" frequency ν and multiply the FID by the "test function" $\sin(2\pi\nu t)$. At each point in time we multiply the value of the FID with the value of the test function, and then we measure the area under the curve of the product:

$$\text{spectrum}(\nu) = \int \text{FID}(t)\sin(2\pi\nu t)dt$$

Suppose, first of all, that we guessed right and the test function has exactly the same frequency as the FID ($\nu = \nu_0 = 2.5$ Hz). The two functions (Fig. 3.28) are completely "in sync": wherever the FID is positive the test function is positive and wherever the FID is negative the test function is also negative. The product of these two functions is thus always positive (*positive* × *positive* = *positive*; *negative* × *negative* = *positive*). For our spectrum,

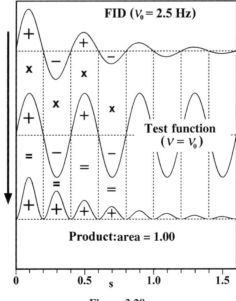

Figure 3.28

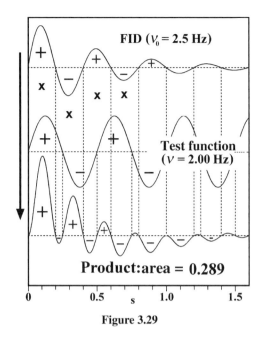

Figure 3.29

we take the area under the curve of this product function (the integral) as the intensity value for the spectrum at frequency $\nu = \nu_o$. This maximum positive intensity (1.00) falls right at the top of our peak in the spectrum. Now consider what happens if we pick a guess frequency that is a bit lower than ν_o: $\nu = 2.0$ Hz (Fig. 3.29). The test function is "slower" than the FID and begins to fall "out of sync" as time progresses, so the product function starts out positive ($p \times p$ or $n \times n$) and then goes negative ($p \times n$ or $n \times p$). As the test function "outruns" the oscillations of the FID, the product function jumps back and forth between positive and negative. Because of the decay of the FID, greater weight is given to the earlier part, and the positive swing outweighs the negative swing, leading to a small positive total area (0.289). In frequency domain, this is down the right side of our peak. An even lower guess frequency (1.66 Hz, Fig. 3.30) leads to a faster oscillation of the product function and better cancelation of the positive and negative areas. This point (intensity 0.124) is farther down the right-hand side of the peak in frequency domain, close to the baseline. Test frequencies still farther from the FID frequency will lead to even more rapid oscillation of the product function and nearly perfect or perfect cancelation of the positive and negative areas: here we are far from the peak in frequency domain, and the intensity of spectrum(ν) is zero.

The real power of the Fourier transform is the linear nature of the calculation. If we have an FID that is a sum of two different pure frequencies (like Fig. 3.16), the spectrum function looks like this:

$$\text{spectrum}(\nu) = \int [\text{FID}_a(t) + \text{FID}_b(t)] \sin(2\pi \nu t) dt$$

We can multiply the terms and separate to obtain

$$\text{spectrum}(\nu) = \int \text{FID}_a(t)\sin(2\pi\nu t)dt + \int \text{FID}_b(t)\sin(2\pi\nu t)dt$$

122   NMR HARDWARE AND SOFTWARE

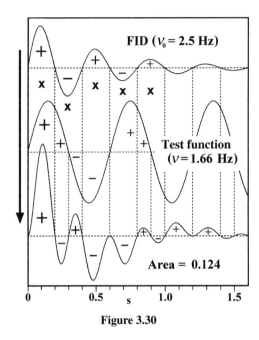

Figure 3.30

Thus, the Fourier transform of the sum of two pure signals is just the sum of the Fourier transforms of the individual signals. The first term above (using $FID_a(t)$) will be nonzero only when the test frequency ν is at or near ν_a (the frequency of FID_a), and the second term will only be nonzero only when the test frequency ν is at or near ν_b (the frequency of FID_b). This is how the Fourier transform "pulls apart" the individual frequencies that are all mixed up in the time-domain data (the FID).

The actual Fourier transform is a digital calculation, so not all frequencies are tested. In fact, the number of frequencies tested is exactly equal to the number of time values sampled in the FID. If we start with 16,384 complex data points in our FID (16,384 real data points and 16,384 imaginary data points), we will end up with 16,384 data points in the real spectrum (the imaginary spectrum is discarded). Another difference from the above description is that the actual Fourier transform algorithm used by computers is much more efficient than the tedious process of multiplying test functions, one by one, and calculating the area under the curve of the product function. This fast Fourier transform (FFT) algorithm makes the whole process vastly more efficient and in fact makes Fourier transform NMR possible.

3.11   DATA MANIPULATION BEFORE THE FOURIER TRANSFORM

3.11.1   Zero Filling

Before performing the FT, there are two things we can do to enhance the quality of the spectrum. First, the size of the data set can be artificially increased by adding zeroes to the end of the list of FID data. This process of zero filling has no effect on the peak positions, intensities, or linewidths of the spectrum, but it does increase the digital resolution (fewer hertz per data point) in the spectrum (Fig. 3.31). This can be useful to give better definition

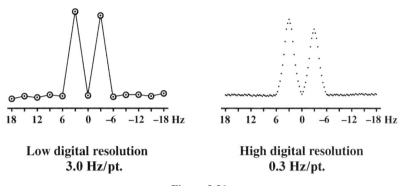

Low digital resolution
3.0 Hz/pt.

High digital resolution
0.3 Hz/pt.

Figure 3.31

of peak shapes for sharp peaks. For example, you might have an FID that contains 3276 total data points (1638 pairs of real, imaginary). If you transform it directly, you will have 1638 points in your spectrum (i.e., the real spectrum). If your spectral width (SW) was 4915 Hz when you acquired the data, your spectrum will have a digital resolution of 4915/1638 or 3.00 Hz per point. A doublet with a splitting of 8 Hz would be described by only three points, so the measurement of the splitting would be very inaccurate due to the "graininess" of the spectrum (Fig. 3.31, left). If, on the other hand, you zero fill the acquired data by adding 14,746 pairs of zeroes to the data list before FT (16,384 total complex pairs), you will get a spectrum with 16,384 (16 K) data points describing the full 4915 Hz spectral window. The digital resolution is much greater ($4915/16{,}384 = 0.300$ Hz per point), and the same doublet would be described by 39 data points (Fig. 3.31, right). Zero filling is accomplished by simply defining the final data size before FT (Bruker *SI*, Varian *FN*) to a larger number than the acquired number of data points (Bruker *TD*, Varian *NP*). In the above example, you would set *TD* (*NP*) to 3276 and *SI* (*FN*) to 16,384.

3.11.2 Weighting or Window Functions

A more common pre-FT massaging of data is the application of a window function or weighting function. The idea is to emphasize ("weight") certain parts of the FID at the expense of others. For example, suppose that your FID signal disappears into the noise after 0.2 s, even though you acquired data up to 1.0 s. The noise from 0.2 to 1.0 s in your FID only increases the noise in your spectrum and does not contribute to the peak height, so your signal-to-noise ratio is reduced. One solution would be to simply set all the data after 0.2 s. to zero, but this introduces a sharp discontinuity in the FID at 0.2 s, which could introduce artifacts into the spectrum. A smoother method is to multiply the FID by an exponential decay function that emphasizes the early data in the FID and deemphasizes the later (mostly noise) data (Fig. 3.32). In the figure, a signal of 1 Hz linewidth is buried in noise after 0.5 s of the 2.6-s acquisition time. The "steepness" of this exponential multiplier (line-broadening parameter LB) can be varied so that it matches the natural decay of the signal (LB = 1 Hz). The net effect on the spectrum (Fig. 3.33) is that the signal-to-noise ratio is increased (from 27.8 to 56.1) and the peak is 1 Hz broader (2 *vs.* 1 Hz), because faster decay of the signal leads to a broader peak. The line broadening actually reduces the absolute peak height, but the reduction in noise level more than compensates for this effect. If the line-broadening effect is not a problem, the increase in S/N is usually worth the price, especially for carbon

124   NMR HARDWARE AND SOFTWARE

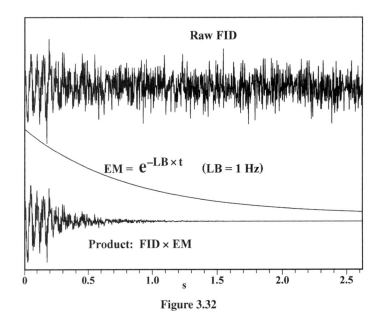

Figure 3.32

spectra where signal is always weak. I usually use an *LB* value of 0.2 Hz for proton spectra and 1.0 Hz for carbon spectra. If you have a very weak signal carbon spectrum and just want to see if there are peaks, you can use an *LB* of 3.0 or 5.0 Hz.

Other window functions can be used for the opposite effect: resolution enhancement (Fig. 3.34). By deemphasizing the beginning of the FID and amplifying the later part, the

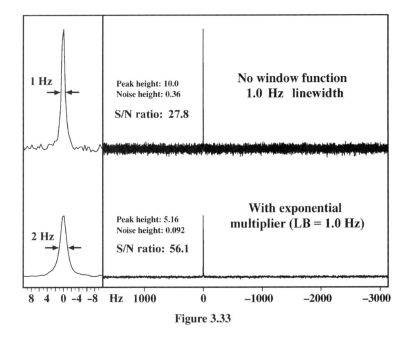

Figure 3.33

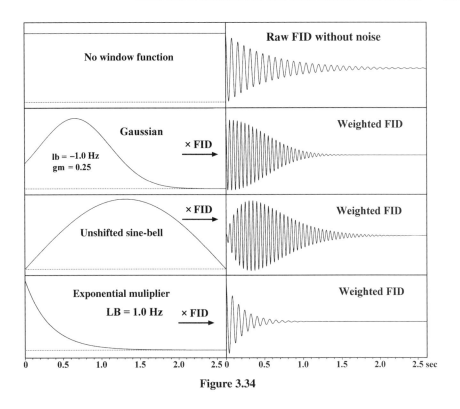

Figure 3.34

natural decay of the FID signal is delayed and the peaks in the spectrum get sharper. This is especially useful for measuring coupling constants. Of course, "there is no free lunch," so you pay a price in poorer signal-to-noise ratio, but with some samples you have more signal than you could ever want. The naïve approach would be to multiply the FID with an exponentially *increasing* function to "slow down" the natural decay of the FID. As with rabbit population and uncontrolled nuclear fission, exponential growth would be disastrous because the end of the FID (dominated by noise) would be huge and then would suddenly drop to zero. But we can rein it in by multiplying by a Gaussian function (the old statistical bell curve):

$$\text{Window} = e^{-\text{LB} \times t} e^{-a(t-\tau)^2}$$

The first exponential is increasing if LB is made negative. The second one, the Gaussian term, reaches a maximum at time $t = \tau$, which can be set to any time during the FID. In Figure 3.34, the parameters for the Gaussian window are set to LB = -1 Hz, with τ adjusted to make the window reach a maximum at one fourth of the way through the FID. The first quarter of the FID is multiplied by an increasing function, slowing down the decay of the FID data, whereas the rest of the window function is decreasing, bringing the noise down.

A very simple window function for resolution enhancement is the sine bell (Fig. 3.34), which is just the function $sin(x)$ for $x = 0$ to $180°$. This function "grows" for the first half of the FID and then brings the signal smoothly to zero during the second half. We saw examples of this window in Chapter 2 (Figs. 2.9 and 2.10). We will see that the sine-bell family of

126 NMR HARDWARE AND SOFTWARE

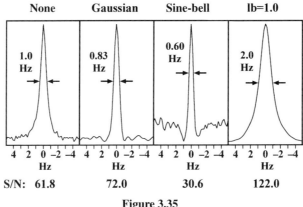

Figure 3.35

window functions is very important for processing 2D NMR data. Figure 3.35 shows the result of Fourier transformation of a single-frequency FID with noise, after multiplication with the window functions of Figure 3.34. The Gaussian window actually increases S/N in this example because it decays during the last three quarters of the FID, but it narrows the peak because it grows during the first quarter. The unshifted sine-bell window narrows the peak even more (from 1.0 to 0.60 Hz), but the peak shape is distorted (prominent negative "ditches" appear on either side) and the S/N is cut in half. The exponential multiplier (LB = 1.0) gives a doubling of signal-to-noise ratio in this example (Fig. 3.35, right). The effect of these window functions on S/N depends greatly on the decay rate of the signal in the raw FID and the acquisition time: If AQ (AT) is long relative to the FID decay, we are acquiring mostly noise in the later part of the FID, and any window that significantly reduces this part of the FID will result in a dramatic S/N improvement. In this case it might be better, however, to just reduce the acquisition time.

Bruker uses the command *EM* (exponential multiplication) to implement the exponential window function, so a typical processing sequence on the Bruker is *EM* followed by *FT* or simply *EF* (EF = EM + FT). Varian uses the general command *wft* (weighted Fourier transform) and allows you to set any of a number of weighting functions (*lb* for exponential multiplication, *sb* for sine bell, *gf* for Gaussian function, etc.). Executing *wft* applies the window function to the FID and then transforms it.

3.12 DATA MANIPULATION AFTER THE FOURIER TRANSFORM

3.12.1 Phase Correction

After you Fourier transform your FID, you get a frequency-domain spectrum with peaks, but the shape of the peaks may not be what you expected. Some peaks may be upside down, whereas others may have a "dispersive" (half up–half down) lineshape (Fig. 3.36). The shape of the peak in the spectrum (+ or − absorptive, + or − dispersive) depends on the starting point of the sine function in the time-domain FID (0° or 180°, 90° or −90°). The starting point of a sinusoidal function is called its "phase." Phase errors come in all possible angles, including those intermediate between absorptive and dispersive (Fig. 3.37). The spectrum has to be phase corrected ("phased") after the Fourier transform to obtain the

DATA MANIPULATION AFTER THE FOURIER TRANSFORM   127

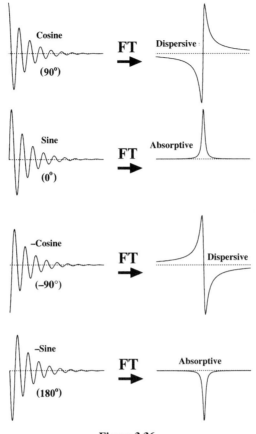

Figure 3.36

desired "absorptive" (0° error) peak shape. Phasing corrects certain unavoidable instrumental errors involved in acquiring the FID.

Recall that the raw NMR data (FID) consists of two numbers for each data point: one real value and one imaginary value. After the Fourier transform, there are also two numbers for each frequency point: one real and one imaginary. In a perfect world, the real spectrum would be in pure absorptive mode (normal peak shape) and the imaginary spectrum would be in pure dispersive (up/down) mode. In reality, each spectrum is a mixture of absorptive and dispersive modes, and the proportions of each can vary with chemical shift (usually in a linear

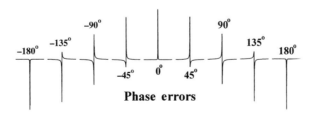

Figure 3.37

Phase correction

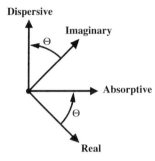

Figure 3.38

fashion). To correct for this, we calculate a linear combination of the real and the imaginary spectra, and use this for our "phased" spectrum (Fig. 3.38). For the mathematically inclined, the actual linear combination is

Absorptive spectrum = (real spectrum) × cos(θ) + (imaginary spectrum) × sin(θ)

The angle θ can be thought of as a rotation of the two mutually perpendicular vectors representing the real and imaginary spectra. The problem of phase correction boils down to finding the correct phase rotation angle θ. Well, actually it is a little more complicated because the phase correction θ is usually a linear function of the chemical shift (δ). Defining the line

$$\theta(\delta) = (m \times \delta) + b$$

requires that you determine two parameters: the intercept b (called the zero-order phase correction) and the slope m (called the first-order phase correction). All phasing routines are based on optimizing these two numbers.

Consider a hypothetical spectrum with six equally spaced peaks (Fig. 3.39, top). There is a chemical shift dependent (linear) phase error that makes the phase error grow by 45° with each peak as we move from left to right. The phase correction process starts with choosing a large peak at one end of the spectrum as the "pivot" peak: this is the peak that is defined as $\delta = 0$ for the purposes of phase correction. In Figure 3.39, we choose the rightmost peak, which has a phase error of 135°. The phase of this pivot peak is optimized by varying the intercept (b) value (the zero-order phase correction) until the pivot peak is perfectly absorptive. This correction applies equally to all of the peaks in the spectrum, regardless of chemical shift, subtracting 135° from the phase error of each of the six peaks. Then another peak is chosen (without moving the pivot) at the other end of the spectrum and its phase is optimized by adjusting the slope (m) parameter (Fig. 3.39, center). The phase correction applied to each peak is determined by the vertical position of the line as it goes through that chemical shift. The key to this method is that changing the slope has no effect on the "pivot" peak because the line goes through zero at this chemical shift. The actual zero of the chemical shift scale is not important—this is just for the purpose of the phase calculation. With both parameters set, the line is defined and all peaks in between should also be correctly phased (Fig. 3.39, bottom). Because the dependence on chemical shift should be linear, correcting both ends of the spectrum should make the whole spectrum

DATA MANIPULATION AFTER THE FOURIER TRANSFORM   129

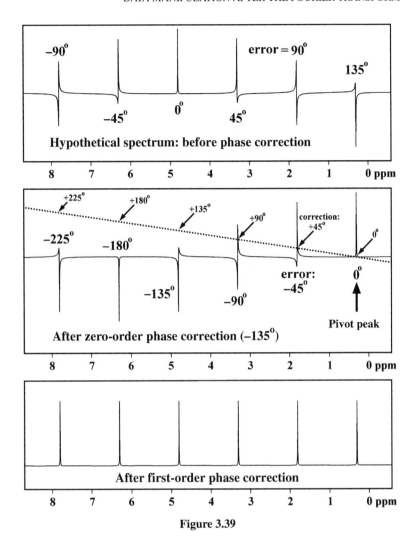

Figure 3.39

"fall into line" with perfect absorptive phase. One notable exception is an aliased (folded) peak, which appears within the spectral window when it really belongs to a very different chemical shift. Its true chemical shift defines its phase error and the method fails for this peak. The two signs of an aliased peak are an uncorrectable phase error and reduced intensity due to attenuation by the audio filter.

There is one situation where any phase correction procedure can fail: It is possible to set m to a large value such that the second (nonpivot) peak is given a phase correction that is 360° too large (Fig. 3.40). This will not affect the shape of the "other" peak, but will introduce a "phase twist" to all the peaks in between. For example, a peak exactly between the pivot peak and the second peak in chemical shift will have a phase error of 180° and thus will appear upside down. Even greater phase twists (720°, 1080°, etc.) can be applied if you are not careful. Using an automatic phasing routine (e.g., Varian *aph* command) can give bizarre values for the first-order phase parameters (slope = m) for spectra with poor signal-to-noise ratios. If you get into this situation, set the first-order phase correction

130   NMR HARDWARE AND SOFTWARE

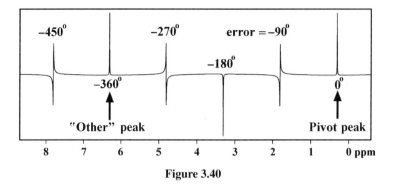

Figure 3.40

(Varian parameter *lp*, Bruker parameter *PHC1*) to zero and start over using manual phase correction. This time adjust the first-order phase correction by looking at a peak that is close to the pivot peak, and then move to peaks farther and farther away.

Whether you are adjusting the *b* (zero-order) or the *m* (first-order) parameter, when you get close to the correct phase setting, focus your attention on the baseline (or noise line) on either side of the peak in question (Fig. 3.41). This should be at the same (vertical) level on each side of the peak. Expand the peak horizontally and increase the vertical scale first so that these baseline differences are greatly exaggerated. If there are distortions to the baseline (curvature), try to imagine a smooth curve between the noise on one side of the peak and the noise on the other side of the peak. Then make the peak blend smoothly into this imaginary curve on both sides of the peak with neither side extending higher over the curve than the other. Sometimes it is necessary to exaggerate the phase error in both directions, especially with noisy data, to clearly see the phase error (one side of the peak extending below the baseline) and then create the same phase error on the other side of the peak. The correct phase setting will then be somewhere near the middle of these two settings.

3.12.2 Setting the Reference

This is a simple procedure whereby a reference peak (e.g., TMS in organic solvents) is selected with a cursor (Bruker uses a triangle or vertical arrow, Varian a vertical red line) and given a specific chemical-shift value. Without this reference, the chemical-shift scale

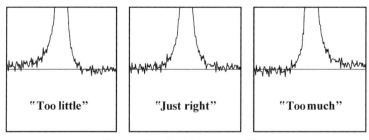

Fine tuning the phase correction

Figure 3.41

of your spectrum will be meaningless. The primary reference for ^{1}H and ^{13}C NMR is TMS at zero ppm, but in many cases this fails or is not practical. Referencing in D_2O or 90% H_2O/10% D_2O presents a challenge because the solvent contains no ^{13}C and TMS is not water soluble. Adding a water-soluble form of TMS ($Me_3SiCD_2CD_2CO_2^-$ Na^+ or TSP) or a small amount of acetonitrile, dioxane, or methanol can give a sharp peak of known chemical shift.

An alternative to the added standards is to use the solvent peak as a chemical-shift reference. If you forgot to add TMS, or the TMS peak is obscured by other peaks, you can use this (residual) solvent peak as the reference peak. This is only valid in dilute solutions where there is only one solvent. Each solvent gives a characteristic *residual ^{1}H peak* in the ^{1}H spectrum due to the 0.2% or so of the solvent molecules that contain ^{1}H, and a *solvent ^{13}C peak* in the ^{13}C spectrum. For example, $CDCl_3$ solvent is typically 99.8% $CDCl_3$ and 0.2% $CHCl_3$, since it is impossible to get 100% incorporation of ^{2}H into the chloroform molecule. In the ^{1}H spectrum one sees a small singlet peak at 7.26 ppm due to the 0.2% of C$\underline{H}$Cl$_3$. This can be used as a chemical-shift reference if the normal reference compound (added TMS) is not present, provided there are no solute peaks at 7.26 ppm. In aqueous solutions (D_2O or H_2O/D_2O) the solvent peak (HOD/H_2O) chemical shift depends on temperature: $\delta(H_2O) = 7.83 - T/96.9$, where T is the absolute temperature in kelvin ($^\circ$C + 273).

In organic solvents the solvent peak is almost always used as the reference in ^{13}C NMR. For ^{13}C spectra in $CDCl_3$ solvent, we observe a "triplet" pattern (1:1:1 intensity ratio) at 77.0 ppm due to the ^{13}C in the $CDCl_3$ solvent. There are three peaks because ^{2}H is a spin-1 nucleus with three spin states possible: spin 1, 0, and -1. Just as a single spin-½ nucleus like ^{1}H will split the NMR signal of a directly bonded ^{13}C into a doublet (1:1 ratio, J ~ 150 Hz), the ^{2}H nucleus splits the ^{13}C signal into three equally spaced peaks (1:1:1 ratio due to the nearly equal populations of the three ^{2}H spin states). Because the magnet strength of the deuterium nucleus is about 1/7 of the strength of the ^{1}H nucleus (γ_H/γ_D ~ 7), the coupling constant is reduced by a factor of 7, and the separation between peaks is around 20 Hz. This solvent ^{13}C peak is usually used for a chemical-shift reference since the tiny amount of TMS added (typically 0.02%) does not give a strong enough peak in the ^{13}C spectrum to be observed over the noise level. You might think that this solvent ^{13}C peak would be enormous compared to the solute peaks due to the preponderance of solvent molecules, but the relaxation of ^{13}C is very slow if it has no ^{1}H atoms attached, so the peak is usually similar in height to the solute peaks.

Solvents with more than one ^{2}H give more complicated patterns in both ^{1}H and ^{13}C spectra. The ^{13}C and residual ^{1}H chemical shifts and coupling patterns of all deuterated solvents can be found in charts provided by the solvent manufacturers (isotope companies). For example, CD_2Cl_2 (d_2-dichloromethane) gives a residual ^{1}H peak (from the 0.2% of $CHDCl_2$ present), which is a 1:1:1 "triplet" (J = 1.1 Hz) at 5.32 ppm, and a solvent ^{13}C peak (from the 99.8% of CD_2Cl_2 present), which is a 1:2:3:2:1 "quintet" (J = 27 Hz) at 54.0 ppm. The "quintet" pattern is due to splitting first by one deuterium into three equally spaced peaks and then splitting each of these by the second deuterium, resulting in a pattern of five peaks. The geminal ^{1}H–^{2}H splitting in the residual ^{1}H solvent peak is quite small due to the reduced magnet strength (γ) of ^{2}H, so that often these splittings are barely resolved or not resolved at all depending on the quality of shimming. Even if poorly resolved, the shape of these peaks can be a dead giveaway in identifying them and using them as a chemical-shift reference, or at least for ignoring them in interpreting the solute ^{1}H spectrum. The ^{13}C solvent peak splitting patterns can be quite complicated. For example, d_6-acetone (CD_3COCD_3) gives a "septet" (1:3:6:7:6:3:1 ratio, J = 19 Hz) at

29.9 ppm and a singlet (C=O peak) at 206.7 ppm. The terms triplet, quintet, and so on are placed in quotations because these are not the classical spin-½ splitting pattern intensities we will concentrate on in this book. One can diagram these splitting patterns as long as each splitting is diagramed as a division into three equally spaced peaks of equal intensities. You can even draw a "spin-1 Pascal's triangle" as follows:

```
No D                  1              (e.g., CHCl₃)
One D              1  1  1           (CDCl₃, C₆D₆)
Two Ds          1  2  3  2  1        (CD₂Cl₂)
Three Ds     1  3  6  7  6  3  1     (d₆–acetone, –DMSO)
```

Note that each number is the sum of three numbers: the number directly above it, the number above it to the right, and the number above it to the left. The long-range (2 or 3 bond) couplings between ^{2}H and ^{13}C are usually not resolved (~1 Hz) so we do not need to worry about these. Keep in mind that the ^{1}H to ^{13}C couplings are not observed in ordinary ^{13}C spectra because we are using ^{1}H decoupling to actively suppress these couplings. Because ^{2}H has a completely different resonant frequency than ^{1}H, the ^{1}H decoupling does not affect the ^{2}H to ^{13}C couplings at all.

3.12.3 Peak Lists

You will often want a printed list of chemical shifts for all the major peaks of your spectrum. First you have to set a threshold intensity (Bruker minimum intensity *MI*, Varian threshold *th*) below which a peak is not included in the list. If you set the threshold too low, you will get a very long list that includes many noise intensities; if you set it too high, you will miss real peaks. Peak lists can be displayed on the screen next to each peak, plotted on the spectrum next to each peak, or printed out as a list on a printer. With a list showing both ppm and hertz values for each peak, simple subtraction gives the *J* values in hertz. Be careful of using subtraction of ppm values to get *J* couplings: these are often not accurate enough. For example, even if ppm values are printed with four digits after the decimal point (e.g., 7.3293 ppm) the precision is 0.0001 ppm or (on a 600 MHz instrument) 0.6 Hz. Subtracting another ppm values increases this error to 1.2 Hz. Much more accurate *J* values can be measured by printing out peak lists in hertz or by using the software to visually position two cursors and compute the separation in hertz. Another common error is to measure *J* couplings directly between peaks when the *J* value is similar to or not much more than the linewidth. If the peaks are not resolved to baseline (intensity dropping to the baseline between the peaks), the distance between peaks is less than the *J* value because one peak "rides up" on the other, skewing the peak shape and shifting the maximum of the peak toward the other peak. In the extreme of a single peak with a slight "notch" at the top, the difference between the two maxima may be a small fraction of the true coupling. In this case, a resolution-enhancing window function (e.g., an unshifted sine bell) can be used to sharpen the peaks, or a nonlinear least squares fit can be performed to extract both the peak width and the *J* coupling independently.

3.12.4 Baseline Correction

The "baseline" is the average of the noise part of your spectrum. Ideally, this would be a straight, horizontal line representing zero intensity. In the real world it can drift, roll, and wiggle like a drunken sailor. These errors generally result from erroneous data that are collected at the very beginning of the FID, when the electronics is still recovering from the shock of the RF excitation pulse. This becomes a problem when you try to measure peak areas (see Section 3.12.5). On the Varian, you indicate where you have peaks and where you have noise in your spectrum by indicating integral regions as part of the integration process. Then the command *bc(1)* fits the noise portions of the spectrum to a smooth function, which is then subtracted from the whole spectrum including peaks. Bruker uses the command *abs* (automatic baseline straightening) to accomplish the same thing. There are also a variety of more sophisticated baseline correction methods, such as mathematically or visually fitting the noise points to polynomial functions.

3.12.5 Integration

To get quantitation of peak areas (numbers of protons), you need to plot an integral. In the old days before Fourier transform NMR, the plotter was set to integral mode and the pen was swept through the peak as the pen level rose with the integrated intensity. For this reason, integrals are still presented as lines that start at the left-hand side of a peak and rise vertically as they pass through the peak. In FT NMR there is often a problem with baseline "wiggle" and this will lead to inaccurate integration of proton peaks (carbon peaks are essentially never integrated because their peak areas are determined more by differences in relaxation rates than by differences in the number of carbons). To get good integrals, you may need to correct the baseline first (see above). Other causes of inaccurate integration include low pulse power (poor excitation of peaks at the edge of the spectral window), "droop" at the edges due to the response of audio filters, and incomplete relaxation due to short relaxation delays. The first and last become more important for higher field instruments because the spectral width (in hertz) and the relaxation times (T_1) increase as the field strength (B_o) increases. Integration generally involves adjusting the display height of the integrals, indicating the start and end points of each peak integral, correcting the drift and curvature so that noise regions give a horizontal line in the integral, and normalizing the peak areas so that the number of protons can be read directly. All processing software allows you to plot the integral area numbers directly on the spectrum next to each integral or to print out a list of integral values. Details of how these steps are accomplished are specific to the software being used and will not be dealt with here.

3.12.6 Plotting

A hardcopy of your spectrum can be obtained using a pen plotter, an inkjet, or laser printer. There are a number of things you can include in your plot:

- spectrum
- integrals
- integral areas (numerical values)
- scale (*x* axis in ppm)
- peak position labels (in ppm)
- parameters
- title
- text describing sample, experiment

134 NMR HARDWARE AND SOFTWARE

Plots can be made on "normal" paper (8.5″ × 11″) or "large" paper (11″ × 17″). Multiple spectra can also be plotted on the same paper, either side-by-side or one above the other (horizontal or vertical "stacked" plots). The plotting procedures for Varian, Bruker, and "third" party software packages are quite different, so these details will not be covered here. Usually the software will also plot to a file, in PostScript format or the more PC-friendly TIFF or JPEG formats. These files can then be introduced into PC drawing programs and annotated with structure diagrams, text, arrows, and lines for use in publications and posters. For example, most of the spectra in this book were processed using Felix software and "plotted" to a graphics file in PostScript format. This text file is converted to a bitmap and imported into a drawing program on a PC.

3.12.7 Archiving (Saving) Your Data

If you had the foresight to save your data on the NMR instrument's hard disk, you will find that these data must be periodically "purged" as the disk gets full. Why save it forever? Someday you will be writing up a paper or thesis and will ask the inevitable question, "what was the coupling constant for that triplet at 3.5 ppm?" Since you probably did not anticipate this question when you plotted your spectrum, you will need to get the data back and reprocess it. At that point you will thank yourself profusely for having the foresight to archive your data. NMR data are generally archived in the raw (FID) form so that you have the maximum flexibility in processing it. Data can be transferred to a PC via the internet or a local network using file transfer protocol (ftp) or the more secure and modern version SSH secure file transfer. From the PC, it can be saved on a CD-ROM or DVD. Bruker and Varian software give data file sizes in terms of the number of data points (Bruker TD, Varian NP). But because each data point uses 2 bytes (Varian with dp = 'n'), 3 bytes (Bruker AM), or 4 bytes (Bruker AMX, DRX or Varian with dp = 'y') of data, the actual file size of a 16 K FID (16,384 data points) can be a little more than 32 or 64 kB on Varian (depending on the dp setting), a little more than 48 kB on the AM or 64 kB on the DRX (the "little more" is for a file header). When you transfer it to a PC (Windows), you will see the file size in actual bytes. On modern UNIX-based NMR instruments, the NMR data file is actually a directory that contains a number of files in addition to the FID binary data file, and even may contain other directories. When transferring these data by ftp or SSH secure file transfer, be sure to specify that directories as well as files are to be transferred. Sometimes long file names will be truncated by Windows when the files are transferred to the PC, and restoring files can cause problems because the names are not correct when they return to the UNIX environment. This problem can be fixed on a case-by-case basis and is never disastrous.

4

CARBON-13 (^{13}C) NMR SPECTROSCOPY

4.1 SENSITIVITY OF ^{13}C

After ^{1}H, the second most important nucleus is ^{13}C because carbon is the building block of all organic molecules, including natural products as well as biopolymers. The ^{13}C nuclear magnet strength is very close to one fourth of that of ^{1}H ($\gamma_C/\gamma_H = 1/4$), leading to a sensitivity of 1/64 (γ^3) of that of ^{1}H. Further bad news is that the natural abundance of ^{13}C on earth is only 1.1%, with nearly all of the remainder being ^{12}C, whose nucleus has no magnetic properties. Thus the overall sensitivity of ^{13}C is about $(1/64) \times (0.011) = 1.72 \times 10^{-4}$ relative to that of ^{1}H, a "hit" of nearly four orders of magnitude. To get the same ^{13}C signal-to-noise ratio as a single-scan proton signal would require 33,850,000 scans because S/N is proportional to the square root of the number of scans! In fact, ^{13}C NMR was not practical until pulsed Fourier transform instruments were available. While a ^{1}H spectrum can be obtained in a single scan for samples of organic molecules as small as 1 mg, a "fat" sample of 30 mg might require 1000 scans or more for a ^{13}C spectrum.

4.2 SPLITTING OF ^{13}C SIGNALS

4.2.1 ^{13}C–^{13}C J Coupling

Although the low natural abundance of ^{13}C carries a big sensitivity disadvantage, it also is a big advantage in that ^{13}C is a "dilute" nucleus: the chances of a ^{13}C being right next to another ^{13}C in a molecule are extremely small ($0.011 \times 0.011 = 1.21 \times 10^{-4}$). For this reason we never see ^{13}C–^{13}C splitting in ^{13}C spectra of natural-abundance samples. Compared to the complexity and wide "footprint" of ^{1}H signals due to ^{1}H–^{1}H splitting, this is an enormous

---

NMR Spectroscopy Explained: Simplified Theory, Applications and Examples for Organic Chemistry and Structural Biology, by Neil E. Jacobsen
Copyright © 2007 John Wiley & Sons, Inc.

simplification of the spectrum. Of course, this is also a loss of information, but we will make up for that later by using ^1H–^{13}C couplings to piece together the carbon skeleton. Thus from the point of view of carbon isotopes, the NMR sample of a pure compound is a complex mixture of isotopomers (molecules of different isotopic composition at specific positions within the molecule). For example, a sample of *n*-propanol (3 carbons) at a concentration of 1 mM actually has the following components, each giving rise to a resonance in the ^{13}C spectrum:

$$^{13}\text{CH}_3-{}^{12}\text{CH}_2-{}^{12}\text{CH}_2-\text{OH} \qquad 11\,\mu\text{M} = 0.011 \times 1\,\text{mM}$$
$$^{12}\text{CH}_3-{}^{13}\text{CH}_2-{}^{12}\text{CH}_2-\text{OH} \qquad 11\,\mu\text{M}$$
$$^{12}\text{CH}_3-{}^{12}\text{CH}_2-{}^{13}\text{CH}_2-\text{OH} \qquad 11\,\mu\text{M}$$

In addition, there are isotopomers with two ^{13}C isotopes in one molecule:

$$^{13}\text{CH}_3-{}^{13}\text{CH}_2-{}^{12}\text{CH}_2-\text{OH} \qquad 121\,\text{nM} = (0.011)^2 \times 1\,\text{mM}$$
$$^{12}\text{CH}_3-{}^{13}\text{CH}_2-{}^{13}\text{CH}_2-\text{OH} \qquad 121\,\text{nM}$$
$$^{13}\text{CH}_3-{}^{12}\text{CH}_2-{}^{13}\text{CH}_2-\text{OH} \qquad 121\,\text{nM}$$

Each of these gives rise to an AB pattern due to ^{13}C–^{13}C splitting ($^1J_{CC}$ for the first two and $^2J_{CC}$ for the third species). These additional ^{13}C signals appear as weak satellite peaks (0.55% of the main peak) around the main peaks from the first three species, and because signal-to-noise ratios are typically much less than 200 for ^{13}C spectra, these signals will be buried in the noise. Finally, there is one isotopomer with three ^{13}C nuclei:

$$^{13}\text{CH}_3-{}^{13}\text{CH}_2-{}^{13}\text{CH}_2-\text{OH} \qquad 1.33\,\text{nM} = (0.011)^3 \times 1\,\text{mM}$$

This species, which can be prepared by uniform isotopic labeling, would give a spectrum even more complex than a ^1H spectrum: the CH$_3$ signal, for example, would be split into a double doublet by $^1J_{CC}$ (large: $\sim$35 Hz) and $^2J_{CC}$ (small: <5 Hz). The central CH$_2$ signal would be split into a double-doublet or triplet by two large $^1J_{CC}$ splittings. In a natural abundance sample we would have to have a signal-to-noise ratio of 33,000:1 to see these signals rising out of the noise!

The remainder of the 1 mM concentration is made up of the predominant isotopomer: the one without any ^{13}C at all:

$$^{12}\text{CH}_3-{}^{12}\text{CH}_2-{}^{12}\text{CH}_2-\text{OH}$$
$$0.96663567\,\text{mM} = 1\,\text{mM} - 3 \times 11\,\mu\text{M} - 3 \times 121\,\text{nM} - 1.33\,\text{nM}$$

This species is invisible to ^{13}C NMR and does not contribute at all to the ^{13}C spectrum.

The advantage of ^{13}C's low natural abundance can be seen clearly in this example: each carbon resonance in the spectrum represents a pure isotopomer with ^{13}C only at that position and ^{12}C at all other positions within the molecule. Any species with two or more ^{13}C atoms in the molecule is present at a concentration of at most two orders of magnitude lower than the one-^{13}C isotopomers, so we will never see any contribution from these species in our ^{13}C spectrum.

4.2.2 Isotopic Enrichment

You may be familiar with the use of carbon-14 as a "tracer" in biosynthetic studies: a metabolic building block such as acetate (CH$_3$–CO$_2^-$) can be prepared with ^{14}C (a radioactive isotope) enriched at one of the carbon positions in the molecule. Any biomolecule that is put together using this building block will end up with ^{14}C in it, which can be detected by measuring radioactivity. We can do the same thing with ^{13}C, without the dangers and cumbersome precautions of working with radioactivity. For example, starting with ^{13}CH$_3$–^{12}CO$_2^-$ (prepared synthetically), an enzyme, cell-free extract, cell culture, or a whole organism can be used to prepare a natural product. This molecule is isolated and purified, and in the ^{13}C NMR spectrum we would see that the peak corresponding to any carbon position that is derived from the methyl group of acetate will be 91 times more intense (abundance 100% *vs.* 1.1%) than the other peaks! The ^{14}C tracer method only tells us whether the building block is incorporated or not, but the ^{13}C NMR method tells us exactly at which position the labeled carbon is incorporated, assuming that the peaks in the spectrum can be assigned to specific carbon positions in the molecular structure.

An even more powerful technique is to label both the positions of a two-carbon building block such as acetate (^{13}CH$_3$–^{13}CO$_2^-$) and mix it equally with natural-abundance molecules (primarily ^{12}CH$_3$–^{12}CO$_2^-$). If the two-carbon "synthon" is incorporated intact, without breaking it apart into two one-carbon pieces, we should see ^{13}C–^{13}C coupling due to $^1J_{CC}$ in all of the final molecules that contain ^{13}C from the acetate building block. If the acetate is broken down first into one-carbon pieces and then joined together in the biosynthesis, there would only be 50% abundance of ^{13}C at each of the two positions, and we would see a normal resonance (singlet) superimposed on a split resonance (doublet) at each position derived from the building block.

^{13}C labels can also be used in metabolic studies to watch the breakdown of biological molecules. This can even be done in suspensions of living cells in an NMR tube, watching the progression of ^{13}C peaks in a starting molecule (such as glucose) moving to ^{13}C signals of breakdown products (such as ethanol). The background of natural abundance ^{13}C is much weaker and usually does not rise above the noise level.

Finally, uniform labeling with ^{13}C is extremely important in biological NMR. Expression of proteins in cell culture can be carried out with uniformly labeled ^{13}C-glucose or ^{13}C-acetate at high enrichment (95–99%) as the only carbon source. Isolation and purification of the overexpressed protein leads to an NMR sample with the potential of measuring and assigning ^{13}C chemical shifts at all positions. We will see in Chapter 12 how ^{13}C–^{13}C and ^{13}C–^{15}N one-bond couplings can be used to build complex and sophisticated biological NMR experiments capable of determining the three-dimensional structure and residue-specific dynamics of very large (e.g., 30 kD) biological molecules.

4.2.3 ^{1}H–^{13}C *J* Coupling

So far we have ignored the effect of protons on the ^{13}C spectrum. The ^{1}H–^{13}C one-bond coupling ($^1J_{CH}$) is very large (~150 Hz), so we can expect to see very wide doublets (for methine, CH), triplets (for methylene, CH$_2$), and quartets (for methyl, CH$_3$) for the ^{13}C resonances in our spectrum. Only the quaternary carbons (C$_q$) would be free of this large coupling. In fact, for all but the simplest molecules, a simple pulse-and-observe ^{13}C experiment (relaxation delay–pulse–acquire FID) will give a forest of overlapping peaks that is very difficult to unravel and analyze. In addition, there are long-range (two-bond and three-bond) couplings between ^{1}H and ^{13}C. Because ^{1}H has essentially 100% natural abundance, any coupling to ^{1}H will show up completely and not as a small satellite. For example, consider *n*-propanol again. The isotopomer that gives rise to the CH$_3$ peak in the

^{13}C spectrum is

$$^{13}CH_3 - {}^{12}CH_2 - {}^{12}CH_2 - OH$$

The ^{13}C signal will be split into a quartet by the three methyl protons ($^1J_{CH} = 125$ Hz) and each line of this quartet will be split into a triplet by the two protons on the adjacent carbon ($^2J_{CH} = 4$–6 Hz). Each of these 12 lines will be further split into a triplet by the two protons on the CH$_2$OH group ($^3J_{CH} \sim 5$ Hz). Thus our ^{13}C chemical shift position will give as many as 36 lines in the spectrum, spreading our already miserable signal-to-noise ratio into a multitude of tiny peaks barely discernible in the noise. A similar cascade of splittings will complicate the other two resonances of *n*-propanol.

Later we will see how these couplings can be exploited in experiments that enhance the sensitivity of ^{13}C spectra (INEPT), measure the number of hydrogens attached to each carbon (APT and DEPT), and correlate ^{13}C chemical shifts with ^{1}H chemical shifts using a second dimension (2D-HETCOR, -HMQC, -HSQC, and -HMBC). But for detecting a simple ^{13}C spectrum, we need a way to suppress these ^{13}C–^{1}H couplings so we can observe a single line (singlet) for each ^{13}C resonance.

4.3 DECOUPLING

For the remainder of this chapter we will be exploring the effects of continuous low-power irradiation of one nucleus on the spectrum of another. Two important phenomena occur as a result of low-power irradiation: *decoupling*, which reduces or eliminates the *J*-coupling (splitting) effect on the observed nucleus and the *nuclear Overhauser effect* (NOE), which enhances the population difference (and hence the signal intensity) of the observed nucleus. Decoupling is accomplished by continuous low-power irradiation during the acquisition of the FID, and the NOE develops during continuous irradiation at even lower power during the relaxation delay.

Decoupling is the process of removing specific kinds of *J*-coupling interactions in order to simplify a spectrum or to identify which pair of nuclei is involved in the *J* coupling. In order to understand how decoupling works, we should review what causes *J* coupling in the first place. As we saw in Section 1.1, a resonance is split into a doublet by a nearby spin-½ nucleus because the tiny magnetic field produced by that nucleus perturbs the B_o field experienced by the nucleus we are observing. If the perturbing nucleus is aligned with the B_o field (α state), we see a shift in the effective field B_{eff} in one sense (increase by our convention), and if the nucleus is aligned against the B_o field (β state) we see a perturbation of B_{eff} in the opposite sense (decrease). These changes in B_{eff} lead to a shift in the Larmor frequency ($v_o = \gamma B_{eff}/2\pi$) by $J/2$ Hz downfield (perturbing nucleus in the α state) or by $J/2$ Hz upfield (perturbing nucleus in the β state). Because the perturbing nucleus has a 50% chance of being in the α state and a 50% chance of being in the β state (actually something like 50.0001 and 49.9999, respectively), we see a doublet with a 1:1 ratio, centered on the chemical shift position and separated by *J* Hz. It is important to recognize that the *J*-coupling effect is transmitted through bonds and not through space. A much larger effect occurs directly through space (with couplings in the order of kHz instead of Hz), but this effect (*dipolar* or *direct* coupling) depends on molecular orientation relative to B_o and is averaged exactly to zero by the rapid isotropic reorientation (tumbling) of molecules in solution. This dipolar interaction is important as a mechanism of relaxation in liquid state NMR, but it shows up as a splitting only in solid state NMR.

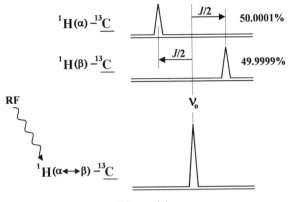

Figure 4.1

A methine carbon (<u>C</u>H) is split into a wide ($^1J_{CH} \sim 150$ Hz) doublet, one line representing the population of molecules with ^{13}C in that position and the attached ^{1}H in the α state and the other line representing the population of molecules with ^{13}C in that same position and the attached ^{1}H in the β state. The C is underlined in <u>C</u>H to indicate that we are observing and discussing the C resonance, not the H resonance. The H is included in the discussion only with respect to its effect on the C resonance.

Decoupling is accomplished by irradiating at the frequency of one nucleus (^{1}H) with continuous low-power RF (Figure 4.1). This irradiation causes the ^{1}H nucleus to "flip" from the lower energy (α or aligned) to the higher energy (β or opposed) state and back again very rapidly. Because the NMR "timescale" or "shutter speed" is relatively slow (in this case on the order of $1/J = 1/150 = 6.67$ ms), the other ^{13}C sees only an average magnetic environment, which is not perturbed at all by the presence of the proton's magnetic field. The two components of the ^{13}C doublet are averaged to a single peak in the center as long as the ^{1}H spins are "flipping" back and forth rapidly enough. If the RF power is not enough to create perfect averaging, the protons will flip back and forth more slowly and we will see a doublet for ^{13}C with a reduced separation or J value. The RF irradiation must go on during the entire process of recording the FID (the acquisition time) in order to eliminate the coupling. If the frequency of the irradiation is not exactly at the resonant frequency of the CH proton, there will still be some decoupling, but it depends on the power of the RF signal and the frequency difference. The larger the frequency difference between the RF signal and the resonant frequency of the proton, the greater the power required to achieve decoupling. Another way of saying that is that a high-power RF signal will decouple a wider range or band of frequencies (chemical shifts) around the frequency of the RF signal. Most of the time this is desirable, but in some cases, where we want to irradiate a specific peak in the ^{1}H spectrum and not any other peaks, higher power is undesirable because it reduces the selectivity of decoupling.

4.4 HETERONUCLEAR DECOUPLING: ^{1}H DECOUPLED ^{13}C SPECTRA

4.4.1 Why Decouple?

There are two main reasons to decouple. The first is to identify which pair of nuclei is involved in the J coupling, and the second is to simplify ^{13}C spectra by removing the

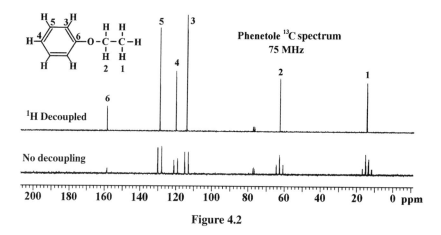

Figure 4.2

^{1}H–^{13}C couplings. The latter application is so routine that most users forget that these large couplings (J up to 180 Hz) even exist. In fact, without ^{1}H decoupling all ^{13}C spectra would show very wide quartets for CH$_3$ carbons, triplets for CH$_2$ carbons, and doublets for CH carbons. This can be useful information, but for molecules of any size and complexity it leads to a tangled forest of multiplets and a costly reduction in signal-to-noise ratio. ^{1}H decoupling gives ^{13}C spectra in which there is only one (singlet) peak for each unique carbon in the molecule. For example, the ^{13}C spectrum of phenetole (ethoxybenzene) is shown with ^{1}H decoupling in Figure 4.2 (top). In the aromatic region we see two large peaks (two carbons each, *ortho* and *meta* to the ethoxy group), one smaller (*para*) and the other quite small quaternary peak (*ipso*, or at the point of attachment of the ethoxy group). In the upfield region of the spectrum we see two peaks (one singly oxygenated sp^3 carbon and one carbon without oxygen). In the ^{13}C spectrum without ^{1}H decoupling (Fig. 4.2, bottom), only the *ipso* aromatic carbon (quaternary) is a singlet. The other aromatic carbons are doublets (CH), and the ethoxy group gives rise to a triplet (CH$_2$) and a quartet (CH$_3$). In Figure 4.3 we see the ^{13}C spectrum of sucrose with and without ^{1}H decoupling. The CH$_2$OH region (60–63 ppm) is particularly crowded with overlapping triplets in the absence of ^{1}H decoupling.

4.4.2 Continuous-Wave Heteronuclear Decoupling

Low-power irradiation at a single frequency tends to excite only a very narrow range of frequencies because a rectangular pulse of duration t_p seconds excites a bandwidth of roughly $1/t_p$ Hz. For a typical ^{13}C acquisition time of 1.0 s, irradiation of protons during the entire acquisition period would correspond to an excitation bandwidth of 1.0 Hz (1/1.0 s) in the proton spectrum. A more precise treatment describes the reduction of the "undecoupled" coupling constant J_o to the observed (reduced) coupling constant J_R, by a continuous ^{1}H irradiation at decoupler field strength B_2 with frequency offset $\Delta \nu$ away from the frequency of the proton being decoupled (Fig. 4.4):

$$\gamma_H B_2/2\pi = \Delta \nu (J_o^2 - J_R^2)^{1/2}/J_R \qquad (4.1)$$

The left-hand side of the equation can be regarded as the *decoupler field strength* in units of hertz. This is the same as describing the main magnetic field, B_o, as $\gamma_H B_o/2\pi$ in hertz. For

HETERONUCLEAR DECOUPLING: ^{1}H DECOUPLED ^{13}C SPECTRA 141

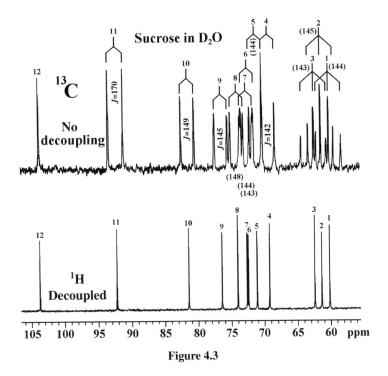

Figure 4.3

example, you might say "we have a 300 MHz instrument," which means that you have a magnetic field strength B_o that gives a resonance frequency of 300 MHz *for protons*. To be precise, it means that $\gamma_H B_o/2\pi$ is 300 MHz, where γ_H is the magnetogyric ratio for protons. Likewise, if you say "we have a decoupler field strength of 10 kHz," this means that in the rotating frame of reference the *proton* magnetization precesses at 10 kHz around the B_2

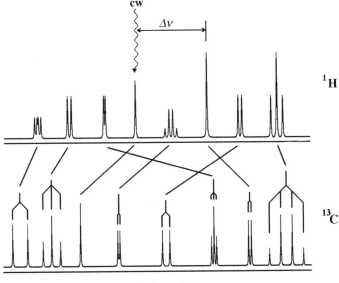

Figure 4.4

field vector, which is in the x'-y' plane. More precisely, it means that $\gamma_H B_2/2\pi$ is 10 kHz, where we use the proton magnetogyric ratio γ_H. We use B_2 to refer to the decoupler and B_1 for the transmitter, but they represent the same thing: the magnetic field due to the radio frequency signal applied to the probe coil, which is a stationary vector in the x'-y' plane when viewed in the rotating frame of reference.

The right-hand side of the equation represents the amount by which the proton frequency is off-resonance ($\Delta \nu$) and the factor by which the apparent ^{13}C–^{1}H coupling constant is reduced. For nice, sharp ^{13}C singlets we would like to have the apparent J value, J_R, be less than the natural ^{13}C linewidth so that it does not even broaden the singlet carbon peak. The equation makes more sense in rearranged form:

$$J_R/(J_o^2 - J_R^2)^{1/2} = \Delta \nu/(\gamma_H B_2/2\pi) \tag{4.1}$$

This says that the residual coupling, J_R, is larger if the proton resonance is farther away from the decoupler frequency (larger $\Delta \nu$) and smaller if we use more decoupler power (larger $\gamma_H B_2/2\pi$). Figure 4.4 shows that peaks near the decoupler frequency in the ^{1}H spectrum (top) have small J_R values (^{1}H–^{13}C splittings) in the ^{13}C spectrum (bottom), and protons that are far away from the decoupler position have wide multiplets in the ^{13}C spectrum for the corresponding ^{13}C directly bound to that ^{1}H. Equation (4.1) can actually be used to calibrate the decoupler field strength B_2 by observing the effect of off-resonance decoupling on the observed J value of a ^{13}C multiplet.

4.4.3 Selective Decoupling

Another reason for decoupling is to identify the coupling "partner" of a particular peak in the spectrum. Irradiation of that peak at its exact frequency using low-power (for selectivity) continuous RF during the acquisition time will "collapse" to a singlet any multiplet patterns that result from the protons in the irradiated peak. For example, you might irradiate a ^{1}H multiplet at 4.68 ppm and find that a ^{1}H double doublet ($J = 12.2, 5.6$ Hz) at 3.24 ppm "collapses" to a doublet ($J = 12.2$ Hz). This means that the multiplet at 4.68 ppm was the source of the 5.6 Hz coupling in the double doublet at 3.24 ppm, the coupling that "disappeared." This is an example of selective *homonuclear* decoupling: the nucleus we are irradiating is of the same type (^{1}H) as the nucleus we are observing. This selective technique can also be used for heteronuclear couplings, so that irradiating a particular proton resonance results in the collapse of a ^{13}C multiplet to a sharp singlet in the ^{13}C spectrum. This is called *selective* heteronuclear decoupling to distinguish it from the broadband nonselective ^{1}H decoupling that is normally used during the acquisition of ^{13}C spectra. As we saw above, not only will we collapse the ^{13}C multiplet corresponding to the carbon directly bound to the proton we are irradiating ($\Delta \nu = 0$), but other ^{13}C multiplets will be narrowed ($J_R < J_o$) depending on the frequency difference ($\Delta \nu$) between the irradiated proton and the other ^{13}C multiplet's proton, and on the decoupler field strength. Figure 4.5 shows the ^{13}C spectrum of phenetole with selective continuous-wave irradiation of the methyl protons (bottom) and the methylene protons (center). In the bottom spectrum, with irradiation of the CH$_3$ proton peak at 1.37 ppm, the CH$_3$ carbon peak is a clean singlet, and the CH$_2$ carbon is a distorted triplet with a reduced coupling $J_R = 109$ Hz (*vs.* $J_o = 140$). In the middle spectrum, with irradiation of the CH$_2$ proton peak at 3.96 ppm, the CH$_2$ carbon peak is a clean singlet, and the CH$_3$ carbon is a distorted quartet with a reduced coupling $J_R = 96$ Hz (*vs.* $J_o = 131$).

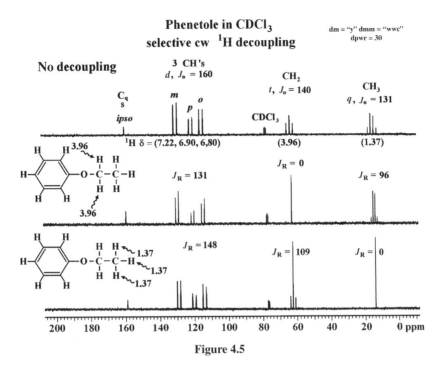

Figure 4.5

The aromatic CH ^{13}C peaks in the bottom spectrum are little affected and show nearly the full coupling (148 vs. 160 Hz) because the aromatic CH *protons* are far from the CH$_3$ protons in the proton spectrum (large Δv). Selective heteronuclear decoupling is rarely used because the two-dimensional (2D) HETCOR and related inverse 2D experiments (HMQC, HSQC, and HMBC) give the same information with far less ambiguity (Chapter 11). In fact, selective *homonuclear* decoupling has all been replaced by 2D-COSY and related variants such as DQF-COSY and COSY-35 (Chapter 9). There are instances, however, where only one or two couplings are ambiguous and a 1D selective decoupling experiment can sort it out quickly.

4.4.4 Broadband Heteronuclear Decoupling

Normally in ^{13}C spectra we want to decouple *all* of the protons from their attached ^{13}C atoms. This means that we cannot irradiate exactly at the frequency of each proton simultaneously. We need "broadband" decoupling that will "cover" the entire range of ^{1}H chemical shifts, which typically range from 0 ppm to 10 ppm, a width of 3000 Hz on a 300 MHz instrument. Because the decoupler frequency cannot be on-resonance for all of the protons in the sample at the same time, it is usually set in the center of the expected range of ^{1}H frequencies. The problem then becomes how to "cover" the entire range of proton chemical shifts with effective decoupling. If we place the ^{1}H decoupler frequency at the center of the ^{1}H spectrum, the worst case would be trying to decouple a ^{1}H signal at the upfield or downfield extremes of the ^{1}H chemical shift range, which could be as much as 5 ppm (1500 Hz on a 300 MHz spectrometer) away from the center. According to equation (4.1), reduction of the observed J value from 150 to 1 Hz with $\Delta v = 1500$ Hz would require a decoupler field strength

144   CARBON-13 (^{13}C) NMR SPECTROSCOPY

($\gamma_H B_2/2\pi$) of 225 kHz. This is an RF field strength corresponding to a 1.1 μs 90° pulse because one cycle of rotation of the sample ^{1}H magnetization takes 1/225,000 s or 4.4 μs. This is ten times the amplitude of a high-power excitation (B_1) pulse, corresponding to 100 times the power: a power level that cannot be achieved without frying the sample and vaporizing the probe coil and the RF amplifiers!

4.4.5 Composite-Pulse Decoupling: Waltz-16

What we need is a method to achieve "broadband" decoupling of protons over the entire chemical shift range (e.g., 0–10 ppm) of the protons, in a very efficient way that uses the lowest possible $\gamma_H B_2/2\pi$ value (i.e., the lowest possible decoupler power). An early solution to this problem was to vary (modulate) the decoupling frequency over a wide range of ^{1}H chemical shifts either by sweeping it back and forth or by random (noise modulated) variation. The currently accepted method to achieve wide decoupling bandwidths at low power levels is to employ repeated pulses of different phase and duration at a single frequency: "composite pulse decoupling." A "composite pulse" is a sandwich of several pulses designed to give an overall rotation that is less dependent on the resonance offset than a single pulse. Later we will see (Chapter 8, Figs. 8.5 and 8.6) that a "sandwich" of $90°_x$–$180°_{-x}$–$270°_x$ (written as 1**2**3 in multiples of 90°, with bold italics indicating a phase of $-x$) gives efficient inversion (overall 180° pulse) over a wide range of chemical shifts ("broadband inversion"). A rapid-fire sequence of repeating 180° pulses would give good decoupling because the spins are inverted ($\alpha \to \beta, \beta \to \alpha$) over and over again very rapidly, averaging the J-coupling effect to zero. By using sandwich pulses in place of simple 180° pulses, the decoupling performance is good over a wide range of chemical shifts around the pulse frequency v_r. To eliminate the accumulation of pulse calibration errors, the pulse phase is reversed (from x to $-x$) at regular intervals in the sequence: using R = 1**2**3, we have RR**RR** for 1**2**31**2**3*1231231*. Moving the beginning "1" ($90°_x$) to the end gives 2**3**1*231231231* or (combining 90° and 270° rotations of the same phase, 31 = 4 and **3***1* = **4**) 2**4**231*24231*. Repeat this with all phases reversed and you have: *242**3**1 242**3**1* 2**4**231 2**4**231. Finally, if we move the ending 1 to the beginning and combine (*1***2** = **3**, 12 = 3) we have *3***4**231**2***4*2**3** *3*42*3*12**4**2**3** (Fig. 4.6). This can be represented as R′**R**′, which when repeated with opposite phase (R′**R**′**R**′R′) gives a "supercycle" called "waltz-16": "waltz" because of the 1**2**3 building block and 16 because it contains 16 of the original 1**2**3 sandwiches. The 36 pulse block is repeated as many times as necessary to cover the entire time of acquiring the FID (Bruker *aq*, Varian *at*). From a hardware perspective, waltz-16 only involves changing the phase of the RF (x or $-x$) at specific times while keeping the amplitude constant (Fig. 4.6). The only parameters you need to set are the RF amplitude (Varian *dpwr*, Bruker *DP* or *pl17*) and the duration of the 90° pulse at that power level (Varian *dmf* = $1/t_{90}$, Bruker *pcpd2*).

With this method we can achieve decoupling of the full ^{1}H chemical shift range with a decoupler power level ($\gamma_H B_2/2\pi$) of less than 2500 Hz, or about one tenth of the amplitude (one percent of the power) used for single-pulse excitation of protons (e.g., $\gamma B_1/2\pi$ = 25,000 Hz for a 10 μs 90° ^{1}H pulse). This decoupling power level corresponds to a 90° ^{1}H pulse width of $1/(4 \times 2500 \text{ Hz})$ = 100 μs, and a reduction in power of 10 log [power ratio] = 10 log [(25,000/2500 Hz)^2] = 10 log [(100/10 μs)^2] = 10 log [100] = 20 dB. The decoupler field strength, expressed in units of Hz, is proportional to the B_1 amplitude, so the relation dB = 20 log [amplitude ratio] = 20 log [90° pulse width ratio] = 10 log [power ratio] can be applied. As power is the square of amplitude, we can also say that the power

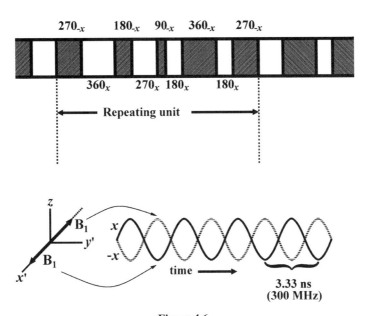

Figure 4.6

level required for decoupling is 100 times (10^2) less than that of hard pulses (typically 0.5 watts for decoupling instead of 50 watts for hard pulses).

Figure 4.7 shows a series of ^{13}C spectra of dioxane (four chemically equivalent CH$_2$ groups) with waltz-16 decoupling, setting the proton decoupler frequency 12 ppm downfield of the ^{1}H peak of dioxane and then repeating the experiment, each time moving the decoupler ^{1}H frequency upfield by 2 ppm (600 Hz on a Varian Unity-300). We see excellent decoupling over a range of 16 ppm, more than sufficient for "covering" the normal range of ^{1}H chemical shifts. At the edges the peak height falls off drastically as the reduced coupling, J_R, begins to show up enough to broaden the singlet line. A lower decoupler power setting would result in a narrower pattern, and higher power a larger range of ^{1}H offsets ($\Delta\nu$) that still give good decoupling. We try to minimize decoupler power because at high-power sample, heating will degrade the field homogeneity by setting up a radial temperature gradient in the sample.

4.5 DECOUPLING HARDWARE

How does a spectrometer deliver this RF irradiation to the probe? Compared to normal excitation pulses, which are very high-power and short ($\sim$10 μs) duration, decoupling requires low-power irradiation for the entire acquisition time (1–2 s). This is usually accomplished by having two separate sources of RF power, a "broadband" *transmitter* that can be

146 CARBON-13 (^{13}C) NMR SPECTROSCOPY

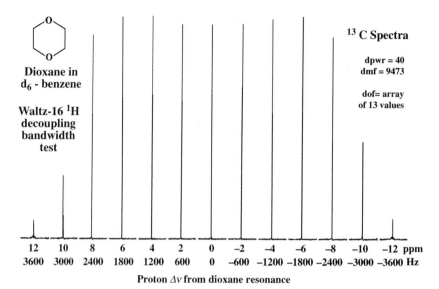

Figure 4.7

operated at a wide range of frequencies (e.g., ^{15}N is 30.4 MHz, ^{13}C is 75.4 MHz, and ^{1}H is 300.0 MHz on a 7.05 T instrument) and a proton *decoupler* that can only produce the proton frequency (e.g., 300.0 MHz). The transmitter is set to the frequency of the nucleus to be observed with a high-power level for pulses, and the decoupler is set to a low-power level for proton decoupling. As soon as the pulse sequence is over and acquisition of the FID begins, the decoupler is turned on for the duration of the acquisition time. Even when ^{1}H is being observed, the proton (high-power) pulses come from the transmitter, and the decoupler is used to deliver the low-power ^{1}H irradiation during the acquisition time. This is necessary in older machines because it takes time (milliseconds!) and requires physical switching of relays to change the power level of the transmitter, so you can not just use a single source of RF to supply high-power pulses and low-power decoupling irradiation.

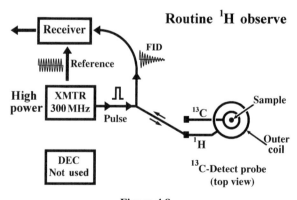

Figure 4.8

DECOUPLING HARDWARE   147

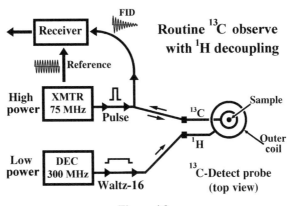

Figure 4.9

With instruments that are a bit more modern than this basic system, the decoupler is not fixed at the ^{1}H frequency. Instead, it is identical to the transmitter in that it can be set to the frequency of any nucleus. Thus, you could decouple ^{13}C while observing ^{1}H, for example (an "inverse" experiment). This arrangement is called "dual broadband" because both RF sources are "broadband"—adjustable over a wide range of frequencies. Still more modern spectrometers can switch power levels in a few microseconds without relays, so that a single "box" can be used for all proton RF, whether it is for high-power pulses or for low-power decoupling. This feature makes the whole concept of "transmitter" and "decoupler" a matter of language rather than real hardware differences. Figures 4.8–4.10 show the configuration of a Varian Unity-300 spectrometer (with direct ^{13}C probe) for routine ^{1}H, for ^{13}C with ^{1}H decoupling, and for ^{1}H with homonuclear decoupling. Note that the *inner* coil of the probe is used for ^{13}C to maximize the sensitivity of detection of this "insensitive" nucleus. The *outer* coil, which is farther from the sample and therefore less sensitive, is used for ^{1}H because it is much easier to detect.

For routine ^{1}H acquisition (Fig. 4.8), the transmitter is set to the ^{1}H frequency (300 MHz) and pulses from the transmitter are directed to the outer coil of the probe, which is tuned to the ^{1}H frequency. The decoupler is not used. After the exciting pulse, the ^{1}H FID is detected on the outer coil of the probe and directed to the receiver, which uses a continuous signal from

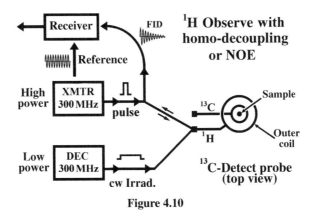

Figure 4.10

the transmitter (300 MHz) as a reference frequency that is "mixed" with (i.e., subtracted from) the FID frequency. For routine ^{13}C acquisition with ^{1}H decoupling (Fig. 4.9), the transmitter is set to the ^{13}C frequency (75 MHz) and pulses from the transmitter are directed to the inner coil of the probe, which is tuned to the ^{13}C frequency. The decoupler operates continuously at low power (with waltz-16 phase modulation) and its output is directed to the outer coil of the probe, which is tuned to the ^{1}H frequency. After the exciting pulse, the ^{13}C FID is detected on the inner coil of the probe and directed to the receiver, which uses a continuous signal from the transmitter (75 MHz) as a reference frequency that is "mixed" with (i.e., subtracted from) the FID frequency. For ^{1}H acquisition with selective ^{1}H decoupling or NOE difference (Fig. 4.10), the transmitter is set to the ^{1}H frequency (300 MHz) and pulses from the transmitter are directed to the outer coil of the probe, which is tuned to the ^{1}H frequency. The decoupler operates continuously at very low power during the relaxation delay (NOE difference) or during the acquisition of the FID (selective homonuclear decoupling) and its output is also directed to the outer (^{1}H) coil of the probe. In either case, the ^{1}H FID is detected on the outer coil of the probe and directed to the receiver, which uses a continuous signal from the transmitter (300 MHz) as a reference frequency that is "mixed" with (i.e., subtracted from) the FID frequency.

Changing from one of these configurations to another simply involves changing the frequency settings of the two channels ("transmitter" and "decoupler") and rerouting the outputs to the probe inputs. This is done by resetting electrical relays, which give a "click" when you issue the command (Varian: *su* or *go*, Bruker: *ii* or *zg*) to set the hardware according to the experimental parameters. Figure 4.11 shows the configuration for an "inverse-mode" experiment (Chapter 11), in which ^{1}H is detected and pulses are delivered to the probe on both the ^{1}H and ^{13}C channels (e.g., 2D HSQC, a ^{1}H-detected 2D ^{13}C–^{1}H correlation experiment). In this case an inverse probe is used, which has the *inner* coil tuned to ^{1}H (the "observe" nucleus) and the *outer* coil tuned to ^{13}C (the "decoupler" nucleus). The transmitter is set to the ^{1}H frequency (300 MHz) and pulses from the transmitter are directed to the inner coil of the probe, which is tuned to the ^{1}H frequency. The decoupler is set to the ^{13}C frequency (75 MHz) and pulses from the decoupler are directed to the outer coil of the probe, which is tuned to the ^{13}C frequency. After the HSQC pulse sequence, the ^{1}H FID is detected on the inner coil of the probe and directed to the receiver, which uses a continuous signal from the transmitter (300 MHz) as a reference frequency that is "mixed" with (i.e., subtracted from) the FID frequency.

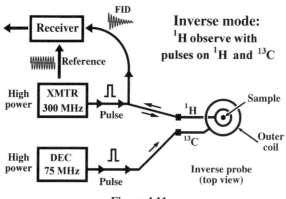

Figure 4.11

Modern NMR spectrometers may have many sources of RF ("channels"). For biological NMR, it is typical to have three channels, usually set to the frequencies of ^{1}H (e.g., 600 MHz), ^{13}C (e.g., 150 MHz), and ^{15}N (e.g., 60 MHz). Bruker refers to these as F1, F2, and F3, whereas Varian uses transmitter, decoupler A, and decoupler B. Recently, a fourth channel has become common, used for decoupling of deuterium (^{2}H) in ^{2}H-labeled proteins and nucleic acids. In these 3-channel and 4-channel spectrometers, the RF for pulses and for the reference frequency is produced by separate fully broadband (zero to ^{1}H frequency) sources, and these low-power (~1 V) signals are fed into power amplifiers to boost them to the high power (50–300 W) needed for pulses fed into the probe. The power amplifiers are not broadband: one is devoted to ^{1}H and the others are for all nuclei except ^{1}H ("X" nuclei). For example, on a 3-channel 600-MHz spectrometer, one power amplifier handles only 600 MHz (^{1}H) pulses, whereas the other two are broadband from 6 to 242 MHz (242 MHz is the ^{31}P frequency, the highest frequency below ^{1}H (except ^{19}F, 570 MHz)). The interface between the flexible low-power RF sources and the more restricted power amplifiers is a kind of switchboard, whose connections depend on which nucleus is being detected and what kind of experiment is being done. The connections in the switchboard are not made by moving cables or switching physical relays, but rather by solid-state switches (PIN diodes) that are controlled by software. These switches can be changed in 1 μs or less and do not have moving parts to wear out like relays.

4.6 DECOUPLING SOFTWARE: PARAMETERS

Most Varian decoupling parameters start with the letter "*d*" to distinguish them from the transmitter parameters, which start with a "*t*." Bruker uses a "1" (F1 channel) to specify the transmitter channel and a "2" (F2 channel) to specify the decoupler channel. The following parameters can be examined by entering *dg* (Varian) or *eda* (Bruker):

Bruker	Varian	
nuc2	dn	*Decoupler nucleus*: H1, C13, N15, and so on. This determines the basic frequency of the decoupler irradiation (300.0, 75.4, 30.4 MHz, etc.)
o2	dof	*Decoupler offset*: This sets the exact frequency (chemical shift) of the decoupler irradiation in hertz.
—	dm	*Decoupler mode*: This determines when the decoupler is on or off during the pulse sequence (e.g., "nny" for on, on, off).
pl17	dpwr	*Decoupler power*: Power level of the decoupler irradiation in decibels (dB) (increasing power from 0 to 63 for Varian, from 120 to −6 for Bruker)
cpdprg2	dmm	*Decoupler modulation*: This defines the decoupling sequence for composite pulse decoupling (e.g., Varian 'w' for waltz-16)
dcpd2	dmf	*Decoupler modulation frequency*. This sets the 90° pulse width at the power level used for decoupling (90° pulse = dcpd2 = 1/dmf)
—	homo	*Homonuclear*: Set to 'y' for homonuclear (^{1}H–^{1}H) decoupling, or 'n' for heteronuclear (e.g., ^{1}H–^{13}C) decoupling.

Bruker has no corresponding parameter for *dm* and *homo* because these options are written into each pulse program.

The Varian parameter *dmm* determines whether the decoupler output is a simple continuous irradiation (*dmm* ="c") or a pulsed waltz-16 modulation (*dmm* ="w"). For nonselective ("broadband") decoupling such as that desired for a 1D ^{13}C spectrum, the waltz-16 mode is used to minimize power requirements and maximize the range ("bandwidth") of chemical shifts decoupled. For selective decoupling, the continuous mode is used to minimize the range of chemical shifts affected. For Bruker the choice between continuous wave (cw) and composite pulse decoupling (cpd) is coded into the pulse program; the parameter *cpdprg2* defines the sequence used for composite pulse decoupling (e.g., waltz-16). The Varian *dmf* (decoupler modulation frequency) parameter is used to set the duration of pulses (e.g., 90° pulse, 270° pulse) in the waltz-16 sequence. It is determined by calibration of the 90° pulse width at the power level *dpwr* and is defined as the reciprocal of the 90° pulse width: $dmf = 1/t_p(90)$ in units of hertz. From the example above (100 μs 90° pulse for waltz-16), we would use $dmf = 1/(100 \mu s) = 10{,}000$ Hz. The 90° pulse is much longer than a hard pulse (~10 μs) because we use much lower power for decoupling. Note that the decoupler field strength in hertz is one fourth of *dmf* (typical $\gamma B_2/2\pi = dmf/4 = 2500$ Hz). Bruker uses the parameter *dcpd2* for the 90° degree pulse width calibrated at power level *pl17* (decoupler power level). The decoupler frequency is set by the Varian parameter *dof* (decoupler offset) and Bruker parameter *o2* (oh-two, offset channel 2), which function just like *tof* (transmitter offset) and *o1* (oh-one, offset channel 1), respectively.

The decoupler power (Varian *dpwr*, Bruker *pl17*) is set according to the desired effect of the decoupler irradiation. Bruker uses a decibel scale for RF power that decreases as power increases (120 to −6 dB), and Varian uses a decibel scale that increases as power increases (0 to 63 dB). For homonuclear (i.e., ^{1}H–^{1}H) NOE experiments, a very low power (5 dB Varian, 58 dB Bruker) is used to maximize selectivity—only a "simmer" is required to equalize populations. For selective decoupling, values of 10–15 (Varian) or 48–53 (Bruker) are typical—small enough to be selective but powerful enough to maintain the "rolling boil" necessary for decoupling. For broadband (nonselective) decoupling (e.g., waltz-16), a power level of 40 (Varian) or 23 (Bruker) is typical, adjusted to obtain good decoupling over the entire range (e.g., 5 ppm ± 6 ppm: −1 to 11 ppm) of proton chemical shifts. For each setting of decoupler power, the 90° pulse must be measured and *dmf* ($1/t_{90}$) or *dcpd2* (t_{90}) set appropriately.

There are more advanced experiments such as DEPT (Chapter 7) that observe ^{13}C and use the decoupler to supply high power, short duration ("hard") pulses at the ^{1}H frequency. This requires full power from the decoupler, but the parameters *dpwr* and *pl17* are avoided for these pulses. Setting decoupler power to the maximum might lead to disastrous mistakes because the decoupler can only deliver full power for short (~10 μs) periods of time without burning up the decoupler, the probe, and the sample. Instead, the parameters *pp* (Varian) and *p3* (Bruker) are used for the 90° pulse width for decoupler hard pulses and *pplvl* (Varian) and *pl2* (Bruker) indicate the power level for short-duration high-power decoupler pulses.

4.7 THE NUCLEAR OVERHAUSER EFFECT (NOE)

The population distribution of a nucleus (difference between populations in the upper spin state and the lower spin state) can be affected by the population distributions of other nuclei that are nearby in space. Experimentally, one can observe an enhancement of the population difference of one nucleus by saturating (equalizing the populations of) a nearby nucleus (Fig. 4.12). In the figure, filled circles represent a slight excess of population ($+\delta$) and open circles represent a slight deficit ($-\delta$). Irradiation of one proton signal (H_a) equalizes its populations across the $\alpha \leftrightarrow \beta$ transition (Fig. 4.12, center), and over a period of time this perturbation "propagates" through space to a nearby proton (H_b), which

THE NUCLEAR OVERHAUSER EFFECT (NOE) 151

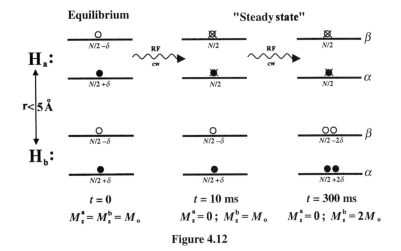

Figure 4.12

experiences a population perturbation in the opposite sense: an increase in population difference (Fig. 4.12, right). An enhanced population difference means a larger net magnetization along the z axis ($M_z > M_o$) and can be observed by applying a 90° pulse to the affected nucleus and observing its NMR signal, which will also be enhanced. The intensity of this effect dies off very quickly with increasing distance between the saturated nucleus and the observed nucleus: the exact dependence is $1/r^6$, where r is the distance between the nuclei. This is extremely important for proton–proton interactions because it allows distances between individual atoms in a molecule to be measured. This strategy has led to the accurate determination of 3D structures of proteins and nucleic acids in aqueous solution, so that NMR now rivals X-ray crystallography as a method for defining the precise conformations of biomolecules.

Heteronuclear NOEs can also be observed; for example, between ^1H and ^{13}C nuclei (Fig. 4.13). Continuous irradiation of the proton signal during the relaxation delay leads over time to an increase in the population difference (and net z-magnetization) of the ^{13}C bonded to that proton. This builds up and finally levels off at a steady state, where relaxation exactly balances the enhancement from the proton. At this point the ^{13}C pulse rotates this enhanced z-magnetization into the x-y plane where it precesses and induces an enhanced signal in the probe coil. After Fourier transformation we have an enhanced peak height in the ^{13}C spectrum (Fig. 4.13). We want all of the ^{13}C signals to get this benefit, so waltz-16 decoupling is used to irradiate the protons, covering the entire range of ^1H chemical shifts. Because saturation of protons is typically carried out during acquisition of ^{13}C signals anyway to eliminate the effects of ^1H-^{13}C J coupling, it is convenient to continue this saturation throughout the whole experiment, including during the relaxation delay. The effect is that a heteronuclear NOE builds up on the ^{13}C nuclei during the relaxation delay, enhancing their z magnetization and giving a stronger signal in the FID after this z magnetization is rotated into the x–y plane by the observe pulse. This gives the ^{13}C signals a much-needed increase in the signal-to-noise ratio. It is not used to measure distances because the majority of the enhancement comes from directly bound protons, and this covalent bond distance is already known.

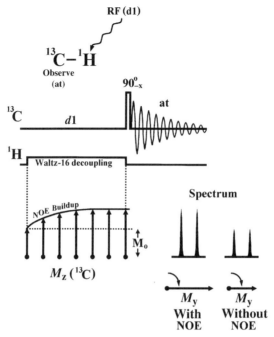

Figure 4.13

4.7.1 Comparison of NOE and Decoupling

It is important to recognize that the power level required for saturation (equalization of populations in the two energy levels) of nuclei, which causes the NOE, is much less than that required for decoupling. Decoupling requires not just equalization of populations but a situation where each ^{1}H nucleus jumps back and forth rapidly between the two levels. It is sort of like simmering the spins *versus* a raging boil. Continuous saturation causes the NOE to "build up" to a steady-state level over a period of time on the order of T_1. The NOE manifests itself as an enhancement of M_z in the target nucleus, and the effect dies off after saturation is discontinued with a time constant on the order of T_1. The irradiation in an NOE experiment occurs *during the relaxation delay* and before the exciting 90° pulse. Decoupling is effective only during the acquisition period because it averages out the effect of the spin state of other nuclei on the precession frequency of the observed nucleus. Thus to decouple the irradiation must occur *during acquisition of the FID*. Decoupling manifests itself as a reduction or elimination of the J coupling, and the effect stops immediately after the decoupler is turned off.

4.8 HETERONUCLEAR DECOUPLER MODES

The standard ^{13}C experiment leaves the decoupler on continuously to take advantage of the NOE enhancement (^{1}H decoupler on at very low power during the relaxation delay) and to get decoupling of J_{CH} (^{1}H decoupler on at low power during the acquisition of the FID). Even when decoupling is not used during acquisition (i.e., when you want to observe fully

HETERONUCLEAR DECOUPLER MODES   153

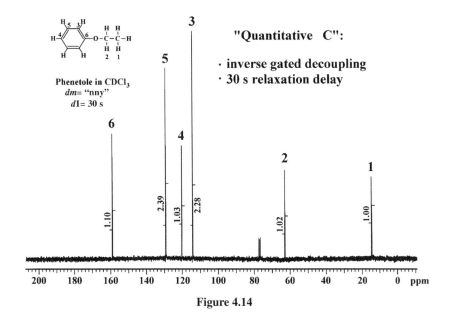

Figure 4.14

coupled ^{13}C multiplets) you should keep the decoupler on during the relaxation delay to get the benefit of the heteronuclear NOE. Decoupling that is applied only during the relaxation delay is sometimes called *gated decoupling* because the decoupler signal is "gated" on and off during each transient. We saw the gated-decoupled ^{13}C spectrum of phenetole in Figure 4.2 (bottom). Occasionally you may want to measure quantitative ^{1}H-decoupled ^{13}C spectra that can be integrated just like proton spectra to determine the number of carbons represented by each line. To get accurate peak areas, you will have to increase the relaxation delay to be at least $5 \times T_1$ for the most slowly relaxing ^{13}C in the sample. You will also want to turn the decoupler off during the relaxation delay to eliminate the heteronuclear NOE, which would enhance some peaks more than others. This experiment is the reverse of gated decoupling so it is called *inverse gated* decoupling. Figure 4.14 shows the inverse-gated ^{13}C spectrum of phenetole with a relaxation delay of 30 s. While it is difficult to integrate these very sharp peaks with a noisy baseline, the integrals and peak heights clearly follow the pattern 1:2:1:2:1:1 for the six resonances, showing that the *ortho* and *meta* positions of the aromatic ring represent two carbons each. Without any ^{1}H decoupling at all the ^{13}C spectrum is very weak and split into complex and overlapping multiplets, so this mode is not used. In fact, the first sign that your ^{1}H decoupler is not working is that your ^{13}C spectra only show the solvent (e.g., CDCl$_3$) peak.

Figure 4.15 shows the ^{13}C spectrum of sucrose with four different decoupling modes: decoupler off, decoupler on during the relaxation delay (A: NOE only), decoupler on during the acquisition of the FID (B: decoupling only), and decoupler on continuously (decoupling and NOE). The effect of the NOE can be seen by comparing the first and second spectrum (none vs. A only), and the effect of decoupling can be seen by comparing the second and fourth spectra (A only vs. A and B). The increase in peak height seen with decoupling is due to the multiplet signals being combined into a single, tall peak. In the third spectrum (B only) we are already seeing some NOE enhancement due to NOE buildup during the acquisition of the FID that does not completely dissipate during the relaxation time.

154 CARBON-13 (^{13}C) NMR SPECTROSCOPY

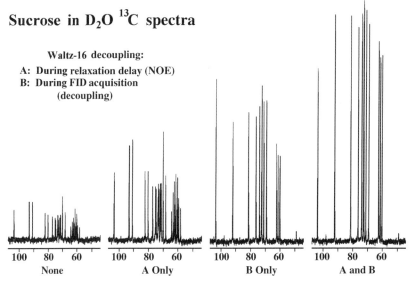

Figure 4.15

4.8.1 Parameters for Decoupler Gating

Varian uses the decoupler mode (*dm*) parameter to determine when the decoupler should be turned on and off during the pulse sequence. Time periods A, B, C, ... are defined during any pulse sequence, and the definition of these time periods can be observed by entering *dps* (display pulse sequence). For the general 1D sequence called "s2pul" (simple 2-pulse), the periods A, B, and C are defined as follows:

```
        d1        -    p1    -    d2    -    pw    -    at
   Relaxation delay    Pulse       Delay      Pulse      Acquisition
|         A         |         B         |         C         |
```

The B period is irrelevant in a simple ^{13}C experiment because *p1* and *d2* are usually set to zero.

These four decoupling modes along with their Bruker equivalents can be defined as follows:

Varian	Bruker	Definition	Result
dm = "yyy"	pulprog: zgdc	Continuous decoupling.	Singlets with enhancement.
dm = "yyn"	pulprog: zggd	Gated decoupling.	Multiplets with enhancement.
dm = "nny"	pulprog: zgig	Inverse-gated decoupling.	Singlets without enhancement.
dm = "nnn"	pulprog: zg	No decoupling.	Multiplets without enhancement.

where enhancement refers to the nuclear Overhauser effect. Bruker uses different pulse programs (parameter *pulprog*) for each application, but Varian uses different settings of the parameter *dm*.

5

NMR RELAXATION—INVERSION-RECOVERY AND THE NUCLEAR OVERHAUSER EFFECT (NOE)

5.1 THE VECTOR MODEL

Before we can understand any experiment more complicated than a simple ^{1}H spectrum, we need to develop some theoretical tools to help us describe a large population of spins and how they respond to RF pulses and delays. The vector model uses a magnetic vector to represent one peak (one NMR line) in the spectrum. The vector model is easy to understand but because it represents a quantum phenomenon in terms of classical physics, it can describe only the simpler NMR experiments. It is important to realize that the vector model is just a convenient way of picturing the NMR phenomenon in our minds and is not really an accurate description of what is going on. As human beings, however, we need a physical picture in our minds and the vector model provides it by analogy to macroscopic objects.

5.2 ONE SPIN IN A MAGNETIC FIELD

The nucleus is viewed as a positively charged sphere that spins on its axis, producing a small magnetic field whose strength (magnetogyric ratio or gamma: γ) is characteristic of the particular isotope (e.g., ^{1}H, ^{13}C, ^{31}P, etc.).

Throughout this book we will treat γ as the "strength of the nuclear magnet," ignoring its units and its definition as a ratio. We will also assume that γ is positive.

The spinning is a fundamental property of the nucleus, so it never stops or changes speed and the magnetic field it produces is a constant. The magnetic field provided by the NMR magnet (B_o) is always shown on the vertical axis (z axis) and the spin axis is represented as forming an angle Θ with the applied magnetic field (Fig. 5.1). Like a compass needle in the earth's magnetic field, the nuclear magnet wants to align with the B_o field, and it experiences

---

NMR Spectroscopy Explained: Simplified Theory, Applications and Examples for Organic Chemistry and Structural Biology, by Neil E. Jacobsen
Copyright © 2007 John Wiley & Sons, Inc.

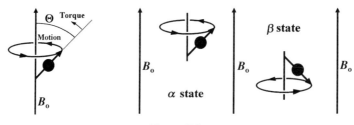

Figure 5.1

a torque pushing the spin axis toward the z axis. Because it is spinning and possesses angular momentum, however, the torque does not change the angle of the spin axis but instead causes the axis to *precess*, describing a circular motion in a plane perpendicular to the B_o field, similar to the motion of a gyroscope (a spinning top) in the earth's gravitational field. Although we are always talking about NMR-active nuclei as "spins," it is not the spinning rate we are interested in but rather the *precession rate*, which is the resonant frequency that forms the basis of nuclear magnetic *resonance*. This precession rate is proportional to the strength of the nuclear magnet (γ), and to the strength of the applied magnetic field B_o:

$$\nu_o = \gamma B_o / 2\pi$$

The precession frequency, also called the *Larmor* frequency, is represented as ν_o ("new-zero," in hertz or cycles per second). The division by 2π is sometimes omitted because the units of γ can be expressed in hertz per tesla rather than radians per seconds per tesla. When the Larmor frequency is represented in units of radians per second it is called the angular velocity or ω_o. These are related by a factor of 2π:

$$\omega_o \text{ (radians/s)} = \gamma B_o = 2\pi \nu_o \text{ (Hz)}$$

For a typical superconducting magnet with a field strength of 7.05 T, protons (^{1}H nuclei) precess at a rate of 300 million revolutions per second (300 MHz). This frequency is in the radio frequency portion of the electromagnetic spectrum, a bit higher than the frequencies on your FM radio dial (88–108 MHz). The important thing is that ν_o depends on the nuclear magnet strength γ, which is a fundamental property of the type of nucleus we are looking at (e.g., ^{1}H or ^{13}C) and never changes, and the field strength B_o, which depends on the NMR instrument ("200 MHz," "500 MHz," etc.) and can also be changed slightly by NMR hardware (lock system Z_o coil, shim coils, or gradient coils) or by the chemical environment of the molecule in which the nucleus finds itself (chemical shift) and the neighboring nuclear magnets (J coupling). Virtually everything we observe in NMR depends on this resonant frequency, and the resonant frequency depends on the effective magnetic field that the nucleus experiences (B_{eff}), which is very close to the NMR magnet's field, B_o. We will return to this fundamental relationship again and again.

In the classical world, all possible angles Θ can be formed with the magnetic field direction, between 0° (+z axis) and 180° (−z axis). To gain some flavor of the quantum world, however, we will allow only two angles corresponding to the two quantum states of a spin-1/2 nucleus. The lower energy or α state will be represented by an angle of 45°, and the higher energy or β state will be represented by an angle of 135° to the positive z axis. Thus the axis of the nuclear spin will sweep out a cone as it precesses around the

z axis, and this cone will be opening upward (toward the positive z axis) for spins in the α state and downward (toward the negative z axis) for spins in the β state (Fig. 5.1). Our vector model combines these two fundamental aspects of the NMR phenomenon: precession (circular motion in the x–y plane that leads to observable NMR signals) and energy levels (the quantum requirement that each spin be in either the α or aligned energy state or the β or opposed energy state).

5.3 A LARGE POPULATION OF IDENTICAL SPINS: NET MAGNETIZATION

People tend to think of NMR as a quantum phenomenon. But quantum mechanics deals with the options available to a single particle (e.g., two options for a spin-1/2 nucleus) and the *probabilities* of being in each of the energy states. In NMR we cannot measure something until we have a very large number of spins, like 10^{20} spins, all behaving statistically in a similar way so their miniscule microscopic magnetic fields add together to generate a macroscopic magnetic signal. Thus NMR is really about *statistical mechanics*, the sociology of large groups of spins, rather than quantum mechanics, the psychology of individual spins. In order to generate this measurable macroscopic signal the individual spins have to be "organized" so that they behave in a coherent manner, with "teamwork" to make sure that their individual signals do not just cancel each other out. This coherence or organization is provided by the radio frequency (RF) pulse.

A large population of identical spins in a sample is called an "ensemble" of spins. This would correspond to a sample with a single compound in solution with only one NMR peak (e.g., the ^{1}H spectrum of chloroform, $CHCl_3$). Each individual nucleus in the sample precesses at its resonant frequency around the external magnetic field B_o, which is along the $+z$ axis. Forget about the location of the each molecule within the volume of the sample solution, or the orientation of that molecule relative to the z axis. The nucleus maintains its orientation with respect to the B_o field (α or β state) even as the atom it belongs to is tumbling with the molecule and moving through the solution. The only thing the nucleus can interact with is a magnetic field; it is not "attached" in any way to the molecule so we can think of the spinning nucleus mounted on frictionless bearings ("gimbles" to the nautically inclined) that allow it to stay oriented to the B_o field regardless of the molecule's position or orientation. The orientation of the spin axis of the nucleus can be represented as a magnetic *vector*, pointing along the spin axis from the South pole of the nuclear magnet to the North pole, with length equal to the magnitude of the spin's magnetic field (γ). If all of the vectors representing the magnetic dipoles of the individual spins are lined up in a row, we have a sort of "chorus line" of spins facing us, and all of them will be precessing in the same direction (counterclockwise) and at exactly the same rate, the Larmor frequency ν_o (Fig. 5.2). But at equilibrium they are all rotating with random phase; that is, at any moment in time if we take a snapshot we see that some are pointing to the right side (y axis), some are pointing to the left side ($-y$), some to the front (x axis) and some to the back ($-x$ axis), and in fact every possible direction around the cone defined by the 45° angle with the $+z$ axis will be represented (Fig. 5.2(a)). For the moment, we consider only the spins in the lower energy (α) state, which are precessing around a cone which opens upward. This is like a bad ballet company: the dancers are spinning around together, but at any moment some are facing the audience, some have their backs to the audience, some face the right side, and some face the left side in a random fashion. In technical terms we

158 NMR RELAXATION—INVERSION-RECOVERY

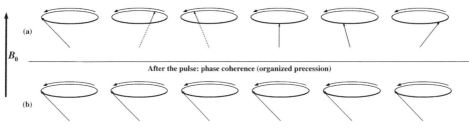

Figure 5.2

say that there is no *phase coherence* in the ensemble. Now we subject the sample to a high power pulse of RF for a precise, short period of time. The effect of the RF pulse is to get the spins "in sync" in terms of phase. After the pulse, all the spins are pointing in the same direction at any point in time (Fig. 5.2(b)). Their precessional motion is now identical and we say that the ensemble has *phase coherence*. This is like a good ballet company: all the dancers are spinning at the same rate and come around to face the audience at exactly the same time. With this organization of the sample spins extending to the bulk level of, say, 10^{20} spins, the individual magnetic vectors add together to give a bulk magnetic vector (the "net magnetization") of the sample that is also rotating counterclockwise at the Larmor frequency around the upper cone. The bulk magnetism is large enough to measure, and by placing an electrical coil next to the sample we can detect a weak voltage oscillating at the Larmor frequency. Thus we can detect the signal (the FID) and measure the Larmor frequency very precisely. This concept of organization or phase coherence created in an ensemble of spins by a pulse is fundamental to the understanding of NMR spectroscopy. Without the pulse there is no coherence and the random orientations of the precessing spins cancel out their motions: there is no measurable signal.

Now that the concept of coherence has been introduced, let us make our model of the ensemble of spins a little more accurate. Instead of lining up the spins in a row, we move their magnetic vectors to the same origin, with the South pole of each vector placed at the same point in space (Fig. 5.3(a)). Furthermore, we need to consider both quantum states, the "up" cone (α or lower energy state) and the "down" cone (β or higher energy state).

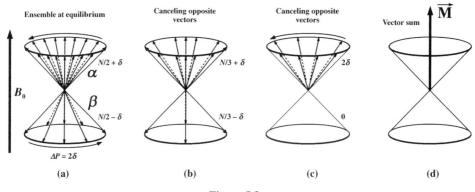

Figure 5.3

For a large population of identical spins, the individual magnetic vectors are all precessing at exactly the same rate (assuming a perfectly homogeneous B_o field) around either the upper cone (α state, "aligned" with B_o) or the lower cone (β state, "opposed" to B_o). The populations will be nearly equal in the two states, with very slightly more spins in the upper cone (lower energy state). In Figure 5.3(a) we have a cartoon representation with 8 spins in the β state and 16 in the α state, so the total population is 24 ($N = 24$) and an even distribution would be P_α (population in the α state) = 12 and P_β = 12. The energy difference creates a slight preference for the α state, which we have enormously exaggerated in the figure: $P_\alpha = N/2 + \delta = 12 + 4 = 16$; $P_\beta = N/2 - \delta = 12 - 4 = 8$, with $\delta = 4$. (In the real world δ is only about 0.001 or 0.0001% of N, not 1/6). Imagine a snapshot at one instant in time: the dipoles are not aligned in any particular direction with respect to the x and y axes, but are spread out evenly over each cone. In other words, the individual spins are all precessing at the same frequency, but their *phase* is random. Notice that for every vector in the upper cone (α state), there is another vector in the lower cone (β state) exactly opposite that vector. Their magnetic vectors will exactly cancel each other so we can erase them in our picture: they contribute nothing to the net magnetization, which is the only thing we can measure. In the cartoon, we first cancel four pairs of opposing spins (Fig. 5.3(b)), then four more pairs (Fig. 5.3(c)), leaving only 8 spins of the original 24, all of which are in the upper cone (α state). Thus we have wiped out nearly all of the N spins in the ensemble and we are dealing with only the population difference, $P_\alpha - P_\beta = (N/2 + \delta) - (N/2 - \delta) = 2\delta = 8$. So the vast majority of the spins cancel each other throughout the NMR experiment and we can only detect about one in 10^5 spins! Remember that in the real world, unlike this example shown in Figure 5.3, δ is much, much less than N. This gives you some idea why NMR is a relatively insensitive experiment, requiring milligrams (mg) of material rather than micrograms or nanograms.

After canceling the opposing spins, we see that the x and y components of the individual vectors cancel when they are combined to form the vector sum because all possible directions are equally represented in the population. The motion of precession is thus not detectable in a sample at equilibrium: no voltage will be induced in the probe coil and no NMR signal will be received by the spectrometer. Now consider the z component of the individual vectors remaining in the upper cone (Fig. 5.3(c)). All of these vectors are pointing upwards at a 45° angle to the z axis, so all have the same positive z component. Adding these together we get the vector sum, which is called the *net magnetization vector* (Fig. 5.3(d)). This vector is a macroscopic property of the sample, and we can view it as a large magnet with potentially measurable properties. At the moment it is stationary because only the x and y components of the individual vectors are moving, and all of these cancel due to their random distribution around the upper cone. We will see, though, how the RF pulse can create from this stationary vector a moving vector that can induce a measurable voltage in the probe coil.

So this net magnetization vector **M**, pointing along the $+z$ axis and stationary at equilibrium, is the starting material for all NMR experiments. Its magnitude (length of the vector) is called M_o, and it is proportional to the population difference, 2δ, and to the length of the individual magnetic vectors, the "strength of the nuclear magnet" or γ.

$$M_o = \text{(constant)} \times (P_\alpha - P_\beta)_{eq} \times \gamma = \text{(constant)} \times 2\delta \times \gamma$$

How big is this population difference at equilibrium? The Boltzmann distribution defines the populations of the two states precisely, and it turns out that the equilibrium population difference ΔP_{eq} is proportional to the energy difference between the two states (α and β)

and inversely proportional to the absolute temperature T (in degrees kelvin):

$$\Delta P_{eq} = (P_\alpha - P_\beta)_{eq} = 2\delta = \text{(constant)} \times N \times \Delta E/T$$

This makes sense because the larger the energy gap between the two states, the greater the preference will be for spins to be in the α state. If we lower the absolute temperature, each individual spin has less thermal energy available so it will be even more likely to prefer the lower energy state. In practice, liquid state NMR cannot benefit much from lowering the absolute temperature because the sample will freeze if we go very far below room temperature, but we can increase the energy gap (ΔE) by getting stronger and stronger NMR magnets. Because the energy difference is proportional to the Larmor frequency:

$$\Delta E = h\nu_o = h\gamma B_o/2\pi$$

the energy gap is proportional to both B_o and to γ. Thus M_o, our NMR "starting material," depends on B_o and γ as follows:

$$M_o = \text{(constant)} \times \Delta P_{eq} \times \gamma = \text{(constant)} \times (N \times \Delta E/T) \times \gamma$$
$$= \text{(constant)} \times (N \times \gamma B_o/T) \times \gamma = \text{(constant)} \times N \times \gamma^2 \times B_o/T$$

So the net magnetization at equilibrium is proportional to the number of identical spins in the sample (i.e., the concentration of molecules), the square of the nuclear magnet strength, and the strength of the NMR magnet, and inversely proportional to the absolute temperature. For example, M_o for ^{1}H is 16 times larger than M_o for ^{13}C because $\gamma_H/\gamma_C = 4$. This net magnetization vector is the material that we mold, transform and measure in all NMR experiments.

Now consider the effect of a 180° pulse on the ensemble of spins represented in Fig. 5.3. The RF pulse is actually a rotation, and we will see in Chapter 6 that this rotation is exactly analogous to the precession of magnetic vectors around the B_o field. The pulse itself can be viewed as a magnetic field (the "B_1" field) oriented in the x–y plane, perpendicular to the B_o field, and for the short period when it is "turned on" it exerts a torque on the individual nuclear magnets that makes them precess counterclockwise around the B_1 field. This is shown in Fig. 5.4. Each magnetic vector is rotated by 180°, so the entire structure of two cones is turned upside down, with the upper cone and all its magnetic vectors turned down to become the lower cone, and the lower cone turned up to become the upper cone. This

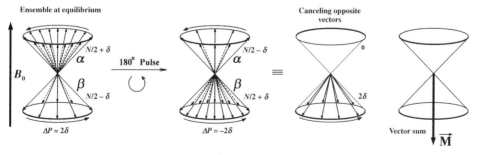

Figure 5.4

means that every spin that was in the α state is now in the β state and every spin that was in the β state is now in the α state. Each individual magnetic vector is still precessing at the Larmor frequency counterclockwise around the z axis (B_o field), and they are still randomly distributed around each of the cones. The only obvious difference is that the population of the upper cone (α or lower energy state) is now $N/2 - \delta$, or a tiny bit less than half of the spins in the ensemble. This is because every one of those spins started out in the lower cone (higher energy state) at equilibrium, before the pulse. The population in the lower cone (β or higher energy state) is now $N/2 + \delta$, or a tiny bit more than half of the spins. We have *inverted* the population distribution, and Herr Dr. Prof. Boltzmann is turning over in his grave, because you are not supposed to have more spins in the higher energy state than in the lower energy state! As before, we cancel the exactly opposing pairs of spins until we have only 2δ spins in the lower cone (β state) and these vectors are then combined by vector addition to give the net magnetization vector, **M** (Fig. 5.4, right). This vector is now pointing along the $-z$ axis, with the same magnitude, M_o, that it had at equilibrium. We see a general pattern here: rotating the ensemble of magnetic vectors by an angle Θ has the effect of rotating the net magnetization vector M by the same angle Θ. This is not surprising because the net magnetization is just the vector sum of all the individual magnetic vectors.

5.4 COHERENCE: NET MAGNETIZATION IN THE X–Y PLANE

The effect of a 90° pulse (i.e., turning on the B_1 field for half of the time that we used for the 180° pulse) is more interesting (Fig. 5.5). Again the entire double-cone structure is rotated counterclockwise by the pulse, this time stopping with the (formerly) upper cone at the left-hand side and the (formerly) lower cone at the right-hand side. Each individual magnetic vector has experienced a 90° rotation around the B_1 vector, which is extending out toward us. As before, we can cancel all of the exactly opposing magnetic vectors, leaving just 2δ spins in the left-hand side cone. Now we have to wave the magic quantum wand because most of the magnetic vectors are violating the rule that they must choose either the upper cone or the lower cone, as defined by the B_o field direction. In other words, they must be either "aligned" (45° angle to B_o) with or "opposed" (135° angle to B_o) to the B_o field in our model. We sort these out by rotating them around the left-hand side cone to the nearest point where they are in a "proper" cone pointing either up or down. Now we have δ spins in the upper cone and δ spins in the lower cone, with a population difference ΔP of zero. The 90° pulse has destroyed the equilibrium population difference. More importantly, all of these magnetic vectors are pointing to the left-hand extreme of their respective cones at the moment the pulse stops. We have created phase coherence because at this instant

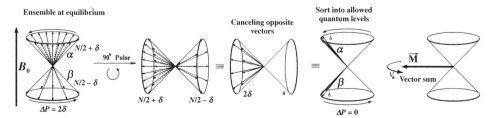

Figure 5.5

in time all of the spins point to the left side and will start their precession around B_o in unison. The net magnetization is the vector sum of all 2δ magnetic vectors, and it points to the left side (Fig. 5.5, right) and has the same magnitude (M_o) as the vector sum formed from the 2δ surviving magnetic vectors in the upper cone at equilibrium. As we saw with the 180° pulse, the effect of the 90° pulse on the net magnetization vector is to rotate it in a counterclockwise direction around the B_1 field (this time by 90°), just as it does to each of the individual magnetic vectors. Hopefully this exercise will give you some confidence in the validity of the vector model, which ignores the individual nuclear spins and deals only with the net magnetization vector **M**. Pulse rotations are applied to the net magnetization vector and we do not need to worry about the complex behavior of the individual spins any more. But the model of the individual nuclear magnetic vectors combining to form the net magnetization is very important to keep in your head because it gives us an understanding of the nature of phase coherence. We will see when we look at NMR relaxation how the individual behavior of spins contributes to the loss of coherence over time.

The simplest NMR experiment is just to apply a 90° pulse to the sample and then record the FID signal. A pulse of high power and short duration radio frequency energy at the Larmor frequency will have the effect of organizing the individual precessing spins into a coherent, in-phase motion so that at any instant in time all 2δ of the "net" spins are oriented in the same direction with respect to the x and y axes. Now, instead of canceling out (random phase), the x and y components of the individual spins add together (phase *coherence*) to form a net magnetization vector in the x–y plane that rotates at the Larmor frequency (Fig. 5.6). Because at equilibrium each of the $N/2 - \delta$ spins in the higher energy state was directly opposed to one of the spins in the lower energy state, only the 2δ spins representing the equilibrium population difference, ΔP, contribute to this vector. This rotating magnetic vector induces a sinusoidal voltage in the probe coil of the spectrometer that can be amplified and detected to give a free induction decay (FID). Fourier transformation of the FID gives a frequency domain spectrum with a single peak at the Larmor frequency.

5.5 RELAXATION

The equilibrium state is characterized by a complete lack of coherence (random phase), a slight excess of population in the α state ($N/2 + \delta$), and a deficit in the β state ($N/2 - \delta$). Anything that perturbs this equilibrium (e.g., an RF pulse) will be followed immediately by a process of relaxation back to the equilibrium state that can take as long as seconds to reestablish. Relaxation is extremely important in NMR because it not only determines how long we have to wait to repeat the data acquisition for signal averaging, but it also determines how quickly the FID decays and how narrow our NMR lines will be in the spectrum. Relaxation is also the basis of the nuclear Overhauser effect (NOE), which can be used to measure distances between nuclei: one of the most important pieces of molecular information we can obtain from NMR.

5.5.1 Relaxation After a 90° Pulse

Immediately after a 90° pulse the net magnetization vector is in the x–y plane. This means that the z component of the net magnetization is zero and that there is no difference in population between the upper (β) and lower (α) energy states. The net magnetization vector will rotate (precess) in the x–y plane at the Larmor frequency, ν_o. The phase coherence

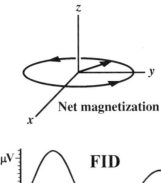
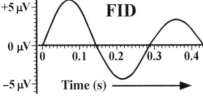
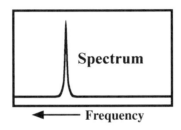

Figure 5.6

dies out with time due to inhomogeneity of the magnetic field and small differences in the local magnetic field experienced by each of the spins, largely due to the presence of nearby nuclear magnets in the tumbling molecule. The individual vectors "fan out" around the cone, and eventually they are randomly oriented again and the net magnetization in the *x* and *y* directions is zero (Fig. 5.7, A→B→C→D). The magnitude of this *x–y* component of the net magnetization vector will decrease exponentially toward zero as the individual magnetic vectors "fan out" over the cone and lose phase coherence (Fig. 5.7, bottom right: M_{xy}). We can represent this magnetization as an exponential decay with time constant T_2 (Fig. 5.7, bottom left: $T_2 = 0.4$ s, $\nu_o = 2$ Hz):

$$M_y = -M_o \cos(2\pi\nu_o t)\, e^{-t/T_2}$$

$$M_x = M_o \sin(2\pi\nu_o t)\, e^{-t/T_2}$$

The cosine and sine functions represent the rotation of the net magnetization vector: it starts on the $-y'$ axis ($-\cos(0) = -1$, $\sin(0) = 0$) and moves toward the $+x'$ axis. After 1/4 counterclockwise rotation ($\nu_o t = 1/4$) we have $-\cos(90°) = 0$, $\sin(90°) = 1$. The decaying exponential function e^{-t/T_2} represents the fanning out of individual vectors and loss of

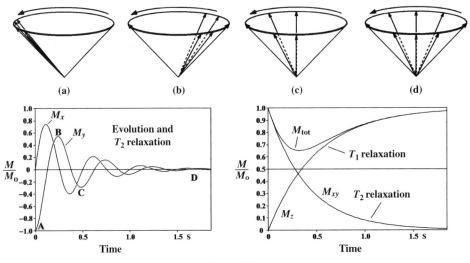

Figure 5.7

coherence. We can define a "half-life" for the process just as you might for radioactive decay:

$$0.5 = e^{-t/T_2}; \ln(0.5) = \ln[e^{-t/T_2}] = -t/T_2; \ln 2 = t/T_2; t_{1/2} = (\ln 2) T_2 = 0.693 T_2$$

So after the period of time $0.693 T_2$ (0.277 s for $T_2 = 0.4$ s) we would see the coherence reduced to 1/2 of its original value; after twice that time ($1.386 T_2 = 0.554$ s) we would see it reduced to 1/4 of its original value, and after three times the half-life ($2.079 T_2 = 0.832$ s) we would see the net magnetization in the x–y plane reduced to one eighth of its original value. T_2 (tea-two; the T is always upper case) is called the *transverse relaxation time* or spin–spin relaxation time, and after this amount of time (T_2) the coherence has decayed to 36.8% (e^{-1}) of its original value right after the 90° pulse. After twice T_2 we have 13.5% (e^{-2}) of the coherence left, and after three times T_2 we have only 5.0% (e^{-3}) left. Sometimes it is more convenient to talk about the rate of loss of coherence, $R_2 = 1/T_2$, which is in units of s^{-1} (Hz) instead of seconds. This is just like a rate constant in chemical kinetics. If you are familiar with exponential decay (e.g., radioactive decay) or, more generally, first-order processes (e.g., heat flow or first order chemical reactions), you will have no trouble understanding NMR relaxation.

At the same time that the individual spins are losing their phase coherence, some of the spins in the higher energy β state are "dropping down" to the α state (Fig. 5.8) as the system moves back to thermal equilibrium (the Boltzmann distribution). In Figure 5.8, we start with equal populations ($P_\alpha = P_\beta = 12$) right after the 90° pulse (Fig. 5.8, left). After a time corresponding to one "half-life" (0.693 times T_1), 2 spins have dropped down, increasing the population difference to $\Delta P = 4$ ($P_\alpha = 14$, $P_\beta = 10$). Eventually a total of δ spins will move from the β state to the α state, decreasing the β state population from $N/2$ to $N/2 - \delta$ and increasing the α state population from $N/2$ to $N/2 + \delta$ (Fig. 5.8, right: $\delta = 4$ and $N/2 = 12$). The reestablishment of the Boltzmann distribution between the spin states will cause a z component of the net magnetization to appear and grow toward the equilibrium magnitude M_o. This is shown in Fig. 5.7 in the graph at the bottom right for $T_1 = 0.5$ s.

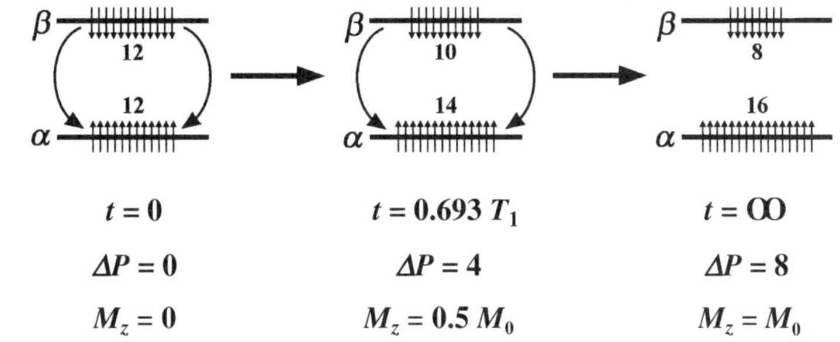

Figure 5.8

$t = 0$	$t = 0.693\,T_1$	$t = \infty$
$\Delta P = 0$	$\Delta P = 4$	$\Delta P = 8$
$M_z = 0$	$M_z = 0.5\,M_0$	$M_z = M_0$

(β: 12, 10, 8; α: 12, 14, 16)

The z magnetization (M_z) grows from zero to M_o with characteristic time $T_1 = 0.5$ s whereas the magnetization in the x–y plane ($M_{xy} = [M_x^2 + M_y^2]^{1/2}$) decreases from M_o to zero with characteristic time $T_2 = 0.4$ s. Note that in Figure 5.7 the magnitude (length) of the net magnetization vector ($M_{tot} = [M_x^2 + M_y^2 + M_z^2]^{1/2}$) drops initially because T_2 is always shorter than T_1; that is, the loss of M_{xy} is faster than the recovery of M_z. The "regrowth" of M_z is an exponential process, characterized by the function e^{-t/T_1}. The mathematical form of M_z is not quite as simple as the loss of transverse (x–y) magnetization because the z component of net magnetization is "growing back" from zero to M_o rather than decaying. What we can say is that the amount of "disequilibrium," defined by the difference between the z magnetization at any point in time, M_z, and the equilibrium value M_o is decaying exponentially:

$$\Delta M_z = M_z - M_o = \Delta M_z(t=0)\, e^{-t/T_1}$$

This is true regardless of the extent or the nature of the perturbation away from the Boltzmann distribution (90° pulse, 180° pulse, saturation, etc.). The z magnetization will always move toward M_o in this way, so that the distance to equilibrium (ΔM_z) is decaying exponentially. For the specific case of a 90° pulse, we can describe the z magnetization by an exponential function that approaches M_o with time constant T_1:

$$\Delta M_z(t) = M_z - M_o = -M_o\, e^{-t/T_1}$$

$$M_z(t) = M_o - M_o\, e^{-t/T_1} = M_o(1 - e^{-t/T_1})$$

since $\Delta M_z(t=0)$ is equal to $(0 - M_o)$ or $-M_o$. Note that $M_z = M_o(1-1) = 0$ at time zero ($e^{-0} = 1$), immediately after the pulse, and after a very long time the exponential term dies away to zero and we have $M_z = M_o$ (Fig. 5.7, bottom right). The populations in the α and β states as a function of time are shown in Figure 5.9. After the 90° pulse, the z magnetization grows from 0 to 63% of M_o after one T_1, to 86% after two times T_1, to 95% after three times T_1, and to 99% of M_o after five times T_1. Often $5T_1$ is used as a rule of thumb for a complete return to equilibrium—apparently 99% is "good enough for government work." Remember that the z component of net magnetization, M_z, is proportional to the difference in population $\Delta P = P_\alpha - P_\beta$, which grows as spins move from the β state to the α state. The rate of spins "dropping down" from β to α is driven by the amount of "disequilibrium" or the deviation from the Boltzmann distribution, so we see the rate get slower and slower as the populations get closer and closer to the equilibrium distribution.

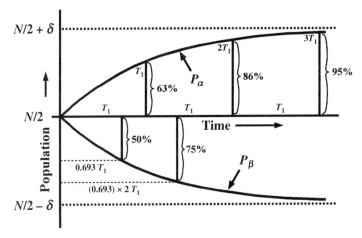

Figure 5.9

5.5.2 Relaxation After a 180° Pulse

At the end of a 180° pulse, the populations are inverted so that there is a slight excess ($N/2 + \delta$) in the upper energy (β) state and a slight deficit ($N/2 - \delta$) in the lower energy (α) state (Fig. 5.10, $T_1 = 0.5$ s). This is twice as far from equilibrium as the situation immediately after a 90° pulse ($\Delta P = -2\delta$ after a 180° pulse, 0 after a 90° pulse, and 2δ at equilibrium). Spins drop down from the β state to the α state, reducing the population in the β state and increasing the population in the α state. After δ spins have dropped down (time = $t_{1/2}$ = $0.693\,T_1 = 0.35$ s), we have reached the point where populations are equal in the two states ($M_z = 0$). This is half of the way to equilibrium. Spins continue to drop down until in all 2δ spins have dropped down from β to α, leaving a population of $N/2 + \delta$ in the α state and $N/2 - \delta$ in the β state. In mathematical terms, $M_z = -M_o$ at the end of the 180° pulse, so

$$\Delta M_z(t=0) = M_z(t=0) - M_o = -M_o - M_o = -2M_o$$

$$\Delta M_z(t) = M_z(t) - M_o = \Delta M_z(t=0)\,e^{-t/T_1} = -2M_o\,e^{-t/T_1}$$

$$M_z(t) = M_o\,(1 - 2\,e^{-t/T_1})$$

Thus M_z starts at $-M_o$ and after $0.693\,T_1$ it equals zero (halfway to equilibrium from the starting point: Fig. 5.10). After two half-lives it equals $1/2\,M_o$ (3/4 of the way to equilibrium) and after $3 \times 0.693 \times T_1$ it equals $3/4\,M_o$. After a long time it equals M_o. It is important to recognize that this return to equilibrium, which moves the net magnetization vector from the $-z$ axis to the $+z$ axis after a 180° pulse, is *not a rotation*. This process of longitudinal or spin-lattice relaxation involves only the process of spins dropping down from the higher energy state to the lower energy state and therefore cannot create magnetization in the x'–y' plane (coherence). After a 180° pulse there is no coherence, so the net magnetization vector simply shrinks in magnitude along the $-z$ axis until it reaches zero magnitude, and then grows along the $+z$ axis until it reaches the magnitude M_o defined by thermal equilibrium.

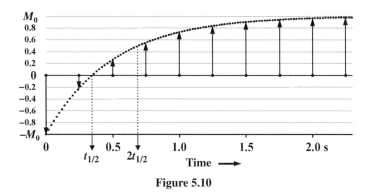

Figure 5.10

5.5.3 Spin Temperature

It is sometimes convenient to compare longitudinal (T_1) relaxation to the process of heat flow (also a first-order process). When we apply a pulse to the spins, we "heat them up," and as they return to equilibrium they "cool down" again to the temperature of their surroundings. We can even define a "spin temperature" (T_s) as the temperature corresponding to a given population difference between the α and β states:

$$P_\beta/P_\alpha = e^{-\Delta E/(kT_s)}$$

where ΔE is the energy gap between the α and β states. This is the same as the Boltzmann relationship, except that the population ratio can assume any value and not just the equilibrium ratio. In Figure 5.8, the equilibrium ratio at room temperature (300 K) was 8/16 = 0.5 or $e^{-0.693}$. If we start to saturate the spins and the ratio increases to 9/15 = 0.714 or $e^{-0.511}$, the spin temperature has increased to 407 K (407 = 300 × 0.693/0.511). Full saturation ($P_\beta/P_\alpha = 1$) corresponds to an infinite spin temperature ($e^{-\Delta E/\infty} = e^{-0} = 1$) and inversion (180° pulse) corresponds to a negative spin temperature of −300 K ($P_\beta/P_\alpha = e^{-\Delta E/(-T)} = e^{+\Delta E/T}$). These are not physically reasonable temperatures, but the concept is still useful that we are "heating up" the spins when we promote spins from the α state to the β state. After we heat up the spins, T_1 relaxation can be viewed as a flow of heat from the spin "container" to the "outside world" of the sample solution (sometimes called the "lattice," a term from solid-state NMR). The amount of energy is very small compared to average thermal energy of the molecules, so the sample temperature increases only very slightly, but the spin temperature goes down and approaches the sample temperature in an exponential manner.

5.5.4 T_1 versus T_2

Because T_2 is always smaller than T_1, *the loss of magnetization in the x–y plane will always be faster than the re-establishment of magnetization along the z axis* (Fig. 5.7, lower right). The equation for M_z is independent of the equations for M_x and M_y. This means we can deal with transverse relaxation (the T_2 process) and longitudinal relaxation (the T_1 process) as completely separate phenomena. Any transverse (x–y plane) component of the magnetization will undergo exponential decay with time constant T_2 and any nonequilibrium longitudinal component (z axis) will approach M_o exponentially with time constant T_1. It is

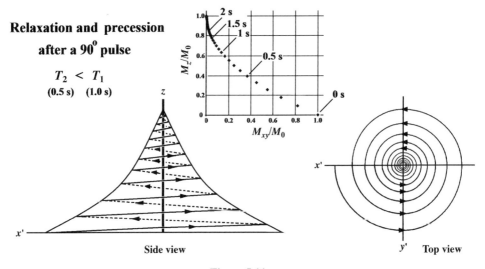

Figure 5.11

important to realize that relaxation is not a rotation back to the z axis: the net magnetization vector only rotates in a plane perpendicular to the x–y plane during an RF pulse. In the absence of pulses, the x and y components decay toward zero with time constant T_2 and the z component recovers toward M_o with the longer time constant T_1. The combination of precession, loss of coherence (T_2 relaxation) and the slower recovery of z magnetization (T_1 relaxation) after a 90° pulse is illustrated in Figure 5.11. The tip of the net magnetization vector describes an inward spiral over the surface of a circular "tent" with a single pole in the center ($+z$ axis). In the extreme case where T_2 is much less than T_1, the x and y components would decay to zero first, and then the z magnetization would "grow back" along the $+z$ axis to the M_o value. Relaxation may seem like a troublesome side issue, sort of like friction in classical mechanics, but we will see later that cross relaxation, the process by which the relaxation of one spin influences the relaxation of a nearby spin, leads to the NOE, which is an important method for measuring distances within a molecule.

5.6 SUMMARY OF THE VECTOR MODEL

The vector model is a way of visualizing the NMR phenomenon that includes some of the requirements of quantum mechanics while retaining a simple visual model. We will jump back and forth between a "classical" spinning top model and a "quantum" energy diagram with populations (filled and open circles) whenever it is convenient. The vector model explains many simple NMR experiments, but to understand more complex phenomena one must use the product operator (Chapter 7) or density matrix (Chapter 10) formalism. We will see how these more abstract and mathematical models grow naturally from a solid understanding of the vector model.

Consider a large population of identical spins—for example the protons in a liquid sample of $^{12}CHCl_3$—in a strong, uniform magnetic field B_o oriented along the positive z axis.

1. Each ^{1}H nucleus can be in one of two states: aligned at a 45° angle to the $+z$ axis ("up") or aligned at a 135° angle to the $+z$ axis ("down").
2. Each ^{1}H nucleus precesses about the z axis, with its spin axis tracing a conical path always at a 45° (or 135°) angle to the $+z$ axis, at a rate equal to the *Larmor frequency* ν_o:

$$\nu_o = \gamma B_o / 2\pi$$

where γ is a measure of the magnet strength of the nuclear magnet. The Larmor frequency is in the radio frequency range of the electromagnetic spectrum.

3. The "up" or α state, which is aligned with the laboratory magnetic field, is lower in energy than the "down" or β state, which is aligned against the field. This energy "gap" is proportional to the magnetic field strength and to the strength of the nuclear magnet:

$$\Delta E = h\nu_o = h\gamma B_o / 2\pi$$

4. At thermal equilibrium, slightly more than half of the population of spins is in the lower energy "up" state and slightly less than half is in the higher energy "down" state. The population difference is on the order of one spin in every 10^5 spins. At equilibrium each spin is randomly oriented around the cone at any moment in time, even though all spins precess at exactly the same frequency. Thus the "phase" of the precessing nuclei is random and spread equally around the two cones.

5. The *net magnetization* is defined as the vector sum of all of the nuclear magnets in the sample. If each vector's origin is moved to the origin of the coordinate system, the vectors can be added together for the whole population. The x and y components of the net magnetization are zero at equilibrium because the spins are spread equally around the two cones at any instant in time. The z components cancel for each pair of one "up" and one "down" spin, but because there is a slightly larger population of spins in the "up" state the net magnetization is a small vector pointing along the positive z axis. This macroscopic equilibrium net magnetization has a magnitude of M_o, and it is the starting point of all NMR experiments.

6. A radio frequency pulse has the effect of "organizing" the phase of the spins in their precession, so that at the end of a pulse all of the excess spins have the same phase. At any moment in time, all of these spins point in the same direction on the cone, so that they add together to make a net magnetization vector that rotates in the x–y plane at the Larmor frequency. We can say that the pulse created "phase coherence" or simply "*coherence*" in the sample. The net magnetization is no longer stationary, and its rotation induces a voltage in the probe coil that oscillates at the Larmor frequency. After the pulse the spins begin to lose the organization imparted by the pulse, and the spins spread out on the cone until they are again randomly oriented on the cone at any moment in time. This loss of coherence causes the induced signal in the probe coil to decay exponentially to zero. The probe coil signal is called the free induction decay (FID). The information contained in the FID signal is the frequency of the oscillating voltage, which corresponds to the chemical shift of the nuclei in the sample, the amplitude of the voltage, which corresponds to the height of the peak, and the phase

of the signal, which corresponds to the phase (absorptive positive, dispersive, etc.) of the NMR line.

7. The z component (M_z) of the net magnetization vector represents the difference in population between the two spin states ("up" and "down"). For example, immediately after a 90° pulse enough spins have been promoted from the lower energy ("up") state to the higher energy ("down") state to equalize the populations. At this moment the z components of the population of spins exactly cancel and the z component of the net magnetization vector is zero. The net magnetization vector is in the x–y plane, rotating at the Larmor frequency.

8. The radio frequency pulse is a very short (tens of microseconds), and a very high power (tens or hundreds of watts) pulse of radio frequency power applied to the probe coil at or very near the Larmor frequency. It has a rectangular envelope: the power turns on and instantly reaches full power, then at the end of its duration it goes instantly to zero. The pulse creates an oscillating magnetic field, which can be represented by a vector (the "B_1 vector") that rotates in the x–y plane at the frequency of the pulse. The length of the B_1 vector is equal to the amplitude of the radio frequency pulse.

9. After a 90° pulse, the phase coherence created by the pulse begins to be lost as the individual spins "fan out" around the cones due to slight local differences in magnetic field. The loss of coherence is exponential and goes to zero with time constant T_2. At the end of the 90° pulse the populations of the two spin states are equal: half of the spins are in the "up" state and half are in the "down" state. Immediately spins begin to drop down from the higher energy ("down" or β) state to the lower energy ("up" or α) state until the equilibrium population difference is reestablished. This process, which leads to an exponential growth of M_z with time constant T_1 until it is equal to the full equilibrium value M_o, is called *longitudinal relaxation*. It is always slower than the loss of coherence, which is called *transverse relaxation*.

10. A 180° pulse rotates the equilibrium sample magnetization to the $-z$ axis. Immediately after the pulse there is no phase coherence (no x or y component to the net magnetization) and no FID can be recorded. The population difference in now reversed: slightly more than half of the spins are in the higher energy ("down" or β) state and slightly less than half are in the lower energy ("up" or α) state. This is the largest deviation from the equilibrium population distribution that can be achieved by an RF pulse. The reversal of populations actually occurs by moving every single spin that was in the "up" state to the "down" state, and every spin that was in the "down" state to the "up" state. After the pulse the slight excess of spins in the higher energy state begin to drop down and the net magnetization vector along the $-z$ axis shrinks, passes through zero, and grows toward M_o along the $+z$ axis in an exponential fashion.

5.7 MOLECULAR TUMBLING AND NMR RELAXATION

What is the mechanism of spins "dropping down" from the β state to the α state and "fanning out" around the two cones, and what determines the rates ($R_1 = 1/T_1$ and $R_2 = 1/T_2$) of NMR relaxation? These processes are intimately tied to the motion of molecules as they tumble ("reorient") in solution in their rapid Brownian motion, and measurement of the NMR relaxation parameters T_1 and T_2 can even give us detailed information about molecular dynamics (motion) from the point of view of each spin in the molecule. A simplified model of these physical processes and their consequences for NMR will help you to understand the effect of molecular size on relaxation and the NOE.

The nucleus cannot simply transfer energy to molecular motions (vibration, rotation, translation, etc.) by collisions because the nucleus is not really "attached" to the rest of the molecule. The tiny nucleus sits in a vacuum very far from the bonding electrons that hold the molecule together, and its only mechanism for interaction with the outside world is through its magnetic properties. Thus only magnetic fields can affect the nucleus and induce transitions from the β to the α state (longitudinal relaxation) or create random differences in the rate of precession (transverse relaxation). For T_1 relaxation, a magnetic field must be oscillating at the frequency of the transition (the Larmor frequency, ν_o) in order to induce a transition from the β to the α state. Molecular motion ("tumbling") can generate these oscillating magnetic fields if the molecule is tumbling at the right frequency. There are several ways this can happen; we will focus on the major one, which is called the *dipole–dipole* interaction, and later mention one other mechanism.

5.7.1 Dipole–Dipole Relaxation

Consider a nucleus such as a ^{13}C nucleus within a molecule. If the carbon in question is a methine (CH) carbon, then it has a hydrogen atom (^{1}H) rigidly attached to it at a distance of 1.1 Å (1 Å = 10^{-10} m). As the molecule tumbles, from the point of view of the ^{13}C nucleus, the ^{1}H nucleus is rotating around it in a circle of radius 1.1 Å at the tumbling rate of the molecule. The ^{1}H nucleus retains its orientation with respect to the external (B_o) magnetic field as the molecule tumbles because the nucleus is not "attached" to the molecule in any way. Precession of the ^{1}H nucleus is not important for this phenomenon, so we can view the proton as a rigid magnet oriented along $+z$ (α state) or $-z$ (β state). As the molecule tumbles, the tiny magnetic field of the ^{1}H nucleus is "felt" by the ^{13}C nucleus at its location in space (Fig. 5.12). When the ^{1}H is above or below the ^{13}C, its field adds to the B_o field at the

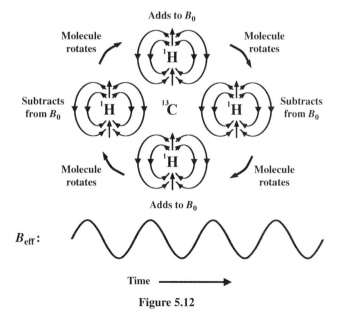

Figure 5.12

location of the ^{13}C nucleus, but when it is to the left or right of the ^{13}C nucleus its magnetic field subtracts a slight amount from the B_o field *at the position of the ^{13}C nucleus*. Thus the effective magnetic field experienced by the ^{13}C nucleus (B_{eff}) is modulated by a sinusoidal variation whose frequency is the rate of tumbling of the molecule and whose amplitude is proportional to the ^{1}H "magnet strength" (γ_H) and to the inverse third power of the distance ($r_{CH} = 1.1$ Å) between the ^{1}H and the ^{13}C. If the frequency of this perturbation is exactly equal to the Larmor frequency for the ^{13}C nucleus ($\gamma_C B_o/2\pi$), then ^{13}C spins in the upper energy state will be stimulated to drop down to the lower energy state in a process similar to stimulated emission (Chapter 1, Section 1.4.7). Unlike stimulated emission, which occurs when we add a radio frequency signal to the sample, there are no photons absorbed or emitted, and the energy is coupled to the rotational motion of the molecule and released as heat. Because the process of relaxation involves interaction of the ^{13}C nuclear magnet's field with the ^{1}H nuclear magnet's field, the rate of stimulated transitions is proportional to both γ_C/r^3 and to γ_H/r^3, so it depends on the inverse sixth power ($1/r^6$) of the distance between the two nuclei. This dipole–dipole interaction is central to the NOE as well, which also exhibits a $1/r^6$ dependence.

5.7.2 The Distribution of Tumbling Rates

All of this requires that the molecule be tumbling at exactly the Larmor frequency of the nucleus that is undergoing relaxation in order to stimulate a transition from β state to α state. In fact, only a very small fraction of molecules is tumbling at this frequency at any one time. We can look at the distribution of tumbling rates as a histogram with tumbling frequency ν (in hertz) on the horizontal axis and number of molecules tumbling at that frequency on the vertical axis. This function, which is a property of molecular size and shape as well as the viscosity of the solvent, is called the spectral density function or $J(\nu)$. A simplified logarithmic plot is shown in Figure 5.13 for five different molecules ranging in molecular weight from 10 to 100,000 Da (100 kD). A typical "organic" molecule (a "small" molecule to the NMR spectroscopist) would be on the order of 100 Da, whereas a peptide, glycopeptide, or oligosaccharide might be in the range of 1000 Da ("medium-sized") and proteins and nucleic acids (DNA and RNA) would range from 10 to 100 kD ("large") in molecular weight. Each molecule can also be characterized by its average tumbling time τ_c (formally known as the rotational correlation time). This is essentially the average time it takes for the molecule to change its orientation with respect to the B_o field (z axis); it is the reciprocal of the average tumbling frequency. Small molecules tumble rapidly and have a short τ_c, whereas large molecules tumble slowly and have a long τ_c. As you can see from the histogram (Fig. 5.13), the distribution of tumbling frequencies for any size molecule is flat over a wide range of frequencies up to a maximum or cutoff value. Above this frequency the number of molecules drops rapidly to zero. The cutoff frequency is very high for small molecules (10–100 Da), lower for the "medium-sized" molecules (1 kD) and quite low for large molecules (10–100 kD). The graph is based on the tumbling of spherical molecules in water at 27 °C, with molecular density typical of proteins.

Because in each case we are dealing with the same number of molecules, as we squeeze the population of molecules into a smaller range of frequencies, the number of molecules at any one frequency within the range increases. This explains the higher level at the left side of the histogram as the molecular size increases. Because of the logarithmic horizontal scale, the differences are much larger than those shown in the plots—note the increase in vertical scaling at lower molecular weights. At 500 MHz, which is the Larmor frequency

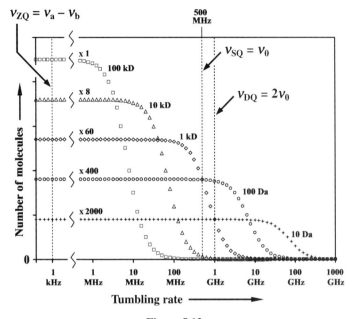

Figure 5.13

for protons in a magnetic field of 11.74 T, the dotted line ($\nu_{SQ} = \nu_0$) shows that the number of molecules tumbling at this rate is greatest for the 1 kD ("medium-sized") molecules (the 100 Da curve is actually only 15% of the height shown, in relation to the 1 kD curve). This would give the fastest T_1 relaxation because a larger proportion of molecules of this size are tumbling at exactly the Larmor frequency. For smaller molecules, the much wider range of tumbling rates means that it is unlikely to find a molecule tumbling at exactly the Larmor frequency, and for larger molecules the Larmor frequency is higher than the "cutoff"; so very, very few molecules tumble this fast. This is consistent with experimental results: As molecular size increases from 10 to 1000 Da, the relaxation rate ($R_1 = 1/T_1$) increases (T_1 decreases), and above this critical size (1000 Da) the relaxation rate falls off (T_1 increases). This relationship is shown in Figure 5.14 for a pair of protons separated in a rigid molecule by a distance of 1.8 Å at a field strength corresponding to a Larmor frequency of 500 MHz. Again, the molecular weight scale is based on a spherical molecule of density 1.42 g/L (typical for a protein) in water at 27 °C. The critical molecular weight, where the reciprocal of the tumbling time ($1/\tau_c$) is close to the Larmor frequency in radians s^{-1} (ω_o), depends on molecular shape as well as solvent viscosity. This "crossover" condition is usually written as: $\omega_o \tau_c \sim 1$. We will see that molecules with this "medium" size are a problem for NOE experiments because the theoretical NOE falls to zero. In the "small molecule" regime ($\omega_o \tau_c \ll 1$) an increase in molecular weight increases the relaxation rate (decreases T_1) and in the large molecule regime ($\omega_o \tau_c \gg 1$) an increase in molecular weight decreases the relaxation rate (increases T_1).

Although T_1 increases with molecular size for large molecules, note that the spin–spin or transverse relaxation rate T_2 continues to decrease as molecular size increases (Fig. 5.14). This is harder to explain, but we can rationalize this effect by considering that the "fanning

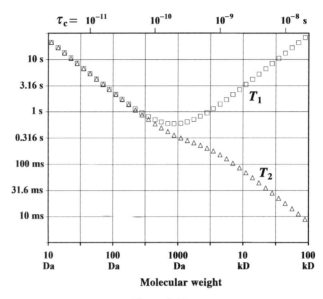

Figure 5.14

out" of individual magnetic vectors in the cones is the result of slightly different effective fields experienced at each of the identical nuclei leading to slightly different precession rates. If the molecule is tumbling rapidly, the oscillation in the magnetic field is too rapid to have any effect on the phase of the individual magnetic vectors as they rotate in the x–y plane. Think of a race with a large number of exactly matched runners, each one speeding up and slowing down repeatedly with the same average speed. If the cycle of speeding up and slowing down is very rapid so that many, many cycles occur during the race, the runners will remain in a "pack" and will all cross the finish line at the same time. If the molecule is tumbling very slowly, however, one molecule might start out with a slightly faster magnetic vector and another with a slightly slower vector, and the slow vector would begin to fall behind the fast one during the recording of the FID. In the extreme case (e.g., for a solid sample) the differences in Larmor frequency do not oscillate at all and each spin is "locked" at one part of the tumbling cycle, either slower or faster than the average precession rate. This leads to very rapid loss of coherence and very short T_2 values, so short that without special techniques we can not observe NMR peaks in the spectrum at all: they are "broadened" out of existence and fall into the noise baseline. Even for slowly tumbling molecules in solution, it makes a difference over the course of the FID what the orientation of the molecule was at the start of the FID and how fast the molecule is tumbling. Because the molecules are all oriented randomly at any moment, these slight oscillations in Larmor frequency are phase incoherent and there is a random distribution of tumbling frequencies at any one time during the FID, leading to loss of phase coherence. Referring back to the histogram of tumbling rates (Fig. 5.13), we see that the number of molecules tumbling at the very slow rates at the left side of the diagram increases monotonically as we go from small molecules to medium-sized molecules to large molecules. Because it is these slowly tumbling molecules that have the most dramatic "fanning out" due to random differences in v_o, this explains why T_2 always gets shorter (faster relaxation) as the size of the molecule increases (Fig. 5.14).

What is the significance for a simple NMR experiment of differences in T_1 and T_2? We are almost always using signal averaging of many FIDs to obtain a good signal to-noise ratio, and we will have to wait until M_z has recovered to near M_o before repeating the acquisition. So T_1 determines how rapidly we can obtain NMR data: long T_1 values will force us to wait longer to repeat the acquisition and will slow down the overall experiment. The best relaxation delay (RD or D1) is determined by the value of T_1—an ideal value would be $5 \times T_1$ for 99% recovery of z magnetization. The T_2 value determines the decay rate of the FID: a short T_2 corresponds to a rapidly decaying FID and a long T_2 value corresponds to a long, ringing FID. The Fourier transform converts time-domain data to frequency-domain data, and because time and frequency are inversely related (s $vs.$ s^{-1}) there are opposite effects in an FID and a spectrum. The shorter the time duration of the FID (short T_2, faster transverse relaxation), the broader the resulting peak in the spectrum after Fourier transformation. Conversely, an FID that decays slowly leads to a very sharp (narrow) peak in the spectrum. Because resolution is very important in NMR spectroscopy, especially as we study more and more complex molecules, we always want the narrowest peaks we can get. With a perfectly homogeneous magnetic field (perfectly shimmed magnet) the decay rate of the FID is determined by T_2, which is determined by the molecular tumbling processes described above. A perfect world for the NMR spectroscopist would be one in which all T_1s are very short and all T_2s are very long. Unfortunately, this cannot happen because for any nucleus T_1 is always longer than T_2. We can see why this is if we consider that T_1 is primarily determined by the number of molecules tumbling at the Larmor frequency (ν_{SQ}), whereas T_2 is primarily determined by the number of molecules tumbling at the low frequencies (ν_{ZQ}, Fig. 5.13). For small molecules, the numbers are nearly equal ($T_1 \sim T_2$) and for large molecules there are far more molecules tumbling at the low ("zero quantum") frequencies than at the Larmor ("single quantum") frequency ($T_1 \gg T_2$).

These arguments only attempt to capture some general trends. The math leads to a precise dependence of T_1 and T_2 on the tumbling rates at three frequencies: ν_{ZQ} (difference between the Larmor frequencies of the two nuclei that are close in the molecule), ν_{SQ} (the Larmor frequency of the nucleus being observed), and ν_{DQ} (the sum of the two Larmor frequencies). For a pair of protons interacting within a molecule, ν_{ZQ} is on the order of Hz or kHz because ν_o differs only slightly due to chemical shift differences, and ν_{DQ} is essentially twice the Larmor frequency (Fig. 5.13). We will come back to this topic and give some more detailed numbers when we look at the NOE and the effect of molecular size on the sign and magnitude of the NOE. Another detail we have ignored is that the term "tumbling rate" implies that molecules rotate at a constant rate in solution as they would in the gas phase and that they behave as rigid bodies. In fact, the ν we use in the spectral density function $J(\nu)$ is really an instantaneous "reorientation rate" describing how rapidly the molecule is changing its orientation with respect to the B_o direction as it is bumped and shoved around by solvent molecules. Futhermore, this reorientation rate is the rate of reorientation of the H–H (or C–H) vector with respect to B_o, which may not be the same as the motion of molecule as a whole. This is an advantage because we can look at local flexibility and conformational change within a large molecule such as a protein by studying relaxation rates at different locations within the molecule. The reorientation of a particular relationship between two nuclei is determined by the motion of the molecule as a whole as well as the sometimes much faster motion of the two nuclei within the molecule. It is this ability to study molecular dynamics at many different timescales (ms, μs, ns, etc.) that makes solution-state NMR a powerful tool in biology.

5.7.3 Other Relaxation Mechanisms: CSA

There are several other ways besides the dipole–dipole mechanism by which spins can be induced to drop down and reestablish the Boltzmann equilibrium, but we will look at only one. Recall that the chemical shift of a spin within a molecule actually depends on the orientation of the molecule with respect to the magnetic field B_o (Chapter 2, Section 2.6.2). In some cases (e.g., aromatic rings or amide bonds) this variation (chemical shift anisotropy or CSA) can be quite large, and in other cases (e.g., a CH group in a saturated hydrocarbon environment) there is very little dependence on orientation. As far as the NMR spectrum is concerned, the rapid tumbling of a molecule in a solution causes this variation to blur so that on the NMR timescale (roughly milliseconds) only a single, sharp peak is observed at a chemical shift that is the average over all orientations. But chemical shift is nothing more than a perturbation of the magnetic field strength experienced by a nucleus (B_{eff}), so as the molecule tumbles and samples various orientations the B_{eff} field at the nucleus is modulated in a sinusoidal fashion at a rate equal to the tumbling rate of the molecule and with an amplitude proportional to the amount of chemical-shift dependence on orientation (the CSA). Like the oscillating magnetic fields produced by the through-space interaction of the magnetic dipoles of a pair of nuclei (dipole–dipole relaxation), the oscillating magnetic field resulting from CSA can also induce transitions and lead to NMR relaxation. The dependence on molecular tumbling rate and molecular size is exactly the same as that described above for the dipole–dipole effect.

5.8 INVERSION-RECOVERY: MEASUREMENT OF T_1 VALUES

The inversion-recovery method is a convenient way to measure T_1 values of both ^1H and ^{13}C nuclei. In a moderately complex molecule (15–30 carbons), the T_1 values of all positions in the molecule can be determined simultaneously, with spectral overlap the only limitation. The method is a multiple-pulse experiment in which net magnetization of the sample nuclei is first inverted with a 180° pulse ("inversion") and then allowed to relax along the z axis with the characteristic time constant T_1 ("recovery"). The effect of the 180° pulse is to interchange all of the spins between the upper and lower energy levels, so that now the higher energy spin state has a slight excess of population and the lower energy spin state has a slightly depleted population. This causes the net magnetization vector to be turned upside-down so that M_z now equals $-M_o$. Recovery begins immediately according to the exponential law, with characteristic rate $R_1 = 1/T_1$. Because z magnetization is not a directly observable quantity, the recovery period is followed by a 90° pulse that "samples" or "reads" the z magnetization by converting it into observable x–y magnetization (Fig. 5.15).

> Notice how we diagram a multiple-pulse NMR experiment: the horizontal axis represents time and the vertical axis represents RF amplitude for pulses. The times and amplitudes are not drawn to scale—they are just cartoon representations. 90° pulses are shown as half the width of 180° pulses, and recording of the FID is shown as a decaying signal. Each RF channel is labeled according to the nucleus being irradiated (pulses) and/or observed (FID).

The magnitude of the FID signal that results from this x–y magnetization (and the peak height in the spectrum) should be directly proportional to the sample's z magnetization just before the 90° pulse. By repeating the experiment with different time delays after the 180° pulse, we can monitor this return of z magnetization to equilibrium and determine the value

INVERSION-RECOVERY: MEASUREMENT OF T_1 VALUES

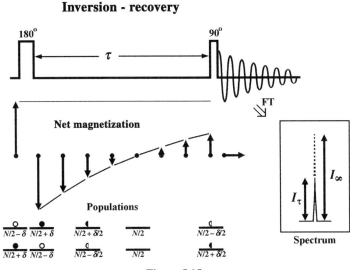

Figure 5.15

of T_1 by curve fitting of the data to an exponential function. The phase correction parameters are first set using a simple 90° pulse acquisition (starting with equilibrium magnetization, along $+z$) and then applied to a series of inversion-recovery spectra acquired with increasing values of the delay τ. For $\tau = 0$ we should see an upside-down spectrum, with each peak at its maximum height but inverted. As the delay is increased each peak will become less intense, pass through zero, and finally become positive. At very long τ delays the spectrum should look just like a normal spectrum. For each signal in the spectrum, corresponding to a unique position in the sample molecule, the recovery of z magnetization will obey the first equation and the signal intensity after the 90° pulse will follow the second equation as a function of τ:

$$Z \text{ magnetization: } M_z(\tau) = M_o + (M_z(0) - M_o)\,e^{-\tau/T_1} = M_o + (-M_o - M_o)\,e^{-\tau/T_1}$$
$$= M_o(1 - 2\,e^{-\tau/T_1})$$

$$\text{Signal intensity (peak height): } I(\tau) = I_{\text{inf}}(1 - 2\,e^{-t/T_1})$$

where I_{inf} is the signal intensity for very long values of τ where M_z has recovered all the way to M_o. We will use peak height as a surrogate for peak area because peak width should not change for a given peak as a function of τ. A simple way to estimate T_1 without making a plot is to look for a null or near-null condition for a given peak:

$$I(\tau_o) = (1 - 2\,e^{-\tau/T_1})\,I_{\text{inf}} = 0;\ \ 1 - 2\,e^{-\tau/T_1} = 0;\ \ 1 = 2\,e^{-\tau/T_1};\ \ 0.5 = e^{-\tau/T_1}$$

$$\ln(0.5) = -\tau_o/T_1;\ \ \ln(2) = \tau_o/T_1;\ \ T_1 = \tau_o/\ln(2) = \tau_o/0.693$$

Here we use τ_o to indicate the delay time at the null point and set the intensity $I(\tau_o)$ to zero. To eliminate the exponential, we take the natural logarithm of both sides of the equation: $\ln(e^x) = x$. This is the same as saying that the half-life of T_1 relaxation occurs at the time 0.693 times T_1. For more accurate T_1 measurement, a line can be fitted to a plot of log(I) versus, time, or even better a nonlinear least-squares fit can be performed directly on the time course data of $I(\tau)$ vs. τ, with I_{inf} and T_1 as parameters to be adjusted.

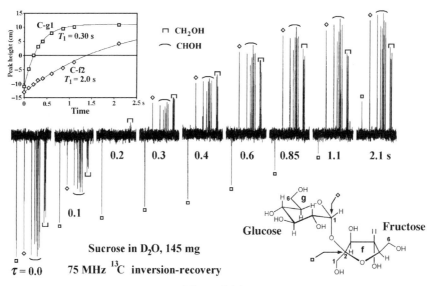

Figure 5.16

The data from a ^{13}C inversion-recovery experiment on sucrose in D$_2$O is shown in Figure 5.16. The experiment was run on a Varian Unity-300, and the data were acquired in an array, a back-to-back series of FIDs acquired one after the other by varying a single parameter—in this case the recovery delay τ. The spectra are plotted side-by-side in a "horizontal stacked plot" with the τ values ranging from 0.0 (left) to 2.1 s (right). In each spectrum we see the anomeric carbons fructose-2 (quaternary, □) and glucose-1 (CH, ◊) furthest downfield, followed by seven peaks representing singly oxygenated CH carbons (CHOH), and at the upfield end three peaks representing singly oxygenated CH$_2$ carbons (CH$_2$OH). We see that the "CHOH" carbons and C-g1 all pass through zero ($M_z = 0$) at $\tau = 0.2$ s, which means that $t_{1/2} = 0.2$ s for these carbons ($M_z = 0$ is halfway between $M_z = -M_0$, the starting value, and $M_z = +M_0$, the equilibrium value). Since $t_{1/2} = 0.693\, T_1$, we have as a rough estimate $T_1 = 0.2/0.693 = 0.29$ s for all of the CH ("methine") carbons. To get more accurate values, the peak heights can be measured and plotted against time, fitting the data to the equation $I = I_{\inf}(1 - 2\,e^{-t/T_1})$. This gives T_1 values of 2.0 s for the quaternary carbon C-f2 and 0.30 s for the anomeric CH carbon C-g1 (Fig. 5.16, inset). There is a very dramatic difference between having no directly bonded protons (C-f2) and having one directly bonded proton (all CH carbons). The directly bonded protons provide a strong oscillating magnetic field to the ^{13}C nucleus as the molecule tumbles, and those molecules tumbling at the Larmor frequency (75 MHz) are stimulated to relax from the β state to the α state. A slight difference in T_1 can be observed between the CH carbons and the CH$_2$ carbons: at 0.2 s the three CH$_2$OH carbons at the upfield end of the spectrum are beginning to rise from the noise as positive peaks, whereas the C-g1 and CHOH carbons are still nulled. This means that $t_{1/2}$ is a bit shorter for the CH$_2$OH carbons, and $T_1 = t_{1/2}/0.693$ is also shorter. This is because two protons at close range are a bit more effective than one at stimulating T_1 relaxation.

Figure 5.17 shows the ^{1}H inversion-recovery experiment for sucrose. Recovery values of 0, 0.1, 0.2, 0.3, 0.5, 0.75, 1.0, and 2.0 s were used, plotting the whole ^{1}H spectrum for each experiment. In the analysis of these data, we use the proton assignments for sucrose derived

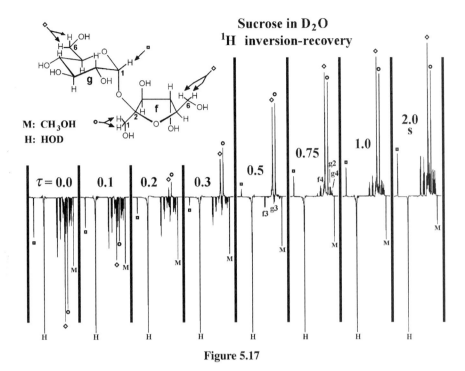

Figure 5.17

from the homonuclear decoupling experiment (see section 5.10 below). The residual water peak (HOD in the D_2O solvent, labeled "H") does not show any recovery at all: at 18 Da it is very small and has a very long T_1. At the right edge of each spectrum is the singlet peak for the added methanol reference (CH_3OH, 32 Da, labeled "M"), which has recovered only about 30% of the way from $-M_o$ to M_o after 2 s, corresponding to a T_1 value of about 6 s ($e^{-2.0/6.0} = 0.72$). The T_1 values for sucrose ($C_{12}H_{22}O_{12}$, 358 Da) are all much shorter, in keeping with the trend for small molecules of decreasing T_1 with increasing molecular size. Site-specific assignments for all of the sucrose resonances are determined below using the homonuclear decoupling experiment. These allow us to look at differences in T_1 relaxation within the sucrose molecule. The fastest relaxing protons are the CH_2OH protons: H-g6 and H-f6 ($\diamond$) and H-f1 (O). These can be seen rising above the baseline in the $\tau = 0.2$ s spectrum as two peaks (H-g6 and H-f6 on the left-hand side, H-f1 on the right-hand side). We can estimate the time they cross zero ($t_{1/2}$) as about 0.18 s, corresponding to a T_1 value of $0.18/0.693 = 0.26$ s. The farthest downfield peak, H-g1 ($\square$), crosses zero between 0.3 and 0.5 s, so if we use $t_{1/2} = 0.4$ s we can estimate $T_1 = 0.4/0.693 = 0.58$ s. In the $\tau = 0.5$ s experiment, we can see that the H-f3 doublet and the H-g3 triplet are negative, whereas the H-f5/g5 multiplet, the H-f4 triplet, the H-g2 double doublet, and the H-g4 triplet are all very close to zero. Thus T_1 can be estimated to be $0.5/0.693 = 0.72$ s for all five of these resonances. H-f3 and H-g3 are the last of the sucrose resonances to cross zero with $t_{1/2}$ of about 0.75 s and $T_1 = 0.75/0.693 = 1.1$ s:

g1	HOD	f3	f4	f5/g5	g6/f6	g3	f1	g2	g4	MeOH
0.58	long	1.1	0.72	0.72	0.26	1.1	0.26	0.72	0.72	6.0 s

Most of these differences can be explained by the number and proximity of nearby protons in the sucrose structure. H-f3 has only one vicinal neighbor, whereas H-g3's two vicinal neighbors are both far away in a fixed *anti* relationship. This may explain their relatively slow relaxation. H-g1 relaxes faster than the "pack" of –CH–O protons (f4, f5, g5, g2, and g4), perhaps due to its proximity to the H-f1 CH_2OH group. This is not evident in the structure diagram, but the strong NOE observed between these two resonances (see below) suggests that a conformation that places them close together may dominate. Finally, the CH_2O protons f1, f6, and g6 relax the fastest because of the very close proximity of a geminal proton in a fixed geometric relationship. In all of these cases, we can see how molecular tumbling rotates the vector between a pair of protons, causing each one to experience an oscillating magnetic field due to the nuclear magnet of the other. For those molecules that happen to be tumbling at the Larmor frequency at any particular moment, transitions are stimulated that allow them to return to the equilibrium (Boltzmann) distribution between the aligned (α) and disaligned (β) states.

What kind of information can we gain from T_1 values? First of all, regardless of any physical interpretation, we need to have some idea of T_1 values to set up any kind of repetitive scanning experiment because the relaxation delay must allow for reestablishing the equilibrium Boltzmann population distribution before starting the next scan. In the real world we usually do not wait as long as $5T_1$, and we get a "steady state" for each spin where it recovers only partially and then gets hit again by the pulse. This means that positions in the molecule with long T_1 values, such as quaternary carbons, will have lower intensity peaks than positions with short T_1. Usually this does not affect proton integration because all protons relax fairly rapidly, but at higher B_o field strengths (600–900 MHz), even the proton T_1 values can be long enough that you need relaxation delays of 2 or 3 s to get accurate integration.

To understand the physical meaning of T_1 values, we need to consider the mechanism of longitudinal relaxation. As we saw above, the relaxation rate depends primarily on: (1) the number, type, and proximity of nearby nuclear magnets (especially the strong ^1H magnets) within the molecular structure and (2) the percentage of molecules tumbling at the Larmor frequency. In a practical sense, a few rules of thumb can be stated. For ^{13}C nuclei, the effect of directly bonded protons far outweighs the effect of any more distant nuclei, so that the relaxation rate depends pretty much on the number of attached protons: CH_3 groups have the shortest T_1 values, followed by CH_2 groups and CH groups. Quaternary carbons, with no directly attached protons, have very long T_1 values because they experience only very weak oscillating magnetic fields and therefore relax very slowly. In fact, the relaxation delay in ^{13}C experiments is dominated by the need to allow time for the quaternary carbons to relax. The majority of protons in organic molecules are bound to ^{12}C rather than ^{13}C, and therefore experience no intense oscillating magnetic fields from directly bound atoms because ^{12}C nuclei are not magnetic. Protons are affected weakly by a large number of other protons in their immediate vicinity. For small molecules, local motions of flexible groups within the molecule can decrease the rotational correlation time, τ_c, leading to longer T_1 values than the rigid portions of the molecule. In a sense, the flexible portions are behaving like smaller molecules (independent pieces) and thus have a wider distribution of reorientation rates and a smaller percentage moving at the Larmor frequency.

Unpaired electrons produce much stronger magnetic fields than nuclei, so in paramagnetic molecules this effect dominates and greatly reduces T_1 values. These strong magnetic fields have a much longer "reach" (20–30 Å), so even paramagnetic ions in solution can increase longitudinal relaxation rates (reduce T_1) of non-paramagnetic molecules. Transition metal ions such as Cr(III) or Cu(II) can be used in this way to shorten ^{13}C T_1 values and speed up data acquisition. In medical NMR imaging, T_1 values of water vary greatly in different tissue types due to differing degrees of association with large biological molecules and membranes, providing contrast in the images. Injectable contrast agents in MRI consist of T_1-increasing paramagnetic ions complexed to ligands with specific affinities for tissues (e.g., tumors).

5.9 CONTINUOUS-WAVE LOW-POWER IRRADIATION OF ONE RESONANCE

The following sections examine three techniques that all use low-power irradiation of a proton resonance in a proton-observe experiment. *Homonuclear decoupling* involves irradiation at a proton peak during the *acquisition* period in order to eliminate a *J* coupling to another proton resonance. This is similar to selective heteronuclear decoupling in that it is used to identify *J* coupling relationships between nuclei. *Presaturation* and *NOE difference* involve irradiation of a signal during the *relaxation delay* period, either to eliminate an unwanted signal (presaturation) or to observe the enhancement of z magnetization (M_z) of protons that are close in space to the irradiated proton (NOE difference). In either case the purpose of the irradiation is "saturation": to equalize the populations of the two energy levels (spin states). All three of these experiments are being replaced by new methods using shaped pulses and pulsed field gradients (Chapter 8): Selective 1D TOCSY for identifying *J* coupling relationships, "Watergate" for solvent suppression, and selective 1D transient NOE for identifying through-space relationships. Later on we will use the two-dimensional (2D) equivalents of the homonuclear decoupling and NOE difference experiments, which are known as COSY and NOESY, respectively (Chapter 9).

We saw in Chapter 1 how continuous low-power irradiation at a single frequency (the exact frequency of one proton resonance in a spectrum) leads to net absorption of energy by those spins. This happens because there are slightly more spins in the lower energy (α) state to absorb "photons" of RF energy and jump up to the higher energy (β) state than there are in the β state to undergo stimulated emission and drop down to the α state. But this absorption of energy rapidly reduces the tiny population difference to zero, at which point the rates of emission and absorption are equal. This state is called *saturation*, and it differs from the situation after a 90° pulse because no coherence is produced and there is no net magnetization at all. Spins are being promoted all the time during the irradiation period and other spins are dropping down at the same rate, unlike the "starting gun" effect of an RF pulse that gets all the spins moving in phase at the same moment. The relatively slow (∼1 s) process of irradiation with low-power (<1 W) RF differs in another way from the rapid and precise rotation that results from a short (tens of μs), high power (50–300 W) pulse of RF energy: the irradiation is fairly selective, affecting only a narrow range of frequencies around the exact frequency of the RF. The lower the RF power level used, the narrower a band of resonances that is saturated. Hence we can use this technique to "wipe out" the population difference (and hence the net magnetization) of just one resonance in the spectrum. Once the saturation condition is established ($\Delta P = 0$), any particular nucleus in the ensemble is cycling between the α and β states, and the average time between transitions depends on the amount of RF power used: for very low RF power, transitions are relatively rare, but for higher power the spins are rapidly bouncing back and forth between the α and β states.

A word about RF power levels would be useful at this point. The amplitude of the RF pulse can be set to a very wide range of values from very weak ("spin tickling") to the high power ("hard") pulses we use to excite all of the spins of a particular type (e.g., ^{1}H) in the sample equally. The 90° pulse "width" is the time required for the pulse to rotate the net magnetization by a 90° angle from its equilibrium position (+z axis) down to the x–y plane. The sample magnetization rotates faster during a pulse of higher amplitude (higher power) than during a "weaker" pulse, so the 90° pulse width depends on the pulse amplitude (the "B_1 amplitude"). One way to talk about power levels, then, is to simply specify the 90° pulse width at that power level. You could say, "set the RF power level of the pulse so that the 90° pulse width for ^{1}H is 100 μs" and anyone in the world on any spectrometer could duplicate that power level, without reference to volts or watts or any electronic measurement. We let the spins do the measuring. Having said this, we can give you an idea of the power levels used for CW irradiation of protons. For NOE difference, a typical power level would give a 30 ms 90° pulse, or 3000 times lower RF pulse (B_1) amplitude than a "hard" (10 μs) 90° pulse. This is very low power, but we will be applying it for much longer (typically 1.0 or 1.5 s) than the hard pulse (10 μs). For homonuclear decoupling we need to have spins bouncing back and forth rapidly between the α and β states, and this typically requires 10 times higher power, corresponding to a 90° pulse width of 3 ms. This is still 300 times lower RF amplitude than the hard pulse. For presaturation of 90% H$_2$O, a typical power level corresponds to a 90° pulse width of 6 ms, or 600 times lower pulse amplitude than the hard pulse. For waltz-16 *heteronuclear* decoupling, which must "cover" the entire ^{1}H chemical shift range (0–10 ppm) rather than just a single peak, and must overcome a much large J value ($^1J_{CH} \sim 150$ Hz), a power level is used that corresponds to a 100 μs 90° pulse, only 10 times lower amplitude (100 times lower power) than the hard pulse.

Experiment	Relative B_1 amplitude	Relative RF power	90° Pulse
"hard" pulse	1	1	10 μs
Hetero-dec. (waltz-16)	1/10	1/100	100 μs
Homo-decoupling	1/300	1/90,000	3 ms
Presaturation (90% H$_2$O)	1/600	1/360,000	6 ms
NOE Difference	1/3000	1/90,000,000	30 ms

The important point is that a much longer duration, much lower power pulse can give selectivity of excitation, affecting only one resonance in the ^{1}H spectrum if it is sufficiently separated from other resonances in chemical shift. Very little RF power is required to saturate a resonance (equalize populations), but a bit higher power is necessary for decoupling because the spins must rapidly shuttle between the two spin states to give the "blurring" effect which makes other spins blind to the J coupling phenomenon. Nonselective or "broadband" decoupling (waltz-16 heteronuclear decoupling) requires much higher power levels, near the limits set by amplifier and sample heating.

5.10 HOMONUCLEAR DECOUPLING

This traditional 1D NMR experiment involves selectively decoupling a proton from its *J* coupling partners. This is accomplished by low-power irradiation at the frequency of the peak of interest in the ^{1}H spectrum during the acquisition of the FID. The power level required is a bit higher because the presaturation and NOE experiments need only enough power to *equalize* populations, whereas the decoupling experiment requires enough power to rapidly (relative to the "NMR timescale" $1/(2.2\,J)$) flip each spin at the selected frequency back and forth between the upper and lower energy levels. As with heteronuclear decoupling, the rapid flipping on the NMR timescale means that other protons that are J coupled to this spin see an average magnetic field that is unaffected by the orientation (α or β state) of the spin being irradiated. The selected spin is thus removed from the coupling network, and any peak in the spectrum that is coupled to it will be simplified by removal of that one splitting. For example, a doublet will become a singlet if the J coupling is from the peak being irradiated, and a double doublet will become a doublet. In the latter case, the coupling that is removed can be unequivocally assigned to the proton being irradiated and the remaining coupling must be due to some other proton in the molecule. In some cases the decoupling power is insufficient to completely remove the coupling and the apparent J value is reduced rather than being eliminated. This is the same as the heteronuclear case where the reduced J value (J_R) is a function of decoupler field strength (Chapter 4, Section 4.4.2). For example, a triplet might be changed into a double doublet with one small coupling due to the proton being irradiated. In general, any change in a peak's coupling pattern can be interpreted as a J coupling to the peak being irradiated.

The results of a homonuclear decoupling experiment on sucrose are shown in Fig. 5.18. The experiment is set up by acquiring a normal ^{1}H spectrum and determining the exact RF frequency of each peak (each resonance) we wish to "test" by CW irradiation during the acquisition of the FID. The desired frequencies are actually offsets from a fundamental decoupler frequency; for example, an offset (Bruker: *o2* for channel 2 offset; Varian: *dof* for decoupler offset) of 132.6 Hz is added to the fundamental frequency (Bruker: *BF2*

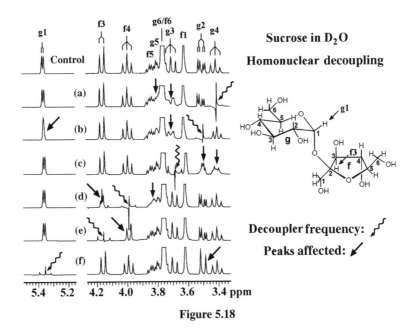

Figure 5.18

for base frequency, channel 2; Varian: *dfrq* for decoupler frequency) of 299.956 MHz to get the exact frequency of the decoupler, 299.9561326 MHz (299.956 + 0.0001326). The experiment is run as a series of ^{1}H acquisitions yielding a series of FIDs, each using a different decoupler offset. In Fig. 5.18, the top spectrum is a control spectrum using a decoupler frequency away from any of the peaks, and the other spectra (a)–(f) are acquired with different peaks selected. On the Varian, the *dof* values are loaded into an array by typing in the values: *dof* = 1536.7, 467.8, 258.4, and so on. On the Bruker, a separate text file is created, a *frequency list*, in which the *o2* values are entered one line at a time.

> From a hardware standpoint, homonuclear decoupling is more challenging than presaturation or NOE difference because we need to do the CW irradiation at the *same time* that we are acquiring FID data. This is accomplished by shutting off ("gating") the decoupler RF for a brief period while each FID data point is being observed and digitized. Thus the send/receiver (or T/R) switch is very busy going back and forth between transmitting the decoupling CW RF signal and "listening" to the FID.

For sucrose, the experiment allows us to assign all of the peaks in the ^{1}H spectrum. As always, we have to start with some prior knowledge, based on a unique chemical shift or coupling pattern. For the glucose part, we have H-g1, the only anomeric proton, which is farthest downfield because it is bonded to a carbon with two bonds to oxygen. For the fructose part, we have H-f3, which is the only doublet peak (besides H-g1) because it has a quaternary carbon on one side (C-f2). From these two pieces of information we begin the process of assignment. Irradiation of H-g1 (Fig. 5.18(f)) converts the double doublet at 3.5 ppm into a doublet. Thus the double doublet represents H-g2, and the remaining large doublet coupling is the *J* coupling between H-g2 and H-g3. The smaller coupling of the H-g2 double doublet is the small (axial-equatorial) coupling observed in the H-g1 doublet (3.8 Hz). Irradiation of H-g2 (Fig. 5.18(b)) causes the triplet at 3.7 ppm to "collapse" into a broad doublet, as well as converting the H-g1 doublet into a broad singlet. Thus the triplet at 3.7 ppm is H-g3. Irradiation at this position in turn causes (Fig. 5.18(c)) the triplet at 3.4 ppm to collapse to a sort of ugly doublet, as well as simplifying the H-g2 double doublet into a narrow doublet (glucose J_{1-2} remains). Now we select H-g4 (Fig. 5.18(a)) and see the H-g3 triplet simplify to a broad doublet (3.7 ppm) and a very subtle change in the overlapped region at 3.8 ppm, which is evident if you compare to the control spectrum directly above it. So we can guess that the H-g5 resonance is buried in overlap at about 3.8 ppm. This is the end of the trail for the glucose part. Now pick the H-f3 peak to irradiate at 4.16 ppm (Fig. 5.18(e)). The triplet at 3.99 ppm collapses very neatly to a doublet, proving that this resonance is H-f4. No other peak is changed, confirming that H-f3 is at the end of the "spin system," next to a quaternary carbon. Irradiating H-f4 (Fig. 5.18(d)) we see a subtle change in the messy region at 3.84 ppm as well as the collapse of the H-f3 doublet to a narrow pattern resembling the superposition of a doublet and a singlet. The H-f5 proton can be assigned to an overlapped peak at 3.84 ppm. The only protons that remain to be assigned are the C$\underline{\text{H}}_2$OH protons H-f6, H-f1, and H-g6. The singlet at 3.62 ppm integrates to 2 protons, so it must be H-f1. H-f6 and H-g6 should have couplings to the proton at position 5. In addition, the NOE difference spectrum (see Fig. 5.30, below) clearly shows an NOE from this singlet peak to H-g1 (across the glycosidic linkage) and to H-f3 within the fructose unit. By process of elimination, the large overlapped peak at 3.76 ppm must be both H-f6 and H-g6.

Homonuclear decoupling is pretty much a historical experiment, as newer experiments such as 1D-TOCSY (Chapter 8) and 2D-COSY (Chapter 9) have replaced it.

5.11 PRESATURATION OF SOLVENT RESONANCE

Often the solvent gives rise to a very large peak in the spectrum that interferes with observation of the solute peaks. For example, many biological samples are run in 90% H_2O/10% D_2O so that exchangeable peptide amide and nucleotide base N–H resonances can be observed (the 10% D_2O is for locking). In this case the water protons are present at a concentration of about 100 M (55 M × 2 × 0.90) and the solute may be at 1 mM or less—a factor of 10^5 difference in concentration. Without some strategy to suppress the solvent signal, the solute peaks in the spectrum will be very difficult to observe. The receiver gain would have to be turned down dramatically so that the FID, which is dominated by the water signal, will not overflow the digitizer. At a low receiver gain setting (low amplification of the FID signal) the signal-to-noise ratio of the solute peaks is greatly reduced. Furthermore, the solute portion of the FID can be lost in the accuracy limits of the digitizer because the solvent signal is now filling the range of the digitizer. For example, with a 12-bit digitizer a signal that is 0.024% ($100\%/2^{12}$) or less of the maximum signal cannot be represented in digital form because the entire signal is less than the least-significant digit (bit) of the digitizer. A 1 mM solution in 90% H_2O corresponds to a 0.001% solute signal, or 24 times smaller than the smallest digital "currency" of the digitizer, if the water signal is just filling the digitizer. A 16-bit digitizer improves the situation, but still provides only two thirds of a bit to digitize the solute signal! Without active suppression of the water peak, you will not see the solute at all.

There are many methods for suppressing a strong solvent signal, but we will consider here only the simplest: presaturation. This technique involves irradiating at the precise solvent frequency with a long (∼1 s) low-power signal to saturate (equalize populations of) the solvent protons. Then a normal high-power (nonselective or "hard") pulse is immediately delivered to excite the solute nuclei and obtain an FID. The solvent protons have no population difference ($M_z = 0$) and no coherence ($M_x = M_y = 0$) at the time of the high-power pulse and therefore produce no magnetization in the x–y plane and no signal in the FID.

Presaturation (as well as homonuclear decoupling and NOE difference) requires that you have two power levels of RF available at the proton frequency: high power (e.g., 30–50 W) for the 90° nonselective "hard" pulse, and low power (<1 W) for continuous irradiation. In newer models of spectrometers, this is done simply by switching the power level (attenuation) of a single RF source from low power (during presaturation) to high power (during the "read" pulse). Older spectrometers cannot switch power levels rapidly, and some even use mechanical relays for power switching. The repeated switching of a relay with every scan would burn it out in the course of a few experiments. For this reason, the traditional way of presaturating the solvent resonance involves using the proton decoupler (which produces low power RF at the ^{1}H frequency during ^{13}C experiments) to deliver the low-power irradiation, and the broadband "transmitter" (the usual source of high-power pulses in both ^{1}H and ^{13}C experiments) to generate the "read" pulse. These two signals are electronically combined and delivered to the probehead. This allows the frequency for presaturation to be set independently from the observe frequency (the pulse and reference frequency—center of the spectral window). In a practical sense, this means that the solvent resonance does not have to be placed at the center of the spectral window. The

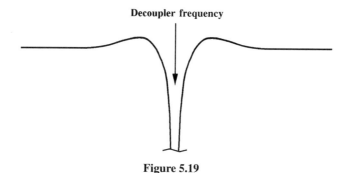

Figure 5.19

same hardware setup is used for NOE difference and homonuclear decoupling experiments. Current spectrometers can switch *both* power level *and* frequency of the ^{1}H channel in a few μs or less without mechanical relays, so there really are no longer any hardware issues for CW irradiation experiments.

Presaturation is very precise and cuts a razor-sharp swath out of the spectrum, but there will be some attenuation of peaks near the solvent peak. For 90% H_2O (10% D_2O) samples typically used in biological NMR, the method is very demanding because if the shimming is not perfect, there will be broad components at the base of the water peak and these will not be removed by the narrow slice of the presaturation. This leads to broad "humps" at the water frequency that can still be very large compared to the solute peaks, usually caused by higher order shim problems (Z^4, Z^5, and Z^6: see Chapter 3, Fig. 3.5). With very good shimming, the water signal appears as a sharp, negative dip between two humps (Fig. 5.19). Since even with the best shimming the FID is dominated by the water signal, the quality of water suppression can be measured by how high the receiver gain can be increased without overloading the digitizer. Better water suppression means a less intense FID signal in the probe coil, and more amplification is possible before reaching the digitizer. Good presaturation should allow a receiver gain of 64 (Bruker) or 22 (Varian). Solute signals, such as amide H_N protons that exchange with water protons, can be strongly attenuated ("bleached") because these protons spend some of their time at the water resonance. Depending on the rate of exchange with water, protons bound to nitrogen may be slightly attenuated or completely wiped out. For this reason, solvent saturation methods involving gradients and selective pulses (e.g., Watergate, Chapter 8) have largely replaced "presat." It is still useful as a very demanding test of the experimental setup: if the presat spectrum looks good, the shimming must be excellent!

Figure 5.20 shows the 500 MHz presaturation ^{1}H spectrum of a cyclic octapeptide in 90% H_2O/10% D_2O. H_2O is used so that the exchangeable amide protons (CO-N$\underline{H}$) can be observed; in D_2O they would be replaced with deuterium. The tiny remaining H_2O peak is seen as a sharp spike at 4.6 ppm with a broad signal around it. The tryptophane NH is seen as a singlet at 10 ppm, and the region from 7.5–9.0 ppm contains most of the amide NH resonances, as well as the histidine NH at 8.4 ppm. The aromatic protons fall in the region of 7.0–7.5 ppm, and the eight $C_\alpha H$ protons and the threonine $C_\beta H$ (C$\underline{H}$(OH)CH$_3$) are near the water at 3.5–4.5 ppm. The 2.2–3.5 ppm region contains the other $C_\beta H$ protons and the lysine $C_\delta H$ (C$\underline{H_2}$-NH$_3^+$), and the farthest upfield region at 0.5–1.5 ppm has the $C_\gamma H$ protons and the methyl groups. The $C_\alpha H$ protons near the water are somewhat attenuated, as are some of the amide NH protons that exchange more rapidly with water.

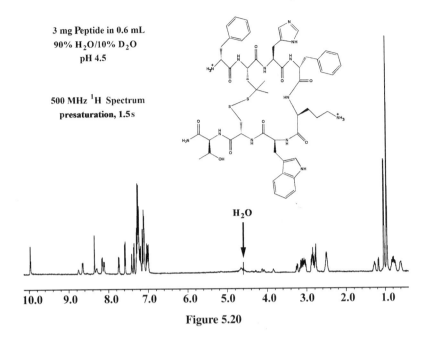

Figure 5.20

5.12 THE HOMONUCLEAR NUCLEAR OVERHAUSER EFFECT (NOE)

In the previous chapter the NOE was introduced but not explained in detail. Why should equalization of the populations of one spin cause an enhancement of the population difference of a nearby spin? The key to understand this effect is to take a close look at the longitudinal (T_1) relaxation process, now in the context of two nearby protons within one molecule. Consider two nonequivalent protons H_a and H_b in an organic molecule, with a distance between them of less than 5 Å. If we saturate nucleus H_a by selective low-power irradiation, it will be out of equilibrium because the populations in the α (lower) and β (upper) energy levels will be equal. From the point of view of H_a, H_b will be rotating around it at the tumbling rate of the molecule. For those molecules that are tumbling at a rate very close to the Larmor frequency (e.g., 300 MHz in a 7.05 T magnet), H_a will experience an oscillating magnetic field that will stimulate it to drop from the β level to the α level. This is the major mechanism for T_1 relaxation; if it were the only mechanism, there would be no nuclear Overhauser effect.

How can relaxation lead to an NOE? There is another pathway for relaxation that involves *both* H_a and H_b changing spin state *simultaneously*. For example, if both H_a and H_b are in the β state (overall state $\beta\beta$), both can flip simultaneously to the α state (overall state $\alpha\alpha$). This is called a *double-quantum* transition, and it is sensitive to magnetic fields oscillating at *twice* the Larmor frequency (600 MHz in our case). Small organic molecules (MW < 1000) tumble rapidly and have significant populations tumbling at both the Larmor frequency and twice the Larmor frequency (Fig. 5.13), so this "cross-relaxation" (simultaneous spin flip) is a significant relaxation pathway. Consider an extreme case where double-quantum relaxation is the *only* pathway available: after saturation of H_a, every time an H_a spin drops down from the β state to the α state it drags an H_b spin (in the same molecule) down with it. This is fine for the H_a spins—they are reestablishing the Boltzmann distribution by increasing

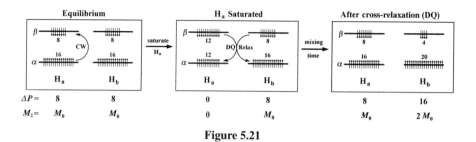

Figure 5.21

P_α and decreasing P_β—but as the H_b spins drop down they are actually "pumping up" the H_b population difference *beyond* the equilibrium difference! In Fig. 5.21 (center) we start the relaxation process ("mixing time") with $\Delta P(H_a) = 0$ and $\Delta P(H_b) = 8$. If four of the H_a spins drop down from β to α we get back to the equilibrium population difference of $\Delta P(H_a) = 8$, but we also drag four of the H_b spins down from the β state to the α state, giving a population difference of $\Delta P(H_b) = 16$, or twice the equilibrium population difference! This means that M_z for the H_b resonance has increased from M_o ($\Delta P = 8$) to $2M_o$ ($\Delta P = 16$) as a result of this cross-relaxation process. If we rotate the net magnetization into the x–y plane with a 90° pulse and record a spectrum at this point, we would see the H_b peak twice as large as it would be if we had not saturated H_a first. This is an exaggerated version of the simple NOE experiment for small molecules ($\omega_o \tau_c < 1$): preirradiation of one resonance leads to enhancement of the ^1H peak for another resonance representing a proton that is nearby (<5 Å) in space.

While we are looking at this figure, consider another extreme where the only route for relaxation is the zero-quantum pathway, whereby H_a drops down from the β state to the α state while "dragging up" an H_b spin from the α state to the β state. This transition is stimulated by molecules tumbling at the zero-quantum frequency, which is just the difference in frequency between the two spins: $\nu_o(H_a) - \nu_o(H_b)$. This frequency is very low in the audio frequency range for two protons, and for large molecules there are significant numbers of molecules tumbling at this rate (Fig. 5.13). Again we start (Fig. 5.22, center) with $\Delta P(H_a) = 0$ (saturation) and $\Delta P(H_b) = 8$ (equilibrium) and allow H_a spins to drop from the β state to the α state, each time *pulling up* an H_b spin from the α state to the β state. By the time H_a has reached equilibrium (Fig. 5.22, right: $\Delta P = 8$), we have destroyed the population difference for H_b ($\Delta P = 0$). The effect of irradiating H_a has been to *reduce* the intensity of the H_b peak in the final spectrum. This is the case for large molecules ($\omega_o \tau_c > 1$). This might seem like a stupid thing to do, but we will see that it is the *difference* in peak intensity

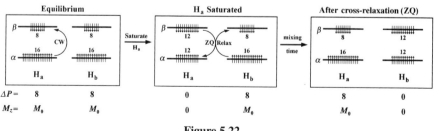

Figure 5.22

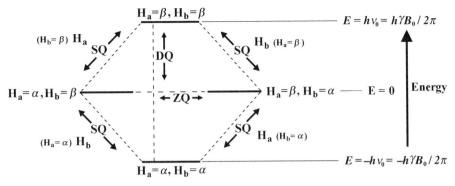

Figure 5.23

compared to a normal ^{1}H spectrum that defines the NOE, and this difference can be either an increase or a decrease. Either way it can be measured and quantified and interpreted in terms of a close approach (<5 Å) of two protons.

Although this is a useful exercise, it is not really accurate to draw two energy diagrams—one for H_a and one for H_b. Because the two spins are interacting in a single molecule, we use an energy level diagram with *four* spin states for the two nuclei H_a and H_b: one with both spins in the α state ("$\alpha\alpha$"), one with both spins in the β state ("$\beta\beta$"), one with H_a in the α state and H_b in the β state ("$\alpha\beta$") and one with H_a in the β state and H_b in the α state ("$\beta\alpha$"). In an energy diagram the $\beta\beta$ state is highest in energy, the $\alpha\alpha$ state is lowest, and the $\alpha\beta$ and $\beta\alpha$ states are in the middle at the same energy level (Fig. 5.23). The transitions can be identified as four "single-quantum" transitions, where only one of the protons changes state while the other remains the same ($\alpha\beta \rightarrow \alpha\alpha$, $\beta\alpha \rightarrow \alpha\alpha$, $\beta\beta \rightarrow \alpha\beta$, $\beta\beta \rightarrow \beta\alpha$), a "double-quantum" transition, where both spins drop down (or move up) simultaneously ($\beta\beta \rightarrow \alpha\alpha$), and a "zero-quantum" transition where one spins drops down and the other spin moves up ($\alpha\beta \rightarrow \beta\alpha$). The single-quantum transitions are observable, so they can be associated with the peaks in the spectrum. The lower right and upper left SQ transitions give rise to the H_a peak in the spectrum because only H_a undergoes a transition and H_b does not change. The lower left and upper right SQ transitions give rise to the H_b peak in the spectrum, with H_b changing its spin state and H_a doing nothing. Note that the energy difference for any SQ transition is just the energy difference for one proton to change state: $\Delta E = h\nu_0 = h\gamma B_0/2\pi$. The energy difference for a DQ transition is twice this amount, and the energy difference for a ZQ transition is zero (actually it is $h[\nu_a - \nu_b]$ due to the slight difference in Larmor frequencies of H_a and H_b). We will be using this energy diagram throughout the book whenever we discuss populations and NOE interactions in a two-spin homonuclear (two proton) system.

If all four energy states had the same population, they would each have N/4 molecules (or N/4 "spin pairs" H_a–H_b). But at thermal equilibrium the upper spin state ($\beta\beta$) will lose a few molecules to the middle states, leaving a slightly depleted population of $N/4 - 2\delta$ (represented by two open circles in Fig. 5.24). The two middle levels ($H_a = \alpha$, $H_b = \beta$ and $H_a = \beta$, $H_b = \alpha$) will initially gain 2δ molecules from $\beta\beta$ but will lose these to the lower state $\alpha\alpha$, leaving them with populations of $N/4$ (represented by the absence of any circles). Finally, the lower spin state ($\alpha\alpha$) will gain 2δ molecules from the middle states to get a slightly augmented population $N/4 + 2\delta$ (represented by two filled circles). The

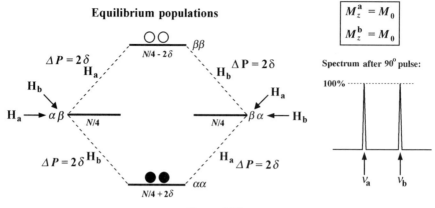

Figure 5.24

populations can be derived from the Boltzmann distribution using the energies: zero for the $\alpha\beta$ and $\beta\alpha$ states, $+\Delta E$ for the $\beta\beta$ state and $-\Delta E$ for the $\alpha\alpha$ state, where ΔE is the energy gap of a single proton transition. But it should make sense that the two middle states ($\alpha\beta$ and $\beta\alpha$) have the same population because they are equal in energy, whereas the upper $\beta\beta$ state is depleted a bit and the lower $\alpha\alpha$ state's population is enhanced by the same amount: 2δ (Fig. 5.24). Note that the population difference across each of the observable transitions is exactly 2δ, with the lower energy level of each transition having the greater population. The H_a transitions are those in which H_a changes from α to β or from β to α without any change in H_b, and the H_b transitions are labeled where H_b changes and H_a remains the same. This is the population diagram corresponding to a sample that has been placed in the magnetic field and been given time to come to thermal equilibrium. Because z magnetization (M_z) is proportional to the population difference and is equal to M_o at equilibrium, we can associate this single-quantum population difference 2δ with a z magnetization of $+M_o$. Thus if we perturb the system at this moment with a 90° pulse and record an FID, the H_a net magnetization vector will be rotated into the x–y plane where it will precess according to its Larmor frequency, ν_a, and induce an oscillating and decaying voltage in the FID. The same thing will happen to the H_b net magnetization, which will induce a decaying voltage corresponding to a slightly different frequency, ν_b, in the FID. Fourier transformation of this FID will give a spectrum with two peaks, one at frequency ν_a and one at frequency ν_b. The height of each of these two peaks will be taken as our "control" experiment: 100% because we started with an equilibrium distribution of populations (Fig. 5.24, right). By executing a 90° pulse, recording the FID and doing the Fourier transform we effectively "read out" the population differences across the single-quantum transitions. Any deviation from the equilibrium population difference (2δ) across any of the SQ transitions will change the z magnetization just before the pulse, which will change the length of the vector in the x–y plane after the pulse and change the height of the peak in the resulting spectrum. For example, if the populations are equal across the H_a transitions ($\Delta P = 0$), the H_a peak will disappear from the resulting spectrum. If, on the contrary, the populations are inverted across the H_b transitions ($\Delta P = -2\delta$), the H_b peak in the spectrum will be upside-down (-100% peak height).

Now let's us have some fun by perturbing this equilibrium population distribution! Selective saturation of the H_a transitions ($\alpha\alpha \rightarrow \beta\alpha$ and $\alpha\beta \rightarrow \beta\beta$) by low-power

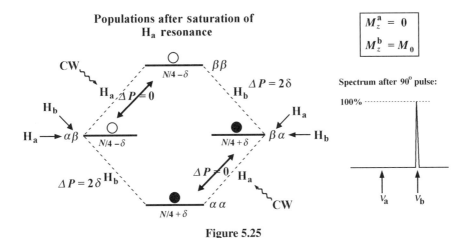

Figure 5.25

continuous irradiation at frequency ν_a (Fig. 5.25) will promote exactly δ H_a spins in each transition, decreasing the $\alpha\alpha$ state population from $N/4 + 2\delta$ to $N/4 + \delta$ (one filled circle) and increasing the $\beta\alpha$ state population from $N/4$ to $N/4 + \delta$ (one filled circle). Likewise for the other H_a transition (upper left) the $\alpha\beta$ state population is decreased from $N/4$ to $N/4 - \delta$ (one open circle) and the $\beta\beta$ state population is increased from $N/4 - 2\delta$ to $N/4 - \delta$ (one open circle) as exactly δ spins are promoted to the higher level. At this point, the population differences across the H_b transitions ($\alpha\alpha \rightarrow \alpha\beta$ and $\beta\alpha \rightarrow \beta\beta$) are still all exactly equal to 2δ, but the H_a transitions have no population difference. If we were to execute a 90° pulse and collect the FID at this point we would see a normal peak for H_b ($\Delta P = 2\delta$ for both of the H_b transitions before the pulse, so $M_z^b = M_o$) and no peak at all for H_a ($\Delta P = 0$ for both of the H_a transitions before the pulse, so $M_z^a = 0$). This makes sense because we have saturated the H_a spins selectively without affecting the H_b spins.

Now suppose that double-quantum relaxation ($\beta\beta \rightarrow \alpha\alpha$) is the only mechanism of relaxation. Of course, simple one-nucleus relaxation is also going on but because it does not lead to an NOE we will ignore it. At this point, the population difference between the $\alpha\alpha$ and the $\beta\beta$ states is just 2δ ($N/4 + \delta$ vs. $N/4 - \delta$). Because the equilibrium population difference between the $\beta\beta$ and $\alpha\alpha$ states is 4δ (Fig. 5.24), we will see δ molecules drop down from $\beta\beta$ to $\alpha\alpha$ to restore the equilibrium. The result is shown in Figure 5.26: the $\beta\beta$ state population has been reduced to $N/4 - 2\delta$ and the $\alpha\alpha$ state population has been increased to $N/4 + 2\delta$. Note that the H_a transitions are now halfway back to their equilibrium distribution (population difference of δ *versus* equilibrium difference of 2δ). This corresponds to a net z magnetization of $M_z^a = M_o/2$. More importantly, the H_b transitions now have a population difference even greater than equilibrium (3δ *versus* an equilibrium population difference of 2δ). The net z magnetization of the H_b spins is therefore $M_z^b = 3M_o/2$, or 50% enhanced from its equilibrium value. Thus the process of saturation of H_a followed by allowing time for H_a to relax, with double-quantum relaxation predominating, leads to a 50% NOE enhancement of H_b. Note that as our model becomes more accurate, the theoretical maximum NOE enhancement drops!

Exercise: Go through the same thought experiment with relaxation occurring to completion only by the zero-quantum pathway ($\alpha\beta \leftrightarrow \beta\alpha$, the dominant pathway for a large molecule).

192 NMR RELAXATION—INVERSION-RECOVERY

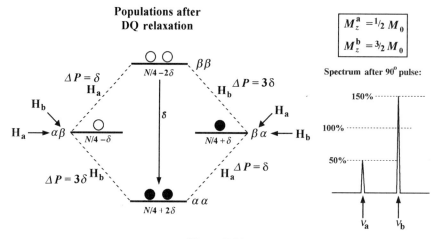

Figure 5.26

What is the equilibrium population difference between $\alpha\beta$ and $\beta\alpha$? After cross-relaxation, what is the percentage change in the z magnetization of H_b, and is it increased (enhanced) or decreased?

Because z magnetization is not observable, we need to convert it to observable (x–y plane) magnetization in order to measure the NOE enhancement. At this point a 90° hard pulse on all protons will yield a spectrum in which the H_a peak is reduced to 50% of its normal intensity and the H_b peak is enhanced by 50% of its normal intensity (Fig. 5.26, right). Note that the amount of z magnetization gained by the H_b nuclei is exactly equal to that lost by the H_a nuclei. We can say that the NOE experiment has *transferred z magnetization* from the H_a to the H_b nuclei. This concept of *magnetization transfer* from one nucleus to another is the key to understanding all 2D NMR experiments. Actual NOEs are quite a bit less than 50% because cross relaxation by this pathway ($\beta\beta \rightarrow \alpha\alpha$) is not the only relaxation pathway available. Also, the 1D NOE difference experiment does not instantly saturate the H_a resonance and then stop irradiation and wait for cross relaxation to happen. Instead, cross relaxation and normal relaxation are happening as the H_a transitions are being irradiated, so that eventually a steady state is reached when all of these processes are going on simultaneously and the population levels are constant. It is this steady-state population distribution that is then sampled by the 90° "read" pulse. The enhancement in the H_b signal observed is referred to as the "steady state NOE," and it is typically in the range of a fraction of a percent to 10%.

Our crude picture of the distribution of tumbling rates explains why zero-quantum relaxation dominates for large molecules because they have negligible populations tumbling at the SQ (single-quantum or T_1 relaxation) frequency (ν_o) or the DQ frequency ($2\nu_o = \nu_a + \nu_b$) and large numbers tumbling at the ZQ frequency ($\nu_a - \nu_b$). Effectively the SQ and DQ frequencies are above the cutoff tumbling frequency for these large molecules (Fig. 5.13). But for small molecules it looks like ZQ, DQ, and SQ would all have the same rates because all three frequencies lie in the flat region of the tumbling rate distribution. But there are inherent differences in the efficiency of ZQ and DQ relaxation relative to SQ relaxation, and a more detailed mathematical analysis gives the following relative rates for ZQ, SQ, and DQ relaxation:

	ZQ ($\alpha\beta \to \beta\alpha$)	SQ ($\beta\alpha \to \alpha\alpha$, etc.)	DQ ($\beta\beta \to \alpha\alpha$)
Small molecule: ($\omega_o\tau_c \ll 1$)	2	3	12
Large molecule: ($\omega_o\tau_c \gg 1$)	2	$\ll 1$	$\ll 1$

Each of these numbers is multiplied by $1/r^6$, reflecting the distance dependence of the dipole–dipole interaction. Now we see that double-quantum relaxation does in fact dominate the dipole–dipole relaxation of small molecules, and our cartoon model of relaxation exclusively by the DQ pathway during the mixing time is not that far off. Likewise, the assumption that only ZQ relaxation occurs for large molecules (see exercise above) is also qualitatively correct.

5.12.1 NOE Difference

The NOE interaction between two protons in a molecule can conveniently be measured by applying a low power, continuous-wave RF irradiation at the exact resonance frequency of one nucleus (i.e., one peak in the spectrum: H_a) for a period of seconds (usually during the entire relaxation delay). During the irradiation period, the NOE builds up at another nucleus that is close (<5 Å) in space (H_b), meaning that its M_z value increases by a few percent beyond M_o (Fig. 5.27). Eventually a steady state is reached and the M_z values of nearby nuclei do not increase any more, whereas the M_z of the irradiated nucleus remains near zero. At the end of the irradiation period (also called the "mixing time"), a 90° high power pulse is applied to excite all of the nuclei in the sample, rotating all z magnetization down to the x–y plane where it precesses and generates signals in the FID. The nucleus being irradiated (H_a) gives essentially no signal because its M_z was near zero at the start of the 90° pulse, and the nuclei that are close to it in space (e.g., H_b) give signals that are enhanced by a few percent over their normal peak intensities. This difference in peak intensity is often difficult

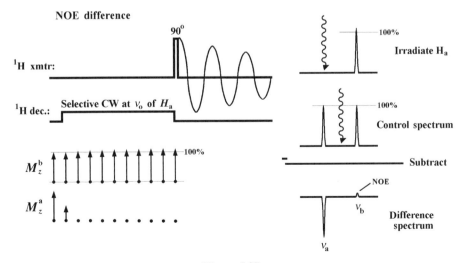

Figure 5.27

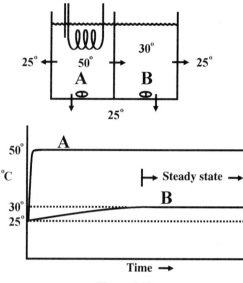

Figure 5.28

to detect directly, so normally a reference spectrum is collected in which the irradiation occurs in a region of noise rather than on any peak in the spectrum. This reference spectrum (which should be identical to a normal 1D spectrum) is then mathematically subtracted from the NOE spectrum at every point to yield a *difference spectrum*. In the difference spectrum, the peak that was irradiated will be upside down (zero minus a normal peak), peaks that show no NOE enhancement will be missing (normal minus normal), and peaks that show an NOE will appear as weak positive peaks (enhanced minus normal). This method is called the *NOE difference* experiment.

An analogy to heat flow helps to explain how the experiment works. Consider a container of water divided into two compartments by a glass partition (Fig. 5.28, top). The left-hand side represents proton H_a and the right-hand side represents a nearby proton H_b. At the start of the experiment (equilibrium), both compartments are at 25 °C, the temperature of the surrounding water bath (the "lattice" temperature). Then we rapidly heat the left-hand compartment (H_b) to 50 °C and hold it constant at that temperature for several minutes. During this time, heat flows from compartment A through the glass sides of the container to the environment (H_a's T_1 relaxation), but we keep adding heat to maintain the temperature at 50 °C (saturation: continuous input of RF energy). Some heat also flows through the partition to compartment B (cross-relaxation) and raises the temperature of compartment B (NOE enhancement of H_b). As the temperature of compartment B rises, heat begins to flow through the sides to the environment (H_b's T_1 relaxation), limiting the rise in temperature of compartment B to a small amount. Eventually we reach a steady state where the temperature in compartment B is no longer rising and the heat flow from compartment A (cross relaxation) exactly equals the heat loss to the environment (H_b's T_1 relaxation). The temperature in compartment B is steady at 30 °C (M_z of H_b is steady at 1.10 M_o). At this point we measure the temperature in compartment B (90° pulse to "read" the H_b z magnetization: the H_b peak is enhanced by 10%).

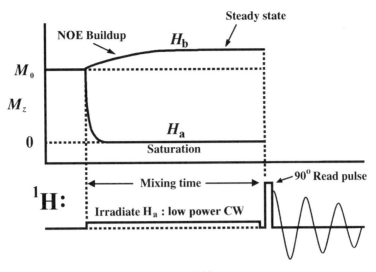

Figure 5.29

The time course of these changes is shown in at the bottom of Fig. 5.28 for the heat flow analogy. The temperature of compartment A rises very quickly from 25 to 50 °C, and then the temperature of compartment B rises slowly to reach its steady-state value of 30 °C (the "mixing time"). From the point of view of z magnetization, the CW irradiation quickly reduces M_z of H_a from M_o to zero (Fig. 5.29). During the mixing time, the M_z of H_b "builds up" from the equilibrium value of M_o to the steady-state enhanced value of $1.10\,M_o$. When the steady state is reached, we sample the z magnetization with a 90° pulse, recording an FID. The spectrum will show no peak for H_a ($M_z^a = 0$ before the pulse), and a 10% taller than normal peak for H_b ($M_z^b = 1.10 M_o$ before the pulse). Note that for small molecules (i.e., molecules for which $\omega_o \tau_c \ll 1$) the effect of "heating up" H_a is actually to "cool down" H_b, moving spins from the β state down to the α state and increasing the population difference above the equilibrium value of 2δ. For this reason the small molecule NOE is often referred to as a "negative" NOE. For large molecules, the effect is opposite: heating up H_a leads to a loss of z magnetization (heating up) for H_b. We call this a "positive" NOE. For "medium-sized" molecules ($\omega_o \tau_c \sim 1$, MW ~ 1000 Da) the sign of the NOE crosses zero and we see no NOE at all! We will see later on that there are other experimental techniques available to get around this problem.

Any change in the experimental parameters, such as temperature, B_o field (lock feedback loop variation), RF power level or phase coming from the amplifiers, or vibration, will degrade the subtraction process because we are looking for very small differences in peak intensities. For this reason, the NOE spectrum and the control spectrum are usually collected in an *interleaved* manner: for example, eight scans with the decoupler frequency (Varian *dof*, Bruker *o2*) set to the resonance of interest, eight scans with the decoupler set to a region of noise, and then repeating this sequence as many times as required for the desired signal-to-noise ratio. Each set of eight scans is added into the appropriate "NOE" or "control" FID. In this way, the two spectra are acquired essentially simultaneously and subtraction artifacts are minimized. The process can be expanded to include each peak in the ^{1}H spectrum and a single control frequency, giving a map of all NOE interactions in the

196 NMR RELAXATION—INVERSION-RECOVERY

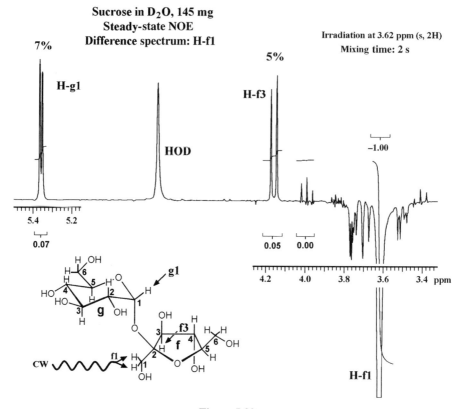

Figure 5.30

molecule, which identifies all of the close approaches of one proton to another (distances over 5 Å are generally too weak to be detected as an NOE). This NOE correlation map can also be obtained by the two-dimensional NOESY experiment (Chapter 10). Artifacts are a big problem with the NOE difference experiment, as with any subtraction experiment, and currently the much cleaner transient NOE experiment is used with selective (shaped) pulses and pulsed field gradients (PFGs). This technique will be discussed in Chapter 8.

Figure 5.30 shows the steady-state NOE difference spectrum of sucrose in D_2O, selecting the H-f1 singlet at 3.62 ppm for saturation. We can see that the crowded region around H-f1 is also affected to some extent, giving small negative peaks for H-g3 (triplet just downfield of H-f1), for H-g6 and H-f6 (a bit further downfield) and for H-g2 (just upfield of H-f1). Because these protons are partially saturated, they will also give NOE enhancement to the peaks representing protons near them in space. This makes the data harder to interpret, so it is best to select peaks that are clearly resolved and far from any other peaks in chemical shift. In this case, the small negative peaks are so weak compared to the negative H-f1 peak that these unwanted NOE interactions are probably not even measurable. We see very strong NOE enhancements to H-f3 (doublet at 4.16 ppm, 5% peak area relative to −100% for H-f1) and to H-g1 (7% peak area). The NOE to H-f3 is due to the *cis*-1,3 relationship within the five-membered fructose ring, and the NOE to H-g1 is across the glycosidic linkage from the fructose ring to the glucose ring. This type of NOE across a glycosidic

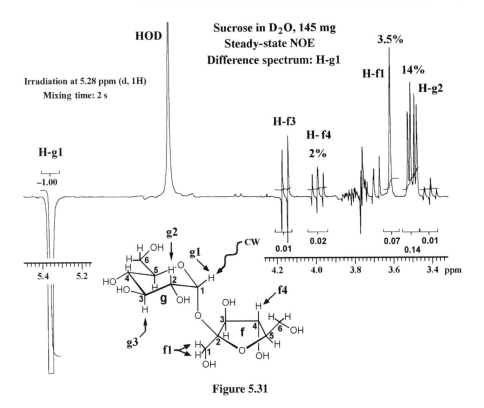

Figure 5.31

linkage is extremely useful in establishing the linkage (connectivity of monosaccharides) within an oligosaccharide. Before NMR was available, this kind of information had to be obtained by a series of chemical protection and degradation steps to find which position on each side was contributing to the glycosidic linkage. This NOE also indicates that our structure diagram (Fig. 5.30) is probably not accurate as drawn: the fructose ring must rotate around the glycosidic linkage to put the CH_2 group of position f1 close to H-g1 in the glucose unit. This illustrates how NOE measurements can be used to determine the conformation of biomolecules.

What if we try the NOE experiment the other way around, irradiating H-g1 on the other side of the glycosidic linkage? Figure 5.31 shows the result of this NOE difference experiment. Note that H-g1 is cleanly selected, with no saturation of any other resonances in the spectrum. Anomeric protons in sugars are useful "handles" for this kind of experiment because they are rare (no more than one per monosaccharide) and shifted to another region of the spectrum (4.5–6 ppm) away from the simple –CH–O protons on singly oxygenated carbons. We see a strong and clean NOE enhancement of the H-f1 singlet at 3.62 ppm (3.5% NOE—we divide by two because there are two protons in the H-f1 resonance), proving that the NOE interaction is a mutual effect: if saturating H_a affects H_b, then saturating H_b should affect H_a to the same extent. We see an even stronger NOE for the cis 1,2-related H-g2 resonance (14% enhancement). There are also some artifacts due to imperfect subtraction: the H-f3 doublet at 4.16 ppm shows a "dispersive" line shape due to imperfect horizontal alignment of the identical doublets in the two raw spectra (NOE and control). Subtraction then gives the "dispersive" appearance. The integral area is zero, so we know it is not a true NOE. The H-f4 triplet at 3.99 ppm looks somewhat dispersive but is more positive than negative. The peak area is positive, representing a 2% NOE enhancement. Thus H-g1 "talks" to both H-f1 and H-f4 across the glycosidic linkage, suggesting that the fructose ring can adopt two different conformations relative to the glucose ring.

5.13 SUMMARY OF THE NUCLEAR OVERHAUSER EFFECT

1. Perturbation of the equilibrium population difference for one nucleus (increased spin temperature) spreads over time to perturb the population difference (increase or decrease the spin temperature) of other nuclei that are nearby in space. For small molecules, increasing the spin temperature of one nucleus will *decrease* the spin temperature of nearby nuclei ("negative NOE"). This leads to an enhancement of peak intensities corresponding to the nearby nuclei. For large molecules, increasing the spin temperature of one nucleus will *increase* the spin temperature of nearby nuclei ("positive NOE"), leading to a reduction in peak intensity.

2. The NOE effect (perturbation of population difference in nearby nuclei) takes time to develop—this process is called the *NOE buildup*. The time allowed for the NOE to build up is called the *mixing time* and is on the order of magnitude of T_1, or typically hundreds of milliseconds (ms) for small molecules.

3. The *heteronuclear NOE*, for example, from ^1H to ^{13}C, is used to enhance the signal-to-noise ratio of ^{13}C peaks in the spectrum. All protons are irradiated equally and simultaneously during the relaxation delay to "pump up" the z magnetization of ^{13}C above the equilibrium value of M_o. Theoretically this can *triple* the M_z of ^{13}C.

4. The *homonuclear NOE*, almost always between two protons, is used to measure distances and determine stereochemical relationships. The NOE intensity (the percent increase or decrease of z magnetization observed at a nearby proton) is proportional to the inverse sixth power of distance between the two protons ($1/r^6$), and is generally too weak to be observed for distances over 5 Å.

5. There are two experimental methods for observing the NOE: *steady-state NOE* and *transient NOE*. The *steady-state* NOE involves a long, continuous-wave irradiation at the resonant frequency of the proton of interest, which equalizes the populations ("saturation"). During this time, the NOE builds up and reaches a steady state with the processes of NOE buildup and relaxation back to equilibrium in balance. The *transient* NOE (Chapter 8) involves a sudden perturbation (usually by a selective 180° pulse) followed by a mixing time with no pulses. During the mixing period, the perturbation propagates to nearby protons, changing their z magnetization. In both cases, at the end of the mixing time a 90° pulse samples the z magnetization of all nuclei, and enhancement or reduction of peak heights is observed in the spectrum.

6. The NOE is caused by dipole–dipole interaction (through-space) of two nuclear magnets, modulated by the tumbling of the molecule in solution. The NOE is an effect of mutual relaxation (or *cross relaxation*) of two nuclei. Mutual relaxation can occur in two ways: zero-quantum (ZQ) relaxation involves a transition from the $\alpha\beta$ state (one spin up and one down) to the $\beta\alpha$ state (one spin down and one up); double-quantum (DQ) relaxation involves a transition from the $\beta\beta$ state (both spins down) to the $\alpha\alpha$ state (both spins up). These transitions are driven by a population difference out of

equilibrium (Boltzmann) for the two states, and are stimulated by molecular tumbling at the frequency of the transition ($\nu_a - \nu_b$ for ZQ and $\nu_a + \nu_b$ for DQ) that leads to an oscillating magnetic field at both nuclei whose amplitude is strongly dependent on the distance between the two nuclei (overall inverse sixth power).

7. Longitudinal (population) relaxation of large molecules is dominated by ZQ relaxation, leading to a positive NOE (reduction of peak intensity). Small-molecule relaxation is dominated by DQ relaxation, leading to a negative NOE (enhancement of peak intensity). This effect can be understood by a thought experiment in which the mixing period of the NOE experiment consists exclusively of ZQ or DQ relaxation, which goes to completion (equilibrium population difference between the two states) before the 90° "read" pulse.

8. Molecules in the transition area of molecular weight (2000–4000 Da depending on molecular shape, rigidity, and solvent viscosity) show little or no NOE. For these molecules an alternative experiment called ROESY (rotating-frame Overhauser effect spectroscopy, Chapters 8 and 10) is effective.

9. Conformational flexibility can lead to loss of NOE interactions because the observed NOE is the weighted average over all conformations. A strong NOE resulting from a close approach of two protons in one conformation may be "diluted" by larger distances in other conformations to the extent that it is not observed at all. Rigid small molecules (e.g., fused ring systems) and tightly folded large molecules (proteins and nucleic acids) give the best NOE information. Flexible molecules, such as lipids, peptides, and oligosaccharides, give few useful NOEs.

10. Long NOE mixing times can lead to *spin diffusion*, in which perturbation of one proton leads to perturbation of a second proton, whose nonequilibrium population now perturbs a third proton. The appearance of an NOE between the first and third proton may be misinterpreted as a close (<5 Å) approach.

11. Organic chemists often misinterpret the NOE experiment by: (a) making distance predictions based on two-dimensional drawings rather than energy-minimized three-dimensional models, (b) testing only one isomer in a pair of stereo- or regioisomers, (c) calculating distances from NOE intensities rather than from initial rates of NOE buildup, or, (d) reading subtraction artifacts as NOE peaks. NOEs are always weak and must be interpreted with great care.

6

THE SPIN ECHO AND THE ATTACHED PROTON TEST (APT)

A major theme of this book is developing a "toolbox" of nuclear magnetic resonance (NMR) pulse sequence building blocks. So far we have described a number of fundamental tools of NMR experiments: pulses, continuous-wave (CW) low-power irradiation, and decoupling sequences such as waltz-16. In this chapter we will see the value of a *delay*, a simple waiting period of precise duration without any pulses, in building up a more complex "building block." The *spin echo* is a combination of pulses and delays that carries out a specific function: It allows us to control the type of changes that occur (due to chemical shifts only, *J* coupling only, or neither) during a precise period of time. We can "plug in" this module anywhere we want in a complex pulse sequence to achieve these predictable effects. Once we have used the vector model to understand what happens during a spin echo, we will not have to go through this analysis again because we will see that the overall effect is predictable in a simple way. Eventually, we will add other fundamental tools (e.g., selective pulses, pulsed field gradients, and spin locks) and continue combining these into more pulse sequence building blocks. As we learn how to predict the effects of these building blocks, eventually even the most complex and advanced NMR experiments can be pulled apart into a series of modules that we understand completely.

In this chapter, we will look at a one-dimensional technique for ^{1}H-decoupled ^{13}C spectra that uses the phase of the ^{13}C signal (positive or negative peaks) as a way to encode information about the number of protons attached to a carbon (C_q, CH, CH_2, or CH_3). We saw that a fully coupled spectrum gives this information, but the sensitivity is very low, and even with a simple molecule like sucrose, the overlapping multiplets are difficult to sort out. By designing experiments that modulate the sign of a single ^{1}H-decoupled carbon peak (positive or negative), we can get this information without overlap, because the peak remains a singlet and does not increase its horizontal "footprint". This "editing" of the ^{13}C spectrum according to the number of attached protons (C_q, CH,

---

NMR Spectroscopy Explained: Simplified Theory, Applications and Examples for Organic Chemistry and Structural Biology, by Neil E. Jacobsen
Copyright © 2007 John Wiley & Sons, Inc.

CH$_2$, or CH$_3$) is achieved with a ubiquitous building block of NMR pulse sequences: the *spin echo*.

To understand these new building blocks, we need to look in detail at the motion of the net magnetization vector during a delay as it precesses in the *x–y* plane. This motion is called "evolution" because the net magnetization changes with time or "evolves" as it rotates.

6.1 THE ROTATING FRAME OF REFERENCE

In a real sample, there will be more than one kind of nucleus of a given type. For example, a sample of ethanol has two different kinds of ^{13}C nuclei: the CH$_3$ carbons and the CH$_2$OH carbons. Due to differences in the amount of nuclear shielding, these two kinds of ^{13}C will have slightly different Larmor frequencies. For example, on a 300-MHz (7.05-T) spectrometer we might have 75.000000 MHz as the resonant frequency (ν_o) of the CH$_3$ carbon and 75.003375 MHz for the resonant frequency of the CH$_2$OH carbon. Because these differences are small (on the order of parts per million of the Larmor frequency), it is convenient to view the NMR experiment from a rotating frame of reference, so that the fundamental resonant frequency (near which all of the sample nuclei precess) is removed from consideration. In our example, all ^{13}C nuclei have a resonant frequency very near to 75 MHz, so we are not really interested in this; it is the differences in resonant frequency, which are thousands of hertz at most, that are important. For example, the reference axes in the rotating frame, referred to as the x' and y' axes, could be rotated counterclockwise (ccw) about the *z* axis at a frequency of 75.000000 MHz. In this case, the ^{13}C nuclei of the CH$_3$ group of ethanol precess at exactly the frequency of rotation of the x' and y' axes, and the net magnetization vector for these nuclei after a pulse will appear to stand still in the x'–y' plane of the rotating frame of reference. The other ^{13}C nuclei, corresponding to the CH$_2$OH group of ethanol, will give rise to a net magnetization vector that appears to rotate slowly counterclockwise in the x'–y' plane at a rate corresponding to the difference between its Larmor frequency and that of the reference frequency (75.003375 − 75.000000 = 0.003375 MHz = 3375 Hz). This rotation will be counterclockwise in this case because the Larmor frequency of the CH$_2$OH group is greater than that of the CH$_3$ group, but in other cases it could be clockwise (cw). The rotation frequency we choose for the rotating frame of reference establishes a reference frequency in the spectrum, which is just the center of the spectral window. This is also the frequency of the radio frequency (RF) pulses we apply to the sample and of the reference signal used in the detector of the NMR receiver. If a certain type of spin (e.g., CH$_3$ of ethanol) has a chemical shift at the exact center of the spectral window, it is said to be "on resonance" and its magnetization vector will stand still in the x'–y' plane. All other nuclei will rotate at a frequency and direction determined by their "offset": the chemical shift (in hertz or radians per second) relative to zero at the center of the spectral window.

Moving from the laboratory frame of reference to the rotating frame of reference is exactly analogous to the detection step in the NMR hardware. If the reference frequency is 75.000000 MHz, the detection step subtracts this frequency from the free induction decay (FID) frequency, which for the CH$_2$OH carbon of ethanol is 75.003375 MHz. The resulting analog signal, 3375 Hz, is the audio signal that we digitize and record as the FID. After Fourier transformation, this leads to a single peak positioned 3375 Hz to the left of the center of the spectral window in the NMR spectrum.

THE SPIN ECHO AND THE ATTACHED PROTON TEST (APT)

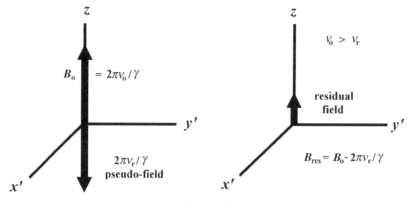

Figure 6.1

For those of you who have studied physics, you may recognize that the rotating frame of reference is an accelerating frame of reference, which is a major no-no if we want the laws of physics to be preserved. A similar situation arises on the earth where all of us are in a rotating frame of reference at the surface of the earth. The laws of physics will not work in this accelerating frame of reference, so we have to invent fictitious forces called Coriolis forces to correct for the discrepancies. For example, it is the Coriolis force that makes storm systems rotate on the earth's surface. In the NMR world, the sins of the accelerating frame can be atoned for by inventing a fictitious magnetic field (or "pseudo-field") that is opposed to the applied magnetic field B_o (Fig. 6.1). This fictitious field has a strength of $2\pi\nu_r/\gamma$, where ν_r is the reference frequency. The faster we spin the x' and y' axes in the rotating frame, the larger is the pseudofield required to maintain the laws of physics. If the spin in question is exactly on-resonance ($2\pi\nu_r/\gamma = 2\pi\nu_o/\gamma = B_o$), the fictitious field precisely cancels the B_o field and there is no magnetic field; hence, any net sample magnetization in the x'–y' plane remains stationary. If the spin is not on-resonance, the B_o field is not perfectly canceled and there remains a small residual field ($B_{res} = B_o - 2\pi\nu_r/\gamma = 2\pi\nu_o/\gamma - 2\pi\nu_r/\gamma = 2\pi(\nu_o - \nu_r)/\gamma$), along either the $+z$ or the $-z$ axis, which is proportional in strength to the difference between the Larmor frequency (chemical shift) and the center of the spectral window. This residual field makes the net magnetization vector rotate in the x'–y' plane the same way the B_o field makes the net magnetization rotate in the x–y plane of the laboratory frame:

$$\nu_o = \gamma B_o/2\pi, \qquad \Delta\nu = \nu_o - \nu_r = \gamma B_{res}/2\pi = \gamma[2\pi(\nu_o - \nu_r)/\gamma]/2\pi$$

Note that we use $\Delta\nu$ to refer to the rotating-frame frequency (sometimes called the resonance offset). This is the difference between the Larmor frequency and the reference frequency: $\nu_o - \nu_r$. The above equation shows that the same physical law expressed in the equation on the left-hand side (precession rate is proportional to γ and to B_o) is operating in the equation on the right-hand side (resonance offset is proportional to γ and to B_{res}) in the rotating frame of reference, as long as we introduce the pseudofield. In the NMR spectrum, $\Delta\nu$ is the distance from the center of the spectral window to the NMR peak (Fig. 6.2), also represented as Ω in units of radians per second. If the peak is in the downfield half (left half) of the spectrum, the Larmor frequency is greater than the reference frequency ($\nu_o > \nu_r$) and we have a positive resonance offset ($\Delta\nu > 0$). This corresponds to the motion of the net magnetization

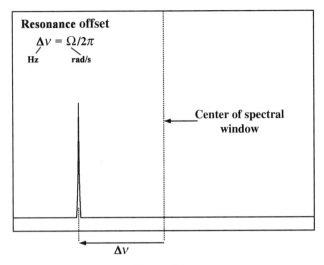

Figure 6.2

vector in a counterclockwise direction in the rotating frame at the frequency Δv. If the NMR peak is in the upfield half (right half) of the spectrum, the Larmor frequency is less than the reference frequency ($v_o < v_r$) and we have a negative resonance offset ($\Delta v < 0$). This corresponds to the motion of the net magnetization vector in a *clockwise* direction in the rotating frame at the frequency $-\Delta v$. Remember that in the NMR spectrum the frequency scale runs from right to left, opposite to every other graphical scale known to man. In discussions of NMR theory, we will ignore the chemical shift scale (δ in ppm) and view the NMR spectrum as a frequency scale (Hz) with zero at the center, negative frequencies on the right-hand side, and positive frequencies on the left-hand side. The resonance offset is the key to understanding what happens to the net magnetization vector during a delay. We call this motion of the net magnetization vector in the x'–y' plane of the rotating frame of reference "evolution." Sometimes to simplify the terms in equations, we will use the upper case Greek letter omega to represent the resonance offset in radians per second:

$$\text{laboratory frame: } \omega_o = 2\pi v_o, \quad \text{rotating frame: } \Omega = 2\pi \Delta v$$

6.2 THE RADIO FREQUENCY (RF) PULSE

The RF pulse is a short (~10 μs) burst of a very high power (50–300 W) RF signal with a specific frequency, amplitude, and phase. The frequency is the same as the reference frequency, v_r. The ideal pulse turns on and off instantly and has constant amplitude (Fig. 6.3), leading to a rectangular shape or envelope ("rectangular pulse"). The duration of the pulse is called the pulse width, usually measured in microseconds. The phase of the pulse is determined by its starting point in the sine function: starting at 0° the amplitude increases at first from zero to maximum; starting at 90° it decreases at first from maximum to zero; starting at 180° it decreases at first from zero to the negative peak; and starting at 270° it increases at first from the negative peak to zero. This can be precisely controlled by the hardware and

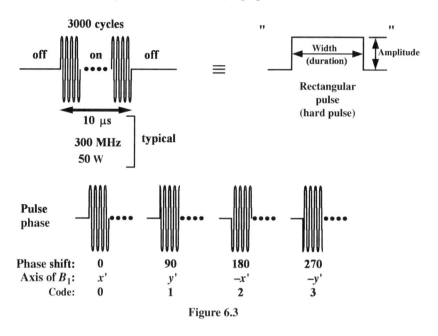

Figure 6.3

is programed into the pulse sequence. In the pulse sequence code that drives the hardware, the pulse phase is referred to as 0, 1, 2, and 3 for 0°, 90°, 180°, and 270°, respectively. In the probe, the pulse is applied to a tuned circuit consisting of the probe coil (an inductance) and a variable capacitance that is used to tune the probe (Fig. 6.4). The actual circuit is much more complex, with at least two variable capacitors, but the simple tuned LC circuit is sufficient to understand how it works. The probe coil is saddle shaped, which is designed to produce a magnetic field oriented perpendicular to the NMR tube axis, that is, to the z-axis. When a pulse is applied to the coil, an oscillating magnetic field (B_1) appears along an axis in the x–y plane. For example, if this is aligned on the x axis, we have B_1 oscillating in time along the x axis: first zero, then growing to a maximum B_1 field oriented on the positive x axis, then decreasing to zero, then growing to a maximum B_1 field along the negative x axis, and then decreasing to zero again. This sequence repeats itself ν_r times per second, where ν_r is the frequency of the pulse (e.g., 300 MHz for ^{1}H excitation on a 7.05-T instrument).

To get interaction with the individual magnetic vectors, which are precessing in a counterclockwise path around the cones at the Larmor frequency, we need a B_1 field that is also rotating in a counterclockwise direction at the same frequency. We can divide the oscillating B_1 field into two components: one that rotates clockwise in the x–y plane at frequency ν_r and the other that rotates counterclockwise in the x–y plane at the same frequency (Fig. 6.5). The vector sum of these two rotating vectors is the B_1 vector, which oscillates in amplitude along the x axis alone. Of the two rotating components, only the counterclockwise one has any effect on the precessing spins; the other one is effectively $2\nu_r$ away from the resonant frequency because it is rotating in the wrong direction. So we ignore this component and from now on we will describe the RF pulse as a magnetic vector of constant magnitude B_1, which rotates counterclockwise in the x–y plane at the frequency ν_r. This explanation is

THE RADIO FREQUENCY (RF) PULSE   205

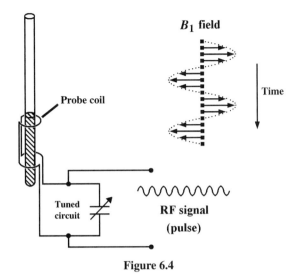

Figure 6.4

formally required to come up with a rotating B_1 vector, but you can forget you ever heard of it if you like because we will always talk about the pulse as a rotating B_1 vector from now on.

In the rotating frame of reference, the B_1 vector always stands still because the reference frequency and the pulse frequency are the same. Changing the pulse phase changes the position of the B_1 vector in the x'–y' plane, so that a 0° phase corresponds to the x' axis, a 90° phase to the y' axis, a 180° phase to the $-x'$ axis, and a 270° phase to the $-y'$ axis. This ability to position the B_1 vector wherever we want in the x'–y' plane through the RF hardware allows us to control precisely the effect of the pulse. We can also change the amplitude of the pulse, which adjusts the length of the B_1 vector and changes the strength of the magnetic field it represents.

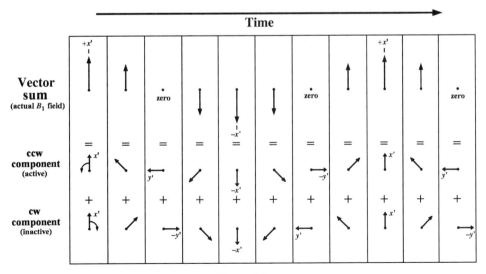

Figure 6.5

6.3 THE EFFECT OF RF PULSES

In the rotating frame of reference, the effect of an RF pulse can be described in terms of the interaction of the stationary magnetic field vector (B_1) of the pulse and the net magnetization of the sample. At equilibrium, the sample net magnetization vector lies along the $+z$ axis. The RF pulse magnetic field, which is referred to as B_1 to distinguish it from the B_o field, exerts a torque on the net magnetization vector that rotates it in a plane perpendicular to the B_1 field vector (Fig. 6.6). For example, an RF pulse with appropriate phase to place the B_1 field vector along the x' axis will rotate the net magnetization vector in a counterclockwise direction from the $+z$ axis toward the $-y'$ axis in the rotating frame. This rotation is the same as the precession under the influence of the B_o field ($\nu_o = \gamma B_o/2\pi$), except that it is much slower because the B_1 field is much weaker than B_o even for the highest power RF pulses. In the rotating frame, for an on-resonance spin, the B_o field goes away and the only field affecting the net magnetization vector is the B_1 field. The net magnetization vector rotates faster if the B_1 amplitude is greater. The rate of precession of the net magnetization vector around B_1 can be written as $\nu_1 = \gamma B_1/2\pi$. Be careful to recognize that ν_1 is not the frequency of the pulse (that would be ν_r), but rather the frequency of rotation of the sample net magnetization around the B_1 vector during the pulse. This rotation rate, $\gamma B_1/2\pi$, is often used as a measure of the amplitude of the pulse because it is in more convenient units of hertz rather than tesla. This is analogous to the way we measure B_o field strength in megahertz ($\gamma_H B_o/2\pi$) rather than tesla.

The extent of precession of the sample net magnetization vector under the influence of the B_1 field depends on the duration of the pulse. The longer we leave the pulse on, the farther the net magnetization vector rotates. A pulse that lasts just long enough to rotate the net magnetization vector by an angle of 90° is called a "90° pulse." A stronger B_1 field (higher RF power during the pulse) will rotate the net magnetization faster and will lead to a shorter duration for the 90° pulse. If we are measuring pulse amplitude in units of hertz ($\gamma B_1/2\pi$), we can calculate the 90° pulse width from the amplitude:

$$t_{360} = \text{time to rotate one cycle around } B_1 = 1/\nu_1 = 1/(\gamma B_1/2\pi)$$

$$t_{90} = \text{time to rotate one-fourth cycle around } B_1 = 1/(4 \times \nu_1) = 1/(4\gamma B_1/2\pi)$$

Likewise if we calibrate the pulse width to give the maximum peak height in the spectrum (90° pulse), we can calculate the pulse amplitude in hertz:

$$\gamma B_1/2\pi = 1/(4 \times t_{90})$$

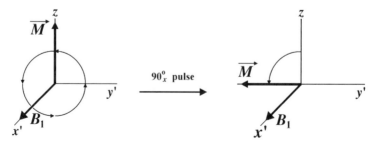

Figure 6.6

THE EFFECT OF RF PULSES   207

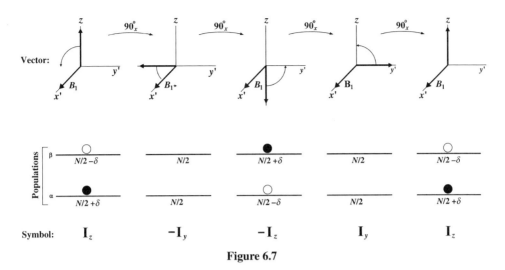

Figure 6.7

All pulse rotations are counterclockwise when viewed with the B_1 vector pointing toward you.

As we look at the effect of pulses on the net magnetization vector, it is useful to keep track of the population difference between the α and β states at the same time. This can be done by drawing the energy diagram (two levels) and using filled circles to represent excess population and open circles to represent population deficits relative to an equal division of spins between the two energy states. For example, at equilibrium we have $N/2 + \delta$ spins in the α state and $N/2 - \delta$ spins in the β state, where δ is a number much smaller than N (e.g., $10^{-5} N$). We can draw an open circle ($-\delta$) in the upper energy level and a closed circle ($+\delta$) in the lower energy state (Fig. 6.7, left). Note that as we are only concerned with the population difference ΔP (in this case $P_\alpha - P_\beta = 2\delta$), we ignore the $N/2$ term in our circle representation.

At the end of the 90° pulse with B_1 on the x' axis, the net magnetization is on the $-y'$ axis, and we have no z component. We will refer to this spin state as $-\mathbf{I}_y$. Because the z component of net magnetization results from the population difference between the α and β states, we can say that there is no population difference at the end of a 90° pulse (Fig. 6.7). With the 90° pulse, we have effectively converted the population difference into coherence. If we record the FID right after this pulse, we would get a normal spectrum with a positive absorptive peak.

A 180° pulse, which lasts twice as long as a 90° pulse, will rotate the net magnetization from the $+z$ axis to the $-z$ axis (Fig. 6.7, center). We can call this spin state $-\mathbf{I}_z$. There is no coherence, and we would not observe any spectrum if we collected an FID at this point. As net magnetization on the $+z$ axis at equilibrium results from the slight excess of spins in the lower energy (α) state, rotating this to the $-z$ axis means that we have inverted the population difference, so that now there is a slight excess of spins in the higher energy (β) state. We have the same population difference but now it is negative: $\Delta P = P_\alpha - P_\beta = -2\delta$. This is represented by a filled circle ($N/2 + \delta$) in the higher energy (β) state and an open circle ($N/2 - \delta$) in the lower energy (α) state. We call this an "inversion" pulse because it inverts the populations in the α and β states. The really bizarre thing is that the 180° pulse does not simply move 2δ spins from the α to the β state; instead, it moves *every single spin* that was in the α state to the β state and *every spin* that was in the β state to the α state. From the point of view of the spins, the entire world is turned upside down! The $N/2 + \delta$ spins that were in the α state are now in the β state, and the $N/2 - \delta$ spins that were in the β state are now all in the α state. This is truly an inversion pulse! This may seem like a trivial point, but it will be very important later on when we talk about *J*-coupling relationships.

At the end of a 270° pulse, the net magnetization vector lands on the $+y'$ axis, opposite to where it would be at the end of a 90° pulse. We call this state $\mathbf{I}_y$. There is no z component, so we have no population difference. If the 90° pulse gives a normal (positive absorptive) peak in the spectrum, a 270° pulse will give an upside down peak. In this case we would call the $-y'$ axis the "reference axis," and any magnetization vector that is on this axis at the start of the FID would give a normal (positive absorptive) peak in the spectrum. If the magnetization vector is on the opposite axis (the $+y'$ axis) at the start of the FID, it will give an upside-down (negative absorptive) peak. Vectors on the $+x'$ or $-x'$ axes would lead to dispersive (up/down or down/up) peaks. We can choose which axis we want to use for the reference axis, and this is referred to in the pulse program (the software that drives the experiment) as the "receiver phase." NMR data processing software can always change the phase of peaks, but it cannot (or at least should not) change the *relative* phase of peaks, so that if the 90° pulse on the x' axis gives an upside-down peak, the 270° pulse on the x' axis will give a normal peak (i.e., we have changed the phase reference to the $+y'$ axis by phase correction in software).

At the end of a 360° pulse, the net magnetization vector has made one complete rotation around the B_1 vector and lands on the $+z$ axis, exactly where it started (Fig. 6.7, right). The spin state is identical to the equilibrium state, $\mathbf{I}_z$. We have the equilibrium (Boltzmann) population distribution, represented with one open circle in the upper state and one filled circle in the lower state. If we collect an FID right after the 360° pulse, we will see no spectrum.

Pulse *calibration* is the process of collecting a series of FIDs, each with the pulse width increased a little bit from the last one. For example, we might acquire 18 ^{13}C spectra starting with a pulse width of 0 and increasing by 3 μs each time (i.e., 0, 3, 6, 9, 12, etc.) (Fig. 6.8). The fifth spectrum (12 μs pulse) gives the maximum positive peak (90° pulse), the ninth (24 μs pulse) gives a very weak negative peak—nearly a null (180° pulse), the 13th spectrum (36 μs pulse) gives a maximum negative peak (270° pulse), and the second null (360° pulse) occurs halfway between the 45 μs pulse and the 48 μs pulse. A very long relaxation delay (70 s) is necessary to make sure we are starting with the equilibrium state ($M_z = +M_0$) each time. It is important to understand that only one peak is shown in the spectrum, and we are repeating the experiment each time with a new pulse width value, plotting it to the right-hand side of the previous spectrum. As it is easier to pin down the null point exactly rather than the maximum spectrum, we usually find the 180° pulse width and divide by 2, or the 360° pulse width and divide by 4, to get the 90° pulse width. In this case, the 360° pulse can be interpolated as 46.5 μs, so the 90° pulse is 11.6 μs (46.5/4). The calibrated 90° pulse is always reported at a particular RF power setting, in this case 60 dB. Because the 90° pulse width depends on the B_1 amplitude (RF power level), we need to specify that value, or the calibration will be meaningless. The B_1 field strength in hertz ($\gamma B_1/2\pi$) is $1/(4 \times t_{90}) = 1/(4 \times 11.6$ μs$) = 1/(4 \times 0.0000116$ s$) = 1,000,000/(4 \times 11.6$ s$) = 1,000,000/(46.4$ s$) = 21,552$ Hz or 21.552 kHz. This is the reciprocal of the 360° pulse: the rate of rotation of the sample net magnetization vector around the B_1 vector during the pulse (ν_1). Note that this is only 0.0287% (100×0.021552 MHz/75 MHz) of the B_0 field strength expressed in hertz for a 7.05-T instrument. Here we are comparing $\gamma B_1/2\pi$ to $\gamma B_0/2\pi$, so we have

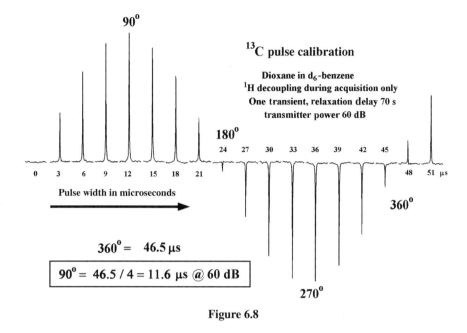

Figure 6.8

to use γ for ^{13}C, which gives us 75 MHz for $\gamma B_0/2\pi$ on a "300-MHz" instrument. The B_1 field (oscillating at 75 MHz and oriented along the x' or y' axis) is very small (short vector) compared to the B_0 field (static and oriented along the $+z$ axis), but we can not get much higher than this without heating the sample and/or burning up the amplifiers.

6.4 QUADRATURE DETECTION, PHASE CYCLING, AND THE RECEIVER PHASE

The real and imaginary channels of the NMR receiver can be considered to record the x' and y' components of the net magnetization vector in the rotating frame: M_x and M_y. Thus, if we use a 90° excitation pulse on the x' axis, the net magnetization vector will rotate to the $-y'$ axis and then undergo chemical shift evolution in a ccw direction (for peaks in the downfield half of the spectral window) or in a cw direction (for peaks in the upfield half). For example, for a positive rotating-frame frequency, the net magnetization vector will move from the $-y'$ axis to the $+x'$, $+y'$, $-x'$ axis and back to the $-y'$ axis, as it rotates. The x' component (M_x) will start at zero, then increase to a positive maximum as the vector passes the x' axis, then decrease to zero as the vector crosses the $+y'$ axis, and pass to a negative maximum as it crosses the $-x'$ axis. This is shown in Figure 6.9, along with the y' component (M_y), which starts at a negative maximum, decreases to zero, increases to a positive maximum, and decreases to zero again in the same time period. If we see these two waveforms, can we deduce the position of the net magnetization vector at the beginning of the FID (the peak phase), as well as its direction of rotation (the sign of the peak frequency)? First consider that we have only the real part of the FID (the M_x component): What can we say about the sample net magnetization? It could be starting on the $-y'$ axis and moving ccw (positive frequency), or it could be starting on the $+y'$ axis and moving cw (negative

210   THE SPIN ECHO AND THE ATTACHED PROTON TEST (APT)

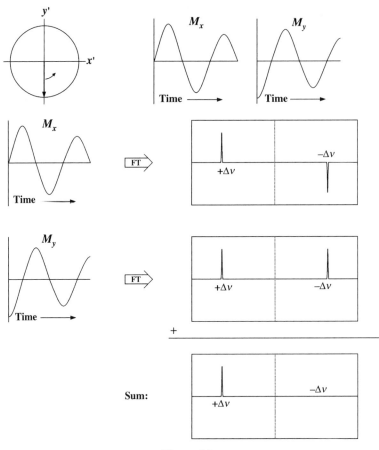

Figure 6.9

frequency). In both cases, we would see the same M_x waveform. So if we use the $-y'$ axis as our reference axis, we could draw the spectrum as a positive absorptive peak with frequency ("chemical shift") equal to $+\Delta\nu$, on the left-hand side of the spectral window, or as a negative absorptive peak with frequency equal to $-\Delta\nu$, on the right-hand side of the spectral window. As both are equally likely given the information available, the Fourier transform includes both peaks with equal intensity in the spectrum (Fig. 6.9). If instead we only have the imaginary part of the FID (M_y), we could say that the net magnetization definitely starts on the $-y'$ axis (positive absorptive peak), but we do not know whether it moves ccw or cw, because both directions would give the same results in the M_y trace: starting at a negative maximum, decreasing to zero, building to a positive maximum, and so forth. So the Fourier transform would give a spectrum with positive peaks of equal intensity at $+\Delta\nu$ and $-\Delta\nu$. If we add these two spectra together, we get an idea of how quadrature detection works: Combining what we know from the real part of the FID with what we know from the imaginary part of the FID gives us a spectrum with a single positive peak at frequency $+\Delta\nu$, and the "quadrature image" peak at $-\Delta\nu$ is canceled out. This gives us a sense of how the mathematics of the complex (i.e., real and imaginary combined) Fourier transform works.

Another term for quadrature detection is "phase-sensitive detection," and we can see how the complex Fourier transform is truly sensitive to the phase of the NMR signal: It can not only distinguish positive frequencies (ccw rotation) from negative frequencies (cw rotation), but also tell on which axis the net magnetization started at the beginning of the FID (corresponding to the phase of the NMR peak). In Figure 6.9 we saw how the ghost image, or quadrature image, at $-\Delta\nu$ in the spectrum was canceled out in the two spectra obtained from M_x and M_y. Sometimes this cancellation is not perfect and we get a small image, either positive or negative on the opposite side of the spectral window from our peak. This is especially likely for a very intense peak in the spectrum. The most likely cause of this artifact is that the two receiver channels, real and imaginary, have unequal gain or amplification so that the negative peak and the positive peak at $-\Delta\nu$ are not equal in intensity. Consider, for example, that the imaginary channel has a gain 10% higher than the real channel. We would then see a signal for M_x equal to $\sin(2\pi\Delta\nu t)$ and a signal for M_y equal to $-1.1\cos(2\pi\Delta\nu t)$, ignoring the decay of the FID. The center spectrum with two positive peaks in Figure 6.9 would be 10% more intense than the upper one with one positive and one negative peak, and in the sum the true peak at $+\Delta\nu$ would have an intensity of 2.1 whereas the quadrature artifact at $-\Delta\nu$ would have an intensity of $+0.1$.

We could spend a lot of time adjusting and balancing the two receiver channels of the spectrometer to get exactly the same gain, but there is a much simpler way to eliminate the "quad" artifacts. We can acquire a four-scan spectrum, but with each scan we advance the phase of the pulse, starting with a pulse on x' (scan 1), moving ccw by 90° to a pulse on y' (scan 2), then ccw again to a pulse on $-x'$ (scan 3), and finally to a pulse on $-y'$ (scan 4). The table below shows the signals that we would observe for M_x and M_y for each of these pulses. You should draw the three axes (x', y' and z) and verify for yourself that the two components will vary as shown as the net magnetization vector rotates ccw in the x'–y' plane after the excitation pulse.

Scan	Pulse phase	Vector starts on	M_x	M_y
1	$+x'$	$-y'$	$\sin(2\pi\Delta\nu t)$	$-1.1\cos(2\pi\Delta\nu t)$
2	$+y'$	$+x'$	$\cos(2\pi\Delta\nu t)$	$1.1\sin(2\pi\Delta\nu t)$
3	$-x'$	$+y'$	$-\sin(2\pi\Delta\nu t)$	$1.1\cos(2\pi\Delta\nu t)$
4	$-y'$	$-x'$	$-\cos(2\pi\Delta\nu t)$	$-1.1\sin(2\pi\Delta\nu t)$

Now we can construct a perfectly balanced real and imaginary FID from these signals by combining all the sine functions into a real FID and all the cosine functions into an imaginary FID:

$$\text{real FID} = M_x(1) \quad +M_y(2) \quad -M_x(3) \quad -M_y(4) = 4.2\sin(2\pi\Delta\nu t)$$
$$\text{imaginary FID} = M_y(1) \quad -M_x(2) \quad -M_y(3) \quad +M_x(4) = -4.2\cos(2\pi\Delta\nu t)$$

Now the two parts of the FID are perfectly balanced, regardless of the matching of gain of the two receiver channels. This technique is an example of *phase cycling*, a general way of eliminating artifacts by subtraction in the sum-to-memory as a number of scans are acquired.

The trick of directing the M_x and M_y components of net magnetization to different data tables in the sum-to-memory (real and imaginary sums) is a way of changing the reference phase (also called the *receiver phase* or the *observe phase*). If we just add M_x to the real sum and M_y to the imaginary sum in the sum-to-memory, we say that the receiver phase is $-y'$, and magnetization that starts on the $-y'$ axis will give rise to a positive absorptive peak in the spectrum. If instead we subtract M_x from the real sum and subtract M_y from the imaginary sum (as in scan 3 above), we say that the receiver phase is $+y'$, and magnetization that starts on the $+y'$ axis will give rise to a positive absorptive peak. But we can also "swap" the channels: If we add M_y to the real sum and subtract M_x from the imaginary sum (as in scan 2 above), we say that the receiver phase is $+x'$, and net magnetization that starts on the $+x'$ axis will give a positive absorptive peak. Finally, if we subtract M_y from the real sum and add M_x to the imaginary sum (as in scan 4 above), we have receiver phase $-x'$, and magnetization that starts on $-x'$ will give a positive absorptive peak. Thus, another way of describing the phase cycle above is

Scan No.	1	2	3	4
Pulse phase	$+x'$	$+y'$	$-x'$	$-y'$
M starts on	$-y'$	x'	y'	$-x'$
Receiver phase	$-y'$	x'	y'	$-x'$

In the phase cycle, the receiver "follows" the starting phase of the sample net magnetization, leading to addition of four positive absorptive peaks. In pulse sequence programing, the $+x'$ axis is given a code of 0, $+y'$ is 1, $-x'$ is 2, and $-y'$ is 3, so we would say the pulse phase is 0 1 2 3 and the receiver phase is 3 0 1 2. The important point is that every time we advance the phase of the pulse by 90°, which advances the starting position of the net magnetization by 90°, we also advance the receiver phase (our point of view) by 90°, so it looks the same in each scan. But we are alternately exercising different physical receiver channels (M_x and M_y) so that any imbalances in the two channels will cancel out and there will be no quadrature artifacts in the spectrum.

6.5 CHEMICAL SHIFT EVOLUTION

We can consider the detection coil as lying along the y' axis of the rotating frame and recording a voltage proportional to M_y, the y' component of net magnetization. Of course, the coil is not rotating at hundreds of megahertz, but the electronics involved in detecting the FID signal and converting it to an audio signal are equivalent to placing the coil at a stationary position in the rotating frame of reference. If we start the experiment with a 90° RF pulse that places the B_1 field along the y' axis, the net magnetization will be rotated to the x' axis and will start to precess in a counterclockwise direction at a rate corresponding to its resonance offset $\Delta \nu$ (or Ω in radians per second) relative to the center of the spectral window (Fig. 6.10). This motion induces an RF signal in the probe coil, which corresponds to a cosine function ($M_x = +M_0, 0, -M_0, 0$, etc., as the vector rotates) for the real audio signal (M_x) and a sine function ($M_y = 0, +M_0, 0, -M_0$, etc.) for the imaginary audio signal (M_y). Fourier transformation of this signal leads to a peak in the spectrum with the normal absorptive lineshape.

Things get interesting if we insert a delay between the end of the 90° pulse and the beginning of the FID. Because of the delay, the net magnetization recorded in the FID will start at a different place in the x'–y' plane due to precession during the delay. This motion is called *chemical shift evolution*. If the delay is just long enough to allow the net magnetization to precess from the x' axis to the y' axis, the real FID signal will be a

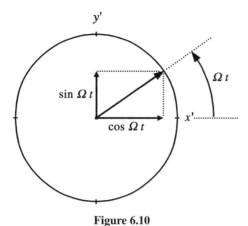

Figure 6.10

negative sine (−sin) function ($M_x = 0, -M_o, 0, +M_o$, etc.) and the peak in the resulting spectrum will have a dispersive lineshape in the absence of phase correction. A delay that is twice as long will allow the magnetization to precess to the $-x'$ axis, which will give a real FID that is a negative cosine (−cos) function ($M_x = -M_o, 0, M_o, 0$, etc.), leading to a negative or upside-down absorptive line. In this case, we can say that the reference axis is $+x$, so that magnetization starting on the $+x$ axis at the beginning of the FID will give a positive absorptive peak, and if we start with the spin state $-\mathbf{I}_x$ we will get an upside-down peak.

The exact amount of rotation that occurs during the delay can be calculated in degrees as $360(\nu_o - \nu_r)t$, where ν_o is the resonant frequency in hertz (Larmor frequency) in the laboratory frame, ν_r is the reference frequency in hertz (frequency of rotation of the axes in the rotating frame, frequency at the center of the spectral window), and t is the length of time of the delay in seconds. If the peak is in the upfield half of the spectral window, $\nu_o < \nu_r$ and the rotating-frame frequency ($\nu_o - \nu_r$) will be negative. The total rotation will be negative, indicating a clockwise rotation looking down from the $+z$ axis. If the peak is in the downfield half, $\nu_o > \nu_r$ and the rotating-frame frequency (or resonance offset) will be positive. Total rotation in the above equation will be positive and the rotation will be counterclockwise. For example, if the NMR peak is 75 Hz downfield of the center of the spectral window ($\Delta\nu = 75$ Hz), to get a rotation of $90°$ we would have to insert a delay τ such that

$$360\,(75\,\text{Hz})\tau = 90;\ \tau = (90/360)/75 = 0.25/75\,\text{Hz} = 0.003333\,\text{s} = 3.333\,\text{ms}$$

Note that as Hz = s^{-1}, when we divide by hertz we get seconds. With this delay, the net magnetization will rotate from the x' axis to the y' axis, and the peak in the spectrum will be dispersive (reference axis = x').

6.6 SCALAR (J) COUPLING EVOLUTION

Now consider a two-spin system which is scalar (J) coupled, such as the ^{1}H–^{13}C pair in chloroform (CHCl$_3$). The ^{13}C nuclei have two different resonant frequencies depending on whether the attached ^{1}H nucleus is in the α or the β spin state. In the absence of proton

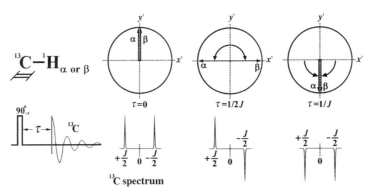

Figure 6.11

decoupling, the ^{13}C spectrum will show a doublet with two peaks separated by the coupling constant J. The population of ^{13}C nuclei in the sample can be divided into two parts. One half of the ^{13}C nuclei are attached to a ^{1}H nucleus in the α state, and the magnetization of these ^{13}C spins can be summed to give one net magnetization vector with a precession rate $\Delta \nu + J/2$ (in hertz). The remaining half of the ^{13}C nuclei are attached to a ^{1}H nucleus in the β state, and they add up to form another net magnetization vector that precesses at a rate $\Delta \nu - J/2$. Of course, there are slightly more ^{13}C nuclei in the H $= \alpha$ group, but the difference is so small as to be insignificant in this analysis. To describe the motion of these two net magnetization vectors, it is convenient to choose the rotation rate of the rotating frame of reference to be ν_0, the chemical shift position of the ^{13}C. This is the equivalent of placing the center of the spectral window exactly between the two components of the ^{13}C doublet in the spectrum, so that $\Delta \nu = 0$ ("on resonance"). In this case, rotating-frame frequencies for the two components of the ^{13}C doublet are $+J/2$ for the H $= \alpha$ peak and $-J/2$ for the H $= \beta$ peak.

Consider the sequence shown in Figure 6.11, which consists of a 90° ^{13}C pulse followed by a variable time delay before the start of acquisition of the (^{13}C) FID. The vector diagram shows the x'–y' plane, viewed from the $+z$ axis (i.e., from above). At equilibrium, both net magnetization vectors will lie along the $+z$ axis. A 90° pulse on the $-x'$ axis of the rotating frame will rotate both vectors to the $+y'$ axis. Both magnetization vectors will precess, but the "β" vector will rotate with a velocity of $-J/2$ Hz (in a clockwise direction toward the x' axis) and the "α" vector will rotate with a velocity of $+J/2$ Hz (in a counterclockwise direction toward the $-x'$ axis). After a delay of time $1/(2J)$ from the end of the 90° pulse, the two vectors will be opposite to each other on the $+x'$ and $-x'$ axes. This state means that the first half of the ^{13}C nuclei, which are attached to protons in the α state, give rise to a net magnetization vector that is 180° out of phase with the vector resulting from the other half, which are attached to protons in the β state. This condition is called *antiphase magnetization*, and it is crucial to many NMR experiments. We will see in the next chapter that this very special antiphase state is a prerequisite for *magnetization transfer*, the process of making the net magnetization "jump" from the ^{13}C nucleus to the attached ^{1}H. Collecting an FID beginning with this antiphase state would yield a spectrum in which both of the components of the ^{13}C doublet are dispersive, but one component peak is opposite in phase relative to the other (up-down vs. down-up). A 90° phase correction of this spectrum (i.e., changing the phase reference from y' to $-x'$) would yield a spectrum in which the H $= \alpha$ component is positive absorptive ("up") and the H = β component is negative absorptive ("down") (Fig. 6.11, center spectrum). A further delay of 1/(2J), for a total delay of 1/J, causes the two vectors to meet again along the −y' axis. This state is called "in-phase" because the two components have the same phase. Starting the FID at this point would yield an upside-down (negative) absorptive doublet using the original phase reference (y' axis). A total delay time of 2/J would allow both magnetization vectors to precess in opposite directions a full rotation, meeting back on the y' axis. This in-phase state would lead to a normal positive absorptive doublet in the spectrum.

It is important to recognize the difference between the terms *absorptive* and *dispersive* on the one hand, and *in-phase* and *antiphase* on the other hand. An antiphase doublet is sometimes confused with a dispersive peak because both have "up" and "down" components. Absorptive and dispersive lineshapes are characteristic of a single resonant frequency or "line" in a spectrum, and they can be interchanged by a 90° zero-order phase correction (i.e., changing the reference axis by 90°). In the vector model, using the y' axis as a phase reference, absorptive and dispersive lineshapes correspond to net magnetization on the y' and x' axes, respectively, at the start of the FID. Thus, they differ only in the phase of the NMR signal resulting from a single magnetization vector. In-phase and antiphase states refer to the relative phase of the two components of a J-coupled doublet system (Fig. 6.12). The antiphase state is one in which the two magnetization vectors of a doublet, which correspond to the two lines in the spectrum (in the above ^{13}C example H = α and H = β), are directly opposite to each other in the x'–y' plane. The in-phase state is one in which the two component vectors are aligned. In the spectrum, an antiphase doublet cannot be phase corrected to look like a normal doublet; if positive absorptive peak shape is achieved for one component, the other will be negative absorptive (upside down). It is quite possible to have a doublet that is in-phase absorptive, in-phase dispersive, antiphase absorptive, or antiphase dispersive (Fig. 6.12).

Evolution (rotation of net magnetization in the x'–y' plane) occurs during delays, and the direction and speed of motion in the x–y plane depend on the resonant frequency of the NMR line relative to the reference frequency ($v_o - v_r$). In general, when the NMR peak is not on-resonance, there are two kinds of evolution. We think of the *chemical shift* as the frequency of the whole resonance or peak due to a nucleus or group of equivalent nuclei,

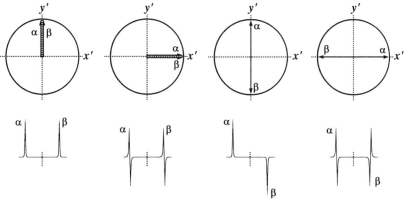

Figure 6.12

which is at the center of a symmetrical multiplet such as a doublet. *Chemical shift evolution* is the movement of this central position in the x'–y' plane during a delay—for a doublet this position is exactly in the middle of the two magnetization vectors representing the α and β components ("lines") of the doublet. *J-coupling evolution* is the divergence of the two vectors in a doublet away from this central position during the delay as they rotate at slightly different frequencies. Eventually, they reach the antiphase state and begin to converge again on the other side.

6.7 EXAMPLES OF J-COUPLING AND CHEMICAL SHIFT EVOLUTION

Consider the example of a ^{13}C–^{1}H pair with $J = 150$ Hz and ^{13}C chemical shift of 51.5 ppm on a Bruker DRX-600 instrument (Fig. 6.13). We know that the ^{13}C Larmor frequency is very close to one fourth of the ^{1}H frequency (600 MHz), so the spectrometer frequency is 150 MHz ($\gamma_C/\gamma_H = 1/4$) and 1 ppm is 150 MHz $\times$ 10^{-6} = 150 Hz. If the center of the ^{13}C spectral window is placed at 50 ppm, the rotating frame chemical shift for the doublet ($\Delta\nu$) is 225 Hz (1.5 ppm downfield of the center, which is 0 Hz). The H = α line of the ^{13}C doublet (in the absence of ^{1}H decoupling) is at 225 + 75 = 300 Hz in the rotating frame, and the H = β line is at 225 − 75 = 150 Hz. Immediately after a 90° pulse on the y' axis, both of the net magnetization vectors are on the x' axis. After a delay of 0.8333 ms (833.3 μs), which corresponds to 1/(8J), we can calculate the amount of rotation each vector has experienced:

$$\Theta(H = \alpha) = 360° \times 300\,\text{Hz} \times 0.0008333\,\text{s} = 90°\,(\text{counterclockwise})$$
$$\Theta(H = \beta) = 360° \times 150\,\text{Hz} \times 0.0008333\,\text{s} = 45°\,(\text{counterclockwise})$$

The angle between the two vectors is 45° and the center position (chemical shift position of the doublet) has rotated 67.5° (360° $\times$ 225 Hz $\times$ 0.0008333 s) to find itself exactly between the two vectors. We will represent this position with a dotted line (Fig. 6.13, $\tau = 1/(8J)$). Note that the H = α vector moves faster than the H = β vector because the NMR line that corresponds to it is farther from the center of the spectral window.

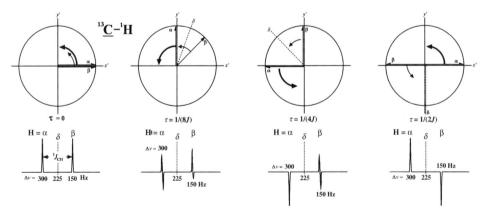

Figure 6.13

Both move counterclockwise, and they diverge from the center line even as the center line itself rotates counterclockwise at a rate of 225 Hz (the "chemical shift" of the doublet). After a total delay of 1.6667 ms (1/(4J)), the H = α vector is on the $-x'$ axis and the H = β vector is on the y' axis, with a 90° angle between them. After a total delay of 3.333 ms (the magic 1/(2J) value), the H = α vector has rotated 360° back to the x' axis and the H = β vector has rotated 180° to the $-x'$ axis. The pair is now in the antiphase relationship, and the chemical shift position (dotted line) is on the $-y'$ axis, having rotated three fourths of a full rotation in the counterclockwise direction. The chemical shift position is not a magnetization vector, it is just a bookkeeping device to keep track of where the vectors would be if they had not undergone J-coupling evolution (i.e., if they had not diverged from each other during the delay). We can draw what the spectrum would look like at each stage of evolution if we started recording an FID at that moment. Typically, we choose the reference axis to be the one that would give positive absorptive peaks if there were no delay: in this case, the x' axis. After a 1/(4J) delay, the H = α line is upside down (its vector is on the $-x'$ axis) and the H = β line is dispersive (its vector is on the y' axis). After a 1/(2J) delay, the H = α peak is positive absorptive and the H = β peak is negative absorptive.

6.7.1 Experimental Example: ^{1}H Observe with J-Coupling Evolution Only

The effect of J-coupling evolution can be observed directly using a sample of methyl iodide (CH$_3$I) in CDCl$_3$, which is enriched in ^{13}C to the level of 60% (Fig. 6.14). Now we are looking at the ^{1}H–^{13}C one-bond coupling from the point of view of the three equivalent protons, rather than from the point of view of ^{13}C. In the ^{1}H spectrum, we see a singlet at the center for the ^{12}CH$_3$I peak (placed on-resonance, $\Delta v = 0$, and representing 40% of peak intensity) and a doublet centered on the same chemical shift for the ^{13}CH$_3$I molecules (60% of the sample, each peak 30% of the total intensity). If we observe ^{13}C, we would see only

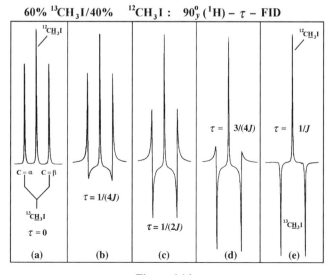

Figure 6.14

the ^{13}CH$_3$I (60%) part of the sample, and the peak would appear as a quartet due to coupling to the three protons, but observing ^{1}H we see a doublet because each of the three equivalent protons is coupled to only one ^{13}C nucleus. If we insert a delay of duration τ between the 90° ^{1}H excitation pulse and the start of the FID, we will see J-coupling evolution for the ^{13}CH$_3$I protons but the ^{12}CH$_3$I protons will not evolve. Let us put the ^{1}H 90° pulse on the y' axis, so the two vectors representing ^{1}H net magnetization (C = α and C = β) will both be rotated by the pulse from the $+z$ axis to the $+x'$ axis. We choose the $+x'$ axis as our reference axis, so the FID with no delay ($\tau = 0$, Fig. 6.14(a)) gives positive absorptive peaks for both components, at $+J/2$ (C = α) and $-J/2$ (C = β) in the spectrum. The net ^{1}H magnetization for ^{12}CH$_3$I is also rotated to the $+x'$ axis and gives a positive absorptive peak in the spectrum at the center of the spectral window ($\Delta \nu = 0$). If we increase the delay from zero to $1/(4J)$, the C = α component rotates counterclockwise (toward the y' axis) by an angle

$$\text{rotation} = \Delta \nu \times \tau = 1/(4J) \times J/2 = 1/8 \text{ cycle} = 45°$$

The C = β vector rotates clockwise (toward the $-y'$ axis) by the same amount. The two vectors have diverged (J-coupling evolution) to an angle of 90° between them but the center position between them has not moved (no chemical shift evolution: the chemical shift is on-resonance). The net magnetization for ^{12}CH$_3$I has not moved because it has a single peak which is on-resonance. We see a positive absorptive peak at the center and the outer peaks show some dispersive character, in opposite directions (Fig. 6.14(b)). Repeating the experiment with a longer delay, $t = 1/(2J)$, the C = α vector has now rotated by 90° and is in the y' axis, whereas the C = β vector has rotated 90° in the opposite direction and is on the $-y'$ axis. As the reference axis is $+x'$, both peaks are completely dispersive, but in the opposite sense. The ^{12}CH$_3$I vector remains on the $+x'$ axis and still gives a positive absorptive peak at the center (Fig. 6.14(c)). This is the *antiphase* state, which we always reach from the in-phase state after J-coupling evolution for a period of time equal to $1/(2J)$. The two vectors (C = α and C = β) have diverged to the maximum angle (180°) and are opposite to each other on the y' axis. Increasing the delay to $3/(4J)$, we have a rotation of 135° for each vector, and they have diverged to an angle of 270° between them. The C = α vector is halfway between the $+y'$ and $-x'$ axes and the C = β vector is halfway between the $-y'$ and $-x'$ axes. In the spectrum, both are nearly upside down, with some dispersive character in the opposite sense (Fig. 6.14(d)). Finally, after a delay of $\tau = 1/J$, each vector has rotated 180° in opposite directions, meeting each other on the $-x'$ axis. The doublet is in-phase on the $-x'$ axis, which is opposite to the reference axis ($+x'$), so the peaks are negative absorptive. The ^{12}CH$_3$I peak is still positive absorptive at the center of the spectrum (Fig. 6.14(e)). The $1/J$ delay has turned the ^{13}CH$_3$I doublet upside down without changing the ^{12}CH$_3$I singlet. We will see that this reversal of sign for a doublet after a $1/J$ delay is the basis of the attached proton test (APT).

The antiphase doublet (Fig. 6.14(c)) is dispersive because J-coupling evolution to the antiphase state moves the vectors by 90°, from the $+x'$ axis to the $+y'$ and $-y'$ axes. This dispersive antiphase doublet can be phase corrected by moving the reference axis from the $+x'$ axis to the $+y'$ axis (90° zero-order phase correction). Now the C = α peak is positive absorptive and the C = β peak is negative absorptive (Fig. 6.15) and the central ^{12}CH$_3$I peak is pure dispersive because the vector is on the $+x'$ axis and the reference axis is now $+y'$ (90° phase error).

EXAMPLES OF J-COUPLING AND CHEMICAL SHIFT EVOLUTION   219

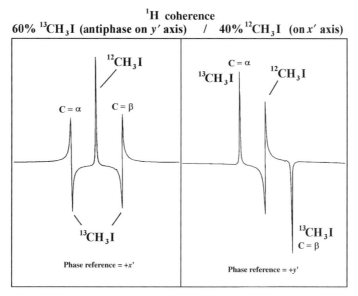

Figure 6.15

We usually ignore relaxation during short delays (e.g., 1/(2J) is usually milliseconds or tens of milliseconds) and consider it only when relaxation is essential to the experiment (e.g., inversion-recovery or nuclear Overhauser effect (NOE) experiments, typically hundreds of milliseconds or seconds for small molecules). Pulses are very short (tens of microseconds), so we do not usually worry about either evolution or relaxation during pulses. Although pulses may look "fat" in pulse sequence diagrams, they are really much shorter than most delays and their duration is not important in terms of evolution. Pulses lead to rotation of the net magnetization vector around the B_1 axis, always in the counterclockwise direction.

6.7.2 Summary of Evolution

In the rotating frame of reference, the x' and y' axes are rotating at the frequency of the pulse relative to the laboratory frame of reference. If the pulse frequency is not exactly equal to the Larmor frequency ("off-resonance pulse"), then the sample magnetization vector will not be stationary in the x'–y' plane after a 90° pulse. The pulse frequency corresponds to the center of the spectral window, so any peak that is not exactly in the center of the spectral window will lead to a magnetization vector that rotates in the x'–y' plane at a rate that is equal to the distance (in hertz) between the peak and the center of the spectral window. A peak in the upfield half of the spectral window will give rise to a magnetization vector that rotates clockwise (negative frequency) in the x'–y' plane, and a peak in the downfield half of the spectral window will give rise to a magnetization vector that rotates counterclockwise (positive frequency) in the x'–y' plane. A peak that is on-resonance (exactly at the center of the spectral window) will give rise to a stationary magnetization vector in the x'–y' plane. The motion of the magnetization vector in the rotating frame is called *evolution*.

6.8 THE ATTACHED PROTON TEST (APT)

APT is a technique for ^{1}H-decoupled ^{13}C spectra, which uses the phase (normal or upside down) of the ^{13}C peaks as a way to encode information about the number of protons attached to a carbon: C_q (quaternary carbon, no protons), CH (methine, one proton), CH_2 (methylene, two), or CH_3 (methyl, three). These spectra are called "edited" because the phase (positive absorptive or negative absorptive) is modified relative to a normal ^{13}C spectrum in order to encode additional information. APT gives all of the information of a normal carbon spectrum with somewhat reduced sensitivity, and it tells you whether the number of attached protons is odd (CH_3 or CH) or even (CH_2 or quaternary).

To illustrate the concept, Figure 6.16 shows the expected results of a normal ^{13}C spectrum and an APT spectrum of 4-hydroxy-3-methyl-2-butanone. The APT spectrum shows all carbons including the quaternary C=O and solvent carbons, and sorts the carbons into categories of CH and CH_3 ("up" peaks) and quaternary and CH_2 ("down" peaks). Note that sometimes APT spectra are presented "upside down" with CH and CH_3 peaks "down" and quaternary and CH_2 peaks "up", but the deuterated solvent peak (no attached protons) tells us how to interpret it.

The APT spectrum of sucrose in D_2O is shown in Figure 6.17. The quaternary carbon (fructose C2) is upside down, as are the three CH_2OH carbons (fructose C1 and C6 and glucose C6). The remaining carbons are all CHOH carbons, with positive absorptive phase. From the chemical shifts we can assign glucose C1 as the most downfield of the CH carbons (anomeric carbon—two oxygens attached) and we can easily distinguish fructose C2 (anomeric and quaternary) from the CH_2OH carbons (typical region 60–70 ppm). Figure 6.18 shows the downfield region of the APT spectrum of cholesterol in $CDCl_3$, aligned with the ^{13}C spectrum acquired with the same number of scans for comparison. The $CDCl_3$ solvent peak (1:1:1 "triplet") is upside down at 77.0 ppm (technically, it is a quaternary carbon because it has no attached protons), as is the quaternary olefinic carbon

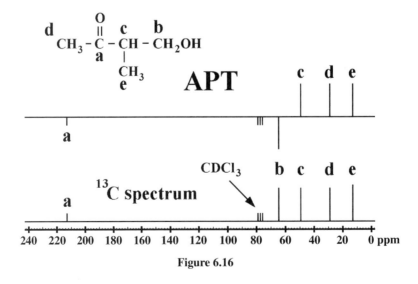

Figure 6.16

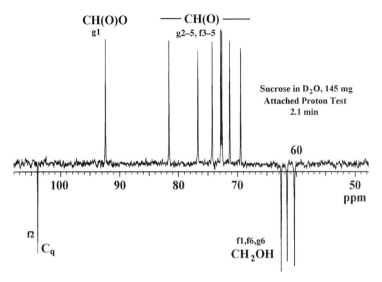

Figure 6.17

(C5, 140.8 ppm). The other olefinic carbon (C6, 122.0 ppm) is a CH and gives a positive APT peak. The single oxygenated carbon (C3, CHOH at 72.0 ppm) is also positive (CH), as are the three downfield CH carbons (C9, C14 and C17, 50–60 ppm) that are adjacent to sp^3-hybridized quaternary carbons (sterically crowded). The upfield region is shown in Figure 6.19, also aligned with the ^{13}C spectrum. The structural assignments (carbon numbers), which are derived from analysis of two-dimensional spectra, are also shown. In the ^{13}C spectrum, there is a peak at 32.0 ppm that represents two overlapped carbons peaks: a CH and a CH_2. Because these have slightly different chemical shifts, the opposite sign in the APT spectrum allows them to be clearly resolved. A broad peak at 42.3 ppm in the ^{13}C spectrum represents two peaks that are not resolved: a CH_2 and a C_q. These are also

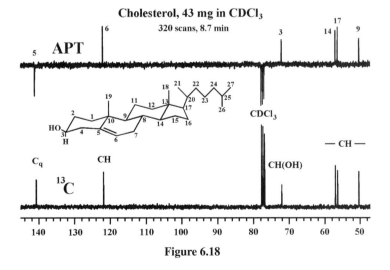

Figure 6.18

222 THE SPIN ECHO AND THE ATTACHED PROTON TEST (APT)

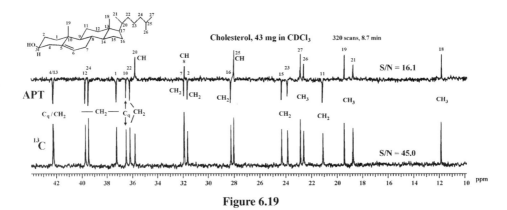

Figure 6.19

not resolved in the APT spectrum because both lead to negative peaks. In all, there are 13 positive peaks and 14 negative peaks (counting as two the broad peak at 42.3 ppm) in the APT spectrum. This is consistent with the structure of cholesterol, which has 3 quaternary carbons and 11 CH_2 carbons (14 negative peaks), and 8 CH carbons and 5 CH_3 carbons (13 positive peaks).

You might wonder why anyone would do a simple ^{13}C spectrum when an APT spectrum gives the same information (chemical shifts and intensities) plus the added distinction of spectral editing (CH and CH_3 opposite in phase to C_q and CH_2). But the signal-to-noise ratio is 45.0 for the simple ^{13}C spectrum and 16.1 for the APT with the same number of scans (Fig. 6.19); to get the same signal-to-noise ratio for the APT would require 2.5 times as long an acquisition. The reduced sensitivity for APT is the result of T_2 relaxation during the long $2/J$ (13.33 ms) delay of the spin echo.

6.8.1 Understanding APT with the Vector Model

APT is a very simple and elegant method to distinguish the number of protons attached to a carbon atom. Recall that in the rotating frame of reference, the net magnetization vector stands still in the x'–y' plane for resonance frequencies exactly at the center of the spectral window ("on resonance"). Off-resonance lines give rise to magnetization vectors that rotate in the x'–y' plane at an angular velocity $\Delta \nu$, where $\Delta \nu$ is the frequency offset (in hertz) from the center of the spectral window. Peaks downfield of the center will rotate with a positive angular velocity (counterclockwise from x' to y', $-x'$, $-y'$, etc.) and peaks in the upfield half of the spectral window will rotate in the opposite direction (clockwise). In the APT experiment, ^{13}C magnetization is rotated into the x–y plane by a 90° ^{13}C pulse, and then allowed to precess (without ^1H decoupling) for a short period of time equal to $1/J$, where J is the one-bond ^1H–^{13}C coupling constant (~150 Hz). The effect of this precession period is shown in Figure 6.20 for a ^{13}C nucleus with a single ^1H attached, assuming that the ^{13}C resonance frequency (center of the doublet) is exactly on-resonance (center of the spectral window). The 90° ^{13}C pulse (on the y' axis) rotates the ^{13}C z magnetization onto the x' axis of the rotating frame of reference. The downfield component of the ^{13}C doublet, which arises from ^{13}C nuclei attached to ^1H nuclei that are in the α state, begins to rotate in the x'–y' plane counterclockwise (ccw) toward the y' axis with angular frequency $J/2$ in hertz. The upfield component of the ^{13}C doublet, which arises from ^{13}C attached to ^1H nuclei

THE ATTACHED PROTON TEST (APT)   223

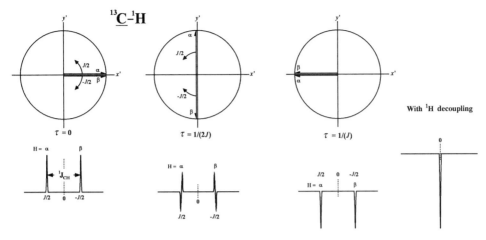

Figure 6.20

that are in the β state, rotates in the opposite direction (cw) in the x'–y' plane toward the $-y'$ axis with angular frequency $-J/2$ in hertz. After a period of time τ equal to $1/(2J)$, the H = α vector is on the y' axis and the H = β vector is on the $-y'$ axis; this is the antiphase state. After a period of time τ equal to $1/J$, both components have rotated exactly $360° \times J/2 \times 1/J = 180°$, meeting at that moment on the $-x'$ axis. If we begin acquisition at this point, the FID will be exactly $180°$ out of phase from a normal FID acquired without the $1/J$ time delay and will yield an upside-down doublet in frequency domain (reference axis $= +x'$). If we apply ^{1}H decoupling during the acquisition of the FID, the two components (H = α and H = β) now have the same resonant frequency and we observe a single upside-down peak at the center of the spectral window. A quaternary carbon (on-resonance) has a single magnetization vector that will not budge from the x' axis during the whole $1/J$ delay period and will give a normal spectrum with a positive peak (Fig. 6.21). A ^{13}CH$_2$ group will give a triplet ^{13}C spectrum (Fig. 6.22). With the central peak of the triplet on-resonance, the downfield component of the triplet (with attached ^{1}H nuclei H$_1$ = α, H$_2$ = α) rotates

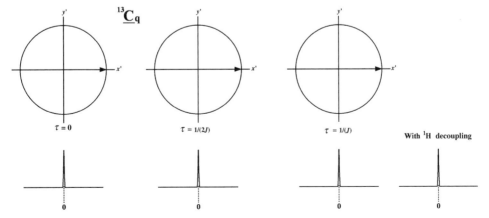

Figure 6.21

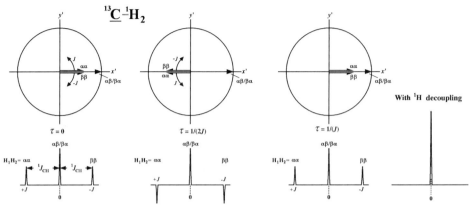

Figure 6.22

with angular frequency J Hz (ccw) whereas the central peak (attached ^{1}H nuclei $\alpha\beta$ or $\beta\alpha$) will remain on the y' axis (on-resonance) and the upfield component (^{1}H nuclei $\beta\beta$) will rotate with angular frequency $-J$ Hz (cw). Note that each attached proton in the α state leads to an increase in the resonant frequency of the ^{13}C nucleus (downfield shift) of $J/2$ Hz, and each attached proton in the β state leads to a decrease in resonant frequency (upfield shift) of $J/2$ Hz. We are assuming that the two protons are equivalent, so the effect of each proton is the same (^{13}C–H$_1$ coupling = ^{13}C–H$_2$ coupling) and they cancel out if one is α and the other is β. After a period of time τ equal to $1/(2J)$, the two outer peaks ($\alpha\alpha$ vector and $\beta\beta$ vector) will have traveled 180° ($J \times 1/(2J) = 1/2$ cycle) in opposite directions to meet on the $-x'$ axis, whereas the inner peak ($\alpha\beta/\beta\alpha$ vector) remained stationary on the x' axis. After a total delay of $1/J$, both off-resonance components will have made a complete rotation (360° and −360°) back to the positive x' axis, so that all three components of the triplet will give positive peaks after acquisition and Fourier transform. With ^{1}H decoupling, the ^{13}C triplet will "collapse" into a single on-resonance positive peak. Similar arguments can be used to show that a ^{13}C quartet (^{13}CH$_3$ group) will end up with all four components on the $-x'$ axis after a period of time $1/J$ (Fig. 6.23). The outer lines $\alpha\alpha\alpha$ (three times $J/2$ downfield shift) and $\beta\beta\beta$ ($3J/2$ upfield shift) rotate 270° ($3J/2 \times 1/(2J) = 3/4$ cycle) in a delay of $1/(2J)$, with the $\alpha\alpha\alpha$ vector rotating ccw from $+x'$ to $+y'$, $-x'$ and finally to $-y'$. The $\beta\beta\beta$ vector rotates cw at the same rate to $-y'$, $-x'$ and ending at $+y'$. After another $1/(2J)$ period, the $\alpha\alpha\alpha$ and $\beta\beta\beta$ vectors rotate another 3/4 turn: $\alpha\alpha\alpha$ from $-y'$ ccw to x', y', and $-x'$ and $\beta\beta\beta$ from $+y'$ to $+x'$, $-y'$, and $-x'$. Thus, after a total delay of $1/J$, both of the "outer line" vectors are on $-x'$. The "inner line" vectors correspond to the line at $+J/2$, composed of all ^{13}C nuclei with two Hs in the α state and one in the β state ($\alpha\alpha\beta$, $\alpha\beta\alpha$, and $\beta\alpha\alpha$), and the line at $-J/2$, composed of all ^{13}C nuclei with one H in the α state and two Hs in the β state ($\beta\beta\alpha$, $\beta\alpha\beta$, and $\alpha\beta\beta$). In each case, the two opposing ^{1}H spins cancel out and only the effect of the remaining one is observed, leading to a shift of $J/2$ Hz from the chemical shift position of ^{13}C. The vector corresponding to the $+J/2$ line behaves just like the H = α vector in the CH case (Fig. 6.20) and moves to the $+y'$ axis after $\tau = 1/(2J)$ and to the $-x'$ axis after $\tau = 1/J$. The vector with frequency $-J/2$ moves to the $-y'$ axis after $\tau = 1/(2J)$ and to the $-x'$ axis after $\tau = 1/J$. Thus, at the end of the $1/J$ delay all four vectors are on the $-x'$ axis, leading to an upside-down quartet in the ^{13}C spectrum. With ^{1}H

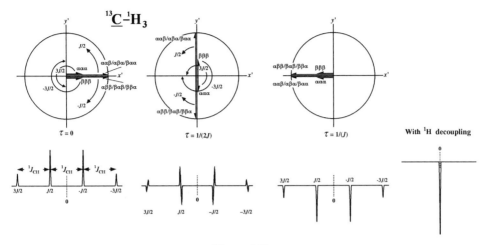

Figure 6.23

decoupling, we have an upside-down (negative absorptive) singlet at the center of the spectral window.

With this simple process, we have encoded information about the number of attached protons into the phase of the ^{13}C peak. Proton decoupling during the acquisition period will give a spectrum with single peaks for each carbon resonance, pointing either up or down according to the number of attached protons.

Group	Pattern	Angular Frequency in hertz	Rotation in cycles after $\tau = 1/J$	Axis	Phase
C (quaternary)	Singlet	0	0	$+x'$	Positive
CH	Doublet	$J/2, -J/2$	$0.5, -0.5$	$-x'$	Negative
CH$_2$	Triplet	$J, 0, -J$	$1, 0, -1$	$+x'$	Positive
CH$_3$	Quartet	$3J/2, J/2, -J/2, -3J/2$	$1.5, 0.5, -0.5, -1.5$	$-x'$	Negative

If we choose the $+x'$ axis as our reference, we will have positive peaks for the C_q and CH$_2$ groups, and negative peaks for the CH and CH$_3$ groups. Usually, the APT spectrum is phased the other way, using the $-x'$ axis for the reference axis, so that the CH and CH$_3$ peaks are positive and the C_q and CH$_2$ peaks are negative.

There is, however, one very important assumption we have made that is not practical: The carbon resonance was assumed to be on-resonance. Obviously, we cannot guarantee this for all carbon resonances in a spectrum because different ^{13}C peaks have different resonant frequencies. What happens if the carbon chemical shift is not at the center of the spectral window? The magnetization, which starts on the y' axis, will precess in the x–y plane at an angular frequency $\Delta \nu$, where $\Delta \nu$ is the position of the ^{13}C resonance relative to the center of the spectral window on a hertz scale. Components of the multiplet will rotate a little faster or a little slower relative to the central component; for example, the three components of a triplet will rotate at angular frequencies $\Delta \nu + J$, $\Delta \nu$, and $\Delta \nu - J$. This additional rotation due to the chemical shift ($\Delta \nu$) will affect the phase of the magnetization when acquisition is started, and the phase information of interest, which is determined by the number of attached protons, will be hopelessly lost in the jumble of chemical shift effects. What we need is a trick that will allow the *J*-coupling precession ("evolution") to proceed but will somehow cancel out the effect of the chemical shift evolution. In other words, we need some control over *what kind* of evolution occurs during a delay. Such a trick exists, of course, and it is called the *spin echo*.

6.9 THE SPIN ECHO

The spin echo is one of the fundamental building blocks of pulse sequences and is used in a variety of 1D and 2D experiments. Consider first a ^{13}C spectrum with three peaks, each one representing a separate ^{13}C position in the molecule (singlet). The first half of the spin-echo sequence is very much like the primitive sequence described in Figure 6.11: a 90° ^{13}C pulse puts the sample magnetization on the y' axis in the rotating frame and a delay of duration τ follows. The magnetization vectors of the three ^{13}C resonances in the sample will "fan out" in both directions away from the y' axis with angular frequencies $\Delta \nu$ depending on the resonance offset $\Delta \nu$ of each NMR peak relative to the center of the spectral window (Fig. 6.24). Then a 180° ^{13}C pulse is applied along the y' axis of the rotating frame. This rotates all three of the magnetization vectors in the *x–y* plane to the opposite side of the y' axis. Essentially, if you view all of the magnetization vectors as lines radiating out from the center of a pancake, you are flipping the pancake along the y' axis. Now a delay of the exact same duration τ follows. Precession continues, with each carbon resonance rotating in the *same direction and velocity* as before, depending on its resonance offset (position of the peak in the spectrum). Because the two delays are of the same length, all of the magnetization vectors will be exactly aligned along the y' axis at the end of the second

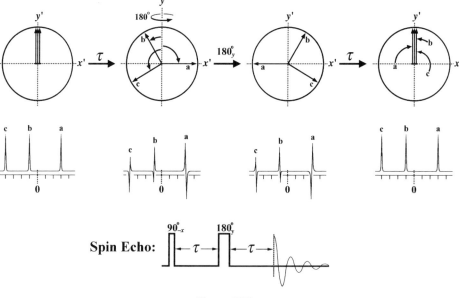

Figure 6.24

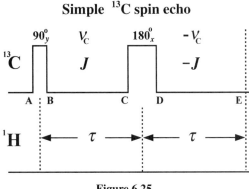

Figure 6.25

delay (Fig. 6.24, right). This is analogous to a foot race around a circular track. The starting gun is fired; the runners fan out according to each runner's characteristic speed. After a certain period of time τ, the gun is fired again and all of the runners turn around and run in the opposite direction. As long as all runners maintain the same characteristic speeds in the second half of the race, all of them will reach the starting line at the same time, exactly 2τ after the start of the race. It is this coalescence of all the runners (or magnetization vectors) at the starting point at the end of the 2τ period that is referred to as the "echo," because it bounces back and reappears after a specific length of time. The spin echo is sometimes described as "time reversal": With the spin echo we have reversed the effects of chemical shift evolution, effectively making time stand still for a period of time 2τ. The second half of the echo (second τ delay) is sometimes called the "refocusing" period because the chemical shift effects are focused back to the starting axis.

Consider the example of Figure 6.13 once again using a simple ^{13}C spin-echo sequence (Fig. 6.25) with the delay τ set to $1/(8J)$. We saw how the initial $90°_y$ pulse on the ^{13}C channel rotates the two ^{13}C net magnetization vectors (H = α and H = β) representing the two components of the ^{13}C doublet from the $+z$ axis (equilibrium) to the $+x'$ axis. After a delay of $1/(8J)$, the H = α vector has rotated one-fourth turn ccw and lies on the $+y'$ axis, whereas the "slower" H = β vector has rotated only $45°$ and lies between the $+x'$ and $+y'$ axes (Fig. 6.26(c)). If we apply the ^{13}C $180°$ pulse now ($\tau = 1/(8J)$) on the $+x'$ axis, the H = α vector is rotated to the $-y'$ axis and the H = β vector ends up between the $+x'$ and $-y'$ axes (Fig. 6.26(d)).

It is very important to note that a $180°$ pulse (sometimes called an "inversion" pulse) does not always move a vector exactly to the opposite side of the $x'-y'$ plane! The H = α vector in this case sweeps out a plane (the $y'-z$ plane) as it rotates up to the $+z$ axis and down to the $-y'$ axis, but the H = β vector sweeps out a cone as it rotates up to halfway between the $+x'$ and z axes and down to halfway between the x' and $-y'$ axes. To get this concept clear you might want to think of the B_1 vector (the pulse) as a physical object, an "axle" or broomstick glued to the net magnetization vector, another broomstick. The pulse "twists" the B_1 vector (the axle) on the x' axis by a counterclockwise rotation of $180°$, whereas the net magnetization vector is dragged along because it is physically attached. If the B_1 vector forms a $90°$ angle with the net magnetization, the twisting moves the net magnetization to the opposite side—in this case from the $+y'$ axis to the $-y'$ axis. But if the net magnetization is at a different angle to B_1—$45°$ in this example—it rotates in a conical path, reversing the sign of its y' axis projection without changing the x' axis projection at all. In the extreme case where the net magnetization is on the

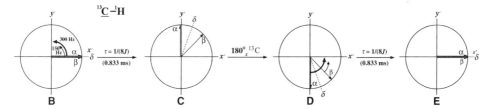

Figure 6.26

same axis as the B_1 vector (in this case, on the $+x'$ or $-x'$ axis), it is not affected at all by the pulse, as we can visualize a broomstick colinear with the B_1 broomstick being twisted but not changing its direction as the B_1 "axle" is rotated. To make this distinction, we refer to a 180° pulse as an "inversion" pulse only when the net magnetization moves from $+z$ to $-z$, and we use the term "refocusing" pulse when the net magnetization starts and ends in the x'–y' plane.

Continuing with the simple spin-echo sequence (Fig. 6.25), during a second $1/(8J)$ delay the H = α vector will rotate another one-fourth turn ccw, ending on the x' axis, and the H = β vector will rotate by 45° ccw, also landing on the x' axis (Fig. 6.26(e)). This is exactly where we started after the initial ^{13}C 90° pulse, with both vectors on the x' axis: the simple spin echo refocuses both the J-coupling evolution (divergence of the α and β components) and the chemical shift evolution (rotation of the center position—the dotted line representing the ^{13}C chemical shift of the doublet). Basically, we have wasted a period of time equal to $1/(4J)$ and nothing has happened! This might seem like a pointless exercise, but later we will see that there might be things you need to do during that delay, and the spin echo is a way to get everything back to where you started.

Applying this to the APT experiment, we could solve the problem of chemical shift differences affecting the final phase by applying a 180° pulse in the middle of the $1/J$ delay period, effectively making a spin echo with delay $\tau = 1/(2J)$. The chemical shift evolution that occurs during the first half is now refocused in the second half, and we do not have to require that the ^{13}C peak be on-resonance. But the desired information would also be lost, because the evolution of magnetization vectors under the influence of J coupling would also be canceled in the second half of the spin echo! Each line of a ^{13}C multiplet (doublet, triplet, or quartet) would undergo the same evolution and "de-evolution" described above and end up on the same axis at the end of the spin echo. We can, however, turn J coupling on and off at will using the proton decoupler. By using a spin-echo delay of $\tau = 1/J$ and turning the decoupler on during the second half of the spin-echo only, we generate the desired phase encoding due to J coupling during the first half of the spin echo and refocus the chemical shift effects during the second half, when each ^{13}C resonance behaves as a single line centered at the chemical shift position (Fig. 6.27). Because the proton decoupler eliminates the J couplings for the second half of the spin echo, the divergence of multiplets that occurred during the first half does not get refocused during the second half. Note that the ^{1}H decoupler is on all of the time except for the first $1/J$ delay of the spin echo: During the relaxation delay, the decoupler is "pumping up" the ^{13}C z magnetization above M_o due to the heteronuclear NOE, and during the acquisition of the FID the decoupler is collapsing the ^{13}C multiplets into singlets so that we see only one line for each ^{13}C resonance.

Often a second, short spin echo is added to this sequence so that pulses shorter than 90° (e.g., 30°) can be used for the first pulse. A starting pulse of less than 90° is desirable to

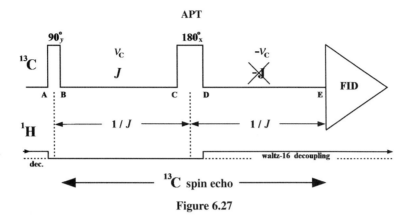

Figure 6.27

allow for shorter relaxation delays, but in the simple APT sequence (Fig. 6.27) the sample magnetization left on the z axis is inverted by the 180° pulse of the spin echo and a long relaxation delay would be required to bring it back from the $-z$ axis to equilibrium by T_1 relaxation. The combination of two spin echoes (two 180° pulses) brings this z magnetization back to the positive z axis and allows for a more rapid return to equilibrium.

6.9.1 Measurement of T_2 Values Using Multiple Spin Echoes: CPMG

We saw in Chapter 5 that in a perfectly homogeneous magnetic field the FID decays to zero in an exponential fashion with characteristic time T_2. In the real world, field inhomogeneity makes the FID decay faster (Fig. 6.28), with a characteristic time we can call T_2^* ($T_2^* < T_2$). The envelope of the FID drops to 63% (e^{-1}) of its original size after this period of time T_2^*. If we look at decay *rate* rather than characteristic time, the intrinsic rate ($R_2 = 1/T_2$) of decay of coherence is added to the rate of decay due to "fanning out" of individual vectors from different physical locations in the sample ($R_i = 1/T_2^i$, the inhomogeneity decay rate) to obtain the experimental rate of decay ($R^* = 1/T_2^*$):

$$R_2^* = R_2 + R_2^i, \quad 1/T_2^* = (1/T_2) + (1/T_2^i)$$

The Fourier transform converts the FID into a Lorentzian peak with absorptive lineshape (after phase correction). The full width of this peak at one half of the peak's height (the "linewidth") is inversely related to the decay time constant of the FID, T_2^*:

$$\text{linewidth} = \Delta v_{1/2} = 1/(\pi T_2^*)$$

So just by measuring the width of the NMR peak we can determine the time constant for decay of the FID, but this is not an interesting number because it depends on shimming. The interesting number, which is a fundamental physical measurement for that particular spin in a specific environment, is the T_2 value. How can we extract the T_2 value from the easily measured T_2^* value, which is a combination of T_2 and the inhomogeneity decay constant T_2^i? Recall that in a spin echo the differences in chemical shift evolution that occur during a delay are "refocused" or removed during the second half of the spin echo following the 180° pulse. These differences may be real differences due to different positions within a molecule, or differences in resonant frequency due to the different locations of identical

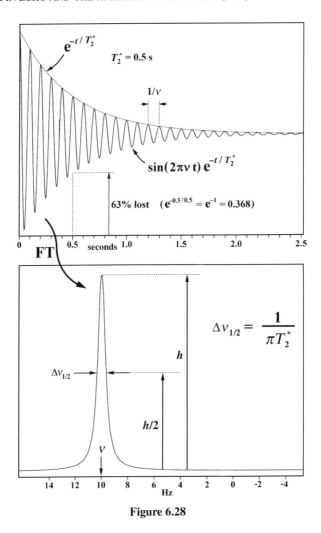

Figure 6.28

spins within the sample volume in an inhomogeneous magnetic field. It does not matter: In either case the vectors that lag behind in precession during the first half of the spin echo have less distance to travel in the second half, and the vectors that pull ahead in the first half have farther to travel in the second half. All of these individual vectors ("isochromats" if you like fancy words) representing different locations within the sample come together at the end of the second delay, creating a maximum in the net magnetization measured throughout the sample (an "echo": Fig. 6.29). In the extreme case, where field homogeneity is very bad ($T_2^i \ll T_2$), we would see the transverse magnetization we are measuring in the FID (e.g., M_x) decay rapidly (e^{-t/T_2^*}) during the first half of the FID as the individual vectors "fan out" in the x–y plane and become evenly distributed in all directions. After the 180° pulse, these vectors begin to gather together and the transverse magnetization grows exponentially until it reaches a maximum when all the vectors cross the "finish line" (the echo). After they cross, the vectors fan out again just as rapidly due to field inhomogeneity and the coherence is lost (Fig. 6.29, top).

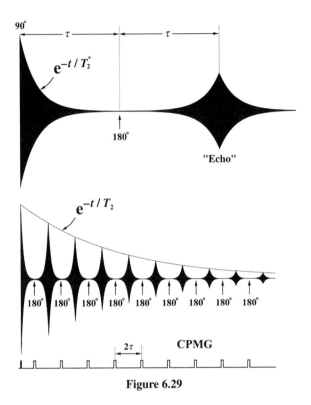

Figure 6.29

The interesting thing is that the maximum intensity of the FID at the "top" of the echo is still less that that at the start of the FID: not all of the coherence is recovered by refocusing in the second half of the spin echo. The part that is lost is the intrinsic decay, the loss of coherence due to pure T_2 relaxation, a fundamental relaxation process. The spin echo simply gets back the losses due to inhomogeneity of the magnetic field (T_2^i losses). This gives us a method to measure T_2: We could repeat the spin-echo experiment a number of times with different echo delays (t values) and start the acquisition of the FID at the "top" of the echo:

$$90° - \tau - 180° - \tau - \text{FID}$$

The signal intensity at the start of the FID is proportional to the peak height after Fourier transformation, so we could make a plot of peak height versus τ delay and fit the exponential decay to a theoretical curve to measure the T_2 value. This is the T_2 equivalent of the inversion-recovery experiment (Section 5.8) for measurement of T_1.

There is still one problem with this method. As we make the τ delay longer and longer, it is possible that some molecules will *diffuse* from one part of the sample to another so that the spins do not experience the same magnetic field (B_o) in the two delays (τ) of the spin echo. The refocusing of the spin echo works only if the inhomogeneity experienced in the first half is identical to that experienced in the second half for each of the identical spins. This diffusion problem will lead to loss of signal at the top of the echo in addition to the intrinsic T_2 loss and will be more pronounced for smaller molecules or in less viscous solvents. One way to avoid this is to keep the echo time (τ) short and use a large number of repeated spin echoes (Fig. 6.29, bottom). Instead of increasing the τ value to explore the T_2 decay curve, we increase the number of repeats of the spin-echo unit (τ–180°–τ) while keeping τ constant. As long as the τ delay is quite short, the loss of signal due to diffusion is kept to a minimum. This method is called CPMG, or Carr–Purcell–Meiboom–Gill, after the four investigators who developed it.

6.10 THE HETERONUCLEAR SPIN ECHO: CONTROLLING *J*-COUPLING EVOLUTION AND CHEMICAL SHIFT EVOLUTION

As we move on to more complex and more powerful pulse sequences for *heteronuclear* (e.g., ^{1}H–^{13}C) experiments, we would like to use the spin echo to refocus only one of the two kinds of evolution, J-coupling evolution *or* chemical shift evolution. This can be done in a very simple and elegant way by adding another 180° pulse to the ^{1}H channel, simultaneous with the ^{13}C 180° pulse. This "heteronuclear" spin-echo sequence is shown in Figure 6.30. Notice that we now are using the ^{1}H channel (lower line) for more than just decoupling: We are delivering high-power pulses of defined duration, calibrated for a specific rotation—in this case, 180°. In the early commercial FT spectrometers this was an advance in hardware capability, because power could not be rapidly and repeatedly switched from high power ("hard" pulses) to low power (waltz-16 decoupling), so two separate sources of ^{1}H RF were required.

Again using the example of Figure 6.26, consider what happens if we include the ^{1}H 180° pulse at the center of the spin echo. At this point in the sequence (time "C" in Fig. 6.30), at the end of the first delay, we have the H = α vector on the +y' axis and the H = β vector halfway between the +x' and +y' axes (Fig. 6.31 C). As before, the ^{13}C 180° pulse on the x' axis rotates the H = α vector to the −y' axis and the H = β vector to a position halfway between the −y' axis and the x' axis. We can consider the effect of the ^{1}H 180 pulse after the ^{13}C 180° pulse, even though they are simultaneous. The ^{1}H 180° pulse does not rotate the vectors because they represent ^{13}C net magnetization and are not affected by a pulse at the ^{1}H frequency. It does, however, affect the ^{1}H nuclei that are attached to (and J-coupled to) to the ^{13}C nuclei. The effect is to change *every* ^{1}H nucleus that was in the α state to the β state and *every* ^{1}H nucleus that was in the β state to the α state. This means that our "H = α" vector, representing the net magnetization of all ^{13}C nuclei whose attached ^{1}H is in the α state, is now an "H = β" vector, because all of those protons are now in the β

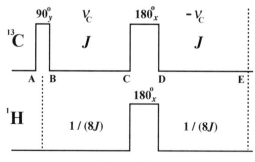

Figure 6.30

THE HETERONUCLEAR SPIN ECHO   233

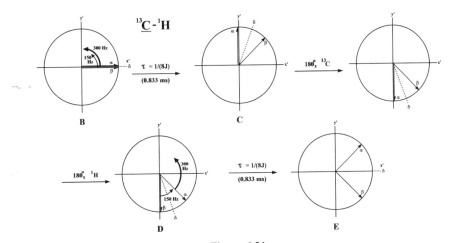

Figure 6.31

state. Likewise, our "H = β" vector can now be called an "H = α" vector because all of the protons attached to those ^{13}C nuclei are now in the α state. In other words, the effect of the ^{1}H 180° pulse is to "swap the labels" on the ^{13}C net magnetization vectors *without moving them*. Now we have the H = α vector halfway between the $+x'$ and $-y'$ axes and the H = β vector on the $-y'$ axis (Fig. 6.31 D). Next we have the second delay, of duration $1/(8J)$, of the spin echo. The H = α vector will rotate one-fourth turn ccw as before, ending up halfway between the $+x'$ and $+y'$ axes, and the H = β vector will rotate 45° ccw as before, ending up halfway between the $+x'$ and $-y'$ axes (Fig. 6.31 E). Note that the only thing that is different in the second half of the spin echo is the *behavior* of the two vectors, because we know that the H = α component of the ^{13}C doublet has a rotating-frame frequency (Δv) of 300 Hz and the H = β component has a rotating-frame frequency of 150 Hz. By flipping the ^{1}H from the α state to the β state, we changed the effective magnetic field experienced by the ^{13}C nucleus, decreasing its Larmor frequency by 150 Hz (J).

So how does the result compare to the simple spin echo (without the ^{1}H 180° pulse: Fig. 6.26)? The two vectors diverged during the first half of the spin echo, ending up with a 45° angle between them: this is J-coupling evolution. The center position between the two vectors rotated 67.5° from the $+x'$ axis to a position three-fourths of the way from the $+x'$ axis toward the $+y'$ axis: this is chemical shift evolution. In the simple spin echo, both of these types of evolution were reversed during the second half, as the two vectors converged toward each other and the center position moved back to the $+x'$ axis. But with the ^{1}H 180° pulse included, the divergence of the two vectors (J-coupling evolution) continues in the second half, resulting in an angle of 90° (twice as large) between them at the end of the delay. J-coupling evolution is "active" throughout the pulse sequence, and we have a divergence of 90° that is the result of J-coupling evolution for a total period of $1/(4J)$, which is the sum of the two delays. Recall that the "magic time" for J-coupling evolution is $1/(2J)$, which takes us from in-phase to antiphase (0° angle between the two vectors of a doublet to 180°), so a delay of $1/(4J)$ should give a divergence of 90°.

We can say that the 180° ^{13}C pulse reverses the J-coupling evolution (Fig. 6.26), but the 180° ^{1}H pulse "reverses the reversal" so that the two vectors continue to diverge during the second half (Fig. 6.31). Another way to look at it is that ^{13}C chemical shift evolution (a result of the

B_o field interacting with the ^{13}C nucleus) is sensitive only to ^{13}C pulses, because only these can rotate the ^{13}C net magnetization vectors. But J-coupling evolution is a mutual interaction between the ^{13}C nucleus and the ^{1}H nucleus, so both of the 180° pulses affect it, and effectively they cancel each other out, just as the product of two negative numbers is a positive number.

What about the chemical shift evolution? During the second half, the center position between the two vectors (dotted line in Fig. 6.31) rotates ccw from a position three fourths of the way between the $+x'$ and $-y'$ axes to end up back on the $+x'$ axis. We have refocused the chemical shift evolution which occurred during the first delay. The chemical shift evolution of the ^{13}C net magnetization "sees" only the ^{13}C 180° pulse, so it is refocused just as it was in the simple spin echo. In Figure 6.30, we represent the overall effect of the spin echo by writing "$+\nu_C$" (^{13}C chemical shift evolution) and "$+J$" (J-coupling evolution) in the space of the first delay, and "$-\nu_C$" (^{13}C chemical shift refocusing) and "$+J$" (J-coupling evolution continuing in the same sense) in the space of the second delay. Overall, we have $+\nu_C$ for time $1/(8J)$ and $-\nu_C$ for time $1/(8J)$, for a net chemical shift evolution of zero. We also have J evolution for $1/(8J)$ and again J evolution in the same sense for $1/(8J)$ for a total J evolution of $1/(4J)$, which leads to a total divergence of the two vectors by J Hz times $1/(4J)$ seconds or one-fourth turn (90°). This way of getting a simple overview of the effects of a pulse sequence building block will be crucial to your understanding of more complex experiments: Once you are very comfortable with the details of vector rotation in the $x'-y'$ plane and the effects of pulses, you no longer will need to consider these details—only the overall effect of the pulse sequence building block is important. You will eventually be able to break up complicated pulse sequences into well-understood building blocks and guess the effect of each building block on the net magnetization.

This is a very powerful pulse sequence building block! We now have control over the two kinds of evolution. We will see how in many heteronuclear experiments we want to get the ^{13}C (or the ^{1}H) into the "magic" antiphase state, but we do not want to complicate things with the chemical shift evolution, which would be different for every peak in the spectrum leading to a great confusion of phases and peak shapes at the end of the experiment. Let's see if we can apply this technique to the APT sequence. Recall that the goal of the APT sequence is to have J-coupling evolution for a period of time $1/J$ to introduce the editing effect (CH and CH_3 phases reversed, C_q and CH_2 phases unaffected). We do not want to mess up all the phases by having chemical shift evolution, so we used a simple spin echo of total duration $2/J$ with the ^{1}H decoupler on for one of the $1/J$ delays (Fig. 6.27). Chemical shift evolution that occurs during the first half ($+\nu_C$) is reversed during the second half ($-\nu_C$) because of the ^{13}C 180° pulse. Whether the ^{1}H decoupler is on or off is irrelevant to ^{13}C shift evolution. We have J-coupling evolution ($+J$) during the first half, when the decoupler is off, and no J-coupling evolution during the second half. So the net J-coupling evolution is $+J$ for a period of time $1/J$, leading to a divergence of the two vectors of a CH group by an angle of J Hz times $1/J$ seconds or one cycle (360°). The two vectors are once again in-phase but on the opposite side of the $x'-y'$ plane, each individual vector having traveled by an angle of $J/2$ times $1/J$ or one half cycle (180°): Each component of the doublet considered alone is $J/2$ Hz away from the center or chemical shift position. Now let's try to achieve the same effect but with an "advanced" heteronuclear spin echo including a ^{1}H 180° pulse at the center along with the ^{13}C 180° pulse. Using the sequence of Figure 6.30, we can set the delays to $1/(2J)$ each so that we get a total J-coupling evolution period of $1/J$ without any chemical shift evolution. That is it! We do not need to play around with the ^{1}H decoupler, except to turn it on during the acquisition of the FID. This "improved" APT

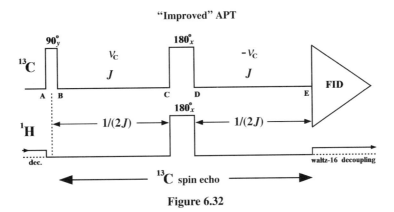

Figure 6.32

sequence is shown in Figure 6.32. In what way is it superior to the original APT sequence? The total time that we have ^{13}C net magnetization in the x'–y' plane is reduced from $2/J$ to $1/J$, cutting in half the loss of NMR signal due to T_2 relaxation. Thus, the loss of sensitivity that we see in the APT compared to a simple ^{13}C spectrum could be cut in half! Hardly anyone uses this sequence, however, because old methods die hard. We will also see in the next chapter that an even better alternative to the APT experiment exists, one that is actually quite a bit *more* sensitive than the simple ^{13}C experiment.

As a final illustration of the power and versatility of the heteronuclear spin echo, let's see if we can design a spin-echo sequence that allows chemical shift evolution but refocuses J-coupling evolution. We start with the simple spin echo, and we remove the ^{13}C 180° pulse because it is responsible for the reversal of chemical shift evolution (changing $+\nu_C$ to $-\nu_C$ in our shortcut notation). But now we do not have any refocusing, just two delays of equal duration. How can we reverse the J-coupling evolution without rotating the ^{13}C magnetization vectors? Remember that because the J-coupling evolution is due to a mutual interaction of the ^{1}H and the ^{13}C nucleus, it can be reversed with a 180° pulse on *either* the ^{1}H channel or the ^{13}C channel, and that 180° pulses on both channels cancel each other in this effect. So we can put our 180° pulse on the ^{1}H channel *only*, where it will reverse the J-coupling evolution without affecting the ^{13}C chemical shift evolution. The pulse sequence is shown in Figure 6.33, with the notations "$+\nu_C$" and "$+J$" in the first delay and "$+\nu_C$" and "$-J$" in the second delay. The overall effect of this pulse sequence is ^{13}C

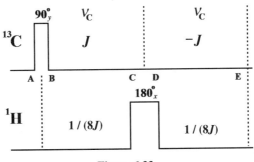

Figure 6.33

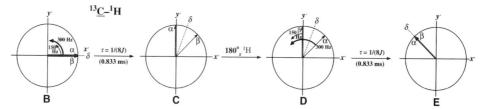

Figure 6.34

chemical shift evolution for a total time of 2τ, leading to a total rotation of $\Delta\nu$ Hz times 2τ seconds, where $\Delta\nu$ is the rotating-frame frequency at the center of the CH ^{13}C doublet. The J-coupling evolution is $+J$ for time τ and $-J$ for time τ, indicating that J-coupling evolution is refocused in the second half and the vectors will still be in-phase, without diverging at all. Figure 6.34 shows the effect of this pulse sequence using vector diagrams for our example. As before, at the end of the first $1/(8J)$ delay we have the H = α vector on the $+y'$ axis and the H = β vector halfway between the $+x'$ and $+y'$ axes. As we saw before, the only effect of the 180° ^{1}H pulse is to reverse the labels on the ^{13}C net magnetization vectors: now the H = β vector is on the $+y'$ axis and the H = α vector is halfway between the $+x'$ and $+y'$ axes. The chemical shift position is in the same place between the two vectors. During the second delay, the chemical shift position continues to rotate counterclockwise. The H = β vector rotates 45° ccw to end up halfway between the $+y'$ and $-x'$ axes, and the faster H = α vector rotates 90° ccw to end up at exactly the same place. The two vectors are in-phase again, and there has been no net J-coupling evolution. But the chemical shift evolution has continued, as if the pair of vectors never diverged, and moved a total of 135° (67.5° times 2) in a ccw direction (225 Hz × 0.833 ms × 2 = 0.375 rotations, 0.375 × 360° = 135°).

One technical question remains: What is the effect of the *phase* of the ^{13}C 180° pulse in these spin-echo sequences? If we place the B_1 field on the $-x'$ axis, the result is exactly the same, because rotation by 180° in one direction is the same as rotation by 180° in the opposite direction. But if we place the ^{13}C 180° pulse on the y' axis, the details of the vector motions will change, and the final result will be different, but the question of evolution remains the same. Do this as an exercise, going through our example for three types of spin echoes: the simple spin echo (^{13}C 180°$_y$ only), the spin echo with 180° pulses on both channels, and the spin echo with 180° pulse on the ^{1}H channel only. You will see that the simple spin echo refocuses J-coupling evolution and chemical shift evolution, but the vectors end up on the $-x'$ axis instead of on the $+x'$ axis. You might have to try a different chemical shift (e.g., $\Delta\nu = 150$ Hz) to convince yourself that the vectors always land on the $-x'$ axis, regardless of the resonance offset of the center of the doublet. The reason is that not only did the ^{13}C 180° pulse refocus the chemical shift and J-coupling evolution, but it also flipped the vectors from the $+x'$ axis to the $-x'$ axis. We can say that the effect of the sequence is the same as it would be if the delay times were set to zero (90°$_y$–180°$_y$–): the two vectors are rotated to the $+x'$ axis by the 90°$_y$ pulse, and then to the $-x'$ axis by the 180° pulse on the y' axis. In the original example, the ^{13}C 180° pulse was on the x' axis, so it had no effect overall on the position of the vectors: If we imagine the delays being zero, we have a 90°$_y$ pulse rotating the vectors to the $+x'$ axis, and a 180°$_x$ pulse having no effect on the two vectors because they are colinear with the B_1 field. Thus, the two vectors end up on the $+x'$ axis. But be careful about saying that the 180° ^{13}C pulse on the x' axis has "no effect": It has the effect of refocusing the chemical shift and J-coupling evolution, and if we leave it out the result will be entirely different. It just has no effect *beyond* the refocusing, whereas a ^{13}C 180° pulse on the y' axis has the additional effect of rotating the vectors from their original position on the $+x'$ axis to the $-x'$ axis.

The same reasoning can be applied to the sequence of Figure 6.30: Overall we have J-coupling evolution only, as if the ^{13}C doublet were on-resonance. The two vectors diverge from each other but the center position does not move; the more downfield H = α line (J/2 Hz downfield of the center of the doublet) leads to a ccw rotation at a rate of J/2, and the more upfield H = β line ($-J/2$ relative to the center of the doublet) leads to a cw rotation at a rate of J/2. The effect of the phase of the 180° pulse is correctly accounted for if we imagine it happening at the beginning of the first delay: The two vectors rotate from the $+x'$ axis to the $-x'$ axis as a result of the 180° pulse on the y' axis, and then they diverge by an total angle of 90° (each one moving J/2 Hz times 1/(4J) or 45°) without changing the position of the center, which remains on the $-x'$ axis. The phase of the 180° ^{1}H pulse in either spin-echo sequence (Fig. 6.30 or Fig. 6.33) is irrelevant because it serves only to convert every ^{1}H from the α state to the β state and vice versa. It does not rotate the ^{13}C magnetization vector, so we do not care which axis the B_1 field is on. We refer to this pulse as an inversion pulse, and like an inversion of net magnetization from $+z$ to $-z$, it does not matter which axis the B_1 field is on.

The overall lesson from this exercise is that we can ignore the details of vector rotation during a spin echo if we understand its overall effect: Which types of evolution are allowed and which are refocused? This is easily determined by looking at the 180° pulses at the center. A 180° pulse on the nucleus that has the net magnetization in the x'–y' plane (i.e., the coherence) will lead to refocusing of chemical shift evolution. A 180° pulse on one of the two nuclei involved in the J coupling will lead to refocusing of J-coupling evolution. The 180° pulses on neither or both of the two nuclei will allow J-coupling evolution to occur. The phase of the 180° pulse on the channel where we have coherence can be accounted for by imagining that the pulse occurs at the beginning of the first delay and accounting for the allowed evolution for the full delay time (2τ) of the spin echo. The phase of any 180° pulse on the other channel, corresponding to the nucleus that is not evolving, is irrelevant. By looking at the spin echo in this overall view we can avoid the tedious analysis of considering the effect of each pulse and delay individually.

7

COHERENCE TRANSFER: INEPT AND DEPT

Now we are ready to see something truly magical in NMR. With our toolbox of pulses and delays, and the more complex "plug in" units of spin echoes, we can make net magnetization "jump" from one nucleus to another across the "bridge" of a J coupling or an nuclear Overhauser effect (NOE) interaction. This strategy allows us to enhance the sensitivity of many experiments by starting with ^{1}H, which has the highest equilibrium population difference, and moving its magnetization to less sensitive nuclei such as ^{13}C. Later, we will see how we can correlate two related spins (actually we correlate their chemical shift positions in the spectrum) to demonstrate the nature (J or NOE) and intensity of the relationship. We will do this first with selective one-dimensional (1D) experiments and then with two-dimensional (2D) experiments. In both cases, the basis of the correlation is making the magnetization "jump" from one nucleus to another by the process of magnetization transfer. But first we need to understand very clearly what this material is that we work with in NMR: the net magnetization. A little bit of review will help.

7.1 NET MAGNETIZATION

Each atom that is NMR-active in a sample (each "spin") contributes its magnetism to the bulk or net magnetization of the sample. This spin produces a tiny magnetic field, and we can represent this little magnet with a vector pointing in the direction of the spin axis (Fig. 7.1(a)). If we move the origin (the magnet's south pole) of each vector to a common point, we will have the spins in the "up" or α state precessing at the Larmor frequency forming a 45° angle to the $+z$ axis, tracing out an upward-facing cone. The spins in the "down" or β state form a 135° angle to the z axis and trace out a downward-facing cone as they precess (Fig. 7.1(b)). To trace the path of each spin in the sample would be an impossibly

---

NMR Spectroscopy Explained: Simplified Theory, Applications and Examples for Organic Chemistry and Structural Biology, by Neil E. Jacobsen
Copyright © 2007 John Wiley & Sons, Inc.

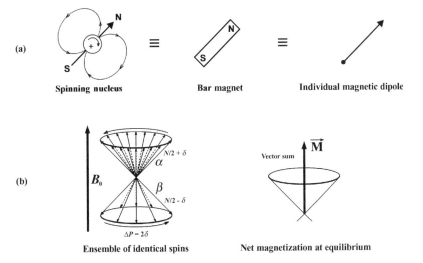

Figure 7.1

complicated task. Instead, we add together all of the individual magnetic dipoles (vectors) to obtain the bulk magnetization of the sample, and we can work with this bulk feature only. At equilibrium, after the spins have been in the applied field for a while, there will be a slight excess of spins in the lower energy (aligned, α, "up") state. Because the energy difference is very small compared to the average thermal energy at room temperature, this population difference is only about one spin in 10^5. For this reason, for nearly every spin in the "up" state, there will be another in the "down" state pointing in exactly the opposite direction. These pairs can be erased from our picture as they do not contribute anything to the net magnetization. After erasing all these pairs, we are left with a much smaller number of spins in the upper cone and none in the lower cone. Of these, the individual spins are precessing at exactly the same frequency but at any moment if we took a snapshot of their orientations we would find that they are equally distributed around the cone; no one orientation is preferred. Thus, the components of magnetism in the x–y plane all cancel each other out and we are left only with the sum of the z components, which add together to give a macroscopic magnetic field oriented along the positive z axis. This equilibrium z magnetization simply means that we have weakly magnetized the sample by giving a slight preference to spins oriented with the B_0 field.

This equilibrium magnetization is the stuff we have to work with in NMR. With pulses of radio frequency energy, we can make this vector move around, "dance," and tell us things of importance about molecular structure. NMR is like radar with a magnet: high-power pulses of radio frequency energy are sent to the sample, and an "echo" is received (the FID). Analysis of this echo provides information about the sample. In particular, we can learn about the relationships between different spins (atoms) within a molecule, both in the sense of the number of bonds separating them and the angles of those bonds, and in the sense of the direct through-space distance between the atoms.

In NMR, there is a very important distinction between the z-axis component of the net magnetization ("z magnetization") and the component of the net magnetization that lies in the x–y plane ("coherence"). z-Magnetization, which is the result of unequal populations in the two spin states ("up" and "down", α and β, aligned and disaligned, lower energy and higher energy), gradually relaxes to its equilibrium value M_o, defined by the Boltzmann distribution of spins between the two energy levels. It is not directly measurable because the net vector is stationary and does not rotate. Net magnetization in the x–y plane is called coherence because it results from the temporary organization (coherence) of the individual spins as they rotate around the cone. Coherence always relaxes to zero as the individual spins gradually get out of phase and lose their "memory" of the organizing pulse. Coherence is measurable because it rotates and creates the FID signal in the probe coil. The RF pulse converts z magnetization into coherence.

The effect of the RF pulse can be viewed more simply if we forget about the individual spins and think only about the net magnetization vector. The RF pulse is a magnetic field that rotates in the x–y plane at the frequency of the pulse. This signal is turned on for a very short time (about 10 μs) and then abruptly turned off. During the pulse, we can view this rotating magnetic field as a vector, the B_1 vector, which rotates in the x–y plane while the B_0 field (about 20,000 times larger) is on the positive z axis, as is the equilibrium net magnetization of the sample. To make the analysis simpler, we rotate the x and y axes at the frequency of the pulse, so that the B_1 vector stands still in this rotating frame of reference (Fig. 7.2(a)). In order to preserve the laws of physics in this artificial rotating frame, we have to remove the B_0 field from our picture. Now we have only the stationary B_1 vector in the x–y plane and the sample net magnetization on the z axis. During the pulse, the sample net magnetization vector, **M**, rotates around the B_1 vector. This is analogous to the rotation of **M** around the B_0 field during the FID. If we place the B_1 vector on the x axis, for example (we can control this by setting the phase of the radio frequency), the **M** vector will rotate counterclockwise from the z axis to the $-y$ axis, then to the $-z$ axis, then to the $+y$ axis

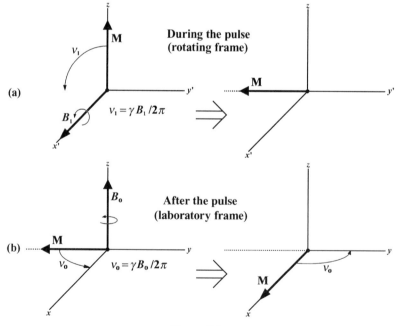

Figure 7.2

and then back to the $+z$ axis. If we time the duration of the pulse correctly, we can rotate **M** by exactly 90° so that it ends up on the $-y$ axis at the moment the pulse is turned off. This pulse is called a 90° pulse.

After a 90° pulse there is no z magnetization: all of it has been converted into x–y magnetization (coherence). This means there is no difference in population between the two spin states. We go back to the laboratory frame at this instant to look at the motion of the net magnetization **M** (Fig. 7.2(b)). It rotates in the x–y plane at the Larmor frequency, inducing the FID signal in the probe coil and gradually decaying (due to loss of coherence) to zero. At a bit slower rate, the population difference is reestablished between the two levels as a small percentage of spins fall down (relax) from the upper energy level to the lower energy level. This causes the z magnetization to grow and eventually return to the equilibrium magnitude, M_o, aligned along the positive z axis. When all coherence is gone and the z magnetization has returned to equilibrium (a few seconds), we can repeat the whole process of pulse, recording the FID and relaxation, adding the new FID to the previous one to obtain a better signal-to-noise ratio.

7.2 MAGNETIZATION TRANSFER

Magnetization transfer is the central process in all advanced NMR experiments, both 1D and 2D. By an appropriate combination of pulses and/or waiting periods ("delays"), we can make net magnetization "jump" from one nucleus in a molecule to another. I will not attempt to explain the details of how this happens, I will just "narrate" the process in general terms and you will have to take it mostly on faith. Just as there are two kinds of net magnetization, z magnetization and coherence, there are two ways to transfer magnetization: NOE (transfer of z magnetization) and INEPT (coherence transfer). The NOE transfer occurs directly through space from one proton in a molecule to a nearby proton. The distance between them must be less than 5 Å and the efficiency of transfer is proportional to the inverse 6th power of the distance between them ($1/r^6$). In this way we can measure distances within a molecule and make conclusions about stereochemistry and conformation. The INEPT (insensitive nuclei enhanced by polarization transfer) transfer occurs via J couplings, which means it is a through-bond effect between atoms that are two or three (occasionally more) bonds apart in the covalent bonding network of a molecule. As J coupling values depend on the dihedral angle for vicinal (three-bond) relationships (Karplus relation), we can learn about conformation as well as covalent connectivity. These are the structural relationships we can discover using NMR, and the key to connecting one spin to another via these relationships is magnetization transfer.

The NOE works like this: if you perturb the z magnetization of one proton in a molecule so that it is no longer at equilibrium (i.e., no longer $+M_o$), this perturbation will propagate over time (0.2–1 s for small molecules) to other protons in the molecule, creating perturbations of their z magnetization away from equilibrium. For small molecules (<1000 Da) the effect is "negative": reducing the z magnetization of one proton will lead to the buildup of an increase in z magnetization of nearby spins. There are several experiments that can be designed to perturb one proton's z magnetization and to convert the transferred z magnetization into a measurable signal (enhanced peak height in the spectrum) and thereby determine the distances between protons. The initial perturbation can be created by a long, low power radio frequency signal at the exact resonant frequency of one proton in the spectrum ("saturation") or by a 180° pulse which rotates the net magnetization of one proton

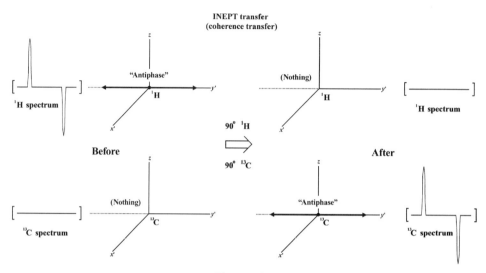

Figure 7.3

in the molecule to the negative z axis. The NOE process can be viewed as magnetization transfer as the z magnetization difference (perturbation from equilibrium) of one proton is transferred to another during the "mixing time" of the experiment, leading to a z magnetization difference in the second proton. Two-dimensional experiments that use NOE as a means of magnetization transfer include the NOESY and ROESY experiments.

INEPT coherence transfer works differently. We create coherence on one nucleus with a 90° pulse and then wait a period of time (equal to $1/(2J)$, where J is the coupling constant). At this point, the two components of the doublet signal are opposite in phase: an FID acquired at this point would give a spectrum of a doublet with one component pointing up and one pointing down (Fig. 7.3, left). This "antiphase" state has a very special property: if we subject it to 90° pulses simultaneously affecting both nuclei in the J-coupled pair, the coherence will "jump" from one nucleus to the other. We will now have antiphase coherence on the J-coupled nucleus and no coherence on the starting nucleus (Fig. 7.3, right). This can be applied to any pair of nuclei that are J coupled: two protons on adjacent carbons (vicinal relationship), a proton and its directly bonded carbon (1-bond heteronuclear coupling), a proton and the carbon next to its own carbon (2-bond heteronuclear J coupling), and so on. The INEPT transfer is used in advanced 1D experiments such as DEPT, as well as in a number of 2D experiments (COSY, DQF-COSY, HETCOR, HSQC, HMBC).

7.3 THE PRODUCT OPERATOR FORMALISM: INTRODUCTION

It is very difficult to describe coherence transfer using the vector model. To understand it we will need to expand our theoretical picture to include product operators. Product operators are a shorthand notation that describes the spin state of a population of spins by dividing it into symbolic components called operators. You might wonder why you would trade in a nice pictorial system for a bunch of equations and symbols. The best reason I can give is that the vector model is useless for describing most of the interesting NMR experiments, and product operators offer a bridge between the familiar vectors and the more formal and mathematical matrix representation. Later on you will see that product operators are just shorthand for the elements of the density matrix.

Using the vector model, when we want to describe the spin state of a particular nucleus, we can draw a vector in three-dimensional space, or we can describe the projection of that vector onto the three axes (the "components" of the vector). For example, a vector of length M_o on the $-x'$ axis could be described as

$$M_x = -M_o$$
$$M_y = 0$$
$$M_z = 0$$

Note that this description requires that we make three statements about the components of the vector. In the product operator formalism, we simply say that the spin state is $-\mathbf{I}_x$. We do not have to say anything about the y and z components because if we do not see $\mathbf{I}_y$ or $\mathbf{I}_z$ in the spin state, we just assume that those components are zero. If we wish to talk about two different spins at the same time in the vector model, we need two different sets of coordinate axes; for example, one for ^{1}H and one for ^{13}C, to keep it straight which one we are talking about. With product operators, we usually use $\mathbf{I}$ for ^{1}H and $\mathbf{S}$ for ^{13}C, so we could describe a spin state as $-\mathbf{I}_z + \mathbf{S}_x$, meaning that the ^{1}H net magnetization is on the $-z$ axis and the ^{13}C net magnetization is on the $+x$ axis. If we want to talk about two protons, we can use either $\mathbf{I}$ and $\mathbf{S}$ or $\mathbf{I}^a$ and $\mathbf{I}^b$ to describe H_a and H_b. As you can see, we are going to move away from the pictorial representation on a coordinate system (vectors) to a symbolic representation using the product operators. These symbols can be easily manipulated using simple math and rules about how they behave with pulses (rotations about the B_1 vector) and delays (rotation about the z axis, also known as "evolution"). The rules all refer back to the vectors, so if we understand how the vectors behave we can manipulate these symbols very easily.

A more complicated case is that of the ^{13}C doublet of a methine (^{13}C–^{1}H) group in the absence of ^{1}H decoupling. In the vector model we draw two vectors: one for the net magnetization of all ^{13}C spins whose ^{1}H coupling partner is in the α state, and the other for the net magnetization of all ^{13}C spins whose ^{1}H coupling partner is in the β state. This corresponds to the two lines in the ^{13}C spectrum (without ^{1}H decoupling) for the CH doublet: the ^{13}C spins with a ^{1}H partner in the α state give rise to the left-hand peak of the doublet, and the ^{13}C spins with a ^{1}H partner in the β state give rise to the right-hand peak of the doublet. If both the "α" vector and the "β" vector are on the $-x$ axis, we would call this state "in-phase magnetization on the $-x$ axis." In the product operator notation, we would call this $-\mathbf{S}_x$, meaning that the ^{13}C magnetization, regardless of the spin state (α or β) of its ^{1}H coupling partner, is on the $-x$ axis (Fig. 7.4). If the "α" vector is on the y axis and the "β" vector is on the $-y$ axis, we call this state "antiphase." In the product operator notation, we represent the spin state with the symbol $2\mathbf{S}_y\mathbf{I}_z$, which we read as "^{13}C magnetization on the y axis, antiphase with respect to its coupling partner ^{1}H." Everyone who sees this for the first time is completely mystified. First of all, the 2 is just a normalization constant, so you can ignore it. It is necessary any time you have two operators multiplied together (hence the name "product" operators). The $\mathbf{S}_y$ means that the ^{13}C magnetization from carbons attached to ^{1}H in the α state is on the $+y$ axis. The multiplication by $\mathbf{I}_z$ is the hard part. This $\mathbf{I}_z$ says nothing about the *net* magnetization of ^{1}H, so do not make the mistake of thinking that the ^{1}H net magnetization vector is on the $+z$ axis. The $\mathbf{I}_z$ multiplier represents the *microscopic* spin state of each individual proton: half are in the α state ($\mathbf{I}_z = +1/2$, nuclear magnet

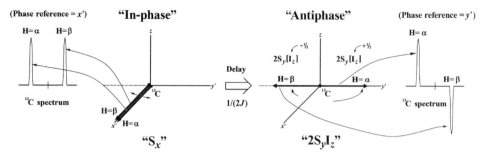

Figure 7.4

pointing "up," along the $+z$ axis) and half are in the β state ($I_z = -1/2$, nuclear magnet pointing "down," along the $-z$ axis), and we are multiplying $2S_y$ by this number. So as we add up the individual nuclear magnetic dipoles (vectors) to yield the net magnetization, we are saying that the individual ^{13}C vectors are aligned with the $+y$ axis if the attached ^{1}H is in the α state (multiply the $2S_y$ by $+1/2$), and the individual ^{13}C vectors are aligned with the $-y$ axis if the attached ^{1}H is in the β state (multiply the $2S_y$ by $-1/2$). We end up with the two net magnetization vectors, one on the $+y$ axis for those ^{13}C nuclei whose attached protons are in the α state, and one on the $-y$ axis for those ^{13}C nuclei whose attached protons are in the β state. Later, when we consider the density matrix representation, we will see that multiplication by I_z is really multiplication by a 4×4 matrix, and all the math works out perfectly to generate the vector model picture of two opposed vectors. For now, you can think of the I_z multiplier as a $+1/2$ (for $\mathbf{I} = \alpha$) or a $-1/2$ for ($\mathbf{I} = \beta$). I_z by itself still represents the ^{1}H net magnetization on the $+z$ axis ($M_z(^1\text{H}) = +M_o$).

7.4 SINGLE SPIN PRODUCT OPERATORS: CHEMICAL SHIFT EVOLUTION

If there is only one NMR line in the spectrum (a population of identical nuclei) there are only three product operators, and they correspond to the three components of the net magnetization vector. A complete description of a population of spins can be given by the spin state σ:

$$\sigma = c_x \mathbf{I}_x + c_y \mathbf{I}_y + c_z \mathbf{I}_z$$

where c_x, c_y, and c_z are coefficients equal to M_x/M_o, M_y/M_o, and M_z/M_o. This is the product operator representation of the spin state of a population of spins. Compare this to the vector representation of the net magnetization (sum of individual spin vectors):

$$\mathbf{M} = M_x \mathbf{i} + M_y \mathbf{j} + M_z \mathbf{k}$$

where $\mathbf{i}$, $\mathbf{j}$, and $\mathbf{k}$ are the unit vectors along the x, y, and, z axes. The product operators $\mathbf{I}_x$, $\mathbf{I}_y$, and $\mathbf{I}_z$ can be viewed as pure spin states. $\mathbf{I}_z$ is the equilibrium state, $\mathbf{I}_y$ is the spin state immediately following a 90° pulse on the $-x'$ axis, and $\mathbf{I}_x$ is the spin state immediately following a 90° pulse on the y' axis. In the literature you will find that pulses are regarded as counterclockwise rotations for product operators, so that a 90° pulse on the x' axis rotates

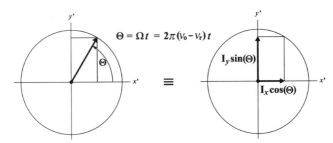

Figure 7.5

the I_z state into the $-I_y$ state. This is opposite to the convention sometimes used for the vector model. For consistency, in this book *all* pulse rotations are counterclockwise.

These simple product operators precess in the x'–y' plane of the rotating frame at a frequency corresponding to the chemical shift in hertz relative to the center of the spectral window (the resonance offset $\Delta \nu = \nu_0 - \nu_r$). The chemical shift frequency $\Delta \nu$ can also be represented as the angular velocity Ω in units of rad/s ($\Omega = 2\pi \Delta \nu$). Using Ω allows us to skip all the 2π terms.

$$\mathbf{I}_x \xrightarrow{\text{delay } \tau} \mathbf{I}_x \cos\Omega\tau + \mathbf{I}_y \sin\Omega\tau$$
$$\mathbf{I}_y \xrightarrow{\text{delay } \tau} \mathbf{I}_y \cos\Omega\tau - \mathbf{I}_x \sin\Omega\tau$$
$$\mathbf{I}_z \xrightarrow{\text{delay } \tau} \mathbf{I}_z \text{ (no precession for } z \text{ magnetization)}$$

The first two changes represent the circular motion described by the components M_x and M_y of the net magnetization vector in the x'–y' plane (Fig. 7.5), a process called *evolution*. The projection of the net magnetization vector on the x axis, relative to M_0, gives the factor in front of $\mathbf{I}_x$ and the projection on the y axis, relative to M_0, gives the factor in front of $\mathbf{I}_y$. Note that in every evolution period, the spin state you start with is multiplied by a cosine term ($\cos\Theta = 1$ for $\Theta = 0$; $\cos\Theta = 0$ for $\Theta = 90°$) and the spin state you are moving toward (by 90° counterclockwise rotation in the x'–y' plane) is multiplied by a sine term ($\sin\Theta = 0$ for $\Theta = 0$, $\sin\Theta = 1$ for $\Theta = 90°$). If the NMR line is upfield of the center of the spectral window ($\nu_0 < \nu_r$, negative value of Ω), the sine terms will start negative and the cosine terms will start out positive, reflecting the clockwise rotation. So we always think of the motion as counterclockwise and let the sign of Ω correct for any clockwise rotations:

{starting state}—τ delay, Ω evolution → {starting state} $\cos\Omega\tau$ + {Next ccw stop} $\sin\Omega\tau$

Any complicated representation of a spin system in terms of $\mathbf{I}_x$, $\mathbf{I}_y$, and $\mathbf{I}_z$ can be described after a delay τ by substituting the corresponding expression on the right side every time one of the terms on the left side ($\mathbf{I}_x$, $\mathbf{I}_y$, or $\mathbf{I}_z$) occurs in the representation. For example, if we start with the spin state

$$\mathbf{I}_x - \mathbf{I}_y + \mathbf{I}_z$$

at time $t = 0$, we will have at time $t = \tau$ as a result of chemical shift evolution

$$\{\mathbf{I}_x \cos\Omega\tau + \mathbf{I}_y \sin\Omega\tau\} + \{-\mathbf{I}_y \cos\Omega\tau + \mathbf{I}_x \sin\Omega\tau\} + \mathbf{I}_z$$
$$= \mathbf{I}_x(\cos\Omega\tau + \sin\Omega\tau) + \mathbf{I}_y(\sin\Omega\tau - \cos\Omega\tau) + \mathbf{I}_z$$

Each operator in the original spin state is replaced by that operator ("where we are starting"), multiplied by the cosine term, plus the next operator we encounter moving counterclockwise in the x–y plane ("where we are headed"), multiplied by the sine term. Because z magnetization is stationary during delays, $\mathbf{I}_z$ does not undergo evolution.

The beauty of the product operators is that we never have to deal with any more than one operator at a time. If we know how $\mathbf{I}_x$, $\mathbf{I}_y$, and $\mathbf{I}_z$ behave for a pulse or delay, we know how any combination of these operators behaves as we just replace each operator with the result of the pulse or delay *on that operator alone*. Instead of the net magnetization vector, which can point anywhere in 3D space, we have reduced the problem to understanding how pulses and delays affect the three simple components: $\mathbf{I}_x$, $\mathbf{I}_y$, $\mathbf{I}_z$.

So far we have not included the relaxation processes (T_1 and T_2), and for many pulse sequences we can leave out this aspect to make the math simpler. We know that relaxation is going on, but in many cases this is merely a technicality and is not essential in understanding the pulse sequence. In general, pulses are on the timescale of microseconds (μs), delays for evolution are on the order of milliseconds (ms), and delays for buildup of NOE can be hundreds of ms. For organic-sized molecules, we can safely ignore relaxation for delays in the μs or ms range. Of course, for some experiments such as NOE, the relaxation process is central to the experiment so we cannot ignore it.

The chemical shift evolution (precession of spins as a result of chemical shift) can be represented as a circle with the rotation rate (in radians per second) written in the center (Fig. 7.6). This is the same as the motion of the net magnetization vector, viewed from the $+z$ axis. Homonuclear product operators (H_a represented by $\mathbf{I}^a$ and H_b represented by $\mathbf{I}^b$) undergo chemical shift evolution in the same way:

$$\mathbf{I}^a_x \xrightarrow{\text{delay } \tau} \mathbf{I}^a_x \cos\Omega_a\tau + \mathbf{I}^a_y \sin\Omega_a\tau$$

$$\mathbf{I}^b_y \xrightarrow{\text{delay } \tau} \mathbf{I}^b_y \cos\Omega_b\tau - \mathbf{I}^b_x \sin\Omega_b\tau$$

Note that the rate (and direction) of precession in the rotating frame is controlled by the chemical shift Ω term, which is specific to each different proton in the molecule (each different resonance or peak in the spectrum).

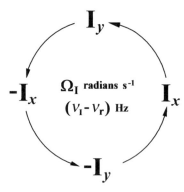

Figure 7.6

7.5 TWO-SPIN OPERATORS: *J*-COUPLING EVOLUTION AND ANTIPHASE COHERENCE

For a system with two kinds of nuclei, the symbols **I** and **S** are used to represent the two nuclei. For example, in a ^{1}H–^{13}C system, the ^{1}H is usually represented as **I** and the ^{13}C as **S**. The six simple spin states $\mathbf{I}_x$, $\mathbf{I}_y$, $\mathbf{I}_z$ and $\mathbf{S}_x$, $\mathbf{S}_y$, $\mathbf{S}_z$ can be represented, but we can also form products of these spin states to represent situations which cannot be described by the vector model. This is where the "product" of product operator comes from. Consider first the product of one nucleus in the x'–y' plane with another nucleus on the z axis:

$2\mathbf{I}_x\mathbf{S}_z, 2\mathbf{I}_y\mathbf{S}_z$:   **I** (^{1}H) magnetization in the x'–y' plane, antiphase with respect to the z orientation of the *J*-coupled spin **S** (^{13}C)

$2\mathbf{S}_x\mathbf{I}_z, 2\mathbf{S}_y\mathbf{I}_z$:   **S** (^{13}C) magnetization in the x'–y' plane, antiphase with respect to the z orientation of the *J*-coupled spin **I** (^{1}H)

The 2 is a normalization factor that will be explained in Chapter 10—it is needed any time we multiply two operators together. In the vector model, these product operators can each be represented by two vectors in the x'–y' plane, 180° apart. For example, $2\mathbf{I}_x\mathbf{S}_z$ represents a spin state where the half of the **I** nuclei that are coupled to an **S** nucleus in the α state add up to form a vector along the $+x'$ axis, whereas the other half of the **I** nuclei that are coupled to an **S** nucleus in the β state add up to form a vector along the $-x'$ axis (Fig. 7.7). We always put the α vector on the axis represented by the first part of the product: for example, for $-2\mathbf{I}_x\mathbf{S}_z$ we put the ^{1}H net magnetization (^{13}C = α) vector on the $-x$ axis and the ^{1}H net magnetization (^{13}C = β) vector on the $+x$ axis.

The chemical shift evolution of these product operators is obtained simply by plugging in the time evolution of the component single-nucleus operators. For example

$$2\mathbf{I}_x\mathbf{S}_z \xrightarrow{\tau \text{ delay}} 2\{\mathbf{I}_x \cos\Omega_I\tau + \mathbf{I}_y \sin\Omega_I\tau\}\{\mathbf{S}_z\} = 2\mathbf{I}_x\mathbf{S}_z \cos\Omega_I\tau + 2\mathbf{I}_y\mathbf{S}_z \sin\Omega_I\tau.$$

This sequence of events can be represented schematically in a circle (Fig. 7.8) with the rotation rate (Ω_I) in the center. This is simply the rotation of the two opposed (antiphase) vectors in the x'–y' plane at the frequency determined by the chemical shift of nucleus **I**. The $\mathbf{S}_z$ part just "goes along for the ride" because the antiphase relationship is retained throughout. Stand up with your arms outstretched at your sides: your right arm represents the ^{1}H net magnetization vector (^{13}C = α) on the $+x$ axis, and your left arm represents the ^{1}H net magnetization vector (^{13}C = β) on the $-x$ axis. The $+y$ axis is in front of you and the $-y$ axis is behind you. Now slowly turn your body counterclockwise (to your left), holding

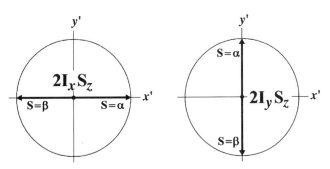

Figure 7.7

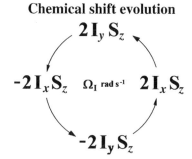

Figure 7.8

your arms apart opposite each other. The "$^{13}C = \alpha$" vector (your right arm, corresponding to the downfield line of the ^1H doublet) moves from the $+x$ axis to the y axis, then the $-x$ axis and then the $-y$ axis, whereas the "$^{13}C = \beta$" vector (your left arm, corresponding to the upfield line of the ^1H doublet) moves from the $-x$ axis to the $-y$ axis, then the $+x$ axis and then the $+y$ axis, always opposite the "$^{13}C = \alpha$" vector. You are turning at a rate corresponding to Ω_I, the chemical shift position of the proton resonance relative to the center of the spectral window. Of course, it requires some special tricks (decoupling, spin echo, etc.) to have a coupled nucleus affected only by its chemical shift and not by the J coupling, but it is easiest to understand these two effects separately.

Evolution under the influence of J coupling *alone* results in refocusing of antiphase magnetization, whereas in-phase magnetization evolves into antiphase:

$$\underset{\text{in-phase}}{\mathbf{I}_x} \rightarrow \underset{\text{antiphase}}{2\mathbf{I}_y\mathbf{S}_z} \rightarrow \underset{\text{in-phase}}{-\mathbf{I}_x} \rightarrow \underset{\text{antiphase}}{-2\mathbf{I}_y\mathbf{S}_z} \rightarrow \underset{\text{in-phase}}{\mathbf{I}_x}$$

Note that the operator in the x–y plane (^1H or $\mathbf{I}$ in this case) evolves, just like chemical shift evolution ($\mathbf{I}_x \rightarrow \mathbf{I}_y \rightarrow -\mathbf{I}_x \rightarrow -\mathbf{I}_y$), a simple counterclockwise rotation in the x–y plane, but with each 90° rotation it alternates between in-phase (omitting the 2 and the $\mathbf{S}_z$) and antiphase (including them). You can do some NMR calisthenics by first putting both arms forward in front of you ($\mathbf{I}_x$, in-phase) and then moving them apart until they are at your sides sticking out ($2\mathbf{I}_y\mathbf{S}_z$, antiphase—your left arm is the $^{13}C = \alpha$ component on the $+y$ axis and your right arm is the $^{13}C = \beta$ component on the $-y$ axis) and then moving them further around to meet in the back ($-\mathbf{I}_x$, in-phase). You cannot go further without hurting yourself, but if you *could* move further your arms would cross and your right arm would point left and your left arm would point right ($-2\mathbf{I}_y\mathbf{S}_z$, antiphase in the opposite sense, with your left arm, $^{13}C = \alpha$, on the right, the $-y$ axis). Further rotation would bring your (broken) arms to the front ($\mathbf{I}_x$, in-phase). This sequence can be represented in a circle (Fig. 7.9, left) with the rotation rate (πJ rad/s or $J/2$ Hz) in the center. If we start with $\mathbf{I}_y$ instead of $\mathbf{I}_x$, we see the same progression of axes for the $\mathbf{I}$ spin (^1H) going counterclockwise from $\mathbf{I}_y$ to $-\mathbf{I}_x$ to $-\mathbf{I}_y$ to $\mathbf{I}_x$, but we start with in-phase on the y axis ($\mathbf{I}_y$) and alternate in-phase and antiphase as we go around: $\mathbf{I}_y \rightarrow -2\mathbf{I}_x\mathbf{S}_z \rightarrow -\mathbf{I}_y \rightarrow 2\mathbf{I}_x\mathbf{S}_z \rightarrow \mathbf{I}_y$ (Fig. 7.9, right).

We can also think about the spectrum that would be observed at each stage of this evolution ("J coupling evolution") *if* we started recording the FID at that point in time. For this purpose, we have to decide on a phase reference (receiver phase): let's use the $+x$ axis as representing a positive absorptive peak in the spectrum. In other words, if a vector is on the $+x$ axis at the start of the FID, it will give a peak in the spectrum that is positive and absorptive. $\mathbf{I}_x$ will give a nice positive absorptive peak for both components (Fig. 7.10) of

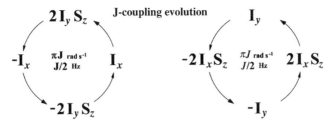

Figure 7.9

the doublet ($^1\underline{H}$–^{13}C system with J coupling $^1J_{CH} \sim 150$ Hz). $2I_yS_z$ will give a dispersive (up/down) peak for the $^{13}C = \alpha$ component on the $+y$ axis and an opposite dispersive peak (down/up) for the $^{13}C = \beta$ component of the doublet on the $-y$ axis. $-I_x$ gives an upside-down (negative absorptive) doublet and $-2I_yS_z$ gives the dispersive antiphase doublet in the opposite sense (down/up, up/down) to $2I_yS_z$. If we use the $+y$ axis as the phase reference, any vector on $+y$ will give a positive absorptive peak, and any vector that leads $+y$ by 90° (i.e., any vector on $-x$, which is 90° counterclockwise from $+y$) will be dispersive up/down. This gives absorptive phase for $2I_yS_z$ and $-2I_yS_z$ (Fig. 7.10, bottom). A real life example of this evolution was shown for ^{13}C labeled methyl iodide ($^{13}CH_3I$) in Chapter 6, Fig. 6.14. It is easier to think about the spectrum if we jump back and forth between the $+x$ axis phase reference (for I_x and $-I_x$) and the $+y$ axis reference (for $2I_yS_z$ and $-2I_yS_z$), avoiding the dispersive lineshape (Fig. 7.10, following the arrows).

We can also use the sine and cosine functions to describe any general rotation, not confined to the four "points of the compass":

$$I_x \xrightarrow{\tau \text{ delay}} I_x \cos(\pi J\tau) + 2I_yS_z \sin(\pi J\tau) \quad \text{(evolution into antiphase: Fig. 7.9, left)}$$

$$I_y \xrightarrow{\tau \text{ delay}} I_y \cos(\pi J\tau) - 2I_xS_z \sin(\pi J\tau) \quad \text{(evolution into antiphase: Fig. 7.9, right)}$$

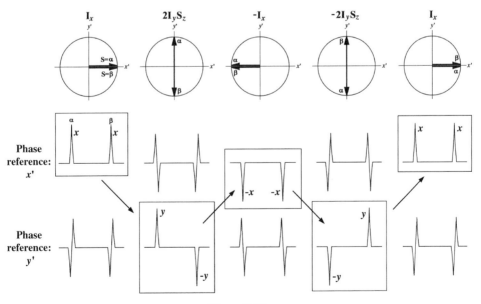

Figure 7.10

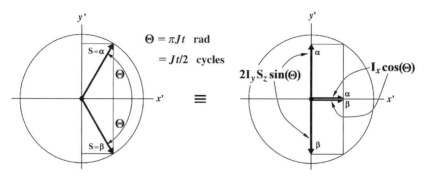

Figure 7.11

$$2\mathbf{I}_x\mathbf{S}_z \xrightarrow{\tau \text{ delay}} 2\mathbf{I}_x\mathbf{S}_z \cos(\pi J\tau) + \mathbf{I}_y \sin(\pi J\tau) \quad \text{(refocusing: Fig. 7.9, right)}$$

$$2\mathbf{I}_y\mathbf{S}_z \xrightarrow{\tau \text{ delay}} 2\mathbf{I}_y\mathbf{S}_z \cos(\pi J\tau) - \mathbf{I}_x \sin(\pi J\tau) \quad \text{(refocusing: Fig. 7.9, left)}$$

As before, the cosine term multiplies the starting spin state of the rotation, and the sine term multiplies the "next stop" on the counterclockwise rotation of J-coupling evolution. For example, for $\mathbf{I}_x$, we know that the next stop for simple chemical shift evolution is $\mathbf{I}_y$ (if you have any doubts, draw a small set of coordinate axes (x, y, and z) and trace the counterclockwise rotation by 90° from the starting spin state). We multiply this by 2 in front and $\mathbf{S}_z$ after the $\mathbf{I}_y$ term (evolution into antiphase) to get $2\mathbf{I}_y\mathbf{S}_z$. Starting from $-2\mathbf{I}_y\mathbf{S}_z$, we move counterclockwise from $-\mathbf{I}_y$ to $\mathbf{I}_x$ and remove the 2 and the $\mathbf{S}_z$ (refocusing) to get $\mathbf{I}_x$. This representation breaks down the vector model of two vectors moving in opposite directions into two components: the in-phase component and the antiphase component (Fig. 7.11). Note that the cosine term always goes with the unchanged product operator and the sine term always goes with the new product operator it evolves into. This makes sense, as the cosine function is 1 and the sine function is 0 at time zero.

Finally, consider the effect of both chemical shift evolution and J coupling. This gets pretty complicated, but we can consider either one of them first and then apply the effect of the other. Let's consider the chemical shift evolution first:

$$\mathbf{I}_x \xrightarrow{\text{delay } \tau} \mathbf{I}_x \cos\Omega_I\tau + \mathbf{I}_y \sin\Omega_I\tau \quad \text{(chemical shift only)}.$$

Now substitute for $\mathbf{I}_x$ and $\mathbf{I}_y$, considering the effect of J-coupling evolution:

$$\mathbf{I}_x \cos\Omega_I\tau + \mathbf{I}_y \sin\Omega_I\tau \xrightarrow{\text{delay } \tau} [\mathbf{I}_x \cos\pi J\tau + 2\mathbf{I}_y\mathbf{S}_z \sin\pi J\tau]\cos\Omega_I\tau$$
$$+ [\mathbf{I}_y \cos\pi J\tau - 2\mathbf{I}_x\mathbf{S}_z \sin\pi J\tau]\sin\Omega_I\tau = \underset{A}{\mathbf{I}_x \cos\pi J\tau\cos\Omega_I\tau} + \underset{B}{2\mathbf{I}_y\mathbf{S}_z \sin\pi J\tau\cos\Omega_I\tau}$$
$$+ \underset{C}{\mathbf{I}_y \cos\pi J\tau\sin\Omega_I\tau} - \underset{D}{2\mathbf{I}_x\mathbf{S}_z \sin\pi J\tau\sin\Omega_I\tau}$$

In the square brackets we have the result of J-coupling evolution starting from $\mathbf{I}_x$ (first term in brackets) and from $\mathbf{I}_y$ (second term in brackets). The result is four separate terms, and we can think of them like this: (A) coherence that underwent neither J-coupling evolution nor chemical shift evolution; (B) coherence that underwent *J*-coupling evolution but not chemical shift evolution; (C) coherence that underwent chemical shift evolution but not *J*-coupling evolution; and (D) coherence that underwent both chemical shift evolution and *J*-coupling evolution. In each case, we use a cosine term if that type of evolution ($\Omega\tau$ for chemical shift evolution and $\pi J\tau$ for *J*-coupling evolution) *did not occur*, and a sine term if it *did occur*. The correct product operators can also be written directly using this type of reasoning. Starting with $\mathbf{I}_x$, we can go directly to the four terms: first, write $\mathbf{I}_x$ with two cosine terms (i.e., no evolution at all); then write $2\mathbf{I}_y\mathbf{S}_z$ (counterclockwise rotation $\mathbf{I}_x \rightarrow \mathbf{I}_y$ plus evolution from in-phase into antiphase) with a $\sin\pi J\tau$ term (*J*-coupling evolution *did* occur) and a $\cos\Omega_I\tau$ term (chemical shift evolution *did not* occur); then write $\mathbf{I}_y$ with a $\cos\pi J\tau$ term (no *J*-coupling evolution) and a $\sin\Omega_I\tau$ term (chemical shift evolution); finally, write $-2\mathbf{I}_x\mathbf{S}_z$ with two sine terms (both kinds of evolution, resulting in a 180° rotation of the I operator *and* evolution into antiphase). If we use s and c for $\sin\pi J\tau$ and $\cos\pi J\tau$, respectively, and s′ and c′ for $\sin\Omega_I\tau$ and $\cos\Omega_I\tau$, respectively, this can be written quickly and simply as follows:

$$\mathbf{I}_x \xrightarrow{\tau\text{ delay}} \mathbf{I}_x \, cc' + 2\mathbf{I}_y\mathbf{S}_z \, sc' + \mathbf{I}_y cs' - 2\mathbf{I}_x\mathbf{S}_z ss'$$

This is pretty complicated, but the advantage is that we can keep track of everything of importance. Any pulse sequence can, in principle, be examined to see what effect it will have on the sample magnetization and what observable signals will remain at the end. Product operator formalism represents the full quantum-mechanical phenomenon of NMR, so that any type of experiment including mysterious things like multiple-quantum coherences (MQCs) can be represented correctly.

7.6 THE EFFECT OF RF PULSES ON PRODUCT OPERATORS

The effect of pulses is very simple: each individual operator is acted on by the pulse, and replaced by the result of that rotation about the B_1 vector. For example, consider the effect of a 90° ^{1}H pulse on the *x* axis:

$$\mathbf{I}_z \rightarrow -\mathbf{I}_y \quad \mathbf{I}_x \rightarrow \mathbf{I}_x \quad \mathbf{I}_y \rightarrow \mathbf{I}_z \quad -\mathbf{I}_y \rightarrow -\mathbf{I}_z \quad \mathbf{S}_y \rightarrow \mathbf{S}_y$$

These are exactly the same as the vector rotations, and you should draw a small set of coordinate axes in the margin of your paper to figure out these rotations as you work with product operators (Fig. 7.12). Note that ^{13}C net magnetization is not affected by a ^{1}H pulse ($\mathbf{S}_y \rightarrow \mathbf{S}_y$). The effect of a 90° ^{13}C pulse on the *y* axis is likewise the same as the vector model predicts:

$$\mathbf{S}_x \rightarrow -\mathbf{S}_z \quad \mathbf{I}_x \rightarrow \mathbf{I}_x \quad \mathbf{S}_y \rightarrow \mathbf{S}_y \quad \mathbf{S}_z \rightarrow \mathbf{S}_x \quad -\mathbf{S}_x \rightarrow \mathbf{S}_z$$

180° pulses lead to inversion ($z \rightarrow -z$ and vice versa) or refocusing ($x \rightarrow -x, y \rightarrow -y$) as long as they are applied on an axis 90° from the axis of the starting magnetization. Any pulse applied on the same axis as the net magnetization has no effect. A 180° ^{13}C pulse on the *y* axis brings about the following rotations (or "non-rotations"):

$$\mathbf{S}_x \rightarrow -\mathbf{S}_x \quad \mathbf{S}_y \rightarrow \mathbf{S}_y \quad -\mathbf{S}_z \rightarrow \mathbf{S}_z \quad \mathbf{I}_x \rightarrow \mathbf{I}_x \quad \mathbf{I}_z \rightarrow \mathbf{I}_z$$

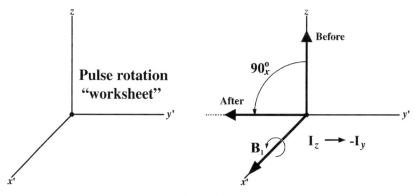

Figure 7.12

For two-spin operators, just figure out the effect of the pulse on each of the operators and replace the each starting operator with the result of the rotation. For example, for a 180° ^{13}C pulse on the x axis

$$2\mathbf{I}_y\mathbf{S}_z \rightarrow 2\mathbf{I}_y(-\mathbf{S}_z) = -2\mathbf{I}_y\mathbf{S}_z \quad \text{(only } \mathbf{S}_z \text{ is affected by the}^{13}\text{C pulse)}$$

$$-2\mathbf{S}_x\mathbf{I}_z \rightarrow -2\mathbf{S}_x\mathbf{I}_z \qquad 2\mathbf{S}_y\mathbf{I}_z \rightarrow 2(-\mathbf{S}_y)\mathbf{I}_z = -2\mathbf{S}_y\mathbf{I}_z$$

For a 90° ^{1}H pulse on the y axis

$$2\mathbf{I}_y\mathbf{S}_z \rightarrow 2\mathbf{I}_y\mathbf{S}_z \quad 2\mathbf{I}_x\mathbf{S}_x \rightarrow 2(-\mathbf{I}_z)\mathbf{S}_x = -2\mathbf{S}_x\mathbf{I}_z \quad 2\mathbf{S}_y\mathbf{I}_z \rightarrow 2\mathbf{S}_y\mathbf{I}_x \quad 2\mathbf{I}_x\mathbf{S}_z \rightarrow -2\mathbf{I}_z\mathbf{S}_z$$

Note that the observable operator (the operator representing coherence or net magnetization in the x–y plane) is always written first in the product. Also, we see above some examples where both operators are in the x–y plane, or both operators are on the z axis! These products represent nonobservable states which are nonetheless very important in NMR experiments. The only observable product operators are those with only one operator in the x–y plane ("single-quantum transitions").

For a homonuclear system (H_a and H_b with coupling constant J) we can do the same kinds of tricks:

$$\mathbf{I}_x^a \xrightarrow{(90_y^H)} -\mathbf{I}_z^a \qquad \mathbf{I}_y^b \xrightarrow{(180_x^H)} -\mathbf{I}_y^b \qquad \mathbf{I}_z^b \xrightarrow{(90_x^H)} -\mathbf{I}_y^b$$

Coupled protons also evolve into antiphase and refocus during delays:

$$\mathbf{I}_y^b - 1/(2J) \rightarrow -2\mathbf{I}_x^b\mathbf{I}_z^a \rightarrow -\mathbf{I}_y^b \rightarrow 2\mathbf{I}_x^b\mathbf{I}_z^a \rightarrow \mathbf{I}_y^b$$

For example, $-2\mathbf{I}_x^b\mathbf{I}_z^a$ is read as "proton H_b net magnetization on the $-x$ axis, antiphase with respect to its coupling partner H_a." Again, we can think of the $\mathbf{I}_z^a$ part as a multiplier, equal to $+1/2$ or $-1/2$ depending on the spin state (α or β) of each individual H_a nucleus. The observable magnetization is on H_b, and as it rotates around the axes ($y \rightarrow -x \rightarrow -y \rightarrow x \rightarrow y$) it alternates between in-phase and antiphase with respect to its coupling partner H_a. The effect of pulses is similar to what we saw for a ^{1}H–^{13}C pair, except that with hard (high power, short duration, nonselective) ^{1}H pulses we cannot deliver a rotation to one of the spins and not the other: all hard pulses affect both H_a and H_b. For example, for a 180° ^1H pulse on the y axis:

$$I_x^a \to -I_x^a \qquad -I_x^b \to I_x^b \qquad 2I_y^a I_z^b \to 2I_y^a(-I_z^b) = -2I_y^a I_z^b$$

$$2I_x^b I_z^a \to 2(-I_x^b)(-I_z^a) = 2I_x^b I_z^a \qquad 2I_y^b I_z^a \to 2(I_y^b)(-I_z^a) = -2I_y^b I_z^a$$

Likewise, a 90° ^1H pulse on the x axis is viewed as a simultaneous pulse on H_a and on H_b:

$$I_y^a \to I_z^a \qquad I_z^b \to -I_y^b \qquad 2I_y^a I_z^b \to 2(I_z^a)(-I_y^b) = -2I_y^b I_z^a$$

$$2I_x^b I_z^a \to 2(I_x^b)(-I_y^a) = -2I_x^b I_y^a \qquad -2I_y^b I_z^a \to 2I_y^a I_z^b$$

In the two examples above on the right, observable magnetization (antiphase coherence) is transferred by the 90° ^1H pulse from H_b to H_a (top) and from H_a to H_b (bottom). This is a key process in all advanced NMR experiments that depend on J couplings. The role of the two operators is reversed as the operator in the x–y plane (the observable net magnetization) rotates to the z axis and the operator on the z axis (the multiplier that represents microscopic z magnetization) rotates to the x–y plane. After the rotations, we reverse the order of the two operators because we always write the observable operator first in the product.

7.7 INEPT AND THE TRANSFER OF MAGNETIZATION FROM ^1H TO ^{13}C

Now that we have the precise tools of product operator notation, we can look at coherence transfer in detail. In the NOE difference experiment (Chapter 5), we saw an example of transfer of z magnetization from one nucleus to another via the through-space interaction of cross-relaxation. It is also possible to transfer magnetization in the x–y plane (observable magnetization or coherence) via the through-bond J-coupling interaction. The simplest form of the INEPT pulse sequence is shown in Figure 7.13 for transfer of ^1H coherence to ^{13}C coherence. Consider a simple case of a single proton bonded to a ^{13}C (e.g., benzene) and assume for the sake of simplicity that both the ^1H and the ^{13}C frequencies are exactly on-resonance. The 90° ^1H pulse (B_1 vector on $-x$) rotates the proton z magnetization onto the y' axis of the rotating frame of reference (Fig. 7.14). The downfield component of the proton doublet, which arises from protons attached to ^{13}C nuclei in the α state, begins to rotate counterclockwise in the x'–y' plane toward the $-x'$ axis with angular frequency $J/2$ Hz (or πJ

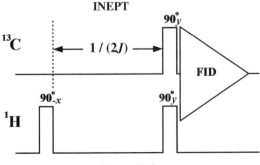

Figure 7.13

254 COHERENCE TRANSFER: INEPT AND DEPT

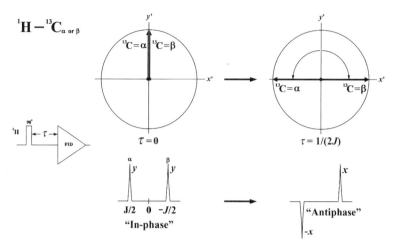

Figure 7.14

radians). The upfield component of the proton doublet, which arises from protons attached to ^{13}C nuclei in the β state, rotates in the opposite direction in the x'–y' plane (clockwise) toward the $+x'$ axis with angular frequency $-J/2$ Hz ($-\pi J$ radians). After a period of time equal to $1/(2J)$, the two components are exactly opposite to each other, with the downfield (H = α) component on the $-x$ axis and the upfield (H = β) component on the $+x$ axis. This special condition is called *antiphase magnetization* and is represented as $-2\mathbf{I}_x\mathbf{S}_z$, where **I** stands for the ^{1}H spins and **S** stands for the ^{13}C spins. This can be read as "**I** spin magnetization along the $-x'$ axis, antiphase with respect to the z orientation of the individual **S** spins." The downfield component of the doublet (which we will always show as the α component) points along the indicated axis ($-x$ from the $-2\mathbf{I}_x$ part of the product) in the vector model, and the upfield (β) component points along the opposite ($+x$) axis (Fig. 7.14).

Acquisition of data at this point would yield a spectrum with a proton doublet, in which the upfield component of the doublet is of opposite phase (upside down) with respect to the downfield component. If we choose $+x$ as our reference axis (receiver phase), the downfield component will be negative absorptive and the upfield component will be positive absorptive in the spectrum (Fig. 7.14). Instead of starting acquisition at this point, however, we deliver two simultaneous $90°$ pulses: one at the ^{1}H frequency and the other at the ^{13}C frequency (Fig. 7.13). This is where the magic happens. The ^{1}H magnetization, which is antiphase with respect to the attached ^{13}C nucleus, is converted into ^{13}C magnetization, which is antiphase with respect to its attached ^{1}H nucleus. The observable (x–y plane) magnetization has jumped from the ^{1}H to the ^{13}C attached to it as a result of the simultaneous $90°$ pulses on the ^{1}H and ^{13}C channels. This *coherence transfer* is possible only if we have a J coupling between the two nuclei, and if the magnetization is in this antiphase state at the time of the pulses. In terms of product operators, the proton $\mathbf{I}_x$ component is rotated by the $90°$ proton pulse on the y' axis onto the $-z$ axis, whereas the ^{13}C $\mathbf{S}_z$ component is rotated by the $90°$ ^{13}C pulse on the y' axis onto the $+x'$ axis:

$\mathbf{I}_x$ component $\rightarrow -\mathbf{I}_z$ by $90°$ ^{1}H pulse on y' axis
$\mathbf{S}_z$ component $\rightarrow \mathbf{S}_x$ by $90°$ ^{13}C pulse on y' axis

$-2\mathbf{I}_x\mathbf{S}_z \rightarrow -2(-\mathbf{I}_z)\mathbf{S}_x = 2\mathbf{S}_x\mathbf{I}_z$ by simultaneous $90°$ ^{1}H and ^{13}C pulses on y' axis

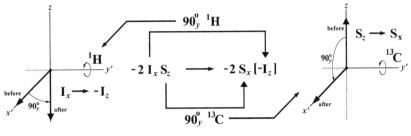

Figure 7.15

This is just following the rules for each individual operator in the product (Fig. 7.15). The resulting product operator, $2S_xI_z$, can be described as ^{13}C magnetization on the x' axis, antiphase with respect to the z magnetization (α or β state) of its attached ^1H nucleus. Acquisition at this time would yield an FID at the ^{13}C frequency, and Fourier transformation with phase reference $+x$ would give a carbon doublet with a downfield component of normal phase and an upfield component of opposite phase (upside-down). This is what you will see in your INEPT spectrum: an antiphase ^{13}C doublet. Later, we will see that this spectrum is actually four times as intense as the normal ^{13}C spectrum: this is the *enhancement* part of INEPT.

As with the APT experiment (Chapter 6), we have the problem that chemical shift differences during the $1/(2J)$ delay will lead to a hopelessly confused pattern of phases in the final spectrum. To eliminate the chemical shift evolution, we use the same strategy we used in the APT experiment: convert the $1/(2J)$ delay into a spin echo. By placing a 180° ^1H pulse in the center of the $1/(2J)$ delay, we refocus the phase shifts due to ^1H chemical shift differences. To make sure that the J coupling effect from the attached ^{13}C nucleus is not refocused as well, a 180° ^{13}C pulse is delivered simultaneously with the 180° ^1H pulse. This is a more sophisticated method than the decoupler switching method used in the APT sequence. In this case to use the APT approach, we would have to decouple the ^{13}C nuclei, which requires nonstandard hardware. The result is the same: the ^1H magnetization evolves into antiphase under the influence of the J coupling from the attached ^{13}C nucleus, but the ^1H chemical shift differences are refocused by the spin echo. The complete INEPT sequence is shown in Figure 7.16.

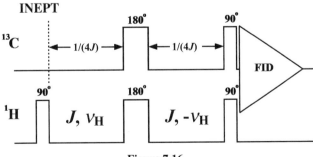

Figure 7.16

There are two important consequences of this method. Both result from the fact that the magnetization that is being observed (^{13}C antiphase doublet) arises from ^{1}H magnetization that is rotated from its equilibrium state along the z axis by the first 90° proton pulse. The first consequence arises because the ^{1}H population difference at equilibrium (sometimes called "polarization") is four times the carbon population difference at equilibrium. This results from the larger energy separation between the α and β states for protons:

$$\Delta P(^1\text{H}) = (\gamma_\text{H} B_o h / 4\pi kT)N \qquad \Delta P(^{13}\text{C}) = (\gamma_\text{C} B_o h / 4\pi kT)N$$

$$\Delta P(^1\text{H})/\Delta P(^{13}\text{C}) = \gamma_\text{H}/\gamma_\text{C} = 3.977$$

where N is the number of identical nuclei in the sample and T is the temperature in kelvin. This means that, compared to a normal ^{13}C experiment in which the observed signal originates from $\Delta P(^{13}\text{C})$, the signal will be four times as large! This increase in sensitivity is extremely important for nuclei such as ^{13}C that have a weak nuclear magnet (small γ). The second consequence is that with repetitive scans it is the ^{1}H T_1 value, and not the ^{13}C T_1 value, which determines how long the relaxation delay needs to be, as we start the experiment with $\Delta P(^1\text{H})$ rather than $\Delta P(^{13}\text{C})$. The ^{13}C z magnetization can be completely saturated and this experiment will still work! Because the ^{1}H T_1 values are often shorter than those of ^{13}C, and may be dramatically shorter than those of more "exotic" heavy nuclei, the reduction in experiment time can be significant.

Figure 7.17 shows the INEPT spectrum of sucrose (top) compared to the nondecoupled ^{13}C spectrum (bottom) acquired with the same number of scans. This is run with the phase cycle to remove all coherence that comes from the ^{13}C z magnetization ($\mathbf{S}_z$), so the result is pure antiphase coherence on ^{13}C. Note that the CH resonances give antiphase doublets with each component four times as intense as the corresponding component in the ^{13}C spectrum. The lone quaternary carbon (C-f2) is missing in the INEPT spectrum, and the CH$_2$ resonances give rise to triplets with the central line missing (1, 0, −1). These CH$_2$ patterns are antiphase with respect to only one of the two protons of the CH$_2$ group (the

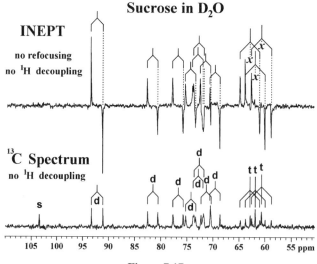

Figure 7.17

one from which the coherence was transferred), so the intensities result from two antiphase doublets (1, −1 and 1, −1) for which the two inner peaks are overlapped and cancel. For example, if the coherence comes from H$_1$ to C, we have $\alpha_1\alpha_2$ (most downfield peak) antiphase with respect to $\beta_1\alpha_2$ (center peak): 1, −1; and $\alpha_1\beta_2$ (center peak) antiphase with respect to $\beta_1\beta_2$ (most upfield peak): 1, −1. These four lines form the triplet and the two inner lines ($\beta_1\alpha_2$ and $\alpha_1\beta_2$) have the same frequency, so the intensities are 1, 0, −1 for the triplet. It is easy to see that as it is, this is not a very practical experiment: without decoupling the peaks are spread out into complex overlapping multiplets (doublets, triplets, quartets), and with decoupling there is no intensity because each multiplet adds up to intensity zero. In Section 7.13, we will see how a refocusing delay brings the antiphase terms back to in-phase so we can apply ^{1}H decoupling and have singlet peaks for each ^{13}C resonance in the spectrum.

7.8 SELECTIVE POPULATION TRANSFER (SPT) AS A WAY OF UNDERSTANDING INEPT COHERENCE TRANSFER

At this point you may understand the math of the product operator "switch" that occurs with the two simultaneous 90° pulses, but you might be feeling unsatisfied and in need of an explanation of what is really happening. Of course, the sooner you accept the product operator math as an explanation the easier the things will be for you, but there *is* a vector model way of explaining INEPT coherence transfer that involves keeping track of populations. Consider a "thought experiment" in which we apply a 180° (inversion) pulse to *only one component* of the ^{1}H doublet. This can actually be done using shaped (selective) pulses, as we will see in Chapter 8. In the vector model, we start with both vectors (representing the two components of the ^{1}H net magnetization in the ^{1}H–^{13}C pair) pointing along the +z axis. This is the equilibrium state. Now we apply a selective 180° pulse to the downfield (^{13}C = α) component of the ^{1}H doublet. The vector labeled "α" will rotate from the +z axis to the −z axis, giving us an antiphase state along the z axis (Fig. 7.18). We can describe this state as $-2\mathbf{I}_z\mathbf{S}_z$, as the $\mathbf{S}_z$ "multiplier" represents +1/2 for the pairs with ^{13}C in the α state and −1/2 for the pairs with ^{13}C in the β state. The "α" vector is on the −z axis ($-2\mathbf{I}_z[1/2] = -\mathbf{I}_z$) and the "$\beta$" vector is on the +z axis ($-2\mathbf{I}_z[-1/2] = \mathbf{I}_z$). We will see in the next section that this state is a useful "intermediate" in coherence transfer.

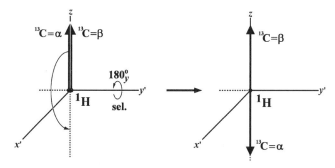

Figure 7.18

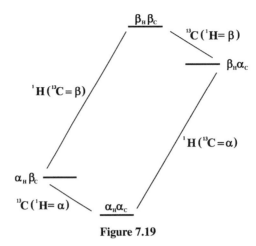

Figure 7.19

Usually when we are talking about nonequilibrium populations, we use the population diagrams (energy states with filled and open circles) to describe the changes. Let's draw an energy diagram for the ^{1}H–^{13}C pair, just like we did for the ^{1}H$_a$–^{1}H$_b$ pair in our discussion of the NOE difference experiment (Chapter 5, Section 5.12). As with the homonuclear case, there will be four energy states: $\alpha_H\alpha_C$, $\alpha_H\beta_C$, $\beta_H\alpha_C$, and $\beta_H\beta_C$. The difference is that the energy gap for the ^{13}C $= \alpha$ to ^{13}C $= \beta$ transition is one-fourth of the energy gap for the ^{1}H $= \alpha$ to ^{1}H $= \beta$ transition because $\gamma_H = 4\gamma_C$. In other words, the much stronger ^{1}H nuclear magnet requires four times more energy to turn it against the B_o field. So we need to draw the energy diagram to reflect this difference in energy gaps. From the center of the diagram ($E = 0$) we go up in energy 4 units for ^{1}H $= \beta$, down in energy 4 units for ^{1}H $= \alpha$, up 1 unit for ^{13}C $= \beta$ and down 1 unit for ^{13}C $= \alpha$. This puts the $\alpha_H\alpha_C$ state at $E = -4 - 1 = -5$, the $\alpha_H\beta_C$ state at $E = -4 + 1 = -3$, the $\beta_H\alpha_C$ state at $E = +4 - 1 = 3$ and the $\beta_H\beta_C$ state at $E = +4 + 1 = +5$ units (Fig. 7.19). The energy unit here is $\Delta E_C/2$, where ΔE_C is the energy gap for a ^{13}C transition. We can identify the ^{1}H transitions as $\alpha\alpha \rightarrow \beta\alpha$ (^{1}H transition with ^{13}C $= \alpha$) and $\alpha\beta \rightarrow \beta\beta$ (^{1}H transition with ^{13}C $= \beta$). The ^{13}C transitions, which have an energy difference one-fourth of the ^{1}H transitions, are $\alpha\alpha \rightarrow \alpha\beta$ (^{13}C transition with ^{1}H $= \alpha$) and $\beta\alpha \rightarrow \beta\beta$ (^{13}C transition with ^{1}H $= \beta$). The two ^{1}H transitions correspond to the two components of the doublet in the ^{1}H spectrum (downfield component is $\alpha\alpha \rightarrow \beta\alpha$, upfield component is $\alpha\beta \rightarrow \beta\beta$) and the two ^{13}C transitions correspond to the two components of the doublet in the ^{13}C spectrum (downfield component is $\alpha\alpha \rightarrow \alpha\beta$, upfield component is $\beta\alpha \rightarrow \beta\beta$).

Now we need to write in the population of each level at equilibrium (Fig. 7.20). According to the Boltzmann distribution, the population of each level will be $N/4$ times the exponential factor:

$$P = (N/4)\,e^{-E/kT} \sim (N/4)\,(1 - E/kT)$$

The deviation in population from an equal distribution between all four states ($N/4$) is thus proportional to the energy. For the $\alpha_H\alpha_C$ state, we draw five filled circles ($E = -5$) to indicate a population of $N/4 + 5\delta$ where δ is $(N/4)(\Delta E_C/8kT)$. For the $\alpha_H\beta_C$ state we draw three filled circles ($E = -3$), for the $\beta_H\alpha_C$ state we draw three open circles ($E = +3$), and for the $\beta_H\beta_C$ state we draw five open circles ($E = +5$) representing a population deficit of 5δ (Fig. 7.20). Now look at the population differences at equilibrium: for the ^{1}H transitions,

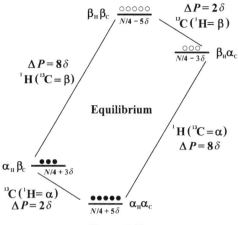

Figure 7.20

we have $\Delta P = 8\delta$ [$+5\delta - (-3\delta)$ for the $^{13}C = \alpha$ transition and $+3\delta - (-5\delta)$ for the $^{13}C = \beta$ transition] and for the ^{13}C transitions we have $\Delta P = 2\delta$ [$+5\delta - (+3\delta)$ for the $^1H = \alpha$ transition and $-3\delta - (-5\delta)$ for the $^1H = \beta$ transition]. Note that population difference is always defined as the population in the lower energy state minus the population in the higher energy state. As expected, the ^1H population differences are four times as large as the ^{13}C population differences (8δ vs. 2δ).

We are finally ready to consider the effect of the selective 180° ^1H pulse on the downfield component of the ^1H doublet ($\alpha_H\alpha_C \rightarrow \beta_H\alpha_C$ transition). We know that the 180° ^1H pulse reverses ("inverts") the spin state of each proton in the sample: every proton in the α state is converted to the β state and every proton in the β state is converted to the α state. Looking at the energy diagram for the ^1H–^{13}C pair, we see that the 180° ^1H pulse swaps the entire population across the ^1H transition: each pair in the $\alpha\alpha$ state is now $\beta\alpha$, and each pair in the $\beta\alpha$ state is now $\alpha\alpha$. The population of the $\alpha\alpha$ state, which was $N/4 + 5\delta$, is now $N/4 - 3\delta$, and the population of the $\alpha\beta$ state, which was $N/4 - 3\delta$, is now $N/4 + 5\delta$ (Fig. 7.21). The redistribution of populations is the result of *every single pair* in the $\beta\alpha$ state moving down to the $\alpha\alpha$ state, and *every single pair* in the $\alpha\alpha$ state moving up to the $\beta\alpha$ state. Consider the analogy of two towns on opposite sides of a river. East Podunk has a population of 51, and West Podunk has a population of 49. If two people move from East Podunk to West Podunk, we now have reversed the populations of the two towns. But this is not analogous to NMR inversion: the correct analogy is that *all 51 residents* of East Podunk move to West Podunk, and *all 49 residents* of West Podunk move to East Podunk, reversing the populations.

The spin state diagrammed in Figure 7.21 is the $-2I_zS_z$ state, an antiphase state on the z axis. It can be represented in a vector diagram with the ^1H net magnetization ($^{13}C = \alpha$) vector pointing down along the $-z$ axis and the ^1H net magnetization ($^{13}C = \beta$) vector pointing up along the $+z$ axis (Fig. 7.18). Note that spin **I** (^1H) and spin **S** (^{13}C) play equal roles in the product; we could also write it as $-2S_zI_z$ because neither operator is observable and the rule that observable operators go first does not apply. Writing it with S_z first implies that we have net **S** spin (^{13}C) magnetization on the z axis, and the I_z term is acting as a multiplier ($+1/2$ for ^1H in the α state and $-1/2$ for ^1H in the β state). Now look at this spin state from the point of view of population differences for the ^{13}C transitions (Fig. 7.21). The $\alpha\alpha \rightarrow \alpha\beta$ transition (downfield or "$^1H = \alpha$" component of the ^{13}C doublet) has a population difference of -6δ ($-3\delta - (+3\delta)$), or -3 times the equilibrium difference of 2δ, and the $\beta\alpha \rightarrow$

260 COHERENCE TRANSFER: INEPT AND DEPT

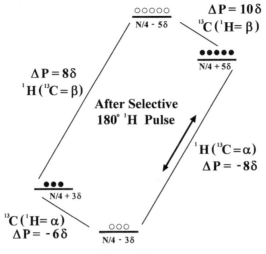

Figure 7.21

$\beta\beta$ transition (upfield or "^{1}H = β" component of the ^{13}C doublet) has a population difference of 10δ $(+5\delta - (-5\delta))$, or 5 times the equilibrium difference of 2δ. In the vector model, we can represent the ^{13}C net magnetization (on a separate coordinate system from the ^{1}H net magnetization) as a vector labeled "^{1}H = α" pointing down along the $-z$ axis with length 3 (three times the normal length for ^{13}C) and another vector labeled "^{1}H = β" pointing up along the $+z$ axis with length 5 (five times the normal length for ^{13}C (Fig. 7.22). We can think of this as the sum of the equilibrium ^{13}C net magnetization (both "α" and "β" vectors on $+z$ with length 1) and the antiphase z magnetization transferred from ^{1}H ("α" vector on $-z$ and "β" vector on $+z$, both four times as long as the normal ^{13}C equilibrium magnetization). The equilibrium ^{13}C z magnetization is still there because we have not perturbed it: only a selective ^{1}H pulse has been delivered to the sample. The transferred magnetization is four times as large as normal ^{13}C magnetization because it came from the equilibrium population difference ("polarization") of the proton, which is four times as large as ^{13}C. This is the origin of the *enhancement by polarization transfer* part (EPT) of the name INEPT. The

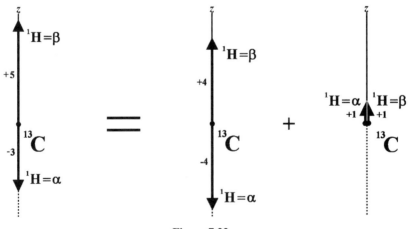

Figure 7.22

proton population difference has been converted into a ^{13}C population difference of the same magnitude. The product operator view of the process thus far is as follows:

$$\mathbf{I}_z + \mathbf{S}_z \xrightarrow{\text{selective }^1\text{H 180° pulse}} -2\mathbf{I}_z\mathbf{S}_z + \mathbf{S}_z = 4[-2\mathbf{S}_z\mathbf{I}_z] + \mathbf{S}_z$$

Our equilibrium starting point is both $\mathbf{I}_z$ (equilibrium proton net magnetization) and $\mathbf{S}_z$ (equilibrium ^{13}C carbon magnetization). Adding these two together ($\mathbf{I}_z + \mathbf{S}_z$) simply means that both are present; it is like a list of the types of net magnetization in the sample. This is very different from multiplying them together ($-2\mathbf{I}_z\mathbf{S}_z$); in this case the first operator is net magnetization and the second is a multiplier that makes it antiphase ($+1/2$ for ^{13}C $= \alpha$, $-1/2$ for ^{13}C $= \beta$). Note that there is no systematic way of calculating the effect of the selective ^{1}H pulse on $\mathbf{I}_z$; we have to reason it out using the vector model and then name the final state $-2\mathbf{I}_z\mathbf{S}_z$ based on our understanding of how to represent antiphase magnetization using the product operators. When we switch the order of operators from $2\mathbf{I}_z\mathbf{S}_z$ to $2\mathbf{S}_z\mathbf{I}_z$ we put in a factor of 4 to indicate that net ^{13}C magnetization is now four times larger than equilibrium ^{13}C magnetization ($\mathbf{S}_z$). Because we are viewing $2\mathbf{S}_z\mathbf{I}_z$ as ^{13}C net magnetization modified by the multiplier $\mathbf{I}_z$, the point of comparison for the magnitude of this magnetization is now $\mathbf{S}_z$ and not $\mathbf{I}_z$.

At this point in order to observe a ^{13}C spectrum, we need to rotate the ^{13}C net magne-

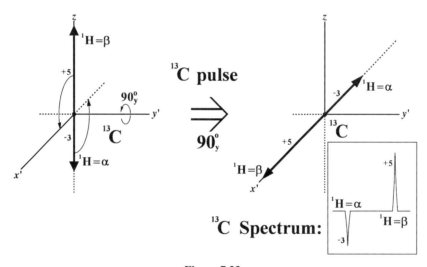

Figure 7.23

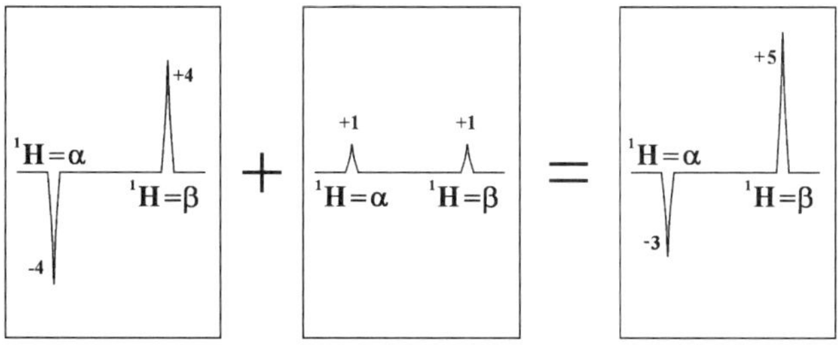

Figure 7.24

tization vectors into the x–y plane with a 90° ^{13}C pulse. If we set the phase of the pulse to y (B_1 field on the y' axis in the rotating frame), the ^{1}H = β vector, which is on the $+z$ axis, rotates 90° ccw to the x' axis and the ^{1}H = α vector, which is on the $-z$ axis, rotates 90° ccw to the $-x'$ axis (Fig. 7.23). If we begin acquiring the FID at this moment, with the x' axis as our phase reference, we will see a ^{13}C doublet with the downfield component (^{1}H = α) upside-down with intensity three times the normal ^{13}C spectrum, and the upfield component (^{1}H = β) positive absorptive with intensity five times the normal ^{13}C spectrum. In product operator terms, we have

$$4[-2\mathbf{S}_z\mathbf{I}_z] + \mathbf{S}_z \xrightarrow{^{13}\text{C}\ 90°\ \text{pulse on }y'\text{ axis}} 4[-2\mathbf{S}_x\mathbf{I}_z] + \mathbf{S}_x$$

The vector representation of the final state is the sum of antiphase ^{13}C magnetization on the $-x$ axis (^{1}H = α vector of length 4 on $-x'$, ^{1}H = β vector of length 4 on x') and in-phase ^{13}C magnetization on the $+x'$ axis (^{1}H = α vector of length 1 on x' and ^{1}H = β vector of length 1 on x'). The sum can also be represented as the sum of two spectra: an antiphase ^{13}C doublet four times as intense as a normal ^{13}C spectrum (downfield component upside-down) plus a normal (in-phase) ^{13}C spectrum (Fig. 7.24). The antiphase part was transferred from ^{1}H equilibrium net magnetization (enhancement by polarization transfer), and the in-phase part is just the normal ^{13}C spectrum derived from ^{13}C equilibrium net magnetization.

What does this "thought experiment" have to do with the INEPT experiment? The same spin state that results from the selective ^{1}H 180° pulse on the downfield component of the ^{1}H doublet ($-2\mathbf{I}_z\mathbf{S}_z + \mathbf{S}_z$) can be obtained in a more practical way by a nonselective 90° ^{1}H pulse followed by a delay of $1/(2J)$ and another nonselective ^{1}H 90° pulse:

$$\mathbf{I}_z + \mathbf{S}_z \xrightarrow{^1\text{H}\ 90°\ \text{pulse on }x'\text{ axis}} -\mathbf{I}_y + \mathbf{S}_z$$

$$-\mathbf{I}_y + \mathbf{S}_z \xrightarrow{1/(2J)\ \text{delay}} 2\mathbf{I}_x\mathbf{S}_z + \mathbf{S}_z \quad (J-\text{coupling evolution})$$

$$2\mathbf{I}_x\mathbf{S}_z + \mathbf{S}_z \xrightarrow{^1\text{H}\ 90°\ \text{pulse on }y'} -2\mathbf{I}_z\mathbf{S}_z + \mathbf{S}_z$$

This may seem more complicated, but it is much more practical because all of the pulses are nonselective ("hard") pulses. If we add the final ^{13}C 90° pulse used in the SPT experiment to bring the magnetization to the x–y plane, we have the INEPT experiment:

$$^1\text{H}\ 90°_x - 1/(2J)\ \text{delay} - {}^1\text{H}\ 90°_y - {}^{13}\text{C}\ 90°_y - \text{FID}$$

The final 90° pulses on ^{1}H and ^{13}C can be applied simultaneously or one immediately after the other, as shown above.

Exercise: The equilibrium ^{13}C z magnetization makes the SPT analysis more complicated than it has to be. Go through the SPT thought experiment again showing the population diagrams and the final ^{13}C spectrum if the ^{13}C resonances are first *saturated* by continuous low-power RF irradiation at both ^{13}C transition frequencies. Draw the population diagram after saturation of the ^{13}C transitions. Then start the SPT experiment with the selective 180° ^{1}H pulse on the ^{1}H (^{13}C = α) component of the ^{1}H doublet.

7.9 PHASE CYCLING IN INEPT

All NMR experiments are beset by *artifacts*: peaks in the spectrum that are not supposed to be there. They may be caused by imperfectly calibrated pulses, by delays (such as $1/(2J)$), that cannot be set perfectly for every resonance in the spectrum, or by hardware imperfections. We can also speak of coherence pathways—different origins and histories of NMR magnetization that eventually reach the receiver during the FID. We saw in the previous section, for example, that observable ^{13}C coherence can come from ^{1}H "polarization" (population difference) via INEPT transfer, or it can come from ^{13}C polarization directly as a result of the final ^{13}C 90° pulse. These two components result in the antiphase (+4, −4) ^{13}C spectrum and the in-phase (+1, +1) ^{13}C spectrum combined in the receiver. If we view the direct (^{13}C → ^{13}C) pathway as an artifact, we can explore methods to remove it. There are two main techniques for removing artifacts: pulsed field gradients (which we will discuss in Chapter 8) and phase cycling. Phase cycling is a subtraction method, which relies on acquiring more than one scan and combining the FIDs (by adding or subtracting) in such a way that the desired signals add together and the artifacts cancel out by subtraction. A phase cycle for INEPT can be designed very easily by looking at the product operator description at each stage of the experiment. By changing the phase of one of the pulses, we can differentiate between the desired operators (the "signal") and the undesired operators (the "artifacts"). Starting from the equilibrium state, we have

$$\mathbf{I}_z + \mathbf{S}_z \xrightarrow{^1\text{H } 90° \text{ pulse on } x' \text{ axis}} -\mathbf{I}_y + \mathbf{S}_z$$

$$-\mathbf{I}_y + \mathbf{S}_z \xrightarrow{1/(2J)\text{delay}} 2\mathbf{I}_x\mathbf{S}_z + \mathbf{S}_z \quad (J-\text{coupling evolution})$$

$$2\mathbf{I}_x\mathbf{S}_z + \mathbf{S}_z \xrightarrow{^1\text{H and }^{13}\text{C } 90° \text{ pulses on } y'} 2[-\mathbf{I}_z][\mathbf{S}_x] + \mathbf{S}_x$$

Note that the conversion $\mathbf{S}_z \rightarrow \mathbf{S}_x$ in the last step is brought about by the 90° ^{13}C pulse only; the 90° ^{1}H pulse has no effect on this conversion. The desired term, $4[-2\mathbf{S}_x\mathbf{I}_z]$, comes from $2\mathbf{I}_x\mathbf{S}_z$ and requires *both* 90° pulses, on ^{1}H and ^{13}C, to be produced. Consider what happens if we change the phase of the ^{1}H 90° pulse from $+y'$ to $-y'$:

$$2\mathbf{I}_x\mathbf{S}_z \xrightarrow{^1\text{H } 90° \text{ on } +y' \text{ and } ^{13}\text{C } 90° \text{ on } +y'} 2[-\mathbf{I}_z][\mathbf{S}_x] = 4[-2\mathbf{S}_x\mathbf{I}_z]$$

$$2\mathbf{I}_x\mathbf{S}_z \xrightarrow{^1\text{H } 90° \text{ on } -y' \text{ and } ^{13}\text{C } 90° \text{ on } +y'} 2[\mathbf{I}_z][\mathbf{S}_x] = 4[2\mathbf{S}_x\mathbf{I}_z]$$

By inverting the phase of the ^{1}H 90° pulse ($+y'$ to $-y'$), we have inverted the phase of the resulting antiphase ^{13}C signal. If we choose the $+x'$ axis as our phase reference, the $-2\mathbf{S}_x\mathbf{I}_z$ term will give an antiphase doublet with the downfield (left) component negative (upside-down) and the upfield (right) component positive (Fig. 7.25, top left). The $2\mathbf{S}_x\mathbf{I}_z$ term will give the opposite spectrum: downfield component positive and upfield component negative (Fig. 7.25, top center). The final $\mathbf{S}_x$ term will be positive and in-phase in

264 COHERENCE TRANSFER: INEPT AND DEPT

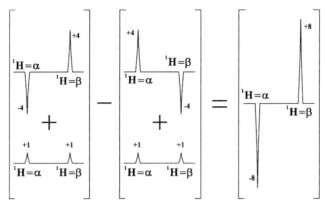

Figure 7.25

either case:

$$S_z \xrightarrow{^1H\,90°\,\text{on}+y'\,\text{and}\,^{13}C\,90°\,\text{on}+y'} S_x$$

$$S_z \xrightarrow{^1H\,90°\,\text{on}+y'\,\text{and}\,^{13}C\,90°\,\text{on}+y'} S_x$$

Only the ^{13}C 90° pulse operates on S_z to rotate it to S_x, so the phase of the ^{1}H 90° pulse is irrelevant (Fig. 7.25, bottom left and bottom center). Now we can apply the phase-cycling strategy: acquire one FID using the ^{1}H 90° pulse on $+y'$ and a second FID with the ^{1}H 90° pulse on $-y'$. Subtract the second FID from the first. The result will be:

$$\text{Difference} = \{4[-2S_xI_z] + S_x\} - \{4[2S_xI_z] + S_x\} = 8[-2S_xI_z]$$

Fourier transformation will give a spectrum corresponding to $8[-2S_xI_z]$: an antiphase ^{13}C doublet with intensities of -8 (^{1}H = α component) and $+8$ (^{1}H = β component) (Fig. 7.25, right). This spectrum is the pure INEPT spectrum, without any contribution from the in-phase ^{13}C doublet (the "artifact"). If you understand this strategy of cancellation, you will understand the use of phase cycling in all NMR experiments: pick a pulse whose phase has a different effect on the desired signal than on the artifact peaks, change its phase over more than one scan, and combine the FIDs by addition or subtraction so as to cancel the artifact signals and sum the desired signals.

Figure 7.26 (left) shows the INEPT spectrum of neat benzene (C_6H_6) using the sequence of Figure 7.16 with no ^{1}H decoupling and no phase cycling. With the final ^{1}H pulse phase set to 1 (y' axis), we see the H = α component upside-down with intensity 3 and the H = β component positive with intensity 5 (spectrum A). With the final ^{1}H pulse phase set to 3 ($-y'$ axis), the antiphase (-4, $+4$) portion of the signal is inverted (spectrum B), giving intensities of $+5$ (H = α) and -3 (H = β). If the FIDs from spectrum A and spectrum B are subtracted, Fourier transformation of the difference FID gives only the portion of the signal that results from coherence transfer from ^{1}H to ^{13}C: an antiphase doublet with intensities of -8 (H = α) and $+8$ (H = β). If instead the two FIDs are added together, Fourier transformation of the sum FID gives only that portion of the signal that results from direct excitation of the ^{13}C z magnetization: an in-phase doublet with intensities of

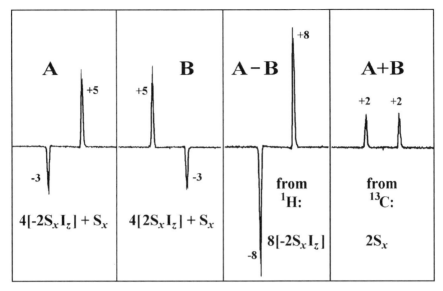

Figure 7.26

+2 (H = α) and +2 (H = β). Note that the intensities add directly from the two scans, but the noise increases as the square root of the number of scans increases, so the noise in the sum and difference spectra (right) is $\sqrt{2}$ times larger (1.414 times larger) than in the individual scan spectra (left).

7.10 INTERMEDIATE STATES IN COHERENCE TRANSFER

In the INEPT experiment, the final step of coherence transfer is the simultaneous 90° pulses on ^{13}C and ^{1}H:

$$2\mathbf{I}_x\mathbf{S}_z \xrightarrow{^1\text{H and }^{13}\text{C 90° pulses on } y'} 2[-\mathbf{I}_z][\mathbf{S}_x] = 4[-2\mathbf{S}_x\mathbf{I}_z]$$

Again, reversing the roles of ^{1}H and ^{13}C operators requires the additional factor of four because our point of comparison is now ^{13}C equilibrium net magnetization. We do not actually need to do the two 90° pulses simultaneously—we can do them in sequence with a small delay in between. First, let's try doing the 90° pulse on ^{1}H, followed by a short delay and a 90° pulse on ^{13}C:

$$2\mathbf{I}_x\mathbf{S}_z \xrightarrow{^1\text{H 90° pulse on } y'} -2\mathbf{I}_z\mathbf{S}_z \xrightarrow{^{13}\text{C 90° pulse on } y'} -2\mathbf{I}_z\mathbf{S}_x = 4[-2\mathbf{S}_x\mathbf{I}_z]$$

We can consider the state $-2\mathbf{I}_z\mathbf{S}_z$ as an intermediate state in coherence transfer, just as we did in the analysis of populations (SPT) above. Often in INEPT-based experiments, this intermediate state is used as a way of "cleaning up" other coherences that are not desired. A pulsed field gradient (PFG, Chapter 8) is a way of temporarily messing up the shims, and this will destroy any magnetization that is in the x–y plane. The intermediate state $-2\mathbf{I}_z\mathbf{S}_z$ is not affected, however, because there is no net magnetization in the x–y plane. After the "spoiler" gradient, we can complete the coherence transfer (with the ^{13}C 90° pulse). The field gradient ("temporary bad shimming") can be regarded as a *filter* that lets only z magnetization and more complicated spin states involving z magnetization get through.

Something even more interesting happens if we reverse the order of the two 90° pulses, starting with the ^{13}C 90° pulse:

$$2\mathbf{I}_x\mathbf{S}_z \xrightarrow{^{13}\text{C 90° pulse on } y'} 2\mathbf{I}_x\mathbf{S}_x \xrightarrow{^{1}\text{H 90° pulse on } y'} -2\mathbf{I}_z\mathbf{S}_x = 4[-2\mathbf{S}_x\mathbf{I}_z]$$

Now in the intermediate state $2\mathbf{I}_x\mathbf{S}_x$, we have both operators in the x–y plane. An operator in the x–y plane corresponds to observable coherence, corresponding to a transition between two energy states in which only one spin changes state (α to β or β to α). This is called a single-quantum (SQ) transition. For example, the $\alpha_H\alpha_C \rightarrow \beta_H\alpha_C$ transition is an SQ transition because only the ^{1}H changes its spin state. The corresponding coherence could be represented as $\mathbf{I}_x$ or $\mathbf{I}_y$, for example. But what happens if two of these x–y plane operators are multiplied together? The product corresponds to transitions in which *both* spins change their spin state: for example, $\alpha_H\alpha_C \rightarrow \beta_H\beta_C$. In this case, because both spins jump up simultaneously from α to β, we call this a double-quantum (DQ) transition. The transition $\alpha_H\beta_C \rightarrow \beta_H\alpha_C$ is called a "zero-quantum" (ZQ) transition because one spin (^{1}H) jumps up and one (^{13}C) falls down, leading to a net change of zero in the total spin quantum number. At this point, all we can say about the product operator $2\mathbf{I}_x\mathbf{S}_x$ is that it represents a mixture of double-quantum coherence (DQC) and zero-quantum coherence (ZQC). We can call it "DQC/ZQC" or simply "MQC" for multiple-quantum coherence. This may seem pretty vague, but we will see later that ZQC and DQC can be very precisely defined and they undergo precession (evolution) and respond to RF pulses in completely predictable ways. Unlike $2\mathbf{I}_z\mathbf{S}_z$, we cannot even draw a vector diagram of the spin state $2\mathbf{I}_x\mathbf{S}_x$, so you can see we have finally left the vector model behind completely. Furthermore, if we turn on the ADC and record an FID at this point in the pulse sequence, starting with $2\mathbf{I}_x\mathbf{S}_x$, we will see nothing at all: no FID and no spectrum. Double-quantum and zero-quantum coherences are not observable. From the point of view of quantum mechanics, there is a "selection rule" that states that observable transitions can only have a change in the total spin quantum number of $+1$ or -1 (1/2 to $-1/2$ or $-1/2$ to 1/2). Double-quantum transitions involve a change of $+2$ or -2, and zero-quantum transitions involve a change of zero. These violate the selection rule and therefore they are not observable in the FID. This "stuff" called DQC/ZQC is looking pretty mysterious: you can't draw a picture of it or see it. Does it really exist? It does because you can change it into observable single-quantum coherence (SQC) by applying an RF pulse:

$$2\mathbf{I}_x\mathbf{S}_x \xrightarrow{^{1}\text{H 90° pulse on } y'} -2\mathbf{I}_z\mathbf{S}_x = 4[-2\mathbf{S}_x\mathbf{I}_z]$$

Not only does this make it observable, but whatever changes might happen to $2\mathbf{I}_x\mathbf{S}_x$ during a delay (evolution, due to chemical shifts or J couplings) will change the observable outcome of the RF pulse, so we can infer changes that happen during the invisible MQC state. This is similar to z magnetization, which changes during the recovery period of an inversion-recovery experiment. We cannot observe z magnetization because it does not undergo precession in the magnetic field, but we can convert it into observable magnetization by "flipping" it into the x–y plane with a 90° pulse. What we observe in the FID after the 90° pulse is affected by what happened to the z magnetization (i.e., relaxation) during the recovery delay.

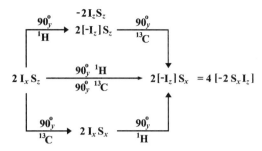

Figure 7.27

The MQC intermediate state in coherence ("INEPT") transfer can also be used to "clean up" the spectrum. In this case, we can apply a double-quantum filter (using either gradients or a phase cycle) to kill all coherences at the intermediate step that are not DQC. We will see the usefulness of this technique in the DQF (double-quantum filtered) COSY experiment (Chapter 10). As with the "spoiler" gradient applied to the $2\mathbf{I}_z\mathbf{S}_z$ intermediate state, a double-quantum filter destroys any unwanted magnetization, leaving only DQC that can then be carried on to observable antiphase magnetization in the second step of INEPT transfer.

In either case, whether we do the ^{1}H 90° pulse first or the ^{13}C 90° pulse first, we are simply choosing the order of the two processes (Fig. 7.27): the ^{1}H operator ($\mathbf{I}_x$) in the product moves from the x–y plane to the z axis (^{1}H 90° pulse) and the ^{13}C operator ($\mathbf{S}_z$) in the product moves from the z axis to the x–y plane (^{13}C 90° pulse). We can "bump up" the ^{1}H operator to the z axis first, resulting in both operators on the z axis, and then "knock down" the ^{13}C operator to the x–y plane. Alternatively, we can first "knock down" the ^{13}C operator from the z axis to the x–y plane, resulting in both operators in the product in the x–y plane (MQC), and then "bump up" the ^{1}H operator from the x–y plane to the z axis.

7.11 ZERO- AND DOUBLE-QUANTUM OPERATORS

Product operators in which both components are in the x'–y' plane represent zero-quantum and double-quantum coherences (collectively called "multiple-quantum" coherences). DQC is a superposition of the spin states $\alpha_I\alpha_S$ and $\beta_I\beta_S$, which involves promotion of both nuclei **I** and **S** simultaneously from the α state to the β state or *vice versa*. ZQC is a superposition of the spin states $\alpha_I\beta_S$ and $\beta_I\alpha_S$, which involves nucleus **I** flipping from α to β while nucleus **S** flips from the β state to the α state, or the reverse process. Neither of these coherences can be directly observed, but we can convert them into observable (single-quantum) coherence and see the effect of evolution during the time spent as zero- and double-quantum coherences. In product operator notation they look like this

$$\tfrac{1}{2}[2\mathbf{I}_x\mathbf{S}_x - 2\mathbf{I}_y\mathbf{S}_y] = \text{Pure DQC along the } x' \text{ axis} = \{\mathbf{DQ}\}_x$$

$$\tfrac{1}{2}[2\mathbf{I}_x\mathbf{S}_y + 2\mathbf{I}_y\mathbf{S}_x] = \text{Pure DQC along the } y' \text{ axis} = \{\mathbf{DQ}\}_y$$

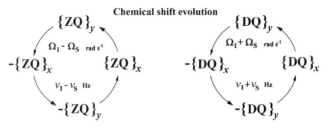

Figure 7.28

$\frac{1}{2}[2\mathbf{I}_x\mathbf{S}_x + 2\mathbf{I}_y\mathbf{S}_y]$ = Pure ZQC along the x' axis = $\{\mathbf{ZQ}\}_x$

$\frac{1}{2}[2\mathbf{I}_x\mathbf{S}_y - 2\mathbf{I}_y\mathbf{S}_x]$ = Pure ZQC along the y' axis = $\{\mathbf{ZQ}\}_y$

DQC precesses under the influence of chemical shifts at a rate determined by the *sum* of the two chemical shifts, whereas ZQC precesses at a rate determined by the *difference* (Fig. 7.28). This can be demonstrated by plugging in $[\mathbf{I}_x \cos(\Omega_I\tau) + \mathbf{I}_y \sin(\Omega_I\tau)]$ for $\mathbf{I}_x$ and $[\mathbf{S}_y \cos(\Omega_S\tau) - \mathbf{S}_x \sin(\Omega_S\tau)]$ for $\mathbf{S}_y$, etc., and multiplying the expressions together. The math is rather messy (although very satisfying) so we will not go through it here.

Neither double-quantum nor zero-quantum coherence undergoes J-coupling evolution due to J_{IS}. This is because the energy of interaction of the two nuclear magnets does not change in the transitions $\alpha\alpha \rightarrow \beta\beta$ and $\alpha\beta \rightarrow \beta\alpha$, as they remain either aligned with each other or against each other. Because this energy is not involved, the J coupling has no effect on the evolution. This can also be see by examining the exact energies of the four quantum states: $\alpha\alpha$, $\alpha\beta$, $\beta\alpha$, and $\beta\beta$. For two protons, we have (all energies are divided by Planck's constant h so we can read out frequencies in hertz directly)

$$\beta_a\beta_b = \nu_0^a/2 + \nu_0^b/2 - J/4$$

$$\alpha_a\beta_b = -\nu_0^a/2 + \nu_0^b/2 + J/4 \qquad \beta_a\alpha_b = \nu_0^a/2 - \nu_0^b/2 + J/4$$

$$\alpha_a\alpha_b = -\nu_0^a/2 - \nu_0^b/2 - J/4$$

Note that the four single-quantum transitions have energy differences corresponding to their exact frequencies in the ^{1}H spectrum:

$$\alpha\beta - \alpha\alpha \text{ (H}_b \text{ transition``}\alpha\text{'')} = \nu_0^b + J/2 \quad \beta\beta - \beta\alpha \text{ (H}_b \text{ transition``}\beta\text{'')} = \nu_0^b - J/2$$

$$\beta\alpha - \alpha\alpha \text{ (H}_a \text{ transition``}\alpha\text{'')} = \nu_0^a + J/2 \quad \beta\beta - \alpha\beta \text{ (H}_a \text{ transition``}\beta\text{'')} = \nu_0^a - J/2$$

These are the four lines of a ^{1}H spectrum with two doublets: one centered on ν_a with splitting J and the other centered on ν_b with splitting J. In each case, the higher frequency (downfield line) of the pair is the transition in which the other spin remains in the α state—the line we label "α" in the doublet. The two states in which the H$_a$ and H$_b$ spins are aligned (both α or both β) are slightly lower in energy (by $J/4$) and the two states in which the H$_a$ and H$_b$ spins are opposed ($\alpha\beta$ and $\beta\alpha$) are slightly higher in energy. This is not universally true, but as we have chosen to always label the downfield component of a doublet the "α" component, we are committed to this relationship.

Now look at the energy differences corresponding to the double-quantum and zero-quantum transitions:

$$\beta\beta - \alpha\alpha \text{ (DQ transition)} = \nu_o^a + \nu_o^b \quad \beta\alpha - \alpha\beta \text{ (ZQ transition)} = \nu_o^a - \nu_o^b$$

The J terms cancel out in both cases because the interaction between H_a and H_b is the same: it remains opposed (ZQ transition) or it remains aligned (DQ transition).

If there is another spin besides **I** and **S**, which is coupled to either **I** or **S**, this "passive coupling" *will* undergo J-coupling evolution. It is only the J coupling between the two nuclei involved in the DQC or ZQC (in this case **I** and **S**), which does not lead to any evolution.

As you can see, the bookkeeping for MQCs gets pretty messy, and we will see later (Chapter 10, Section 10.4) that another kind of operator called **spherical operators** is neater and easier to visualize for MQCs.

7.12 SUMMARY OF TWO-SPIN OPERATORS

For a system of two kinds of spins (**I** and **S** for heteronuclear, or $\mathbf{I}^a$ and $\mathbf{I}^b$ for homonuclear systems), there are 16 product operators, formed from the 16 matrix elements of the density matrix. So far we have discussed 14 of them:

$\mathbf{I}_z, \mathbf{S}_z$	z magnetization (population difference)
$\mathbf{I}_x, \mathbf{I}_y, \mathbf{S}_x, \mathbf{S}_y$	in-phase magnetization in the $x'-y'$ plane
$2\mathbf{I}_x\mathbf{S}_z, 2\mathbf{I}_y\mathbf{S}_z, 2\mathbf{S}_x\mathbf{I}_z, 2\mathbf{S}_y\mathbf{I}_z$	antiphase magnetization in the $x'-y'$ plane
$2\mathbf{I}_x\mathbf{S}_y, 2\mathbf{I}_y\mathbf{S}_x, 2\mathbf{I}_x\mathbf{S}_x, 2\mathbf{I}_y\mathbf{S}_y$	zero- and double-quantum coherences (not observable)

The remaining two are the longitudinal spin order, which results when the macroscopic z magnetization of one nucleus (e.g., ^{1}H) is opposite depending on the microscopic z magnetization (α or β) state of the other nucleus (e.g., ^{13}C), and the identity (**1**) operator, which simply represents the vast majority of spins that cancel each other out and play no role in NMR experiments. Longitudinal spin order can be viewed as an intermediate state in coherence transfer: $2\mathbf{I}_x\mathbf{S}_z \rightarrow 2\mathbf{I}_z\mathbf{S}_z \rightarrow 2\mathbf{I}_z\mathbf{S}_x = 2\mathbf{S}_x\mathbf{I}_z$. Like z magnetization, it is not affected by gradients. The identity operator is usually ignored because we are interested only in population differences.

$2\mathbf{I}_z\mathbf{S}_z$	longitudinal spin order
1	the identity operator (equal populations in all four states)

Together, these 16 product operators describe the 16 matrix elements in the 4×4 density matrix representation of a two-spin system (Chapter 10). In the matrix, each element represents coherence between (or superposition of) two spin states. As there are four spin states for a two-spin system ($\alpha_I\alpha_S$, $\alpha_I\beta_S$, $\beta_I\alpha_S$, and $\beta_I\beta_S$), there are 16 possible pairs of states, which can be superimposed or share coherence. The product operators are closer to the visually and geometrically concrete vector model representations, so in most cases they are preferable to writing down the 16 elements of the density matrix, especially as only a few of the elements are nonzero in most of the examples we discuss.

These operators and the rules that govern chemical shift and *J*-coupling evolution in time can be used to describe any combination of RF pulses and delays, giving a prediction of the observable magnetization (and thus the spectrum) at the end of the sequence. This gives us the means of understanding all of the 1D and 2D NMR experiments. By comparison, the vector model can explain only a few of the 1D experiments.

7.13 REFOCUSED INEPT: ADDING SPECTRAL EDITING

The INEPT sequence is not very useful because we cannot apply ^{1}H decoupling to the antiphase signals observed in the FID. For example, the antiphase ^{13}C doublet of a CH group (1:−1) would, with ^{1}H decoupling, collapse to a single frequency with the positive peak right on top of the negative peak. These would exactly cancel, and we would see no peak at all. The same is true for a CH$_2$ antiphase triplet (1:0:−1) and a CH$_3$ antiphase quartet (1:1:−1:−1); all of these have zero net signal if they are collapsed by ^{1}H decoupling into a single frequency. In order to apply ^{1}H decoupling, we need to add a refocusing period to allow antiphase magnetization to evolve back into in-phase magnetization. For the CH group, this is very simple: we add a 1/(2*J*) delay to go from antiphase to in-phase: $2\mathbf{S}_x\mathbf{I}_z$ —1/(2*J*)→ $\mathbf{S}_y$. To prevent phase twisting by chemical shift evolution, we need to add simultaneous ^{1}H and ^{13}C pulses in the center of the 1/(2*J*) delay, just as we did in the first ("defocusing") 1/(2*J*) delay. Figure 7.29 shows the full sequence with a general refocusing time of 2τ. For the CH group, we would set 2τ = 1/(2*J*). In the first spin echo, we have *J*-coupling evolution only from in-phase to antiphase: $\mathbf{I}_y$ —1/(2*J*)→ $-2\mathbf{I}_x\mathbf{S}_y$. We think of the 180° ^{1}H pulse on the *y*′ axis as occurring at the start of the first delay, where it has no effect on $\mathbf{I}_y$. The simultaneous 90° pulses on ^{1}H and ^{13}C then lead to transfer of coherence: $-2\mathbf{I}_x\mathbf{S}_z \to -2\,[-\mathbf{I}_z]\,[\mathbf{S}_x] = 4[2\mathbf{S}_x\mathbf{I}_z]$. Finally, the refocusing delay 1/(2*J*) brings us from antiphase ^{13}C coherence back to in-phase: $4[2\mathbf{S}_x\mathbf{I}_z] \to 4\mathbf{S}_y$. The 180° ^{13}C pulse on the *x*′ axis has no effect because we imagine it at the beginning of the refocusing period, where ^{13}C coherence is on the *x*′ axis. For a general refocusing delay Δ = 2τ, we have

$$2\mathbf{S}_x\mathbf{I}_z \to 2\mathbf{S}_x\mathbf{I}_z \cos(\pi J\Delta) + \mathbf{S}_y \sin(\pi J\Delta)$$

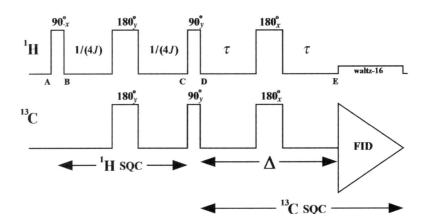

Figure 7.29

REFOCUSED INEPT: ADDING SPECTRAL EDITING 271

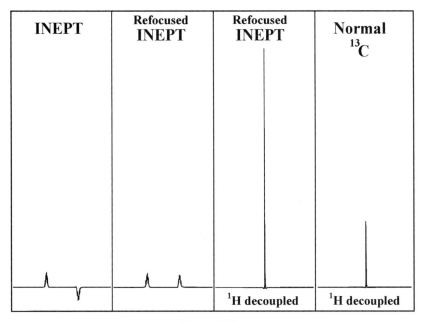

Figure 7.30

With ^{1}H decoupling, the antiphase term is not observed, so we see only the in-phase signal with an intensity of sin Θ, where the "angle" Θ (in radians) is equal to $\pi J\Delta$ or $2\pi J\tau$. The maximum signal occurs with $\Theta = \pi/2$ (90°) or $\Delta = 1/(2J)$. We will see that while this is ideal for the CH group, we would see no signal at all for CH$_2$ or CH$_3$ groups! This can be useful because it allows us to get a ^{13}C spectrum with only the CH peaks, so we can distinguish between CH and CH$_3$ peaks—something that the APT experiment cannot do.

Figure 7.30 shows the INEPT spectrum of neat benzene, C$_6$H$_6$, acquired with the sequence of Figure 7.16. On the left is the spectrum without refocusing or ^{1}H decoupling: we see an antiphase doublet with complex long-range ($^2J_{CH}$ and $^3J_{CH}$) couplings making the components "ragged." With refocusing (sequence of Fig. 7.29, $\Delta = 1/(2J)$, where J is $^1J_{CH}$) we see an in-phase doublet, still showing the long-range couplings. When ^{1}H decoupling is applied during acquisition of the FID, we see a sharp singlet. This singlet peak is about four times the height of the ^{13}C singlet obtained in a simple ^{13}C spectrum with ^{1}H decoupling (Fig. 7.30, right) because of the enhancement coming from coherence transfer from the ^{1}H.

Now we need to look at the refocusing step in general—for all three types of ^{13}C nuclei that are coupled to ^{1}H. The defocusing step (first $1/(2J)$ period) was simple because any proton, whether it is a part of a CH, CH$_2$, or CH$_3$ group, is still connected to only one ^{13}C, so it can be looked at as a doublet. In the refocusing step, however, we have ^{13}C coherence and it behaves differently depending on whether it is a doublet (CH), triplet (CH$_2$), or quartet (CH$_3$). We will have to define two ^{1}H product operators — I^1 and — I^2 for the CH$_2$ group and three — I^1, I^2 and I^3 — for the CH$_3$ group. As soon as we have ^{13}C magnetization in the x'–y' plane, we can have an in-phase or antiphase relationship to any of the attached protons. Thus, for pure ^{13}C SQC with a CH$_2$ group, we can have product operators like

$$\mathbf{S}_x \quad 2\mathbf{S}_x\mathbf{I}_z^1, 2\mathbf{S}_x\mathbf{I}_z^2 \quad 4\mathbf{S}_x\mathbf{I}_z^1\mathbf{I}_z^2$$

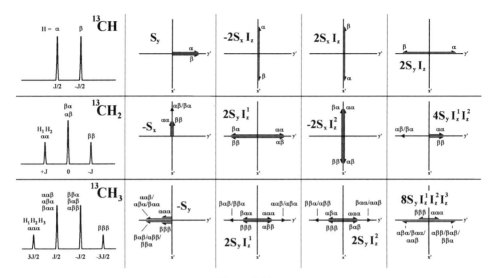

Figure 7.31

S_x is in-phase with respect to both protons, $2S_xI_z^1$ is antiphase with respect to H_1 and in-phase with respect to H_2, and $4S_xI_z^1I_z^2$ is antiphase with respect to both attached protons. In each case, the multiplication by I_z^1 means that the ^{13}C coherence is multiplied by 1/2 if $H_1 = \alpha$ and by $-1/2$ if $H_1 = \beta$. We can draw vector diagrams for any of these states, but we have to be more specific about the vector labels. We are already familiar with the in-phase and antiphase states for a CH group (Fig. 7.31, top). For a CH_2 group (triplet in the nondecoupled ^{13}C spectrum), we have four states for the two protons (Fig. 7.31, center). Both can be in the α state ($\alpha\alpha$, giving rise to the downfield peak of intensity 1), one can be α and the other β ($\alpha\beta$ or $\beta\alpha$, giving rise to the central peak of intensity 2), or both can be β ($\beta\beta$, giving rise to the upfield peak of intensity 1). For the operator product $2S_yI_z^1$, we have four net magnetization vectors, each representing one-fourth of the ^{13}C nuclei, differentiated by the spin state of the two attached protons: $\alpha\alpha$, $\alpha\beta$, $\beta\alpha$, or $\beta\beta$. Multiplication by I_z^1 corresponds to multiplication by $+1/2$ if H_1 is in the α state and by $-1/2$ if H_1 is in the β state, so the $\alpha_1\alpha_2$ and $\alpha_1\beta_2$ vectors lie on the $+y'$ axis (α_1) and the $\beta_1\alpha_2$ and $\beta_1\beta_2$ vectors lie on the $-y'$ axis (β_1, Fig. 7.31, center). In other words, it does not matter what state H_2 is in, but the H_1 state (α or β) determines whether the vector is on $+y'$ or $-y'$. Even though they are on the same axis, we cannot combine the $\alpha\alpha$ and $\alpha\beta$ vectors into a single vector because they have different rotating frame frequencies: $\alpha\alpha$ rotates at $+J$ and $\alpha\beta$ rotates at 0 Hz for an on-resonance triplet. The product $-2S_xI_z^2$ has the $\alpha\alpha$ and $\beta\alpha$ vectors on $-x$ and the $\alpha\beta$ and $\beta\beta$ vectors on $+x'$: in this case only the H_2 spin state affects the direction of the vector. The product $4S_yI_z^1I_z^2$ represents ^{13}C coherence that is antiphase with respect to both attached protons. The $\alpha\alpha$ and $\beta\beta$ vectors lie on the $+y'$ axis (multiplication by $+1/2$ and $+1/2$ for $\alpha\alpha$ and by $-1/2$ and $-1/2$ for $\beta\beta$) and the $\alpha\beta$ and $\beta\alpha$ vectors lie on the $-y'$ axis (multiplication by $+1/2$ and $-1/2$). These last two can be combined into one vector of twice the length as $H_1H_2 = \alpha\beta$ and $H_1H_2 = \beta\alpha$ have the same rotating-frame frequency (i.e., the same behavior during delays). Note that the normalization factor of 4 is needed to take care of the two factors of 1/2 coming from the I_z terms.

For a CH$_3$ group (Fig. 7.31, bottom), we have eight different vectors: $\alpha\alpha\alpha$ (frequency = $3J/2$); $\alpha\alpha\beta$, $\alpha\beta\alpha$, and $\beta\alpha\alpha$ (frequency = $J/2$); $\alpha\beta\beta$, $\beta\alpha\beta$, and $\beta\beta\alpha$ (frequency = $-J/2$); and $\beta\beta\beta$ (frequency = $-3J/2$). The product $2\mathbf{S}_y\mathbf{I}_z^1$ can be drawn with all the ^{1}H states starting with α on the $+y'$ axis and all the ^{1}H states starting with β on the $-y'$ axis. The $\alpha\alpha\beta$ and $\alpha\beta\alpha$ vectors can be combined, as can the $\beta\alpha\beta$ and $\beta\beta\alpha$ vectors. Notice that the normalization factor of 2 is introduced for *each* antiphase relationship. Thus, for a CH$_3$ group, we can have $8\mathbf{S}_y\mathbf{I}_z^1\mathbf{I}_z^2\mathbf{I}_z^3$, which could be represented by a vector on the $+y'$ axis for the H$_1$H$_2$H$_3$ = $\alpha\alpha\alpha$ state (length 1), another vector on the $+y'$ axis for the $\alpha\beta\beta$, $\beta\beta\alpha$, and $\beta\alpha\beta$ states (length 3), a vector on the $-y'$ axis for the $\alpha\alpha\beta$, $\alpha\beta\alpha$, and $\beta\alpha\alpha$ states (length 3), and a fourth vector on the $-y'$ axis for the $\beta\beta\beta$ state (length 1). In all cases where the vector is on $+y'$, we have an even number of protons in the β state (so the multiplication by $-1/2$ is cancelled), and in all cases where the vector lies on the $-y'$ axis, we have an odd number of protons in the β state (so the multiplication by $-1/2$ takes effect). Again, we can group together vectors that are on the same axis and have the same rotation frequency. This product operator represents ^{13}C coherence antiphase with respect to all three of the attached protons. A normalization factor of 8 is required because of three factors of $\pm 1/2$.

During a delay, all antiphase relationships evolve toward the in-phase relationship and all in-phase relationships evolve toward the antiphase relationship. For example, for a CH$_2$ group during a delay Δ we have

$$2\mathbf{S}_x\mathbf{I}_z^1 \xrightarrow{J_1} 2\mathbf{S}_x\mathbf{I}_z^1 \cos(\pi J\Delta) + \mathbf{S}_y \sin(\pi J\Delta) \qquad H_1 \text{ J-coupling evolution}$$

$$\xrightarrow{J_2} 2\mathbf{S}_x\mathbf{I}_z^1 \cos^2(\pi J\Delta) + 4\mathbf{S}_y\mathbf{I}_z^1\mathbf{I}_z^2 \cos(\pi J\Delta)\sin(\pi J\Delta)$$

$$+ \mathbf{S}_y \sin(\pi J\Delta)\cos(\pi J\Delta) - 2\mathbf{S}_x\mathbf{I}_z^2 \sin^2(\pi J\Delta) \qquad H_2 \text{ J-coupling evolution}$$

Here we are assuming that both J couplings (^{13}C–H$_1$ and ^{13}C–H$_2$) are the same. For simplicity, we are treating the Δ delay as two separate J-coupling evolution periods—the first for J-coupling evolution with respect to H$_1$ only and the second for J-coupling evolution with respect to H$_2$ only. In the second step, the first two terms (in the upper line) come from the $2\mathbf{S}_x\mathbf{I}_z^1\cos(\pi J\Delta)$ term and the last two terms (in the lower line) come from the $\mathbf{S}_y \sin(\pi J\Delta)$ term. So we see that, starting with ^{13}C coherence antiphase with respect to H$_1$, we get the unchanged product $2\mathbf{S}_x\mathbf{I}_z^1$ times the cosine term plus the in-phase state with respect to H$_1$, rotated 90° ccw in the x'–y' plane ($\mathbf{S}_y$) times the sine term. This last term represents refocusing. The effect of H$_2$ J-coupling evolution is to bring both of these terms toward the antiphase state with respect to H$_2$ (defocusing) with a cosine term for the starting state and a sine term for the "destination" state.

Now we have the theoretical tools to look at the refocusing for CH$_2$ and CH$_3$ groups in the refocused INEPT experiment. For a CH$_2$ group, we can start the experiment with either $\mathbf{I}_z^1$ or $\mathbf{I}_z^2$, the result is the same. So we will start with $\mathbf{I}_z^1$ and multiply the final result by 2, as the ^{13}C coherence is coming equally from two different attached protons. Up until the refocusing delay, the product operator analysis is the same, resulting in the product $2\mathbf{S}_x\mathbf{I}_z^1$ that represents coherence transfer from H$_1$ to ^{13}C. The refocusing process was already described above for this product, and yields

$$2\mathbf{S}_x\mathbf{I}_z^1 \cos^2(\pi J\Delta) + 4\mathbf{S}_y\mathbf{I}_z^1\mathbf{I}_z^2 \cos(\pi J\Delta)\sin(\pi J\Delta)$$

$$+ \mathbf{S}_y \sin(\pi J\Delta)\cos(\pi J\Delta) - 2\mathbf{S}_x\mathbf{I}_z^2 \sin^2(\pi J\Delta)$$

274 COHERENCE TRANSFER: INEPT AND DEPT

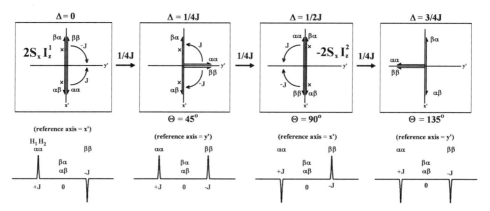

Figure 7.32

With ^{1}H decoupling, all antiphase terms will give no signal, so we are interested only in the in-phase term S_y. Its intensity is modulated by the factor $\sin(\pi J\Delta)\cos(\pi J\Delta)$, which when we consider an equal amount of ^{13}C coherence coming from H$_2$ can be increased to $2\sin(\pi J\Delta)\cos(\pi J\Delta)$. If we think of the product $\pi J\Delta$ as an angle Θ, we would get a relative intensity of 1 for $\Theta = 45°$ ($\Delta = 1/(4J)$), zero for $90°$ ($\Delta = 1/(2J)$), and -1 for $135°$ ($\Delta = 3/(4J)$). Considering that $2\sin\Theta\cos\Theta = \sin 2\Theta$, we can see that the maximum intensity occurs at $\Theta = 45°$ or $135°$ ($2\Theta = 90°$ or $270°$).

We can also look at the refocusing of the CH$_2$ group using vectors (Fig. 7.32). The vector representation of $2S_xI_z^1$ has the $\alpha\alpha$ and $\alpha\beta$ vectors on the $+x'$ axis and the $\beta\alpha$ and $\beta\beta$ vectors on the $-x'$ axis. With ^{1}H decoupling, this operator gives zero intensity, as all of the vectors have the same frequency and will always cancel each other out. Even without decoupling, the central peak ($\alpha\beta/\beta\alpha$) is gone because the two vectors are exactly opposed (a small x in Fig. 7.32 indicates cancellation). During a total refocusing delay of $1/(4J)$ ($\Theta = 45°$), the $\alpha\alpha$ vector rotates $90°$ ccw and the $\beta\beta$ vector rotates $90°$ cw, bringing the two together on the $+y'$ axis. The $\alpha\beta$ and $\beta\alpha$ vectors are frozen because their rotating frame frequencies are both zero relative to the chemical shift position. They will always cancel each other and we can never get that half of the intensity back! So the intensity of 1 that we observe ($2\sin\Theta\cos\Theta = 1$) is one-half of the intensity of an in-phase triplet getting its coherence from two protons. After another delay of $1/(4J)$, the $\alpha\alpha$ vector continues to rotate ccw and the $\beta\beta$ vector rotates cw until they are again opposite each other, with $\alpha\alpha$ on the $-x'$ axis and $\beta\beta$ on the $+x'$ axis. At this point (total refocusing delay = $1/(2J)$, $\Theta = 90°$), we have no intensity with ^{1}H decoupling, and the spin state can be represented as $-2S_xI_z^2$. Note that all states in which H$_2 = \alpha$ are on the $-x'$ axis and all states in which H$_2 = \beta$ are on the $+x'$ axis, as the spin state is on $-x'$ and antiphase with respect to H$_2$. In the product operator expression, we see that $\pi J\Delta = \pi J/(2J) = \pi/2$, so each sine term is 1 and each cosine term is 0, leaving only the one $\sin^2$ term $-2S_xI_z^2$. This is the null point where only CH peaks are observed in the refocused INEPT spectrum. After a third delay of $1/(4J)$, for a total delay of $3/(4J)$, the $\alpha\alpha$ vector rotates another $90°$ ccw and the $\beta\beta$ vector rotates another $90°$ cw, so both land on the $-y'$ axis. Again we have one-half of the full intensity we would have with all four vectors aligned, and the observable intensity is opposite in sign (on the $-y'$ axis). With $\Theta = 135°$, the factor $2\sin\Theta\cos\Theta$ equals -1, and we see upside-down peaks in the ^{1}H-decoupled spectrum for all CH$_2$ groups.

REFOCUSED INEPT: ADDING SPECTRAL EDITING   275

Finally, let's look at the CH$_3$ group. We will start with $\mathbf{I}_z^1$ and assume that starting with $\mathbf{I}_z^2$ or $\mathbf{I}_z^3$ would give the same result, so the final result will be increased by a factor of 3 to represent coherence transfer from all three attached protons. We now know that only the in-phase ^{13}C coherence will be observable with ^{1}H decoupling, so we can ignore any terms that will lead to antiphase terms at the end of the refocusing period. To get an in-phase operator, we need to "unwind" the antiphase relationship with H$_1$ and avoid "winding up" the antiphase state with respect to H$_2$ or H$_3$. We can therefore ignore any term that does not do what we want, while considering the three coupling relationships in three separate delays of length Δ:

$$2\mathbf{S}_x\mathbf{I}_z^1 \rightarrow \mathbf{S}_y \sin(\pi J\Delta) \quad \text{full } J\text{-coupling evolution with respect to H}_1$$

$$\rightarrow \mathbf{S}_y \sin(\pi J\Delta)\cos(\pi J\Delta) \quad \text{no } J\text{-coupling evolution with respect to H}_2$$

$$\rightarrow \mathbf{S}_y \sin(\pi J\Delta)\cos(\pi J\Delta)\cos(\pi J\Delta) \quad \text{no } J\text{-coupling evolution with respect to H}_3$$

Putting in the factor of 3, we have a final observable spin state of $\mathbf{S}_y [3\sin\Theta\cos^2\Theta]$. This factor has a value of 1.061 for $\Theta = 45°$, 0 for $\Theta = 90°$, and 1.061 for $\Theta = 135°$, compared to 3 if it were possible to fully refocus all coherence coming from the three protons. Just for fun, the vector diagram is shown in Figure 7.33. For both $\Theta = 45°$ and $\Theta = 135°$, the "short" vectors $\alpha\alpha\alpha$, $\beta\beta\beta$, $\alpha\beta\beta$, and $\beta\alpha\alpha$ are at four opposite corners, canceling out, and the two "long" vectors $\beta\alpha\beta/\beta\beta\alpha$ and $\alpha\alpha\beta/\alpha\beta\alpha$ are halfway between the $+x'$ and $+y'$ axes and halfway between the $-x'$ and $+y'$ axes. Each of the "long" vectors represents one-fourth of the intensity of a fully in-phase CH$_3$ peak coming from one proton, so the vector sum has magnitude $\sqrt{2}/4 = 0.3536$. Multiplying by 3 for the three protons, we get 1.061 for the peak intensity. For $\Theta = 90°$, we have the two long vectors on the $+y'$ axis and the four short vectors on the $-y'$ axis, leading to complete cancellation and a peak intensity of zero. The maximum value of the function $3\sin\Theta\cos^2\Theta$ occurs at $\Theta = 35.26°$ (or $144.74°$), corresponding to a Δ delay of about $1/(5J)$ and an in-phase intensity of 1.155.

So you see that refocusing of ^{13}C is always complicated by the fact that the optimal refocusing time is different for a CH, CH$_2$, or CH$_3$ group. Of course, we can exploit these differences to get spectral editing (making peak phase—up, down, or missing—tells us about

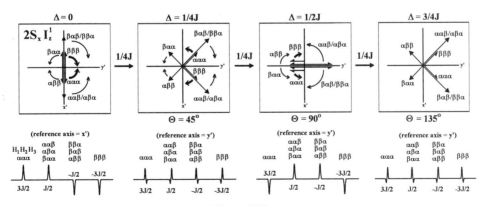

Figure 7.33

the number of attached protons), which is even more informative than the APT experiment. If we do three refocused INEPT experiments using the three delay times $\Delta = 1/(4J)$, $1/(2J)$, and $3/(4J)$, we get the following in-phase intensities:

Δ	$= 1/(4J)$	$1/(2J)$	$3/(4J)$
Θ	$= 45°$	$90°$	$135°$
CH: $\sin\Theta$	$= 0.707$	1	0.707
CH$_2$: $2\sin\Theta\cos\Theta$	$= 1$	0	-1
CH$_3$: $3\sin\Theta\cos^2\Theta$	$= 1.061$	0	1.061

Thus, an "INEPT-45" experiment will give all positive ^{13}C peaks, an "INEPT-90" will give only CH peaks, and an "INEPT-135" will give positive peaks for CH and CH$_3$ and negative peaks for CH$_2$. These three experiments (actually just the last two) give us a complete identification of each peak as CH, CH$_2$, or CH$_3$ while maintaining the ^{1}H-decoupled, singlet character of each peak. We will see how another experiment, called DEPT, achieves exactly the same results with a simpler pulse sequence by varying the width of a *pulse* rather than a delay time, using 45°, 90°, and 135° pulses.

7.14 DEPT: DISTORTIONLESS ENHANCEMENT BY POLARIZATION TRANSFER

The INEPT method can be used to distinguish CH, CH$_2$, and CH$_3$ carbons in addition to enhancing the signal-to-noise ratio, but a related method called DEPT is superior in many ways. The DEPT pulse sequence is shown in Figure 7.34. We will discuss how it works later; for now, just note that if we forget about the two 180° pulses and the second $1/(2J)$ delay, and we set $\Theta = 90°$, we have a refocused INEPT that would work well for the CH group. DEPT combines the *coherence transfer* technique of INEPT (90° on ^{1}H, delay of $1/(2J)$, simultaneous 90° pulses on ^{1}H and ^{13}C) with the *spin-echo* technique of APT ($1/(2J)$ delay, 180° pulse, $1/(2J)$ delay) to achieve both signal enhancement and spectral editing

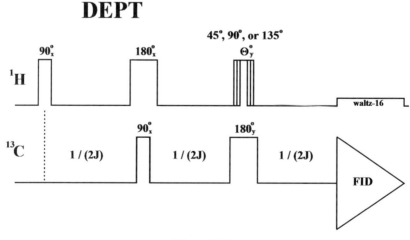

Figure 7.34

(distinction between CH, CH$_2$, and CH$_3$ carbons by phase labeling). The three "flavors" of the DEPT experiment are called "DEPT-45", "DEPT-90", and "DEPT-135" depending on the pulse width (45°, 90°, and 135°, respectively) of the final ^{1}H pulse in the sequence. We need only to perform a normal ^{13}C spectrum, a DEPT-90, and a DEPT-135 spectrum in order to assign each carbon in the molecule to quaternary (C$_q$), methine (CH), methylene (CH$_2$), or methyl (CH$_3$) based on its exact number of attached protons. The expected behavior is as follows: quaternary carbons (C$_q$) will be present in the ^{13}C spectrum but absent in all DEPT spectra; methine carbons (CH) will be positive in the DEPT-90 and in the DEPT-135; methylene carbons (CH$_2$) will be negative in the DEPT-135 and absent in the DEPT-90; and methyl carbons (CH$_3$) will be positive in the DEPT-135 and absent in the DEPT-90. Because the ^{1}H frequency is being used for "hard" pulses as well as for decoupling in this sequence, we need either a ^{1}H decoupler separate from the ^{1}H and ^{13}C pulse transmitters, or a means of rapidly switching the power level from high power (for ^{1}H pulses) to low power (for ^{1}H decoupling) and back again. Modern spectrometers use the latter approach, with power level changing in a matter of microseconds without relays.

The expected results for the ^{13}C and DEPT experiments are diagrammed in Figure 7.35 for 4-hydroxy-3-methyl-2-butanone, which has one quaternary carbon, one methine, one methylene, and two methyl groups. Note that the solvent resonance (CDCl$_3$ in this case) is absent in the DEPT spectra because solvent carbon is effectively quaternary (no attached hydrogens) and cannot undergo INEPT transfer from ^{1}H to ^{13}C. This presents a problem for referencing the DEPT spectra; usually the exact chemical shift of an easily identified peak in the referenced ^{13}C spectrum is used to reference the DEPT spectra. The ketone carbonyl peak, which appears at 213 ppm in the ^{13}C and APT spectra, is missing in the DEPT spectra because it has no attached proton. This identifies it as a quaternary carbon, C$_q$. The next peak, at 64 ppm, is negative in the DEPT-135, so it must be the

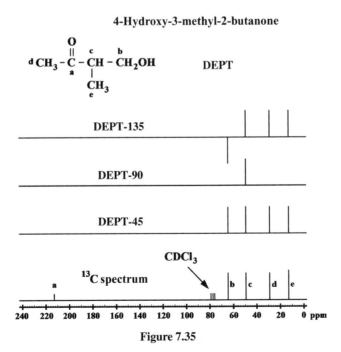

Figure 7.35

CH$_2$ carbon (position b, CH$_2$OH). The peak at 50 ppm is the only peak in the DEPT-90 spectrum, so it can be assigned to the CH carbon (position c) in the molecule. The two most upfield peaks are positive in the DEPT-135 and missing in the DEPT-90, so they must be the two CH$_3$ carbons (positions d and e in the structure). Note that the DEPT-135 is similar to an APT spectrum (Chapter 6, Fig. 6.16): there is an alternation of sign as we go from CH (positive) to CH$_2$ (negative) to CH$_3$ (positive). But in the APT spectrum, we would also see the carbonyl carbon peak (negative) and the solvent resonance (CDCl$_3$, negative), and the sensitivity would be worse than that of a ^{13}C spectrum (which has about one-fourth of the sensitivity of DEPT) because there is no coherence transfer involved.

The width of the final ^{1}H pulse in the DEPT sequence (Θ) affects the intensity and phase (positive or negative) of the ^{13}C peaks in a way very similar to the length of the refocusing delay Δ in refocused INEPT, according to the relation $\Theta = \pi J\Delta$. The intensities are shown below for the three types of carbon in DEPT:

	Scans	CH peak intensity	CH$_2$ peak intensity	CH$_3$ peak intensity	Relative noise level
$\Theta°$ ^{1}H pulse:		$\sin\Theta$	$2\sin\Theta\cos\Theta$	$3\sin\Theta\cos^2\Theta$	
s1. 45° ^{1}H pulse:	n	0.707	1.0	1.060	1.0
s2. 90° ^{1}H pulse:	$2n$	2.0	0	0	1.414
s3. 135° ^{1}H pulse:	n	0.707	−1.0	1.060	1.0

If all three experiments are run on the same sample, simple linear combinations of the three spectra (s1, s2, and s3) can be generated, which are "pure subspectra": one containing only CH carbon peaks, another containing only CH$_2$ carbon peaks, and a third containing only CH$_3$ carbon peaks. Generally, the 90° pulse experiment (spectrum 2) is run with twice the number of transients as the others (or simply run twice and added together) so that its signal-to-noise ratio will be comparable to the others. The precise weightings are shown below:

(A) 0.354*(s1 + s3) − 0.25*(s2) = CH$_3$ peaks only
(B) 0.5*(s1 − s3) = CH$_2$ peaks only
(C) 0.5*(s2) = CH peaks only
(D) 0.854*(s1) + 0.25*(s2) − 0.146*(s3) = all three types together

The predicted signal intensities and noise levels can be calculated from the weighting in the above equations, giving equal signal-to-noise in the three pure subspectra:

	CH peak intensity	CH$_2$ peak intensity	CH$_3$ peak intensity	Relative noise level	S/N
CH$_3$ only	0	0	1.225	0.612	2
CH$_2$ only	0	1.414	0	0.707	2
CH only	1.414	0	0	0.707	2
All protonated	1.069	1.069	0.802	0.936	
(S/N)	(1.142)	(1.142)	(0.857)		

DEPT: DISTORTIONLESS ENHANCEMENT BY POLARIZATION TRANSFER

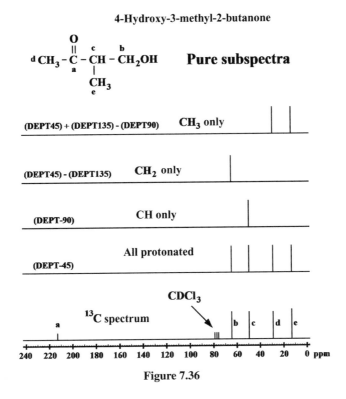

Figure 7.36

The pure subspectra are diagramed for 4-hydroxy-3-methyl-2-butanone in Figure 7.36, with simplified linear combinations. We can directly read off the assignments for peaks a–e from the pure subspectra: a is C_q (missing in all DEPT spectra), c is CH, b is CH_2, and d and e are CH_3. Varian sets up the full DEPT experiment by defining the parameter *mult* as 0.5, 1.0, or 1.5 and using the parameter *pp* for the proton 90° pulse. The actual final ^1H pulse used is then *mult*pp*, which will be a 45°, 90°, or 135° pulse. To run all three spectra in succession as a single experiment, an array is created with mult = 0.5, 1.0, 1.0, 1.5. This will acquire four FIDs with final ^1H pulses of 45°, 90°, 90°, and 135°. Bruker uses three different pulse programs called "dept45", "dept90," and "dept135" in which *p1* is the transmitter 90° ^1H pulse and *p0* is the decoupler ^1H 90° pulse. These three can be set up in consecutive experiment numbers and run in queued fashion using the "multizg" command.

In the DEPT-90 experiment (Fig. 7.35), if the pulse is slightly too long (e.g., $\Theta = 100°$), the DEPT-90 spectrum will contain a little bit of the DEPT-135: there will be weak negative peaks for the CH_2 carbons and weak positive peaks for the CH_3 carbons. If the pulse is slightly too short (e.g., $\Theta = 80°$), the DEPT-90 spectrum will contain a little bit of the DEPT-45: there will be weak positive peaks for both the CH_2 carbons and the CH_3 carbons. Thus, the DEPT-90 experiment, using a concentrated sample of a simple molecule like menthol, is the best way to calibrate the decoupler 90° ^1H pulse. Simply calibrating the ^1H 90° pulse by observing the ^1H spectrum may give a different result. In older instruments, the ^{13}C pulses come from the *transmitter* (a "broadband" RF source that can be set to any frequency up to the proton frequency) and the ^1H pulses come from the *decoupler*

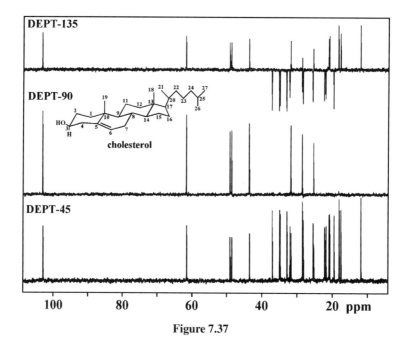

Figure 7.37

(a different RF source that is often limited to proton frequency only). The normal method for calibrating ^{1}H pulses uses pulses from the transmitter, which does not necessarily put out the same power level as the decoupler.

Figure 7.37 shows the three DEPT spectra of cholesterol in CDCl$_3$, run on a Bruker AM-250 (62.9 MHz ^{13}C). The DEPT-90 spectrum was acquired with twice the number of scans of the other two. The DEPT-45 spectrum (bottom) shows all of the "protonated" (nonquaternary) carbon peaks as normal, positive peaks. The three quaternary carbons in cholesterol—C5 in the C=C group, C10 at the A-B ring juncture, and C13 at the C-D ring juncture—are missing from the DEPT spectra. The DEPT-90 spectrum (middle) clearly shows eight methine (CH) carbons: one olefinic (C6 at 103 ppm), one "alcohol" (C3 at 61 ppm), three "crowded aliphatic" (C9, C14, and C17 between 40 and 50 ppm) and three "normal aliphatic" (C8, C20, and C25 between 24 and 34 ppm). From the DEPT-135 spectrum (top), we can count 11 negative peaks, corresponding to the 11 methylene (CH$_2$) groups in cholesterol: C1, 2, 4, 7, 11, 12, 15, 16, 22–24. Those peaks that are missing from the DEPT-90 and positive in the DEPT-135 are CH$_3$ groups: the five most upfield positive peaks in the DEPT-135 can be assigned to C18, 19, 21, 26, and 27. Note that the CDCl$_3$ peak (1:1:1 "triplet" at 77 ppm) is missing from all DEPT spectra.

Figure 7.38 shows the pure subspectra derived from the "raw" DEPT data in Figure 7.37 by linear combination. Although the interpretation of the "raw" DEPT spectra is straightforward, this presentation is especially simple and beautiful to look at. The eight CH carbons, 11 CH$_2$ carbons, and five CH$_3$ carbons can directly be read from the three subspectra. Keep in mind, however, that these are not in fact "spectra": they are the results of adding and subtracting the "raw" DEPT spectra.

When you have a very small amount of compound, a full DEPT analysis is a luxury. All you really need is a ^{13}C spectrum, a DEPT-90, and a DEPT-135 to assign all of the

DEPT: DISTORTIONLESS ENHANCEMENT BY POLARIZATION TRANSFER

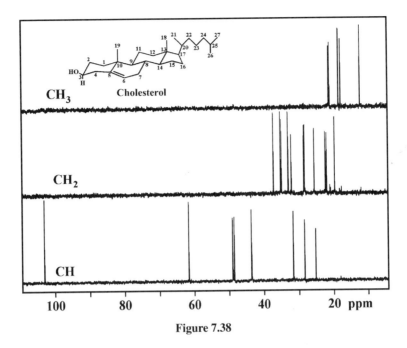

Figure 7.38

carbons to one of the four categories: C_q, CH, CH_2, or CH_3. Figure 7.39 shows a stacked plot of spectra obtained on a 0.19 mg sample of a testosterone (steroid) metabolite dissolved in CD_3OD. The data were acquired in an overnight run on a Bruker DRX-500, using about three-fourths of the time for the ^{13}C and the remaining one-fourth divided equally between the DEPT-90 and the DEPT-135 spectra. Because of the theoretical advantage of four times in sensitivity, the DEPT spectra take much less time than the ^{13}C spectrum. The

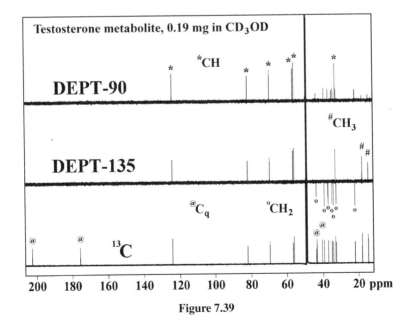

Figure 7.39

$^1H \rightarrow {}^{13}C$ NOE (up to 3× enhanced signal) observed in the ^{13}C spectrum reduces the difference in sensitivity somewhat (without NOE it would take 16 times longer than the DEPT to acquire a ^{13}C spectrum), but as we need to see the slowly relaxing quaternary carbons in the ^{13}C spectrum, most of the time is allotted to this experiment. The ^{13}C spectrum (bottom) includes the intense CD_3OD solvent peak at 49.15 ppm (seven peaks, 1:3:6:7:6:3:1 ratio), which goes off-scale at the top of the spectrum. The solvent peak is intense because we are comparing it to the very weak ^{13}C peaks of the dilute sample. The DEPT-90 spectrum (top) clearly shows six strong peaks (*), but a number of weak peaks result from incorrect calibration of the final 90° ^1H pulse. Because all of these are positive, including those that show up negative in the DEPT-135, it is clear that the final ^1H pulse was less than 90°, so that we have some DEPT-45 mixed in with the DEPT-90. It is impossible to calibrate the pulse on such a dilute sample, but it would have been possible to calibrate on a concentrated sample of menthol or cholesterol dissolved in CD_3OD before starting the experiment. This was not done, however, so this is what we have to work with. It is easy to see that the weak peaks in the DEPT-90 are a combination of CH_2 peaks (negative in the DEPT-135) and CH_3 peaks (positive in the DEPT-135), so we can ignore them. We can see two quaternary carbons (@) at 202.59 and 175.61 ppm (ketone C=O and β carbon of α,β unsaturated C=O, respectively), and six CH (*) carbons (olefinic at 124.24, two alcohols at 81.97 and 69.58, two "crowded" aliphatic and one "normal" aliphatic). We count seven CH_2 carbon peaks (O) in the DEPT-135 (negative peaks) and two CH_3 (#) peaks (positive in DEPT-135 and weak in DEPT-90) at the upfield end of the spectrum. Two more quaternary (@) carbons, the bridgehead C10 and C13 carbons, are seen in an expanded plot (Fig. 7.40) at 43.47 and 40.38 ppm. From this data alone we can infer that one of the eight CH_2 groups of testosterone was converted into a CH group bearing an alcohol (OH). Testosterone has only one alcohol, and there are two "alcohol" carbons in the metabolite. To find out which position in the steroid bears the "new" alcohol functionality would require 2D experiments.

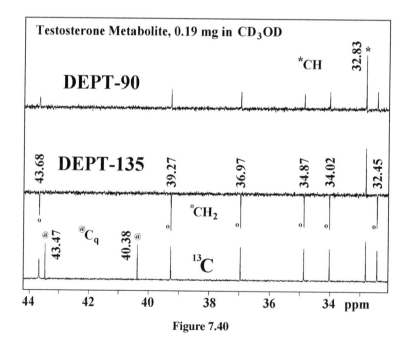

Figure 7.40

7.15 PRODUCT OPERATOR ANALYSIS OF THE DEPT EXPERIMENT

The DEPT pulse sequence is more difficult to understand than refocused INEPT, but it accomplishes the same thing. Coherence is transferred from ^{1}H to ^{13}C via the one-bond J_{CH} coupling, and the ^{13}C signal is edited according to the number of hydrogens attached. We saw in the refocused INEPT that the refocusing period Δ can be used to modify the editing: $\Delta = 1/(4J)$ gives all positive peaks, $\Delta = 1/(2J)$ gives only the CH peak, and $\Delta = 3/(4J)$ gives CH and CH$_3$ carbons positive and CH$_2$ carbons negative. This is due to the effect of refocusing time on the in-phase signal intensity: $\sin(\pi J\Delta)$ for CH, $2\sin(\pi J\Delta)\cos(\pi J\Delta)$ for CH$_2$, and $3\sin(\pi J\Delta)\cos(\pi J\Delta)\cos(\pi J\Delta)$ for CH$_3$. DEPT accomplishes the same control over spectral editing by maintaining the refocusing period constant at $1/(2J)$ and varying the pulse width of the final ^{1}H pulse, Θ. The in-phase intensity of the ^{13}C signal in the FID is $\sin\Theta$ for CH, $2\sin(\Theta)\cos(\Theta)$ for CH$_2$, and $3\sin(\Theta)\cos(\Theta)\cos(\Theta)$ for CH$_3$. Thus, for a 45° pulse ("DEPT-45": $\sin\Theta = \cos\Theta = 0.707$), we get positive peaks for CH, CH$_2$, and CH$_3$ (CH:CH$_2$:CH$_3$ = 0.707:1:1.060), for a 90° pulse ("DEPT-90": $\sin\Theta = 1$; $\cos\Theta = 0$) we see only the CH peaks (CH:CH$_2$:CH$_3$ = 1:0:0), and for a 135° pulse ("DEPT-135": $\sin\Theta = 0.707$; $\cos\Theta = -0.707$) we see positive peaks for CH and CH$_3$ and negative peaks for CH$_2$ (CH:CH$_2$:CH$_3$ = 0.707:−1:1.060).

To understand the pulse sequence, we will try to get an overview of what is happening and then look at some simplified product operator analysis. Consider first the CH case in the DEPT-90 experiment. Ignoring the 180° pulses, the DEPT-90 sequence can be viewed as an INEPT sequence in which the coherence transfer is split up into two steps (Fig. 7.41): the two 90° pulses are no longer simultaneous and between them we have an intermediate state in coherence transfer: multiple-quantum coherence (ZQC and DQC).

$$\mathbf{I}_z \underset{A}{\xrightarrow{\,^1\text{H}\,90^\circ_x\,}} \underset{B}{-\mathbf{I}_y} \xrightarrow{1/(2J)} \underset{C}{2\mathbf{I}_x\mathbf{S}_z} \underset{}{\xrightarrow{\,^{13}\text{C}\,90^\circ_x\,}} \underset{D}{-2\mathbf{I}_x\mathbf{S}_y}$$

The 90°$_x$ ^{1}H pulse puts the ^{1}H magnetization on the $-y'$ axis, and J-coupling evolution for a period of exactly $1/(2J)$ allows this in-phase magnetization to evolve into antiphase. For simplicity, we assume that the ^{13}C and ^{1}H are on-resonance so we can ignore chemical shift evolution during the delays. The ^{13}C 90° pulse then converts this to a mixture of ZQ and

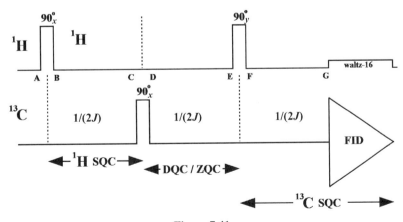

Figure 7.41

284 COHERENCE TRANSFER: INEPT AND DEPT

DQ ($-2\mathbf{I}_x\mathbf{S}_y = \mathbf{ZQ}_y - \mathbf{DQ}_y$). Both operators in the product are in the x'-y' plane, so this can be thought of as an intermediate state in coherence transfer.

$$-2\mathbf{I}_x\mathbf{S}_y \xrightarrow{1/(2J)} -2\mathbf{I}_x\mathbf{S}_y \xrightarrow{^1\text{H }90°_y} 2\mathbf{S}_y\mathbf{I}_z \xrightarrow{1/(2J)} -\mathbf{S}_x$$
$$\quad\quad\quad\quad\text{D}\quad\quad\quad\quad\quad\text{E}\quad\quad\quad\quad\quad\quad\quad\text{F}\quad\quad\quad\quad\quad\text{G}$$

During the second $1/(2J)$ delay, we have no J-coupling evolution because ZQC and DQC do not undergo evolution of the active (^{1}H–^{13}C) coupling of the MQC. Because both ^{1}H and ^{13}C are on-resonance, we can ignore chemical shift evolution as well. The next pulse, the second 90° pulse on ^{1}H, completes the coherence transfer by moving the ^{1}H operator from the x'-y' plane to the z axis, resulting in antiphase ^{13}C coherence ($2\mathbf{S}_y\mathbf{I}_z$). The final $1/(2J)$ delay is for refocusing: the antiphase ^{13}C coherence undergoes J-coupling evolution back to the in-phase state and we can observe the FID with ^{1}H decoupling. The result is exactly the same as a refocused INEPT with refocusing delay set for observing the CH ^{13}C signal. All we have done is pull apart the simultaneous ^{1}H and ^{13}C 90° pulses and insert a delay of $1/(2J)$ between them.

The next step in understanding the DEPT-90 sequence is to insert the 180° pulses and look at their effect on chemical shift evolution for "real" ^{1}H and ^{13}C peaks that are not on-resonance. For this it is best to consider what kind of evolution is going on at each stage of the pulse sequence. In the first delay, we have ^{1}H coherence that is undergoing J-coupling evolution (in-phase to antiphase) as well as ^{1}H chemical shift evolution, so we can write "J" and "ν_H" in this space. The 180° ^{1}H pulse at the end of this delay reverses the ^{1}H chemical shift evolution so that after this we have "$-\nu_\text{H}$." But now we have ZQC and DQC, so the chemical shift evolution that occurs in this second delay is "$-\nu_\text{H} - \nu_\text{C}$" for DQC and "$-\nu_\text{H} + \nu_\text{C}$" for ZQC. For simplicity let's consider just the DQC part: "$-\nu_\text{H} - \nu_\text{C}$." Notice that the ^{1}H chemical shift evolution that occurred in the first $1/(2J)$ delay is now refocused by the opposite ^{1}H chemical shift evolution in the second delay. We can think of the first two $1/(2J)$ delays as a ^{1}H spin echo, with the ^{1}H 180° pulse in the center (Fig. 7.42). There is no J-coupling evolution during this second delay because ZQC and DQC are not affected by the active J coupling. At the end of this delay, the ^{1}H 90° pulse

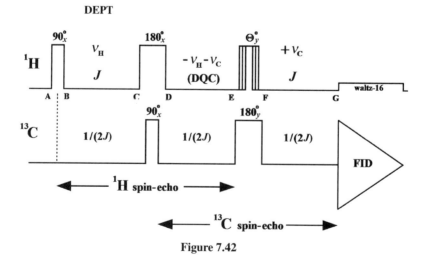

Figure 7.42

rotates the ^{1}H magnetization to the z axis and the ^{13}C 180° pulse reverses the ^{13}C chemical shift evolution that occurred during the central $1/(2J)$ delay. So during the final $1/(2J)$ delay, we have "$+\nu_C$" for chemical shift evolution and "J" for J-coupling evolution (from ^{13}C antiphase to ^{13}C in-phase coherence). We see that the last two $1/(2J)$ delays form a ^{13}C spin echo with a ^{13}C 180° pulse in the center. Thus, all chemical shift evolution, ^{1}H and ^{13}C, is refocused in this sequence and we have only the necessary J-coupling evolution to move ^{1}H from in-phase to antiphase before the coherence transfer and to refocus ^{13}C from antiphase to in-phase after the coherence transfer.

Next, we need to make the pulse width of the final ^{1}H pulse variable, with a rotation angle of Θ ($\Theta = 45°, 90°,$ or $135°$). For the CH case, we have already discussed the full coherence transfer, so in the final pulse we have:

$$-2\mathbf{I}_x\mathbf{S}_y \xrightarrow{^1\text{H}\,\Theta°_y/^{13}\text{C}\,180_y} -2\mathbf{I}_x\mathbf{S}_y\cos\Theta + 2\mathbf{S}_y\mathbf{I}_z\sin\Theta$$
$$\phantom{-2\mathbf{I}_x\mathbf{S}_y}\text{E} \text{F}$$

Note that the $180°_y$ pulse on the ^{13}C channel has no effect on $\mathbf{S}_y$. The cosine term is just the product operator we started with, unaffected by the ^{1}H pulse, and the sine term is the operator we would get with a full $90°$ ^{1}H pulse. Note that rotation of the $\mathbf{I}_x$ magnetization vector by a ^{1}H B_1 field on the y' axis goes from x to $-z$ to $-x$ to $+z$ as Θ is incremented from $0°$ to $90°$ to $180°$ to $270°$ in the trigonometric expression. The first term is DQC/ZQC, which will not be observable in the FID—there are no more pulses in the sequence to convert it to observable magnetization. Only the second term represents full coherence transfer to antiphase ^{13}C coherence, which will refocus during the final $1/(2J)$ delay into in-phase ^{13}C coherence:

$$2\mathbf{S}_y\mathbf{I}_z\sin\Theta \xrightarrow{1/(2J)} -\mathbf{S}_x\sin\Theta$$
$$\phantom{2\mathbf{S}_y\mathbf{I}_z\sin\Theta}\text{F} \text{G}$$

Thus, the intensity of the in-phase coherence varies as $\sin\Theta$ with the pulse width Θ of the final ^{1}H pulse for a CH group. This gives intensities of 0.707, 1, and 0.707 for the DEPT-45, DEPT-90, and DEPT-135 experiments, respectively.

Now we can understand all aspects of DEPT clearly, at least for a CH group. The final step is to understand the spectral editing aspect of the DEPT experiment, and here we will have to look at the complexities of CH$_2$ and CH$_3$ groups. For the CH$_2$ group, we can start with $\mathbf{I}_z^1$ and work our way through the pulse sequence. Although starting with $\mathbf{I}_z^2$ would give the same result (coherence transfer to the same ^{13}C), we can just multiply by 2 when we are finished to reflect the fact that coherence is transferred to the ^{13}C from each of the two attached protons. Things get more complicated after the ^{13}C 90° pulse. The DQC/ZQC term $-2\mathbf{I}_x^1\mathbf{S}_y$ represents a multiple-quantum "dance" between the ^{13}C nucleus and one of the attached protons (H$_1$), and the other proton (H$_2$) is not involved. While it is true that J coupling is not involved in the evolution of ZQC/DQC, this is true only for the active coupling, in this case the ^{13}C–^{1}H$_1$ coupling. The other coupling, ^{13}C–^{1}H$_2$, is a passive coupling not involved in the MQC and we will see J-coupling evolution with respect to this proton during the second $1/(2J)$ delay. Because of the spin-echo effects of the two 180° pulses, we can continue to ignore chemical shift evolution but we have to consider the J-coupling evolution due to these passive couplings during the second delay:

$$-2\mathbf{I}_x^1[\mathbf{S}_y] \xrightarrow{1/(2J)} -2\mathbf{I}_x^1[-2\mathbf{S}_x\mathbf{I}_z^2] = 4\mathbf{I}_x^1\mathbf{S}_x\mathbf{I}_z^2$$
$$\phantom{-2\mathbf{I}_x^1[\mathbf{S}_y]}\text{D} \text{E}$$

Note that the $\mathbf{S}_y$ operator is replaced by the $-2\mathbf{S}_x\mathbf{I}_z^2$ operator because it undergoes J-coupling evolution with respect to H_2 for a period of exactly $1/(2J)$, arriving at the antiphase state on the $-x'$ axis. We can view this product operator as a mixture of ZQC and DQC involving ^{13}C and H_1 (more precisely, $\{DQ\}_x^1 + \{ZQ\}_x^1$), which is antiphase with respect to the spin state of H_2. That is, if H_2 is in the α state, we multiply by $1/2$ (to get $\{DQ\}_x^1 + \{ZQ\}_x^1$) and if H_2 is in the β state, we multiply by $-1/2$ (to get $-\{DQ\}_x^1 - \{ZQ\}_x^1$).

The effect of the ^1H pulse of width Θ can be calculated using the sine and cosine terms for each of the proton operators:

$$\underset{E}{4\mathbf{I}_x^1\mathbf{S}_x\mathbf{I}_z^2} \xrightarrow{^1H\,\Theta_y^\circ / ^{13}C\,180_y^\circ} \underset{F}{4(\mathbf{I}_x^1\cos\Theta - \mathbf{I}_z^1\sin\Theta)[-\mathbf{S}_x](\mathbf{I}_z^2\cos\Theta + \mathbf{I}_x^2\sin\Theta)}$$

Note that the 180° ^{13}C pulse on the y' axis inverts $\mathbf{S}_x$. The ^1H pulse rotates $\mathbf{I}_x^1$ from $+x'$ toward $-z$ by the angle Θ and rotates $\mathbf{I}_z^2$ from $+z$ toward $+x'$ by the angle Θ. Any term with either H_1 or H_2 in the $x'-y'$ plane will represent MQC, as ^{13}C is already in the $x'-y'$ plane, so we can ignore these terms because they are not observable in the FID. The only term of interest is the doubly antiphase term:

$$4(-\mathbf{I}_z^1\sin\Theta)[-\mathbf{S}_x](\mathbf{I}_z^2\cos\Theta) = \underset{F}{4\mathbf{S}_x\mathbf{I}_z^1\mathbf{I}_z^2\sin\Theta\cos\Theta}$$

During the final $1/(2J)$ delay both of the antiphase relationships will evolve into an in-phase relationship due to J-coupling evolution:

$$\underset{F}{4\mathbf{S}_x\mathbf{I}_z^1\mathbf{I}_z^2\sin\Theta\cos\Theta} \xrightarrow{1/(2J)} \underset{G}{-\mathbf{S}_x\sin\Theta\cos\Theta}$$

Each J-coupling evolution from antiphase to in-phase involves a 90° rotation in the $x'-y'$ plane. We can view the refocusing as two steps, first J-coupling evolution with respect to H_1 and then J-coupling evolution with respect to H_2:

$$2[2\mathbf{S}_x\mathbf{I}_z^1]\mathbf{I}_z^2\sin\Theta\cos\Theta \xrightarrow{1/(2J)\,H_1} 2[\mathbf{S}_y]\mathbf{I}_z^2\sin\Theta\cos\Theta$$

$$2\mathbf{S}_y\mathbf{I}_z^2\sin\Theta\cos\Theta \xrightarrow{1/(2J)\,H_2} -\mathbf{S}_x\sin\Theta\cos\Theta$$

Because we could do the same thing starting with $\mathbf{I}_z^2$ and get $\sin\Theta\cos\Theta$ again, we have a final spin state of $-\mathbf{S}_x\,[2\sin\Theta\cos\Theta]$, and we can say that the CH_2 group ^{13}C resonance is edited by a factor of $2\sin\Theta\cos\Theta$, which works out to 1 for DEPT-45, zero for DEPT-90, and -1 for DEPT-135.

Now for a real challenge, let's look at the CH_3 group. Again, we start with the z magnetization of one proton, $\mathbf{I}_z^1$, and at the end we will multiply by 3 to account for the coherence transfer from all three attached protons. Everything is the same until the second delay, where we have multiple-quantum coherence (ZQC/DQC) between ^{13}C and H_1, which during the delay undergoes J coupling evolution with respect to both of the passive couplings: $^{13}C-^1H_2$ and $^{13}C-^1H_3$. As before, we can consider J-coupling evolution with respect to H_2 first and then as a second step consider the *J*-coupling evolution with respect to H$_3$:

$$-2\mathbf{I}_x^1[\mathbf{S}_y] \xrightarrow{1/(2J)\,\text{H}_2} -2\mathbf{I}_x^1[-2\mathbf{S}_x\mathbf{I}_z^2] = 4\mathbf{I}_x^1\mathbf{S}_x\mathbf{I}_z^2$$
$$\phantom{-2\mathbf{I}_x^1[\mathbf{S}_y]}_{\text{D}}$$

$$4\mathbf{I}_x^1[\mathbf{S}_x]\mathbf{I}_z^2 \xrightarrow{1/(2J)\,\text{H}_3} 4\mathbf{I}_x^1[2\mathbf{S}_y\mathbf{I}_z^3]\mathbf{I}_z^2 = 8\mathbf{I}_x^1\mathbf{S}_y\mathbf{I}_z^2\mathbf{I}_z^3$$
$$\phantom{4\mathbf{I}_x^1[\mathbf{S}_x]\mathbf{I}_z^2 \xrightarrow{1/(2J)\,\text{H}_3} 4\mathbf{I}_x^1[2\mathbf{S}_y\mathbf{I}_z^3]\mathbf{I}_z^2 =}_{\text{E}}$$

The final ^{1}H pulse rotates all three of the ^{1}H operators by the angle Θ:

$$8\mathbf{I}_x^1\mathbf{S}_y\mathbf{I}_z^2\mathbf{I}_z^3 \xrightarrow{{}^1\text{H}\,\Theta_y^\circ/{}^{13}\text{C}\,180_y^\circ}$$
$$\underset{\text{E}}{\phantom{8\mathbf{I}_x^1\mathbf{S}_y\mathbf{I}_z^2\mathbf{I}_z^3}}$$

$$\underbrace{8(\mathbf{I}_x^1\cos\Theta - \mathbf{I}_z^1\sin\Theta) \times \mathbf{S}_y \times (\mathbf{I}_z^2\cos\Theta + \mathbf{I}_x^2\sin\Theta) \times (\mathbf{I}_z^3\cos\Theta + \mathbf{I}_x^3\sin\Theta)}_{\text{F}}$$

As in the CH case, the $\mathbf{S}_y$ term is not affected by the ^{13}C 180° pulse on the y' axis. The only term of interest is the one with only ^{13}C coherence in the x'–y' plane:

$$8(-\mathbf{I}_z^1\sin\Theta)\mathbf{S}_y(\mathbf{I}_z^2\cos\Theta)(\mathbf{I}_z^3\cos\Theta) = -8\mathbf{S}_y\mathbf{I}_z^1\mathbf{I}_z^2\mathbf{I}_z^3\sin\Theta\cos\Theta\cos\Theta$$
$$_{\text{F}}$$

During the final 1/(2*J*) delay, all three antiphase relationships will evolve under *J*-coupling evolution into in-phase relationships. We can view this process in three steps:

$$-4[2\mathbf{S}_y\mathbf{I}_z^1]\mathbf{I}_z^2\mathbf{I}_z^3\sin\Theta\cos^2\Theta \xrightarrow{1/(2J)\,\text{H}_1} -4[-\mathbf{S}_x]\mathbf{I}_z^2\mathbf{I}_z^3\sin\Theta\cos^2\Theta$$
$$\phantom{-4[2\mathbf{S}_y\mathbf{I}_z^1]\mathbf{I}_z^2\mathbf{I}_z^3\sin\Theta\cos^2\Theta}_{\text{F}}$$

$$2[2\mathbf{S}_x\mathbf{I}_z^2]\mathbf{I}_z^3\sin\Theta\cos^2\Theta \xrightarrow{1/(2J)\,\text{H}_2} 2[\mathbf{S}_y]\mathbf{I}_z^3\sin\Theta\cos^2\Theta$$

$$2\mathbf{S}_y\mathbf{I}_z^3\sin\Theta\cos^2\Theta \xrightarrow{1/(2J)\,\text{H}_3} -\mathbf{S}_x\sin\Theta\cos^2\Theta$$
$$\phantom{2\mathbf{S}_y\mathbf{I}_z^3\sin\Theta\cos^2\Theta \xrightarrow{1/(2J)\,\text{H}_3} -\mathbf{S}_x\sin\Theta\cos^2\Theta}_{\text{G}}$$

When we combine this with the other two identical terms that come from $\mathbf{I}_z^2$ and $\mathbf{I}_z^3$, we have $-\mathbf{S}_x\,[3\sin\Theta\cos^2\Theta]$, so we can say that the intensity of in-phase ^{13}C coherence in the FID is edited by the factor $3\sin\Theta\cos^2\Theta$, which works out to 1.060, 0, and 1.060 for DEPT-45, DEPT-90, and DEPT-135, respectively.

This analysis assumes that both the ^{13}C peak and the ^{1}H peak are on-resonance with respect to their reference frequencies (pulse frequencies), but we tried to show at least conceptually how this is not important because all chemical shift evolution, ^{1}H or ^{13}C, is refocused by the two overlapping spin echoes. The result of this analysis is that the observable (i.e., SQC) magnetization at the beginning of the FID will be

$$\begin{aligned}
\text{CH group:} \quad & -\mathbf{S}_x\,[\sin\Theta] \\
\text{CH}_2\text{ group:} \quad & -\mathbf{S}_x\,[2\sin\Theta\cos\Theta] \\
\text{CH}_3\text{ group:} \quad & -\mathbf{S}_x\,[3\sin\Theta\cos^2\Theta]
\end{aligned}$$

These editing factors are the same we observe for the refocused INEPT experiment, if we replace Θ with $\pi J\Delta$, where Δ is variable refocusing delay.

Further Reading

1. Hennel JW, Klinowski J. Fundamentals of NMR. Longman Scientific and Technical; 1993.
2. Shaw D. Fourier Transform NMR Spectroscopy. 2nd ed. Elsevier; 1984.
3. Shriver J. Product operators and coherence transfer in multiple-pulse NMR experiments. *Concepts Magn. Reson.* 1992;**4**:1–33.
4. Sørensen OW, Eich GW, Levitt MH, Bodenhausen G, Ernst RR. Product operator formalism for the description of NMR pulse experiments. *Prog. NMR Spectrosc.* 1983;**16**:163–192.

8

SHAPED PULSES, PULSED FIELD GRADIENTS, AND SPIN LOCKS: SELECTIVE 1D NOE AND 1D TOCSY

8.1 INTRODUCING THREE NEW PULSE SEQUENCE TOOLS

In many nuclear magnetic resonance (NMR) experiments, we wish to excite only one resonance or peak in the spectrum, corresponding to a specific position within the molecule. Once this resonance has been "selected," we can transfer magnetization via J couplings (through bond) or via NOE (nuclear Overhauser effect) transfer (through space) to other positions nearby in the molecule. The new, transferred magnetization can be observed in the free induction decay (FID) and identified by its chemical shift, allowing us to "connect" or "correlate" two resonances in the spectrum (two different chemical shifts). This process of establishing a relationship (through bond or through space) between two spins is central to all advanced NMR experiments.

So far, the only way we know to select a resonance is through saturation: a long, very low power radio frequency (RF) pulse set to the exact frequency of that resonance (Section 5.9), equalizing the populations and destroying any net magnetization on that proton. This technique is used by the 1D NOE difference experiment (Section 5.12), which allows us to select a single resonance in the spectrum and observe an enhancement of the peak intensities of any resonance corresponding to a proton close in space to the selected proton. New technology in NMR hardware began to be commonly available in the 1990s allowing not just saturation but specific pulse excitation (e.g., 90° pulse or 180° pulse) with specific phase (e.g., x', y', $-x'$, or $-y'$) of any resonance in the spectrum, without affecting the other spins in the molecule in any way. This is much more powerful than saturation because now we can create coherence (90° pulse) on a single spin (a single position within the molecule) or invert (180° pulse) specifically just one resonance in the spectrum.

Another new tool in our arsenal comes from the technology of NMR imaging (MRI). In MRI, there is only one chemical shift (that of H_2O) and the chemical shift scale is used

---

NMR Spectroscopy Explained: Simplified Theory, Applications and Examples for Organic Chemistry and Structural Biology, by Neil E. Jacobsen
Copyright © 2007 John Wiley & Sons, Inc.

instead for indicating the physical location of a spin within the sample volume. By "bending" the homogeneity of the magnetic field during the acquisition of the FID, the B_o field becomes dependent on the position within the sample. This intentionally nonhomogeneous field is called a "field gradient." Because the NMR resonance frequency is directly proportional to the field strength ($\nu_o = \gamma B_o/2\pi$), this means that the resonant frequency of each spin now depends only on its position within the sample. The ^{1}H NMR spectrum becomes a physical "map" (or image) of where the spins are located.

In NMR *spectroscopy* (as opposed to imaging), we do not use field gradients during the acquisition of the FID, but the gradient technology can be used for another purpose: for destroying coherences that we do not want to see. Gradients are applied for a brief period of time (typically 1–2 ms) and then removed, returning the magnetic field to its very high degree of homogeneity. The gradient affects coherence because the precession frequency changes during the gradient (again, $\nu_o = \gamma B_o/2\pi$), and this makes all the "identical" spins have different precession frequencies depending on their position within the sample tube. The end result after a millisecond or two of this chaos is that the phase of the coherence is now scrambled throughout the sample and no longer adds together to make measurable net magnetization. The technique of "pulsing" the gradient on and then off again rapidly is called "pulsed field gradients" (PFGs), and it has become an integral part of all modern NMR experiments. We can think of PFGs as the janitorial service of NMR, sweeping up and discarding all of the signals we do not want to see (the "artifacts") and leaving the clean spectrum we are interested in.

A third new technique or pulse sequence tool will be introduced in this chapter: the *spin lock*. A spin lock is just a long RF pulse, applied to give a very large number of rotations of the net magnetization: hundreds or thousands rather than the one-fourth (90° pulse) or half (180° pulse) we usually think of for pulses. Because we can not shim it, the B_1 (RF) field is notoriously inhomogeneous compared to the B_o field. This means that depending on where you are in the sample, a well-calibrated 90° pulse might give a rotation of 88° or 91° rather than 90°. This does not create a big problem for most pulse sequences, but imagine what happens if you rotate 100 times (400 times the duration of a 90° pulse): the 88° pulse rotates 97.777 cycles and the 91° pulse rotates 101.111 cycles. The integer number of cycles does not matter much, but the fraction is 280° (0.777 times 360°) for the 88° pulse and 40° for the 91° pulse. Depending on where you are in the sample, you will see every possible rotation of the sample magnetization, and the total net magnetization throughout the sample will be zero. Again, as with the pulsed field gradient, we have destroyed the magnetization by scrambling it as a function of location in space.

But we have been assuming that the sample net magnetization is forming a 90° angle to the B_1 field. What if the magnetization is colinear with the B_1 field? We have seen with simple pulse rotations that the B_1 field has no effect on magnetization that is aligned with it; for example, a 90°$_y$ pulse does not change $\mathbf{I}_y$. The same is true for a spin lock: All magnetization perpendicular to the B_1 field is scrambled and all magnetization aligned with it is *preserved*. Not only is this magnetization preserved, but it is also *locked* to the axis of the B_1 field for the duration of this long pulse, preventing it from moving. Some very interesting things happen to this magnetization while it is locked on the B_1 axis: Magnetization can transfer from one spin to a nearby spin either by NOE (through space) or by J coupling (through bond). These processes are complex and even with the powerful theoretical tools we have developed, we will only get a glimpse of how they work. But they are the basis of two extremely useful NMR experiments: TOCSY (total correlation spectroscopy) and ROESY (rotating-frame Overhauser effect spectroscopy). We will attempt to at least get a feel for what is going on in the spin lock and how we can get transfer of magnetization.

To understand selective (shaped) pulses and the spin lock, we need to look in detail at the effect of pulses on spins as a function of their resonant frequency, ν_o, that is to say the position of a resonance within the spectral window.

8.2 THE EFFECT OF OFF-RESONANCE PULSES ON NET MAGNETIZATION

In the acquisition of a simple 1D spectrum, our goal is to excite all of the spins of a certain type (e.g., ^{1}H) in the sample, regardless of chemical shift, at the same time. This requires a radio frequency pulse of very high power and short duration. The frequency of the pulse is adjusted to correspond to the resonance frequency at the center of the spectral window, so that it will be close to the resonance frequency of all of the spins in the sample.

8.2.1 On-Resonance Pulses

For a spin whose chemical shift is exactly at the center of the spectral window, we call the pulse an "on-resonance" pulse because the pulse (or "carrier") frequency is exactly equal to the resonant frequency (precession frequency or Larmor frequency ν_o) of the spin. During the pulse, we can use the vector model to show the B_1 field (the pulse) as stationary in the rotating frame of reference, because the x' and y' axes are rotating about the z axis at exactly the frequency of the pulse. The position of the B_1 field in the x'–y' plane depends on the phase of the pulse, which is just the place in the sine function (0–360°) where the radio frequency oscillation starts at the beginning of the pulse. This can be controlled by the spectrometer and is written into the pulse sequence by the user:

Code	Phase shift	Function	B_1 vector
0	0°	$\sin(2\pi\nu_r t)$	x' axis
1	90°	$\cos(2\pi\nu_r t)$	y' axis
2	180°	$-\sin(2\pi\nu_r t)$	$-x'$ axis
3	270°	$-\cos(2\pi\nu_r t)$	$-y'$ axis

where ν_r is the frequency of the pulse. In the rotating frame of reference, we compensate for the physical violation of using an accelerating (rotating) frame of reference by including a fictitious magnetic field, oriented along the $-z$ axis with magnitude $2\pi\nu_r/\gamma$, where ν_r is the rate of rotation (in hertz) of the x' and y' axes of the rotating frame of reference (Fig. 8.1(a)). If the spins are on-resonance, then the pulse frequency, ν_r, is equal to the Larmor frequency, ν_o. In this case, the fictitious field strength is $2\pi\nu_o/\gamma$, which is equal to B_o (because $\nu_o = \gamma B_o/2\pi$). The fictitious field, which is oriented along $-z$, exactly cancels the real field B_o, which is oriented along $+z$, and there is no field at all in the absence of a pulse (Fig. 8.1(b)). During the pulse, the only magnetic field experienced by the spins is the B_1 field, which is in the x'–y' plane.

Thus for an on-resonance pulse, the B_o field does not exist in the rotating frame and the effective field experienced by the spins is just the B_1 field. The magnitude of the effective

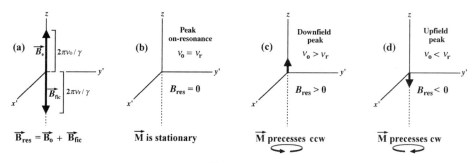

Figure 8.1

field is $B_{\text{eff}} = B_1$, and the $\boldsymbol{B}_{\text{eff}}$ vector is oriented in the x'–y' plane at a position determined by the pulse phase. The net magnetization vector **M** has magnitude M_0 and, starting from its equilibrium orientation along the $+z$ axis, precesses counterclockwise (ccw) about the $\boldsymbol{B}_1$ vector at a rate $\nu_1 = \gamma B_1/2\pi$. If the pulse duration, t_p, is adjusted so that **M** precesses exactly one fourth of a complete rotation (90° pulse: $\nu_1 t_p = 1/4$; $t_p = 1/(4\nu_1)$), then the **M** vector ends up in the x'–y' plane at the end of the pulse. This is the picture we have been using so far for all pulses.

8.2.2 Off-Resonance Pulses

What happens if a resonance peak is not exactly in the center of the spectrum? In this case, the pulse is off-resonance ($\nu_r \neq \nu_0$). As always, we choose axes x' and y' that rotate around the z axis at a rate equal to ν_r, the frequency of the pulse (corresponding to the radio frequency at the center of the spectral window). As before, the B_1 field can be described by a vector that is stationary in the x'–y' plane, but now the fictitious field, which is required to correct for the accelerating frame of reference, no longer perfectly cancels the B_0 field. If the resonance peak is downfield (higher frequency) of the center of the spectral window, then $\nu_0 > \nu_r$ and the fictitious field ($2\pi\nu_r/\gamma$) is lower in magnitude than the B_0 field ($2\pi\nu_0/\gamma$). Because the fictitious field is oriented along the $-z$ axis and the slightly stronger B_0 field is oriented along the $+z$ axis, the result is a small residual field (B_{res}) oriented along the $+z$ axis (Fig. 8.1(c)).

During the pulse, the spins do not experience the B_1 field alone, but rather an effective field $\boldsymbol{B}_{\text{eff}}$, which is the vector sum of the small residual field along the $+z$ axis (B_{res}) and the $\boldsymbol{B}_1$ field in the x'–y' plane (Fig. 8.2, left). If the resonance is not far from the center of the spectral window, the $\boldsymbol{B}_{\text{eff}}$ vector will "tilt" slightly out of the x'–y' plane and get slightly longer than $\boldsymbol{B}_1$.

If the resonance peak is upfield (lower frequency) of the center of the spectral window, then $\nu_0 < \nu_r$ and the fictitious field along $-z$ "wins out" over the B_0 field along $+z$, leaving a small residual field (B_{res}) along $-z$ (Fig. 8.1(d)). Now the spins experience an effective field vector $\boldsymbol{B}_{\text{eff}}$ that is tilted slightly below the x'–y' plane during the pulse (Fig. 8.2, right). The exact angle of tilt can be calculated using simple trigonometry ($\tan\Theta = B_{\text{res}}/B_1$) and the magnitude of $\boldsymbol{B}_{\text{eff}}$ comes from Pythagoras ($B_{\text{eff}}^2 = B_{\text{res}}^2 + B_1^2$), but we are concerned here only with a qualitative understanding: the $\boldsymbol{B}_{\text{eff}}$ vector tilts out of the x'–y' plane and gets slightly longer than $\boldsymbol{B}_1$, and this effect depends on the relative magnitudes of $\boldsymbol{B}_1$ and $\boldsymbol{B}_{\text{res}}$, that is, on how far we are off-resonance and how powerful the pulse is. A very high

THE EFFECT OF OFF-RESONANCE PULSES ON NET MAGNETIZATION 293

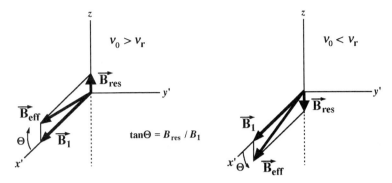

Figure 8.2

power pulse can "resist" and minimize the off-resonance tilting, but a weak RF pulse will be more sensitive to resonance offset.

If the B_1 field is weak (short B_1 vector) or the pulse is far off-resonance (long B_{res} vector), the effective field vector B_{eff} will tilt significantly up or down, out of the x'–y' plane. In this case, a 90° rotation about B_{eff} ("90° pulse") will not really put the net magnetization **M** into the x'–y' plane and a 180° rotation ("180° pulse") will not really put the net magnetization on the $-z$ axis. We can easily compare the magnitudes of B_1 and B_{res} if we think of B_1 amplitude in units of hertz: $\nu_1 = \gamma B_1/2\pi$. This is analogous to expressing the field strength of a magnet in terms of its proton resonant frequency: $\nu_o = \gamma B_o/2\pi$. Thus, we talk about a "300-MHz magnet" rather than a magnet with $B_o = 7.4$ T, and we can talk about a "25-kHz B_1 field" for an RF power setting, which gives a 10-μs 90° pulse. In this B_1 field, the ^{1}H net magnetization rotates one full rotation in 40 μs, so we can say that the net magnetization vector rotates around the B_1 vector at a rate of $\nu_1 = 1/(0.000040 \text{ s}) = 25,000$ Hz.

> Be careful to note that ν_1 is not the frequency of the pulse. We call that frequency ν_r (reference frequency); it is very close to ν_o, or 300 MHz on a "300-MHz" NMR spectrometer. Think of ν_1 as a measure of the pulse *amplitude* rather than frequency: it is the rate at which the sample magnetization rotates around the B_1 vector, a measure of the *effect* of the pulse (length of the B_1 vector).

Now we can directly compare B_1 (the magnetic field of the RF pulse) to B_{res} (what is left of B_o after subtracting out the pseudofield correction for the rotating frame of reference). If a resonance in the ^{1}H spectrum is at 10.0 ppm and the center of the spectral window is at 5.0 ppm on a 300-MHz instrument, we have $\nu_o - \nu_r = (10 - 5)\,300 = 1500$ Hz. This is how far the pulse is off-resonance and it is proportional to B_{res}. If the 90° pulse width is 10 μs, we can describe the B_1 field strength in hertz ($\nu_1 = \gamma B_1/2\pi$) as $1/(4 \times t_p) = 25,000$ Hz. This is proportional to B_1 in the same way that $\nu_o - \nu_r$ is proportional to B_{res}. Thus, the B_1 vector is 16.67 times longer (25,000/1500) than the B_{res} vector, and the tilt will be insignificant. We can say that this pulse is strong enough to "cover" a spectral window 10 ppm wide (0–10 ppm) without any significant loss of effectiveness at the edges. The exact amount of tilt is 3.4° and the B_{eff} vector is 0.18% longer than B_1. Clearly, we can use the same simple vector model for all resonances within the 10 ppm (3000 Hz) wide spectral

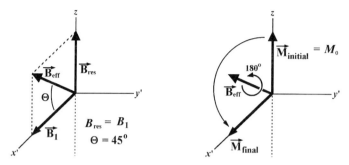

Figure 8.3

window. We will have problems when B_{res} becomes comparable to B_1:

$$B_1 \sim B_{res} \quad \text{or} \quad \nu_1 \sim (\nu_o - \nu_r)$$

Thus, we can avoid off-resonance effects by making B_1 as strong as possible (highest power pulse possible, shortest duration) and by making $\nu_o - \nu_r$ as small as possible (avoid having resonances peaks very far from the center of the spectral window).

As an example of how bad it can get, consider a case where $B_1 = B_{res}$ (i.e., $\nu_1 = \nu_o - \nu_r$). In this case, if we apply the pulse on the x' axis the resultant vector $\boldsymbol{B}_{eff}$ is tilted 45° out of the x'–y' plane toward the z axis (Fig. 8.3, left). The magnitude of the $\boldsymbol{B}_{eff}$ vector is 1.414 (square root of 2) times B_1 so that the net magnetization vector **M** will rotate about $\boldsymbol{B}_{eff}$ about 41% faster than it would for an on-resonance pulse. Suppose we want to apply a 180° pulse to this off-resonance peak. We could compensate for the larger magnitude of $\boldsymbol{B}_{eff}$ by using a pulse that is shorter in duration by a factor of 1.414. This would rotate the net magnetization **M** exactly 180° around the $\boldsymbol{B}_{eff}$ vector (Fig. 8.3, right). Because $\boldsymbol{B}_{eff}$ is tilted 45° up in the x'–z plane, the **M** vector rotates around it maintaining the 45° angle to $\boldsymbol{B}_{eff}$ at all times, tracing out a conical path, and landing after a 180° rotation right on the x' axis! Thus, our "180° pulse" delivered to an off-resonance peak is really only a 90° pulse in our simple vector model.

This can be a real problem for ^{13}C pulses at high field strengths. Consider a 600-MHz spectrometer with a ^{13}C spectral window stretching from 0 to 220 ppm. The center of the spectral window is at 110 ppm and a ketone carbonyl resonance is at 200 ppm. The strongest ^{13}C pulse you can muster is a 16-μs 90° pulse, corresponding to a B_1 field strength of 15.63 kHz (1/0.000064 s). The resonance is 90 ppm or 13.5 kHz (90 × 150 Hz) from the center of the spectral window. B_{res} (13.5 kHz) is nearly equal to B_1 (15.6 kHz). You've got problems!

8.2.3 Composite (Sandwich) Pulses

There are many tricks to get around the problem, such as sandwich 180° pulses (e.g., 90_x–180_y–90_x) and "broadband" shaped pulses. Figure 8.4 (top) shows the inversion profile for a simple 180° pulse at the highest available power ($t_p = 28.4$ μs, $\gamma B_1/2\pi = 17.6$ kHz). The profile is obtained using an inversion-recovery sequence ($180°_x - \tau - 90°_y$) with recovery time $\tau = 0$. The final 90° pulse frequency and the ^{13}C peak (^{13}CH$_3$I) are both at the center of the spectral window, but the frequency of the 180° pulse is moved in 10 ppm (1500 Hz)

THE EFFECT OF OFF-RESONANCE PULSES ON NET MAGNETIZATION

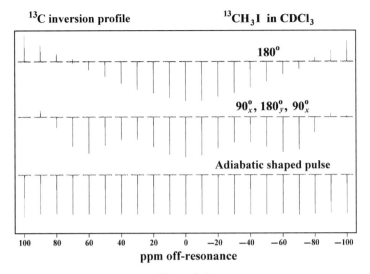

Figure 8.4

steps away from the center right and left, each time printing the spectrum to the left or right of the previous spectrum. On-resonance the peak is upside down and has maximum intensity ($S_z \rightarrow -S_z \rightarrow -S_x$) but as we move off-resonance the intensity diminishes and reaches zero at about 70 ppm off-resonance. At this point, we are getting a 90° pulse rather than a 180° pulse. Beyond this we actually see positive peaks, indicating that the z component of net magnetization after the "180°" pulse is positive.

A more "robust" way to invert the sample magnetization is the sequence $90°_y$–$180°_x$–$90°_y$, with no space in between the pulses. This is called a composite pulse or sandwich pulse because a number of pulses are lined up right next to each other, like slices of cheese and meat in a sandwich. Suppose that what we *think* is a 90° pulse is really an 85° pulse due to miscalibration. The first "90°" pulse on y' rotates the sample magnetization ccw by only 85°, leaving it in the x'–z plane just 5° short of the x' axis (Fig. 8.5).

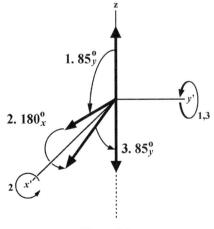

Figure 8.5

The 180°_x pulse rotates the M vector around the x' axis in a very sharp cone, landing at a point 5° *below* the x' axis. The final pulse (85°_y) rotates M precisely down to the $-z$ axis. So we have a perfect 180° pulse even though it was miscalibrated by over 5%. This is not entirely true because the 180° pulse in the middle of the sandwich is really a 170° pulse, but this introduces a very small error as the sample magnetization is so close to the x' axis and rotates in a very narrow cone. The effects of off-resonance pulses are more complex, but one can see in Figure 8.4 (middle) that this sandwich pulse does a better job of inversion than the simple 180° pulse. The effective "bandwidth" or coverage of the pulse is 150 ppm compared to about 80 for the 180° pulse alone. This is still not wide enough for the entire range of ^{13}C shifts, which extends from 5 to around 220 ppm, requiring a 215 ppm bandwidth.

We will see that the major application of shaped pulses is to select a narrow region of the spectrum, thus displaying a narrow bandwidth. But there are also shaped pulses designed to do just the opposite — to give *even* excitation over a very wide range of frequencies. These "broadband" shaped pulses are specialized for inversion ($S_z \rightarrow -S_z$) or refocusing ($S_x \rightarrow -S_x$). Figure 8.4 (bottom) shows the inversion profile of an "adiabatic" inversion shaped pulse with maximum B_1 field strength of 13.7 kHz ($\gamma B_1/2\pi$), average B_1 field strength of 8 kHz and total duration 546 μs. No discernible "droop" in inversion efficiency is seen over a range of 200 ppm, and even over a bandwidth of 260 ppm, only a 15% loss in efficiency is observed at the edges. This is accomplished with an average B_1 field strength of less than half of that used for the simple 180° pulse. We will discuss how this works later in this chapter after we gain an understanding of shaped pulses and spin locks.

8.2.4 Precession in the Rotating Frame

What happens *after* the pulse for an off-resonance spin? Suppose that right after the pulse the net magnetization M of the sample is on the x' axis. The B_1 field is now turned off, so that in the rotating frame of reference the only field experienced by the spins is the residual field along the z axis ($B_{\text{res}} = B_o - B_{\text{fict}}$). If the resonance is downfield of the center of the spectral window ($\nu_o > \nu_r$), then $B_o > B_f$ and the residual field is a small field oriented along the positive z axis of magnitude $2\pi(\nu_o - \nu_r)/\gamma$. The net magnetization vector M will precess about the effective field, which is now B_{res}, in the counterclockwise direction (viewed from the $+z$ axis) at a rate equal to $\gamma B_{\text{res}}/2\pi$, which is simply $\nu_o - \nu_r$ (Fig. 8.1(c), bottom). We can think of the B_{res} vector just like the B_1 vector during the pulse or the B_o vector in the laboratory frame: Any magnetic field rotates the sample magnetization counterclockwise about the field axis. Precession is like a z-axis pulse!

This result is not surprising as the rotating frame of reference simply subtracts ν_r, the rate of rotation of the x and y axes in the rotating frame, from the Larmor frequency ν_o. The same thing goes on electronically in the receiver of the NMR instrument, where the FID (decaying signal of frequency ν_o) is *mixed* (mathematically multiplied at each time point using an analog device) with a reference frequency ν_r, which is the same frequency as the RF pulse. The result of this mixing consists of two signals added together, one with frequency $\nu_o + \nu_r$ and another with frequency $\nu_o - \nu_r$. The first signal is a radio frequency (close to twice the Larmor frequency) and the second is an audio frequency (in the range 0–10 kHz), so it is easy to block the high frequency and pass the audio frequency with an analog filter. The result is the audio FID containing the frequency $\nu_o - \nu_r$, which is the precession frequency in the rotating frame of reference.

If the resonance is upfield of the center of the spectral window ($\nu_o < \nu_r$), then the fictitious field is stronger than the B_o field ($B_o < B_{\text{fic}}$) and the residual field $\boldsymbol{B}_{\text{res}}$ is oriented along the negative z axis. Under the influence of this effective field, the sample magnetization $\boldsymbol{M}$ rotates clockwise (viewed from above) at a frequency $\nu_r - \nu_o$ (Fig. 8.1(d), bottom). In the rotating frame of reference, we consider this a negative frequency as the rotating-frame frequency is always defined as $\nu_o - \nu_r$, corresponding to precession in the counterclockwise direction. Some books use the uppercase omega (Ω) to represent the rotating-frame *angular velocity*: $\Omega = \omega_o - \omega_r$. Angular velocity is just frequency times 2π: $\omega = 2\pi\nu$. This angular velocity (in radians per second) is often referred to as the chemical shift, even though it has no relation to the δ scale in parts per million. The rotating-frame frequencies ($\nu_o - \nu_r$ in Hz) or the rotating-frame angular velocities (Ω in radians per second) depend on the spectrometer field strength B_o and have zero value at the center of the spectral window. The chemical shift (δ in parts per million) is independent of B_o and is zero at the resonance position of tetramethylsilane (TMS).

If the peak is on-resonance ($\nu_o = \nu_r$), there is no residual field and the magnetization vector stands still in the x'–y' plane after the pulse (Fig. 8.1(b), bottom). This is to be expected as in the laboratory frame the magnetization vector is precessing counterclockwise at frequency ν_o, and in the rotating frame we are rotating the axes at exactly that rate ($\nu_r = \nu_o$), so relative to the rotating x' and y' axes the vector is not moving.

8.3 THE EXCITATION PROFILE FOR RECTANGULAR PULSES

A simpler way to look at the effectiveness of an off-resonance pulse is to calculate the excitation profile, which is just a graph of the effective pulse rotation delivered by a pulse as a function of the resonance frequency in the rotating frame ($\nu_o - \nu_r$). This graph can be superimposed on the spectral window to view the effectiveness of the pulse with respect to each of the peaks in the spectrum (Fig. 8.6). It is actually quite simple to calculate this function, as the excitation profile is just the Fourier transform of the pulse. The pulse shape is rectangular as it is simply turned on at the beginning and held on for a time t_p, which is the duration of the pulse (also called the pulse width), and then turned off. The height of this rectangular function is the pulse amplitude B_1 (square root of the pulse power). We can also describe pulse amplitude using the effect of the pulse: the rotation rate of the sample magnetization around the $\boldsymbol{B}_1$ vector ($\nu_1 = \gamma B_1/2\pi$). This is expressed in hertz and is equal to $1/(4 \times t_p(90°))$. Using the B_1 "field strength" in hertz as the height of the pulse, we see that the area of the pulse (height × width) is just the pulse rotation in cycles (0.25 for a 90° pulse). All three pulses in Figure 8.6 are 90° pulses, with the same area but different pulse widths t_p.

The Fourier transform of a rectangular function turns out to be something called the "sinc" function: $\sin(x)/x$. This function reaches a maximum at $x = 0$ and goes to zero on either side, with "wiggles" gradually dying out sort of like an FID in both the $+x$ and $-x$ directions (Fig. 8.6). Notice how the experimental result for inversion efficiency (Fig. 8.4, top) closely resembles this sinc function. The function first passes through zero at rotating-frame frequencies of $-1/(2t_p)$ and $+1/(2t_p)$, so we can say that the pulse excites a range

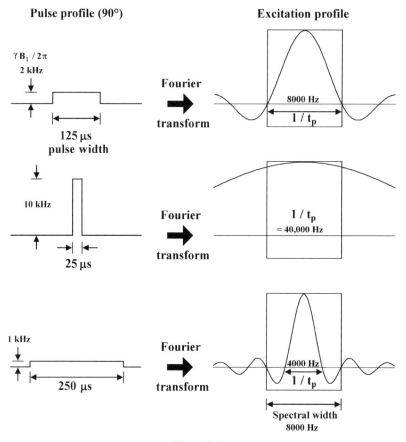

Figure 8.6

of frequencies corresponding to a "bandwidth" of $1/t_p$. Of course, if we superimpose this function on our spectral window with $1/t_p$ = sw (spectral width), we will be getting very poor excitation (rotation much less than 90° for a 90° pulse) near the edge of our spectral window (Fig. 8.6, top), going down to zero at the edge ($\nu_o - \nu_r = +$sw/2 or $-$sw/2). We could take this even further by using a longer pulse with smaller B_1 amplitude (Fig. 8.6, bottom) to get "selective excitation" of the peaks at the center of the spectral window. This is just the opposite strategy, using low-power ("soft") pulses to rotate only the spins corresponding to one peak in the spectrum.

To get nearly equal excitation all across our spectral window, we need to expand the excitation profile horizontally so that the zero points are far outside the spectral window and the function "droops" only slightly at the edges of the window (Fig. 8.6, center). This can be accomplished by using a shorter duration pulse, as the frequency domain has an inverse relationship to the time domain: squeezing the pulse in (shorter t_p) has the effect of horizontally expanding the excitation profile (wider coverage or bandwidth $1/t_p$). We would like to have the "bandwidth" $1/t_p$ much greater than the spectral width sw. Of course, if we use a shorter duration pulse, we must compensate by using a higher amplitude pulse so that the pulse rotation is not changed. For example, if we cut the duration t_p of a 90° pulse in half, we will get a 45° rotation of the sample magnetization M, unless we also double the pulse amplitude (four times the pulse power) to compensate. To get nearly "flat" coverage of the entire spectral window, we want a very short duration pulse (on the order of tens of microseconds) with very high power (on the order of 50–300 W). This is a lot of radio frequency power to put into the small volume (about 300 μl) of the NMR sample covered by the probe coil, so that we must be very careful to limit the pulse width to the microsecond range. Pulses of hundreds of milliseconds to seconds at this power level will boil the sample, fry the probe, and burn out the power amplifiers of the spectrometer. Pulse amplitude, which is the square root of pulse power, is limited not only by the maximum power output of the amplifiers, but also by the tendency of the probe coil to spark or "arc" at very high RF amplitudes. It is usually the arcing limit that sets a maximum on the B_1 amplitude we can use.

A typical 90° pulse might have a duration of 10 μs. Thus, $1/t_p$, the width of the main peak of the sinc function excitation profile, is 1/(10 μs) = 100,000/(1 s) = 100 kHz. A typical spectral width for proton is 12 ppm or 12 × 300 = 3600 Hz on a 300-MHz (7.05-T) instrument. Thus, the "bandwidth" is 28 times (100,000/3600) wider than the spectral window, and we will have minimal "droop" of the excitation profile between the center and the edges of the spectral window. This pulse will deliver very close to a 90° pulse to all of the peaks in the spectrum. Problems arise with low-γ nuclei because the rotation generated by the pulse is much slower ($v_1 = \gamma B_1/2\pi$) and the 90° pulse width is, therefore, much longer even at the highest B_1 amplitude (highest power) available. The longer pulse width corresponds to a narrower "coverage" ($1/t_p$) in the frequency domain. This is compounded by the fact that many low-γ nuclei have very wide ranges of chemical shifts, thus requiring very wide spectral windows. For example, ^{57}Fe has a γ that is 3.2% of γ_H (i.e., its nuclear magnet is only 3.2% of the strength of the proton nuclear magnet), and its range of chemical shifts is around 30,000 ppm. In terms of spectral width in hertz, this is about 1000 times wider than the typical proton spectral window. In these cases, it is often necessary to acquire several spectra with adjoining spectral windows in order to "cover" the entire range of chemical shifts.

8.4 SELECTIVE PULSES AND SHAPED PULSES

Because the excitation profile is the Fourier transform of the time course of the pulse, and because of the inverse relationship between time domain and frequency domain, a long enough pulse will lead to a very narrow sinc function. Figure 8.7 shows actual FT calculations done on rectangular pulse shapes. If we use a 90° pulse that is very long (e.g., 35 ms) and has very low power (3500 times lower amplitude or 12.3×10^6 times lower power than a 10 μs 90° pulse), we will get a very narrow excitation profile ($1/t_p = 28.6$ Hz). If we adjust the reference frequency so that one peak of interest in the spectrum is on-resonance ($v_o = v_r$), we could excite only the spins corresponding to this peak without affecting any of the other spins in the sample. This is called a selective pulse. The problem with the sinc function excitation profile is that there are many "wiggles" in the function that extend out quite far from the center of the spectral window. If another peak in the spectrum falls on the maximum of one of these wiggles, it too will be excited by the pulse, although the excitation will be weak.

300 SHAPED PULSES, PULSED FIELD GRADIENTS, AND SPIN LOCKS

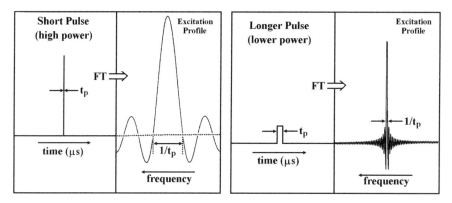

Figure 8.7

How can we eliminate the wiggles? We could try other functions for the pulse other than the rectangular shape and think about what the Fourier transform is for these functions. A Gaussian function (general form e^{-x^2}) has the useful property that its Fourier transform is also a Gaussian function (Fig. 8.8, top). The Gaussian is symmetrical and goes smoothly to zero quickly without wiggles, so it is an ideal shape for a selective pulse. We can adjust the selectivity of the Gaussian pulse by adjusting its pulse duration t_p, just as we do with rectangular pulses. A long, low-power Gaussian pulse corresponds to a narrow (highly selective) excitation profile and a faster, higher-power Gaussian pulse leads to a wider Gaussian excitation profile. We can even create a rectangular excitation profile, exciting a precise region of the spectrum with flat response throughout the region, by using a sinc function for our pulse shape (Fig. 8.8, bottom). Just as the Fourier transform of a rectangular pulse is a sinc function excitation profile, the Fourier transform of a sinc-shaped pulse is a rectangular excitation profile.

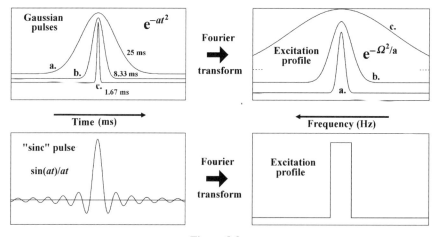

Figure 8.8

The ability to produce pulses with amplitude variation during the pulse according to a precise mathematical function became commonly available in the 1990s as a result of new hardware technology, called waveform generators (Varian) or amplitude setting units (Bruker). These nonrectangular pulses are called shaped pulses, and they are put together not with a continuous function but rather as a "sandwich" of short rectangular pulses. For example, a 35-ms Gaussian pulse might be put together by executing a long string of 350 rectangular pulses, each one 0.1 ms (100 μs) in duration. The amplitudes are set from a list of amplitudes calculated from the mathematical Gaussian function. This list can also contain RF phases (0–360°) that also vary in a precise predetermined fashion during the course of the long pulse.

Because nonselective pulses use high power, they are sometimes called "hard" pulses, whereas the low-power selective pulses are called "soft" pulses. Thus, for selectivity we use low-power (soft), long-duration shaped pulses and when we want to excite all of the signals in the spectral window equally, we use high-power (hard), short-duration rectangular pulses. Pulse power can vary over an enormous range, so we use a logarithmic scale to measure it. In the decibel scale, the pulse power in decibels is ten times the logarithm (base 10) of the pulse power:

$$dB = 10 \log(\text{power}) = 10 \log[(\text{amplitude})^2] = 20 \log(\text{amplitude})$$

To double the pulse power, simply increase the power by 3 dB, as $\log(2) = 0.3$. Because pulse power is the square of pulse amplitude B_1, to double the amplitude we need to multiply pulse power by a factor of 4, which corresponds to increasing power by 6 dB, as $\log(4) = 0.6$. This leads to a simple rule of thumb: Every time you increase the pulse power by 6 dB, you will cut the 90° pulse (t_p) in half (because B_1 is doubled). Likewise, each 6 dB decrease in pulse power will double the 90° pulse width. This is a good rule of thumb, but as the actual power settings are not precise, you will normally have to calibrate the 90° pulse at the new power setting to be sure. To make matters worse, Bruker uses the dB scale to describe power *attenuation* rather than power itself, so that the higher the dB value the *lower* the power. This is the opposite of Varian's system. Be careful whenever you are setting power levels! If you get it wrong, you can burn up the probe, the amplifiers, and your sample!

As good as shaped pulses sound, they have some unpleasant features. The phase properties of the pulse are often less than ideal, leading to phase distortions of the resonance peak being excited. These problems can be eliminated by using pulsed field gradients (PFGs) to "clean up" the selective excitation. PFGs can effectively scramble any undesired excitation, leaving only the absolutely clean pure-phase excitation at the desired resonance in the spectrum. When these two new technologies, shaped pulses and pulsed field gradients, pair up, we get a truly powerful new way to pick apart the spectrum and establish connectivities within a molecule through space and through bonds.

8.5 PULSED FIELD GRADIENTS

Field gradients were developed for magnetic resonance imaging, which is an NMR technique that encodes spatial information (x, y, and z axes) rather than chemical shift information in the FID. The resulting pictures of "slices" of the human body have provided a revolutionary new tool for medicine. More recently, this technology has been applied to NMR spectroscopy, with dramatic results. Pulsed field gradients make it possible to automatically optimize all of the shims simultaneously in a few minutes (Section 12.3). More importantly, with adequate sample concentration the total time required for a 2D experiment can be reduced from many hours to 15–45 min. Gradients make possible water suppression in 90% H_2O that is far superior to the old presaturation method. Finally, many artifacts in 2D spectra can be eliminated and sensitivity can be further improved by selecting only the signals that you are interested in and suppressing all others.

8.5.1 What is a Gradient?

Normally, we go to great efforts to assure that the magnetic field is homogeneous throughout the sample volume. This means that the strength of the magnetic field, or B_o, is exactly the same everywhere in the sample leading to sharp peaks for each resonance in the spectrum. The gradient intentionally destroys this homogeneity in a linear and predictable way. For example, a z-axis gradient alters the magnetic field so that the magnetic field strength is reduced in the lower part of the sample and increased in the upper part in a linear fashion. In other words, the magnetic field strength is now a function of the position of a molecule in the NMR tube along the z axis:

$$B_g(z) = B_o + G_z \times z$$

where B_g is the magnetic field strength with the gradient turned on, G_z is the strength of the field gradient (usually given in gauss per centimeter, where 1 G is 10^{-4} T) and z is the position of the molecule along the z axis. We choose the zero of the z axis to be at the center of the sample, so molecules above the center experience a slightly increased magnetic field and molecules below the center experience a slightly decreased magnetic field. The relative magnitude of this change is very small; for example, a maximum gradient strength of 50 G/cm in a B_o field of 117,440 G (11.744 T, 500 MHz ^{1}H). The gradient can be turned on and off very rapidly, so that typically the gradient is "pulsed" on for a period of 1–2 ms and then turned off.

8.5.2 Effect of Gradients on NMR Signals

What happens to the sample magnetization during a pulsed field gradient? Because the resonance frequency of a nucleus is always proportional to the magnetic field ($v_o = \gamma B/2\pi$), if we have a net magnetization vector in the x'–y' plane (e.g., after a 90° pulse), the magnetization vector will rotate in the x'–y' plane at a different rate depending on the molecule's position in the NMR tube (we can assume that diffusion is slow, so the molecule's position does not change). Magnetization in the upper part of the tube will precess faster than normal, and magnetization in the lower part will precess slower than normal. The result is that a "twist" or helix of magnetization will exist in the sample, so that at the end of the gradient period the magnetization rotates as a function of z coordinate throughout the sample (Fig. 8.9).

Assuming an on-resonance peak, spins at the center of the tube are unaffected by the gradient and do not move ($B_g = B_o$). Spins above the center experience a stronger magnetic field during the gradient ($B_g > B_o$), so they rotate counterclockwise in the x'–y' plane. As we move further up in the tube, this rotation is faster and the total angle of rotation during the gradient pulse is proportionally greater. To show this "twisting," we use an

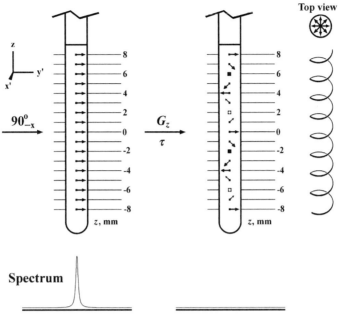

Figure 8.9

open square to indicate magnetization pointing back into the page ($-x'$ axis) and a filled square to indicate magnetization pointing out toward us ($+x'$ axis). Moving up from the center, the net magnetization at each level progresses from $+y$ to $-x$, to $-y$, to $+x$, and back to $+y$. All of the spins underwent precession in the $x'-y'$ plane for the same amount of time (τ) but at different rates ($\gamma G_z z/2\pi$) depending on their vertical distance from the center. The rotation is counterclockwise and is more pronounced as we move up. Moving *down* from the center, the field during the gradient pulse is slightly weaker ($B_g < B_o$) and the on-resonance spins now begin to fall behind the rotating frame, precessing in the clockwise direction. At the end of the gradient pulse, we have net magnetization on $+y$, $+x$, $-y$, $-x$, and back to $+y$ as we move down. The result is a "phase twist" or a helix of coherence. This is shown in cartoon fashion as a spring or spiral of coherence (Fig. 8.9, right).

There may be many hundreds of revolutions of the vector in the full vertical distance of the sample volume. If we try to acquire a spectrum at this point, after the gradient is turned off, we will not have any observable signal because the vectors point in all possible directions in the $x'-y'$ plane equally throughout the sample volume and the net magnetization vector is zero (Fig. 8.9, bottom). When viewed from the top, we see magnetization vectors pointing equally in all directions in the $x'-y'$ plane, leading to a net magnetization of zero throughout the whole sample. Figure 8.10(a) shows a ^{1}H spectrum of sucrose in D_2O at 500 MHz, and Figure 8.10(b) shows the same spectrum with a 1 ms gradient applied between the 90° excitation pulse and the start of the FID. So, a gradient can completely annihilate an NMR signal! Why would we want to do that? The signal might be an artifact, a solvent signal, or some other feature of the spectrum that we do not want to see. In this way, gradients can be used to "clean up" or remove unwanted NMR signals. Compared to the older method

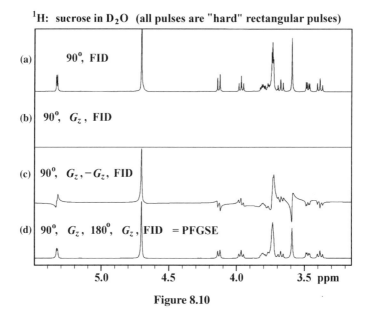

Figure 8.10

of removing unwanted signals, subtraction using a phase cycle, the gradient technique is far superior because it accomplishes the cleanup in one scan. The receiver never sees the artifacts so we do not have to turn down the receiver gain, and we are not dependent on perfect stability to give good subtraction.

8.5.3 Refocusing with Gradients—The Gradient Echo

We can not only kill coherence but also bring it back from the dead! The twisted magnetization in the sample can be "untwisted" by applying another gradient pulse of the same magnitude and duration but of opposite sign. This gradient decreases the magnetic field strength above the center of the sample and increases it below the center. During the second gradient pulse, the magnetization vectors rotate in the x'–y' plane clockwise in the upper part of the sample and counterclockwise in the lower part. The vectors which rotated counterclockwise in the first gradient pulse are now rotating clockwise at the same rate, and *vice versa*, so that at the end of the second gradient pulse all of the magnetization vectors are lined up again throughout the sample (Fig. 8.11). If we start the acquisition of the FID at this point, we will get a normal NMR spectrum, except for the phase "twist" that results from chemical shift evolution during the gradients. This result is shown for sucrose in Figure 8.10(c). Another way of saying this is that the first gradient pulse encoded the position of each molecule into its magnetization, scrambling the net magnetization of the whole sample, and the second gradient pulse decoded this information, unscrambling the net magnetization. So we can destroy with gradients, but we can also reverse the process and regenerate signals that were completely destroyed! All of this assumes that the molecules do not change their "level" in the tube between the time of the first gradient and the time of the second gradient. To the extent that the molecules undergo diffusion, which is faster for smaller molecules, there will be a loss of some signal. In fact, the gradient echo can be used as a way

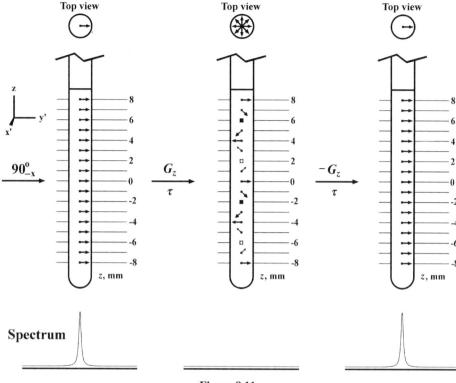

Figure 8.11

of measuring diffusion rates or to distinguish between small molecules and large molecules. In order to minimize "stirring" of the sample, we do not use spinning during gradient experiments.

If this sounds a lot like a spin echo, you are right. In the spin echo, various factors affect the precession rate of the spins in a sample: chemical shift differences for nonequivalent spins or differences in Larmor frequencies for identical spins in a nonhomogeneous magnetic field (bad shimming). In either case, a time delay leads to a "fanning out" of phases in the $x'-y'$ plane as they do not precess at exactly the same rates. A 180° pulse "flips" all the spins to the opposite side of the $x'-y'$ plane, and as long as they continue at the same frequency for the second half, they will all line up again at the end. Each spin "remembers" its precession frequency, either due to its position within a molecule (chemical shift) or due to its physical location in the NMR tube (inhomogeneous field), and by repeating this behavior exactly in the second half, it ends up back where it started. In the gradient echo, we create the differences in frequency by applying a gradient pulse. This is just an inhomogeneous magnetic field. The spins "fan out" in phase during the time of the gradient, depending on where they are physically located within the NMR tube. The second gradient actually *reverses the inhomogeneity of the field*, so that the accumulated error in phase during the first gradient is exactly reversed during the second for each spin in the sample.

If the spins are not on-resonance, they will still undergo chemical shift evolution in the gradient echo. For a resonance in the downfield half of the spectral window, the spins in the upper part of the sample will precess *faster* in the counterclockwise direction and the spins in the lower part will precess *slower* during the first gradient. During the second gradient, the spins in the upper part will precess *slower* and the spins in lower part will precess *faster*. At the end of the second gradient, these two perturbations will exactly cancel out for each spin at each level in the sample, and all the magnetization vectors will point in the same direction, as if the gradients had just been simple delays.

We can look at this more precisely using the product operator formalism, even though it is more important to focus on the conceptual picture rather than the math. For a resonance with Larmor frequency v_0, we have during the first gradient

$$v'_0 = \gamma(B_0 + zG_z)/2\pi = \gamma B_0/2\pi + \gamma zG_z/2\pi = v_0 + v_g$$

where v_g is the change in precession frequency due to the gradient. If the peak is on-resonance ($v_0 = v_r$), we have a rotating-frame precession frequency of $v'_0 - v_r = v_0 + v_g - v_r = v_g$. Starting with magnetization on the y' axis, we have at the end of the first gradient

$$\mathbf{I}_y \rightarrow \mathbf{I}_y \cos(2\pi v_g \tau) - \mathbf{I}_x \sin(2\pi v_g \tau)$$

As v_g is proportional to the z coordinate of the spin within the sample tube ($v_g = \gamma zG_z/2\pi$), we see that this is a helical coherence spinning around in the x'–y' plane as we move up or down the z axis. The net magnetization throughout the whole sample, summed over a large range of values of z, is zero. The second gradient will modify the Larmor frequency to $v'_g = -\gamma zG_z/2\pi = -v_g$, and precession at this frequency for a time τ will convert the pure $\mathbf{I}_y$ and $\mathbf{I}_x$ operators to

$$\mathbf{I}_y \rightarrow \mathbf{I}_y \cos(-2\pi v_g \tau) - \mathbf{I}_x \sin(-2\pi v_g \tau) = \mathbf{I}_y \cos(2\pi v_g \tau) + \mathbf{I}_x \sin(2\pi v_g \tau)$$

$$\mathbf{I}_x \rightarrow \mathbf{I}_x \cos(-2\pi v_g \tau) + \mathbf{I}_y \sin(-2\pi v_g \tau) = \mathbf{I}_x \cos(2\pi v_g \tau) - \mathbf{I}_y \sin(2\pi v_g \tau)$$

Plugging these expressions for $\mathbf{I}_x$ and $\mathbf{I}_y$ into the product operator representation for the spin state at the end of the first gradient, we get

$$\mathbf{I}_y \cos(2\pi v_g \tau) - \mathbf{I}_x \sin(2\pi v_g \tau) \rightarrow [\mathbf{I}_y \cos(2\pi v_g \tau) + \mathbf{I}_x \sin(2\pi v_g \tau)]\cos(2\pi v_g \tau)$$
$$- [\mathbf{I}_x \cos(2\pi v_g \tau) - \mathbf{I}_y \sin(2\pi v_g \tau)]\sin(2\pi v_g \tau)$$
$$= \mathbf{I}_y \cos^2\Theta + \mathbf{I}_x \sin\Theta\cos\Theta - \mathbf{I}_x \cos\Theta\sin\Theta + \mathbf{I}_y \sin^2\Theta$$
$$= \mathbf{I}_y(\cos^2\Theta + \sin^2\Theta) = \mathbf{I}_y$$

where $\Theta = 2\pi v_g \tau$ and we make use of the trigonometric identity $\cos^2\Theta + \sin^2\Theta = 1$. We see that the spin state starts as $\mathbf{I}_y$ and ends as $\mathbf{I}_y$ regardless of the position (z) of the spin in the sample. The math is considerably more complicated if we do not assume that the peak is on-resonance, but the conclusion is that we would have the same result as if the gradients were just simple delays: chemical shift evolution for a period of time 2τ (Fig. 8.10(c)).

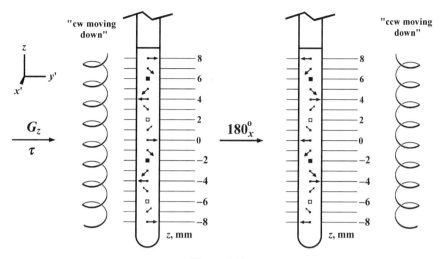

Figure 8.12

8.5.4 The Pulsed Field Gradient Spin Echo (PFGSE)

We saw in Figure 8.10(c) the phase "twist" that results from chemical shift evolution, during the relatively long (ms) time of the two gradients. To refocus this chemical shift evolution, we place a 180° pulse at the center of the sequence, between the two gradients. This makes the gradient echo into a spin echo (τ–180°–τ) with gradients during the two delay times. But we have to consider what the 180° refocusing pulse does to the helix (the "twist") of coherence created by the first gradient. Figure 8.12 shows the effect that a 180° pulse on the x' axis has on the coherence helix: The sense of the twist is reversed, giving the mirror image of the original helix. Before the 180° pulse, we have coherence that moves from x to $-y$ to $-x$ to y (clockwise viewed from above) as we move down in the tube, but after the pulse the coherence moves the from x to y to $-x$ to $-y$ (counterclockwise viewed from above) as we move down. What kind of gradient do we need to untwist this coherence? We need a gradient *identical to the first one*, which will rotate coherence ccw in the upper part of the tube and cw in the lower part, exactly canceling the twist imparted by the first

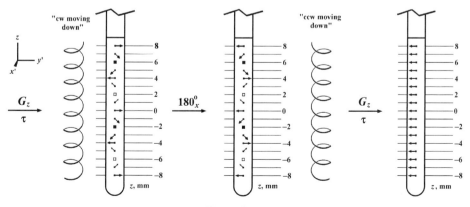

Figure 8.13

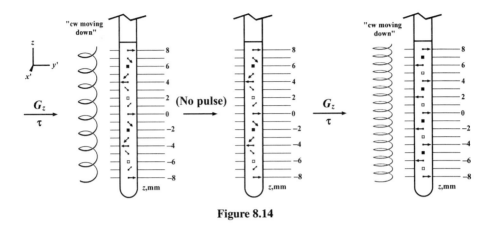

Figure 8.14

gradient (Fig. 8.13). Without the 180° pulse, this second identical gradient would simply reinforce the effect of the first, twisting the helix twice as tightly (Fig. 8.14). This sequence (G_z–180°–G_z) is called PFGSE, and we will see that it forms the basis of many selective excitation experiments. The key to its utility is in the 180° pulse: If it truly flips the sample magnetization to the opposite side of the x'–y' plane, reversing the sense of the helix twist, the sample magnetization is lined up at the end (Fig. 8.10(d)). If the pulse does not give a 180° rotation, the sample magnetization is completely destroyed. We will see how this reinforces the selectivity of a shaped pulse in the next section.

8.6 COMBINING SHAPED PULSES AND PULSED FIELD GRADIENTS: "EXCITATION SCULPTING"

We saw that the PFGSE acts as a spin echo if the central pulse is a 180° pulse, and as a gradient-based coherence annihilator if the central 180° pulse is absent. What happens if we put a 180° *shaped* pulse at the center of the PFGSE (G_z–180°(sel.)–G_z)? This pulse should deliver a 180° pulse to the selected resonance (peak) in the spectrum and have no effect (0° pulse) on all the other peaks. At the end of this sequence, we expect to have aligned coherence for the selected spins (Fig. 8.13) and completely scrambled coherence for all of the other spins (Fig. 8.14). As we started with a 90° pulse, we have no z magnetization and for the nonselected spins we end up with no net coherence either. So overall we have excited the selected resonance with a 90° pulse and we have *destroyed all net magnetization* on *all* other resonances in the spectrum. This is a very radical kind of selectivity, as nothing is left at all but the net magnetization of the desired spins in the x'–y' plane. A 90° shaped pulse will rotate the selected spins into the x'–y' plane, but the other spins will still have their full equilibrium net magnetization on the +z axis. This strategy was developed by A.J. Shaka, who dubbed it "excitation sculpting" because we start by exciting all the spins equally and then we cut away all the magnetization we do not want using the gradients, just as a sculptor reveals the desired shape by cutting away marble from a formless block.

Figure 8.15(b) shows the spectrum of sucrose with a 90° Gaussian pulse applied to the triplet at 3.99 ppm, compared to a normal ^{1}H spectrum (Fig. 8.15(a)). This is done by moving the reference frequency to place the 3.99 ppm triplet at the center of the spectral window (on-resonance), which is the center of the Gaussian-shaped excitation profile

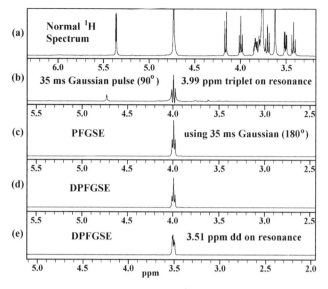

Figure 8.15

resulting from the shaped pulse. We can see some distortion of the peak shape as well as some undesired excitation, particularly of the two strong singlets (CH$_2$OH peaks) and the HOD peak at 4.73 ppm. In Figure 8.15(c), we see the same spectrum using a PFGSE with a 180° Gaussian pulse at the center. The peak shape is improved and we see absolutely none of the nonselective peak intensity. This can be improved even further by repeating the PFGSE with a different gradient strength ($G_z^a - 180$(sel.) $- G_z^a - G_z^b - 180$(sel.) $- G_z^b$) for an overall *double* pulsed field gradient spin-echo or DPFGSE (Fig. 8.15(d)). We can select other peaks in the spectrum by simply moving the reference frequency (ν_r) to place the desired peak on-resonance. For example, with the double doublet at 3.51 ppm on-resonance we see only this peak in the DPFGSE spectrum (Fig. 8.15(e)). The normal spectrum (Fig. 8.15(a)) has narrower lines because the sample is spinning.

We can actually measure the excitation profile of the Gaussian pulse in a DPFGSE sequence by selecting a peak (putting it on-resonance) and then repeating the experiment with the peak moved off-resonance in equal steps both upfield and downfield. The spectra are superimposed to give a series of peaks that map out the shape of the excitation. This is shown in Figure 8.16 for the HOD peak of sucrose in D$_2$O, changing the reference and pulse frequency (ν_r) by 6 Hz for each successive spectrum. We see that the profile from a Gaussian-shaped pulse is indeed Gaussian, with a bandwidth at half-height of about 36 Hz. The bandwidth is inversely proportional to the pulse width (duration of the shaped pulse), so if we used a 70-ms Gaussian pulse (with half the maximum B_1 field strength to maintain a 180° rotation), we would see a Gaussian excitation profile with a bandwidth of 18 Hz at half-height. Stretching the pulse squeezes the excitation profile and *vice versa*.

8.6.1 Frequency-Shifted Laminar Pulses

It is rather tedious to move the spectral window every time we want to select a peak with a shaped pulse, but it is necessary as the center of the Gaussian excitation profile is at

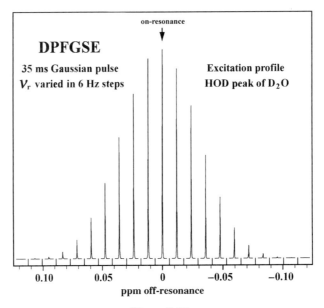

Figure 8.16

the center of the spectral window. One way to get around this is to change the *phase* of the individual rectangular pulses that make up the shaped pulse. If we consider a 35-ms Gaussian pulse made up of 35 rectangular pulses of 1 ms each, we could increase the phase of each pulse relative to the last one by an angle of 10°. This is easy to do because the shaped pulse is created from a list of 35 lines, each line specifying a pulse amplitude and a pulse phase. The first pulse would be delivered with B_1 on the x' axis, the 10th pulse with B_1 on the y' axis, the 19th pulse with B_1 on the $-x'$ axis, the 28th pulse with B_1 on the $-y'$ axis, and so on (Fig. 8.17). We see that the B_1 vector is no longer stationary in the rotating frame of reference—it is moving counterclockwise (in jerks) at a rate of one cycle every 36 ms, which is a frequency of 1/(0.036 s) or 27.78 Hz. The effective frequency of the pulse is 27.78 Hz higher than ν_r, its nominal frequency, so the center of the excitation profile is shifted downfield by 27.78 Hz from the center of the spectral window. Until now, the pulse frequency has always been the same as the reference frequency, at the center of the spectral window. We are using the phase "ramp" as a way of tricking the spins into seeing the excitation pulse at a different position within the spectral window. Now we can place the Gaussian excitation

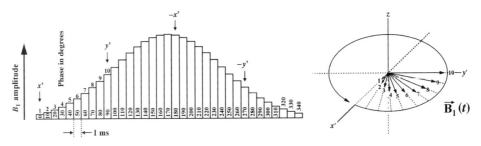

Figure 8.17

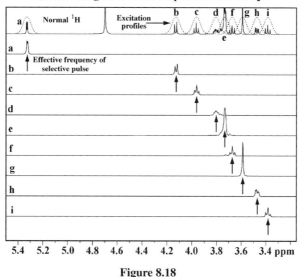

Figure 8.18

profile anywhere in the spectrum we want. Figure 8.18 shows a stacked plot of spectra of sucrose in D$_2$O with different effective frequencies of the shaped pulse, using the phase ramp (frequency-shifted laminar pulse) to move the center of the excitation profile. The reference frequency and pulse frequency (v_r) are the same in all of these spectra. We can cleanly select any of the resolved peaks in the spectrum, with no excitation of other peaks. Only in the case of crowded regions (d–f) do we see any excitation of neighboring peaks.

8.6.2 Selective Annihilation: Watergate

We have seen that the key to selectivity in the PFGSE is whether the spins in question receive a 180° pulse at the center of the spin echo. If they do not, all magnetization (including z magnetization) is destroyed. This can be applied as a strategy for getting rid of unwanted peaks in a spectrum. The most unwanted peak in all of NMR spectroscopy is the water peak in a 90% H$_2$O/10% D$_2$O sample. We saw in Chapter 5 how presaturation can be used to selectively saturate the water protons with a long, low-power irradiation at exactly the water resonance frequency. The problem with presaturation is that these saturated H$_2$O protons can exchange with amide NH positions, carrying their lack of magnetization along with them. This "bleaches" these signals, reducing or even removing them from the spectrum. What if we had a shaped pulse that provides a 180° pulse everywhere *except* the center of the spectral window, where the water peak is positioned? All the peaks of the spectrum will survive the PFGSE, but the water peak will be destroyed by the gradients because its coherence helix is not reversed in the middle of the spin echo. This is better than presaturation because the water magnetization is destroyed quickly at the end of the pulse sequence, just before the start of the FID. Furthermore, the water at any level of the NMR tube still has its full net magnetization—it is only at the level of summing the water magnetization at all levels of the tube that we get cancellation and a zero net magnetization.

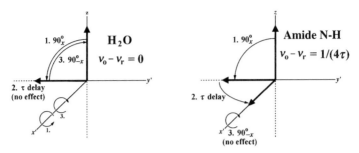

Figure 8.19

The only problem is that we need to find this magic selective pulse that delivers a 180° rotation everywhere but the center of the spectral window. There are shaped pulses that do this, but it turns out that the simplest solution is a series of six hard pulses separated by equal delays. If we divide the 90° rotation into 13 small rotations of equal angle, the sequence is

$$3 - \tau - 9 - \tau - 19 - \tau - \overline{19} - \tau - \overline{9} - \tau - \overline{3}$$

where the numbers are multiples of 90/13 = 6.92° and the bar over the number means that the pulse phase is reversed from that of the first three pulses (e.g., $-x$ instead of x). The actual pulse rotations are 20.77° ("3"), 62.31° ("9"), and 131.54° ("19"), and their durations are calculated from the calibrated 90° pulse width.

To understand this sequence, let's start with a very simple set of two pulses separated by a delay. The sequence $90°_x$–τ–$90°_{-x}$ is called a "jump-return" or $1 - \overline{1}$ sequence and can be used as a selective 90° pulse on everything but the water. The water resonance, which is placed at the center of the spectral window, does not undergo chemical shift evolution during the τ delay. So it is rotated from $+z$ to $-y$, sits motionless on the $-y$ axis during the τ delay, and then returns to $+z$. It receives no excitation at all (Fig. 8.19). Now consider a resonance with an offset ($\nu_o - \nu_r$) of $1/(4\tau)$ Hz. Just like the water magnetization, it moves from $+z$ to $-y$ during the first pulse. But during the τ delay, it precesses ccw in the x'–y' plane by an angle $1/(4\tau) \times \tau = 1/4$ cycle or 90°, from the $-y'$ axis to the x' axis. The final 90° pulse has no effect, as the magnetization is on the same axis as the pulse. At the end of the sequence, we have delivered an overall 90° excitation pulse to the spins at this offset. In general, for an offset of Ω rad/s ($\Omega = 2\pi(\nu_o - \nu_r)$) we have

$$I_z \xrightarrow{90°_x} -I_y \; -\tau \rightarrow \; -I_y \cos\Omega\tau + I_x \sin\Omega\tau \xrightarrow{90°_{-x}} I_z \cos\Omega\tau + I_x \sin\Omega\tau$$

The excitation profile is just a sinc function, with positive peaks to the left of the center of the spectral window rising to a peak at an offset of $\Omega = \pi/(2\tau)$ and then falling to another null at $\Omega = \pi/\tau$. To the right of the center of the spectral window, we see the same thing except that the peaks are negative. At the center ($\Omega = 0$), there is no excitation. This is quite a radical distortion of our spectrum, a high price to pay for destroying the water signal.

Now let's return to the "3-9-19" sequence. For the water peak, it's simple. As there is no evolution during the τ delays, it is just a sequence of six pulses whose rotation is exactly balanced between ccw rotations (3-9-19) and cw rotations ($\overline{19} - \overline{9} - \overline{3}$). The net rotation is zero and the water magnetization ends up on the $+z$ axis, where it started. Water is not affected by the pulse train. The same is true if the offset is $\nu_o - \nu_r = 1/\tau$

COMBINING SHAPED PULSES AND PULSED FIELD GRADIENTS   313

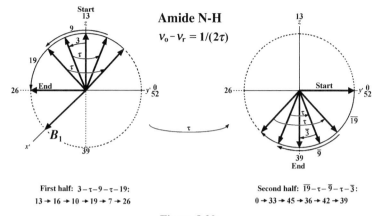

First half: $3-\tau-9-\tau-19$:
$13 \to 16 \to 10 \to 19 \to 7 \to 26$

Second half: $\overline{19}-\tau-\overline{9}-\tau-\overline{3}$:
$0 \to 33 \to 45 \to 36 \to 42 \to 39$

Figure 8.20

or $-1/\tau$ ($\Omega = 2\pi/\tau$ or $-2\pi/\tau$) as the evolution of the x–y component will be 360° during each τ delay, returning the vector to where it started before the τ delay. What if the offset is right between these two extremes, at $\nu_o - \nu_r = 1/(2\tau)$? The x–y component of the magnetization vector will rotate exactly 180° during the τ delay. If the pulses are on the x' axis (or the $-x'$ axis), the net magnetization stays in the y'–z plane and each delay flips it to the opposite side (reversing the y' component). The chemical shift evolution can be thought of as a 180° rotation around the z axis: a "$+z$ pulse." If we divide the 360° rotation around the x' axis in the y'–z plane into 52 equal angles (13 for each 90° rotation), we can describe the position of the net magnetization by a number between 0 and 52 (Fig. 8.20). Now we can "narrate" the effect of the 3–9–19 sequence. Starting from position 13 (the $+z$ axis)

$$13 \xrightarrow{3} 16 \xrightarrow{\tau} 10 \xrightarrow{9} 19 \xrightarrow{\tau} 7 \xrightarrow{19} 26$$

At the center of the sequence the net magnetization is on the $-y'$ axis, and the central τ delay rotates it 180° in the x'–y' plane to the $+y'$ axis: $26 \xrightarrow{\tau} 0$. For the second half:

$$0 \xrightarrow{\overline{19}} 33 \xrightarrow{\tau} 45 \xrightarrow{\overline{9}} 36 \xrightarrow{\tau} 42 \xrightarrow{\overline{3}} 39$$

Note that rotations about $+x'$ simply add to the number, whereas rotations about $-x'$ (equivalent to cw rotations about $+x'$) subtract from the number. The τ delays are 180° rotations about $+z$ (13) or $-z$ (39), so they move the vector to the opposite side ($13 - x$ becomes $13 + x$, $39 + x$ becomes $39 - x$, etc.). At the end of all this gyrating, we end up at the $-z$ axis, for a net rotation of 180° around the x' axis! This is our magic pulse: It gives no rotation to water and a 180° rotation to the peaks we are interested in. If we start at $-y$ (26), we end up on $-y$ (26):

$$26 \xrightarrow{3} 29 \xrightarrow{\tau} 49 \xrightarrow{9} 6 \xrightarrow{\tau} 20 \xrightarrow{19} 39 \xrightarrow{\tau} 39 \xrightarrow{\overline{19}} 20 \xrightarrow{\tau} 6 \xrightarrow{\overline{9}} 49 \xrightarrow{\tau} 29 \xrightarrow{\overline{3}} 26$$

and if we start on $+y$ (0) we end up on $+y$ (0):

$$0 \to 3 \to 23 \to 32 \to 46 \to 13 \quad \to \quad 13 \to 46 \to 32 \to 23 \to 3 \to 0$$

What if we start on $+x$? The pulses have no effect, but each τ delay rotates 180° in the x–y plane from $+x$ to $-x$ and *vice versa*:

$$x \to x \xrightarrow{\tau} -x \to -x \xrightarrow{\tau} x \to x \xrightarrow{\tau} -x \to -x \xrightarrow{\tau} x \to x \xrightarrow{\tau} -x \to -x$$

As there is a odd number (5) of τ delays, we end up on the opposite axis. If we start on $-x$, we end up on $+x$:

$$-x \to -x \xrightarrow{\tau} x \to x \xrightarrow{\tau} -x \to -x \xrightarrow{\tau} x \to x \xrightarrow{\tau} -x \to -x \xrightarrow{\tau} x \to x$$

So for that exact offset ($1/2\tau$ in hertz) we have a true 180° pulse on the y' axis:

$$\mathbf{I}_z \to -\mathbf{I}_z, \quad \mathbf{I}_x \to -\mathbf{I}_x, \quad \mathbf{I}_y \to \mathbf{I}_y$$

The same is true for the opposite side of the spectral window ($-1/2\tau$) as a 180° rotation gives the same result whether it is cw or ccw. If we put this 6-pulse sequence at the center of our PFGSE, it will reverse the sense of the coherence helix for the resonance $1/2\tau$ away from the center, and it will maintain the sense of the coherence helix for the on-resonance water peak and for peaks $1/\tau$ away from the center. The resonance $1/2\tau$ away will be "unwound" under the influence of the second gradient, whereas the on-resonance (water) peak will be wound twice as tightly, leading to zero net magnetization when summed over the whole sample (Fig. 8.21).

What happens between these two extremes? With so many pulses and delays it becomes impossible to draw simple diagrams, and we need to do some calculations. Pulse rotations are simple sine and cosine calculations and they can be simulated on a simple spreadsheet. Figure 8.22 shows the simulated "extinction profile" of the Watergate sequence using the 3–9–19 strategy. For the simulation, the delay τ is set to 217.4 μs, which gives a maximum signal (180° rotation during τ) at an offset of ±2300 Hz (3.833 ppm on a 600-MHz

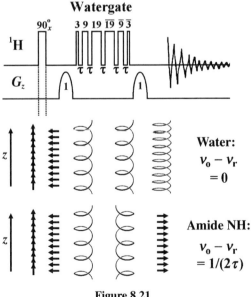

Figure 8.21

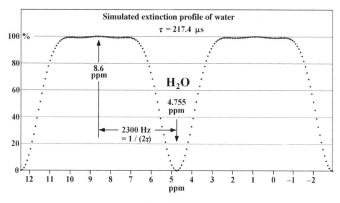

Figure 8.22

spectrometer) from the center. If we put the center of the spectral window on the water peak (4.755 ppm), we see a maximum at $4.755 + 3.833 = 8.588$ ppm and at $4.755 - 3.833 = 0.922$ ppm. The nulls occur at the water frequency (4.755 ppm) and at twice the optimal offset: $4.755 + 2(3.833) = 12.421$ ppm and $4.755 - 2(3.833) = -2.911$ ppm. The advantage of Watergate is the flatness of the curve around the optimal frequency (7–11 ppm and −1 to 3 ppm), where there is little or no loss of signal. The 7–11 ppm region for proteins and peptides corresponds to the amide NH region and most of the aromatic region, and the −1 to 3 ppm region covers most of the aliphatic side chain resonances. The problem is the H_α proton region, which gets really "slammed" in the process of knocking down the water signal. Compared to presaturation, Watergate cuts a very wide "swath" around the water peak and there is a lot of collateral damage done to the H_α resonances. Generally, for experiments that focus on the amide NH resonances, we use Watergate because it avoids "bleaching" these signals, and for looking at the H_α resonances we use presaturation or prepare a sample in D_2O, which removes the amide NH peaks but makes water suppression much easier. We will come back to the unique properties of 90% H_2O and other methods to suppress water in Chapter 12.

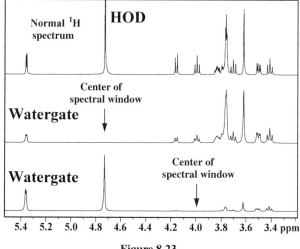

Figure 8.23

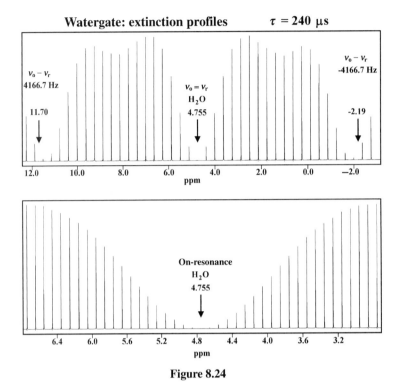

Figure 8.24

Figure 8.23 shows the Watergate sequence applied to a sample of sucrose in D$_2$O. When the HOD peak is put on-resonance (Fig. 8.23, center) the HOD peak disappears with some loss of intensity for the peaks closest to HOD. When we move the reference frequency to the triplet at 3.99 ppm, we see that nearby peaks such as the doublet at 4.15 ppm are almost completely destroyed, whereas the faraway g1 peak at 5.36 ppm is restored to its original intensity because it is now in the "plateau" region of the Watergate extinction profile (Fig. 8.22). We can map out the extinction profile of Watergate by applying it to a sample of H$_2$O alone and moving the reference frequency in small steps (0.1 ppm), repeating the experiment and plotting the spectra side by side. Figure 8.24 shows the results using a τ delay of 240 μs. The two additional nulls occur close to the calculated null points of 11.70 and -2.19 ppm ($1/\tau = 4166.7$ Hz $= 6.94$ ppm), and we see the wide swath cut out around the water resonance (3.4–6.2 ppm). We have a relatively flat response from 6 to 10 ppm, which includes the majority of amide NH protons and aromatic protons.

8.7 COHERENCE ORDER: USING GRADIENTS TO SELECT A COHERENCE PATHWAY

How sensitive is a particular spin state to being twisted by a pulsed field gradient? For example, I_z is completely unaffected by a PFG because it has no coherence. Without magnetization in the x'–y' plane, there is no precession and the gradient has nothing to "twist." The same goes for $2I_zS_z$, as neither I_z nor S_z is affected by a gradient. In fact, this is a common strategy for "cleaning up" coherence transfer by INEPT: The INEPT

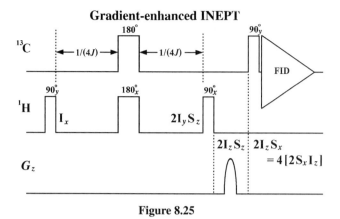

Figure 8.25

transfer is separated into two parts with $2I_zS_z$ as an intermediate state and between the two pulses a gradient is applied to "scramble" any magnetization that is in the x'–y' plane (Fig. 8.25):

$$2I_yS_z \xrightarrow{90^\circ_x(^1H)} 2I_zS_z \xrightarrow{G_z\tau} 2I_zS_z \xrightarrow{90^\circ_y(^{13}C)} 2I_zS_x = 4[2S_xI_z]$$

Any coherences that did not make it to the $2I_zS_z$ state will be killed by the gradient.

We know that I_y is twisted by a gradient. How much is S_y ($I = {}^1H$, $S = {}^{13}C$) twisted? The change in Larmor frequency during the gradient is $\nu_g = \gamma z G_z/2\pi$, and the amount of "twist" or phase change at any given level of the gradient is determined by ν_g times the duration of the gradient, τ. So the twist resulting from a given gradient strength and duration is proportional to γ, the strength of the nuclear magnet. We know that γ_H is about four times as large as γ_C, so we can say that ^1H single-quantum coherence (SQC) is about four times more sensitive to twisting by a gradient than ^{13}C SQC. We could exploit this difference in sensitivity to gradients in an INEPT experiment by using a gradient to "twist" the ^1H SQC spin state $2I_yS_z$ before the coherence transfer, and then using a gradient of opposite sign and *four times the magnitude* to "untwist" the ^{13}C SQC ($2S_xI_z$) after the coherence transfer (Fig. 8.26). This "gradient selection" would destroy any coherence that is not ^1H SQC before the transfer and ^{13}C SQC after the transfer! We can have an amazingly "clean" INEPT transfer using gradients to enforce the pathway ^1H SQC → ^{13}C SQC. Twisting of antiphase coherences is just like chemical shift evolution (Fig. 7.8): The double arrow pointing in opposite directions in the x'–y' plane just rotates as a unit, without changing the 180° angle between them. For example, $2I_yS_z$ is twisted by a gradient of intensity G_z and duration τ_g into $2I_yS_z \cos(\gamma_H z G_z \tau_g) - 2I_xS_z \sin(\gamma_H z G_z \tau_g)$, ignoring the chemical shift evolution that would occur during τ if the I (proton) peak is not on-resonance. Just like with chemical shift evolution, the S_z part is not affected because it is on the z axis.

In general, the sensitivity to twisting of a particular spin state can be classified by something called its *coherence order*. Thus an ordinary magnetization vector in the x'–y' plane has coherence order of 1 (single-quantum coherence, $\mathbf{p} = 1$) and a magnetization vector along the z axis has coherence order zero ($\mathbf{p} = 0$). Only the coherence order of 1

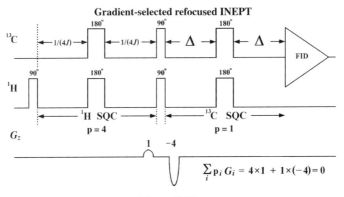

Figure 8.26

can be observed during the FID. There is also double-quantum coherence, which corresponds to a transition in a *J*-coupled system of two spins (e.g., H_a and H_b), where both spins flip together in the same direction: $\alpha\alpha$ to $\beta\beta$ ($H_a^\alpha H_b^\alpha$ to $H_a^\beta H_b^\beta$) or $\beta\beta$ to $\alpha\alpha$. This coherence for the homonuclear (two protons) system has a coherence order of 2 ($\mathbf{p} = 2$). Zero-quantum coherence results from a transition where both spins flip together but in opposite directions (e.g., $\alpha\beta$ to $\beta\alpha$ or $\beta\alpha$ to $\alpha\beta$). Like magnetization along the z axis, it has a coherence order of zero ($\mathbf{p} = 0$) for a homonuclear system. It turns out that the "twisting" effect of a gradient pulse depends precisely on the coherence order. For example, during a gradient a double-quantum coherence ($\mathbf{p} = 2$) rotates twice as fast in the x'–y' plane as a single-quantum coherence ($\mathbf{p} = 1$) and will acquire twice as many "turns" of twist during the gradient. We saw in Chapter 7 that DQC precesses in the x'–y' plane at a rate equal to the sum of the two offsets ($\Omega_I + \Omega_S$); this applies equally to twisting in a gradient—the twist is equal to the sum of the twists that would result for each of the nuclei alone. For z-magnetization and homonuclear zero-quantum coherence (both $\mathbf{p} = 0$), the gradient has no effect.

In *heteronuclear* experiments, we need to consider that different types of nuclei have different "magnet strengths" or magnetogyric ratios γ. For example, the magnetogyric ratio of proton (^{1}H) is about four times as large as the magnetogyric ratio of carbon (^{13}C). This means that in a gradient the proton magnetization rotates (and accumulates a helix twist in the x–y plane) four times faster than the carbon magnetization under the influence of the same gradient. This is extremely useful when we want to select only proton or only carbon coherence (SQC) at a particular point in a pulse sequence. We can put all of this together by including the magnetogyric ratio as part of the coherence order. Thus, for single-quantum coherence we can use $\mathbf{p} = 1$ for ^{13}C and $\mathbf{p} = 4$ for ^{1}H. For heteronuclear double-quantum coherence (^{1}H, ^{13}C pair), we have $\mathbf{p} = 5$ ($p_H + p_C$) and for zero-quantum coherence we have $\mathbf{p} = 3$ ($p_H - p_C$). This means that in addition to z magnetization ($\mathbf{p} = 0$), there are four separate things we can select with a gradient pulse.

A simple way to view a pulsed field gradient experiment is to add up the "twist" acquired by the sample magnetization in each gradient pulse and make sure they add up to zero for the desired pathway. If the "twist" is not zero at the beginning of acquisition of the FID, there will be no observable signal. For example, in the INEPT experiment (Fig. 8.26) we have

$$\begin{array}{cc} \mathbf{p}\,G_z & \mathbf{p}\,G_z \\ +4(1) & +1(-4) = 0; \sum(\mathbf{p}_i \times G_i) = 0 \\ {}^1\text{H SQC} & {}^{13}\text{C SQC} \end{array}$$

In Section 7.9, we saw how phase cycling can be used to remove the ^{13}C coherence that comes from the original ^{13}C z magnetization ($\mathbf{S}_z$), so that only the coherence transferred from ^{1}H z magnetization ($\mathbf{I}_z$) is observed. This is a subtraction process that requires more than one scan to accomplish. With gradients we can do it in one scan alone:

$$2\mathbf{I}_y\mathbf{S}_z + \mathbf{S}_z \xrightarrow{90^\circ_x {}^1\text{H}/90^\circ_y {}^{13}\text{C}} 2[\mathbf{I}_z][\mathbf{S}_x] + \mathbf{S}_x$$

The pathway $\mathbf{S}_z \to \mathbf{S}_x$ is unaffected by the first gradient (Fig. 8.26) because z magnetization does not precess, so the ^{13}C SQC ($\mathbf{S}_x$) is only "twisted" by the second gradient and arrives at the FID in a coherence helix that adds to zero over the whole sample. There is no need to subtract it out—it never reaches the receiver. We can add up the "twists" imparted by the two gradients using the fact that coherence order ($\mathbf{p}$) equals zero for z magnetization:

$$\begin{array}{cc} \mathbf{p}\,G_z & \mathbf{p}\,G_z \\ 0(1) & +1(-4) = -4; \sum(\mathbf{p}_i \times G_i) = -4 \\ {}^{13}\text{C }z\text{ magnetization} & {}^{13}\text{C SQC} \end{array}$$

Because the sum is not equal to zero, we end up with twisted coherence and no signal in the receiver. We call this a "gradient-selected" experiment because the gradients are being used to specifically refocus coherence in the desired coherence transfer pathway (^{1}H SQC $\to$ ^{13}C SQC) and to reject all others. In Chapter 10, we will develop the idea of coherence order in a more precise manner, and we will see that coherence order can be either positive or negative.

8.8 PRACTICAL ASPECTS OF PULSED FIELD GRADIENTS AND SHAPED PULSES

8.8.1 Gradient Hardware

The gradient coils are located in the NMR probe, surrounding the sample. Only the RF send–receive coil is closer to the sample. Gradient amplifiers in the console provide direct currents up to 10 A to create the gradient magnetic field. A heavy cable attached to the probe delivers these currents to the gradient coils. It is important to keep in mind that the gradients are driven by direct currents (DC), not the MHz oscillating signals (RF) that we use for pulses. When not actually producing the gradient, the amplifiers must be either "blanked" (blocked from introducing any current into the gradient coils) or adjusted to a zero current value with very low noise. Some spectrometers have the capability to deliver pulsed field gradients in all three directions: x, y, and z (the NMR tube and the B_0 field are aligned with the z axis). This is achieved with three separate gradient coils in the probe, each driven by a separate current source. This is standard in MRI, but for NMR the three-axis gradient capability is used mainly for improved water suppression in DQF-COSY of biological (proteins and nucleic acids) samples, and the crowded three-coil design actually sacrifices some sensitivity.

The gradient not only "twists" the magnetization of the observed nucleus (^{1}H, ^{13}C, etc.), but also twists the ^{2}H magnetization of the lock channel. You will see that the lock signal drops sharply when a gradient pulse is executed, and then recovers gradually to its former level. This is evident with the dipping down and bouncing back of the lock level on the meter in the Varian remote status unit or the graphic display in the lock window of the Bruker. This gives a convenient "heartbeat" of the gradient experiment, so you know what is going on. The Bruker lock system has a "sample-and-hold" feature that allows it to sample the lock signal before the gradient pulse and then hold onto this value until the lock has recovered fully. During a gradient experiment, you will see the message "Lock Sample and Hold Activated" on the LED display of the shim keyboard.

8.8.2 Gradient Parameters

Gradient pulses can be simply turned on and off like high-power RF pulses ("rectangular pulses") or they can be shaped so that they turn on and off more gradually. The shaped gradient pulses are less troublesome because they do not create a big transient response from the sharp rise and fall times of rectangular gradient pulses. Bruker uses almost exclusively the "sine"-shaped gradient pulse, which has the shape of the first 180° of the sine function. Varian generally uses rectangular (simple on/off) gradients. Parameters related to pulsed field gradients include the (z axis) gradient strength (Bruker *gpz1*, *gpz2*, etc., in percent of maximum gradient current, and Varian *gzlvl1*, *gzlvl2*, etc., in arbitrary units $-32,768$ to $+32,767$), the time duration of the gradient pulse (*gt1*, *gt2*, etc. for Varian and *p16* for Bruker, typically 1–5 ms), the shape of the gradient pulse (Bruker *gpnam1*, *gpnam2*), and the duration of the recovery delay, which allows the magnetic field to go back to homogeneous after the gradient pulse (Varian *gstab*, Bruker *d16*, typically 200 μs).

8.8.3 Shaped Pulse Hardware and Software

Shaped pulses are created from text files that have a line-by-line description of the amplitude and phase of each of the component rectangular pulses. These files are created by software that calculates from a mathematical shape and a frequency shift (to create the phase ramp). There are hundreds of shapes available, with names like "Wurst", "Sneeze", "Iburp", and so on, specialized for all sorts of applications (inversion, excitation, broadband, selective, decoupling, peak suppression, band selective, etc.). The software sets the maximum RF power level of the shape at the top of the curve, so that the *area* under the curve will correspond to the approximately correct pulse rotation desired (90°, 180°, etc.). When an experiment is started, this list is loaded into the memory of the waveform generator (Varian) or amplitude setting unit (Bruker), and when a shaped pulse is called for in the pulse sequence, the amplitudes and phases are set in real time as the individual rectangular pulses are executed.

For each shaped pulse you must select the pulse width (duration in μs: Bruker *p12*, *p13*, etc., or Varian *selpw*), the name of the text file that contains the shape function (Bruker *spnam1*, *spnam2*, etc., Varian *selshape*), the maximum power (B_1 amplitude) at the top of the pulse shape (Bruker *sp1*, *sp2*, etc., or Varian *selpwr*), and the offset frequency in hertz if you want to excite a peak that is not at the center of the spectral window (Bruker *spoffs1*, *spoffs2*, etc., in hertz relative to the center of the spectral window, or Varian *selfrq* in hertz).

8.9 1D TRANSIENT NOE USING DPFGSE

We are finally able to apply our fancy selective excitation building block, DPFGSE, to a real experiment that can give us structural information. Remember that a 90° nonselective (hard) pulse followed by the DPFGSE (containing 180° shaped pulses) is just a 90° excitation pulse for the resonance we are selecting. For all other peaks in the spectrum, it totally destroys all net magnetization. How can we use this to make an NOE experiment? We will add a 90° hard pulse at the end of the DPFGSE to flip the selected resonance magnetization down from the x'–y' plane to the $-z$ axis. This is the largest perturbation of populations (z magnetization) possible, and if we now wait a while (the mixing time τ_m), we will see that this perturbation from equilibrium propagates to nearby (<5 Å away) protons in the molecule. As all the other protons in the molecule have no net magnetization at all at the end of the DPFGSE, they are unaffected by the 90° hard pulse and have no z magnetization at all at the start of the mixing time. Any z magnetization that is transferred from the selected spin via NOE during the mixing time will then show up as positive z magnetization ($-\mathbf{I}_z^a \rightarrow \mathbf{I}_z^b$) and we can "read" it out with a final 90° hard pulse at the end of the mixing time. The full sequence is shown in Figure 8.27 (omitting the repeat of the PFGSE) with the magnetization vectors shown at various levels within the NMR tube for the selected and nonselected spins. The spectrum will show a very large upside-down peak for the inverted peak, which is selected (H_a), and very small in-phase positive peaks for the resonances which receive z magnetization transfer from the selected peak. We have done an NOE experiment in a single scan, with no subtraction of spectra!

This technique differs from the old NOE difference experiment we looked at in Chapter 5, where we selectively saturate one resonance (make $M_z = 0$) over a long period of time (the mixing time) whereas other spins are perturbed and reach a steady-state level of

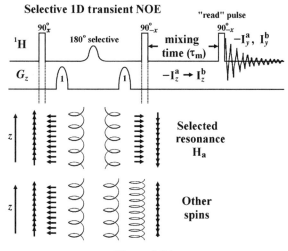

Figure 8.27

z magnetization, which is slightly enhanced over the equilibrium value M_o. The PFGSE method introduces a *sudden* perturbation (inversion of the selected resonance) and then waits for this perturbation to propagate to nearby protons. During the mixing time there is no RF; we just are waiting for the NOE to develop. Furthermore, because the PFGSE kills all magnetization on the nonselected spins, any z magnetization that we "read" with the final 90° pulse has to be an NOE, transferred from the selected proton.

8.9.1 Transient NOE

The rapid perturbation method is called the "transient NOE." Let's look at the process in detail. At thermal equilibrium in a strong magnetic field, there is a slight excess of population of nuclei in the lower energy (aligned with the magnetic field) state and a slight depletion of nuclei in the higher energy state (opposed to the magnetic field). If this equilibrium is perturbed for one group of nuclei (corresponding to a peak in the ^{1}H spectrum), this perturbation is propagated to nearby nuclei in the molecule due to the NOE. Because the intensity of a peak in an NMR spectrum is directly proportional to this population difference, the perturbation can be measured by simply recording a spectrum.

The traditional 1D NOE experiment (Section 5.12) involves irradiating with low-power radio frequency at the resonant frequency of one peak in the ^{1}H spectrum in order to equalize the populations of the two states ("saturation"). This saturated state is maintained by continued irradiation until the perturbation of populations of nearby nuclei in the molecule reaches a steady state and does not change any further. Then a 90° pulse is applied and an FID is recorded to measure the amount of perturbation on the nearby nuclei. As the enhancement of signals is quite small (a few percent), it is necessary to record a control spectrum with irradiation away from any peaks in the spectrum, and then subtract the control spectrum from the NOE spectrum. There are a number of disadvantages to this approach:

1. In any difference spectrum, the conditions (temperature, RF power, sensitivity, magnetic field, and vibration) must be identical in the two experiments in order to get perfect subtraction of the signals that are not affected. This subtraction is always imperfect as the two spectra are recorded at different times, so there are always big subtraction artifacts in the difference spectrum.
2. The magnitude of the NOE is proportional to the inverse sixth power of the distance between two nuclei only for very short times between the perturbation and the measurement of the effect on other nuclei. The magnitude of the steady-state NOE is dependent on many other competing relaxation processes, so it cannot be used as an accurate measure of distance. To accurately measure distances, you need to measure the transient NOE with a number of different times between perturbation and measurement ("mixing times") and measure the initial slope of the curve as the effect increases with time.
3. The selectivity of continuous-wave (CW) irradiation is limited, and in crowded regions of the spectrum nearby peaks are also affected. This sometimes makes the results ambiguous.

With shaped (selective) pulses, we specifically invert (overall 180° pulse) a single peak in the spectrum. This is the most dramatic perturbation you can create, as the excess population in the lower energy level is now in the higher energy level and the depleted population is now in the lower energy level. If we then wait a short time for this perturbation to propagate to nearby nuclei, a 90° pulse will "read" the effect on the other nuclei in the form of a spectrum with enhanced peak areas. We can avoid having to subtract two spectra, as the gradients in the PFGSE kill all magnetization on the other nuclei at the same time that we invert the desired peak. Thus, the only thing that will be detected with the 90° "read" pulse is the perturbation due to the NOE (i.e., the transferred magnetization).

8.9.2 Populations After a Selective Inversion Pulse

We saw the effect of cross-relaxation (DQ relaxation for small molecules) after a selective *saturation* of one resonance in Chapter 5 (Figs. 5.24–5.26) by analyzing the four-state population diagrams. A selective *inversion* of one resonance (H_a) is twice the perturbation of saturation, reducing M_z from M_0 (equilibrium) to $-M_0$ (inverted) rather than to zero (saturated). The resulting population diagram is shown in Figure 8.28, with all of the $N/4 + 2\delta$ spins originally in the $\alpha\alpha$ state (Fig. 5.24) now in the $\beta\alpha$ state ($H_a = \beta$, $H_b = \alpha$) and all of the $N/4$ spins originally in the $\beta\alpha$ state now in the $\alpha\alpha$ state (Fig. 8.28, lower right). Remember that inversion (a 180° pulse) affects every single spin in the ensemble (in this case all of the H_a spins), switching spins in the α state to β and spins in the β state to α. The other H_a transition is affected in the same way, moving the $N/2 - 2\delta$ spins in the $\beta\beta$ state to the $\alpha\beta$ state and the $N/4$ spins in the $\alpha\beta$ state to the $\beta\beta$ state (Fig. 8.28, upper left). After the selective 180° pulse, the population difference across the H_b transitions is unaffected ($\Delta P = 2\delta$) and the population difference across the H_a transition is inverted ($\Delta P = -2\delta$). We can say that the z-magnetization of H_b is M_0 (equilibrium) and the z-magnetization of H_a is $-M_0$ (inverted). If we acquire a spectrum at this point (90° pulse and FID), we will see a normal peak for H_b and an upside-down peak for H_a, both with 100% of normal peak height (Fig. 8.28, right).

Instead of acquiring an FID, we wait for a period of time τ_m (the mixing time) and allow relaxation to occur, dominated (for small molecules) by the DQ relaxation pathway: $\beta\beta \rightarrow \alpha\alpha$. What is the equilibrium population difference between these two states? From Figure 5.24, we see that $\Delta P = (N/2 + 2\delta) - (N/2 - 2\delta) = 4\delta$, or counting the circles

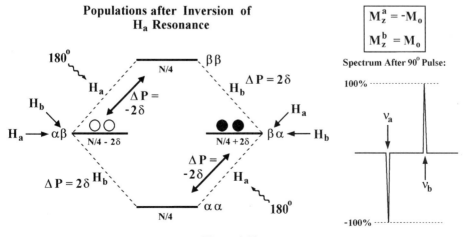

Figure 8.28

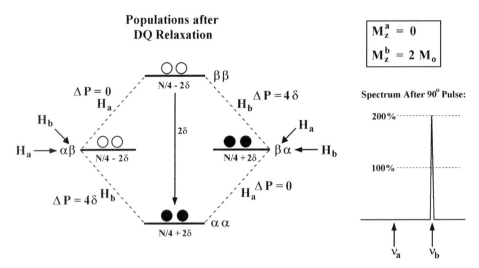

Figure 8.29

we have $2 - (-2) = 4$. This makes sense because the energy difference is twice that of a single-quantum transition, so the population difference should be twice as large according to the Boltzmann distribution. After the selective inversion pulse (Fig. 8.28), there is no population difference between $\beta\beta$ and $\alpha\alpha$, so the molecules want to drop down across that transition. If we let δ molecules drop down, we will have one open circle ($N/2 - \delta$) in the $\beta\beta$ state and one closed circle ($N/2 + \delta$) in the $\alpha\alpha$ state, for a population difference of 2δ (two circles). To reach the equilibrium population difference, we let another δ molecules drop down, leaving two open circles in $\beta\beta$ and two filled circles in $\alpha\alpha$ (Fig. 8.29). This is the equilibrium population difference, so no more molecules will drop down. In the real world, you would not allow enough time to reach the equilibrium population difference, and you would also have competition from other relaxation pathways: ZQ ($\alpha\beta \leftrightarrow \beta\alpha$) and SQ ($H_a$ and H_b transitions). For simplicity, we are allowing relaxation to proceed to completion via the DQ pathway, and we are blocking any relaxation by any other pathway—this will give a greatly exaggerated NOE effect but will simplify the explanation. At this point, we can take stock of the population differences (Fig. 8.29): for the H_a transitions we have $\Delta P = 0$, corresponding to $M_z = 0$, and for the H_b transitions we have $\Delta P = 4\delta$ ($2 - (-2) = 4$ circles), corresponding to $M_z = 2M_0$. We have enhanced the z magnetization on the H_b spins by a factor of 2 (100% NOE), increasing it from the equilibrium value of M_0 to $2M_0$. At this point, a 90° nonselective pulse will lead to no peak at all for H_a and a peak of twice the normal height for H_b (Fig. 8.29, right). Overall, the effect of reducing H_a's z magnetization by $2M_0$ (from M_0 to $-M_0$) has increased H_b's magnetization by M_0 (from M_0 to $2M_0$). Using the product operator notation, we can describe the experiment as follows:

$$\mathbf{I}_z^a + \mathbf{I}_z^b \xrightarrow{\text{(selective 180° on } H_a)} -\mathbf{I}_z^a + \mathbf{I}_z^b \xrightarrow{(\tau_m)} \mathbf{I}_z^b + \mathbf{I}_z^b = 2\mathbf{I}_z^b$$

By writing the result, $2\mathbf{I}_z^b$, as $\mathbf{I}_z^b + \mathbf{I}_z^b$ we can see what happened during the mixing period: $-\mathbf{I}_z^a$ was converted into $\mathbf{I}_z^b$, a *transfer of magnetization*! The cross-relaxation that occurs in an NOE experiment can be described as transfer of z magnetization from one spin (H_a) to another (H_b). Note that the sign changes when magnetization transfers: this is characteristic of small molecules, a direct result of the dominant pathway being DQ relaxation. Although the effect on H_b is enhancement of its z magnetization, the NOE can be described as negative because the effect on H_b (increase in M_z) is opposite to the original perturbation of H_a (decrease of M_z). The "negative" NOE is clearly seen in the product operator representation of cross-relaxation: $\mathbf{I}_z^a \rightarrow -\mathbf{I}_z^b$.

Exercise: Go through the same thought experiment for large molecules, allowing only ZQ relaxation ($\alpha\beta \leftrightarrow \beta\alpha$) during the mixing time. What is the equilibrium population difference between these two states? Allow complete relaxation to this difference during τ_m. What is the effect on the final spectrum? Describe the experiment using product operators and show the net effect of cross-relaxation (transfer of z magnetization) in terms of $\mathbf{I}_z^a$ and $\mathbf{I}_z^b$. Would you call this a positive or negative NOE? Now try the experiment for large molecules and for small molecules with relaxation during the mixing time (ZQ or DQ, respectively) proceeding only *half of the way* to the equilibrium population difference. How does this affect the percent change in z magnetization at the end? This is a slightly more realistic thought experiment.

8.9.3 The Heat Flow Analogy

As we did for the steady-state NOE (Chapter 5), we can look at the transient NOE using the heat flow analogy (Fig. 8.30). As before, we have two beakers filled with water, immersed in a tub of water at 25°C. The beaker on the left (A) represents the selected proton, and the beaker on the right (B) is a nearby proton. A shared glass wall between them allows heat flow (NOE transfer of z magnetization) between the two beakers. The transient perturbation (inversion of H_a) is represented by dropping a hot stone into beaker A. The temperature immediately rises from 25 to 50°C. Heat begins to flow out of beaker A into the surrounding tub of water at 25°C (self-relaxation of H_a through T_1 relaxation), bringing the temperature back

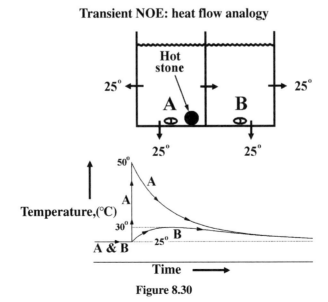

Figure 8.30

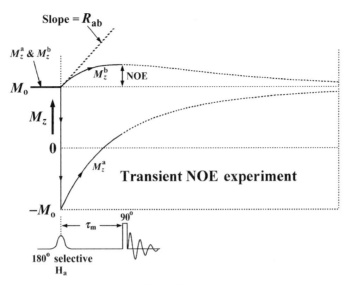

Figure 8.31

toward the equilibrium value (25°C). But some heat flows through the partition to beaker B, and its temperature begins to rise slightly, in a linear fashion at first because the heat flow is constant. But as beaker A begins to cool down, the rate of heat flow to beaker B slows and the rise in temperature begins to occur at a slower rate. In addition, as beaker B is now above the temperature of the surroundings, it begins to lose heat to the tub of water (self-relaxation of H_b through T_1 relaxation). At some point, the temperature in beaker B reaches a maximum (30°C) and begins to fall as the heat flow out to the environment exceeds the heat flow in from beaker A. After a long time, both beakers return to the temperature of the surroundings (25°C).

From the standpoint of NMR z magnetization, the experiment is diagrammed in Figure 8.31. For simplicity, we use a 180° shaped pulse alone, rather than a PFGSE. At equilibrium $M_z = M_o$ for both H_a and H_b. Immediately after a selective inversion of H_a, $M_z = -M_o$ for H_a and M_o for H_b. The z magnetization of H_a recovers in an exponential fashion with a time constant slower than T_1 as it is donating some of its z magnetization to H_b, at the same time it is gaining it by T_1 relaxation. The z magnetization of H_b is *increasing* above M_o because magnetization transfer by NOE for small molecules is in the opposite sense of the heat analogy: $\mathbf{I}_z^a \rightarrow -\mathbf{I}_z^b$. "Heating up" H_a leads to a "cooling down" of H_b. This happens in a linear fashion at first, and the slope of this line is the rate of z-magnetization transfer from H_a, which is proportional to $1/r^6$, where r is the distance between the two protons. This rate of increase falls off until M_z reaches a maximum for H_b and begins to fall. The optimal mixing time depends on what you are trying to measure: for accurate distance measurements you will need to repeat the experiment a number of times with various short values of τ_m in order to measure the initial rate of increase of M_z for H_b. This "NOE buildup study" is the only way to get accurate distances from an NOE experiment. More commonly, however, we just want to know if there is a NOE or not, and the mixing time is set to correspond to the maximum of the NOE buildup curve.

8.9.4 Simulation of the Transient NOE Experiment

Both the heat flow analogy and the NMR experiment can be represented by a pair of linked differential equations:

$$\mathrm{d}\Delta M_z^\mathrm{a}/\mathrm{d}t = -R_\mathrm{aa}\Delta M_z^\mathrm{a} - R_\mathrm{ab}\Delta M_z^\mathrm{b}$$

$$\mathrm{d}\Delta M_z^\mathrm{b}/\mathrm{d}t = -R_\mathrm{ab}\Delta M_z^\mathrm{a} - R_\mathrm{bb}\Delta M_z^\mathrm{b}$$

where $\Delta M_z^\mathrm{a} = M_z^\mathrm{a} - M_\mathrm{o}$ is the "disequilibrium" or perturbation of the z magnetization of H_a from equilibrium and $\Delta M_z^\mathrm{b} = M_z^\mathrm{b} - M_\mathrm{o}$ is the "disequilibrium" of H_b. R_aa is the self-relaxation rate of H_a and R_bb is the self-relaxation rate of H_b, whereas R_ab is the cross-relaxation rate or the rate at which H_a "disequilibrium" is propagated to H_b and *vice versa*. R_ab is actually proportional to $1/r^6$, the "Holy Grail" of NOE experiments. So all this says is that the rate of change of H_b's "disequilibrium" depends on its own "disequilibrium" (the self-relaxation or T_1 process) and on the "disequilibrium" of H_a (the NOE process). It will pick up z magnetization from H_a to the extent that H_a is out of equilibrium, and it will tend to recover from any disequilibrium of its own and move back to M_o.

Figure 8.32 shows a simulation of these equations starting right after the selective 180° pulse that inverts H_a: at this moment $M_z^\mathrm{a} = -M_\mathrm{o}$ and $M_z^\mathrm{b} = M_\mathrm{o}$. The T_1 value is set to 0.7 s and the rates are $R_\mathrm{aa} = R_\mathrm{bb} = 0.95\ \mathrm{s}^{-1}$ and $R_\mathrm{ab} = -0.45\ \mathrm{s}^{-1}$ (negative because it is a small molecule). The behavior is very much like the cartoon of Figure 8.31; M_z^b rises at an initial linear rate of 0.7 M_o/s, which is just R_ab times the initial perturbation $(-2M_\mathrm{o})$ of H_a: $-0.45\ \mathrm{s}^{-1} \times -2M_\mathrm{o}$. This eventually slows down and M_z^b reaches a maximum after 1.125 s with an NOE enhancement of 36.7%. M_z^a passes through zero after 0.8 s, slower than predicted by its T_1 value alone ($t_{1/2} = \ln 2 T_1 = 0.693 T_1 = 0.485$ s). Note that the sum of M_z^a and M_z^b recovers from 0 $(-M_\mathrm{o} + M_\mathrm{o})$ to $2M_\mathrm{o}$ $(M_\mathrm{o} + M_\mathrm{o})$ in the same way as it would if there were no cross-relaxation: a simple exponential curve based on a T_1 value of 0.7 s ($2M_\mathrm{o}\,\mathrm{e}^{-t/T_1}$:□ symbols in Fig. 8.32), passing through the halfway point (M_o) at

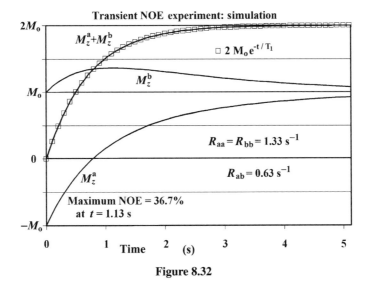

Figure 8.32

exactly 0.485 s. Any z magnetization transferred from H$_a$ to H$_b$ is invisible in the sum $M_z^a + M_z^b$; the slower recovery of M_z^a is exactly compensated by the growth of the NOE for M_z^b. Transfer of magnetization can be accounted for precisely: whatever z-magnetization H$_a$ loses is exactly gained by H$_b$.

8.9.5 DPFGSE-NOE

Instead of using a simple 180° shaped pulse for inversion of the selected resonance, we will use a double PFGSE to excite the selected spins and "destroy" the others. The pulse sequence is as follows (Fig. 8.27): A nonselective 90° pulse rotates all of the sample magnetization onto the −y axis. Then a gradient "twists" the magnetization into a helix. The selective (shaped) 180° pulse is applied to invert the magnetization of the peak of interest, so that its "twist" is now in the reverse direction. A second gradient of equal intensity and duration to the first now unwinds the twist for the peak of interest. But all the other peaks in the spectrum are just twisted twice as far, as their magnetization helix was not reversed by the selective 180° pulse. This destroys this magnetization and leaves only one thing in the sample: the peak of interest with its magnetization aligned along the y axis. A second nonselective 90° pulse is now applied to rotate this magnetization from the y axis to the −z axis. Thus, we have accomplished two things: the peak of interest has been inverted (population inversion) and the rest of the peaks have been destroyed. During the mixing time, the perturbation of populations for the selected resonance (inverted at the start of mixing) propagates to nearby nuclei and perturbs their populations (enhancement of z magnetization, $M_z > 0$). Finally, a 90° pulse rotates this transferred magnetization into the x–y plane where it precesses and is recorded as an FID. Any signal other than the selected one is an NOE.

8.9.6 Transient NOE of Fumarate–Cyclopentadiene Adduct

Figure 8.33 shows the ^{1}H spectrum and a series of transient NOE spectra for a rigid bicyclo[2.2.1] system formed in a Diels–Alder reaction of dimethyl fumarate and cyclopentadiene. Assignment of the ^{1}H spectrum depends primarily on through-space (NOE) interactions. The molecule is chiral (racemic) and has no symmetry elements, so all the protons are unique. In this discussion, we will refer to positions in the structure by number and to peaks in the spectrum by letters. In the ^{1}H spectrum (Fig. 8.33, top), we can immediately assign the olefinic protons (H6 and H7) to the two downfield resonances at 6.03 (H$_i$) and 6.24 ppm (H$_j$) from their chemical shifts, but we do not yet know which is which. The two three-proton singlets at 3.62 (H$_g$) and 3.68 (H$_h$) can be assigned to the two methyl groups H10 and H11. The most upfield signals are due to the protons farthest away from the olefin and ester functional groups: H$_a$ and H$_b$ form an AB system (with one additional small coupling to H$_a$), corresponding to the geminal pair of protons on C2. By chemical shift arguments, we can tentatively assign the upfield peak H$_a$ to the proton lying directly above the olefin (H2o): The area above and below the olefin is *shielded* by the "ring current" of the loosely attached π electrons of the double bond. The remaining four peaks, H$_c$–H$_f$, are intermediate in chemical shift and will have to be assigned by looking at *J* couplings and NOEs.

Although this is strictly an NOE experiment, we see strong *J*-coupling artifacts. Selection of H$_j$, for example, gives a strong antiphase peak at the H$_i$ resonance due to the vicinal H6–H7 coupling in the olefin functional group. This "zero-quantum" artifact comes from

1D TRANSIENT NOE USING DPFGSE   329

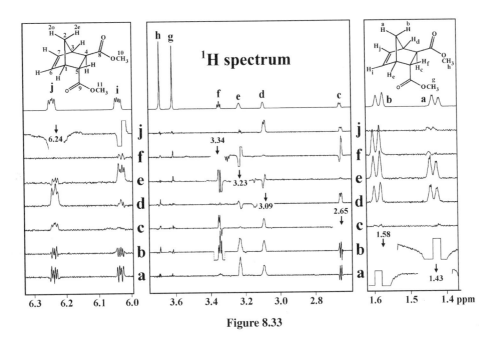

Figure 8.33

coherence transfer via the intermediate ZQC state:

$$I_z \xrightarrow{90°} I_x \xrightarrow{J\ evol.} 2I_yS_z \xrightarrow{90°} 2I_yS_y(DQ/ZQ) \xrightarrow{90°} 2S_yI_z$$

where **I** represents the selected proton and **S** is a proton J-coupled to **I**. A gradient can be used during the mixing time to kill the DQC portion (coherence order $p = 2$) of $2I_yS_y$, but the ZQC part ($p = 0$) is insensitive to gradients and contributes to the final antiphase state. These ZQ artifacts are common in 2D NOE ("NOESY") experiments as well. Because INEPT transfer is very efficient and NOE transfer occurs only to the extent of a few percent, these J-coupling artifacts appear very strong next to the NOE peaks (in Fig. 8.33 they are cut off to avoid messing up the stack of spectra).

Looking at the right-hand side of Figure 8.33, we see that selection of either H_d or H_e gives equally strong NOEs to H_a and H_b, the geminal pair at C2. This identifies H_d and H_e as the bridgehead positions H1 and H3. In contrast, selection of H_f gives an NOE to H_b only and selection of H_c does not give an NOE to either H_a or H_b. Looking at the structure, we see that the C2 proton that points toward the ester side (H2e) is close to H5 ("up") and farther from H4 ("down"). Thus, we can assign H_f as the H5 proton that points "up", toward H2e (H_b), and H_c as the H4 proton that points "down", away from H2e. Remember that H_a was assigned by chemical shift arguments to the C2 proton (H2o) that lies over the olefin and away from H4 and H5.

Although both H_d and H_e are close to the bridgehead protons H_a and H_b, how can we tell which one corresponds to H1 and which to H3 in the structure (Fig. 8.33, upper left)? Note that selection of H_e gives a strong ZQ artifact (antiphase peak) at H_f, and the reverse is also true. This places H_e in a vicinal relationship to H_f, so we can assign it to H1. The other bridgehead proton, H_d, can then be assigned to H3. It is interesting that no such coupling is observed between H_d and H_c (no antiphase peak at H_d when H_c is selected, nor *vice versa*).

Instead, we see a strong mutual NOE. Apparently, the H3–C3–C4–H4 dihedral angle is very near the minimum in the Karplus curve (90° angle) so that the vicinal coupling constant is very small.

Now we can assign the olefinic pair H_i and H_j and the methyl singlets H_g and H_h. The bridgehead proton H3 (H_d) shows a strong NOE to H_j and not to H_i, so we can assign H_j to the olefinic proton H7, next to H3. The other bridgehead proton H1 (H_e) gives a strong NOE to H_i but not to H_j. Selection of H_j gives NOEs to H_d (H3) and H_c (H4) on the same side of the molecule as H7, as well as to H_a (H2o). Likewise, H_c (H4) "talks" to H_j (H7), H_f (H5), and H_d (H3). Finally, weak NOEs can be used to assign the H_g/H_h pair. Selection of H_b (H2e) or H_d (H3) gives a very weak NOE to H_h, but only a subtraction artifact at the H_g chemical shift, and selection of H_h gives a very weak NOE to H_j (not shown). This identifies H_h as H10. Likewise, selection of H_e (H1) or H_f (H5) "lights up" the H_g singlet and not the H_h singlet, so we can assign H_g to H11. This completes the assignments, which are shown on the structure at the upper right-hand side in Figure 8.33.

In contrast to the NOE evidence, the J couplings are rather confusing. H_f appears as a triplet, coupled to H_e and H_c, but H_c appears as a broad doublet, with resolved coupling only to H_f. The absence of a vicinal H_c–H_d coupling was already noted above. Although we see mostly NOE to H_i and H_j when selecting H_a (H2o), selection of H_b (H2e) gives ZQ artifacts to both olefinic protons, suggesting a long-range J coupling ("W" coupling). Likewise, four-bond "W" couplings can be deduced from ZQ artifacts between H_d and H_e, H_d and H_i, H_e and H_j and between H_f and H_i.

8.9.7 NOE Buildup Curve for Sucrose

A study of the NOE intensity as a function of mixing time is called an NOE buildup experiment. The NOE should build up initially at a constant rate (Figs. 8.31 and 8.32) and then level off and eventually decrease to zero as the mixing time is increased. In Chapter 5, we saw the effect of steady-state irradiation of the fructose-1 (CH$_2$OH singlet at 3.62 ppm) resonance of sucrose (Fig. 5.30): strong NOEs are observed to H-g1 (5.36 ppm) and to H-f3 (4.15 ppm). Figure 8.34 shows the NOE buildup curve for *selective transient NOE* (DPFGSE) of sucrose, selecting the fructose-1 resonance. The upper curve (Δ) shows the

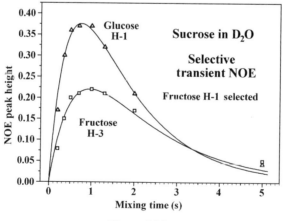

Figure 8.34

peak height of the H-g1 peak, and the lower curve (□) shows the peak height of the H-f3 peak as a function of mixing time. The solid curves are simulations of the transient NOE. An accurate measure of the linear buildup rate in the initial phase (proportional to $1/r^6$) would require a number of data points in the 0–300 ms range of τ_m. The maximum NOE would be obtained at about 0.7 s mixing for H-g1 and 0.9 s for H-f3. Usually, we set the mixing time of an NOE experiment based on the size of the molecule: longer for smaller molecules and shorter for larger molecules. Because the NOE is a relaxation experiment, the T_1 value can give us a rough estimate of the optimal mixing time. The T_1 values for sucrose can be estimated from the ^{1}H inversion-recovery experiment (Fig. 5.17), which gives $T_1 = 120$ ms for H-f1, 280 ms for H-g1, and 1.08 s for H-f3. The range of T_1 values is very large for these three protons, but the order of magnitude (0.1–1 s) is not far off for setting the mixing time of the transient NOE experiment (Fig. 8.34). As a first guess, use an NOE mixing time of 350 ms for small molecules (200–400 Da), 200 ms for "medium-sized" organic molecules (400–1000 Da), and 100 ms for "large" molecules (1–10 kDa).

8.9.8 A Demonstration of Selectivity: Cholesterol

To show the selectivity of the DPFGSE-NOE experiment, consider the H4$_{ax}$ and H4$_{eq}$ protons of cholesterol (Fig. 8.35). Because C4 is flanked on both sides by downfield-shifting functional groups (C3–OH and C5=C6), the two H4 protons are pulled downfield to 2.2–2.4 ppm, away from the "pack" of overlapped resonances in the ^{1}H spectrum. At 500 MHz, the H4$_{ax}$ and H4$_{eq}$ protons are just barely resolved from each other, with H4$_{eq}$ (downfield) appearing as a "doublet" (plus two small couplings) and H4$_{ax}$ (upfield) appearing as a "triplet" (actually a double doublet plus three small, nearly equal long-range couplings to H6, H7$_{eq}$, and H7$_{ax}$). If we focus on the large couplings only, we see that H4$_{eq}$ has only one: the geminal coupling to H4$_{ax}$. This gives it the "doublet" appearance. H4$_{ax}$ has two large couplings: the geminal coupling back to H4$_{eq}$ and the axial–axial coupling to H3. This gives it the "triplet" appearance. The "doublet" and "triplet" lean strongly toward each other due to their strong coupling (Δv in hertz similar in magnitude to J). H4$_{ax}$ is close in space to the angular methyl group (C19) at the A–B ring juncture, and H4$_{eq}$ is close to the H6 olefinic proton in the equatorial plane. So if we could selectively excite mostly H4$_{ax}$, we would expect a strong NOE (H4$_{ax}$ to H19 = 2.39 Å) to the H19 methyl peak (singlet)

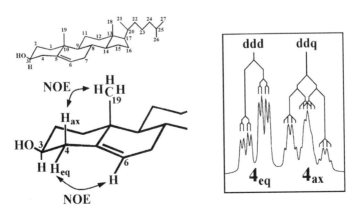

Figure 8.35

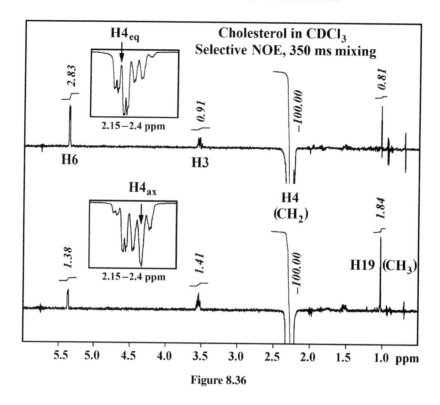

Figure 8.36

and a weak NOE (H4$_{ax}$ to H6 = 3.32 Å) to the H6 peak. Likewise, selective excitation of mostly H4$_{eq}$ would give a strong NOE to the H6 peak (2.32 Å) and a weak NOE to the H19 methyl peak (3.89 Å).

> These distances come from an X-ray crystal structure of cholesterol hydrate, with hydrogen positions added. Because cholesterol is a rigid molecule with the four rings locked in place by the *trans* ring junctures, energy minimized model structures also give fairly accurate distances. The distance to H19 is measured to the nearest of the three hydrogens in the C19 methyl group. The NOE intensity will actually be the sum of the NOEs from the three protons of the methyl group, but because of the 1/r^6 dependence it will be dominated by the closest proton.

Figure 8.36 shows the results of these two experiments, using a mixing time of 350 ms. In the insets, we see the inverted H4 peaks: in the top spectrum, we have excited mostly the H4$_{eq}$ peaks ("doublet") along with the downfield part of the H4$_{ax}$ peak, and in the bottom spectrum we see mostly the H4$_{ax}$ peaks ("triplet") with some intensity due to the upfield half of the H4$_{eq}$ peak ("doublet"). This is not bad for selectivity considering how close the two chemical shifts are to each other. In the rest of the spectrum we see the NOE peaks, integrated relative to an integral value of −100 for the inverted H4 peak. Selecting H4$_{eq}$ gives NOEs of 2.83% for H6 and 0.27% for H19-Me, whereas selecting H4$_{ax}$ gives values of 1.38% for H6 and 0.88% for H19-Me (integral values are divided by the number of protons represented by each peak). These numbers are strictly qualitative, but they are consistent with our expectations based on the structure. They also confirm our assignments of H4$_{ax}$ and H4$_{eq}$, and allow us to assign which of the two CH$_3$ singlet peaks is H19-Me.

8.9.9 Details, Details, Details

The DPFGSE-NOE experiment is a very elegant demonstration of excitation sculpting using the combined power of shaped pulses and gradients. The DPFGSE allows us to destroy all magnetization on the other, nonselected spin, so that any signal that we observe in the spectrum *has to* derive from NOE transfer from the selected spin. In the NOE difference experiment, our result is the difference of two very similar numbers: 103% minus 100%, for example. In the transient NOE experiment using DPFGSE, we see only the 3%. This is however, only a first approximation and now it is time to face up to the nitty-gritty details.

First of all, although it is true that at the end of the DPFGSE sequence there is no overall net magnetization on the nonselected spins, if we look at the sample in detail, we see that the net magnetization alternates between $\mathbf{I}_z$ and $-\mathbf{I}_z$ as we move up in the tube (Fig. 8.27, bottom). At this moment they all cancel perfectly, but during the mixing time the $-\mathbf{I}_z$ levels begin to recover whereas the $\mathbf{I}_z$ levels remain at equilibrium. They no longer cancel and we begin to see net magnetization ($\mathbf{I}_z$) overall for the nonselected spins. So at the end of the mixing time, our 90° "read" pulse will rotate this recovered z magnetization into the x–y plane, producing peaks in the spectrum that have nothing to do with the NOE. Furthermore, although the selected peak is perturbed radically by inversion ($\mathbf{I}_z \rightarrow -\mathbf{I}_z$), the nonselected peaks are perturbed half as much on average by inversion for half of the levels. So we would expect the levels that were inverted to generate NOEs to their nearest neighbors, although the unperturbed levels would not generate any NOEs. This would create a whole bunch of signals in the final spectrum, coming from the nonselected spins (one-half as strong) and from the selected spins.

The solution to both problems—recovery of z magnetization of the nonselected spins and NOEs developed from these same spins—is to phase-cycle ($-x, x$) the 90° pulse at the end of the DPFGSE, the one that flips the selected spin's magnetization down to the $-z$ axis. If we reverse the phase of this pulse, it flips the selected spin's magnetization *up*, back to $+z$. There will be no NOE from the selected spin, but the nonselected spins will experience exactly the same perturbations, with the levels that were previously inverted now at equilibrium and the levels that were previously at equilibrium now inverted. Overall, the same artifact signals (recovery and NOE) will be generated from the nonselected spins. If we alternate the phase of this pulse and alternate the receiver phase with it (add, subtract, add, subtract, ...), we are essentially running a control experiment on every other scan and subtracting out any signals that come from nonselected spins, either from recovery or from NOE. The nonselected spins behave the same either way (half are inverted and half are unaffected), so any signals they give directly will subtract out. Thus, the only signals we see will be NOE signals deriving from the selected spin. Are we fibbing then when we say it is not a difference experiment? Technically, yes, but the signals we are subtracting out are of similar magnitude (actually smaller) than the ones we end up with, so the errors of subtraction are negligibly small. There are a lot of details involved in optimizing this experiment, but the results are absolutely stunning in terms of clarity and lack of artifacts. This is important as NOEs are generally weak and can be ambiguous if the experiment is not really clean. Anyone still doing the old steady-state difference NOE experiment is living in the dark ages!

8.10 THE SPIN LOCK

A spin lock is a relatively long (1–400 ms), low-power (usually 12–33% of the B_1 amplitude of a hard pulse) radio frequency pulse applied on the same axis as the desired sample

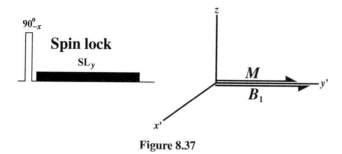

Figure 8.37

magnetization. It can be either continuous wave or a long series ("train") of pulses of varying length and phase. It can be used to remove artifacts, to "grab and drag" the sample magnetization, and to get transfer of magnetization (either by NOE or by J couplings). First, we will look at the continuous-wave spin lock.

8.10.1 Locking the Sample Magnetization

In the rotating frame of reference for an on-resonance peak, the B_o field is exactly canceled by a fictitious field created by the rotation of the axes, so that for nuclei that are on-resonance the only field present is the B_1 field during the spin lock ($B_{eff} = B_1$). If we place the sample magnetization on the y' axis of the rotating frame with a 90° hard pulse (phase $-x$), the spin lock can be placed on the y' axis (phase y). While the spin lock is on, the sample magnetization is "locked" on the y axis and will not undergo precession, as the only field present is the B_1 field and the sample magnetization is on the same axis as the B_1 field (Fig. 8.37).

8.10.2 Fate of Magnetization Perpendicular to the Spin Lock: Purge Pulses

If instead we start by putting the sample magnetization on the x axis (90° hard pulse on y) and then apply the spin lock on the y axis, the sample magnetization will rotate around the spin lock axis (y axis) at the rate $v_1 = \gamma B_1/2\pi$. For typical spin-lock power levels this rate is between 3000 and 9000 Hz. The rotation occurs in the x–z plane (from x to $-z$ to $-x$ to z, and back to x). As the vector rotates, the individual spins that contribute to it began to "dephase" because different parts of the sample experience different B_1 amplitudes and the sample magnetization from each region rotates at a slightly different rate. Although the B_o field is carefully shimmed to be homogeneous to *parts per billion* (10^{-9}) variation throughout the sample, the B_1 field is quite inhomogeneous and varies significantly in amplitude in different parts of the sample. So this "fanning out" occurs rapidly and all components of the net magnetization that are not on the spin-lock axis rapidly decay to zero. We can see this "fanning out" effect by doing a pulse calibration and continuing far beyond the 360° point (Fig. 8.38). This is a measure of B_1 field homogeneity, usually expressed as the ratio of signal intensity for an 810° pulse to the intensity for a 90° pulse. Even though probe designers strive for the best B_1 homogeneity possible, you can see that after 100 or 200 cycles there will be no more signal left. If the pulse calibration data are fit to an exponential decay (Fig. 8.38, inset) we get a half-life of 116 μs for the magnetization rotating around the spin-lock axis. This means that after a 1 ms spin lock at this power level (high power), the net magnetization has rotated 31.25 times around B_1 (1000 μs/(4 × 8 μs)) and has been cut in half 8.6 times (1000 μs/116 μs). After eight half-lives the net

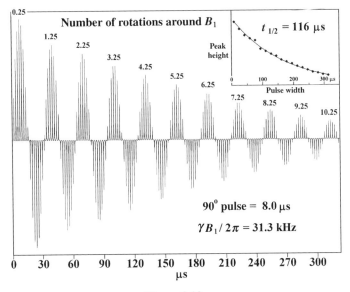

Figure 8.38

magnetization vector is 1/256 of its original magnitude, and after nine it is 1/512. A "trim pulse" or "purge pulse" is a short (1–2 ms) spin lock with high power (~25 kHz, the same as a hard pulse) placed on the axis of the desired magnetization to destroy all magnetization that is not on that axis. A simple application would be the INEPT sequence, where we allow in-phase ^{1}H magnetization to undergo J-coupling evolution into antiphase: $\mathbf{I}_x \to 2\mathbf{I}_y\mathbf{S}_z$. This conversion is perfect only if the delay is exactly equal to $1/(2J)$. Because we just set the delay to our best guess, there will be residual in-phase signals at the end of the delay:

$$\mathbf{I}_x \to \mathbf{I}_x \cos(\pi J\tau) + 2\mathbf{I}_y\mathbf{S}_z \sin(\pi J\tau) \qquad \pi J\tau \neq \pi/2$$

At this point, a purge spin lock on the y axis would preserve the antiphase term $2\mathbf{I}_y\mathbf{S}_z$ and destroy the in-phase term $\mathbf{I}_x$. We then proceed to the coherence transfer step (simultaneous 90° pulses on ^{13}C and ^{1}H) with pure antiphase ^{1}H coherence.

8.10.3 Effect of the Spin Lock on Locked Magnetization

The locked magnetization is parallel to the only magnetic field that is present in the rotating frame: the B_1 field. This is analogous to z magnetization in the B_0 field in the laboratory frame during a delay, when there is no B_1 field. If we follow this analogy further, we see that the spin lock axis is like the z axis and the B_1 field is like the B_0 field, except very much weaker (e.g., 8 kHz vs. 600 MHz, or 75,000 times smaller!). Thus, we can think of the spin lock as a way of temporarily "turning down" the B_0 field to a vastly lower value. This has two effects: First, the tumbling rates required to stimulate SQ and DQ relaxation are very low (on the order of ν_1 and $2\nu_1$ instead of ν_0 and $2\nu_0$), and second, the "chemical shift" differences ($\Delta\nu_1$ instead of $\Delta\nu_0$) are extremely small when compared to the J values. The first effect means that the NOE in the spin-lock world (the "rotating-frame" NOE) will always be dominated by DQ cross-relaxation, which leads to negative NOEs (NOE enhancement), regardless of the size of the molecule. In other words, "all molecules are small molecules" in the spin lock. This is because the DQ and ZQ frequencies are now accessible to even the slowly tumbling biological molecules through dipole–dipole interactions. This is the basis of the ROESY experiment, which transfers magnetization along the spin-lock axis via through-space (NOE) effects, while making all molecules behave like small molecules in terms of their NOE behavior. The second effect means that the chemical shift differences between different protons within a molecule are almost completely eliminated, leading to "strong coupling" and "virtual coupling" within a spin system. In other words, the protons within a spin system behave like each one is J coupled to every other one, even if there is no direct J coupling between them. This is the basis of the TOCSY experiment, which transfers magnetization along the spin-lock axis from one proton to all other protons in the same spin system.

8.10.4 Off-Resonance Effects

So far we have assumed that the spin-locked nucleus is on-resonance or at the center of the spectral window ($\nu_o = \nu_r$). If the nucleus is off-resonance, the effective field in the rotating frame, $\boldsymbol{B}_{\text{eff}}$, is the vector sum of the $\boldsymbol{B}_1$ field vector along the axis of the spin lock (e.g., y') and the residual field along the z axis ($B_{\text{res}} = 2\pi\,(\nu_o - \nu_r)/\gamma = B_o - 2\pi\nu_r/\gamma$). This means that the spin-lock axis tilts out of the x'–y' plane by an angle that increases as ν_o moves farther away from the reference frequency, or as the B_1 field strength is decreased. Each proton in the molecule thus has a different spin-lock axis and must be considered separately. The length of the $\boldsymbol{B}_{\text{eff}}$ vector is greater than B_1 due to the vector sum: B_{eff} (the magnitude of the $\boldsymbol{B}_{\text{eff}}$ vector) is equal to $(B_1^2 + B_{\text{res}}^2)^{1/2}$. The rate of precession about the spin lock axis is $\nu_1 = \gamma B_{\text{eff}}/2\pi$, so the rotation of any magnetization that is not on the spin-lock axis around it becomes faster as we move off-resonance.

What does this mean for the effect of the spin lock on sample magnetization? If the sample magnetization starts on the y' axis, for example, the tilted spin-lock axis will destroy the component that is perpendicular to the spin-lock axis and retain the component that is on the spin-lock axis. This preserved component is "locked" because it is on the axis of the effective field and has no "reason" to precess around the z axis. So even if the spin is off-resonance, its magnetization does not precess around the z axis during the spin-lock period. Instead, the component that is *not* on the tilted spin-lock axis precesses around the spin-lock axis until it is destroyed by B_1 inhomogeneity, and the component that is *on* the spin-lock axis is retained.

8.10.5 Moving the Spin-Lock Axis

What would happen to the spin-locked sample magnetization if we moved the spin-lock axis? Would the spins follow, as the name "spin lock" implies, or would the magnetization vector be left behind, rotating around the spin-lock axis until the B_1 field inhomogeneity "spins it out of existence"? The answer is: It depends on how fast we move the spin-lock axis. If we move it slowly enough (the "adiabatic" condition), the sample magnetization vector will get dragged along with it without any loss of intensity. How do we move the spin-lock axis? We can change the pulse frequency, moving it away from the resonance so that, in the rotating frame defined by the pulse frequency, the spins are off-resonance. This makes the $\boldsymbol{B}_{\text{eff}}$ vector tilt out of the x'–y' plane. By a combination of adjusting the pulse amplitude (B_1) and the resonance offset ($\nu_o - \nu_r$), we can put the $\boldsymbol{B}_{\text{eff}}$ field vector anywhere

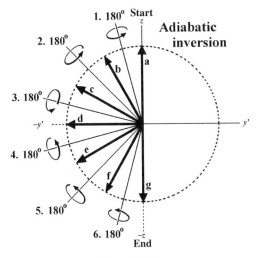

Figure 8.39

we want and give it any magnitude we want. Suppose, for example, that we start with the B_{eff} field vector tilted 15° away from the $+z$ axis in the y'–z plane and turn it on long enough to get a 180° rotation of the sample net magnetization, which starts on $+z$ (a, Fig. 8.39). The magnetization will rotate to a point 30° away from the $+z$ axis (b, Fig. 8.39). Now we move the B_{eff} field vector to an angle of 45° with the $+z$ axis and apply another 180° rotation. The sample magnetization will move around the B_{eff} vector to a point 60° away from the $+z$ axis (c). We could continue this process, placing the B_{eff} vector at angles of 75°, 105°, 135°, and 165° from the $+z$ axis until the last 180° rotation moves the sample magnetization down to the $-z$ axis (g). We have inverted the sample magnetization by picking it up at $+z$ and "shepherding" it around in a series of steps down to $-z$, always keeping the B_{eff} vector close to the M vector. Imagine now that we decrease the increment of angle of the B_{eff} vector and do the inversion in many more steps. Eventually, we would have a continuous RF irradiation with the B_{eff} vector moving smoothly from $+z$ to $-z$. What we have is a spin lock that "grabs" the sample net magnetization at $+z$ and "drags" it down to $-z$ physically, without doing any finite rotations. This can be accomplished by constructing a shaped pulse with a phase ramp that moves rapidly at first (large frequency shift in one direction) and slows down and stops at the center of the pulse, and then reversing the direction of the phase ramp, speeding it up continuously until the end of the pulse. The effective frequency of the pulse starts way downfield of the Larmor frequency, moves upfield until it equals the Larmor frequency at the center of the pulse, and then moves off upfield to a point far upfield of the Larmor frequency. At the same time, the magnitude of the B_1 field is adjusted to be small at the start and end, to allow the B_{res} field (on $+z$ or $-z$) to dominate and tilt the B_{eff} vector up close to $+z$ (start) or $-z$ (end), and maximal at the center of the pulse to give a strong spin lock with no tilt out of the x–y plane. The classical adiabatic-inversion shaped pulsed is called "WURST" and there are many, many variations with equally cute names. Why go to all this trouble? If we consider a series of resonances widely spaced throughout a wide spectral window (like a ^{13}C spectrum on a high-field spectrometer), as the effective frequency of the pulse "sweeps" from far downfield to far upfield, we will see the spins invert one at a time because the spin-lock axis depends on how far away the pulse effective frequency is from the Larmor frequency. As the pulse frequency passes through each peak in the spectrum, these spins are spin locked in the x–y plane halfway through their journey from $+z$ to $-z$. It really does not matter where the spins are in the wide spectral window; they will be picked up on $+z$ and dragged down to $-z$ when their turn comes, like dominos falling in a row. If we sweep from downfield to upfield, the only difference will be that the spins on the downfield edge of the spectrum will be inverted a little earlier than the spins on the upfield edge of the spectrum. We saw an impressive example of this for inversion of the $^{13}CH_3I$ ^{13}C signal using a "cawurst" adiabatic inversion pulse called "ad180" (Fig. 8.4, bottom). The bandwidth of this shaped pulse is far superior to hard pulses or hard pulse sandwiches. Decoupling schemes based on WURST are very effective, covering the wide bandwidth needed for ^{13}C decoupling with much lower power than rectangular pulse trains like waltz-16. The spin-lock field is just swept back and forth across the spectral window, inverting the spins over and over again as the B_{eff} vector shuttles from $+z$ to $-z$ and back to $+z$. Rapid and continuous inversion gives good decoupling because the coupling partner (e.g., ^1H) sees a spin (e.g., ^{13}C) that is moving rapidly back and forth between the α state and the β state, blurring the difference in magnetic field experienced by the coupling partner into a single, constant field on the NMR time scale.

8.11 SELECTIVE 1D ROESY AND 1D TOCSY

The use of a relatively long, low-power spin lock to effect transfer of magnetization is the basis of two modern two-dimensional experiments: ROESY for through-space transfer and TOCSY for through-bond transfer. We will deal with these 2D experiments in Chapters 9 (TOCSY) and 10 (ROESY), but for now our interest is in using the spin lock in a selective 1D experiment. Using the DPFGSE, we will put the magnetization of one selected resonance in the x'–y' plane, apply a spin lock on the same axis for a long enough time to get magnetization transfer to other spins, and then record the FID. First, we need to understand what goes on during the long spin-lock period and how we can get transfer of magnetization. For the first time, a thorough theoretical understanding of the process is out of reach for the level of this book and we will have to resort to giving a "feel" for the process. By using analogy and looking at it from a number of points of view, we will try to give the experimental result some plausibility and make sense of this remarkable phenomenon.

8.11.1 ROESY Mixing

The transfer of magnetization within a spin lock by NOE (through-space) interaction is called ROESY mixing. We are already familiar with NOE transfer of magnetization; the only difference is that ROESY transfer happens on the spin-lock axis ($\mathbf{I}_y^a \rightarrow -\mathbf{I}_y^b$) viewed in the rotating frame, rather than in a simple delay ($\mathbf{I}_z^a \rightarrow -\mathbf{I}_z^b$) viewed in the laboratory frame. In either case, transfer occurs between protons that are close in space (<5 Å), with the "disequilibrium" of one proton gradually transferring to create disequilibrium in the opposite sense on a nearby proton. In fact, the same pair of linked differential equations in Section 8.9.4 that governs z-magnetization transfer also applies to transfer of spin-locked magnetization in the x'–y' plane, except that the "disequilibrium" is defined as $\Delta M_y^a = M_y^a$. This is because the equilibrium state for magnetization in the x'–y' plane is always zero (T_2 relaxation), so any nonzero spin-locked net magnetization is "out of equilibrium" and will lead to NOE transfer to nearby protons. The rate of cross-relaxation in the spin lock is more closely related to T_2 than to T_1. In fact, the self-relaxation of spin-locked magnetization (to zero) is governed by the time constant $T_{1\rho}$, where the Greek letter ρ (rho) refers to the rotating frame. When the spins are on-resonance, $T_{1\rho} = T_2$ because the spin-lock axis is in the $x'-y'$ plane. As the spin-lock axis tilts out of the $x-y$ plane for off-resonance spins, T_1 begins to contribute to $T_{1\rho}$. As with NOE transfer on the z axis, the ROE is an inefficient transfer, with ROEs of a few percent typical for nearby protons. The ROE builds up twice as fast as the NOE, so ROESY mixing times are usually set at half the value of optimal NOE mixing times.

8.11.2 Spin Systems and TOCSY Magnetization Transfer

Transfer of spin-locked magnetization by J-coupling interaction (through bonds) is called TOCSY mixing. We have seen in the INEPT experiment how a single transfer can be achieved from one spin to its J-coupling partner. TOCSY mixing can achieve multiple transfers through J couplings. For example, in a linear string of coupled protons $CH_a-CH_b-CH_c-CH_d-CH_e$, selective excitation of the H_a resonance followed by the INEPT sequence $1/(2J) - 90°$ would lead to transfer of coherence (antiphase to antiphase) from H_a to H_b. In this homonuclear INEPT experiment, the 90° pulse hits both H_a and H_b equally, so we can call it a "simultaneous 90° pulse" on both spins, meeting the requirement for INEPT transfer. In the spectrum, we would see the peak for the selected resonance, H_a, as well as the peak for H_b. We could go through the spectrum, selecting each of the peaks in turn (H_a, H_b, H_c, etc.) and observing which peaks are J-coupled to the selected peak. But if we use a TOCSY spin lock instead of the INEPT sequence for transfer of magnetization, we will see *multiple jumps* through J couplings. Magnetization from the selected resonance (H_a) transfers to H_b (in-phase to in-phase transfer) via the vicinal coupling J_{ab}, and then some of the H_b magnetization transfers to H_c via J_{bc}, and so on. During the course of the spin lock (typical mixing time 70 ms), the magnetization gradually "diffuses" or "smears out" through the string of protons until eventually we have some magnetization on all of the spins in the "spin system." If there are no other J couplings to other parts of the molecule, magnetization cannot escape from this system. At the end of the mixing period, we record the FID and in the spectrum we will see the H_a, H_b, H_c, H_d, and H_e peaks in decreasing intensity according to their order along the carbon backbone of the molecule. The rest of the ^1H spectrum, representing other parts of the molecule, will be absent. We have a spectrum of this one "spin system" alone.

A spin system is a group of spins (usually ^1H) in a molecule that are connected together by J couplings. Specifically, each member of the group has at least one J coupling with another member of the group. For example, n-propyl benzoate (Fig. 8.40) has two proton spin systems: the CH_3 and two CH_2 groups form one spin system and the five aromatic protons form the other. As there is no J coupling between the CH_3 or CH_2 protons and any of

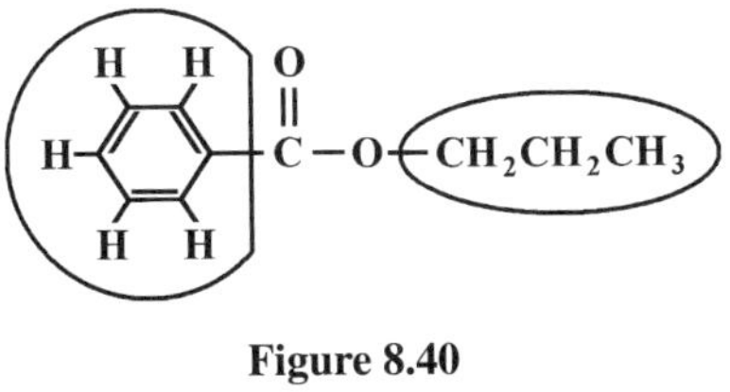

Figure 8.40

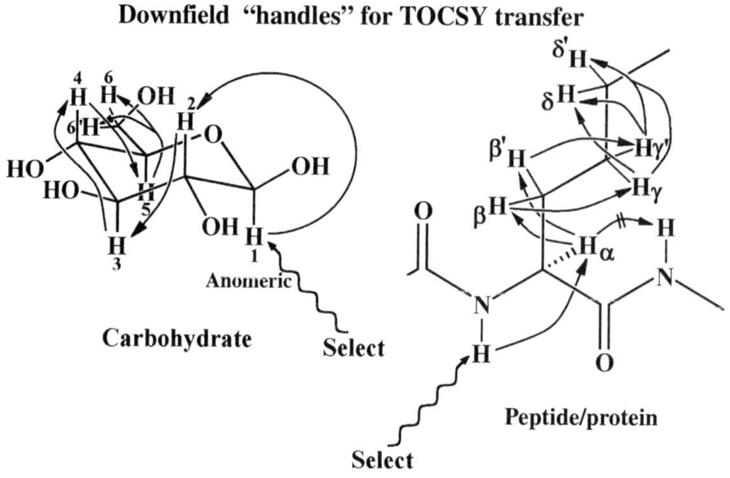

Downfield "handles" for TOCSY transfer

Figure 8.41

the aromatic protons, these are two distinct spin systems. TOCSY is a technique that spreads NMR magnetization from any one member of a spin system to all the other members of a spin system. Thus, if you selectively excite one member of the spin system (e.g., the CH$_3$ protons), the TOCSY mixing sequence will transfer that magnetization first to its J-coupled partners (the next CH$_2$ group) and then to *their* J-coupled partners (the CH$_2$ group next to oxygen) until all of the members of the spin system are excited in the same way. If the FID is recorded at this time, a spectrum will be observed that contains only the peaks due to the spin system that was excited (in this case, the CH$_3$–CH$_2$–CH$_2$ group). Peaks from other spin systems in the molecule (the aromatic protons) would not appear in this selective spectrum.

For biological molecules, the definition of a spin system is simple: Each residue of a peptide, protein, nucleic acid or oligosaccharide is a separate spin system. There is no ^{1}H–^{1}H coupling through a peptide bond, phosphate ester, or glycosidic linkage, so TOCSY transfer will not occur between residues. In peptides and proteins, the amide NH is the beginning of a continuous series of J couplings to the H$_\alpha$ proton, to the H$_\beta$ and H$_{\beta'}$ protons, to the H$_\gamma$ and H$_{\gamma'}$ protons, to the H$_\delta$ and H$_{\delta'}$ protons, and so on, of the side chain (Fig. 8.41, right). The aromatic protons in the side chains of aromatic amino acids are separate spin systems from the H$_N$ and aliphatic (H$_\alpha$, H$_\beta$, H$_{\beta'}$) protons. In oligosaccharides, the anomeric proton (proton bound to carbon with two bonds to oxygen) is the start of a spin system that includes the whole monosaccharide unit: C^{1}H–C^{2}H–C^{3}H–C^{4}H–C^{5}H–C^{6}H$_2$ for pyranose sugars (Fig. 8.41, left). For selective 1D TOCSY experiments on relatively simple peptides and oligosaccharides, we can access these spin systems by selective excitation of the downfield "handles": amide NH protons (7–11 ppm) for peptides, and anomeric protons (5–6 ppm) for sugars. Because there is only one of these protons per residue and they are in a different chemical shift range from all the other protons, it is common to find them well resolved as single peaks.

8.11.3 TOCSY Mixing

How do we get these multiple jumps of magnetization within a spin system? We could use a series of INEPT transfers: $1/(2J)$–$90°$–$1/(2J)$–$90°$–$1/(2J)$–$90°\cdots$, but this would be a very long mixing sequence as for a typical ^{1}H–^{1}H J coupling (7 Hz) the delay $1/(2J)$ is 71 ms. We could shorten the delays, and if we continue this process, we eventually end up with a continuous string of RF pulses. This is essentially what the TOCSY mixing sequence is: A continuous string of pulses, but with varying pulse width and phase in a precise repeating program.

Like ROESY mixing, the TOCSY mixing sequence is a spin lock. The choice of NOE or J-coupling transfer depends on various practical aspects of the spin lock: a low-power ($\nu_1 = \gamma B_1/2\pi \sim 3000$ Hz) continuous-wave spin lock favors magnetization transfer via NOE (through-space) interactions, whereas a higher-power (~ 8000 Hz) *pulsed* spin lock with a specific optimized sequence of pulse durations and phases favors magnetization transfer via J-coupling (through-bond) interactions within an entire spin system. For example, the MLEV-17 TOCSY mixing sequence consists of a repeating pattern of pulses based on the building blocks $90°_x$–$180°_y$–$90°_x$ (A) and $90°_{-x}$–$180°_{-y}$–$90°_{-x}$ (B) repeated in the series ABBA–BBAA–BAAB–AABB, followed by a $60°_y$ pulse. Note that A and B are just our sandwich pulses designed to give improved bandwidth for inversion (Fig. 8.4, center). This "train" of 17 pulses is then repeated as many times as necessary to complete the mixing time of 35–75 ms. A more efficient sequence than MLEV-17 is the DIPSI family of mixing sequences. The basic repeating unit of DIPSI is

$$320°_x - 420°_{-x} - 290°_x - 285°_{-x} - 30°_x - 245°_{-x} - 375°_x - 265°_{-x} - 370°_x.$$

There is no obvious theoretical basis for this sequence; it was arrived at by computer simulation and found to give the most efficient through-bond transfer of magnetization with the least sample heating. The goal of these complex strings of pulses is to create an environment in which the chemical shift differences of the protons within a spin system disappear completely, leaving only the J-coupling interactions. This ultimate extreme of "strong coupling" and "virtual coupling" ($\Delta \nu \ll J$) gives rise to magnetization transfer to all members of the spin system. Computer simulation has been used to improve these mixing sequences even further with the goal of a perfect "isotropic mixing" (J coupling only) sequence.

8.11.4 The Hartmann–Hahn Match

If we used a continuous-wave spin lock (ROESY mixing scheme) for TOCSY mixing, the only way efficient transfer could occur via J couplings within a spin system would be if the two protons have opposite resonance offsets ($\nu_o - \nu_r$), where ν_r is the carrier or reference frequency. This is a special case of the Hartmann–Hahn match:

$$\gamma_A B_{\text{eff}}(A) = \gamma_B B_{\text{eff}}(B)$$

where B_{eff} is the magnitude of the effective field vector $\boldsymbol{B}_{\text{eff}}$ in the rotating frame ($B_{\text{eff}} = [B_1^2 + B_{\text{res}}^2]^{1/2}$, $B_{\text{res}} = 2\pi(\nu_r - \nu_o)/\gamma$) and A and B are any two spins, for example, ^{1}H and ^{13}C. This was originally developed by Hartmann and Hahn for heteronuclear coherence transfer in the solid state, with simultaneous continuous-wave irradiation at both the ^{1}H frequency and the ^{13}C frequency, with the intensity ratio tuned precisely to give the "match." When TOCSY was originally developed for ^{1}H–^{1}H coherence transfer, it was called "HOHAHA" or homonuclear Hartmann–Hahn. When spins A and B are both protons, the Hartmann–Hahn match requires that $B_{\text{eff}}(A) = B_{\text{eff}}(B)$, as $\gamma_A = \gamma_B$. Because B_1

is the same for A and B (they are both protons), this means that both A and B have the same resonance offset ($B_{res}(A) = B_{res}(B)$) or opposite resonance offsets ($B_{res}(A) = -B_{res}(B)$) as B_{res} depends on resonance offset ($B_{res} = 2\pi(\nu_r - \nu_o)/\gamma$). The cw spin lock is not effective for TOCSY transfer because transfer would only occur between pairs of resonances equally disposed on opposite sides of the center of the spectral window ($\nu_A - \nu_r = -(\nu_B - \nu_r)$). This can produce artifacts in 2D-ROESY spectra because transfer by J coupling (TOCSY transfer) does occur in this special case. The purpose of the complex TOCSY pulse trains at higher power is to generalize the TOCSY transfer to include *any* pair of resonance offsets within the spectral window, including where the Hartmann–Hahn condition is *not* met.

In both ROESY mixing and TOCSY mixing, the transfer is from in-phase net magnetization on the spin-lock axis to in-phase net magnetization on the spin-lock axis; for example, $I_y^a \to I_y^b$ (TOCSY) or $I_y^a \to -I_y^b$ (ROESY). This is analogous to the NOESY transfer, $I_z^a \to -I_z^b$ for small molecules and in contrast to the INEPT transfer $2I_x^a I_z^b \to 2I_x^b I_z^a$, which is antiphase to antiphase.

8.11.5 Strong Coupling, Virtual Coupling, and TOCSY Transfer

TOCSY mixing is achieved by creating an environment where chemical shift differences are reduced to zero or near zero and J-coupling interactions are preserved. For liquid samples, these two interactions define the energy of a pair of spins in NMR: the interaction of each spin's magnet with the external magnetic field (the "Zeeman" energy giving rise to the α, β energy gap) and the interaction of the two spins' magnets with each other (J coupling). The first energy term is dependent on the strength of the external field and small, structure-dependent differences in this energy define the chemical shift. The second term is independent of field strength and so we measure it in units of hertz. If both spins are aligned in the same way with respect to the external field ($\alpha\alpha$ or $\beta\beta$ state), the energy is slightly lower, and if they are aligned opposite to each other ($\alpha\beta$ or $\beta\alpha$ state), the energy is slightly higher. The goal of the TOCSY mixing sequence is to eliminate the chemical shift (Zeeman) term and leave only the J-coupling interaction. This would be like turning off the B_0 field completely. We know that a simple homonuclear spin echo can eliminate the chemical shift evolution while allowing the J-coupling evolution, so it is not surprising that the right kind of spin lock can also do it, as the spin lock is similar to a very rapid series of spin echoes: $[\tau-180°-\tau]_n$. This ideal situation where the only interaction left is J coupling is called "isotropic mixing."

Consider an intermediate case where chemical shift differences are not eliminated completely, but are small compared to the J couplings. This is the so called "non-first order" or "strong coupling" situation, which is commonly encountered in low-field (e.g., 60 MHz) instruments but can show up at any field strength. In Chapter 2 (Section 2.8.1), we saw that even weak coupling to a group of protons that are strongly coupled to each other leads to "virtual coupling"—the appearance of coupling to *all* the members of the strongly coupled group, including those too far away to be directly J coupled. This sounds a little like TOCSY mixing: coherence transfer among a "spin system"—a group of protons connected by J couplings—as if each proton in the spin system were coupled to *all* of the other members of the group.

Now consider what happens during the ideal TOCSY mixing scheme. There are no chemical shift differences but the J couplings are still active, so *all* protons in a spin system are *extremely* strongly coupled, like the CH_2 groups in a long, straight-chain hydrocarbon.

Any one proton behaves as if it is coupled to all protons of the spin system as it is coupled to at least one member of the group and all the other protons are strongly coupled to each other. This way of understanding the TOCSY scheme comes closest to the actual theoretical explanation, so hopefully it will give you a feeling for what goes on in the TOCSY experiment.

8.12 SELECTIVE 1D TOCSY USING DPFGSE

A selective TOCSY experiment starts with putting the net magnetization of just one resonance in the x'–y' plane and locking it with the TOCSY mixing spin lock. After an appropriate mixing time, the spin lock field is turned off and we simply start acquiring the FID. These steps can be summarized as follows:

1. *Preparation:* Selectively excite the peak of interest to place its magnetization on the y axis.
2. *Mixing:* Apply the TOCSY mixing sequence (spin lock) on the y axis for a period of time between 30 and 85 ms.
3. *Detection:* Record the FID.

Any coherence that we observe in the FID must have come from the selected resonance by TOCSY transfer and therefore must belong to the same spin system. The best selective 90° pulse is DPFGSE, the same building block used as the front end of the selective NOE experiment: A 90° hard pulse followed by a gradient, and then a selective 180° pulse followed by another gradient of the same magnitude and sign as the first (Fig. 8.42). The first gradient twists all the sample magnetization and the shaped 180° pulse refocuses only the selected spins, whereas all other spins are unaffected. The second gradient "unwinds" the coherence helix of the selected spins and doubles the twist of all the other spins. At the end, we have only the selected net magnetization aligned on the y' axis and all other

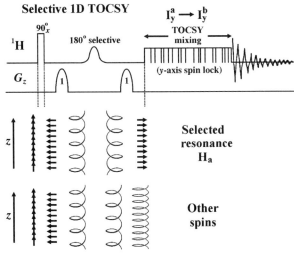

Figure 8.42

magnetization is destroyed. The PFGSE sequence is repeated to reinforce the selectivity (not shown in Fig. 8.42), and then the spin-lock mixing sequence is started with phase on the y' axis. Unlike the NOE experiment, we do not need a 90° pulse at the end of the DPFGSE because we already have the selected resonance's net magnetization right where we want it: in the $x'-y'$ plane. At the end of mixing we have positive, in-phase magnetization on y' for the selected spin and all other spins in the spin system that received the transferred magnetization. As it is already in the $x'-y'$ plane, we just turn on the ADC and collect the FID data. The 1D proton spectrum will have in-phase peaks for the selected resonance and for any other resonance to which magnetization was transferred during the spin lock. The intensity of each peak (integral area) will depend on the efficiency of TOCSY transfer from the selected spins. A small coupling constant (e.g., a fixed *gauche* relationship for vicinal couplings or a long-range (>3 bond) coupling) will lead to a "bottleneck" in TOCSY transfer, reducing the intensity of the destination peak and the peaks of all other protons after it in a linear spin system.

8.12.1 Factors That Affect TOCSY Mixing Efficiency

In the real world, complete mixing throughout the spin system is not observed. Magnetization transfer is a stepwise process starting from the selected proton, so protons near the selected proton in the spin system usually give more intense signals, and protons farther away give weaker peaks in the 1D spectrum. The simplest case, where there are only two protons (H_a and H_b coupled with J_{ab}) in the spin system, has been analyzed precisely. If we start with magnetization on H_a along the y' axis at the beginning of the TOCSY mixing sequence, the magnetization will oscillate between H_a and H_b:

$$\mathbf{I}_y^a \rightarrow [\mathbf{I}_y^a(1 + \cos(2\pi J\tau)) + \mathbf{I}_y^b(1 - \cos(2\pi J\tau))]/2$$

where τ is the mixing time (Fig. 8.43). Note that at time zero we have pure $\mathbf{I}_y^a$, at time $\tau = 1/(2J_{ab})$ we have pure $\mathbf{I}_y^b$ and none of the starting magnetization (100% transfer), and at time $\tau = 1/J_{ab}$ we are back to pure $\mathbf{I}_y^a$ (no transfer). So we can conclude that when magnetization hits the end of a spin system, it "bounces back" and we see *oscillatory* behavior as a function of mixing time. For a typical vicinal coupling ($J = 7.0$ Hz), a mixing

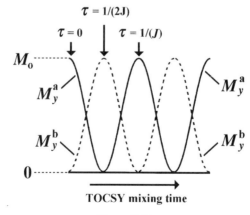

Figure 8.43

time of 71 ms gives complete transfer. For a small coupling constant such as $J_{ab} = 2.0$ Hz, you would get only 4% transfer and with $J_{ab} = 0.5$ (long range) you would get 0.24% transfer. This is why small coupling constants cause a "bottleneck" in transfer of magnetization and most of the magnetization remains on the starting nucleus.

In long spin systems such as flexible chains of CH_2 groups, magnetization transfer is more of a *diffusion*-like process. The signal is strongest on the starting spin and weaker as you move farther away along the chain. With longer mixing times, magnetization spreads farther along the chain. For example, with a mixing time of 70 ms it is usually possible to reach the ε position of the lysine side chain (five jumps) starting with magnetization on the amide NH in a peptide or protein:

$$HN-CH^\alpha-CH_2^\beta-CH_2^\gamma-CH_2^\delta-CH_2^\varepsilon-NH_3^+$$

Short mixing times (e.g., 30 ms) lead to INEPT-type spectra (or COSY-type 2D spectra), where transfer is mostly limited to a single jump over one J coupling. Unlike INEPT and COSY, however, the transfer results in an in-phase rather than antiphase signal. This is a significant advantage as the peaks have the same shape and pattern as they do in a 1D spectrum.

Figure 8.44 shows two selective 1D TOCSY spectra of sucrose in D_2O, with 70 ms of MLEV-17 mixing. Selecting the anomeric H-g1 (glucose) proton as a downfield "handle," we can see the spin system of the glucose unit, without any peaks from the fructose unit (Fig. 8.44, center). Because of the α-glycosidic linkage, H-g1 is in an equatorial position and has a small (3.8 Hz) coupling to H-g2. This "bottleneck" accounts for the inefficient transfer to H-g2 and H-g3. The triplet at 3.71 ppm is much larger than the distorted triplet

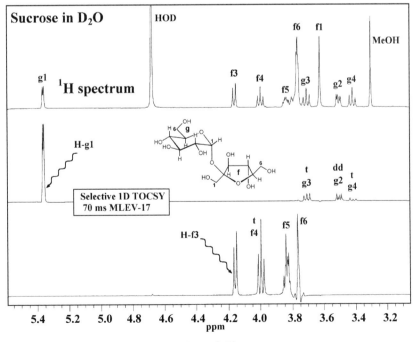

Figure 8.44

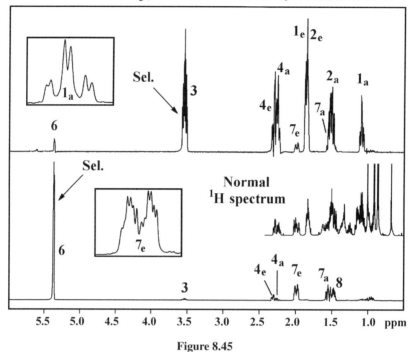

Figure 8.45

at 3.42 ppm, so we can assign the former to H-g3 and the latter to H-g4. No transfer is observed beyond H-g4 (three jumps). Selecting the H-f3 doublet (fructose unit: Fig. 8.44, bottom), we see the fructose spin system in a 1D proton spectrum, with none of the glucose peaks. The triplet at 3.99 must be H-f4 as we expect more complex splitting for H-f5 and H-f6. The multiplet at 3.83 ppm can be assigned to H-f5 (three couplings) and the tall peak at 3.77 ppm is H-f6. We can assign the two-proton singlet at 3.62 to H-f1, so the entire fructose system is assigned. In this way, we can "light up" one unit (residue) of a biological polymer and see only the peaks due to spins in that residue.

Figure 8.45 shows two selective 1D TOCSY spectra of cholesterol. Between the two spectra at the right-hand side is shown the very crowded and heavily overlapped upfield region of the ^{1}H spectrum of cholesterol for comparison. Selecting the H3 multiplet at 3.54 ppm (Fig. 8.45, top), we can follow the spin system clockwise around the A ring to H2$_{ax}$ and H2$_{eq}$ and on to H1$_{ax}$ and H1$_{eq}$ (Fig. 8.46). The completely resolved H1$_{ax}$ peak is shown in the inset (Fig. 8.45, top); this peak is hopelessly overlapped in the normal ^{1}H spectrum. The dt coupling pattern is due to large coupling constants to H1$_{eq}$ (geminal) and H2$_{ax}$ (axial–axial) and a small coupling to H2$_{eq}$ (axial–equatorial). We are now at a dead end because we run into the quaternary carbon, C10, at the A–B ring juncture. Moving counterclockwise around the A ring from H3 we see H4$_{ax}$ and H4$_{eq}$, the peaks we selected in the 1D NOE experiment (Fig. 8.35), and then there is a jump through a long-range ("allylic") coupling to H6 (Fig. 8.46). The H6 peak is very small due to this TOCSY transfer "bottleneck" of a small long-range coupling. Then we see transfer from H6 to H7$_{ax}$ and H7$_{eq}$. These peaks are small primarily because the peak they derive from, H6, is small.

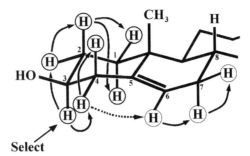

Figure 8.46

Selecting the H6 peak at 5.35 ppm (Fig. 8.45, bottom) gives a relatively weak transfer to H7$_{ax}$ and H7$_{eq}$, and these efficiently give magnetization to H8 (Fig. 8.47). The inset (Fig. 8.45, bottom) shows the resolved H7$_{eq}$ "doublet" peak, which has one large coupling (geminal coupling to H7$_{ax}$) and a number of small couplings (to H8, H6, and H4$_{ax}$). In the other direction (Fig. 8.47), we see weak transfer from H6 to H4$_{ax}$ and H4$_{eq}$ and from these we are just beginning to get transfer to H3 (Fig. 8.47). Note that technically cholesterol is a single spin system (if we count the allylic H4 to H6 coupling), but in reality magnetization does not spread indefinitely. The bottleneck (H4 to H6) confines magnetization to some extent to the A ring, but even without this the spread of magnetization seldom goes more than five jumps (J couplings) in the extreme. Eventually, we "run out of time" for magnetization transfer, and the spins farther away from the selected spin give very weak peaks or none at all. The peak assignments in Figure 8.45 are derived from 2D NMR analysis and are not obvious, based only on the 1D selective TOCSY experiment shown here. We can get some ideas of axial and equatorial from the number of large couplings ("doublet" vs. "triplet" patterns) and we might be able to sort it out completely using a short mixing time ($\tau_m = $ 30 ms) to see the direct "one-jump" relationships only.

Comparing Figure 8.45 (TOCSY) to Figure 8.36 (NOE), we can see that TOCSY transfer is very efficient. Compared to the selected peak, the transfer peaks are of comparable intensity. NOE transfer is very inefficient and, for small molecules, of opposite sign. In the NOE spectrum, the selected peak is negative and enormous compared to the very small (around 1%) positive transfer peaks. This is an important factor in the design of NMR

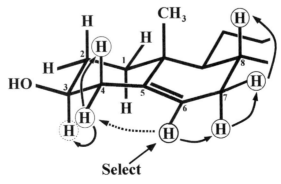

Figure 8.47

348 SHAPED PULSES, PULSED FIELD GRADIENTS, AND SPIN LOCKS

experiments: Through-space transfer is very inefficient, whereas *J*-coupling transfer is very efficient. TOCSY transfer can be so efficient (even 100%, see Fig. 8.43) that the selected peak may be small or even missing in the spectrum!

8.13 RF POWER LEVELS FOR SHAPED PULSES AND SPIN LOCKS

The amplitude of an RF pulse can be expressed in units of telsa (B_1). This corresponds to the magnitude (length) of the $\boldsymbol{B}_1$ vector in a rotating-frame vector diagram. Pulse amplitude is most commonly expressed in terms of the frequency of rotation of sample magnetization as it precesses around the $\boldsymbol{B}_1$ vector (for on-resonance pulses) during the pulse.

$$\text{Pulse amplitude} = B_1(\text{tesla}) \quad \propto \quad \gamma B_1/2\pi (\text{Hz}) = 1/(4 \times t_{90})$$

The inverse of this frequency of rotation ($2\pi/\gamma B_1$ in seconds) is the time it takes for the sample magnetization to rotate one full cycle under the influence of the B_1 field. This is simply the duration of a 360° pulse, and one fourth of this time is the 90° pulse duration, t_{90}.

For example, a 10 μs hard pulse at a B_1 field strength of 25 kHz will rotate the sample magnetization by $\Theta = 10 \times 10^{-6}$ s $\times 25 \times 10^3$ cycle/s $= 0.25$ cycle $= 90°$. So this pulse is a 90° pulse. A 10-ms soft pulse at a B_1 field strength of 25 Hz will rotate the sample magnetization by $\Theta = 10 \times 10^{-3}$ s $\times 25$ cycle/s $= 0.25$ cycle $= 90°$. So this is also a 90° pulse (Fig. 8.48). The "area" of the rectangular pulses is the same:

$$\text{"Area"} = \text{width} \times \text{height} = 10 \,\mu\text{s} \times 25 \,\text{kHz} = 10 \,\text{ms} \times 25 \,\text{Hz} = 0.25$$

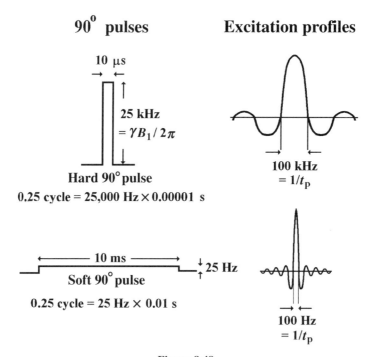

Figure 8.48

The soft pulse is 1000 times lower in amplitude compared to the hard pulse, but it is 1000 times longer than the hard pulse. Although both pulses deliver a 90° rotation when on-resonance, they have very different behavior off-resonance. The excitation profile of both pulses is a "sinc" function:

$$\text{sinc}(x) = \sin(x)/x$$

but the width of the main "hump" of the sinc function is different:

$$\text{Excitation bandwidth} = 1/t_p = 1/10\,\mu s = 100\,\text{kHz for the hard pulse}$$

$$\text{Excitation bandwidth} = 1/t_p = 1/10\,\text{ms} = 100\,\text{Hz for the soft pulse}$$

Pulse *power* is the square of the pulse amplitude:

$$\text{Power} = (\text{amplitude})^2 \propto (\gamma B_1/2\pi)^2 = (1/(4 \times t_p))^2$$

In the laboratory we can measure RF power in watts, but when we set up NMR experiments we use a relative power scale that is logarithmic: the decibel scale. For comparison of power levels, we compare to a standard power level P_o that corresponds to zero on the decibel scale:

$$\text{Power in decibels} = 10\log(P/P_o)$$

Every time the power is increased by a factor of 2, we are *adding* 3 dB to the power level in decibels: $\log(2) = 0.301$, $10\log(2) = 3.01$. Because the decibel scale is logarithmic, multiplying power by a factor corresponds to adding to or subtracting from the power level in decibels.

We can also compare pulse amplitudes using the decibel scale, as we know that power is the square of pulse amplitude:

$$\Delta\text{dB} = 10\log(P_a/P_b) = 10\log(B_1^a/B_1^b)^2 = 20\log(B_1^a/B_1^b)$$

where B_1^a is the B_1 amplitude at one power setting and B_1^b is the amplitude at another setting, ΔdB decibel units lower in power. As the 90° pulse width is inversely proportional to B_1 amplitude, $\Delta\text{dB} = 20\log(t_{90}^b/t_{90}^a)$, where t_{90}^a is the 90° pulse width at one power level and t_{90}^b is the 90° pulse width at a power level ΔdB decibel units lower in power. Thus to cut the 90° pulse width in half, we need to double the B_1 amplitude (quadruple the pulse power), which requires a 6 dB increase in power: $20\log(2) = 6.021$ dB. Likewise, to double the 90° pulse width would require a 6 dB *decrease* in pulse power. This "6 dB rule" is very useful to keep in mind.

Bruker and Varian not only use different zero points for their decibel scales, but also use the opposite sign: Varian considers decibel to be a power level as described above, but Bruker sees the decibel setting as an attenuation—higher decibel values correspond to lower power. As long as you know this, you will not have any problem, but be very careful because setting the wrong power level can fry equipment!

Vendor	Parameters	Minimum	Maximum	Definition
Bruker	*pl1, pl2*	120 dB	−6 dB	"dB of attenuation"
Varian	*tpwr, dpwr*	0 dB	63 dB	"dB of power"

When you calibrate a 90° "hard" ^{1}H pulse, you can estimate the power levels for other uses. The most convenient way to express a power level is by the duration of the 90° pulse at that power level:

Application	90°	$\gamma B_1/2\pi$
TOCSY mixing (MLEV-17)	30 μs	8333 Hz
ROESY mixing (cw)	75 μs	3333 Hz
^{1}H decoupling (waltz-16)	90 μs	2778 Hz

You can then use the decibel scale to estimate power settings. For example, suppose you calibrated the 90° pulse on a Bruker 500 to be 17.6 μs for ^{1}H at a power setting of 3 dB, and you want to know the power setting that will give a 30 μs 90° pulse ($\gamma B_1/2\pi = 1/(4 \times 30\ \mu s) = 8333$ Hz). Just plug in the ratio of pulse widths:

$$\Delta dB = 20 \log(t_{90}^b / t_{90}^a) = 20 \log(30/17.6) = 4.6\ dB$$

As our point of comparison (t_{90}^a) was at 3 dB, we add this number ΔdB to 3 to get the correct power setting: 7.6 dB. This calculation gives us an estimate of the power setting; to get an accurate value you would have to calibrate the 90° pulse (on resonance) using this value as a starting point. Because in this case we want a 90° pulse of 30 μs, you would start with a 60 μs pulse and adjust the pulse power (Bruker parameter *pl1*) until you get a null (180° pulse). When you are calibrating pulse widths and pulse power, at low power, it is extremely important to be on-resonance for the peak you are observing during the calibration. When $\gamma B_1/2\pi$ is small, the effect of being off-resonance by even a small amount can be dramatic. For example, for a 75-μs 90° pulse, $\gamma B_1/2\pi$ is 3333 Hz and on a 600-MHz instrument you would tilt the B_{eff} vector out of the x'–y' plane by 45° if you are off-resonance by the same amount (3333 Hz = 5.56 ppm for protons). Also, keep in mind that near the maximum setting of pulse power, the dB settings do not give as much power as you expect: they begin to "droop" in a process called "amplifier compression." This occurs in the top 6 dB or so of available pulse power. The dB calculations work much better below this range.

Exercise: Estimate the Varian power level settings for TOCSY mixing (8333 Hz), for ROESY mixing (3333 Hz), and for ^{1}H decoupling (2778 Hz) if the 90° pulse is 21.3 μs at a power setting of 59 dB.

8.13.1 Calibrating a Shaped Pulse

Starting from a "first guess" of power level, a shaped pulse should be calibrated to get exactly the correct pulse rotation. Calibration of a rectangular pulse involves changing the pulse duration (pulse width) while maintaining the power level (pulse height) constant with a peak on-resonance. We look for a null in the spectrum at the 180° or 360° pulse width. For a shaped pulse, the selectivity depends on the pulse width, so we keep that constant and adjust the pulse power, increasing or decreasing the vertical scale of the pulse shape to change the area under the curve. When the maximum amplitude is changed, the shape of the

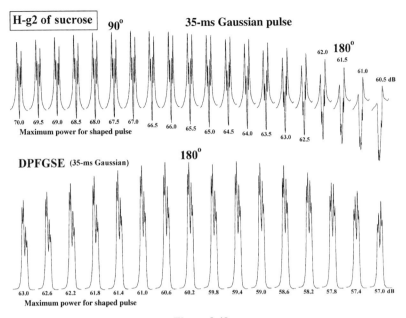

Figure 8.49

pulse is maintained so that all the short pulses that make up the shaped pulse are adjusted in amplitude according to the same ratio. Because pulse power is set using a logarithmic scale (dB), the envelope of a pulse calibration will not be a simple sine wave as it is for varying the duration of rectangular pulses. For example, for a Gaussian pulse we see a maximum for the 90° pulse power and then decreasing intensity to a null for the 180° pulse power (Fig. 8.49, top). When the null point is located (180° pulse at 61.5 dB), we can decrease the maximum power by 6 dB (67.5 dB on Bruker) to get a 90° pulse, rather than dividing the pulse duration by two as we do for hard pulses. These power levels represent the power of the highest amplitude in the shape, which is at the center of a Gaussian pulse. Often it is better to calibrate a shaped pulse in the context of how it is used in the pulse sequence. For example, a 180° Gaussian pulse used in a PFGSE is calibrated for the strongest signal of the selected peak using a PFGSE sequence (Fig. 8.49, bottom). Even though we think of a 180° pulse producing a null, in the context of a PFGSE it produces the maximum refocusing and allows the second gradient to perfectly unscramble the twisted coherence produced by the first.

To estimate a starting point for the maximum power of a shaped pulse, we need to come up with a rectangular pulse that has the same "area" as the shaped pulse. This can be done by mathematically integrating the function used for the pulse shape, for example, the Gaussian function. The Bruker software does this in the PulseTool program, and Varian does it in Pandora's Box (PBox). For example, a 35-ms Gaussian 180° pulse that is truncated at 5% of the maximum amplitude takes up an area that is 50.4556% of the corresponding rectangular pulse with the same duration and the same maximum amplitude as the shaped pulse (Fig. 8.50). Thus, a rectangular pulse of duration 17.66 ms (0.504556 × 35 ms) with the same amplitude as the maximum of the shaped pulse would rotate the sample magnetization the same amount (180°) as the Gaussian pulse, if both pulses are on-resonance. This corresponds to a full rotation in 2 × 17.66 = 35.32 ms, and a $\gamma B_1/2\pi$ of $1/(35.32 \text{ ms}) = 28.31$ Hz. This is a very weak pulse! Suppose we have calibrated the 90° hard pulse at 3 dB

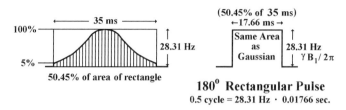

Figure 8.50

on the Bruker to be 15 μs, corresponding to a 180° pulse width of 30 μs. We need to know the pulse power in decibels that will give us a 180° rectangular pulse of duration 17.66 ms:

$$\Delta dB = 20 \log(17660 \mu s / 30 \mu s) = 55.40 \, dB$$

We use common sense to find the correct power level: We know we want lower power, and for Bruker that means a larger number. So we add this to 3 dB to get a power setting of 58.4 dB. As this power level corresponds to the maximum power of the Gaussian shaped pulse, we can set this power level for our shaped pulse and get a 180° rotation. This would be the starting point for the pulse calibration.

Further Reading

1. Bauer C, Freeman R, Frenkiel T, Keeler J, Shaka, AJ. Gaussian pulses. *J. Magn. Reson.* 1984;**58**:442–457.
2. Bothner-By AA, Stephens RL, Lee J-M. Structure determination of a tetrasaccharide: transient nuclear Overhauser effects in the rotating frame. *J. Am. Chem. Soc.* 1984;**106**:811–813.
3. Braunschweiler L, Ernst, RR. Coherence transfer by isotropic mixing: application to proton correlation spectroscopy. *J. Magn. Reson.* 1983;**53**:521–528.
4. Hurd RE. Gradient-enhanced spectroscopy. *J. Magn. Reson.* 1990;**87**:422–428.
5. Sklenář V, Piotto M, Leppik R, Saudek V. Gradient-tailored water suppression for ^1H–^{15}N HSQC experiments optimized to retain full sensitivity. *J. Magn. Reson. A* 1993;**102**:241–245.
6. Stott K, Stonehouse J, Keeler J, Hwang T-L, Shaka AJ. Excitation sculpting in high-resolution nuclear magnetic resonance spectroscopy: application to selective NOE experiments. *J. Am. Chem. Soc.* 1995;**117**:4199–4200.

9

TWO-DIMENSIONAL NMR SPECTROSCOPY: HETCOR, COSY, AND TOCSY

9.1 INTRODUCTION TO TWO-DIMENSIONAL NMR

We are now ready to discuss the most exciting and most powerful technique in NMR, a simple idea that led to an explosion of NMR applications, extending the power of NMR beyond organic chemistry to become a powerful tool in structural biology. Until now, we have looked at an NMR spectrum as a simple graph of intensity (vertical scale) versus frequency (horizontal chemical-shift scale). In a simple one-dimensional (1D) NMR spectrum, we get structural information from the chemical shift and peak area, and in ^{1}H spectra we can learn about near neighbors in the bonding network by examining the coupling patterns of resolved multiplets. We can gain more specific information about interactions between protons (NOE and J coupling) by using low-power irradiation or shaped-pulse selective excitation of specific resonances in the 1D spectrum. But we are severely limited by the chemical shift "space" available for resolution of many resonances in a single frequency scale and by the need to examine relationships one at a time in separate selective experiments. In 2D NMR we have *two* frequency scales: the familiar direct measurement of frequency by Fourier transformation of the FID on the horizontal scale and an *indirect* second frequency scale on the vertical scale. This second dimension is created by recording a series of hundreds of 1D spectra, each time lengthening a delay in the pulse sequence, just as we sample the FID at discreet intervals in real time using the analog-to-digital converter. This incremented delay becomes a second (indirect) time domain, and Fourier transformation of this time domain yields the second frequency scale. Intensity is a third axis coming out of the paper, usually represented by color coding (intensity plot) or a contour (topographic) map. Depending on the way we do the experiment, we can map out specific kinds of interactions between spins. For example, if we are looking at NOE interactions, we will see a "spot" of intensity at the frequency of H_a on the horizontal chemical-shift scale and the frequency of H_b on the

---

NMR Spectroscopy Explained: Simplified Theory, Applications and Examples for Organic Chemistry and Structural Biology, by Neil E. Jacobsen
Copyright © 2007 John Wiley & Sons, Inc.

vertical chemical-shift scale, if H_a and H_b are less than 5 Å apart. All of the information for all possible NOE interactions is contained in this single "map" produced by a single NMR experiment! 2D NMR essentially allows us to selectively excite each of the chemical shifts in one experiment and gives us a matrix or two-dimensional map of all of the nuclei affected by each perturbation.

Let's look at the overall strategy for a 2D pulse sequence. There are four steps to any 2D experiment:

1. *Preparation*: Excite nucleus **A**, creating magnetization in the x–y plane.
2. *Evolution*: Indirectly measure the chemical shift of nucleus **A**.
3. *Mixing*: Transfer magnetization from nucleus **A** to nucleus **B** (via J or NOE).
4. *Detection*: Measure the chemical shift of nucleus **B**.

Of course, all possible pairs of nuclei in the sample go through this process at the same time. *Preparation* is usually just a 90° pulse that excites all of the sample nuclei of a given type (e.g., ^{1}H, ^{13}C, etc.) simultaneously. *Detection* is simply recording an FID and finding the frequency of nucleus **B** by Fourier transformation. A simple 1D spectrum is just steps 1 and 4. To get a second dimension, we have to measure the chemical shift of nucleus **A** *before* it passes its magnetization to nucleus **B**. This is accomplished by simply waiting for a period of time (called t_1, the *evolution* period) and letting the nucleus **A** coherence rotate in the x–y plane. The experiment is repeated many times (e.g., 512 times), recording the FID each time with the delay t_1 incremented each time by a fixed amount. The time course of motion of the nucleus **A** magnetization as a function of t_1 (determined by its effect on the final FID) is unraveled by a second Fourier transform, defining how fast it rotates during the t_1 delay and giving us its chemical shift. *Mixing* is a combination of RF pulses and/or delay periods that induce the magnetization to jump from **A** to **B** as a result of either a J coupling or an NOE interaction (close proximity in space). Different 2D experiments (e.g., NOESY, COSY, HETCOR, etc.) differ primarily in the mixing sequence because in each one we are trying to define the relationship between **A** and **B** within the molecule in a different way. We now have quite an array of tools for transfer of magnetization: transient NOE (z-magnetization transfer via NOE), INEPT transfer (antiphase to antiphase coherence transfer via J-coupling), TOCSY transfer (multiple in-phase to in-phase coherence transfers via J coupling), and ROESY transfer (NOE transfer in the x'–y' plane in a spin lock). Each of these can be applied to create a specific 2D experiment (COSY, HETCOR, HSQC, HMBC, NOESY, ROESY, TOCSY, etc.).

9.2 HETCOR: A 2D EXPERIMENT CREATED FROM THE 1D INEPT EXPERIMENT

Let's pick a concrete example we are familiar with: the INEPT transfer from ^{1}H to ^{13}C. The simplest INEPT sequence is just $90°(^1\text{H}) - \tau - 90°(^1\text{H})/90°(^{13}\text{C})$, where the τ delay is for J-coupling evolution of the ^{1}H doublet into antiphase and the simultaneous 90° pulses give coherence transfer from ^{1}H to ^{13}C. Normally, we would set this delay to $1/(2\,J)$ to get complete conversion from in-phase to antiphase ^{1}H coherence ($\mathbf{I}_x \rightarrow 2\mathbf{I}_y\mathbf{S}_z$), and we would put simultaneous 180° pulses on ^{1}H and ^{13}C in the center of the delay to refocus ^{1}H chemical-shift evolution. But for a 2D experiment we *want* ^{1}H chemical-shift evolution, so we make

HETCOR: A 2D EXPERIMENT CREATED FROM THE 1D INEPT EXPERIMENT   355

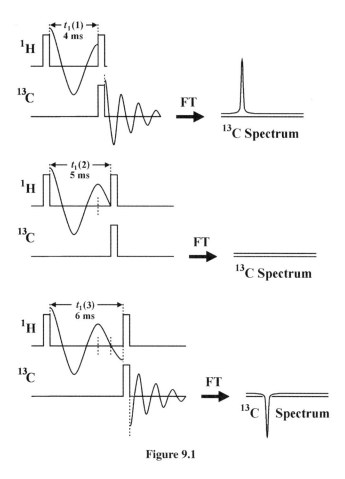

Figure 9.1

the delay τ into the simple evolution delay t_1. During the t_1 delay, the required antiphase component $2\mathbf{I}_y\mathbf{S}_z$ is oscillating due to chemical-shift evolution of the $\mathbf{I}_y$ part (Fig. 9.1). The frequency of this oscillation is just the offset of this proton (H_a) in the ^{1}H spectrum: Ω_a. At some point in this oscillation, we execute the simultaneous 90° pulses and transfer this magnetization to antiphase ^{13}C coherence ($2\mathbf{S}_y\mathbf{I}_z$), which oscillates at the frequency in the ^{13}C spectrum of the ^{13}C (C_b) that is bonded to H_a: Ω_b. This oscillation is recorded in the ^{13}C FID, and Fourier transformation gives a ^{13}C spectrum with a peak at frequency Ω_b. At this particular value of t_1, the ^{1}H coherence is at a positive maximum, and coherence transfer starts the ^{13}C FID at a positive maximum, leading to a positive peak of maximum intensity in the ^{13}C spectrum (Fig. 9.1, top). Now we repeat the experiment with a slightly longer value of t_1, so that the moment of coherence transfer happens when the ^{1}H coherence is zero. No ^{13}C coherence is produced, and the FID is just noise. Fourier transform gives no ^{13}C peak in the ^{13}C spectrum (Fig. 9.1, middle). A third experiment is done with the t_1 value incremented a bit further, and this time the ^{1}H coherence is at a negative maximum. Coherence transfer gives a negative maximum of ^{13}C coherence at the start of the ^{13}C FID, and Fourier transformation gives an upside-down ^{13}C peak in the ^{13}C spectrum (Fig. 9.1, bottom). In this way, the peak intensity in the ^{13}C spectrum oscillates in a way that exactly follows the oscillation of ^{1}H coherence during the evolution (t_1) delay. The intensity of the

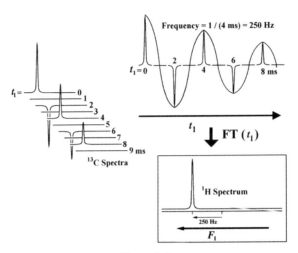

Figure 9.2

^{13}C peak can be plotted as a ^{1}H FID with t_1 as the timescale (the "indirect time domain"). Understanding this is the core of understanding the genius of the 2D idea—creating a "fake" time domain with a delay that allows us to generate a second "indirect" frequency scale. Fourier transformation of this ^{1}H FID (Fig. 9.2, put together from peak heights of a large number of ^{13}C spectra) gives a ^{1}H spectrum with a peak at the frequency of the original ^{1}H (Ω_a). The second Fourier transform traces backward from the intensity variation of the ^{13}C peak to the history of the ^{1}H as it undergoes chemical-shift evolution during t_1. We can say that during t_1 we "encoded" the chemical shift of H$_a$, and this encoded information turns up in the peak height dependence of the C$_b$ peak on the value of the incremented delay t_1. The second Fourier transform "decodes" this information and gives us the chemical shift of H$_a$. The experiment provides a *correlation* between the chemical shift of H$_a$ (Ω_a) and the chemical shift of C$_b$ (Ω_b), proving that they are directly bonded to each other (*J* coupled). In 2D NMR, we do not correlate spins or positions within a molecule; we can only make connections between chemical shifts (frequencies). It is our job in interpreting the 2D spectrum to try to convert this information into structural conclusions.

In Figure 9.1 we show only one component of the proton magnetization ($2\mathbf{I}_y\mathbf{S}_z$) undergoing chemical-shift evolution during the t_1 delay, and it is the magnitude of this component at the moment of magnetization transfer that determines the magnitude of the ^{13}C FID obtained at the end. This is true because magnetization transfer always selects only one component of magnetization. For INEPT transfer we can ignore in-phase ^{1}H coherence ($\mathbf{I}_x$, $\mathbf{I}_y$) because it cannot undergo coherence transfer, so let's look only at the chemical-shift evolution of $2\mathbf{I}_y\mathbf{S}_z$:

$$2\mathbf{I}_y\mathbf{S}_z \xrightarrow{t_1} 2\mathbf{I}_y\mathbf{S}_z\cos(\Omega_a t_1) - 2\mathbf{I}_x\mathbf{S}_z\sin(\Omega_a t_1)$$

The simultaneous 90° pulses on ^{1}H and ^{13}C give INEPT transfer only for the first term:

$$2\mathbf{I}_y\mathbf{S}_z\cos(\Omega_a t_1) - 2\mathbf{I}_x\mathbf{S}_z\sin(\Omega_a t_1) \xrightarrow{90^\circ_x(^1\text{H})/90^\circ_y(^{13}\text{C})}$$
$$2\mathbf{S}_x\mathbf{I}_z\cos(\Omega_a t_1) - 2\mathbf{I}_x\mathbf{S}_x\sin(\Omega_a t_1)$$

HETCOR: A 2D EXPERIMENT CREATED FROM THE 1D INEPT EXPERIMENT

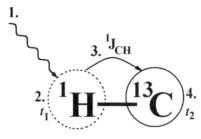

Figure 9.3

The second term gives $2I_xS_x$, a mixture of DQC and ZQC which is not observable in the FID. The INEPT transfer term, $2S_xI_z$, is multiplied by $\cos(\Omega_a t_1)$, which is the amplitude modulation of our ^{13}C FID according the "history" of the ^{1}H (**I** spin) chemical-shift evolution during the t_1 delay. This is how the "encoding" of the chemical shift of the proton comes out in precise mathematical terms.

We can diagram this experiment in a way that summarizes the flow of magnetization (Fig. 9.3). First draw the two nuclei that are being correlated, and indicate the relationship that will lead to a crosspeak: a ^{1}H bonded to a ^{13}C. The "relationship" is a single bond, which leads to a large J coupling (~150 Hz). We start the flow with excitation of the proton (step 1: preparation). Next we measure the chemical shift of the proton indirectly during the t_1 delay (step 2: evolution, shown by a dotted circle around the ^{1}H and the label "t_1"). Then we transfer the coherence from ^{1}H to ^{13}C via INEPT transfer (step 3: mixing, indicated by an arrow from ^{1}H to ^{13}C labeled "$^1J_{CH}$"). Finally, we measure the chemical shift of the ^{13}C directly by recording a ^{13}C FID (step 4: detection, shown by a solid circle around the ^{13}C labeled "t_2"). We call the time domain of the directly detected FID t_2 because it comes after t_1 in the sequence. This leads to the name F_2 for the directly detected frequency domain (horizontal axis of the 2D spectrum) and F_1 for the indirectly detected frequency domain (vertical axis in the 2D spectrum). We will use these diagrams throughout our discussion of 2D NMR as a quick way of showing how the experiment works.

Now let's consider a situation where we have three different carbon resonances in our ^{13}C spectrum, each with one proton attached: -C$_1$H$_1$–C$_2$H$_2$–C$_3$H$_3$-. We do the same INEPT experiment with the variable delay t_1 and record a series of ^{13}C spectra. Starting with the first FID, obtained with $t_1 = 0$, we Fourier-transform each FID and load the resulting ^{13}C spectra (with three peaks) into successive rows moving up in a 2D matrix of data. This gives us a stack of ^{13}C spectra starting with the $t_1 = 0$ spectrum at the bottom and moving upwards as we increment t_1: 1, 2, 3 ms, and so on (Fig. 9.4). This intermediate 2D matrix, after the F_2 Fourier transform, is sometimes called an *interferogram*. The height of each ^{13}C peak oscillates (+, 0, −, 0, etc.) at the frequency Ω_H of its attached proton, and decays due to T_2 relaxation of the attached proton, as we look at successive spectra with increasing t_1 delay values (moving up in the data matrix). Carbon C$_1$ oscillates at a high frequency because proton H$_1$ has a downfield chemical shift. Carbon C$_3$ oscillates slowly in t_1 because its attached proton, H$_3$, has an upfield (lower frequency) chemical shift. In Figure 9.4 the trace of data in each of the three columns is shown to the side of the data column—in each case we have a decaying sinusoidal signal: a t_1 FID. It may have taken several minutes in

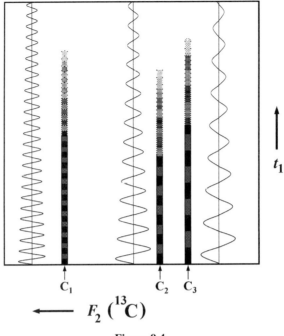

Figure 9.4

real time to acquire the FID for one point of this column (e.g., 16 scans of ^{13}C acquisition), but in the reconstructed time domain of t_1 the time difference is only the t_1 increment (e.g., 1 ms) between successive data rows.

The second Fourier transform is performed on each of the columns of the data matrix, starting with the first one on the left side and moving to the right side. Most of the columns in Figure 9.4 contain noise, but when we reach the column at the chemical shift of C_1 we load the t_1 FID (shown to the left side of the data column) and Fourier-transform it to obtain a ^{1}H spectrum with a peak at the frequency of H_1. This spectrum is put back into the data column, replacing the t_1 FID. Likewise, when we pass through the ^{13}C shift of C_2 we transform its t_1 FID and obtain the ^{1}H spectrum of H_2, and so forth. The completed data matrix (the 2D spectrum) is shown in Figure 9.5. Now for each data column corresponding to a peak in the ^{13}C spectrum we have a ^{1}H spectrum (shown to the side of the data column) of just the one proton attached to that carbon. We have a two-dimensional map that correlates the ^{1}H spectrum (on the vertical or F_1 axis) with the ^{13}C spectrum (on the horizontal or F_2 axis).

Figure 9.6 shows a 2D HETCOR spectrum with the 1D ^{13}C spectrum displayed at the top (horizontal or F_2 dimension) and the 1D ^{1}H spectrum displayed vertically on the left side (vertical or F_1 dimension). From any peak (resonance) in the ^{1}H spectrum, we can follow a horizontal line until we encounter a spot or blob of intensity in the 2D data matrix. These clusters of intensity represent correlations in the 2D spectrum and are called crosspeaks. From the crosspeak we move up along a vertical line and run into the ^{13}C peak in the 1D ^{13}C spectrum corresponding to that proton's personal carbon atom (the one it is directly bonded to). In this way we can pair up each proton peak in the ^{1}H spectrum with a carbon peak in the ^{13}C spectrum—a process called *chemical-shift correlation*.

HETCOR: A 2D EXPERIMENT CREATED FROM THE 1D INEPT EXPERIMENT 359

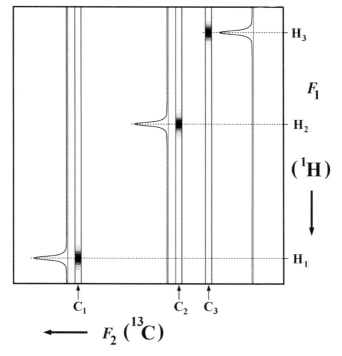

Figure 9.5

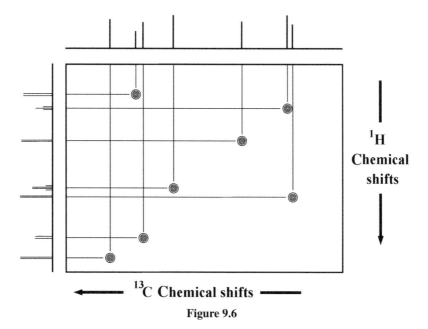

Figure 9.6

360 TWO-DIMENSIONAL NMR SPECTROSCOPY: HETCOR, COSY, AND TOCSY

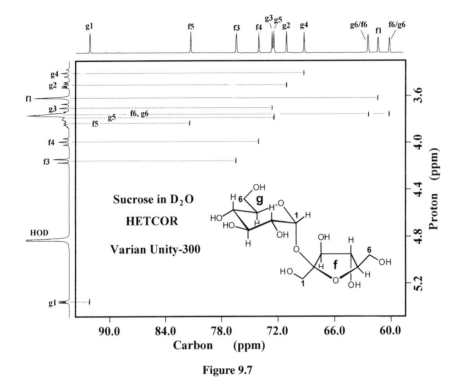

Figure 9.7

Figure 9.7 shows the HETCOR spectrum of sucrose. In Chapter 8 we were able to assign nearly all of the ^{1}H resonances of sucrose using the selective 1D TOCSY experiment, using the H-g1 and H-f3 doublets as unambiguous starting points. Now we can "transfer" these ^{1}H assignments to the ^{13}C spectrum using the one-bond correlations mapped out in the HETCOR spectrum. All 10 protonated (nonquaternary) carbons give clearly resolved crosspeaks, leading us from each ^{13}C peak to a precise chemical shift in the ^{1}H spectrum, even if the ^{1}H resonance is overlapped in the 1D ^{1}H spectrum. Although the C-g3 and C-g5 peaks are very close to each other in the ^{13}C spectrum, we can clearly connect the downfield peak of the pair to the H-g3 triplet. Because all of the other protons except H-g5 have been assigned, the upfield peak of the pair must "point" to H-g5, an overlapped resonance between the H-f5 multiplet and the large, broad singlet representing H-g6 and H-f6. All three CH$_2$OH crosspeaks (f1, f6, and g6) appear in a tight triangle at the right side, consistent with the upfield location of CH$_2$ groups relative to CH groups with the same inductive (electron-withdrawing) factors. The steric effect on ^{13}C shifts gives us general locations of 50–60 ppm for CH$_3$–O, 60–70 for CH$_2$–O, 70–80 for CH–O and 80–90 for C$_q$–O. The top peak of the triangle can be assigned to f1, since the H-f1 singlet (integral area 2) was identified by an NOE across the glycosidic linkage from H-g1. The two other peaks have nearly identical ^{1}H shifts (the peak on the left is slightly lower, or downfield on the ^{1}H shift scale) and we cannot assign them unambiguously. Note that the C-f2 quaternary carbon does not give any crosspeak in the HETCOR spectrum because it is not directly connected to a proton. Only one-bond relationships between ^{1}H and ^{13}C lead to a correlation in this experiment (Fig. 9.3).

9.2.1 Designing a HETCOR Pulse Sequence

Let's apply the general design principles of 2D NMR to the one-bond correlation of ^{13}C (in F_2) to ^{1}H (in F_1). The specific steps for the HETCOR experiment are as follows:

1. *Preparation*: a 90° nonselective ^{1}H pulse rotates ^{1}H z magnetization into the x–y plane.
2. *Evolution*: the ^{1}H magnetization precesses in the x–y plane for a period t_1, encoding its chemical shift as a function of t_1.
3. *Mixing*: an INEPT sequence converts the ^{1}H magnetization into antiphase magnetization with respect to its attached ^{13}C nucleus, and then transfers the ^{1}H magnetization to ^{13}C magnetization by simultaneous 90° pulses on both ^{1}H and ^{13}C channels.
4. *Detection*: The ^{13}C FID is recorded.

The simplest possible pulse sequence would involve a 90° ^{1}H pulse followed by a t_1 delay, and then an INEPT sequence (Fig. 9.8). Now we need to make refinements, thinking carefully about what we want to happen during various delays and what we want to prevent or suppress. During the evolution (t_1) period, we only want chemical-shift evolution. We would like to refocus the J-coupling ($^1J_{CH}$) evolution, and it would be nice if we could also refocus the homonuclear (^{1}H–^{1}H) J-coupling evolution. That way the only information that will be encoded during the evolution period is the information that we want to show up in the F_1 (Ω_H) dimension of the 2D spectrum: the ^{1}H chemical shift. Suppressing the $^1J_{CH}$ coupling is accomplished simply by inserting a ^{13}C 180° pulse into the center of

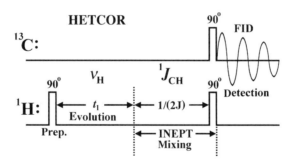

Figure 9.8

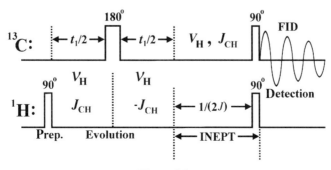

Figure 9.9

the t_1 period (Fig. 9.9). This reverses the J-coupling evolution so that it refocuses during the second half of t_1. The ^{1}H chemical shifts continue to evolve because there is no ^{1}H 180° pulse. This is the reverse of the strategy used in Chapter 6 to refocus J-coupling evolution during a ^{13}C evolution period (Fig. 6.33). Refocusing of homonuclear (^{1}H–^{1}H) J-coupling evolution is a bit more complicated and will be discussed last.

Next, we need to examine the $1/(2J)$ period of the INEPT sequence. In an INEPT experiment we usually allow only the J-coupling evolution, which is required to generate antiphase magnetization, to occur during the $1/(2J)$ delay. The chemical-shift evolution is suppressed so that the phase of detected peaks is not screwed up by off-resonance chemical-shift effects. We could accomplish this by inserting simultaneous 180° pulses on ^{1}H and ^{13}C in the center of the $1/(2J)$ delay. But in this experiment, we will not worry about phase. The data will be looked at in "magnitude mode," which calculates a single number for each data point from the real and imaginary parts of the spectrum:

$$\text{magnitude} = \sqrt{(\text{real})^2 + (\text{imag})^2}$$

Because chemical-shift evolution just rotates the magnetization vector in the x'–y' (real–imaginary) plane without affecting its magnitude, the phase of the detected ^{13}C magnetization is not important. So we can live with a simple $1/(2J)$ delay.

As our sequence stands so far, the detected signals will be fully coupled antiphase peaks. This could be messy since CH groups will appear as antiphase doublets (intensity ratio 1, −1), CH_2 groups as antiphase triplets (1, 0, −1), and CH_3 groups will appear as antiphase quartets (intensity ratio 1, 1, −1, −1) in the F_2 (^{13}C) dimension. We need to use proton decoupling during the FID acquisition ("detection") period. This will collapse all of the ^{13}C multiplets into single peaks. But as we saw in the simple INEPT experiment (Chapter 7), the intensities of the antiphase multiplets add up to zero. So turning on the proton decoupler will lead to a complete loss of all ^{13}C signals in F_2. This is not good. The solution is to allow the antiphase magnetization to undergo J-coupling evolution for an additional delay period so that the individual multiplet components come back together into normal (in-phase) multiplets. The optimal refocusing period is $1/(2J)$ for the CH carbons, $1/(4J)$ for the CH_2 carbons, and about $1/(5J)$ for the CH_3 carbons. The best compromise that allows observation of all three kinds of carbons (Fig. 9.10) is a delay of $1/(3J)$. Again, we do not worry about ^{13}C chemical-shift evolution during this delay because we will display the data in magnitude mode.

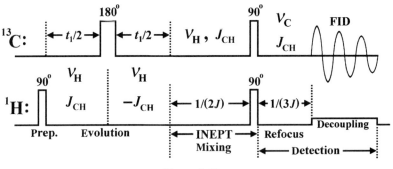

Figure 9.10

HETCOR: A 2D EXPERIMENT CREATED FROM THE 1D INEPT EXPERIMENT

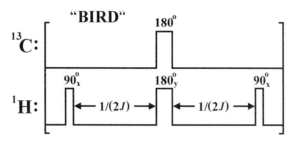

Figure 9.11

Finally, we will deal with the question of ^{1}H–^{1}H coupling during the evolution period. This is frosting on the cake, and I only bring it up to show the beautiful things you can accomplish with pulse sequence building blocks. The only protons we are interested in are those bonded to a ^{13}C, because these are the only ones that can transfer magnetization to ^{13}C. If we consider the other ^{1}H nuclei that have homonuclear coupling to the (^{13}C-bound) ^{1}H that is undergoing chemical-shift evolution during the t_1 period, we see that they have a 99% chance of being bonded to a ^{12}C, since the abundance of ^{13}C is only 1%. The only exception is the case of geminal coupling ($^2J_{HH}$), where the proton is attached to the same carbon, which has to be a ^{13}C. The trick is to apply a "magic" selective 180° pulse that only affects protons bonded to ^{12}C, at the same time as the ^{13}C 180° pulse in the middle of the t_1 period (Fig. 9.10). From the point of view of the proton we are observing in F_1, it is undergoing chemical-shift evolution during the t_1 period. The 180° ^{13}C pulse in the center of t_1 reverses the $^1J_{CH}$ coupling evolution from its directly bound ^{13}C, and the 180° ^{1}H "magic pulse" on the ^{12}C-bound protons reverses its J-coupling evolution from the vicinal ^{1}H coupling.

How do you generate a 180° ^{1}H pulse that only hits ^{12}C-bound protons? This magic is accomplished by a spin-echo sequence called bilinear rotation decoupling (BIRD), which takes advantage of the different J-coupling evolution of the ^{13}C-bound protons and the ^{12}C-bound protons (Fig. 9.11). For the ^{12}C-bound ^{1}H magnetization, this is just a 180° inversion pulse (two 90° pulses with a spin echo sandwiched in between). The spin-echo part is effectively invisible because the 180° ^{13}C pulse has no effect and the 1/J delay does not lead to any J-coupling evolution (J refers to $^1J_{CH}$ as before). Starting with $\mathbf{I}_z$, we rotate the magnetization to $-y'$, and the spin-echo returns it to $-y'$, reversing any chemical-shift evolution. The final $90°_x$ pulse rotates it from $-y'$ down to $-z$. This is just what we want because we wish to reverse any J-coupling evolution due to this proton. For the ^{13}C-bound ^{1}H magnetization, however, the spin echo leads to $^1J_{CH}$ evolution for a total of 1/J s, rotating the magnetization from $-y'$ to $+y'$, just as it does in the APT experiment (Chapter 6). The final $90°_x$ pulse rotates it back up to $+z$, so the ^{13}C-bound ^{1}H nuclei are not affected by this sequence. We can use product operators to verify that for the ^{13}C-bound protons the BIRD sequence is equivalent to a simple 180° ^{13}C pulse, just like the center of the t_1 delay in Figure 9.10:

	$-90°_x(^1H)\to$		$-1/(2J)\to$		$-180°_y(^1H)/180°(^{13}C)\to$		$-1/(2J)\to$		$-90°_x(^1H)\to$	
I_z		$-I_y$		$2I_xS_z$		$2I_xS_z$		I_y		I_z
I_x	$\to$	I_x	$\to$	$2I_yS_z$	$\to$	$-2I_yS_z$	$\to$	I_x	$\to$	I_x
I_y	$\to$	I_z	$\to$	I_z	$\to$	$-I_z$	$\to$	$-I_z$	$\to$	I_y
$2I_xS_z$	$\to$	$2I_xS_z$	$\to$	I_y	$\to$	I_y	$\to$	$-2I_xS_z$	$\to$	$-2I_xS_z$
$2I_yS_z$	$\to$	$2I_zS_z$	$\to$	$2I_zS_z$	$\to$	$2I_zS_z$	$\to$	$2I_zS_z$	$\to$	$-2I_yS_z$

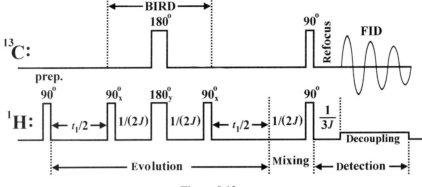

Figure 9.12

Every possible component of ^{13}C-bound ^{1}H net magnetization is affected the same way it would be affected by a simple 180° ^{13}C inversion pulse. For the ^{12}C-bound net magnetization:

I_z	→	$-I_y$	→	$-I_y$	→	$-I_y$	→	$-I_y$	→	$-I_z$
I_x	→	I_x	→	I_x	→	$-I_x$	→	$-I_x$	→	$-I_x$
I_y	→	I_z	→	I_z	→	$-I_z$	→	$-I_z$	→	I_y

The sequence is exactly like a 180°$_y$ pulse on the ^{1}H channel. Throughout we are ignoring ^{1}H chemical-shift evolution during the $1/(2J)$ delays, because the 180° proton pulse in the center will refocus it. We will see other variants of this strategy later. By reversing the phase of the final ^{1}H 90° pulse, we can do just the opposite: deliver a 180° ^{1}H pulse to the ^{13}C-bound protons while leaving the ^{12}C-bound protons alone. If we change the 90° pulses to 45° pulses, we can selectively deliver an overall 90° ^{1}H pulse to the ^{13}C-bound protons alone or to the ^{12}C-bound protons alone. This variant is called TANGO.

So the BIRD sequence is effectively a selective 180° ^{1}H pulse that applies only to ^{12}C-bound protons. The final version of our HETCOR pulse sequence is shown in Figure 9.12. This is beginning to look pretty complicated, but most of the details just have to deal with controlling what is refocused and what is allowed to evolve during the t_1 period. The resulting spectrum will show only single peaks in F_1 for each proton chemical shift, and single peaks in F_2 for each ^{13}C chemical shift (Fig. 9.7). This leads to a remarkable simplification of heavily overlapped regions of the proton spectrum. We will see in Chapter 11 that the HETCOR experiment is largely obsolete, replaced by the more sensitive inverse experiment HSQC, which correlates ^{1}H to directly bound ^{13}C by transferring in the reverse direction, from ^{13}C (F_1) to ^{1}H (F_2).

9.3 A GENERAL OVERVIEW OF 2D NMR EXPERIMENTS

9.3.1 Display of 2D NMR Data

The 2D data matrix is just an array of numbers (intensities) arranged in rows and columns: 2048 columns and 1024 rows is typical. The numbers themselves can be positive or negative

A GENERAL OVERVIEW OF 2D NMR EXPERIMENTS 365

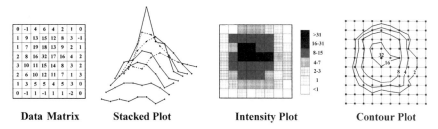

Data Matrix **Stacked Plot** **Intensity Plot** **Contour Plot**

Figure 9.13

and are typically very large (e.g., -2×10^9 to $+2 \times 10^9$). How can we represent this "third dimension"—the intensity values—on a two-dimensional piece of paper or computer screen? We could use a stacked plot of 1D spectra (Fig. 9.13), with a horizontal offset to keep peaks from one spectrum from falling on top of the same peak in the next spectrum. Software can even "whitewash" the spectra behind a peak to make the peak stand out like a three-dimensional object. The nice thing about a stacked plot is that we can see the noise level, but for any degree of complexity it is not practical. An intensity plot is just a color code of each data element ("pixel") according to a color key at the side of the spectrum. Typically red and yellow colors are used for positive values and blue colors are used for negative values. This is very fast for a computer to display (Varian *dconi* command) but when we look at the details of a crosspeak it is difficult to see the fine structure. The standard for display of 2D spectra today is the contour map or contour plot. If you are a hiker, you are familiar with the topographic maps that show elevation as a "third dimension" of the map. If you imagine that the world is flooded to the level of 1000 ft. above sea level, for example, the shoreline is drawn on the map as the 1000 ft. contour. A mountain is represented as a series of concentric circles, successive shorelines around the "island" that would be left if the area were flooded. We can do the same thing to describe an NMR crosspeak, with contours showing positions of equal intensity. For example, we might connect all "pixels" or data cells that have intensities of 2 with the "2" contour line. All data values of 4 would be connected to create the next higher contour, and data values of 8 would be used for the next (Fig. 9.13). Because data values are digitized in discreet "boxes," we need to interpolate if we cannot find the exact value we are looking for. In topographic maps we use even intervals of elevations (e.g., 40 or 80 ft contour intervals), but in NMR we use a geometric series (e.g., 2, 4, 8, 16, 32) because it fits the Lorentzian lineshape better. A Lorentzian mountain would be a real challenge for hikers! For an NMR contour plot, we need to decide on a contour *threshold* (everything below this value is ignored), a number of contours and a contour interval or multiplier. If the threshold is 100 and the multiplier is 1.5, we have contour levels of 100, 150, 225, 338, 506, and so on, up to the number of contours. For some 2D data, negative values are important, so we might choose to show both negative and positive contours. On the computer screen, a typical display consists of 10 positive and 10 negative contours, with a multiplier of 1.20 between levels. If the threshold is 1000, we have contour levels of:

In *white*:	1000	1200	1440	1728	2074	2488	2986	3583	4300	5160
In *red*:	−1000	−1200	−1440	−1728	−2074	−2488	−2986	−3583	−4300	−5160

A contour plot hides many evils. Because we can set the threshold as high as we want, we can eliminate impurities and artifacts without being accused of fraud! To be honest (and avoid missing important details) it is best to move the threshold down until a little bit of noise is visible. It's like trying to see a boat in a choppy ocean: a supertanker is easy to see without showing any of the waves, but if you want to find a canoe you'd better bring your threshold down until you see just the tops of the waves. Beginners in the world of 2D NMR always want to print out their 2D data on paper, but this is difficult because you really need to look at each part of the spectrum at several different contour thresholds, first displaying only the most intense peaks and then moving down until the noise is visible. Aligning peaks and comparing different 2D experiments is extremely difficult on paper. I remember a protein NMR lab in 1990 that had a room full of full-sized drafting tables, with researchers working on table-sized printouts of 2D NMR data using very sharp pencils to align crosspeaks. Soon afterward all of this was replaced with computer programs with sophisticated graphic displays that allow side-by-side display of different regions of the same 2D spectrum or corresponding regions of two different experiments, with correlated crosshairs that move simultaneously in both spectra under mouse control. Although the NMR instrument manufacturers all have software for 2D NMR data processing and analysis, a number of "third party" (neither Bruker nor Varian) programs are available that are far superior to software on the spectrometers: Felix (Accelerys, Inc.), MestRec, NUTS (Acorn NMR), and NMRpipe/NMRview (freeware) are examples.

9.3.2 2D Data Acquisition and Processing in General

The raw data from a 2D experiment consist of a series of FIDs, each acquired with a slightly longer t_1 delay than the previous one. Varian creates an array of FIDs, with the t_1 delay (parameter $d2$) arrayed (e.g., $d2 = 0$, 0.001, 0.002, 0.003, and so on). Bruker puts the FIDs together in a "serial file" (filename *ser*) and uses the parameter $d0$ (*d*-zero) for the t_1 delay in the pulse sequence, increasing it by the increment $in0$ with the pulse program command $id0$ (increment *d*-zero). Keep in mind that the heart of the 2D experiment is the transfer of magnetization from nucleus A to nucleus B during the mixing step. The first step in processing a 2D dataset is to Fourier-transform each of the FIDs in the array. The resulting spectra are loaded into a data matrix (like a spreadsheet) with the rows representing individual spectra in order of t_1 value, with the smallest t_1 value as the bottom row. The horizontal axis is labeled F_2, which is the chemical shift observed directly in each FID, and the vertical axis is t_1, the evolution delay. Each row in the 2D matrix represents a spectrum acquired with a different t_1 delay, and each column in the matrix represents either noise (if the F_2 value of that column is in a noise region of the spectrum) or, if F_2 is the frequency of nucleus B (Ω_b), the column is a t_1 "FID" with maximum intensity at the bottom and oscillating in a decaying fashion as we move up to higher t_1 values (Fig. 9.14). The frequency of this oscillation is just the chemical shift (Ω_a) of nucleus A. Of course, a real sample has more than one peak in its spectrum, so there would be other columns containing different t_1 FIDs.

The second step in processing the 2D data is to perform a second Fourier transform on each of the *columns* of the matrix. Most of columns will represent noise, but when we reach a column which falls on an F_2 peak, transformation of the t_1 FID gives a spectrum in F_1, with a peak at the chemical shift of nucleus A (Fig. 9.15). The final 2D spectrum is a matrix of numerical values that has a pocket of intensity at the intersection of the horizontal line $F_1 = \Omega_a$ and the vertical line $F_2 = \Omega_b$ and has an overall intensity determined by the efficiency of transfer of magnetization from nucleus A to nucleus B. This efficiency tells us something about the relationship (J value or NOE intensity) between the two nuclei

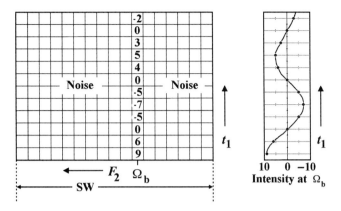

Figure 9.14

within the molecule. Simultaneously with the process we described, other pairs of nuclei are undergoing the same evolution, mixing and detection process resulting in other crosspeaks at the intersections of the appropriate chemical-shift lines and with characteristic intensities representing the efficiency of transfer of magnetization. The 2D spectrum thus represents a complete map of all interactions that lead to magnetization transfer, with the participants in the interaction addressed by their chemical shifts.

The important concept here is the "labeling" of the magnetization with the chemical shift of nucleus A during the evolution period, and the subsequent unraveling of this information to link nucleus **A** to nucleus **B**. Let's look at this process in more detail. Each FID of the 2D experiment samples a single point in the *indirect* time domain t_1, in the same way that the ordinary 1D FID is sampled at discreet, evenly spaced time points by the analog to digital converter (ADC) during a *direct* (real-time) acquisition. During the evolution (t_1) period, the x' component of the nucleus **A** magnetization is a cosine or sine function with amplitude A and angular frequency Ω_a that decays due to dephasing of individual nuclei with time constant T_2^*:

$$M_x^a(t) = A \cos(\Omega_a t) \times \exp(-t/T_2^*)$$

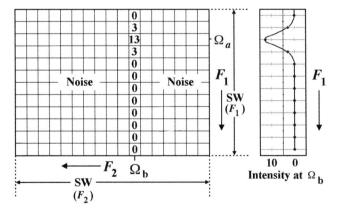

Figure 9.15

Now assume that the x component of nucleus **A** magnetization is transferred to nucleus **B** by the mixing sequence (most mixing schemes can only transfer one component of the magnetization). The transferred component might be z magnetization or antiphase magnetization or even DQC, but for simplicity we will use M_x. Magnetization transfer gives **B** magnetization whose intensity is modulated by the factor $M_x^a(t_1)$ that depends on the offset Ω_a of nucleus **A** and the value of t_1 at the moment of magnetization transfer. This **B** magnetization then precesses during the directly observed FID, inducing a signal in the probe coil. This signal is the normal FID of nucleus **B**, multiplied by the modulation (nucleus **A**) factor and the efficiency of transfer factor:

$$\text{FID} = M_x^a(t_1) \times G_{ab} \times \text{FID(B)}$$
$$\text{FID} = A \cos(\Omega_a t_1) \exp(-t_1/T_2^a) \times G_{ab} \times \cos(\Omega_b t_2) \exp(-t_2/T_2^b)$$

where G_{ab} is the efficiency of magnetization transfer (a function of J_{ab} or r_{ab}, depending on the type of experiment), and t_2 is the direct time domain of the FID. It is this function of Ω_a multiplying the directly observed FID that "labels" the nucleus **B** information with the chemical shift of the original nucleus, **A**. If we sample the t_1 values in the same way we sample the t_2 values, starting with zero and incrementing by a "dwell time" short enough to distinguish all of the frequencies expected, we have all the information needed to determine Ω_a and Ω_b, the chemical shifts of nuclei **A** and **B**. This is all of the information we can expect to obtain from a 2D NMR experiment: the chemical shifts of each pair of nuclei involved and the intensity of their interaction. To get the information out we need to do *two* Fourier transforms: the first in t_2, the second in t_1.

Each individual FID is an oscillating and decaying function of t_2, with the first two terms above equal to a constant. Fourier transformation gives a spectrum of **B** multiplied by the same constant:

$$\text{Spectrum}(t_1, F_2) = A \cos(\Omega_a t_1) \exp(-t_1/T_2^a) \times G_{ab} \times \text{Spectrum}_b(F_2)$$

We have a different spectrum of **B** for each t_1 value, differing only in the value of the first term. For each column in the data matrix, we have a function of t_1 for a fixed value of F_2 (Fig. 9.14, right). Now the first term is the variable (function of t_1) and the last term is a constant. Fourier transformation of the column converts the t_1 FID into a spectrum of **A** in the indirect frequency domain F_1:

$$\text{Spectrum}(F_1, F_2) = \text{Spectrum}_a(F_1) \times G_{ab} \times \text{Spectrum}_b(F_2)$$

For the data matrix, this means pulling out each column of the matrix in succession, treating it as an FID in t_1, performing a Fourier transform and putting the resulting F_1 spectrum back into the matrix at the same column position. When all the columns of the matrix (all the t_1 FIDs indexed by the frequency F_2) have been transformed into **A** spectra, we have a 2D spectrum, which is a function in F_1 and F_2. If we fix F_2 at the offset (chemical shift) of **B** ($F_2 = \Omega_b$), we are looking at a vertical slice through the crosspeak, which is Spectrum$_a$ (F_1) (Fig. 9.15, right). If we fix F_1 at the offset of **A** ($F_1 = \Omega_a$), we have a horizontal slice through the crosspeak, which is Spectrum$_B$ (F_2). The intensity of the crosspeak is determined by the efficiency of magnetization transfer G_{ab}. Depending on the mixing scheme used (i.e., sequence of pulses and/or delays between evolution and detection) the selected relationship (interaction) might be a proximity in space (NOE) or a small number of bonds separating the two nuclei (*J* coupling).

9.3.3 Taxonomy of 2D NMR Experiments

By now you should be in the habit of thinking of a 2D NMR experiment as a transfer ("jump") of magnetization from nucleus **A** (the F_1 frequency on the vertical axis) to nucleus **B** (the F_2 frequency on the horizontal axis). All of the 2D experiments in use can be classified by the two types of nuclei detected in the direct (F_2) and indirect (F_1) dimensions and the criteria for magnetization transfer during the mixing step. The mixing pulse sequences are designed to select for certain types of interactions between nuclei and can be divided into two categories: magnetization transfer based on a *J*-coupling interaction and magnetization transfer based on an NOE interaction. We can further divide the 2D experiments into homonuclear (experiments that transfer magnetization from one nucleus to another nucleus of the same type, usually ^{1}H to ^{1}H) and heteronuclear (experiments that transfer magnetization between two different types of nucleus, e.g., ^{1}H and ^{13}C).

Homonuclear Experiments (^1H–^1H)

Name	F_1 Nucleus	Mixing	F_2 Nucleus
COSY	^{1}H	J (single transfer)	^{1}H
TOCSY	^{1}H	J, J, J, ... (multiple transfers)	^{1}H
NOESY	^{1}H	NOE	^{1}H
ROESY	^{1}H	ROE (spin-lock NOE)	^{1}H

Heteronuclear Experiments (^1H–^{13}C)

Name	F_1 Nucleus	Mixing	F_2 Nucleus
HETCOR	^{1}H	$^1J_{CH}$	^{13}C
HSQC	^{13}C	$^1J_{CH}$ (^{13}C SQC during t_1)	^{1}H
HMQC	^{13}C	$^1J_{CH}$ (^{1}H–^{13}C MQC during t_1)	^{1}H
HMBC	^{13}C	$^{2,3}J_{CH}$ (long-range couplings)	^{1}H

Homonuclear experiments are characterized by a *diagonal* defined by $F_1 = F_2$ and by pairs of crosspeaks at symmetrical positions across the diagonal: $F_1 = \Omega_a$, $F_2 = \Omega_b$ and $F_1 = \Omega_b$, $F_2 = \Omega_a$. This is because both magnetization transfer pathways, ^{1}H$_a \rightarrow {}^1$H$_b$ and ^{1}H$_b \rightarrow {}^1$H$_a$, can be observed. Heteronuclear experiments lack a diagonal and diagonal symmetry. The range of *J* values can be selected in mixing schemes: for ^{1}H–^{13}C couplings, $^1J_{CH}$ (the one-bond or direct coupling) is very large (125–180 Hz) whereas the long-range $^{2,3}J_{CH}$ (two and three-bond coupling) is small (2–12 Hz). Some mixing sequences can allow multiple "jumps" of magnetization: the TOCSY experiment allows for many jumps based on *J* coupling—the first from the F_1 nucleus (**A**) to an intermediate (undetected) nucleus (**C**). Other jumps may transfer the magnetization to other undetected nuclei (e.g., **D** and **E**) and a final jump carries it to the F_2 nucleus (**B**), thus spreading the magnetization out over an entire "spin system" or group of protons interconnected by *J* couplings.

From these basic experiments have grown many variants. The COSY has been extended to DQF-COSY (reduced diagonal intensity and improved phase properties) and COSY-35 (simplified crosspeak structure for *J*-value determination). A common variant of NOESY is the ROESY, which gives better results for molecules in the size range of peptides, oligosaccharides and large natural products. HSQC gives the same results as HMQC but has better relaxation properties for large molecules such as proteins. All 2D experiments and even 3D and higher dimensional experiments are based on the above list of basic 2D experiments.

In all this complexity of acronyms it is easy to forget that all 2D experiments do the same thing: they allow you to correlate two atoms (nuclei) in a molecule based on an interaction that is either through bond (*J* coupling) or through space (NOE). The two nuclei are identified by their chemical shifts, and the correlation appears in the 2D spectrum as a crosspeak at the F_1 chemical shift ("*y* coordinate") of the nucleus where magnetization starts and the F_2 chemical shift ("*x* coordinate") of the nucleus to which the magnetization is transferred. Thus the basis of all 2D experiments is the "jump" or transfer of magnetization. The information (*J* value or NOE intensity) can be used to define structural relationships (dihedral angle or distance) but is only useful if we can unambiguously assign the two chemical shifts (F_1 and F_2) to specific positions within the molecule.

9.4 2D CORRELATION SPECTROSCOPY (COSY)

COSY is the first and the simplest 2D experiment. It correlates one proton (H_a) to another (H_b) via a single *J* coupling that may be 2-bond (geminal), 3-bond (vicinal) or in rare cases 4-bond or 5-bond (long range). The pulse sequence is simply $90° - t_1 - 90° -$ FID (Fig. 9.16). Consider the interaction of two *J*-coupled protons, H_a and H_b (Fig. 9.17). The preparation pulse rotates the H_a magnetization from the *z* axis into the *x*–*y* plane. During the evolution (t_1) period, H_a magnetization precesses in the rotating frame at a rate dependent upon its chemical-shift offset, Ω_a. At the same time, *J*-coupling evolution occurs to produce H_a magnetization that is antiphase with respect to its *J* coupling with H_b. As with the INEPT transfer in the HETCOR experiment, simultaneous 90° pulses applied to H_a and H_b transfer antiphase H_a magnetization to antiphase H_b magnetization (this is actually accomplished with a single nonselective 90° ^{1}H pulse). During the detection period (FID), H_b magnetization precesses at its characteristic rate (Ω_b) in the rotating frame, inducing a voltage in the probe coil, which is digitized as the FID. Fourier transformation in F_2 and then in F_1 leads to a 2D data matrix with a crosspeak at $F_1 = \Omega_a, F_2 = \Omega_b$ (Fig. 9.18). This basic COSY sequence is not used much any more—everyone uses the double-quantum filtered COSY or DQF-COSY sequence. Until we get to the actual DQF-COSY pulse sequence and understand how it works, however, we will treat these two experiments as equivalent.

The appearance of a homonuclear 2D spectrum is different from what you have seen for HETCOR, the ^{1}H–^{13}C correlation. Because both frequency scales, F_2 and F_1, are ^{1}H chemical-shift scales, we can observe the transfer of coherence from H_a to H_b as a crosspeak at $F_1 = \Omega_a, F_2 = \Omega_b$ (Fig. 9.18, lower right), as well as the opposite sense of transfer from H_b to H_a, a symmetrically disposed crosspeak at $F_1 = \Omega_b, F_2 = \Omega_a$ (upper left crosspeak). In HETCOR we observe ^{13}C coherence (in F_2) that was transferred from the attached proton, whose chemical shift appears in F_1. Transfer in the opposite sense (^{13}C to ^{1}H) is not possible

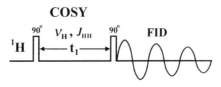

Figure 9.16

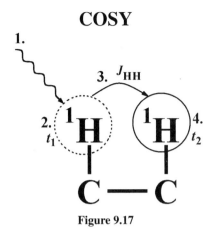

Figure 9.17

because the pulse sequence starts with ^{1}H excitation and the detector is set up to observe only signals at the radio frequency of ^{13}C. Furthermore, if transfer of magnetization "fails" (i.e., if some coherence remains on the ^{1}H after the mixing step), the ^{1}H coherence is not detected and cannot be displayed in the 2D spectrum. But in a homonuclear experiment like COSY, "failed" coherence transfer from H$_a$ means that we encode the frequency of H$_a$ during t_1 and then observe the same frequency in t_2. The FID has the form:

$$\cos(\Omega_a t_1) \times \cos(\Omega_a t_2)$$

which leads to a peak in the 2D spectrum at $F_1 = \Omega_a$ and $F_2 = \Omega_a$. We call this a *diagonal peak* because it falls on the diagonal line defined by $F_1 = F_2$, running from the lower left corner to the upper right corner of the 2D data matrix (Fig. 9.18). Because coherence transfer is never 100% complete at the end of the mixing sequence, we will always see a diagonal peak for each resonance in the ^{1}H spectrum. That means that if we trace along the

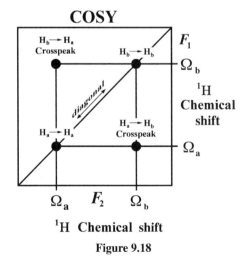

Figure 9.18

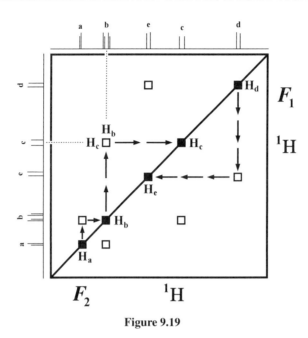

Figure 9.19

diagonal line from lower left to upper right we will trace out the 1D proton spectrum. Keep this in mind because you do not always have the luxury of a ^{1}H spectrum displayed along the top and side of the 2D spectrum. After a while you can wean yourself away from this crutch and begin to view the diagonal as your 1D spectrum.

Figure 9.19 shows a COSY spectrum of a molecule with two simple spin systems separated by a quaternary carbon: $-CH_a-CH_b-CH_c-C_q-CH_d-CH_e-$. One can "walk" through each spin system by moving from diagonal to cross-peak vertically, back to the diagonal horizontally, and repeating this process. Note that either of the two symmetrical crosspeaks can be used for this "walk". The H_b-H_c crosspeak (upper left) shows how the H_b peak in the 1D spectrum at the top is correlated to the H_c peak in the 1D spectrum on the left side. This crosspeak represents coherence on H_c that transferred to H_b in the mixing step. We will always label crosspeaks as shown, with the F_2 assignment on the top or bottom and the F_1 assignment to the left or right of the crosspeak. This process of "walking" (diagonal–crosspeak–diagonal) is especially useful when each carbon has only one proton, or group of equivalent protons, so that each crosspeak is a jump from a proton on one carbon to a proton on the next carbon (a vicinal coupling), a common occurrence in carbohydrates. The walk sometimes doubles back on itself, since the chemical shift changes can reverse direction as we move through a spin system (Fig. 9.20). In this case we walk from the crosspeak to the diagonal peak, passing over the crosspeak for the next jump. The biggest problem occurs when two adjacent (vicinal) protons have identical or nearly identical chemical shifts: then the crosspeak connecting them is on the diagonal or very near the diagonal. Consider the molecular fragment $-CH^1-CH^2-CH^3-C_q^4-CH^5-CH^6-CH^7-$ with two spin systems separated by the quaternary carbon C4. Suppose that the H_a resonance can be assigned to H1 and the H_f resonance can be assigned to H7 (Fig. 9.21) based on chemical shifts or coupling constants. Using the COSY data, we can connect both H_a (H1) and H_f (H7) to the overlapped two-proton peak $H_{d/e}$. So we know that this peak contains

2D CORRELATION SPECTROSCOPY (COSY) 373

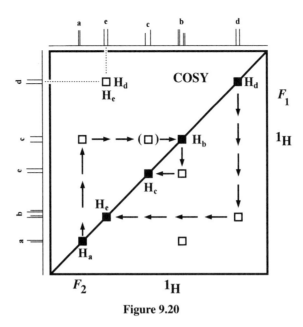

Figure 9.20

H2 and H6. But where do we go from here? The $H_{d/e}$ peak connects to H_b and to H_c, but we cannot tell which is H3 and which is H5 because we lost the specific trails of the two spin systems when they "crossed" at $H_{d/e}$.

We can resolve this ambiguity with a TOCSY spectrum. This is just a homonuclear 2D (^1H–^1H) experiment with the TOCSY spin lock as the mixing portion of the pulse sequence. With the fragment CH^1–CH^2–CH^3–C_q^4–CH^5–CH^6–CH^7 we expect to see H1 correlated

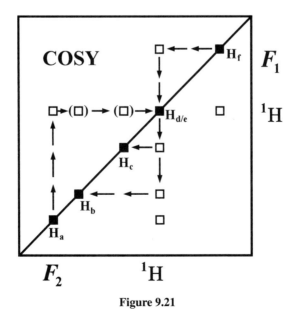

Figure 9.21

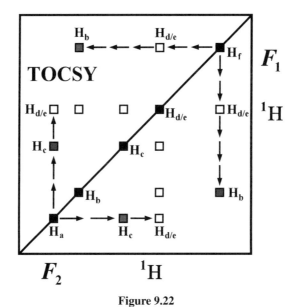

Figure 9.22

to H2 as it was in the COSY spectrum, but we will also see H1 correlated to H3 because the TOCSY spin lock transfers coherence in multiple *J*-coupling jumps. Likewise, H7 will correlate to H6 but also to H5 (Fig. 9.22). Thus there is no ambiguity and we can directly "read off" the two spin systems. Starting with H_a on the diagonal at the lower left, we can extend a line up or to the right and pass through the H_c and the $H_{d/e}$ crosspeaks. Starting with H_f on the diagonal at the upper right, we can move down or to the left and pass through the $H_{d/e}$ and H_b crosspeaks. We can assign peak c to H3 and peak b to H5. The "new" crosspeaks that represent multiple *J*-coupling jumps are shown in gray. The disadvantage is that we lose the information about the order of protons in the spin system: without the COSY data, we don't know if the order is CH_a–$CH_{d/e}$–CH_c or CH_a–CH_c–$CH_{d/e}$. We saw in Chapter 8 that to some extent the intensity of TOCSY transfer peaks can help us to put them in order because in long spin systems the "smearing" of coherence tends to be a diffusion-like process, giving less intensity to peaks that result from larger numbers of jumps. But to be certain of the exact order of protons in a spin system requires a COSY spectrum, which is limited to single "jumps" through *J* couplings.

9.4.1 Examples of 2D COSY Spectra

Figure 9.23 shows the 600 MHz DQF-COSY spectrum of 3-heptanone. As a starting point, we will use the two CH_2 groups closest to the ketone carbonyl. These protons should be shifted downfield to the region of 2–2.5 ppm (compare to acetone at 2.1 ppm). There are two overlapped peaks on the diagonal (lower left) that can be identified as separate peaks because their crosspeaks do not line up: the downfield peak lines up vertically and horizontally with crosspeaks at 0.90 ppm, whereas the upfield peak lines up with crosspeaks at 1.42 ppm. Following the spin system from the downfield diagonal peak (2.30 ppm), we can move to the right and up (or up and to the right) to the diagonal peak at 0.90 ppm. This "box" (dotted lines) is a dead end: there are only two resonances in the spin system. So we can

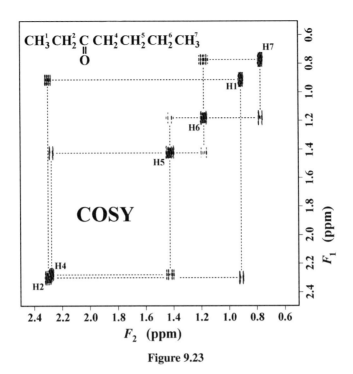

Figure 9.23

assign it to H2 (2.30 ppm) and H1 (0.90 ppm). This spin system represents the ethyl group: CH_2–CH_3. By process of elimination, the upfield peak of the overlapped pair (2.27 ppm) must be H4 on the other side of the ketone carbonyl. Following the crosspeaks on either side of the diagonal (dotted lines) we move to H5 (1.42 ppm), then to H6 (1.18 ppm), and finally to the methyl group H7 (0.77 ppm). The dotted lines show how this spin system of four resonances (*n*-butyl group) is distinct from the spin system of two resonances (ethyl group). Note how the methyl groups are the most upfield, both because they are farthest from the carbonyl group and because there is an inherent upfield shift as we move from CH to CH_2 to CH_3 (δ1.6, 1.2, and 0.85 in a saturated hydrocarbon).

For comparison, the 600 MHz TOCSY spectrum of 3-heptanone is shown in Figure 9.24. The first thing you notice is that there are a lot more crosspeaks! This does lead to more clutter but now instead of having to "walk" from diagonal to crosspeak to diagonal, you can move horizontally or vertically and spell out the entire spin system along one line. The top line, starting at the right with the H7 peak on the diagonal, moves left through crosspeaks for H6, H5, and H4 (it is only coincidence that they are in order). Because the crosspeaks are in-phase, we can even read off their splitting patterns: triplet for H7, quartet for H6, quintet for H5, and triplet for H4. This fine structure is only visible in the F_2 (horizontal) direction; in F_1 the resolution is poor because we collect far fewer data points in the t_1 FID. Along this top horizontal line the magnetization started with the H7 methyl group ($F_1 = 0.77$ ppm), transferring in one *J*-coupling "jump" to give the crosspeak at F_2 = H6, two "jumps" of TOCSY transfer to give the H5 crosspeak, and three "jumps" to reach H4.

There are typically two kinds of artifacts in COSY spectra: t_1 noise (vertical streaks) and DQ artifacts. The t_1 streaks extend up and down from the most intense and sharpest peaks along the diagonal. These result from any instability that can cause random variations

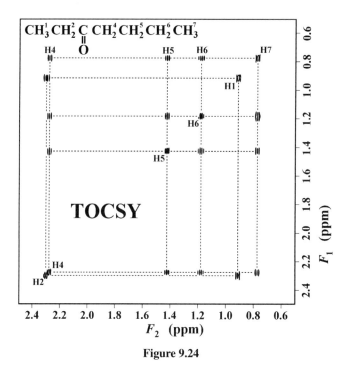

Figure 9.24

between different FIDs in the 2D acquisition (different t_1 values). If after the first Fourier transform the peak heights in the interferogram (Fig. 9.2) are not exactly reproducible from spectrum to spectrum, this introduces random "noise" in the t_1 FID and after the second Fourier transform we have noise in the F_1 spectrum which is inserted back into the data column. This "noise" only shows up in the columns which correspond to peaks in the spectrum, so it looks like vertical streaks in the 2D spectrum. Because they are truly random variations in intensity, however, there are no "tricks" of smoothing or subtraction that can remove them. We can only try to reduce any source of variation (unstable electronics, temperature variation, sample spinning, building vibration, etc.) during the 2D data acquisition. For this reason, we never spin the sample during a 2D acquisition. The DQ artifacts show up along two lines extending from the lower left and lower right corners of the 2D spectrum to a point on the top in the exact center in F_2 (Fig. 9.25), at the F_2 frequency of each peak on the diagonal. Again, the stronger peaks in the diagonal tend to give the most intense artifacts. These can be distinguished from crosspeaks because they are not symmetrically disposed about the diagonal. If you have any doubt about a crosspeak, always check for its partner on the other side of the diagonal.

A portion of the 600 MHz DQF-COSY spectrum of 15-β-hydroxytestosterone (Fig. 9.26) is shown in Figure 9.27. As before, F_1 assignments are written to the left side or right side of a crosspeak and F_2 assignments are written above or below the crosspeak. The amount of information packed into this single experiment is amazing! The assignments were obtained by using this data along with other 2D experiments that connect ^{1}H to ^{13}C (Chapter 11). There are several overlapped groups on the diagonal; for example, at the lower left side we see two overlapped peaks on the diagonal. The downfield one is H6β alone and the larger upfield one is H2β and H16α. We can see that H2β is slightly downfield of H16α by looking

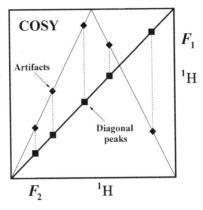

Figure 9.25

at the H2β – H1α crosspeak (center left), which is slightly to the left of the H16α – H16β crosspeak just above it. You can also see that the H2β resonance is wider and appears as a triplet in F_2, whereas the H16α resonance is narrower. In this way, we can see resolved crosspeaks even when they are badly overlapped or even precisely the same chemical shift on the diagonal. This is one of the huge advantages of 2D NMR.

For resolved peaks we can easily move along a horizontal or vertical line to identify all of its coupling partners. Starting at the right with the H8 resonance ($F_1 = 2.03$ ppm), we can see COSY magnetization transfer (moving left) to H14, H9, H7α and then, passing through the diagonal, to H7β. All of these are vicinal relationships involving a single J coupling to H8. Starting at the center left side we can follow the $F_1 = $ H1α line to the right, passing the H2β, H2α, and H1β crosspeaks, then going through the diagonal peak to a very narrow crosspeak with H19, the angular methyl group singlet. How can we have a COSY crosspeak to a singlet? There is a small "W" coupling between the axial H1α proton and the H19 methyl group due to the *anti* relationship of H-1α and C19 (Fig. 9.26, bottom). At least one of the C19 methyl protons is in a "W" relationship with H1α, and this leads to a COSY crosspeak even though the coupling is not resolved in the 1D ^{1}H spectrum.

Figure 9.28 shows the 300 MHz DQF-COSY spectrum of sucrose in D_2O. For convenience, the ^{1}H spectrum is shown at the top and on the left side. To analyze the COSY spectrum you always need a starting point: a ^{1}H resonance that is resolved and can be

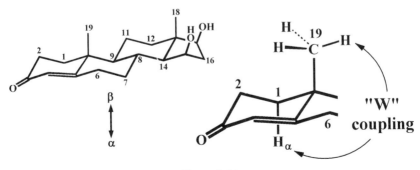

Figure 9.26

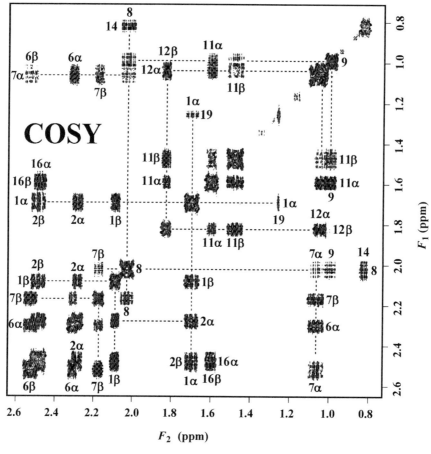

Figure 9.27

unambiguously assigned on the basis of its coupling pattern and/or chemical shift. The sucrose structure (Chapter 8, Fig. 8.44) predicts only two doublet resonances: H-g1 because it is coupled only to H-g2, and H-f3 because it is coupled only to H-f4. H-g1 is bonded to an anomeric carbon, with two bonds to oxygen, so it should be downfield of H-f3. So we can assign H-g1 to the doublet at 5.36 ppm ($J = 3.8$ Hz) and H-f3 to the doublet at 4.2 ppm ($J = 8.8$). Starting at 5.36 ppm (d, $J = 3.8$) with H-g1 on the diagonal (Fig. 9.28, lower left) we can move to the right to a prominent crosspeak (lower right) and then straight up passing through a crosspeak (rectangular box) and returning to the diagonal at 3.5 ppm (dd, $J = 10.0, 3.8$). This diagonal peak can be assigned to H-g2. Moving either left or back down from the H-g2 diagonal peak we encounter a crosspeak (rectangular boxes) that leads us back to the diagonal at 3.7 ppm (t, 10 Hz). This diagonal peak corresponds to H-g3. Before continuing our walk, let's take a closer look at the H-g1 to H-g2 crosspeak.

9.4.2 Fine Structure of COSY Crosspeaks

So far we have looked at the COSY crosspeaks as "blobs" of intensity which correlate one resonance (H_a in F_1) with another (H_b in F_2) at the intersection of their chemical-shift

2D CORRELATION SPECTROSCOPY (COSY)   379

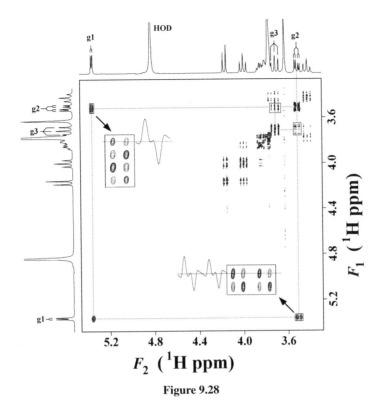

Figure 9.28

positions ($F_1 = \Omega_a$ and $F_2 = \Omega_b$). But we know that the resonances H_a and H_b are not single lines: at the very least we would have doublets for the H_a and H_b resonances if they are coupled only to each other. If we extend the four individual lines of the two doublets into the 2D spectrum, we see that there will be four peaks in each crosspeak and four peaks in each diagonal peak (Fig. 9.29). Positive intensities are shown as closed circles and negative intensities are shown as open circles. The crosspeak arises from a transfer of antiphase coherence to antiphase coherence by the mixing sequence (a single 90° pulse), so the peak intensities are antiphase (+/−) in each dimension. The upper left crosspeak ($F_2 = \Omega_a, F_1 = \Omega_b$) gives an antiphase doublet if we display a horizontal slice or a vertical slice through the peak. It often happens that the crosspeaks are "out in the open"—not overlapped with other crosspeaks—whereas the corresponding peaks in the 1D spectrum are overlapped and cannot be analyzed to extract J couplings. By making a slice through the crosspeak, we can measure the J coupling and assign it clearly to the coupling between the two resonances that give rise to the crosspeak (in this case, J_{ab} between H_a with frequency Ω_a and H_b with frequency Ω_b). When analyzing a 1D spectrum one can determine J couplings accurately in resolved multiplets, but it is not always clear which of the other peaks in the spectrum is the coupling partner for each of the couplings. The 2D COSY clearly shows us which resonances are coupled, and the fine structure of the crosspeak can give us the exact coupling constant value.

In the inset of the sucrose COSY spectrum (Fig. 9.28, upper left), the H-g2 (F_1) to H-g1 (F_2) crosspeak is enlarged to show its fine structure. In the 1D spectrum the H-g1 peak is a doublet ($J = 3.8$ Hz) because it is the anomeric proton (at the "end of the line") and

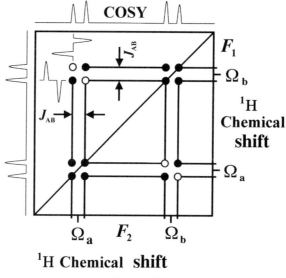

Figure 9.29

only has one coupling: to H-g2. If we make a horizontal slice of this crosspeak we get an antiphase doublet with $J = 3.8$ Hz. The symmetry-related crosspeak (Fig. 9.28, lower right inset) represents transfer of magnetization from H-g1 (F_1) to H-g2 (F_2). The H-g2 resonance in the 1D spectrum is a double doublet ($J = 10.0, 3.8$) because it is coupled to H-g1 (3.8 Hz, axial–equatorial) and to H-g3 (10.0 Hz, axial–axial). The horizontal (F_2) slice through this crosspeak shows the double-doublet structure, but it is antiphase (+,−) with respect to the 3.8 Hz coupling and in-phase (+,+ or −,−) with respect to the 10.0 Hz coupling. The antiphase H-g1 coherence, $2\mathbf{I}_y^1\mathbf{I}_z^2$, undergoes coherence transfer to give the antiphase H-g2 coherence, $2\mathbf{I}_y^2\mathbf{I}_z^1$, but because H-g2 is also coupled to H-g3 we see this coupling as well. Any coupling that does not appear as a multiplier ($\times \mathbf{I}_z$) in the operator product must be in-phase. This is recorded in the FID and after the F_2 Fourier transform we see that any pair of lines separated by the "passive" coupling J_{23} will have the same sign (+,+ or −,−) and any pair of lines separated by the "active" coupling J_{12} will have the opposite sign (+,− or −,+). The "active" coupling is simply the coupling that gave rise to the crosspeak. If the crosspeak is located at the intersection of the H-g1 chemical shift and the H-g2 chemical shift, the active coupling in its fine structure will be J_{12}, the coupling that led to the H-g1/H-g2 antiphase state and allowed coherence transfer to occur. The passive coupling (J_{23}) does not appear at all in the F_1 dimension of this crosspeak because the F_1 resonance is H-g1, which is only coupled to H-g2.

Figure 9.30 shows the upfield region of the 300 MHz DQF-COSY spectrum of sucrose. Starting at the upper left, we follow horizontally from the H-g2 (F_1)/H-g1 (F_2) crosspeak (from Fig. 9.28, upper left) to the right along the dotted line, passing through a crosspeak and on to the H-g2 diagonal peak. From here we go straight down to a crosspeak and then to the left back to the diagonal: this is the H-g3 diagonal peak. From here we reverse direction, moving horizontally to the right and passing the H-g2/H-g3 crosspeak to stop at another crosspeak, enclosed in a rectangular box. From here we move straight up, returning to the diagonal at the most upfield peak of the spectrum. We can assign this peak to H-g4 (a triplet

2D CORRELATION SPECTROSCOPY (COSY) 381

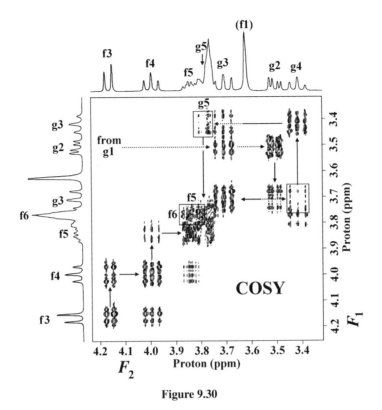

Figure 9.30

in the 1D spectrum). From here we move to the left, passing the H-g3/H-g4 crosspeak to a faint crosspeak just beyond it (rectangular box labeled "g5"). Moving up to the ^{1}H spectrum at the top of the COSY we can pinpoint the chemical shift of H-g5 in the overlapped region. This completes our walk through the glucose spin system. H-g6 (two protons) is too close in chemical shift to H-g5 to provide any useful crosspeak, but we can guess that its resonance is in the tall peak just upfield of H-g5.

Starting again with the H-f3 diagonal peak (Fig. 9.30, lower left), we move up and then right to the H-f4 diagonal peak, and then up again and right to the H-f5 diagonal peak ("X" shape). If we strain our eyes a bit, we can see a blob of intensity above the H-f5 diagonal peak (rectangular box) that leads us to the right side to the H-f6 diagonal peak. This corresponds to the same tall, overlapped peak in the 1D spectrum that we assigned to H-g6. This confirms the assignments we made in Chapter 8 based on selective 1D TOCSY experiments, and it also confirms our ^{13}C assignments made through the HETCOR correlations. Sucrose is assigned!

In the 300 MHz DQF-COSY spectrum of sucrose the crosspeaks appear very large because the chemical-shift range is small, particularly for the nonanomeric protons. One way to "shrink" the size of the crosspeaks is to move to a higher-field instrument. We saw in Chapter 2 how the "footprint" of a ^{1}H resonance (multiplet) gets smaller on the ppm scale as we increase the field strength. Figure 9.31 shows the upfield region of the 600 MHz DQF-COSY spectrum of sucrose. Comparing to Figure 9.30, all of the splittings have been cut in half and the crosspeaks and diagonal peaks are one half the size in both dimensions.

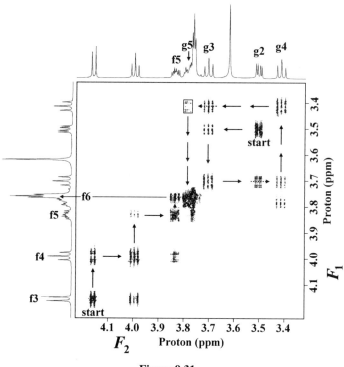

Figure 9.31

This is because 1 ppm now corresponds to 600 Hz rather than 300 Hz and the coupling constants in hertz have not changed. It is now much easier to follow the spin system from H-g2 to H-g5 and from H-f3 to H-f6 (Fig. 9.31, arrows). At 600 MHz the H-f5 resonance is completely resolved (a ddd coupled to H-f4 and the two H-f6 protons) and the H-f5 to H-f6 crosspeak is clearly visible.

In Chapter 5 we looked at the presaturation ^{1}H spectrum of a cyclic peptide in 90% H_2O/10% D_2O (Fig. 5.20). Figure 9.32 shows a portion of the DQF-COSY of the same cyclic peptide in D_2O, also using presaturation of the HOD resonance. The H_α resonances are seen on the diagonal in the region 3.9–5.4 ppm, with connections to the H_β protons shown by dotted lines. One H_α resonance in particular is labeled with crosspeaks to two β protons. This type of spin system is typical of a large number of amino acids and is called "three-spin" or "AMX": ND–CH$^\alpha$–CH$_2^\beta$–Y, where Y is either a heteroatom or a quaternary carbon (aromatic ring or carbonyl). Note that the amide NH is exchanged with deuterium in D_2O so it is no longer part of the spin system. Another AMX spin system is found at H_α = 4.77 ppm, slightly downfield of the HOD streak. Notice that the H_α peak is missing on the diagonal for this spin system, and there are no symmetry-related $H_\alpha \rightarrow H_\beta$ crosspeaks below the diagonal. In this case, the presaturation of HOD "wiped out" (saturated) the H_α proton so it could not transfer magnetization to H_β or $H_{\beta'}$. This kills the H_α diagonal peak (failed transfer from H_α) and the $F_2 = H_\beta$, $H_{\beta'}$ crosspeaks (transfer from H_α). But the F_1 = H_β, $H_{\beta'}$ crosspeaks are fine because they are far from HOD and are not affected by the presaturation. They transfer magnetization to H_α ($F_2 = H_\alpha$), which is observed in the FID.

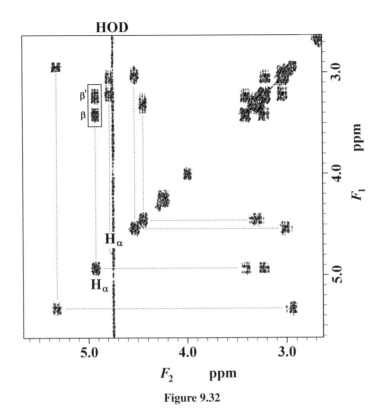

Figure 9.32

It might be possible to measure the side-chain dihedral angle χ_1 (chi-1), defined by the N–C_α–C_β–Y angle, if we can measure the H_α–H_β J-coupling constants. In the AMX spin system, there are three protons and each appears as a double doublet: H_α is split by H_β and $H_{\beta'}$; H_β is split by $H_{\beta'}$ and H_α; and $H_{\beta'}$ is split by H_β and H_α. Assuming that H_β and $H_{\beta'}$ have significantly different chemical shifts, we can diagram the expected crosspeak fine structure in the COSY spectrum (Fig. 9.33, left). For example, in F_2 we have the H_α resonance with a double doublet defined by $J_{\alpha\beta}$ and $J_{\alpha\beta'}$. The H_α–H_β crosspeak (bottom) will have the $J_{\alpha\beta}$ couplings antiphase (active coupling) and the $J_{\alpha\beta'}$ couplings in-phase (passive coupling). The H_α–$H_{\beta'}$ crosspeak (top) will have the $J_{\alpha\beta'}$ couplings antiphase (active coupling) and the $J_{\alpha\beta}$ couplings in-phase (passive coupling). In all, each of the two crosspeaks will have 16 peaks in the fine structure (double doublet in F_1 and dd in F_2). We could make a horizontal (F_2) slice through one of the rows and try to extract the active and passive couplings, but usually this is difficult due to crowding and the complexity of the pattern. A modified COSY experiment, called COSY-35, greatly reduces the intensity of every other peak in the fine structure, leaving only eight peaks in each crosspeak (Fig. 9.33, right). This is accomplished by simply reducing the pulse width of the final pulse in the COSY sequence from 90° to 35° (Fig. 9.34), reducing 8 of the peaks to an intensity of 10% $((1 - \cos\Theta)/(1 + \cos\Theta))$ of the other 8. Now an F_2 slice gives a simple antiphase doublet pattern representing the active coupling, or a direct measurement in F_2 can be made between peaks of the same sign (+ to +, or − to −) to measure the passive coupling (Fig. 9.33, right). Figure 9.35 shows an expansion of the $F_1 = H_\beta$, $H_{\beta'}/F_2 = H_\alpha$ crosspeak highlighted with a rectangle in Figure 9.32. The left side is the DQF-COSY spectrum, with 16 peaks in each crosspeak,

384 TWO-DIMENSIONAL NMR SPECTROSCOPY: HETCOR, COSY, AND TOCSY

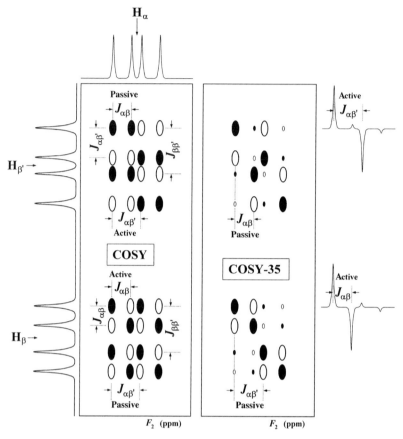

Figure 9.33

and the right side is the COSY-35 spectrum, with only eight peaks in each crosspeak. The contour threshold is set high enough to completely reject the other eight peaks that are much lower in intensity. An F_2 slice of the upper ($H_{\beta'}$) crosspeak yields the active coupling $J_{\alpha\beta'}$, which can be accurately measured by simulation and curve fitting. An F_2 slice of the lower (H_β) crosspeak shows an antiphase doublet with the active coupling $J_{\alpha\beta}$. Alternatively, direct measurement from peaks of like sign gives us the passive coupling $J_{\alpha\beta}$ from the upper ($H_{\beta'}$) crosspeak and the passive coupling $J_{\alpha\beta'}$ from the lower (H_β) crosspeak.

Figure 9.36 shows Newman projections of the three low-energy conformers for the C_α–C_β bond of an "AMX" amino acid residue within a peptide or protein. The *gauche* relationship should give a small coupling (<6 Hz) and the *anti* relationship should give a

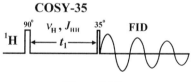

Figure 9.34

2D CORRELATION SPECTROSCOPY (COSY)   385

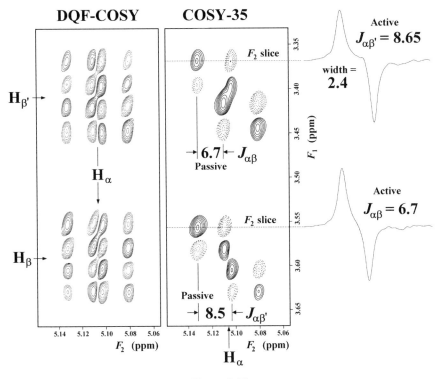

Figure 9.35

large coupling (>8 Hz). If both couplings ($J_{\alpha\beta}$ and $J_{\alpha\beta'}$) are small, we have conformer (a). If one is large and one is small, we have either (b) or (c). Since we do not know which β proton resonance is the pro-R and which is the pro-S (i.e., the β and β' labels could be swapped in Fig. 9.36), we cannot distinguish between these two possibilities. If the couplings are intermediate (6–8 Hz) we have a flexible molecule with averaging over more than one conformation. The example in Figure 9.35 is averaged J couplings ($J_{\alpha\beta'} = 8.65$, $J_{\alpha\beta} = 6.7$ Hz) with a slight preference for conformations (b) and (c).

When measuring J couplings, one should keep in mind that the simple distance between peak maxima for a doublet is not always an accurate measure of the J coupling, especially if the peak width is similar to the J coupling. In-phase doublets "creep" together as peak

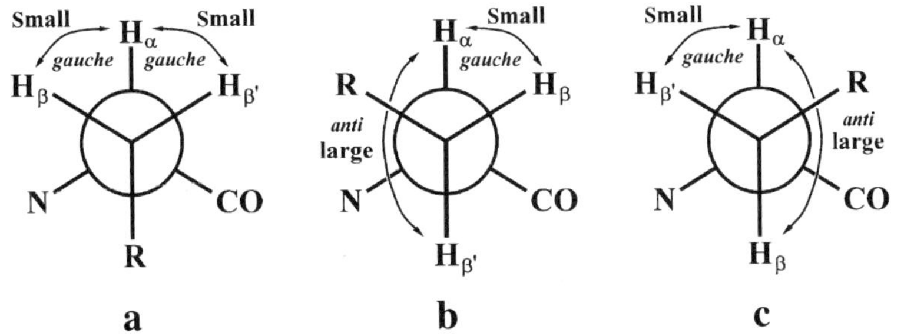

Figure 9.36

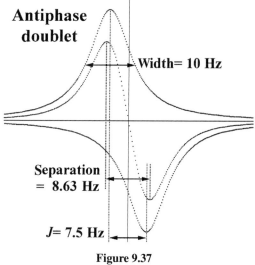

Figure 9.37

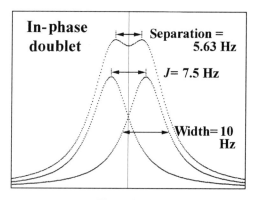

Figure 9.38

width increases, leading to an underestimate of the J value, whereas antiphase doublets move apart, leading to an overestimate. Figure 9.37 shows two calculated Lorenztian peaks with linewidth of 10 Hz, separated by a coupling of 7.5 Hz with one peak inverted. The sum is an antiphase doublet with a separation of 8.63 Hz, which would be mistakenly interpreted as an *anti* relationship (>8 Hz) instead of an average of conformers (actual $J = 7.5$ Hz). An in-phase doublet with the same 10 Hz linewidth and 7.5 Hz J value would give a separation of 5.63 Hz (Fig. 9.38).

9.5 UNDERSTANDING COSY WITH PRODUCT OPERATORS

9.5.1 Analyzing the Homonuclear "Front End"

Most homonuclear 2D (^{1}H–^{1}H) experiments have the same "front end": they start with the sequence $90° - t_1 - 90°$. Let's follow the net magnetization of a single proton, H_a, with offset Ω_a and a single J coupling to H_b. The preparation pulse is just a 90° pulse on the x' axis, which rotates the H_a z magnetization onto the $-y'$ axis. This magnetization undergoes both chemical shift and J-coupling evolution during the t_1 delay. This is a complex motion that involves the two components of H_a magnetization ($H_b = \alpha$ and $H_b = \beta$) separating, spreading until they are opposite each other (antiphase), and then coming back together again (in phase). At the same time, the center between the two components is rotating with angular frequency Ω_a (the chemical-shift offset, corresponding to the center of the H_a doublet). The two components may rotate into antiphase and back again many times during the t_1 period, and where they end up at the end of the t_1 period will vary with t_1, consisting in general of some fraction of the magnetization being in-phase and some fraction antiphase. In addition, the chemical shift Ω_a is encoded into the phase of the magnetization at the end of the t_1 period. As we saw in the INEPT experiment, antiphase magnetization can be transferred to the other J-coupled nucleus by simultaneous 90° pulses on both nuclei. This is just what the second 90° pulse in the COSY sequence does, since it is a nonselective pulse that affects both H_a and H_b equally. The portion of the H_a magnetization that is antiphase at the end of the t_1 delay is transferred to H_b antiphase magnetization and contributes to a crosspeak at $F_1 = \Omega_a$, $F_2 = \Omega_b$. The portion of the H_a magnetization that is in-phase does not undergo transfer, and contributes to a diagonal peak at $F_1 = F_2 = \Omega_a$.

In product operator notation:

$$\mathbf{I}_z^a \;-90_x^\circ \to\; -\mathbf{I}_y^a$$

During the evolution (t_1) period, the H_a magnetization rotates with angular frequency Ω_a in the x'–y' plane, and the doublet components ($H_b = \alpha$ and $H_b = \beta$) separate from this center position with angular frequency $J_{ab}/2$ in Hz, or πJ_{ab} in radians. In contrast to the INEPT experiment, we have no control over these two kinds of evolution and both will happen at the same time. A spin echo will not help because pulses do not distinguish between H_a and H_b: both would receive a 180° pulse and we would have no chemical-shift evolution, only J-coupling evolution. Without chemical-shift evolution we cannot create a second dimension!

This is a complicated motion to describe with vectors, but with product operators it is relatively simple, if you are not afraid of a little algebra and trigonometry. First we consider the chemical-shift evolution, which causes the H_a magnetization to rotate through an angle $\Theta = \Omega_a t_1$ radians after t_1 s:

$$-\mathbf{I}_y^a \to -\mathbf{I}_y^a \cos(\Omega_a t_1) + \mathbf{I}_x^a \sin(\Omega_a t_1)$$

Since counterclockwise rotation leads from the $-y'$ axis to the $+x'$ axis, the x component has a plus sign. Now consider the effect of the coupling J_{ab}. The pure x' and y' magnetization will rotate into and out of the antiphase condition with angular frequency πJ:

$$\mathbf{I}_x^a \to \mathbf{I}_x^a \cos(\pi J t_1) + 2\mathbf{I}_y^a \mathbf{I}_z^b \sin(\pi J t_1)$$

$$\mathbf{I}_y^a \to \mathbf{I}_y^a \cos(\pi J t_1) - 2\mathbf{I}_x^a \mathbf{I}_z^b \sin(\pi J t_1)$$

Verify for yourself that this makes sense: at $t_1 = 0$ you have the starting magnetizations. At $t_1 = 1/(2J_{ab})$ you have the pure antiphase magnetization. At $t_1 = 1/J_{ab}$ you have the starting in-phase magnetizations with the opposite sign, that is, rotated by 180° in the x'–y' plane.

At $t_1 = 2/J_{ab}$ you are back to the starting magnetization. With product operators we do not need to draw vector diagrams, because we treat the vectors as pure in-phase ($\mathbf{I}_x^a$, $\mathbf{I}_y^b$) and antiphase ($2\mathbf{I}_x^a\mathbf{I}_z^b$, $2\mathbf{I}_y^a\mathbf{I}_z^b$) components. Now plug these results in wherever you see $\mathbf{I}_x^a$ or $\mathbf{I}_y^a$ as a result of chemical-shift evolution:

$$-\mathbf{I}_y^a \xrightarrow{\text{(chemical-shift evolution)}} -\mathbf{I}_y^a \cos(\Omega_a t_1) + \mathbf{I}_x^a \sin(\Omega_a t_1)$$

$$\xrightarrow{\text{(J-coupling evolution)}} [-\mathbf{I}_y^a \cos(\pi J t_1) + 2\mathbf{I}_x^a\mathbf{I}_z^b \sin(\pi J t_1)] \cos(\Omega_a t_1)$$
$$+[\mathbf{I}_x^a \cos(\pi J t_1) + 2\mathbf{I}_y^a\mathbf{I}_z^b \sin(\pi J t_1)] \sin(\Omega_a t_1)$$
$$= -\mathbf{I}_y^a \cos(\Omega_a t_1)\cos(\pi J t_1) + 2\mathbf{I}_x^a\mathbf{I}_z^b \cos(\Omega_a t_1)\sin(\pi J t_1)$$
$$+\mathbf{I}_x^a \sin(\Omega_a t_1)\cos(\pi J t_1) + 2\mathbf{I}_y^a\mathbf{I}_z^b \sin(\Omega_a t_1)\sin(\pi J t_1)$$

We can abbreviate a bit by using **s** and **c** for $\sin(\Omega_a t_1)$ and $\cos(\Omega_a t_1)$, respectively, and **s'** and **c'** for $\sin(\pi J t_1)$ and $\cos(\pi J t_1)$, respectively:

$$-\mathbf{I}_y^a \mathbf{c}\,\mathbf{c}' + 2\mathbf{I}_x^a\mathbf{I}_z^b \mathbf{c}\,\mathbf{s}' + \mathbf{I}_x^a \mathbf{s}\,\mathbf{c}' + 2\mathbf{I}_y^a\mathbf{I}_z^b \mathbf{s}\,\mathbf{s}'$$

It still looks pretty messy, but there are only four terms: the x' and y' components of H_a in-phase magnetization (first and third terms), and the x' and y' components of H_a magnetization that is antiphase with respect to the spin state (α or β) of the coupled H_b spin (second and fourth terms). This is a full description of what happens to the H_a magnetization that started on the $-y'$ axis at the beginning of the t_1 period, and it can be applied to all homonuclear 2D NMR experiments.

There is only one more thing to do: consider the effect of the mixing portion of the 2D experiment, which is just a 90° pulse applied along the x' axis of the rotating frame. Each component of the product operators can be treated separately and rotated just as the vectors rotate under the influence of a pulse. All of the individual components behave as follows:

$$\mathbf{I}_y^a \xrightarrow{(90°\text{pulse on }x'\text{axis})} \mathbf{I}_z^a$$
$$\mathbf{I}_x^a \xrightarrow{(90°\text{pulse on }x'\text{axis})} \mathbf{I}_x^a \quad \text{(not affected)}$$
$$\mathbf{I}_z^b \xrightarrow{(90°\text{pulse on }x'\text{axis})} -\mathbf{I}_y^b$$

Applying these results to the products in our four terms:

$$-\mathbf{I}_y^a \xrightarrow{(90°\text{pulse on }x'\text{axis})} -\mathbf{I}_z^a$$
$$2\mathbf{I}_x^a\mathbf{I}_z^b \xrightarrow{(90°\text{pulse on }x'\text{axis})} 2(\mathbf{I}_x^a)(-\mathbf{I}_y^b) = -2\mathbf{I}_x^a\mathbf{I}_y^b$$
$$\mathbf{I}_x^a \xrightarrow{(90°\text{pulse on }x'\text{axis})} \mathbf{I}_x^a$$
$$2\mathbf{I}_y^a\mathbf{I}_z^b \xrightarrow{(90°\text{pulse on }x'\text{axis})} 2(\mathbf{I}_z^a)(-\mathbf{I}_y^b) = -2\mathbf{I}_y^b\mathbf{I}_z^a \quad \text{(coherence transfer!)}$$

So the final expression for the detected magnetization is:

$$-\mathbf{I}_z^a \mathbf{c}\,\mathbf{c}' - 2\mathbf{I}_x^a\mathbf{I}_y^b \mathbf{c}\,\mathbf{s}' + \mathbf{I}_x^a \mathbf{s}\,\mathbf{c}' - 2\mathbf{I}_y^b\mathbf{I}_z^a \mathbf{s}\,\mathbf{s}'$$

The first term represents inverted z magnetization, the starting point for a transient NOE experiment. All we have to do is add a mixing period (simple delay τ_m) and a 90° read pulse, and we have a 2D NOE experiment: NOESY. For the COSY experiment, we start acquiring the FID immediately and I_z is not observable, so this term is not important. The second term represents a mixture of ZQC and DQC, which is also not observable. We will use this term in the double-quantum filtered (DQF) COSY experiment as an intermediate state in coherence (INEPT) transfer, but in the simple COSY we can ignore it. The third term represents in-phase H_a magnetization that is labeled with the chemical shift of H_a ($\mathbf{s} = \sin(\Omega_a t_1)$) and with the J coupling ($\mathbf{c'} = \cos(\pi J t_1)$). This will be our diagonal peak in the 2D spectrum: since it is H_a coherence that will be observed in the FID, it will give an in-phase doublet at chemical-shift position Ω_a in F_2. Since it is labeled with the H_a chemical shift in t_1, the second Fourier transform will place it at chemical-shift position Ω_a in F_1. This is the H_a peak on the diagonal. The fourth term is the important one: it represents coherence transfer (INEPT transfer) from H_a to H_b. It will be observed in the FID as H_b antiphase coherence, leading to an antiphase peak at the position Ω_b in F_2. But this peak is modulated in F_1 according to the chemical shift of H_a ($\mathbf{s} = \sin(\Omega_a J t_1)$), so it carries along the information about the spin it came from. The second Fourier transform places the peak at the chemical shift Ω_a in F_1, so it is the crosspeak at $F_1 = \Omega_a$, $F_2 = \Omega_b$. Now we see in detail how 2D NMR works!

9.5.2 Untangling the J-Coupling Patterns in F_1

Understanding how the chemical shift and J-coupling modulation in t_1 works ($\mathbf{s\,s'}$ for the crosspeak and $\mathbf{s\,c'}$ for the diagonal peak) takes a bit of mathematical manipulation. The $\sin(\Omega_a t_1)\sin(\pi J t_1)$ term ($\mathbf{s\,s'}$) can be written as a sum rather than a product using the trigonometric identity $\cos(\alpha+\beta) = \cos\alpha\,\cos\beta - \sin\alpha\,\sin\beta$:

$$\cos((\Omega_a + \pi J)t_1) = \cos(\Omega_a t_1)\cos(\pi J t_1) - \sin(\Omega_a t_1)\sin(\pi J t_1)$$
$$\cos((\Omega_a - \pi J)t_1) = \cos(\Omega_a t_1)\cos(-\pi J t_1) - \sin(\Omega_a t_1)\sin(-\pi J t_1)$$
$$= \cos(\Omega_a t_1)\cos(\pi J t_1) + \sin(\Omega_a t_1)\sin(\pi J t_1)$$

Subtracting the second equation from the first,

$$\cos((\Omega_a + \pi J)t_1) - \cos((\Omega_a - \pi J)t_1) = -2\sin(\Omega_a t_1)\sin(\pi J t_1)$$

$$\mathbf{s\,s'} = \sin(\Omega_a t_1)\sin(\pi J t_1) = 0.5\,[-\cos((\Omega_a + \pi J)t_1) + \cos((\Omega_a - \pi J)t_1)]$$

Thus Fourier transformation in F_1 will yield two peaks: a positive peak at $F_1 = \Omega_a - \pi J$ and a negative peak at $F_1 = \Omega_a + \pi J$. This is an antiphase doublet in F_1 centered at frequency Ω_a and separated by $2\pi J$ rad or J Hz. So we have a crosspeak that is an antiphase doublet in F_2 ($-2I_y^b I_z^a$ observed in the FID) and an antiphase doublet in F_1 ($\sin(\Omega_a t_1)\sin(\pi J t_1)$), with both doublets showing a separation of J_{ab}. This is the crosspeak fine structure shown in Figure 9.29.

For the diagonal peak F_1 fine structure, we have the t_1 modulation $\mathbf{s\,c'} = \sin(\Omega_a t_1)\cos(\pi J t_1)$. We start with the trigonometric identity: $\sin(\alpha + \beta) = \sin\alpha\,\cos\beta + \cos\alpha\,\sin\beta$, applied to the sum and difference frequencies:

$$\sin((\Omega_a + \pi J)t_1) = \sin(\Omega_a t_1)\cos(\pi J t_1) + \cos(\Omega_a t_1)\sin(\pi J t_1)$$
$$\sin((\Omega_a - \pi J)t_1) = \sin(\Omega_a t_1)\cos(-\pi J t_1) + \cos(\Omega_a t_1)\sin(-\pi J t_1)$$
$$= \sin(\Omega_a t_1)\cos(\pi J t_1) - \cos(\Omega_a t_1)\sin(\pi J t_1)$$

Since we are looking for **s c′**, we add the two equations together to get:

$$\sin((\Omega_a + \pi J)t_1) + \sin((\Omega_a - \pi J)t_1) = 2\sin(\Omega_a t_1)\cos(\pi J t_1)$$
$$\mathbf{s\ c'} = \sin(\Omega_a t_1)\cos(\pi J t_1) = 0.5\,[\sin((\Omega_a + \pi J)t_1) + \sin((\Omega_a - \pi J)t_1)]$$

Fourier transformation in F_1 will yield two peaks, both of them positive: one at $F_1 = \Omega_a - \pi J$ and one at $F_1 = \Omega_a + \pi J$. This is an in-phase doublet in F_1 centered at frequency Ω_a and separated by $2\pi J$ rad or J Hz. So we have a 2D peak that is an in-phase doublet at frequency Ω_a in F_2 ($+\mathbf{I}_x^a$ observed in the FID) and an in-phase doublet at frequency Ω_a in F_1 ($\mathbf{s\ c'} = \sin(\Omega_a t_1)\cos(\pi J t_1)$), with both doublets showing a separation of J_{ab}. The diagonal peak has a fine structure of four peaks in a square pattern, all with the same phase. How does its phase compare to the crosspeak? It is 90° out of phase with the crosspeak in F_2 ($\mathbf{I}_x^a$ vs. $-2\mathbf{I}_y^b\mathbf{I}_z^a$) and it is 90° out of phase with the crosspeak in F_1 ($0.5\,[-\cos((\Omega_a + \pi J)t_1) + \cos((\Omega_a - \pi J)t_1)]$ for the diagonal peak vs. $0.5\,[\sin((\Omega_a + \pi J)t_1) + \sin((\Omega_a - \pi J)t_1)]$ for the crosspeak. Note that a sine function in time is always 90° out of phase with a cosine function. This is a significant conclusion because it means we cannot phase correct the whole 2D spectrum: either we have absorptive crosspeaks and dispersive diagonal peaks or *vice-versa*. The dispersive lineshape does not go to zero quickly as we move away from the center of the resonance (Fig. 9.39) as the absorptive shape does, so this will lead to large streaks extending above and below and to the left and right of the diagonal peaks.

An exponentially decaying FID gives a Lorentzian lineshape upon Fourier transformation. The general form of the absorptive Lorentzian line is $I_{abs} = 1/(1 + \nu^2)$, whereas the dispersive line has the form $I_{disp} = \nu/(1 + \nu^2)$, where I is the intensity at each point in the spectrum. Far from the peak maximum ($\nu^2 \gg 1$), we have $I_{abs} \sim 1/\nu^2$ and $I_{disp} \sim 1/\nu$. This is the reason that the dispersive lineshape extends much further from the peak maximum.

The COSY phase differences can be eliminated by presenting the data in "magnitude mode," as we did for the HETCOR spectrum. But we lose useful information such as the distinction between active and passive couplings, and more importantly the magnitude mode peaks are broader and extend farther outward at their bases (Fig. 9.39), again leading

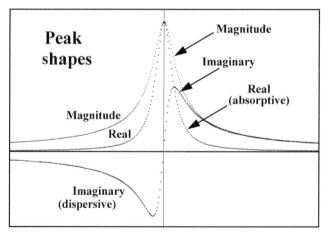

Figure 9.39

to streaks extending up and down, and to the left and right from each diagonal peak or crosspeak. We will see that the DQF-COSY experiment avoids this problem since the phase of the diagonal is the same as that of the crosspeaks.

The magnitude mode lineshape is just:

$$I_{\text{magn}} = \sqrt{I_{\text{abs}}^2 + I_{\text{disp}}^2} = \sqrt{1/(1+\nu^2)^2 + \nu^2/(1+\nu^2)^2}$$
$$= \sqrt{(1+\nu^2)/(1+\nu^2)^2} = 1/\sqrt{1+\nu^2}$$

Far from the peak maximum ($\nu^2 \gg 1$) we have $I_{\text{magn}} \sim 1/\nu$, just like the dispersive lineshape.

9.5.3 COSY-35: Simplifying Crosspeak Fine Structure

What about the COSY-35 experiment (Fig. 9.34)? We can now show with product operators why it simplifies the crosspeak fine structure. Consider again the AMX spin system of a peptide residue in D_2O: $ND-CH_\alpha-CH_\beta H_{\beta'}-Y$. For the crosspeaks shown in Figure 9.33 (left) let's focus on the lower one: $F_1 = H_\beta/F_2 = H_\alpha$ ($H_\beta \rightarrow H_\alpha$ coherence transfer). We start the t_1 period with $-\mathbf{I}_y^\beta$ and write down the terms that result from J coupling, keeping in mind that there are two J couplings affecting H_β: $J_{\alpha\beta}$ and $J_{\beta\beta'}$:

$$-\mathbf{I}_y^\beta \xrightarrow{t_1} \underbrace{-\mathbf{I}_y^\beta cc'}_{A} \quad \underbrace{+2\mathbf{I}_x^\beta \mathbf{I}_z^\alpha sc'}_{B} \quad \underbrace{+2\mathbf{I}_x^\beta \mathbf{I}_z^{\beta'} cs'}_{C} \quad \underbrace{+4\mathbf{I}_y^\beta \mathbf{I}_z^\alpha \mathbf{I}_z^{\beta'} ss'}_{D} \quad (J \text{ evolution})$$

where c and s are cosine and sine of $\pi J_{\alpha\beta} t_1$ and c' and s' are cosine and sine of $\pi J_{\beta\beta'} t_1$. As before, we advance the phase by 90°, multiply by $2\mathbf{I}_z$ and change the cosine term to sine for each coupling that undergoes evolution from in-phase to antiphase. Only terms B and D above represent H_β coherence that is antiphase with respect to H_α, so we can ignore the A and C terms because they cannot give us coherence transfer from H_β to H_α. Now consider the effect on terms B and D of chemical-shift evolution of H_β coherence, which multiplies the starting terms by $\cos(\Omega_\beta t_1)$ and then adds new terms with the phase of H_β coherence advanced by 90° and multiplying by $\sin(\Omega_\beta t_1)$:

$$-t_1 \rightarrow \underbrace{2\mathbf{I}_x^\beta \mathbf{I}_z^\alpha s c' c''}_{E} \quad \underbrace{+4\mathbf{I}_y^\beta \mathbf{I}_z^\alpha \mathbf{I}_z^{\beta'} s s' c''}_{F} \quad \underbrace{+2\mathbf{I}_y^\beta \mathbf{I}_z^\alpha s c' s''}_{G} \quad \underbrace{-4\mathbf{I}_x^\beta \mathbf{I}_z^\alpha \mathbf{I}_z^{\beta'} s s' s''}_{H} \quad \text{(shift evolution)}$$

where s'' or c'' refer to $\sin(\Omega_\beta t_1)$ or $\cos(\Omega_\beta t_1)$, respectively. Terms E and G above come from the singly antiphase term B and terms F and H above come from the double antiphase term D. The final pulse is on x', so the E and H terms (H_β coherence on x') cannot give us coherence transfer. We only need to consider the F and G terms:

$$\underbrace{4\mathbf{I}_y^\beta \mathbf{I}_z^\alpha \mathbf{I}_z^{\beta'} s s' c''}_{F} \quad \underbrace{+2\mathbf{I}_y^\beta \mathbf{I}_z^\alpha s c' s''}_{G} \quad -\Theta \text{ pulse on } x' \rightarrow$$

Now we do the coherence transfer with our final 90° pulse on x', but we allow the pulse to be any rotation angle Θ (90° for COSY, 35° for COSY-35). This would generate two terms for each operator, for a total of 12 terms (!), but we need to worry only about those terms that represent coherence transfer from H_β to H_α. For this to happen, we need to have

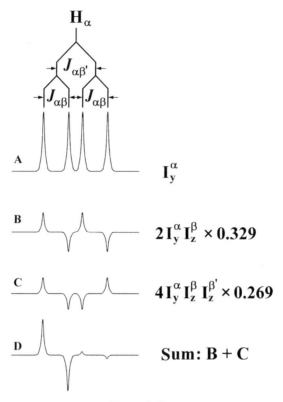

Figure 9.40

I_β move from y' to the z axis and I_α move from z to the x'–y' plane (antiphase to antiphase INEPT transfer). The "Θ" pulse on x' converts I_z to $[I_z \cos \Theta - I_y \sin \Theta]$ and I_y to $[I_y \cos \Theta + I_z \sin \Theta]$. Just like with evolution, the cosine term goes with the unchanged operator and the sine term goes with the $\Theta = 90°$ result. Each term in the product is affected by the pulse, but we need only consider the results that have the I_α operator in the x'–y' plane and all others on z. Any terms with more than one operator in the x'–y' plane represent unobservable ZQC/DQC. Term G gives the standard INEPT coherence transfer result:

$$2I_y^\beta I_z^\alpha \xrightarrow{\Theta_x} 2(I_y^\beta \cos \Theta + I_z^\beta \sin \Theta)(I_z^\alpha \cos \Theta - I_y^\alpha \sin \Theta)$$
$$= -2I_y^\alpha I_z^\beta \sin^2 \Theta + 3 \text{ other terms}$$

Only the $\sin^2 \Theta$ term represents coherence transfer; the three others can be ignored. The F term can give coherence transfer to H_α also, as long as I^α is the only operator in the x'–y' plane:

$$4I_y^\beta I_z^\alpha I_z^{\beta'} \xrightarrow{\Theta_x} 4(I_z^\beta \sin \Theta)(-I_y^\alpha \sin \Theta)(I_z^{\beta'} \cos \Theta) + 8 \text{ other terms}$$
$$= -4I_y^\alpha I_z^\beta I_z^{\beta'} \sin^2 \Theta \cos \Theta$$

None of the other eight terms represents $H_\beta \to H_\alpha$ coherence transfer. We also ignore the s s'c" and s c's" multipliers since they carry the F_1 chemical shift and J-coupling information—we are only interested in the F_2 slice. What do these two terms look like in F_2? Remember that the H_α resonance ($\mathbf{I}_y^\alpha$) is a doublet of doublets (Fig. 9.40 A). $2\mathbf{I}_y^\alpha \mathbf{I}_z^\beta$ is antiphase with respect to H_β only, so we get the pattern 1, −1, 1, −1 (if $J_{\alpha\beta} < J_{\alpha\beta'}$) (Fig. 9.40 B). $4\mathbf{I}_y^\alpha \mathbf{I}_z^\beta \mathbf{I}_z^{\beta'}$ is antiphase with respect to both H_β and $H_{\beta'}$, so we get an antiphase doublet on the left side (antiphase with respect to $J_{\alpha\beta}$) and another antiphase doublet on the right side that is opposite in phase (antiphase with respect to $J_{\alpha\beta'}$): the pattern is 1, −1, −1, 1 (Fig. 9.40 C). Now if we consider that the first pattern ($2\mathbf{I}_y^\alpha \mathbf{I}_z^\beta$) is multiplied by $\sin^2 \Theta$ (0.329 for $\Theta = 35°$) and the second pattern ($4\mathbf{I}_y^\alpha \mathbf{I}_z^\beta \mathbf{I}_z^{\beta'}$) is multiplied by $\sin^2 \Theta \cos \Theta$ (0.269 for $\Theta = 35°$), we see that when we add them together there are two kinds of lines (Fig. 9.40 D): the first and second where the two add together ($0.329 + 0.269 = 0.598$) and the third and fourth where they are subtracted ($0.329 − 0.269 = 0.060$): an absolute value of $\sin^2 \Theta + \sin^2 \Theta \cos \Theta$ for lines 1 and 2 and $\sin^2 \Theta − \sin^2 \Theta \cos \Theta$ for lines 3 and 4. If we look at the intensity ratio between these two types of lines we get:

$$\text{Ratio} = (\sin^2\Theta - \sin^2\Theta \cos\Theta)/(\sin^2\Theta + \sin^2\Theta \cos\Theta) = (1 - \cos\Theta)/(1 + \cos\Theta)$$

For $\Theta = 90°$, this ratio is 1 and we see all four lines equally in each row (16 peaks in all in the crosspeak, Fig. 9.33, lower left). For $\Theta = 35°$, the ratio is 0.1 (1 to 10): lines 3 and 4 are only 10% of the intensity of lines 1 and 2 (eight intense peaks in all in the crosspeak, Fig. 9.33 lower right). Some people use a 45° pulse ("COSY-45"), for a ratio of 0.17 (1 to 5.8). As we make Θ smaller, we pay a price in overall intensity since both types of line are multiplied by $\sin^2 \Theta$; although the ratio gets better as the overall sensitivity goes down. This analysis illustrates the power of product operators as well as the need to look ahead and anticipate which terms will be important to avoid an explosion of complexity.

9.6 2D TOCSY (TOTAL CORRELATION SPECTROSCOPY)

In Chapter 8 we saw how the TOCSY spin lock, a continuous string of medium-power pulses with carefully designed widths and phases, can transfer coherence in multiple jumps along a chain of J-coupled protons: a "spin system". In the selective 1D TOCSY experiment, the DPFGSE (a combination of shaped pulses and gradients) is used to selectively excite one resonance in the spectrum with the equivalent of a 90° pulse, and this coherence is then transferred to other protons in the spin system with the TOCSY mixing sequence (pulsed spin lock). To make this sequence (Fig. 8.42) into a 2D TOCSY (Fig. 9.41), we simply replace the selective 90° pulse (the DPFGSE) with a non-selective 90° pulse and insert a t_1 delay (evolution period) between this preparation pulse and the mixing sequence (Fig. 9.42). We already know how the $90° - t_1$ sequence produces four terms:

$$-\mathbf{I}_y^a cc' + 2\mathbf{I}_x^a \mathbf{I}_z^b cs' + \mathbf{I}_x^a sc' + 2\mathbf{I}_y^a \mathbf{I}_z^b ss'$$

If we apply the TOCSY spin lock at this point on the x' axis, we will destroy the first and fourth terms (B_1 field inhomogeneity) and "lock" the second and third terms. Because the TOCSY mixing sequence transfers coherence from in-phase to in-phase, only the third

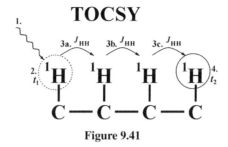

Figure 9.41

term, $\mathbf{I}_x^a \mathbf{sc}'$, leads to a crosspeak at $F_1 = \Omega_a/F_2 = \Omega_b$:

$$\mathbf{I}_x^a \mathbf{sc}' - \text{TOCSY spin lock} \rightarrow \mathbf{I}_x^b \mathbf{s}\,\mathbf{c}'$$

We also saw that the $\mathbf{s}\,\mathbf{c}'$ encoding in t_1 ($\sin(\Omega_a t_1)\cos(\pi J t_1)$) represents an in-phase doublet at the H_a chemical shift in F_1. So the crosspeak is in-phase in both dimensions.

We saw in Chapter 8 that a continuous-wave spin lock is not effective for TOCSY transfer, giving efficient transfer only when the Hartmann–Hahn match is satisfied: $\Omega_a = \pm\Omega_b$. This corresponds to the diagonal and the "antidiagonal"—a line extending from the lower right corner of the 2D matrix to the upper left corner. In fact, TOCSY transfer crosspeaks do appear as artifacts in 2D ROESY spectra along the antidiagonal. To get efficient TOCSY transfer we use a specific sequence of medium-power pulses such as MLEV-17 or DIPSI-2. The holy grail of TOCSY mixing sequences is the "ideal isotropic mixing" sequence that completely eliminates the chemical shifts and leaves only the J-coupling interactions, just as if the B_0 field were reduced to zero. In this ideal case if we start with H_a magnetization on the x' axis, we get conversion to H_b magnetization on the x' axis as follows:

$$\mathbf{I}_x^a \rightarrow 0.5\mathbf{I}_x^a(1+\mathbf{cos}) + 0.5\mathbf{I}_x^b(1-\mathbf{cos}) + 0.5(2\mathbf{I}_y^a\mathbf{I}_z^b - 2\mathbf{I}_y^b\mathbf{I}_z^a)\,\mathbf{sin}$$

where $\mathbf{cos} = \cos(2\pi J \tau_m)$, $\mathbf{sin} = \sin(2\pi J \tau_m)$, and τ_m is the mixing time. The derivation of this result will be shown in Chapter 10. The first two terms represent the transfer of in-phase coherence from H_a to H_b: after a time $\tau_m = 1/(4J)$ the cosine term equals zero and we have 0.5 $\mathbf{I}_x^a$ and 0.5 $\mathbf{I}_x^b$, i.e., 50% conversion. But we also have a combination of antiphase H_a coherence and antiphase H_b coherence ($\mathbf{sin} = 1$)! This term decreases again and becomes zero when $\tau_m = 1/(2J)$. At this time we have complete conversion of $\mathbf{I}_x^a$ to $\mathbf{I}_x^b$ since the cosine term equals -1. We saw this oscillation in Figure 8.43 but the antiphase terms were ignored at that time. These terms lead to distortion of peak shape unless the

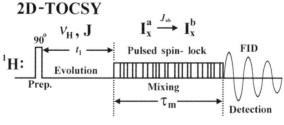

Figure 9.42

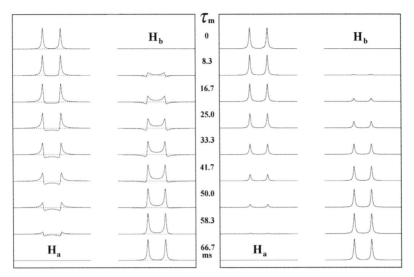

Figure 9.43

sine term is zero ($\tau_m = 0$, $1/(2J)$, $1/J$, etc.). This distortion is shown in Figure 9.43 (left) for a simulation of the $H_a - H_b$ spin system with $J_{ab} = 7.5$ Hz. You might think that the antiphase terms would disappear since they are perpendicular to the spin-lock axis and B_1 field inhomogeneity should cause them to "fan out" over time. But these terms are actually immune to pulses on the x' axis:

$$2I_y^a I_z^b - 2I_y^b I_z^a - 90_x^\circ \rightarrow 2I_z^a(-I_y^b) - 2I_z^b(-I_y^a) = 2I_y^a I_z^b - 2I_y^b I_z^a$$

$$2I_y^a I_z^b - 2I_y^b I_z^a - 180_x^\circ \rightarrow 2(-I_y^a)(-I_z^b) - 2(-I_y^b)(-I_z^a) = 2I_y^a I_z^b - 2I_y^b I_z^a$$

Many tricks have been applied (trim pulses, z filters, gradients, etc.) to remove these antiphase terms, leaving only the pure phase I_x^a and I_x^b terms (Fig. 9.43, right). For example, a "z filter" is a $90_{-y}^\circ - \Delta - 90_y^\circ$ sequence that puts the desired magnetization on the z axis and then allows a bit of evolution to occur for the undesired terms:

$$-90_{-y}^\circ \rightarrow 0.5I_z^a(1 + \cos) + 0.5I_z^b(1 - \cos) + 0.5(-2I_y^a I_x^b + 2I_y^b I_x^a)\sin$$

The last term is now ZQC on the y' axis, which undergoes chemical-shift evolution during the delay Δ at a rate of $\Omega_a - \Omega_b$. The second 90° pulse puts the desired terms back on x' and returns the undesired term to antiphase SQC on y'. If we repeat the acquisition with different values of the delay Δ the antiphase terms will tend to cancel out due to different amounts of evolution during Δ. A variable delay for evolution of undesired ZQC terms can also be used in NOE mixing to remove ZQ artifacts. The strategy of "storing" desired terms on the z axis while taking care of other components of magnetization is also a common strategy we will encounter again later. Still, distortion of peak shape is commonly encountered in both 1D and 2D TOCSY spectra. In 2D TOCSY this appears as negative intensity in the center of a crosspeak or negative "ditches" on the sides of a crosspeak.

The MLEV-17 mixing sequence falls far short of achieving the "holy grail" of isotropic mixing, and the more complex DIPSI-2 sequence (Chapter 8) is superior in its tolerance of large resonance offsets ($v_o - v_r$) while avoiding high power and sample heating. But most people still use MLEV-17 out of blind tradition. For large biological molecules, there is a "clean" or "relaxation compensated" version of the DIPSI sequence (DIPSI-2rc) in which there are short delays separating all of the pulses. The magnetization is on the z axis for the delays, and this allows an NOE to develop that is opposite in sign to the ROE that develops in the spin lock along with TOCSY transfer. This positive NOE (for large molecules) cancels the negative ROE and leaves pure coherence transfer (TOCSY) mixing.

We saw a 2D TOCSY spectrum in Figure 9.22 and compared it to a COSY spectrum: in the TOCSY spectrum, we have more peaks because starting from any proton in the spin system we can see correlations to all other members of the spin system, not just to the protons connected by a single J coupling. We saw the same thing in a real example by comparing the COSY spectrum of 3-heptanone (Fig. 9.23) with the TOCSY spectrum of the same sample (Fig. 9.24).

9.6.1 Examples of 2D TOCSY

The 600 MHz 2D TOCSY spectrum of cholesterol (Chapter 8, Fig. 8.35) is shown in Figure 9.44. Note that all peaks (diagonal and crosspeaks) are positive (black) and in-phase.

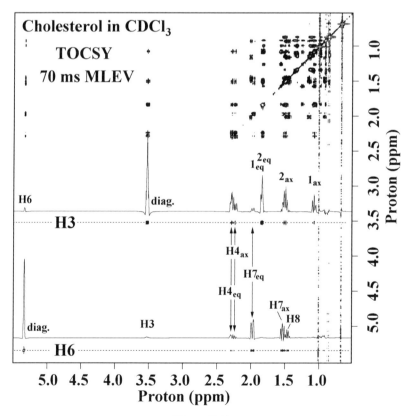

Figure 9.44

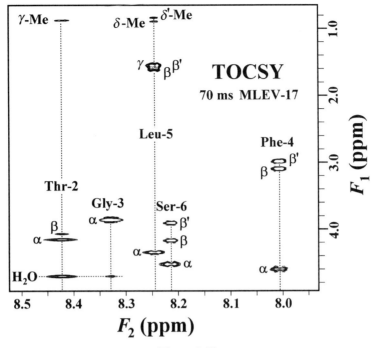

Figure 9.45

The t_1 noise streaks can be seen extending downward from the two singlet methyl peaks on the diagonal (H-18 and H-19) and to a lesser extent from the methyl doublet peaks between them. An F_2 slice at the H6 resonance (5.35 ppm) on the diagonal shows efficient TOCSY transfer to H7$_{ax}$ and H7$_{eq}$ and then on to H8, and weak transfer to H4$_{ax}$ and H4$_{eq}$ (long-range coupling) and on to H3. Another F_2 slice at the H3 resonance (3.5 ppm) on the diagonal shows transfer to H4$_{ax}$, H4$_{eq}$ and then on weakly to H6 and H7$_{eq}$, as well as strong crosspeaks to H2$_{ax}$, H2$_{eq}$, H1$_{ax}$, and H1$_{eq}$. These F_2 slices are almost identical to the two separate selective 1D TOCSY experiments shown in Figure 8.45. The advantage of 2D TOCSY is that we only have to do one experiment to get all possible TOCSY correlations, and we do not rely on selecting resolved peaks in the 1D spectrum.

Figure 9.45 shows the amide NH region of the TOCSY spectrum of the glycopeptide Tyr-Thr-Gly-Phe-Leu-Ser(O-Lactose) in 90% H_2O/10% D_2O. In F_2, we see the amide proton (H_N) resonances in the range of 8–8.5 ppm, and in F_1 we see the entire region from the water resonance upfield. We can see the entire spin system of each amino acid residue stretching upwards from its H_N chemical-shift position on the horizontal (F_2) ppm scale. Coherence was transferred to the H_N proton by TOCSY mixing from the H_α proton, the H_β protons, the H_γ protons, and so on, and each of these was labeled during t_1 with its proton chemical shift. Just by looking at the patterns of chemical shifts in each vertical line, we can identify the amino acid or at least narrow it down to a group of amino acids. The farthest left spin system (H_N = 8.42 ppm) is a threonine (side chain $CH^\beta OH$–$CH^\gamma{}_3$), since the H_α and H_β shifts are close together and the γ-methyl is far upfield ($\sim$1.0 ppm). The next residue (H_N = 8.33 ppm) has only one crosspeak, so it must be a glycine (no side chain) with only the H_α peak. Note that these two H_N protons exchange more rapidly with water, leading to an exchange crosspeak at the H$_2$O resonance (F_1 = 4.6 ppm). Moving to the right, the next system (H$_N$ = 8.25 ppm) is a leucine (side chain CH$^\beta_2$–CH$^\gamma$(CH$^\delta_3$)$_2$). The H$_\gamma$ signal is overlapped with the H$_{\beta'}$, and we see two different δ-methyl crosspeaks due to the chiral environment. Just upfield of this system is a serine (side chain CH$^\beta_2$–OH), with both H$_\beta$ and H$_{\beta'}$ close to H$_\alpha$ in chemical shift. So far every one of these patterns of crosspeaks is unique, leading to the identification of a single amino acid among the 20 naturally occurring possibilities. The farthest upfield H$_N$ (8.01 ppm) is a classic AMX ("three-spin") system: H$_\alpha$ plus two H$_\beta$ signals in the 2.5–3.5 ppm region. There are many possible amino acids that give this pattern (side chain CH$_\beta$H$_{\beta'}$–Y): all of the aromatic amino acids (Phe, His, Trp, Tyr) plus Cys, Asp, and Asn. In this case, however, there are only two residues left: Phe and Tyr. Because the tyrosine is at the unprotected N terminus, there is no amide H$_N$. The H$_3$N$^+$ protons at the amino terminus are exchanging so rapidly with water that they are never observed in NMR. That leaves phenylalanine (side chain CH$^\beta_2$–C$_6$H$_5$) for the most upfield H$_N$. Note that TOCSY mixing does not penetrate the aromatic ring because there is no J coupling between the H$_\beta$ protons and the aromatic ring protons: these are two separate spin systems.

9.7 DATA SAMPLING IN T_1 AND THE 2D SPECTRAL WINDOW

The hardware and data-processing details of 1D NMR data were discussed in Chapter 3: data sampling in the ADC, quadrature detection, the spectral window, weighting (window) functions, and phase correction. We will have to revisit each of these topics in the second (t_1, F_1) dimension and some of them will take on added significance.

The evolution delay (t_1) is usually started at zero for the first FID and increased by the same amount, the t_1 increment Δt_1, for each successive FID. We are trying to describe the evolution of the nucleus A magnetization as it precesses during t_1, so the same digital sampling limitations (Nyquist theorem) apply as they do in direct (analog-to-digital converter) sampling of the t_2 FID. The rule is that we need to have a minimum of two samples per cycle to define a frequency: we can think of it as a sample in each crest and a sample in each trough of the wave. This fundamental limitation defines the maximum frequency that we can observe without aliasing. Before the advent of quadrature detection (real and imaginary FIDs), the audio frequency scale ran from zero on the right side edge to the maximum frequency (*sw*) on the left side edge of the spectral window. The spectral window is still defined by this maximum frequency, which is determined by the sampling rate. The sampling delay Δt_1 is half of the time of one full cycle (1/*sw*) of the maximum frequency since we have to have two samples per cycle:

$$\Delta t_1 = 1/(2 \times sw1) \quad \text{where } sw1 = \text{the spectral width in } F_1 \text{ in Hz}$$

In 2D NMR the spectral window is now a rectangle (Fig. 9.46), with horizontal width *sw* defined by the sampling rate in t_2 (the dwell time of the ADC, Δt_2) and the vertical "width" defined by the sampling rate in t_1 (the t_1 increment Δt_1). Any F_1 frequency greater than *sw1* (above the upper edge or below the lower edge of the rectangle) will alias, folding back vertically into the rectangle.

Quadrature detection in t_2 gives us two FIDs (real and imaginary) by sampling both the M_x component and the M_y component of the net magnetization as it precesses. This allows us to put zero audio frequency in the center of the spectral window and defines the left side

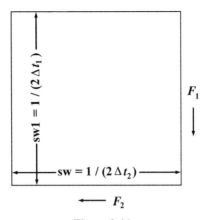

Figure 9.46

edge of the window as *sw*/2 and the right side edge as −*sw*/2. The maximum detectable frequency is now *sw*/2, but the width of the spectral window is still *sw*. Recall that there are two ways to sample the real (M_x component) and imaginary (M_y component) audio channels: the "Bruker" or alternating method (real – Δt_2 – imag. – Δt_2 – real – Δt_2 – imag. – Δt_2, etc.) or the "Varian" or simultaneous method (real & imag. – $2\Delta t_2$ – real & imag. – $2\Delta t_2$, etc.). As long as we define the sampling delay in this way (acquisition time divided by the total number of samples) we can say in either case that the width of the spectral window is $1/(2 \Delta t_2)$, based on the need for 2 samples per cycle. The radio frequency center of the spectral window in F_2 is the reference frequency ν_r, which is subtracted out by analog mixing in the detector of the NMR receiver to give zero audio frequency at the center of the spectral window.

9.7.1 Phase-Sensitive 2D NMR: Quadrature Detection in F_1

In phase-sensitive 2D NMR, the same kind of strategy is used. To create an imaginary data point in t_1, the phase of the preparation pulse is advanced by 90° and the FID is recorded again. This means that the magnetization component of interest, the one that will be transferred in the mixing step, is evolving during t_1 according to a sine function instead of a cosine function. For example, if only the y' component $\mathbf{I}_y$ can be transferred (e.g., in a 2D TOCSY with the spin lock on y'), we have:

"real" FID : $\mathbf{I}_z - 90°_x \rightarrow -\mathbf{I}_y \rightarrow -\mathbf{I}_y\cos(\Omega_a t_1) + \mathbf{I}_x\sin(\Omega_a t_1) \rightarrow \rightarrow -\cos(\Omega_a t_1) \times \text{FID}_b(t_2)$

"imag." FID : $\mathbf{I}_z - 90°_y \rightarrow \mathbf{I}_x \rightarrow \mathbf{I}_x\cos(\Omega_a t_1) + \mathbf{I}_y\sin(\Omega_a t_1) \rightarrow \rightarrow \sin(\Omega_a t_1) \times \text{FID}_b(t_2)$

Both FIDs are acquired with the same t_1 value, and both are encoded with the same frequency Ω_a in t_1, but they are 90° out of phase (cosine vs. sine modulation in t_1), just as the real and imaginary channels of the receiver (M_x and M_y) are 90° out of phase. This gives us our quadrature detection in F_1, allowing us to put zero F_1 audio frequency in the center of the F_1 spectral window.

There are two methods of encoding the phase information, just as there are for 1D spectra in t_2—alternating and simultaneous—except that the alternating method is done a little

400 TWO-DIMENSIONAL NMR SPECTROSCOPY: HETCOR, COSY, AND TOCSY

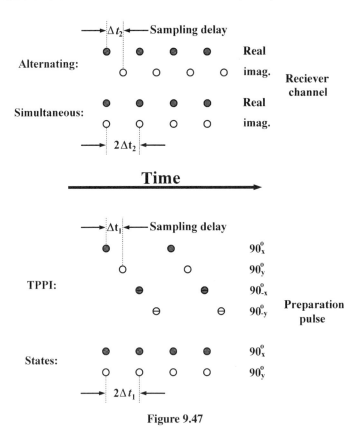

Figure 9.47

differently in the indirect dimension (Fig. 9.47). The "simultaneous" method is called *States* or *States-Haberkorn* after the inventor(s), and the "sequential" or "alternating" method is called *TPPI*, for time proportional phase incrementation. Instead of an ADC choosing between sampling of the real receiver channel or the imaginary receiver channel, the real FID is created with a preparation pulse with phase "x" and the imaginary FID is created with a preparation pulse with phase "y". There is a difference between TPPI and the alternating 1D method: the second pair of data points is recorded with opposite sign of the pulse phase. This means that the preparation pulse has phase 0°, 90°, 180°, 270° (x', y', $-x'$, $-y'$) for the first four t_1 values, and then repeats this pattern. TPPI data is processed in the F_1 dimension with another Fourier transform algorithm called the "real Fourier transform." The end result is the same, and peaks which alias ("fold") in F_1 will alias (vertically) from the same side of the spectral window, just as they do in a 1D alternating ("Bruker") spectrum. The *States* method is not really "simultaneous" in t_1, since t_1 is not a direct or "real-time" variable. You simply repeat the acquisition with the preparation pulse on the y' axis using the same t_1 value, and record this FID as an "imaginary" FID in t_1. Then you increment t_1 by twice the sampling delay ($2\Delta t_1$, where $\Delta t_1 = 1/(2 \times sw1)$) and repeat the process, first with an x' pulse and then with a y' pulse. The data is processed with a standard complex Fourier transform, just like 1D simultaneous ("Varian") data, and peaks outside the spectral window in F_1 will alias vertically from the opposite side of the 2D spectrum.

Although these methods are not spectrometer specific, Varian's software (VNMR) is biased toward States mode. The States method is implemented in VNMR by setting the parameter *phase* to an array of two integers: 1,2. This means that for each value of t_1 you acquire two FIDs, one with the preparation pulse applied along the x' axis (*phase* = 1) and one with the pulse applied along the y' axis (*phase* = 2). This array is in addition to the t_1 array (using *d2* as the t_1 delay), which has *ni* values starting with zero and incrementing by 1/*sw1* (or $2 \times \Delta t_1$) each time. This can be confusing, because the actual number of FIDs acquired will be twice the value of *ni* in States mode. If you want to acquire 512 FIDs, for example, you will set *ni* to 256 if *phase* is set to 1,2. TPPI mode is accomplished by setting *phase* to the single value of 3. Bruker uses the parameter *MC2* to define the phase encoding method in F_1: it can be set to States or TPPI. In either case, the number of FIDs acquired will be equal to $td(F_1)$ so there is no confusion.

Bruker uses an odd parameter called *nd0* (number of d-zeroes) to calculate the t_1 increment Δt_1. Technically, *nd0* is 1 if t_1 is a single delay and 2 if t_1 is split into two delays of $t_1/2$ each (usually by a 180° pulse in the middle of t_1). But if we use TPPI mode these numbers are 2 and 4, respectively, so that the increment in *d0* can be calculated as:

$$in0 = 1/(nd0 \times swh(F_1)), \text{ where } swh(F_1) \text{ is the } F_1 \text{ spectral width in Hz}$$

Thus if there is in fact only one *d0* in the pulse sequence ($d0 = t_1$), we have an increment of $1/swh(F_1)$ for States and $1/(2 \times swh(F_1))$ for TPPI. If there are two *d0*'s in the sequence ($d0 = t_1/2$) the increment of *d0* is cut in half. Setting *nd0* wrong will completely mess up the experiment!

Because this "sampling" is just the incrementation of a delay and changing the phase of a pulse, we have complete control at the software level of how we want to sample the real and imaginary data in t_1.

9.7.2 Weighting (Window) Functions and Zero-Filling in t_1

Massaging the FID with multiplier functions and increasing the digital resolution with zero filling take on much more importance in 2D NMR because we typically sample the FID for a much shorter time (the acquisition time). In t_2, acquisition times are usually cut from 1–2 s for 1D spectra to 100–400 ms for 2D spectra to limit total experiment time and file size (we are collecting hundreds of FIDs) and because resolution is not as important in a crosspeak "blob." In t_1, we are even more parsimonious with acquisition time because each data point in the t_1 FID represents a complete 1D acquisition, which may involve many scans (transients). Consider a 2D COSY with a ^{1}H spectral width in F_1 of 6 ppm on a 600 MHz spectrometer: to cover the spectral width of 3600 Hz we need a sampling delay Δt_1 of 139 μs ($1/(2 \times sw1)$), and if we acquire 512 FIDs the final t_1 value will be 512×139 μs = 71 ms. If we consider that T_2 for a typical proton might be 0.5 s, we have only lost about 13% of our FID intensity by the time we stop collecting data (Fig. 9.48). If we do a Fourier transform of this FID, we will have two big problems. First, the digital resolution of our spectrum will be very low. With only 512 data points in the FID, we have 256 data points in our real spectrum and 256 data points in our imaginary spectrum. After phase correction, we discard the imaginary spectrum and keep the real (absorptive) spectrum, which now has only 256 data points to cover a range of 6 ppm (3600 Hz). That is one data point every 14.1 Hz or 0.023 ppm. The details in the spectrum will be lost since even a large *J* coupling is smaller than the distance between two data points. To solve this problem we simply extend

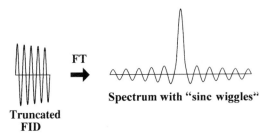

Figure 9.48

the FID by adding zeros at the end (sort of like extending soup by adding water). If we extend our FID of 256 complex data points to 1024 by adding 768 pairs of zeros, we will have 1024 complex pairs in our spectrum after the Fourier transform and 1024 data points in our real spectrum after phase correction. The digital resolution is now 3.5 Hz (0.006 ppm) per data point; not as good as a 1D spectrum but plenty of detail for showing crosspeaks and even some of the larger J couplings. The amount of zero filling is determined by the Varian parameters *fn* (Fourier number) and *fn1* (Fourier number in F_1) and the Bruker parameters $si(F_2)$ and $si(F_1)$ for "size." In either case, the FID is zero filled from the original number of complex data points acquired (Bruker: $td/2$; Varian: $np/2$ in F_2 and ni in F_1) to the final matrix size (Bruker: $si/2$; Varian *fn*/2 in F_2 and *fn1*/2 in F_1). By now you can see that Bruker usually uses the same parameter names for F_2 and F_1 and identifies them by placing them in different columns of a parameter display; Varian adds a "1" to the F_2 parameter name to generate the F_1 parameter name.

The second problem is that the t_1 FID makes a large "jump" from a finite signal at 71 ms to zero because we ran out of time (or patience) sampling it. This sudden discontinuity in the time domain data can be viewed as multiplying our FID by a rectangular window function that is one while we are sampling the FID and falls suddenly to zero after 71 ms. The effect on the spectrum, in frequency domain, is that our peaks get "wiggles" at the base that extend far out in either direction, upfield and downfield from the peak. In a 2D spectrum, these wiggles will appear as intense streaks of alternating positive and negative intensity extending above and below the crosspeaks and diagonal peaks (Fig. 9.49). The sinc artifacts extend far away from the crosspeak because the sinc function (sin ν/ν) decays as $1/\nu$, just like the dispersive peak and the magnitude mode peak. The Fourier transform does not like sudden and radical changes in time domain!

We can understand this effect precisely by applying the *convolution theorem*, which says that multiplying the FID by a function has the effect of "convoluting" the spectrum with the Fourier transform of that function. Convolution is the process of moving a multiplier function from left to right through a digital dataset, stopping at each alignment of the data points, multiplying the data by the multiplier function and adding up all the products to get a single number at each stop. This set of numbers is the result of the calculation: the "convolution" of the multiplier function and the data (Fig. 9.50). The multiplier function in this case is the Fourier transform of a rectangular "pulse" function ("on" from $t_1 = 0$ to $t_1 = 71$ ms). We saw in Chapter 8 that the result is a "sinc" function (sin ν/ν) that has a separation of $1/0.071$ s $= 14$ Hz at the base of the central peak. Now we slide this function by our spectrum, which might be a single NMR peak. As the wiggles pass by the peak, we will get alternating positive and negative intensities that increase as the central peak of

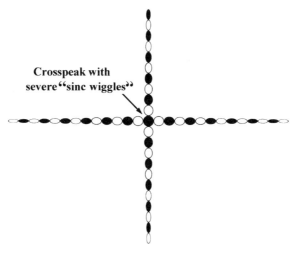

Figure 9.49

the sinc function approaches the NMR peak. As it passes through the NMR peak we get a large positive intensity, and after that we get alternating positive and negative intensity of decreasing amplitude (Fig. 9.48).

What we need to do is to smooth the transition from a finite FID to zero, which will have the effect of "calming down" the wiggles in frequency domain. For this purpose we need a multiplier function that goes smoothly to zero at the end of the FID data. Two commonly used window functions that accomplish this are the *sine-bell* and the *cosine-bell* functions (Fig. 9.51). The *cosine-bell* (or "90°-shifted sine-bell") function starts at the maximum (sine of 90°) at the beginning of the FID and goes smoothly to zero (sine of 180°) at the end of the acquired data. This window function is commonly used for 2D experiments with low-intensity crosspeaks that require *sensitivity enhancement* such as NOESY and ROESY. Because the function gives greater weight to the beginning of the FID where the signal-to-noise ratio is greater, the sensitivity is enhanced at the expense of the resolution. The *sine-bell* (or "unshifted sine-bell") function starts at the zero point of the sine function (sine of 0°) at the start of the FID data, reaches a maximum halfway through the FID (sine of 90°) and falls back smoothly to zero by the end of the acquired FID data (sine of

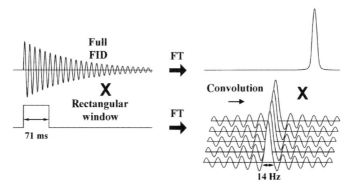

Figure 9.50

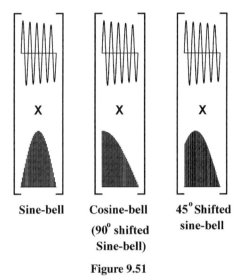

Figure 9.51

180°). Because data later in the FID (in the center) is emphasized over data early on in the FID, this window function leads to *resolution enhancement* at the expense of sensitivity (signal-to-noise ratio). It is commonly used for COSY data where the peaks are antiphase and will "self-cancel" if they are broad (Fig. 9.37). This radical resolution enhancement was encountered in Chapter 2 for 1D spectra, where we saw the resulting "ditches" on either side of the peaks (Fig. 2.9). In the DQF-COSY spectrum of sucrose these ditches are clearly visible in the F_2 slices (Fig. 9.28). In Figure 9.35 the DQF-COSY spectrum was processed with an unshifted sine-bell (left side), but the COSY-35 was processed with a simple exponential multiplier to facilitate accurate curve-fitting (right side). The difference in peak resolution is clearly visible in the 2D spectra, and in the F_2 slices of the COSY-35 we see no ditches. Less radical resolution enhancement can be achieved by shifting the sine bell by 30° or 45° (Fig. 9.51), always keeping the 180° point of the sine function at the end of the FID. These window functions are commonly used for 2D experiments with strong crosspeaks (efficient magnetization transfer) such as 2D TOCSY.

The size of the window must be carefully fit to the FID being processed. Varian uses the parameter *sb* to describe the width (in seconds) of the sine-bell window from the 0° point to the 90° point. Thus for an unshifted sine-bell function, we want the 0° to 180° portion of the sine function (2 *sb*) to just fit over the time duration of the FID (*at*). This is accomplished by setting the value of *sb* to one-half the acquisition time: $sb = at/2$. Since the sine-bell is not shifted, the "sine-bell shift" (*sbs*) is set to zero. For a cosine-bell or 90° shifted sine-bell window, we want the portion of the sine function from 90° to 180° (or *sb*, since the 0° to 90° portion is of the same duration as the 90° to 180° portion) to just fit over the FID (duration *at*): $sb = at$. In addition, the whole sine function is shifted to the left side by the duration of the FID, so we set the parameter *sbs* (sine-bell shift) equal to $-at$ (left shift corresponds to a negative number). In F_1 we do not have a parameter for acquisition time (*at*) in t_1, but we know that the maximum t_1 value is just the number of data points times the sampling delay:

$$t_1(\max) = (ni \times 2) \times \Delta t_1 = (ni \times 2) \times (1/(2 \times sw1)) = ni/sw1$$

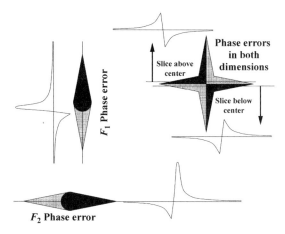

Figure 9.52

So you can just set $sb1 = ni/sw1$ and $sbs1 = -sb1$ for a 90°-shifted sine-bell, and $sb1 = ni/(2 \times sw1)$ and $sbs1 = 0$ for an unshifted sine-bell. Bruker uses the parameter wdw (in both F_1 and F_2) to set the window function (SINE = sine-bell, QSINE = sine-squared, etc.) and ssb for the sine-bell shift. For example, if $ssb = 2$, the sine function is shifted 90° (180°/ssb) and we get a simple cosine-bell window. For an unshifted sine-bell, use $ssb = 0$.

9.7.3 Phase Correction in Two Dimensions

Phase errors appear in 2D spectra as "streaks" with negative intensity on one side and positive intensity on the other side. Vertical streaks correspond to F_1 phase errors and horizontal streaks to F_2 phase errors (Fig. 9.52). For example, if positive intensity is color-coded red and negative intensity blue, an F_2 phase error will appear as a crosspeak or diagonal peak with a red streak extending out to the left side and a blue streak extending out to the right side, or *vice versa*. Sometimes there are severe phase errors in both dimensions, leading to a pattern of horizontal and vertical streaks (Fig. 9.52, upper right).

The complex FID, consisting of real and imaginary parts, is converted by a complex Fourier transform to a complex spectrum, consisting of a real spectrum and an imaginary spectrum. Phase correction involves "rotating" the complex spectrum in the complex plane until the real spectrum is absorptive and the imaginary spectrum is dispersive (Chapter 3, Fig. 3.38). The imaginary spectrum is then discarded and we use the real (absorptive) spectrum. In 2D processing there are two Fourier transforms: one in t_2/F_2 and one in t_1/F_1. Each one generates two spectra, so we can potentially end up with four 2D matrices (Fig. 9.53). Phasing a 2D matrix would then involve forming a linear combination of all four final 2D spectra to get absorptive lineshape in both dimensions. Regardless of the software you are using, you are looking for four numbers: the phase correction parameters in F_2 (zero-order and first-order) and the phase correction parameters in F_1 (zero-order and first-order). The zero-order correction is applied equally to all peaks in the spectrum and the first-order parameter is a linear function of chemical shift, going through zero at the "pivot peak." The process is based on phase correction of 1D "slices": make an F_2 (horizontal) slice through a peak (diagonal or crosspeak) in the 2D spectrum and phase correct it as a 1D spectrum to generate the F_2 phase correction parameters (Fig. 9.52). Then make an F_1 (vertical slice)

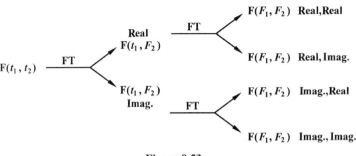

Figure 9.53

through a peak and phase correct that 1D spectrum to get the F_1 phase correction parameters. This works well for 2D experiments with strong crosspeak intensity, such as TOCSY, but for others we may see only one peak in a given 1D slice. How do you come up with a chemical-shift dependent (first-order) phase correction of a spectrum with only one peak? The solution is to use more than one slice to generate multiple 1D spectra with peaks in different parts of the spectrum. Bruker software uses a display with three 1D windows, all controlled by the same two phase parameters (*phc0* and *phc1*). Three different slices of the same type (e.g., rows) are loaded into the windows, the pivot peak is selected in one of the windows and the two phase parameters are adjusted, with all three 1D slices responding in real time to the adjustments. When the optimal parameters are found, this phase correction is applied to all rows of the data matrix. Varian software will only do horizontal slices ("traces") so the matrix has to be turned on its side (command: *trace = "f1"*) in order to load an F_1 slice. The phase correction parameters (*rp* and *lp*) are determined by treating the slice as a 1D spectrum. If you need more than one slice, you can adjust *rp* with one slice (with a peak on the right) and *lp* with another (with a peak on the left side). Other software packages (e.g., Felix) construct a 1D spectrum as a sum of several slices from the 2D matrix. For example, three columns (F_1 slices) can be selected and summed to give a 1D spectrum with three peaks, which is then phase corrected and the parameters are applied to all the columns of the 2D matrix. Regardless of the software used, it can be tricky if there are significant phase errors in both dimensions because the F_1 phase errors can "flip" peaks upside down in the F_2 slices. When making multiple F_2 slices (rows) of the 2D matrix, select a row just above or below the center of a peak (Fig. 9.52, upper right) if there is a significant F_1 phase error. Try to be consistent in making all slices on the same side (e.g., all rows just above the center or all columns just to the left of the center) to keep the phase errors in the other dimension from interfering.

2D spectra with very weak crosspeak intensities (e.g., NOESY) require very careful phase correction because dispersive "tails" (Fig. 9.39) can extend far outwards from the intense diagonal peaks to obscure weak crosspeaks. Sometimes you will also have to "flatten" the baseline to be able to lower the contour threshold enough to see the weak crosspeaks. If there is curvature in the baseline (in two dimensions that would be like a rug held up at the corners and sagging in the middle) you cannot see the weak peaks because the threshold plane cuts through the noise in some places. Baseline errors appear as streaks extending in both directions (up/down or left/right) from a peak with the same sign (same color code) on both sides. There are many techniques to do 1D baseline flattening. The most common method is to generate a 1D spectrum and manually set up a series of baseline "points" which represent specific chemical-shift positions where there is only baseline noise and no peaks.

The program fits the intensity at these point values to a polynomial (up to 5th order) function and then subtracts the polynomial function from the whole dataset. This is repeated for each 1D slice (row or column) of the 2D data matrix. More sophisticated methods calculate the baseline points automatically and use functions other than polynomials. For example, a program called FLATT (by Kurt Wüthrich) is very effective at removing horizontal or vertical streaks resulting from baseline curvature in rows or columns of the data matrix. Especially with NOESY and ROESY data baseline correction is essential to getting "clean" 2D displays and plots.

The following table summarizes the 2D parameters for Bruker and Varian.

Direct (F_2)		Indirect (F_1)		
Varian	Bruker	Varian	Bruker	
	Acquisition Parameters:			
np	td	ni × 2	td	Number of data points (real + complex).
sw	swh	sw1	swh	Spectral width in Hz.
at	aq	ni/sw1	td/(2swh)	Acquisition time in seconds.
tof	o1	tof	o1	Transmitter offset (homonuclear).
tof	o1	dof	o2	Transmitter offset (heteronuclear).
	Processing Parameters:			
fn/2	si/2	fn1/2	si/2	Number of complex pairs after zero-filling.
sb	—	sb1	—	Span of sine-bell function in seconds (0° to 90°).
sbs	—	sbs1	—	Amount of right shift of sine-bell function in seconds.
—	wdw	—	wdw	Window function (SINE, QSINE, etc).
—	sbs	—	sbs	Sine-bell shift: degrees = 180/*sbs* (zero if sbs = 0).

10

ADVANCED NMR THEORY: NOESY AND DQF-COSY

The purpose of this book is to provide a deep and satisfying understanding of how NMR experiments work, while maintaining the practical perspective of an NMR user (organic chemist or structural biologist) rather than a theoretical approach, which would be more appropriate for an analytical chemist, physical chemist, or physicist. The simple vector model for net magnetization, along with the energy diagrams with open and filled circles to represent population differences, has provided a strong conceptual basis for understanding the NOE (nuclear Overhauser enhancement) and many of the complex tricks of evolution of single-quantum coherence (spin echoes, APT, BIRD, TANGO, gradients, etc.). In order to describe coherence transfer (INEPT), we needed the additional theoretical tools of the product operators ($\mathbf{I}_x$, $2\mathbf{I}_y\mathbf{S}_z$, etc.): a simple mathematical approach that is firmly tied to the visual representation of the vector model. We have touched upon the idea of multiple quantum coherences (ZQC and DQC) and defined them in terms of product operators ($2\mathbf{I}_x\mathbf{S}_x$, $2\mathbf{I}_y\mathbf{S}_x$, etc.), but many things had to be taken on faith. Why we multiply the operators together ($2\mathbf{I}_y\mathbf{S}_z$, $2\mathbf{I}_y\mathbf{S}_x$, etc.) in a particular way to represent certain coherences (DQC, antiphase SQC, etc.) has been until now just a set of rules.

In this chapter, we will introduce a new level of theoretical tools—the density matrix—and show by a bit of matrix algebra what the product operators actually represent. The qualitative picture of population changes in the NOE will be made more exact, the precise basis of cross-relaxation will be revealed, and a new phenomenon of cross-relaxation—chemical exchange—will be introduced. With these expanded tools, it will be possible to understand the 2D NOESY (nuclear Overhauser and exchange spectroscopy) and DQF-COSY experiments in detail.

The tricks of selecting desired coherences and rejecting unwanted (artifact) peaks by phase cycling or gradients will be formalized by introducing the spherical product operators and defining the coherence order precisely. This gives us a very simple way of describing an

---

NMR Spectroscopy Explained: Simplified Theory, Applications and Examples for Organic Chemistry and Structural Biology, by Neil E. Jacobsen
Copyright © 2007 John Wiley & Sons, Inc.

NMR pulse sequence without getting tied up in the details of pulse phases and a mountain of sine and cosine terms: only the essential elements of the sample net magnetization will be described at each point. Finally, the formal Hamiltonian description of solution-state NMR will be described and applied to explain two related phenomena: strong coupling ("leaning" of multiplets) and TOCSY mixing (the "isotropic" mixing sequence).

10.1 SPIN KINETICS: DERIVATION OF THE RATE EQUATION FOR CROSS-RELAXATION

In Chapter 5, we demonstrated qualitatively how DQ relaxation alone ($\beta\beta \to \alpha\alpha$) leads to a negative NOE (saturation or inversion of H_a leads to enhancement of H_b's z magnetization), and ZQ relaxation alone ($\alpha\beta \leftrightarrow \beta\alpha$) leads to a positive NOE (reduction of H_b's z magnetization). We also showed how the distribution of tumbling rates changes as molecular size is increased, leading to a change from relaxation dominated by DQ transitions (small molecules) to relaxation dominated by ZQ transitions (large molecules). We will now consider more quantitatively how this happens by looking at the "kinetics" of molecules (or proton pairs H_a and H_b) moving among the four homonuclear spin states $\alpha\alpha$, $\alpha\beta$, $\beta\alpha$, and $\beta\beta$.

For simplicity, we assume that H_a and H_b are close to each other in a molecule ($r_{ab} <$ 5 Å) but have no J coupling. We will use $P_{\beta\beta}$, $P_{\alpha\beta}$, $P_{\beta\alpha}$, and $P_{\alpha\alpha}$ to represent the populations of the four spin states $\beta\beta$ ($H_a = \beta$, $H_b = \beta$), $\alpha\beta$ ($H_a = \alpha$, $H_b = \beta$), $\beta\alpha$ ($H_a = \beta$, $H_b = \alpha$), and $\alpha\alpha$ ($H_a = \alpha$, $H_b = \alpha$), respectively. At equilibrium, $P_{\beta\beta} = N/4 - 2\delta$, $P_{\alpha\beta} = N/4$, $P_{\beta\alpha} = N/4$, and $P_{\alpha\alpha} = N/4 + 2\delta$, where N is the total number of molecules and δ is a very small fraction of N determined by the Boltzmann condition (Fig. 10.1).

Because z magnetization is the result of population differences between spin states, we can equate z magnetization with population difference (actually it is proportional, but for simplicity the proportionality constant is omitted):

$$M_z^a = P_{\alpha\beta} - P_{\beta\beta} = P_{\alpha\alpha} - P_{\beta\alpha} \quad (H_a(2) \text{ and } H_a(1) \text{ transitions})$$

$$M_z^b = P_{\alpha\alpha} - P_{\alpha\beta} = P_{\beta\alpha} - P_{\beta\beta} \quad (H_b(1) \text{ and } H_b(2) \text{ transitions})$$

$$M_z^a + M_z^b = (P_{\alpha\beta} - P_{\beta\beta}) + (P_{\alpha\alpha} - P_{\alpha\beta}) = P_{\alpha\alpha} - P_{\beta\beta} \quad (\text{DQ transition})$$

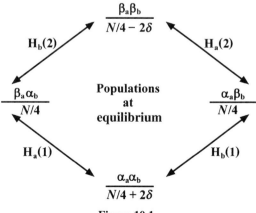

Figure 10.1

$$M_z^a - M_z^b = (P_{\alpha\beta} - P_{\beta\beta}) - (P_{\beta\alpha} - P_{\beta\beta}) = P_{\alpha\beta} - P_{\beta\alpha} \quad \text{(ZQ transiton)}$$

$$M_o = 2\delta \quad \text{(equilibrium)}$$

Note that in each case we subtract the population of the higher energy (less populated at equilibrium) state from the population of the lower energy (more populated at equilibrium) state. We can also define the amount of "disequilibrium" as the difference between the actual z magnetization and the equilibrium z magnetization.

$$\Delta M_z^a = M_z^a - M_o = P_{\alpha\beta} - P_{\beta\beta} - 2\delta = P_{\alpha\alpha} - P_{\beta\alpha} - 2\delta$$

$$\Delta M_z^b = M_z^b - M_o = P_{\alpha\alpha} - P_{\alpha\beta} - 2\delta = P_{\beta\alpha} - P_{\beta\beta} - 2\delta$$

Note that ΔM_z^a and ΔM_z^b both tend toward zero ($\Delta P = 2\delta$) as the nuclei relax.

Now consider the "kinetics" of the flow of spins between spin states. For each pair of spin states, we calculate the difference in population and compare it to the equilibrium difference. If these two are not the same, there will be a flow of spins from the "overpopulated" state to the "underpopulated" state at a rate that is proportional to the "rate constant" for that transition and to the amount by which the transition is out of equilibrium. The rate constants, or relaxation rates, for each transition are determined by

1. the distance **r** between protons H_a and H_b in the molecule ($1/r^6$ effect);
2. the number of molecules in solution that are tumbling at a rate corresponding to the frequency (ν) of that transition: ν_a and ν_b for the SQ transitions, $\nu_a + \nu_b$ for the DQ transition, and $\nu_a - \nu_b$ for the ZQ transition.

The four single-quantum transitions relax at a rate W_1, or more specifically W_1^a for the H_a transitions and W_1^b for the H_b transitions. The double-quantum transition ($\alpha\alpha \leftrightarrow \beta\beta$) relaxes with rate W_2 and the zero-quantum transition ($\alpha\beta \leftrightarrow \beta\alpha$) relaxes at a rate W_o (Fig. 10.2). For small organic molecules in nonviscous solvents (e.g., cholesterol in $CDCl_3$), the molecule tumbles rapidly compared to the Larmor frequency and the ratio of $W_1:W_o:W_2$ is about

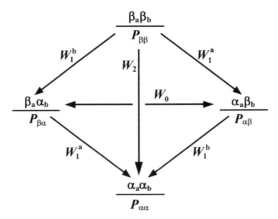

Figure 10.2

3:2:12, meaning that double-quantum relaxation is the fastest pathway. For a protein with a molecular weight of 13,690 Da (Ribonuclease A) in H_2O, the ratio $W_1:W_0:W_2$ is 1:28:1 on a 500-MHz spectrometer. Thus, for large molecules, the zero-quantum pathway is fastest.

Consider first the single-quantum transition between the $\alpha\beta$ and $\beta\beta$ states ($H_a(2)$ transition, Fig. 10.1). This is an H_a transition with relaxation rate W_1^a. The equilibrium difference in population for this transition is $P_{\alpha\beta} - P_{\beta\beta} = 2\delta$. If this equality does not hold, then the "overpopulation" of the $\beta\beta$ state is given by $P_{\beta\beta} - P_{\alpha\beta} + 2\delta$, and the rate of spins dropping down from the $\beta\beta$ state to the $\alpha\beta$ state is $W_1^a(P_{\beta\beta} - P_{\alpha\beta} + 2\delta)$. If this were the only transition available (i.e., if there were no double-quantum or zero-quantum pathways), we could write down the rate of change of population as

$$dP_{\alpha\beta}/dt = -dP_{\beta\beta}/dt = W_1^a(P_{\beta\beta} - P_{\alpha\beta} + 2\delta)$$

We can calculate from this the rate equation for the relaxation of the H_a spins:

$$d\Delta M_z^a/dt = d(M_z^a - M_o)/dt = dM_z^a/dt = d(P_{\alpha\beta} - P_{\beta\beta})/dt$$
$$= dP_{\alpha\beta}/dt - dP_{\beta\beta}/dt = 2W_1^a(P_{\beta\beta} - P_{\alpha\beta} + 2\delta) = -2W_1^a \Delta M_z^a$$

The overall equation is a simple first-order decay with rate constant $2W_1^a$. This corresponds to the longitudinal relaxation rate for H_a in the absence of cross-relaxation: $R_1^a = 1/T_1^a$. Looking at the other H_a transition, $\alpha\alpha$ state to $\beta\alpha$ state, gives the same result, and the H_b transitions yield the analogous result $R_1^b = 1/T_1^b = 2W_1^b$. These self-relaxation times, T_1^a and T_1^b, will decrease when we introduce the cross-relaxation pathways (DQ and ZQ relaxation).

Now let's look at the more interesting situation where the cross-relaxation pathways (single quantum and double quantum) are available. Spins in the $\beta\beta$ state can relax by any of three pathways: they can drop down to the $\alpha\beta$ state (rate W_1^a), drop down to the $\beta\alpha$ state (rate W_1^b), or follow the double-quantum pathway down to the $\alpha\alpha$ state (rate W_2). All of these pathways will contribute to the change in population of the $\beta\beta$ state as a function of time (Fig. 10.3). Considering all three pathways leading away from the $\beta\beta$ state, we can

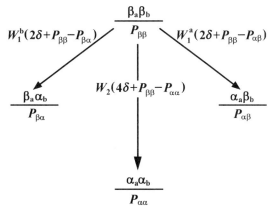

Figure 10.3

write the three terms contributing to the rate of loss of spins from this state:

$$W_1^a(P_{\beta\beta} - P_{\alpha\beta} + 2\delta) \quad \text{via the } H_a \text{ SQ (single quantum) transition}$$

$$W_1^b(P_{\beta\beta} - P_{\beta\alpha} + 2\delta) \quad \text{via the } H_b \text{ SQ transition}$$

$$W_2(P_{\beta\beta} - P_{\alpha\alpha} + 4\delta) \quad \text{via the DQ (double quantum) transition}$$

Note that the equilibrium difference across the DQ transition ($P_{\alpha\alpha} - P_{\beta\beta}$) is 4δ because the energy separation is twice that of an SQ transition. Combining all three terms,

$$dP_{\beta\beta}/dt = -W_1^a(P_{\beta\beta} - P_{\alpha\beta} + 2\delta) - W_1^b(P_{\beta\beta} - P_{\beta\alpha} + 2\delta) - W_2(P_{\beta\beta} - P_{\alpha\alpha} + 4\delta)$$

The minus signs reflect the fact that all three pathways *remove* spins from the $\beta\beta$ state when the spins flow "downhill" to the more stable states $\alpha\beta$, $\beta\alpha$, and $\alpha\alpha$. Likewise for the other three spin states

$$dP_{\alpha\beta}/dt = W_1^a(P_{\beta\beta} - P_{\alpha\beta} + 2\delta) - W_1^b(P_{\alpha\beta} - P_{\alpha\alpha} + 2\delta) - W_0(P_{\alpha\beta} - P_{\beta\alpha})$$

$$dP_{\beta\alpha}/dt = W_1^b(P_{\beta\beta} - P_{\beta\alpha} + 2\delta) - W_1^a(P_{\beta\alpha} - P_{\alpha\alpha} + 2\delta) + W_0(P_{\alpha\beta} - P_{\beta\alpha})$$

$$dP_{\alpha\alpha}/dt = W_1^b(P_{\alpha\beta} - P_{\alpha\alpha} + 2\delta) + W_1^a(P_{\beta\alpha} - P_{\alpha\alpha} + 2\delta) + W_2(P_{\beta\beta} - P_{\alpha\alpha} + 4\delta)$$

Note that the equilibrium population difference for the zero-quantum transition is zero because the two states have (essentially) the same energy. Now we can substitute the (indirectly) measurable quantities M_z^a and M_z^b for the population differences. For the "disequilibrium" of M_z^a we have

$$d\Delta M_z^a/dt = d(M_z^a - M_0)/dt = dM_z^a/dt = d(P_{\alpha\beta} - P_{\beta\beta})/dt = dP_{\alpha\beta}/dt - dP_{\beta\beta}/dt$$

Substituting the expressions above for the "flow" toward and away from the $\alpha\beta$ and $\beta\beta$ states

$$\begin{aligned}
d\Delta M_z^a/dt &= W_1^a(P_{\beta\beta} - P_{\alpha\beta} + 2\delta) - W_1^b(P_{\alpha\beta} - P_{\alpha\alpha} + 2\delta) - W_0(P_{\alpha\beta} - P_{\beta\alpha}) \\
&\quad + W_1^a(P_{\beta\beta} - P_{\alpha\beta} + 2\delta) + W_1^b(P_{\beta\beta} - P_{\beta\alpha} + 2\delta) + W_2(P_{\beta\beta} - P_{\alpha\alpha} + 4\delta) \\
&= 2W_1^a(P_{\beta\beta} - P_{\alpha\beta} + 2\delta) - W_1^b(P_{\alpha\beta} - P_{\alpha\alpha} - P_{\beta\beta} + P_{\beta\alpha}) \\
&\quad + W_2(P_{\beta\beta} - P_{\alpha\alpha} + 4\delta) - W_0(P_{\alpha\beta} - P_{\beta\alpha}) \\
&= 2W_1^a(-M_z^a + M_0) - W_1^b(-M_z^b + M_z^b) \\
&\quad + W_2(-M_z^a - M_z^b + 2M_0) - W_0(M_z^a - M_z^b) \\
&= -2W_1^a\Delta M_z^a - W_2(\Delta M_z^a + \Delta M_z^b) - W_0(\Delta M_z^a - \Delta M_z^b) \\
&= -\Delta M_z^a(2W_1^a + W_2 + W_0) - \Delta M_z^b(W_2 - W_0) \quad (10.1)
\end{aligned}$$

Likewise for ΔM_z^b,

$$d\Delta M_z^b/dt = d(M_z^b - M_o)/dt = dM_z^b/dt = d(P_{\beta\alpha} - P_{\beta\beta})/dt = dP_{\beta\alpha}/dt - dP_{\beta\beta}/dt$$

Substituting the expressions above for the "flow" toward and away from the $\beta\alpha$ and $\beta\beta$ states

$$\begin{aligned}
d\Delta M_z^b/dt &= W_1^b(P_{\beta\beta} - P_{\beta\alpha} + 2\delta) - W_1^a(P_{\beta\alpha} - P_{\alpha\alpha} + 2\delta) + W_0(P_{\alpha\beta} - P_{\beta\alpha}) \\
&\quad + W_1^a(P_{\beta\beta} - P_{\alpha\beta} + 2\delta) + W_1^b(P_{\beta\beta} - P_{\beta\alpha} + 2\delta) + W_2(P_{\beta\beta} - P_{\alpha\alpha} + 4\delta) \\
&= 2W_1^b(P_{\beta\beta} - P_{\beta\alpha} + 2\delta) - W_1^a(P_{\beta\alpha} - P_{\alpha\alpha} - P_{\beta\beta} + P_{\alpha\beta}) \\
&\quad + W_2(P_{\beta\beta} - P_{\alpha\alpha} + 4\delta) + W_0(P_{\alpha\beta} - P_{\beta\alpha}) \\
&= 2W_1^b(-M_z^b + M_o) - W_1^a(-M_z^a + M_z^a) \\
&\quad + W_2(-M_z^a - M_z^b + 2M_o) + W_0(M_z^a - M_z^b) \\
&= -2W_1^b \Delta M_z^b - W_2(\Delta M_z^a + \Delta M_z^b) + W_0(\Delta M_z^a - \Delta M_z^b) \\
&= -\Delta M_z^b(2W_1^b + W_2 + W_0) - \Delta M_z^a(W_2 - W_0) \quad (10.2)
\end{aligned}$$

The two results can be written together as a system of two linked (coupled) first-order differential equations. This means that the return of M_z^a to equilibrium depends on how far M_z^b is from equilibrium, and *vice versa*.

$$d\Delta M_z^a/dt = -R_{aa}\Delta M_z^a - R_{ab}\Delta M_z^b \qquad \text{from (10.1)}$$

$$d\Delta M_z^b/dt = -R_{ab}\Delta M_z^a - R_{bb}\Delta M_z^b \qquad \text{from (10.2)}$$

where the self-relaxation rates R_{aa} and R_{bb} are the longitudinal relaxation rates for H_a and H_b, respectively, and the cross-relaxation rate R_{ab} depends on the competition of the DQ and ZQ pathways:

$$R_{aa} = 2W_1^a + W_2 + W_0 \quad \text{self-relaxation for } H_a$$

$$R_{bb} = 2W_1^b + W_2 + W_0 \quad \text{self-relaxation for } H_b$$

$$R_{ab} = W_2 - W_0 \quad \text{cross-relaxation}$$

For a small organic molecule, W_2 is about six times as fast as W_0, so R_{ab} is positive. This means that if we start at equilibrium ($\Delta M_z^b = 0$) and saturate H_a ($\Delta M_z^a = -M_o$) the z magnetization of H_b is enhanced:

$$d\Delta M_z^b/dt = -R_{ab}\underset{\text{positive}}{\Delta M_z^a} - R_{bb}\Delta M_z^b = -R_{ab}(-M_o) \quad \text{(initial)}$$

This is because the cross-relaxation term $(-R_{ab}\Delta M_z^a)$ becomes positive and makes M_z^b grow initially and become greater than M_o. This leads to enhancement of the H_b signal in a 1D NOE experiment, and the initial rate of growth as a function of time ("mixing time") is a direct measure of R_{ab}, which is proportional to $1/\mathbf{r}^6$. This is the "classical" NOE familiar to the organic chemist. Note that inversion of the H_a spins ($\Delta M_z^a = -2M_o$) has the same effect but twice as strong. We call this a *negative* NOE because decreasing M_z^a has the effect of increasing M_z^b; for this reason the crosspeaks in a 2D NOESY spectrum of a small organic molecule are negative with respect to the positive diagonal.

For a large molecule such as a protein, W_o is much faster than W_2, so R_{ab} is negative. This means that if we saturate H_a ($\Delta M_z^a = -M_o$) when the H_b spins are at equilibrium ($\Delta M_z^b = 0$), the z magnetization on H_b will begin to decrease:

$$d\Delta M_z^b/dt = -R_{ab}\Delta M_z^a - R_{bb}\Delta M_z^b = \underbrace{-R_{ab}(-M_o)}_{\text{negative}}$$

This is because the cross-relaxation term ($-R_{ab}\Delta M_z^a$) becomes negative ($-$negative $\times$ negative) and makes M_z^b decrease initially below M_o. This leads to reduction of the H_b signal in a 1D NOE experiment, and an NOE "buildup" study using a series of different mixing times can be used to measure R_{ab} as the initial rate. We call this a *positive* NOE because decreasing ΔM_z^a has the effect of decreasing ΔM_z^b; for this reason the crosspeaks in a 2D NOESY spectrum of a large molecule are positive with respect to the positive diagonal.

There is a molecular size in between these extremes for which W_2 and W_o are the same. In this case $R_{ab} = 0$ and there is no NOE. For an ideal spherical protein in water at 27 °C on a 500-MHz instrument, this occurs at a molecular weight of 2370 Da, or a typical 20-residue peptide. Fortunately, there is another 2D experiment called ROESY (*r*otating frame *o*verhauser *e*ffect *s*pectroscopy) that carries out the NOE transfer in a weak B_1 field (typically $\gamma B_1/2\pi \sim 3000$ Hz) rather than in the B_o field (e.g., $\gamma B_o/2\pi = 500$ MHz). Under these conditions, the SQ and ZQ transition frequencies are so low (3000 and 6000 Hz) that even large molecules such as proteins have significant populations tumbling at these rates. This means that all molecules, regardless of size, have $W_2 > W_o$ and a positive R_{ab}; that is, all molecules behave like small molecules. For small and large molecules alike the NOE effect ("ROE") is an enhancement of M_z, and the crosspeaks in the 2D ROESY experiment are negative with respect to the positive diagonal.

10.2 DYNAMIC PROCESSES AND CHEMICAL EXCHANGE IN NMR

So far we have assumed that a nucleus exists in a stable chemical environment; that is, that its chemical shift does not change with time. The chemical shift is a result of the effective field at the nucleus, and this is sensitive to inductive effects (electron withdrawal or donation through bonds), through-space effects (magnetic anisotropy due to nearby π-bonds), and steric effects. A nucleus can change its chemical environment, and therefore its chemical shift, by a chemical reaction (bond breaking and bond breaking) or by a conformational change (bond rotation). The effect this has on the NMR spectrum depends on the rate of the exchange process, compared to a time commonly referred to as the *NMR timescale*. There are actually a number of different timescales that can be studied using NMR, but the commonly used NMR timescale is that of direct observation of chemical shifts—the timescale of recording an NMR spectrum. The NMR timescale is essentially the "shutter speed" of taking an NMR picture of the molecule, and is of the order of magnitude of

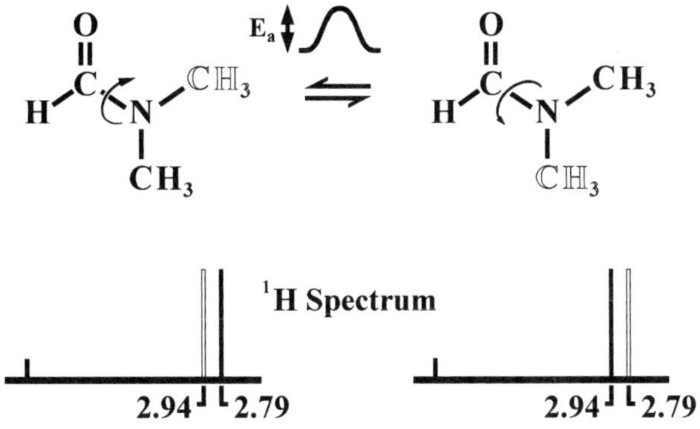

Figure 10.4

milliseconds. This is a very slow timescale compared to optical spectroscopies, which operate on a nanosecond to picosecond or faster timescale. In fact, most molecular motions (molecular tumbling, bond rotation of methyl groups, etc.) and many chemical reactions (e.g., acid–base reactions) are much faster than the NMR "shutter" and we see only a single chemical shift that is the time average of the chemical shifts of the individual environments that the nucleus visits.

The simplest effect occurs when a given nucleus in a molecule changes its magnetic environment, and thus its chemical shift, as a result of a simple molecular motion. For example, the methyl groups in N,N-dimethylformamide (DMF) change places as a result of the relatively slow rotation about the amide bond (Fig. 10.4). The protons of the methyl group closer to the carbonyl oxygen have a larger chemical shift (2.94 ppm) than the other site (2.79 ppm) so that the resonant frequency of a given nucleus is bouncing back and forth between these two chemical shifts as the bond rotates. A "shutter time" can be defined for the NMR experiment, which is inversely proportional to the difference in chemical shift (in Hz!) between the two environments. On a 200-MHz instrument:

$$\text{"shutter time"} = \sqrt{2}/(\pi \Delta \nu) = 1/(2.22 \Delta \nu) = 1/(2.22 \times 0.15 \text{ ppm} \times 200 \text{ Hz/ppm})$$
$$= 1/(2.22 \times 30 \text{ Hz}) = 0.015 \text{ s} = 15 \text{ ms}$$

If the average lifetime in one state (τ_{ex}) is longer than the shutter speed, we will see two distinct peaks in the spectrum. If the average lifetime is shorter than the shutter time we will only see one averaged peak. This shutter time is formally called the coalescence time, τ_c, for the exchange process, or simply the "NMR timescale".

Slow exchange ($\tau_{ex} \gg \tau_c$) means that each nucleus is, on average, entirely in one environment during the shutter time, so that the motion is "frozen" and two sharp peaks are observed for different nuclei in the two environments (Fig. 10.5, top). Heating the sample speeds up the exchange so that a blur is observed (Fig. 10.5, center) as nuclei move back and forth between chemical environments during the shutter time ($\tau_{ex} \sim \tau_c$). At even higher temperature, the average nucleus moves back and forth so many times during the shutter time that a single sharp peak is observed (Fig. 10.5, bottom) at the average of the two chemical shifts (*fast exchange*, $\tau_{ex} \ll \tau_c$). Study of this behavior as a function of temperature allows determination of the rate constant and the energy barrier for the bond rotation.

416   ADVANCED NMR THEORY: NOESY AND DQF-COSY

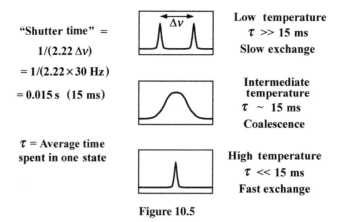

"Shutter time" =
1/(2.22 Δv)
= 1/(2.22 × 30 Hz)
= 0.015 s (15 ms)

τ = Average time spent in one state

Low temperature
$\tau \gg$ 15 ms
Slow exchange

Intermediate temperature
$\tau \sim$ 15 ms
Coalescence

High temperature
$\tau \ll$ 15 ms
Fast exchange

Figure 10.5

The DMF bond rotation can be considered as a dynamic equilibrium with equilibrium constant of 1 ($k_1 = k_{-1}$):

$$H_a \underset{k_{-1}}{\overset{k_1}{\rightleftharpoons}} H_b$$

The average time spent in the H_a state ($\delta(CH_3) = 2.94$ ppm)) before jumping to H_b is $1/k_1$, as the k_1 rate defines the end of its lifetime in the H_a state. The average time spent in the H_b state is likewise equal to $1/k_{-1}$, as the k_{-1} process defines the end of its life. Figure 10.6 shows a simulation of the DMF bond rotation starting with slow exchange (top) and raising the temperature to increase the average time spent in each state ($\tau_{ex} = 1/k_1 = 1/k_{-1} = 1.5$ s, 0.15 s, 15 ms, 1.5 ms, and 0.15 ms). In the simulation, an equal amount of noise has been added to each spectrum and the spectra are scaled vertically to the tallest peak. Comparing the average lifetime τ_{ex} to the coalescence lifetime (shutter time) τ_c, we see sharp lines at the two chemical shift positions when τ_{ex} (1.5 s) is 100 times longer than τ_c (Fig. 10.6, top).

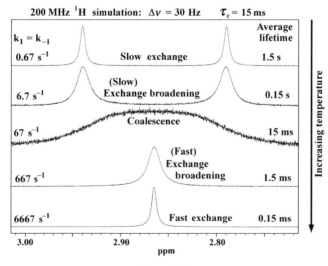

Figure 10.6

The exchange process is much slower than the "shutter time," which means that the nucleus spends much more time on average than the NMR timescale in each environment. In this *slow exchange* limit, the intensity (integral area) of each peak will reflect the fraction of time that the nucleus spends in that environment ($1/(K_{eq} + 1)$ for H_a, $K_{eq}/(K_{eq} + 1)$ for H_b), in this case a 1:1 ratio ($K_{eq} = 1$). When the temperature is increased to the point where τ_{ex} (0.15 s) is only 10 times longer than τ_c, the peaks begin to broaden and the peak maxima begin to "creep" inwards each other. This "exchange broadening" can be used to measure the rate constants k_1 and k_{-1} (see below). When the temperature is raised further until τ_{ex} is equal to the coalescence time (Fig. 10.6, center), there is no longer a "dip" between the two peaks and we see one single, very broad peak. Because the same peak area (two CH_3 groups) is now spread over a very broad peak, the peak height is much lower and the noise appears larger in comparison. It is not uncommon for peaks to be broadened out of existence (into the noise) by exchange!

Continuing to raise the temperature, we arrive at a point where τ_{ex} (1.5 ms) is 10 times shorter than τ_c. The single peak is now much sharper but still broadened relative to the natural linewidth in the absence of exchange. This is a case where we might see one peak in the spectrum broader and shorter than the others and start thinking about the possibility of an exchange process. Finally, at the high temperature limit where τ_{ex} is 100 times or more shorter than τ_c, we see a single sharp peak at the average chemical shift position (Fig. 10.6, bottom), with no broadening and a peak area representing the total of both environments (in this case 6H).

Note that on a 200-MHz instrument, $\Delta \nu$ is 30 Hz (0.15 ppm × 200 Hz/ppm) and the NMR timescale τ_c is 15 ms ($1/(2.22 \times 30)$), but the "shutter speed" is faster as we go to higher field instruments because the chemical shift difference $\Delta \nu$ is measured in hertz, not ppm. Thus, moving to higher field shortens the shutter time τ_c in a way that is inversely proportional to B_o. Figure 10.7 shows simulated spectra of the DMF sample at the same temperature that gives $\tau_{ex} = 15$ ms, analyzed on three different spectrometers with $\gamma_H B_o/2\pi = 60$, 200, and 600 MHz. At 60 MHz (top), we have an exchange-broadened fast

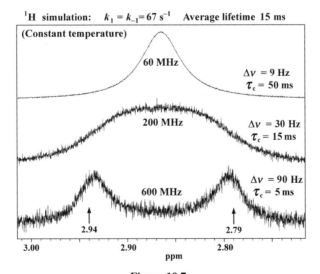

Figure 10.7

exchange spectrum ($\tau_{ex} = \tau_c/3.33$); at 200 MHz (middle), we are at the coalescence point ($\tau_{ex} = \tau_c$); and at 600 MHz (bottom), we have an exchange-broadened slow exchange spectrum ($\tau_{ex} = 3\,\tau_c$). In terms of its effect on the spectrum, going to higher field is like cooling down the sample and going to lower field is like heating it up. If you simply want to verify if a broadened peak is due to exchange it might be much simpler to try a different field strength instead of doing a variable temperature study.

If the nucleus is farther away in the molecule from the site of chemical change, its chemical shift may be affected less so that $\Delta\nu$ is small or even zero. We can have as many NMR timescales as there are distinct nuclei within a molecule. Thus, the NMR timescale is not an absolute time, but rather it depends on field strength and on the significance of the chemical exchange in terms of its effects on the chemical shift of a particular nucleus.

10.2.1 Slow Exchange

Let's consider in more detail what happens in slow exchange as we increase the exchange rate and the single NMR peak begins to broaden. A simple 1D NMR spectrum is recorded by rotating the sample magnetization into the x–y plane with a pulse and then observing the precession of this sample magnetization in the x–y plane as it induces a sinusoidal signal in the probe coil. As the sample magnetization rotates during the recording of the FID, individual H_a spins become H_b spins, keeping the same spin state they had before, but changing their resonant frequency (chemical shift) from ν_a to ν_b. Each individual spin undergoes the jump from one chemical environment to another at random intervals, so that the phase coherence is rapidly lost as it switches its precession frequency from ν_a to ν_b and back again (Fig. 10.8). The process is random ("stochastic") from the point of view of any one nucleus, but we can say that on average the nucleus will spend half its time in the H_a state and half its time in the H_b state (if $K_{eq} = 1$). We can also say that on average a nucleus will remain in the H_a state for $\tau_a = 1/k_1$ s and it will remain in the H_b state for $\tau_b = 1/k_{-1}$ s. Phase coherence is lost because after one spin changes to a different frequency for a brief period, it cannot jump back into the original frequency at the same position in the x–y plane (the same phase) that it *would have occupied* if it had stayed at the original frequency for the whole time. So now it has lost phase coherence with other spins that did not make the "jump." This randomization of individual phases leads to a "fanning out" of the individual vectors that make up the H_a net magnetization, and the net magnetization vector for the population of H_a spins rapidly decays to zero magnitude as it rotates. This is similar to the T_2 relaxation process, and it adds to the loss of coherence resulting from T_2 relaxation:

$$H_a \text{ slow exchange linewidth} = \Delta\nu_{1/2} = 1/(\pi T_2) + 1/(\pi T_2^i) + 1/(\pi \tau_a)$$

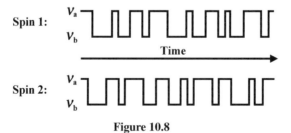

Figure 10.8

where T_2^i is the line broadening due to magnetic field inhomogeneity (Chapter 6, Section 6.9) and τ_a is the average lifetime in the H_a chemical environment ($\tau_a = 1/k_1$, $\tau_b = 1/k_{-1}$). Note that the lifetime in the H_a environment is not affected by the reverse rate constant k_{-1}; once we have a nucleus in environment H_a, the clock starts and the only thing that removes this nucleus from the H_a environment is the rate process represented by k_1. The k_{-1} process produces a different nucleus in the H_a environment and simply restarts the clock for that nucleus. Likewise the lifetime of a particular nucleus in the H_b environment is affected only by the reverse rate constant k_{-1}, which puts an end to its life as H_b. This equation for linewidth applies only in the slow exchange regime, where the time constant for exchange is longer than the coalescence time τ_c. In this regime, we have separate peaks in the spectrum for each chemical environment and the rate constants can be obtained from measuring the linewidths and comparing to "pure" linewidths that are not broadened by exchange. The integrated areas of these peaks give the relative amounts of the different species involved in the exchange. For example, in the simple two-site example ($H_a \leftrightarrow H_b$), the area of the H_b peak divided by the area of the H_a peak gives the equilibrium constant for the exchange, K_{ex}.

The anomeric carbon of reducing sugars undergoes ring opening and reclosing, interchanging the α and β diastereomers (anomers) slowly at room temperature (Chapter 1, Fig. 1.12). Figure 10.9 shows the structure of lactose, a disaccharide that exists as a mixture of two anomeric forms, differing in the stereochemistry of the anomeric position (1) of the glucose ring. In the ^{1}H spectrum (Fig. 10.10, top), separate, sharp peaks are observed for the glucose-1 proton of α-lactose (5.11 ppm, doublet, $J = 3.8$ Hz) and the glucose-1 proton of β-lactose (4.54 ppm, doublet, $J = 7.5$ Hz). The smaller coupling is due to the equatorial orientation of the glu-1 proton in α-lactose. The equilibrium constant for the conversion α-lactose $\rightleftharpoons$ β-lactose can be calculated from the ratio of integral areas: $K_{eq} = 0.657/0.363 = 1.81$. The broad triplet at 3.15 ppm ($J = 7.7$ Hz) has an integral area equal to that of the β-glu-1 peak, so it must come from β-lactose. The doublet at 4.32 ppm ($J = 7.7$ Hz) is an anomeric proton with an integral area equal to the sum of the α-glu-1 area and the β-glu-1 area ($0.363 + 0.657 \cong 1.00$). The large coupling suggests a β-orientation of the anomeric oxygen, with the anomeric proton in an axial position. This is the galactose H-1 proton, which is not affected by the exchange process because it is far enough away in

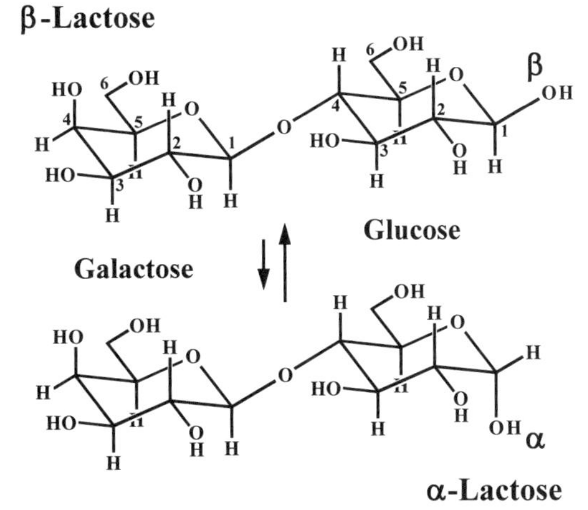

Figure 10.9

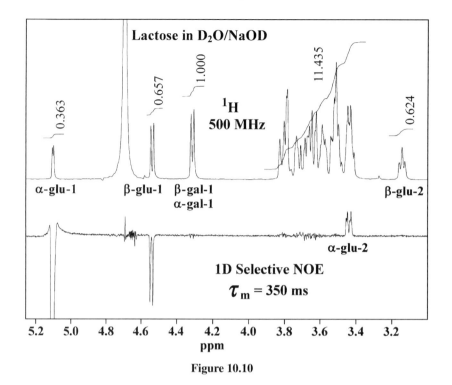

Figure 10.10

the molecule that both forms (α-lactose and β-lactose) give identical peaks. The area of complex overlapped peaks from 3.3–3.9 ppm must contain 12 protons from α-lactose and 11 from β-lactose, as there are 14 carbon-bound protons in all in each form. We predict the area of this region to be 0.363 (12) + 0.624 (11) = 11.22, which is close to the measured area of 11.44. Because the two forms are present in different concentrations, an area of 0.363 corresponds to one proton for α-lactose and an area of 0.624 corresponds to one proton for β-lactose. The OH protons are in fast exchange with residual HOD (large peak at 4.7 ppm) and resonate at the average chemical shift, which is the HOD position because the vast majority of protons in this large "reservoir" are on the much more concentrated HOD at any one time.

Even with slow exchange it is possible to detect the exchange by an NOE experiment. During the mixing time of the NOE, the perturbed z magnetization on H_a becomes perturbed z magnetization on H_b in an exchange event:

$$-\mathbf{I}_z^a \rightarrow -\mathbf{I}_z^b$$

The spin state has not changed; we just changed the label on the proton when it passed into a new chemical environment with a different Larmor frequency.

The formal analysis of this exchange process as it affects the longitudinal relaxation of two resonances H_a and H_b leads to the following equations, similar to those for cross-relaxation by the NOE:

$$d\Delta M_z^a/dt = (-R_{aa} - k_1)\Delta M_z^a + k_{-1}\Delta M_z^b$$
$$d\Delta M_z^b/dt = k_1 \Delta M_z^a + (-R_{bb} - k_{-1})\Delta M_z^b$$

Consider the first equation: every time an H_a spin becomes an H_b spin, it takes its disequilibrium away from the H_a category with a rate of k_1, speeding up the self-relaxation of H_a. More importantly, every time an H_b spin becomes an H_a spin, it carries its disequilibrium (ΔM_z^b) into the H_a category, with a rate of k_{-1}. This is the term that replaces the cross-relaxation term R_{ab} in the corresponding equations for the NOE. For H_b, the rate constant k_1 describes the rate at which H_a's disequilibrium (ΔM_z^a) is converted into disequilibrium of the z magnetization of H_b. In a transient NOE experiment, if we invert H_a (180° selective pulse) while leaving H_b at equilibrium, the rate equation for H_b becomes

$$d\Delta M_z^b/dt = dM_z^b/dt = k_1 \Delta M_z^a = k_1(-2M_o)$$

for the initial rate. Note that the z magnetization of H_b is *decreasing*, opposite to the "enhancement" of z magnetization observed for small-molecule NOEs. Solving for k_1, we have

$$k_1 = -[dM_z^b/dt]/2M_o$$

This can be measured by equating the (negative) peak area of the inverted H_a peak at $\tau_m = 0$ with $-M_o$ and measuring M_z^b as the peak area of the H_b peak relative to "$-M_o$".

Figure 10.10 (bottom) shows the result of a 1D selective NOE (DPFGSE) experiment on the mixture of lactose anomers, selecting the glucose-1 peak of α-lactose at 5.11 ppm. We see a strong exchange peak at 4.54 ppm representing the glucose-1 peak of β-lactose. Note that this peak is the same sign as the selected peak (negative), as exchange does not change the spin state. A strong NOE peak is observed at 3.44 ppm for the glucose-2 proton of α-lactose. This peak is positive, opposite in sign to the selected peak, as for small molecules the NOE is negative:

$$-\mathbf{I}_z^a \rightarrow \mathbf{I}_z^b$$

Exchange peaks correlate one molecule with a different molecule, whereas NOE peaks correlate through space within the same molecule. The NMR timescale for this exchange ($\Delta\delta = 5.11 - 4.54 = 0.57$ ppm; $\Delta\nu = 0.57 \times 500 = 285$ Hz) can be calculated as $1/(2.22 \times \Delta\nu) = 1.58$ ms. Because we see sharp peaks for the glucose-1 resonances of both forms of lactose, the lifetime of each state must be much longer than 1.58 ms.

Figure 10.11 shows the results of a buildup study, varying the mixing time from 5 to 350 ms and measuring the relative areas of the NOE and exchange peaks. The vertical scale is peak area in percent of the inverted α-glu-1 peak at $\tau_m = 0$. We already know that the lifetime of either state (α-lactose or β-lactose) is much greater than 1.58 ms, but we can see that as the mixing time is increased, the number of molecules "jumping across" from α-lactose to β-lactose during the time "window" of τ_m increases steadily. The NOE experiment gives us another timescale to observe exchange: the much longer timescale of the mixing time, which is limited only by T_1 ($\sim$2 s) relaxation. In this case, the rate of conversion α-lactose $\rightarrow$ β-lactose can be estimated from the initial rate (1.4% in 150 ms) of exchange buildup: $k_1 = 0.046$ s^{-1}. The average lifetime in the α-lactose state is $1/k_1 = 21.6$ s! This is extremely slow, 13,600 times longer than the coalescence lifetime (chemical shift time scale) of 1.58 ms. As we know the equilibrium constant, we can calculate the reverse rate (β-lactose $\rightarrow$ α-lactose): $k_{-1} = k_1/K_{eq} = 0.046$ s$^{-1}/1.81 = 0.026$ s^{-1}. The average lifetime in the β-lactose state is $1/k_{-1} = 39.1$ s.

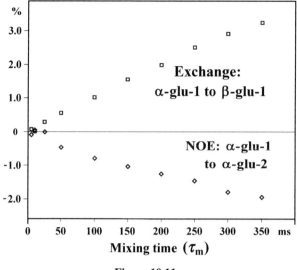

Figure 10.11

There are many other examples of slow exchange at room temperature, including hindered rotation about single bonds, where the energy barrier may be due to partial double-bond character (amide N–CO rotation, aniline N-aromatic ring rotation) or to steric crowding. Amide bonds in peptides (except proline) do not usually show exchange, but simple amides used for protecting groups (benzoyl, benzyloxycarbonyl, etc.) can give doubling of many peaks. Some sigmatropic rearrangements can bring about reversible carbon–carbon bond breaking and formation that is slow at room temperature and fast at high temperatures. There are many examples of exchange in inorganic complexes, which can adopt different geometries. For example, pentavalent phosphorus interchanges the two axial ligands with the three equatorial ligands in a process called pseudorotation, which has been studied extensively by ^{31}P and ^{19}F NMR. In many cases, the process of coalescence from slow exchange to fast exchange can be studied without changing the temperature, by the addition of different concentrations of a catalyst.

10.2.2 Fast Exchange

In the fast exchange regime, where $k_{ex} \gg k_c$ ($\tau_{ex} \ll \tau_c$), we have only one peak for all of the chemical environments ("sites") involved in the exchange process. The chemical shift position of this peak is the weighted average of the individual site chemical shifts, weighted by the "population" (equilibrium concentration) of each site. In the simplest case of two sites with an equilibrium constant of 1 (e.g., N,N-dimethylformamide), the chemical shift will be the simple average $(\nu_a + \nu_b)/2$. If the equilibrium constant is not 1, we have the weighted average

$$\nu_{av} = (\nu_a n_a + \nu_b n_b)/(n_a + n_b)$$

where n_a and n_b are the populations (or concentrations) of the H_a and H_b sites, respectively, with $K_{eq} = n_b/n_a$. Exchange can cause line broadening in this regime without splitting the resonance into more than one peak. This can show up as a broad peak in a spectrum of otherwise sharp peaks, as $\Delta \nu$ may be small for nuclei that are farther away from the focus of a conformational change. This may show up in the ^{13}C spectrum but not in the ^{1}H spectrum, as ^{13}C chemical shifts are more sensitive to differences in the steric environment and generally have a larger range of chemical shifts in the ppm scale, even though the $\Delta \nu$ in Hz is four times larger for a proton spectrum if the chemical shift difference is the same in ppm. Sometimes peaks in a spectrum can be so broad as to be missing altogether. As a peak becomes very broad, its integral area remains the same and the peak height must become very low, sometimes lower than the noise level. The same applies for crosspeaks in a 2D spectrum: a peak that is strongly affected by exchange may disappear entirely whereas the other peaks, for which $\Delta \nu$ is much less than the exchange rate, are sharp and unaffected. The missing or broad peaks can be verified as exchange effects by raising the temperature (faster exchange) or lowering the B_o field (slower shutter speed) until the peak becomes sharp. Alternatively, one can lower the temperature or go to higher field in order to observe splitting of the peak into more than one peak (slow exchange).

Typical systems that show fast exchange at room temperature include most rotations around single bonds in acyclic molecules (methyl groups always show a single chemical shift with sharp peaks), acid–base reactions, and cyclohexane chair interconversions. Protons on electronegative atoms, such as alcohol oxygens, are typically in fast exchange with other OH groups in the sample due to the presence of trace amounts of acid or base catalysts:

$$\text{ROH}_a + \text{H}_b^+ \rightleftharpoons \text{RO}^+(\text{H}_a)\text{H}_b \rightleftharpoons \text{ROH}_b + \text{H}_a^+$$

The effect of this fast exchange is to decouple the OH proton from other spins in the molecule. If an OH proton is in the α state (spin "up") on a particular molecule, it will soon exchange with another proton in the β state (spin "down"), so that other spins in the molecule will see a rapid "blur" of α and β state protons on the oxygen. The NMR timescale is just $1/(2.22 J)$ (J is difference in resonant frequency, $\Delta \nu$, between the two states for the spin that is coupled to the OH proton) and the exchange is usually much faster than this. Coupling to an OH proton is observed only with aprotic solvents that are hydrogen bond acceptors (DMSO-d_6) or in cases where the molecule itself provides an intramolecular hydrogen bond acceptor. Hydrogen bonding greatly reduces the exchange rate and allows the J coupling to be observed. The NH protons of an amine or aniline are also in fast exchange, but in an amide (peptide) bond the NH exchanges slowly enough to allow the observation of J coupling. This amide NH exchange is pH dependent in aqueous solution (e.g., 90% H$_2$O/10% D$_2$O) and reaches a minimum exchange rate at around pH 3. Decoupling itself is an exchange process, in which we repeatedly invert one spin (e.g., ^{1}H) very rapidly by RF irradiation, leading to averaging of the Larmor frequencies for the coupled spin (e.g., ^{13}C) between the H = α and H = β chemical shift positions. Thus, the doublet is collapsed to a sharp singlet as long as the protons are jumping back and forth between α and β states at a rate much faster than $1/(2.22\ ^1J_{CH})$ during the acquisition of the FID.

Acid–base reactions are of particular interest because one can titrate the NMR sample and observe a titration curve of chemical shifts. This is because the two forms (protonated and unprotonated) usually show chemical shift differences at one or more sites near the basic atom:

$$B: + H^+ \rightleftharpoons BH^+$$
$$\,\nu_a \,\nu_b$$

For example, the imidazole ring of a histidine residue in a protein shows significant pH dependence of the ^{13}C chemical shift at all three of the carbon positions. Although the fast-exchange chemical shift observed is the weighted average of the two chemical shifts (B and BH$^+$), the basic form chemical shift is observed at high pH and the acidic form chemical shift is observed at low pH. As the pH is gradually decreased (by removing the NMR tube from the magnet, adding aliquots of acid and replacing it), the chemical shift changes from the basic form shift value to the acid form shift value in the form of a classical titration curve, with a midpoint at the pK_a value for the acid–base equilibrium. The peaks in the NMR spectrum are sharp at each step of the titration because the acid–base reaction is very fast on the NMR timescale at all pH values. This technique has been used to identify the pK_a values of specific active-site amino acid side chains in proteins, using 2D NMR to determine the chemical shift values at each pH.

10.2.3 Variable Temperature Operation

Most spectrometers control the sample temperature by passing a constant flow of gas (air or nitrogen) by a heater and then past the NMR tube, with a thermocouple just below the NMR tube to sense the temperature of the gas. A feedback loop compares the gas temperature to the set temperature and adjusts the current in the heater coil accordingly to maintain a constant temperature (Fig. 10.12). The gas supply must be 10 °C or more colder than the set temperature to get good regulation. Air chillers are commonly used to access temperatures of 0–25 °C. Below 0°C requires more drastic cooling, usually with liquid nitrogen (77 K, −196 °C). Varian uses a heat exchanger that passes room temperature nitrogen gas through a coil of tubing in a bucket of liquid nitrogen, then into the probe. Bruker generates the nitrogen gas directly from the liquid nitrogen by placing a heater coil in the dewar and boiling off the nitrogen directly into the probe. For high temperature work, you may need to

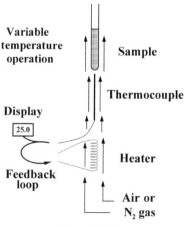

Figure 10.12

use nitrogen for prolonged experiments to avoid oxidation of the probe electronics. Every probe has a rated temperature range, so check the probe manual! Prolonged low temperature work can actually cool and shrink the rubber O-ring vacuum seals at the bottom of the magnet, leading to a quench (violent loss of superconductivity). To avoid this, a heater can be placed between the probehead and the bottom of the magnet. Most VT units include a safety feature that shuts down the heater if there is any interruption of the gas flow. This is important because without gas flow the heat produced in the heater is not transferred to the thermocouple and the heater never turns off—the probe can melt! The VT unit senses gas flow in the console only, so if the gas does not reach the probe for any reason, such as a disconnected air line at the probe, the same hazardous situation arises.

One practical aspect that may seem obvious, but many people overlook with disastrous results: you must operate in the temperature range between the freezing point and the boiling point of the solvent! Freezing can break the NMR tube and boiling can send your sample spewing all over the probe and the bore. At low temperatures you can also encounter highly viscous solvents, which will lead to slow molecular tumbling and broad peaks due to short T_2. Some solvents are ideal for low temperature because of their low boiling points and low viscosity: for example, dichloromethane (mp $-95\,°C$). Keep in mind that the temperature reading on an NMR variable temperature unit can be quite inaccurate; you should calibrate using an internal standard of chemical shift difference (methanol for low temperature and ethylene glycol for high temperature) or a thermocouple directly introduced into solvent in an NMR tube. The variable temperature unit measures the air (or nitrogen) temperature in the stream just before it reaches the NMR tube, not the temperature of the sample, so you need to allow ample time for equilibration even after the temperature reading has stabilized.

10.3   2D NOESY AND 2D ROESY

NOESY and ROESY (Fig. 10.13) correlate protons with other protons via their homonuclear NOE interactions. A NOESY spectrum looks very much like a COSY, except that the crosspeaks correspond to pairs of protons that are close in space (<5 Å) and not necessarily close in the bonding network. The intensities of crosspeaks are roughly proportional to $1/r^6$, where **r** is the direct through-space distance between the two protons correlated by the crosspeak.

10.3.1   The Transient Nuclear Overhauser Effect

In Chapter 5 we observed NOE interactions by 1D NOE difference, measuring the *steady-state* NOE resulting from a long (several seconds), low-power continuous-wave irradiation of one nucleus. The modern selective (DPFGSE) 1D NOE experiment

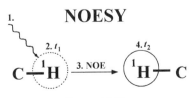

Figure 10.13

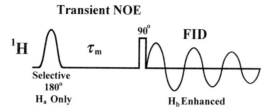

Figure 10.14

(Chapter 8) uses the *transient* NOE, in which a selective 180° pulse is applied to one nucleus and then the enhancement of z magnetization on a nearby nucleus is measured after a mixing delay (Fig. 10.14: the DPFGSE—90° sequence is replaced by a single selective 180° pulse for simplicity). For two nuclei H_a and H_b that are nearby in space, interaction of their magnetic dipoles leads to the phenomenon of cross-relaxation. This means that any sudden perturbation of the H_a populations away from the Boltzmann distribution will lead to a relaxation process that perturbs the H_b populations away from the Boltzmann distribution temporarily. The effect will build up during the relaxation process, but eventually a Boltzmann distribution for both nuclei is reestablished and the effect on H_b goes away. For "small" molecules, this effect enhances the z magnetization of H_b up to a few percent above M_o at the optimum mixing time.

In product operator terms, we can say that the inverted z magnetization on H_a leads to the generation of additional z magnetization on H_b:

$$-\mathbf{I}_z^a \xrightarrow{(\tau_m \text{delay})} \mathbf{I}_z^b$$

Consider now the common "front end" of the homonuclear 2D experiments: 90°—t_1—90°. If we put it in place of the selective 180° pulse of the transient NOE experiment (Fig. 10.14), it will give us the following terms:

$$-\mathbf{I}_z^a \cos(\Omega_a t_1) + \mathbf{I}_x^a \sin(\Omega_a t_1) \qquad \text{(if } H_a \text{ and } H_b \text{ are not } J \text{ coupled)}$$

The first term, which is not observable in the COSY experiment, is now exactly what we need for a transient NOE experiment. We have "inverted" the H_a magnetization in a way that carries the information of its chemical shift encoded in the $\cos(\Omega_a t_1)$ term. Depending on the value of t_1, sometimes H_a will be completely inverted (cosine $= 1$), leading to a maximum NOE transfer to H_b, and sometimes it will not be inverted at all (cosine $= -1$), leading to no NOE transfer to H_b. Thus, the transferred magnetization will also carry the chemical shift information of H_a:

$$-\mathbf{I}_z^a \cos(\Omega_a t_1) \xrightarrow{(\tau_m \text{delay})} \mathbf{I}_z^b \cos(\Omega_a t_1)$$

The final "read" pulse rotates the H_b z magnetization into the x'–y' plane and the FID is recorded with the frequency Ω_b. Fourier transformation of the FIDs gives in each one a peak at $F_2 = \Omega_b$ whose amplitude is oscillating as a function of t_1 at the frequency Ω_a. Fourier transformation of the t_1 FID gives a crosspeak at $F_2 = \Omega_b$, $F_1 = \Omega_a$. Because for small molecules, the transferred z magnetization is opposite in sign from the original

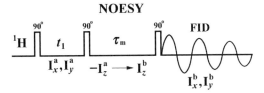

Figure 10.15

perturbation of H_a ($-I_z^a \rightarrow I_z^b$), the crosspeaks will be of negative intensity if we phase the diagonal peaks (coming from untransferred $-I_z^a$) to positive intensity.

10.3.2 The NOESY Pulse Sequence

The NOESY is a simple extension of the COSY pulse sequence, with one additional delay and one additional 90° pulse added (Fig. 10.15). The mixing part of the 2D pulse sequence now consists of two 90° pulses separated by a delay. The first 90° pulse converts magnetization in the x'–y' plane into z magnetization (population difference). To the extent that this z magnetization differs from M_o, it will undergo cross relaxation during the mixing delay τ_m, altering the population difference and thus the z magnetization of nearby nuclei. The final pulse converts the transferred z magnetization into observable x'–y' magnetization on the nearby nuclei, which precesses during t_2 and induces the FID signal in the probe coil. As with all 2D experiments, magnetization transfer is the basis for the appearance of crosspeaks in the spectrum (cf. Chapter 9, efficiency of transfer G_{ab}), but in this case it is z magnetization that is transferred and the intensity of crosspeaks will depend on the cross-relaxation rate for that pair of nuclei. As in the transient NOE experiment, the intensity of the crosspeak will increase with increasing mixing time τ_m, but will eventually reach a maximum and then drop off to zero.

A simple way to gradient enhance the NOESY experiment is to add a single gradient during the mixing delay (Fig. 10.16). This will destroy any SQC present during the mixing time ($\mathbf{p} = 1$) as well as any DQC ($\mathbf{p} = 2$), as there is no other gradient to "untwist" the coherence. z magnetization and ZQC ($\mathbf{p} = 0$) are not affected. Phase cycling is also used to remove artifacts. Because we are only interested in z magnetization during the mixing delay, we can use any phase we want for the final pulse as long as the receiver phase is the same as the pulse phase. So usually this final pulse is cycled as $x, y, -x, -y$ with the receiver making the same cycle: $x, y, -x, -y$. We can also phase cycle the second 90° pulse ($x, -x$), which will reverse the perturbation of z magnetization at the start of the mixing delay: $-I_z^a \cos(\Omega_a t_1)$ for an x' pulse and $+I_z^a \cos(\Omega_a t_1)$ for a $-x'$ pulse. This reverses the sign of the final detected H_b magnetization:

$$-I_z^a \cos(\Omega_a t_1) \stackrel{\text{NOE}}{\rightarrow} I_z^b \cos(\Omega_a t_1) \stackrel{90_x^\circ}{\rightarrow} -I_y^b \cos(\Omega_a t_1)$$

$$-I_z^a [-\cos(\Omega_a t_1)] \stackrel{\text{NOE}}{\rightarrow} I_z^b [-\cos(\Omega_a t_1)] \stackrel{90_x^\circ}{\rightarrow} -I_y^b [-\cos(\Omega_a t_1)]$$

If we then reverse the phase of the receiver we will get addition of the desired H_b magnetization with each scan and cancellation of any magnetization that was not affected by that pulse.

Gradient-enhanced NOESY

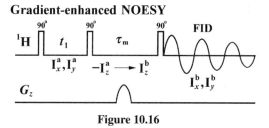

Figure 10.16

10.3.3 The 2D NOESY Spectrum

The NOESY spectrum (Fig. 10.17) looks very much like the COSY spectrum, except that we will see additional crosspeaks that are not present in the COSY or TOCSY spectra. These are the interesting ones—pairs of protons that are close in space but not close enough in the bonding network to be J coupled. Classic examples of this are 1,3-diaxial relationships in rigid cylcohexane chairs, 1,3-diaxial relationships between a proton and a methyl group (e.g., $H4_{ax}$ and H19 in cholesterol: Chapter 8, Fig. 8.35), CH–O–CH across a glycosidic linkage, CH_α–CO–NH (observed in β sheets and β turns), and NH–C_α–CO–NH (observed in an α-helix) across a peptide bond. When there is a large J coupling between two protons, we can see zero-quantum artifacts, just as we noticed in the selective 1D NOE experiment. These result from ZQC that is produced by the "front end" sequence $90°_x$–t_1–$90°_x$:

$$\underset{Crosspeak}{-I_z^a\, cc'} \quad \underset{ZQ\,artifact}{-2I_x^a I_y^b\, cs'} \quad \underset{Diagonal}{+I_x^a\, sc'} \quad \underset{Not\,observed}{-2I_y^b I_z^a\, ss'}$$

where **c**, **s**, **c'** and **s'** are as defined in Chapter 9. The second term is a mixture of ZQC and DQC. The DQC part can be removed by phase cycling or by gradients, but there is no simple way to remove ZQC because it has coherence order zero, just like z magnetization. During the mixing delay τ_m, it undergoes chemical-shift evolution at a rate determined by the chemical-shift difference $\Omega_a - \Omega_b$, and the third 90° pulse completes the coherence

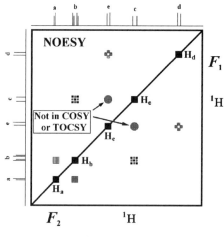

Figure 10.17

transfer from H_a to H_b:

$$-2I_x^a I_y^b \, c \, s' \, c'' \xrightarrow{90_y^\circ} 2I_y^b I_z^a \, c \, s' \, c''$$

where c'' is $\cos((\Omega_a - \Omega_b)\tau_m)$. So this is a COSY-like antiphase crosspeak resulting from antiphase-to-antiphase INEPT coherence transfer with an intermediate ZQC state. The mixing time τ_m can be randomly varied to try to average the artifacts to zero, taking advantage of the cosine dependence on τ_m, but this will also introduce t_1 noise. The artifacts are easily recognized in the 2D spectrum because they have equal amounts of positive and negative intensities, usually in a star-like pattern (H_b–H_c, Fig. 10.17). Sometimes you will see NOE crosspeaks that are distorted in shape because the in-phase negative intensity of the NOE crosspeak is added to the twisted antiphase shape of the ZQ artifact at the same position (H_d–H_e, Fig. 10.17).

Pure NOE crosspeaks in a NOESY spectrum are *in-phase* (normal multiplets) where there is J coupling. The transferred magnetization is on the z axis, and the "read" (third) $90°$ pulse produces in-phase magnetization in the x'–y' plane at the start of the acquisition period. The lack of a $\sin(\pi J t_1)$ term in the observed crosspeak magnetization means that peaks will also be in-phase in the F_1 dimension. Integration of NOESY crosspeaks will give nonzero peak areas (actually volumes, as we are dealing with 2D crosspeaks) that are representative of the intensity of the NOE interaction. The crosspeaks in a phase-sensitive COSY experiment are antiphase and have a zero net volume because equal areas are found in the positive and negative components. Thus, the NOESY and TOCSY experiments lead to *net* transfer of magnetization and the COSY experiment does not.

The sign of NOESY crosspeaks relative to the diagonal depends on the sign of the NOE interaction. Magnetization that does not transfer during the mixing period (τ_m) will have opposite sign to transferred magnetization as long as the cross-relaxation rate R_{ab} is a positive number:

$$-I_z^a \cos(\Omega_a t_1) \xrightarrow{(\tau_m)} \underbrace{-I_z^a \cos(\Omega_a t_1)}_{\text{Diagonal peak}} + \underbrace{R_{ab}\tau_m I_z^b \cos(\Omega_a t_1)}_{\text{Crosspeak}}$$

where R_{ab} is the cross-relaxation rate and we are looking at short mixing times where the NOE buildup is still linear as a function of mixing time. This is true for small organic molecules (molecular weight less than about 2000 Da) in nonviscous solvents for which $\omega\tau_c \ll 1$, where ω is the Larmor frequency in the laboratory frame (e.g., $2\pi \times 600$ MHz) and τ_c is the correlation time for tumbling of the molecule. The double-quantum relaxation pathway dominates and we see an increase in z magnetization on nearby spins. If the diagonal peaks are phased to be positive absorptive peaks, the crosspeaks will be negative (upside down) absorptive peaks. This is called a "negative NOE" because the cause (perturbation of H_a's z magnetization by reducing it) is opposite in sign to the effect (enhancement of H_b's z magnetization). For large molecules, such as proteins, the tumbling rate is much slower (i.e., τ_c is long) and $\omega\tau_c \gg 1$. In this case, the zero-quantum relaxation pathway dominates over the double-quantum pathway, so R_{ab} ($W_2 - W_0$) becomes negative. The crosspeaks will then be the same sign as the diagonal peaks. As the initial perturbation on H_a (reducing its z magnetization) results in the same type of perturbation on H_b during the mixing period (reduction of z magnetization), we call this a "positive NOE."

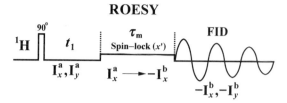

Figure 10.18

10.3.4 ROESY: Rotating-Frame Overhauser Effect Spectroscopy

For "medium-sized" molecules where $\omega\tau_c \cong 1$, zero-quantum and double-quantum relaxation rates are nearly the same and the NOE cross-relaxation rate approaches zero. This size depends on the field strength of the NMR instrument and the viscosity of the solvent: it is around 2000 Da for spherical polypeptides in water at 500 MHz. This can happen for peptides, oligosaccharides, and large natural products. Even if the NOE is not zero, it can be too small to conveniently measure with a NOESY experiment. In these cases we have an alternative experiment, originally called CAMELSPIN, which gives negative crosspeaks regardless of molecular size. Instead of the NOESY mixing sequence, which consists of putting the magnetization on the z axis and waiting a period of time for the z magnetization perturbations to propagate to nearby nuclei, ROESY puts the magnetization on a specific axis in the $x'-y'$ plane and "locks" it there for a period of time (the mixing time) using a continous-wave spin lock. The pulse sequence (Fig. 10.18) is almost identical to the 2D TOCSY: We start with the standard homonuclear $90°_x-t_1$ sequence and then select the in-phase coherence on the x' axis by executing a long, low-power radio frequency pulse.

$$I_x^a \text{ s c}' \xrightarrow{\text{ROESY spin lock}(x')} -I_x^b \text{ s c}'$$

We saw in Chapter 8 that the spin lock causes magnetization to transfer in a through-space manner very similar to the NOE, except that now it is x' magnetization on H_a transferring to x' magnetization on H_b (if the spin lock is applied on the x' axis): $I_x^a \rightarrow -I_x^b$. The main difference is that the effective field felt by the spins is reduced from the static field (B_o) to the radio frequency field strength (B_{eff}), which is typically five orders of magnitude (10^{-5} times) lower. It is as if we could use our normal magnetic field strength (e.g., 500 MHz) for the preparation, evolution, and detection periods, but switch the field to a very low field strength (e.g., 3300 Hz) for the mixing period. In this low field environment, the SQ frequency is 3300 Hz and the DQ frequency is 6600 Hz. The dominant pathway for relaxation is now $H_a(\beta)H_b(\beta) \rightarrow H_a(\alpha)H_b(\alpha)$ (double quantum relaxation), regardless of the molecular size. So we see negative NOE crosspeaks relative to the positive diagonal for large, medium, *and* small molecules.

The ROESY used to be a bit difficult to set up because the low power spin-lock RF had to come from a different source than the hard pulse RF. Now rapid solid-state power switching is so routine that all ^{1}H RF comes from the same source, with no variation of phase or frequency. ROESY tends to replace the NOESY experiment for NOE measurements, especially for small molecules where T_2 is relatively long. Because the NOE builds up about twice as fast in the $x'-y'$ plane as it does on the z axis, ROESY mixing times are set to about half of what would be the NOESY mixing time. There is one additional parameter to set up: the power level ("B_1 field strength") of the spin lock pulse. This is typically

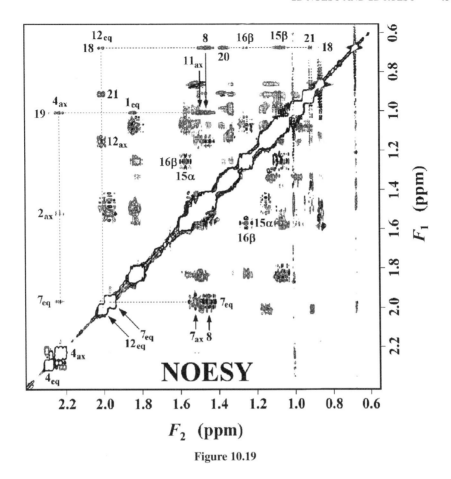

Figure 10.19

calibrated to give a ^{1}H 90° pulse of 75 μs ($\gamma B_1/2\pi = 3333$ Hz) to "cover" a spectral width of about twice that amount (6666 Hz). At this power level, the spin-lock axis is tilted out of the x'–y' plane by an angle of 45° for peaks at the edges of the spectral window.

10.3.5 Examples of NOESY and ROESY

The upfield region of the 600-MHz NOESY spectrum of cholesterol ($\tau_m = 300$ ms) is shown in Figure 10.19. Positive contours are shown in black and negative contours in gray. Note that the diagonal peaks are positive and extremely strong at the contour threshold used to show the weak negative crosspeaks. Streaks of t_1 noise can be seen extending down from the singlet methyl peaks (H18 and H19) on the diagonal. The cholesterol structure is shown in Figure 10.20 with some of the NOE interactions indicated with double arrows. The distances in angstrom, taken from the X-ray crystal structure, are shown in Figure 10.20 next to the arrows. Along the NOESY diagonal at the lower left side are the H4$_{eq}$ (doublet) and H4$_{ax}$ (triplet) peaks around 2.2 ppm, followed (moving up and to the right side) by the H12$_{eq}$ and H7$_{eq}$ diagonal peaks (Fig. 10.19). Moving up vertically from the H4$_{ax}$ peak on the diagonal, we see a ZQ artifact at $F_1 = $ H7$_{eq}$ (long range J-coupling CH–C=C–CH) and NOE peaks at $F_1 = $ H2$_{ax}$ and $F_1 = $ H19 methyl. Both of these are 1,3-diaxial relationships

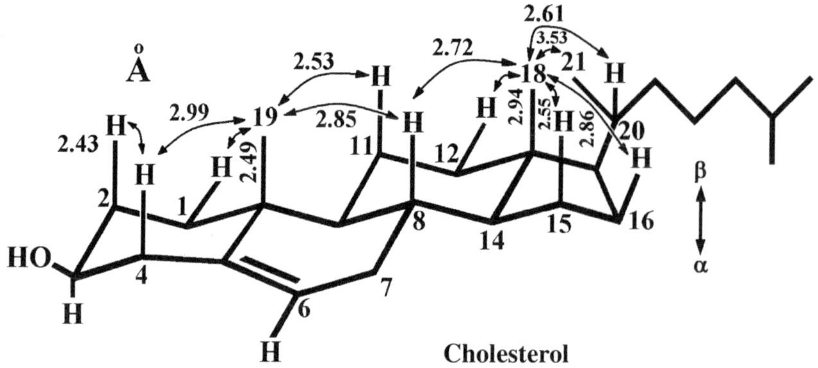

432   ADVANCED NMR THEORY: NOESY AND DQF-COSY

Figure 10.20 Cholesterol

in the A ring of the steroid. Moving to the right on the F_1 = H19 methyl line, we encounter H1$_{eq}$ (1,2-*cis* relationship) and then H11$_{ax}$ and H8 (1,3-diaxial). Moving up from the H12$_{eq}$ peak on the diagonal, we encounter H12$_{ax}$ (ZQ artifact), H21 methyl of the C-17 side chain, and the angular methyl group H18 (1,2-*cis*). Moving to the right on the F_1 = H18 methyl line, we see a number of NOE interactions: H8 (1,3-diaxial), H20 of the C-17 side chain, H15β and H16β in the five-membered D ring, and H21 methyl of the C-17 side chain. The NOE crosspeaks from the angular methyl groups are partially obscured by t_1 noise streaks below the diagonal. As t_1 noise is always a vertical streak (along the F_1 dimension), if a crosspeak is obscured by the streak, we can always look to the other side of the diagonal to find an equivalent crosspeak that is not in the path of a t_1 noise streak.

Moving to the right from the H7$_{eq}$ diagonal peak, we see strong ZQ artifacts at the F_2 = H7$_{ax}$ and F_2 = H8 positions. An especially strong pair of ZQ artifacts appears (center) due to the H15α–H16β coupling. These can be seen clearly as star-shaped antiphase peaks in the expanded region shown in Figure 10.21. The center of the peak as well as four spots extending diagonally from the center are negative; four spots above, below, right, and left of the center are positive.

The 600 MHz ROESY spectrum of cholesterol is shown in Figure 10.22 (τ_m = 200 ms, spin-lock $\gamma B_1/2\pi$ = 3333 Hz). The strong positive diagonal, weak negative crosspeaks, and t_1 noise streaks coming down from the methyl diagonal peaks are all similar to the NOESY spectrum (Fig. 10.19), but the spectrum is cleaner overall with fewer ZQ artifacts. Crosspeaks can be identified from the olefinic proton H6 to H4$_{eq}$ (strong) and H4$_{ax}$ (weak) as well as to H7$_{eq}$ and H7$_{ax}$ (Fig. 10.22). From H3 we see crosspeaks to H1$_{ax}$ (1,3-diaxial relationship) and to the *J*-coupled protons H2$_{eq}$, H4$_{ax}$, and H4$_{eq}$. The expanded upfield region of the ROESY spectrum (Fig. 10.23) can be directly compared to the NOESY (Fig. 10.19) with most of the same correlations identified corresponding to the distances shown on the structure (Fig. 10.20). Figure 10.24 shows three enlarged strips of the ROESY with F_2 slices at F_1 = 0.68 (H18), 1.01 (H19), and 5.35 (H6) ppm. In the slices you can clearly see the "triplet" structure of H4$_{ax}$ (middle slice) and the "doublet" structure of H4$_{eq}$ (bottom slice). Even the smaller coupling from H4$_{eq}$ to H3 is resolved in the bottom slice. The H4$_{ax}$ crosspeak dominates in the F_1 = H19 slice, whereas the H4$_{eq}$ crosspeak is much larger in the F_1 = H6 slice, similar to the results we saw in the selective 1D NOE experiment (Chapter 8, Fig. 8.36). In the F_1 = H6 slice (bottom), the H7$_{eq}$ crosspeak clearly has doublet structure and the H7$_{ax}$ crosspeak has double-doublet structure. These multiplets are low resolution

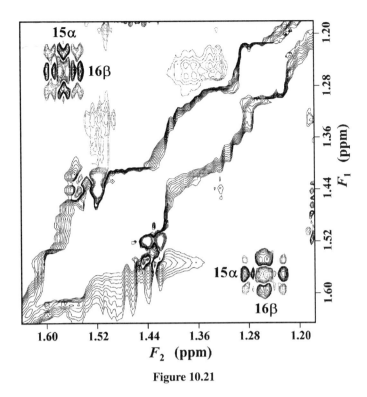

Figure 10.21

and show only the large couplings—in general, only the large axial–axial and geminal ($^2J_{HH}$) couplings are well resolved. From the H18 methyl group (top slice), we see a strong ROE to H8 but only a weak ROE to H11$_{ax}$; from the other angular methyl group H19 (middle slice), we see strong ROEs to both of these β-axial protons. The methyl groups also "see" equatorial protons H12$_{eq}$ (from H18) and H1$_{eq}$ (from H19), which appear as doublets. From H18 we see a "quartet" structure for H16β and a sharp doublet for the H21 methyl group. A crosspeak to H18 in the $F_1 = $ H19 slice may be due to spin diffusion through H8 or H11$_{ax}$.

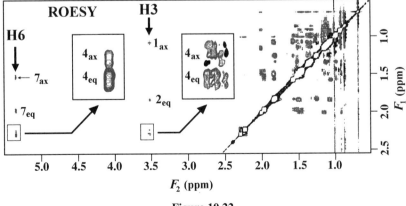

Figure 10.22

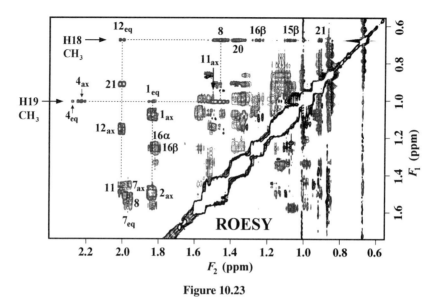

Figure 10.23

In cholesterol we can see nearly all of the 1,3-diaxial and 1,2 axial–equatorial NOE interactions, whether they are H–H or H–CH$_3$ relationships. One should take great care, however, in establishing regiochemistry or stereochemistry by NOE experiments. Even in a small ring (4–6 members), the difference in the H–H distance between a *cis* 1,2 (vicinal) relationship and a *trans* relationship is small. For an ideal cyclohexane chair conformation, for example, the vicinal distances are 2.54 Å for equatorial–equatorial (*trans*), 2.48 Å for

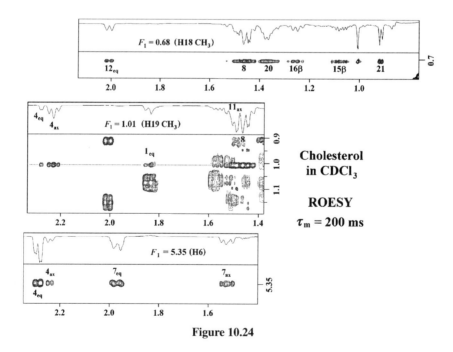

Figure 10.24

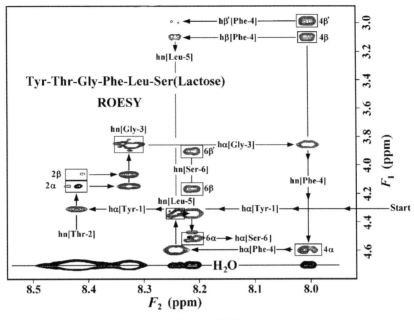

Figure 10.25

equatorial–axial (*cis*), and 3.09 Å for axial–axial (*trans*). The close 1,3 diaxial (*cis*) distance is characteristic for cyclohexane: 2.77 Å compared to 3.89 for ax–eq and 4.35 for eq–eq. NOE interactions across the ring (1,4) are rarely observed, with distances of 4.27 (*cis*), 4.15 (*trans* diaxial), and 5.04 (*trans* diequatorial).

The 200-ms 600 MHz ROESY spectrum of the glycopeptide Tyr-Thr-Gly-Phe-Leu-Ser(Lactose) in 90% H_2O/10% D_2O is shown in Figure 10.25. Crosspeaks that are also found in the TOCSY spectrum (i.e., which are within the same amino acid residue: Fig. 9.45) are enclosed in rectangles. Negative peaks are shown in gray, and positive peaks in black. We saw in Chapter 9 how the spin systems corresponding to each amino acid residue can be identified in the 2D TOCSY spectrum from the H_N resonance in F_2. Now looking at the same region of the ROESY spectrum, we can see numerous NOE connections between one residue and its neighbor in the primary sequence. These are called *sequential* (or $i \rightarrow i+1$) NOEs because they connect across one peptide bond to the next residue in the sequence.

Starting with the F_1 chemical shift of the Tyr-1 H_α proton (Fig. 10.25, right side, F_1 = 4.32 ppm), we can "walk" through the peptide backbone by using the proximity of the H_α proton of residue i to the H_N proton of residue $i+1$: $CH_\alpha^i - CO^i - NH^{i+1}$. There is no H_N chemical shift for Tyr-1 because it is the N-terminal residue and its *amine group* protons (not an amide) are in very fast exchange with water. Moving all the way to the left side on the $F_1 = H_\alpha$ of Tyr-1 line, we encounter a negative crosspeak that is not found in the TOCSY spectrum: a sequential NOE to H_N of Thr-2. Moving up from the NOE crosspeak leads to two weak ZQ artifacts at the same positions as the H_α and H_β shifts of Thr-2 in the TOCSY spectrum (rectangles). Moving to the right from these crosspeaks leads to two strong sequential NOE crosspeaks at the F_2 chemical shift of the H_N of Gly-3. In this case, *both* the H_α proton *and* the H_β proton of residue 2 are close to the H_N of residue 3. Moving up, we come to a messy ZQ artifact at the H_α shift of Gly-3 in F_1. From here, we move all the way to the right to a nice, fat sequential NOE peak at $F_2 = 8.01$ ppm, the H_N proton of Phe-4. Below this is the ZQ arifact-distorted H_α peak and above this are the H_β and $H_{\beta'}$ peaks of the Phe-4 spin system. From any of these three "intraresidue" crosspeaks, one can move to the left and run into a sequential NOE peak at the H_N chemical shift of Leu-5 in F_2. Again, not just the H_α proton of Phe-4 but also the H_β and $H_{\beta'}$ protons are all close enough in space to the H_N of Leu-5 to give ROESY crosspeaks. Looking along this vertical line ($F_2 = H_N$ of Leu-5), we find the ZQ artifact corresponding to $F_1 = H_\alpha$ of Leu-5 (rectangle) and moving to the right a very short distance there is a strong sequential NOE crosspeak at the chemical shift of H_N of Ser-6 in F_2. Above this crosspeak are the intraresidue $F_1 = H_\beta$ and $F_1 = H_{\beta'}$ of Ser-6 crosspeaks, and below this is the $F_1 = H_\alpha$ intraresidue crosspeak of Ser-6. This completes our "walk" along the backbone. In this case, the assignments can be made from the TOCSY alone because each spin system has a unique pattern of chemical shifts that makes it easy to identify in the known primary sequence. But in more complex peptides and proteins, this "walk" is a way to make sequence-specific assignments even when there may be several examples of each of the 20 amino acids in the sequence.

10.3.6 Distance Measurement From NOESY

Ideally, the *initial* rate of increase of the NOE intensity as a function of mixing time is directly proportional to $1/r_{ab}^6$, where r_{ab} is the *distance* between H_a and H_b. There are a number of *caveat*s for those who wish to measure distances using crosspeak intensities in a NOESY spectrum. First, the crosspeak intensities (volumes) are in arbitrary units so that we must calibrate them with a pair (or better yet several pairs) of protons with an accurately known distance within the molecule. This can be done with a geminal pair, a vicinal pair on an aromatic ring, or a 1,3-diaxial pair in a clearly-defined cyclohexane chair structure. Second, the above discussion assumed that there are only two nuclei, H_a and H_b, related by an NOE interaction. This is called the *isolated spin-pair hypothesis*. In reality, it is very rare to find two protons that have no other neighbors within 5 Å. Usually this distance includes other protons, which are related to still others by NOE interactions. The result is a process called *spin diffusion*, where perturbation of the populations (z magnetization) of one nucleus affects the populations of nearby nuclei, which in turn perturb the populations of their neighbors in an expanding process. Spin diffusion can lead to crosspeaks between pairs of nuclei that are not directly related by an NOE interaction and may be considerably more than 5 Å apart in space. Because for small molecules the NOE transfer leads to a reversal of sign of the z magnetization, spin diffusion involving two steps will give crosspeaks that are positive with respect to the diagonal, so that these can be spotted easily. The best way to avoid spin diffusion, however, is to choose as short a mixing time (τ_m) as possible. This also increases the accuracy of distance measurements because only for short mixing times is the crosspeak intensity truly proportional to the cross-relaxation rate and thus to $1/r^6$. Sometimes a series of NOESY spectra is acquired using a number of different mixing times, and the crosspeak intensities are plotted as function of τ_m. This *NOESY buildup* study allows the initial slope of the NOE buildup curve to be measured directly as the cross-relaxation *rate*. Spin-diffusion crosspeaks can be identified in a buildup study because they will always show a lag (sigmoidal shape) before crosspeak intensity starts to climb.

All of these considerations would indicate that a short mixing time is good. Unfortunately, short mixing times also mean weak crosspeaks that are near the noise level. Larger numbers of transients will be required for short mixing times, which means longer overall experiment times. If you are not doing quantitative distance measurements, the mixing time is usually set somewhere near the T_1 value (characteristic time for simple self-relaxation). For these routine experiments, we are shooting for a maximum NOE intensity so that intensity is not strictly proportional to $1/r^6$ and spin diffusion has to be considered as a possible complicating factor in the interpretation of crosspeaks. In this case, we should classify NOE crosspeaks as "weak," "medium," or "strong" and not make any attempt to measure accurate distances.

Because of all of the above pitfalls, NOE is probably the most misinterpreted experiment in organic chemistry. In my experience, J-coupling measurements, both homonuclear and heteronuclear, give far more reliable information than NOE measurements in the determination of small-molecule stereochemistry. To use NOE measurements for stereochemical determinations, it is always best to do the NOESY experiment on *both* isomers and compare the crosspeak intensities (relative to the diagonal peak intensities) and measure distances on both isomers using an energy-minimized computer model of the structures. If the differences in distance and NOE intensity are small between the two isomers, the experiment cannot be conclusive.

10.3.7 Baseline Correction

Just as a rolling *baseline* will affect the integral *area* of a peak in a 1D spectrum, baseline errors (or "baseplane errors") in a 2D spectrum will render crosspeak *volumes* inaccurate. For any quantitative distance determinations, the baseline must be very flat with no streaks or artifacts passing through the crosspeak. One way to check for baseline errors is to measure the volume of two crosspeaks that are symmetry-related across the diagonal—they should be of similar intensity. Another way is to measure the volume within a rectangle in the noise (with no crosspeaks) and compare it to a similar rectangle ("footprint") on a crosspeak. The noise volume should be very small compared to the crosspeak volume. Another way is to display a 1D slice (row or column) from the 2D data matrix and display a zero data set on top of it in a contrasting color. When the vertical scale is increased, you will see where the baseline (noise level where there are no peaks) deviates from the zero line.

10.3.8 EXSY: Chemical Exchange in 2D NMR

Crosspeaks in a NOESY spectrum can also arise from exchange. Even if exchange is slow on the NMR timescale ($\tau_c = \sqrt{2}/(\pi \Delta \nu)$), we can observe it in a longer timescale, the mixing time of the NOESY experiment. If you see positive crosspeaks (relative to a positive diagonal) in the NOESY spectrum of a small molecule (as always, "small" means $\omega \tau_c \ll 1$), these crosspeaks represent exchange rather than an NOE interaction. Exchange means that a nucleus in one environment physically moves to another environment with a different chemical shift. If the rate is much less than the coalescence time $\tau_c = 1/(2.22 \times \Delta \nu)$, two sharp lines will still be observed in the 1D spectrum at the two chemical-shift positions ("slow exchange"). If, for example, half of the H_a nuclei in a 2D NOESY experiment undergo exchange to the H_b position during the mixing time τ_m, we have

$$-\mathbf{I}_z^a \cos(\Omega_a t_1) \xrightarrow{(\tau_m)} \underbrace{-0.5\, \mathbf{I}_z^a \cos(\Omega_a t_1)}_{\text{Diagonal}} \underbrace{-0.5\, \mathbf{I}_z^b \cos(\Omega_a t_1)}_{\text{Crosspeak}}$$

where $\mathbf{I}_z^a$ and $\mathbf{I}_z^b$ represent z magnetization in the two environments. Effectively, the H_a nucleus has changed into an H_b nucleus (changed its chemical shift from Ω_a to Ω_b), and because the chemistry did not affect the nuclear spin, it will have exactly the same nuclear spin state when it becomes H_b magnetization. When a H_a nucleus "turns into" an H_b nucleus, it "drags" its spin state over from chemical shift Ω_a to chemical shift Ω_b. In the slow exchange case, we may still be able to observe the exchange in a NOESY spectrum because the proton changes its environment during the much longer timescale of the mixing time (typically hundreds of milliseconds versus a few milliseconds for τ_c). This term will lead to a crosspeak at $F_1 = \Omega_a, F_2 = \Omega_b$, which has the same sign as the diagonal. For small molecules, exchange peaks stand out clearly because they are opposite in sign to the NOE peaks. Because both NOE and exchange crosspeaks show up in this experiment, NOESY was named nuclear Overhauser and exchange spectroscopy. Sometimes if a NOESY experiment is run solely for the purpose of studying exchange, it is called "EXSY" in the literature (Exchange Spectroscopy), but the experiment is identical to a NOESY experiment. For large molecules, exchange will not stand out in the NOESY spectrum as NOE crosspeaks are also positive. In this case, a ROESY spectrum will allow us to differentiate exchange from NOE because the NOE crosspeaks are all negative in a ROESY spectrum and the exchange crosspeaks are positive.

Figure 10.26 shows the 300-ms NOESY spectrum of lactose (Fig. 10.9) in D_2O/NaOD. Negative contours are shown in gray and positive contours in black; the diagonal is phased

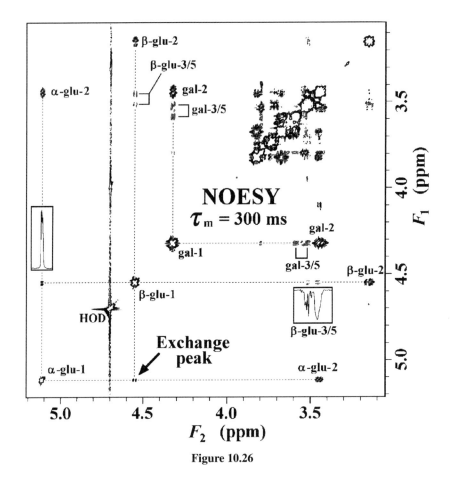

Figure 10.26

to positive intensity. In the lower left corner, we see the diagonal peak for the H-1 proton of the glucose portion of α-lactose (α-glu-1). Moving up on the vertical line, we encounter at $F_1 = \beta$-glu-1 (4.54 ppm) a positive, in-phase crosspeak. This is the exchange peak for conversion of the β-glu-1 proton to an α-glu-1 proton during the mixing time. The inset above the crosspeak shows the 1D horizontal slice through this crosspeak. Moving to the right along the $F_1 = \beta$-glu-1 horizontal line, we pass the diagonal peak and move to the right to a pair of negative (NOE) crosspeaks around $F_2 = 3.5$ ppm. These are NOE peaks from β-glu-1 (axial proton) to H-3 and H-5 of the glucose portion of β-lactose (Fig. 10.9), representing 1,3-diaxial relationships. The inset below the crosspeaks shows a horizontal slice: the peaks are clearly in-phase and negative. Moving farther to the right on the $F_1 = \beta$-glu-1 line we come to the β-glu-2 crosspeak ($F_2 = 3.15$ ppm), which is distorted by the ZQ (J-coupling) artifact. Similar NOE crosspeaks are seen starting from gal-1 on the diagonal, but no exchange peak is observed because the galactose anomeric position is locked in the β-orientation by the glycosidic linkage to the glucose-4 position. On the $F_1 = \alpha$-glu-1 horizontal line, we see the positive exchange peak with β-glu-1 and the (ZQ artifact-distorted) NOE crosspeak with α-glu-2. Because the α-glu-1 proton is equatorial, we do not see any NOE crosspeaks to α-glu-3 or α-glu-5.

10.4 EXPANDING OUR VIEW OF COHERENCE: QUANTUM MECHANICS AND SPHERICAL OPERATORS

10.4.1 Double-Quantum and Zero-Quantum Coherence

Single-quantum coherence (SQC) can be easily defined in terms of the vector model. In a population of identical spins, each individual spin precesses in the laboratory magnetic field at the same frequency, the Larmor frequency ν_o. At equilibrium, the orientation of the spins on the "cone" of precession is random: they are spread out evenly around the cone at any instant in time. Thus the x and y components of their magnetization cancel perfectly leaving no net magnetization in the x–y plane: $M_x = 0$ and $M_y = 0$. After a 90° pulse, the spins become "organized" such that they are now rotating "in phase"—at any instant in time they are all at the same orientation on the cone. Because their phases are coherent instead of random, we call this *phase coherence* or simply *coherence*. Now the individual x and y components of magnetization add together instead of canceling, resulting in a net magnetization in the x–y plane that rotates at the Larmor frequency. It is this net magnetization (coherence) that induces a sinusoidal voltage in the probe coil, which is recorded by the spectrometer as the FID.

There are, however, other kinds of coherence that play an important role in many NMR experiments, such as DEPT, DQF-COSY, HMQC, and HMBC. Coherences other than SQC are called multiple-quantum coherences (MQC), including zero-quantum and double-quantum coherences (ZQC and DQC). When we are talking about SQC, we are referring to NMR transitions that involve only one spin, changing its orientation with respect to the B_o field. For example, in a two-spin system (^{1}H–^{13}C) we can talk about a transition from $\alpha_H \alpha_C$ to $\beta_H \alpha_C$ in which the proton changes from the α to the β state whereas the carbon remains in the α state. This transition corresponds to SQC, as only the proton undergoes a change in orientation. It is an observable transition, giving rise to one of the two components of the proton doublet in the proton spectrum.

Suppose that we are talking about a double-quantum transition in which both the proton and carbon change from the α state to the β state. This transition is thus from the $\alpha_H\alpha_C$ state to the $\beta_H\beta_C$ state of the two-spin, four-state system. This transition corresponds to DQC. Likewise, if the proton flips from β to α while the carbon simultaneously flips from α to β, we have a zero-quantum transition ($\beta_H\alpha_C$ to $\alpha_H\beta_C$) because the total number of spins in the excited (β) state has not changed. This transition corresponds to ZQC. What can we say about these mysterious coherences? In Section 7.10, we encountered ZQC and DQC as intermediate states in coherence transfer, created with pulses from antiphase SQC:

$$2\mathbf{I}_x\mathbf{S}_z \xrightarrow{90^\circ_x \text{ on }^{13}\text{C}} -2\mathbf{I}_x\mathbf{S}_y$$

Note that in the product operator $-2\mathbf{I}_x\mathbf{S}_y$ we have both spins, $\mathbf{I}$ and $\mathbf{S}$ (^{1}H and ^{13}C) in the x–y plane, with the operators multiplied together. This means that both spins are undergoing transitions at the same time, so we have ZQC and DQC. We can convert ZQC and DQC back into observable SQC with a second pulse

$$-2\mathbf{I}_x\mathbf{S}_y \xrightarrow{90^\circ_y \text{ on }^1\text{H}} 2\mathbf{I}_z\mathbf{S}_y = 4[2\mathbf{S}_y\mathbf{I}_z]$$

(The factor of 4 reflects the change from observing ^{1}H to observing ^{13}C, a change in our standard of comparison for magnitude). The net effect of these two steps is to convert antiphase proton SQC into antiphase carbon SQC, an overall coherence transfer with ZQC/DQC as an intermediate state. In Section 7.11 the product operator representations of pure ZQC and DQC were introduced, and we saw that pure DQC rotates in the x–y plane just like SQC, but at a frequency that is the sum of the frequencies of the two spins involved:

$$\{DQ\}_x \to \{DQ\}_y \to -\{DQ\}_x \to -\{DQ\}_y \to \{DQ\}_x$$

frequency of precession $= \nu_H - \nu_C$

For example, on a 600-MHz spectrometer ^{1}H SQC precesses at 600 MHz, ^{13}C SQC precesses at 150 MHz, and $\{^1$H–^{13}C$\}$ DQC precesses at 750 MHz. This makes sense because we are talking about a transition in which both ^{1}H and ^{13}C change from the α to the β state. Zero quantum coherence behaves in a similar way, but the precession rate is the difference between the two SQ precession frequencies:

$$\{ZQ\}_x \to \{ZQ\}_y \to -\{ZQ\}_x \to -\{ZQ\}_y \to \{ZQ\}_x$$

frequency of precession $= \nu_H - \nu_C$

Thus on the same 600-MHz spectrometer, $\{^1$H–^{13}C$\}$ ZQC precesses at 450 MHz (600–150). The good news is that there is no J-coupling evolution due to the active ^{1}H–^{13}C J coupling. Because both spins are undergoing transitions, the interaction between the magnetic dipoles of the two spins does not change: If they are aligned ($\alpha\alpha$ or $\beta\beta$), they remain aligned in DQC; if they are opposite ($\alpha\beta$ or $\beta\alpha$), they remain opposite in ZQC. Passive couplings (coupling to nuclei other than these two, e.g., $\mathbf{I}'$ or $\mathbf{S}'$) will lead to J-coupling evolution and multiplication by $\mathbf{I}_z'$ or $\mathbf{S}_z'$ operators.

The pairwise products of the operators $\mathbf{I}_x$, $\mathbf{I}_y$, $\mathbf{S}_x$, and $\mathbf{S}_y$ can be expressed in terms of pure ZQC and DQC as follows:

$$2\mathbf{I}_x\mathbf{S}_x = 0.5(\{DQ\}_x + \{ZQ\}_x) \quad 2\mathbf{I}_y\mathbf{S}_y = 0.5(\{ZQ\}_x - \{DQ\}_x)$$

$$2\mathbf{I}_x\mathbf{S}_y = 0.5(\{DQ\}_y + \{ZQ\}_y) \quad 2\mathbf{I}_y\mathbf{S}_x = 0.5(\{DQ\}_y - \{ZQ\}_y)$$

Now you can see why the conversion of antiphase ^{1}H SQC with a 90° pulse results in a mixture of DQC and ZQC:

$$2\mathbf{I}_x\mathbf{S}_z \xrightarrow{90°_x \text{ on }^{13}\text{C}} -2\mathbf{I}_x\mathbf{S}_y = -0.5(\{DQ\}_y + \{ZQ\}_y)$$

These multiple quantum coherences cannot be visualized with the vector system, even though we can talk about them as being on the x axis or the y axis. So what do they represent and how can we think about them?

10.4.2 Coherence: The Quantum View

To understand any coherence other than SQC, we need a new and more general definition of coherence. Coherence arises from the quantum mechanical mixing or overlap of spin states ("superposition"). In the two spin system (I, S = ^{1}H, ^{13}C) we have four spin states ($\alpha\alpha$, $\alpha\beta$, $\beta\alpha$, and $\beta\beta$), which are all stable states of defined energy. Let's talk about a single ^{1}H–^{13}C pair (one molecule). It is possible for this pair to be in any one of the four energy states, but it is also possible for the pair to be in a mixture or overlap or superposition of two states. This is one of the fundamental tenets of quantum mechanics: Sometimes you cannot be sure which energy state a particle is in. Let's say that this particular pair is in a mixture of states $\alpha\alpha$ and $\beta\beta$:

$$\Psi = c_1\alpha\alpha + c_4\beta\beta$$

The spin state (or "wave function") of this pair is a linear combination of the states $\alpha\alpha$ and $\beta\beta$, with coefficients c_1 and c_2. These coefficients are actually complex numbers, with real parts (a) and imaginary parts (b):

$$c_1 = a_1 + b_1\mathbf{i} \quad c_4 = a_4 + b_4\mathbf{i}$$

where $\mathbf{i}$ is the square root of -1 ($\mathbf{i}^2 = -1$). We cannot say which energy state the pair is in, but we *can* talk about probabilities. The probability of the pair being in the $\alpha\alpha$ state is

$$P(\alpha\alpha) = c_1^*c_1 = (a_1 - b_1\mathbf{i})(a_1 + b_1\mathbf{i}) = a_1^2 - b_1^2\mathbf{i}^2 = a_1^2 + b_1^2$$

where c_1^* is the complex conjugate of c_1. Note that $P(\alpha\alpha)$ is a positive real number, and if the coefficients are properly normalized, it will be between 0 and 1. Likewise, the probability of this spin pair being in the $\beta\beta$ state is

$$P(\beta\beta) = c_4^*c_4 = (a_4 - b_4\mathbf{i})(a_4 + b_4\mathbf{i}) = a_4^2 - b_4^2\mathbf{i}^2 = a_4^2 + b_4^2$$

which is also a positive real number between 0 and 1, such that $P(\alpha\alpha) + P(\beta\beta) = 1$. That is, it has to be in either the $\alpha\alpha$ or the $\beta\beta$ state because the coefficients for the other states ($\alpha\beta$ and $\beta\alpha$) are zero. If we had a enormous number N of spin pairs (like a mole) in the exact same spin state Ψ, we could say that there are $N \cdot P(\alpha\alpha)$ spin pairs in the $\alpha\alpha$ state and $N \cdot P(\beta\beta)$ spin pairs in the $\beta\beta$ state, but we still could not be sure which state any individual spin pair is in.

Now we get to the interesting part that leads to our expanded definition of coherence. We would like to describe the *degree of overlap* of the $\alpha\alpha$ and $\beta\beta$ spin states for a particular individual spin pair. Consider the product

$$c_4^* c_1 = (a_4 - b_4 i)(a_1 + b_1 i) = \text{degree of mixing or overlap}$$

Note that if $c_1 = 1$ and $c_4 = 0$ (pure $\alpha\alpha$ state), this product is zero; and if $c_1 = 0$ and $c_4 = 1$ (pure $\beta\beta$ state), the product is also zero. So only if there is mixing of the two states for this spin pair will the product be nonzero. Consider the real and imaginary parts of this product, and let's call them x and y:

$$c_4^* c_1 = x + y i = (a_4 a_1 + b_4 b_1) + (a_4 b_1 - a_4 b_1) i$$

If we associate the real part with the x axis and the imaginary part with the y axis, we can think of the "degree of mixing" parameter ($c_4^* c_1$) as a vector with the property of *phase*: It points in a particular direction in the x–y plane. In fact, this is not a stable ("stationary") energy state and it "precesses" in the x–y plane (the "complex plane") at a rate of $\nu_{DQ} = \nu_H + \nu_C$. If we average together the degree of mixing between the $\alpha\alpha$ and $\beta\beta$ states for all of the identical spin pairs in the sample, we have *coherence*:

$$<c_4^* c_1>_{av} = \text{Coherence between the } \beta\beta \text{ and the } \alpha\alpha \text{ states} = \text{DQC}$$

This is analogous to adding up the individual magnetic dipole vectors of individual single spins to get the net magnetization; the part of this net magnetization that is in the x–y plane is what we called coherence, actually SQC. Just like with SQC, if the phase of the individual degree of mixing parameters ($c_4^* c_1$) is random over the population of spin pairs, they will average to zero and there will be no coherence (no DQC). It is only when these phases are *coherent* or tend to have the same phase over the entire population of spin pairs that we have coherence (DQC in this case).

So, in general, in order to have coherence the individual spins or spin pairs have to have mixing or overlap or superposition of two energy states. Because two different energy states define a *transition*, we can say that coherence is associated with a particular NMR transition between energy states. Furthermore, in order to have coherence this mixing or overlap has to add together in a phase coherent manner over all of the spins or spin systems in the sample. The coherence is this sum (actually an average) of the degree of overlap of two energy states over the entire sample. As long as we associate the real and imaginary parts of the overlap product ($c_i^* c_j$) with the x and y axes, we can talk about the phase of this product and whether these phases are coherent in a population ("ensemble") of spin systems.

Just for completeness, let's define all of the coherences for a two-spin (^{1}H–^{13}C) system. If we consider the most general spin state for this system

$$\Psi = c_1 \alpha_H \alpha_C + c_2 \alpha_H \beta_C + c_3 \beta_H \alpha_C + c_4 \beta_H \beta_C$$

We can describe the populations of the four stable energy states:

$N(c_1^* c_1)$ = Population in $\alpha\alpha$ state $N(c_2^* c_2)$ = Population in $\alpha\beta$ state

$N(c_3^* c_3)$ = Population in $\beta\alpha$ state $N(c_4^* c_4)$ = Population in $\beta\beta$ state

The coherences are as follows:

$<c_3^* c_1>_{av}$ = ^{1}H SQC(^{13}C = α) $<c_4^* c_2>_{av}$ = ^{1}H SQC(^{13}C = β)

$<c_2^* c_1>_{av}$ = ^{13}C SQC(^{1}H = α) $<c_4^* c_3>_{av}$ = ^{13}C SQC(^{1}H = β)

$<c_4^* c_1>_{av}$ = DQC $<c_3^* c_2>_{av}$ = ZQC

Note that each of these coherences corresponds to a transition between two energy levels (e.g., $c_3^* c_1$ corresponds to the $\beta_H\alpha_C$ to $\alpha_H\alpha_C$ transition) in the two-spin energy diagram. The four SQCs correspond to the four peaks in the ^{13}C and ^{1}H spectra (two doublets), and the MQCs are not observable. Later, we will see that all of these numbers (the four populations, the four single-quantum coherences, and the ZQC and DQC) can be fit into a 4 × 4 matrix that provides a succinct summary of everything we could ever want to know about the spin state of this ensemble of N spin pairs. This matrix is called the *density matrix*.

10.4.3 Raising and Lowering Operators

The "Cartesian" operators ($\mathbf{I}_x$, $\mathbf{I}_y$, $\mathbf{I}_z$, $\mathbf{S}_x$, $\mathbf{S}_y$, $\mathbf{S}_z$) are easy to use because they relate to the vector model and we can easily figure out what happens to them with a pulse of a particular phase or a delay. For example, $\mathbf{I}_y$ will become $\mathbf{I}_z$ under the influence of a 90° pulse on the x' axis on the ^{1}H channel, and $\mathbf{I}_y$ will become $-\mathbf{I}_x$ after chemical-shift evolution for $\tau = \pi/(2\,\Omega_H)$. But when we talk about DQC and ZQC, the Cartesian operators become more cumbersome:

$$\{ZQ\}_y = 2\mathbf{I}_x\mathbf{S}_y - 2\mathbf{I}_y\mathbf{S}_x \xrightarrow{90^\circ_x(^1H)} 2\mathbf{I}_x\mathbf{S}_y - 2\mathbf{I}_z\mathbf{S}_x = 0.5(\{DQ\}_y + \{ZQ\}_y) - 2\mathbf{S}_x\mathbf{I}_z$$

$$\{ZQ\}_y = 2\mathbf{I}_x\mathbf{S}_y - 2\mathbf{I}_y\mathbf{S}_x \xrightarrow{\tau\text{delay}} 2(\mathbf{I}_x\cos + \mathbf{I}_y\sin)(\mathbf{S}_y\cos' - \mathbf{S}_x\sin')$$
$$-2(\mathbf{I}_y\cos - \mathbf{I}_x\sin)(\mathbf{S}_x\cos' + \mathbf{S}_y\sin') \to \text{so on.}$$

Furthermore, the coherence order (sensitivity to twisting of a coherence by a gradient) is ambiguous, at least with respect to the sign

$\mathbf{I}_x, \mathbf{I}_y$ $p = \pm 4$ $\mathbf{S}_x, \mathbf{S}_y$ $p = \pm 1$
$\{DQ\}_x, \{DQ\}_y$ $p = \pm 5$ $(ZQ)_x, (ZQ)_y$ $p = \pm 3$

We can define another type of operator that describes the spin state without reference to the x and y (Cartesian) axes. These are the raising and lowering operators that refer to coherence in terms of the transitions between spin states. For example, $\mathbf{I}^+$ refers to the transition of spin $\mathbf{I}$ from the α to the β state, while the $\mathbf{S}$ spin does not change state. For the ^{1}H, ^{13}C spin system, this is a transition from the $\alpha_H\alpha_C$ state to the $\beta_H\alpha_C$ state, or from the $\alpha_H\beta_C$ state to the $\beta_H\beta_C$ state (Fig. 10.27). Likewise, the $\mathbf{S}^-$ coherence refers to the transition from

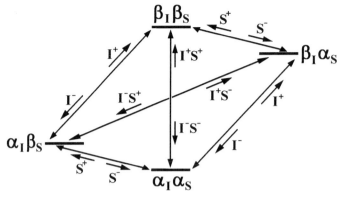

Figure 10.27

$\alpha_H\beta_C$ to $\alpha_H\alpha_C$ or from $\beta_H\beta_C$ to $\beta_H\alpha_C$. These are the S (^{13}C) single-quantum transitions in the energy diagram. The double-quantum transitions give rise to the coherences $\mathbf{I}^+\mathbf{S}^+$ ($\alpha\alpha$ to $\beta\beta$) and $\mathbf{I}^-\mathbf{S}^-$ ($\beta\beta$ to $\alpha\alpha$), and the zero-quantum transitions give rise to the coherences $\mathbf{I}^-\mathbf{S}^+$ ($\beta_H\alpha_C$ to $\alpha_H\beta_C$) and $\mathbf{I}^+\mathbf{S}^-$ ($\alpha_H\beta_C$ to $\beta_H\alpha_C$).

> The effect of the spherical operators on individual spin states is actually opposite to this: for example $\mathbf{I}^+\ |\beta> \rightarrow |\alpha>$. It is the magnetic quantum number that is "raised" by the operator: $-1/2$ (β state) to $+1/2$ (α state). In this book, we will reverse the definition for convenience so that the operators make intuitive sense: $\mathbf{I}^+$ "raises" the spin state from α to β.

Note that these are product operators just like $2\mathbf{I}_x\mathbf{S}_z$; they are two single-spin operators multiplied together. We can also have antiphase SQC states such as $\mathbf{I}^+\mathbf{S}_z, \mathbf{S}^-\mathbf{I}_z$, and so on. The formal definition of $\mathbf{I}^+$ and $\mathbf{I}^-$ is

$$\mathbf{I}^+ = \mathbf{I}_x + i\mathbf{I}_y \qquad \mathbf{I}^- = \mathbf{I}_x - i\mathbf{I}_y$$

By adding and subtracting them, we get the definition of $\mathbf{I}_x$ and $\mathbf{I}_y$ in terms of the raising and lowering operators

$$\mathbf{I}_x = \tfrac{1}{2}(\mathbf{I}^+ + \mathbf{I}^-) \qquad \mathbf{I}_y = \tfrac{1}{2i}(\mathbf{I}^+ - \mathbf{I}^-)$$

These are extremely useful for the math but they do not give us any feel for their physical meaning. We are sacrificing the comfort level of visualizing operators in terms of the vector model in order to focus on the transitions that are associated with a coherence. The phase of pulses and careful tracking of the location of vectors in the x–y plane will be ignored completely, freeing us to focus on the overall processes of coherence transfer, evolution, and so on.

The most important thing about the raising and lowering (or "spherical") operators is the way they react to gradients, which is to say their coherence order. The coherence order is no longer ambiguous. For the heteronuclear system

$$\mathbf{I}^+: p = 4; \qquad \mathbf{I}^-: p = -4; \qquad \mathbf{S}^+: p = 1; \qquad \mathbf{S}^-: p = -1$$
$$\mathbf{I}^+\mathbf{S}^+: p = +5; \quad \mathbf{I}^-\mathbf{S}^-: p = -5; \quad \mathbf{I}^+\mathbf{S}^-: p = +3; \quad \mathbf{I}^-\mathbf{S}^+: p = -3$$

Note that the coherence orders simply add in the product operators: $\mathbf{I^+S^-}$, which represents heteronuclear ZQC, has a coherence order of 3 ($+4\ -1$). The ^{1}H SQC corresponds to $\mathbf{p} = \pm 4$ because the ^{1}H nuclear magnet is four times as strong as the ^{13}C nuclear magnet ($\gamma_H/\gamma_C = 4$) and this makes its precession rate four times as sensitive to changes in the magnetic field ($\Delta v_o = \gamma \Delta B_o$). As gradients are simply a position-dependent change in the magnetic field ($B_g = B_o + G_z \cdot z$), this means that ^{1}H coherence is twisted four times as much as ^{13}C coherence by the same strength and duration of gradient. For a *homonuclear* two-spin system ($\mathbf{I} = H_a, \mathbf{S} = H_b$), we use $\mathbf{p} = \pm 1$ for SQC for both of the protons because they are equally sensitive to gradients. Thus, the coherence order is context dependent and the ^{1}H SQC coherence order reflects the relative magnetogyric ratio (γ_H/γ_X) in comparison to the other nucleus (X) in the spin system.

The effect of a 180° pulse on spherical operators is very simple: it reverses the coherence order of the affected spin. For example, a ^{1}H 180° pulse converts $\mathbf{I^+}$ to $\mathbf{I^-}$, $\mathbf{I^-}$ to $\mathbf{I^+}$, and $\mathbf{I^+S^+}$ to $\mathbf{I^-S^+}$ ($\mathbf{S} = {}^{13}\text{C}$). This is easy to prove using the definitions

$$\mathbf{I^+} = \mathbf{I}_x + i\mathbf{I}_y \xrightarrow{180_x^\circ} \mathbf{I}_x - i\mathbf{I}_y = \mathbf{I^-}$$

$$\mathbf{I^+} = \mathbf{I}_x + i\mathbf{I}_y \xrightarrow{180_y^\circ} -\mathbf{I}_x + i\mathbf{I}_y = -\mathbf{I^-}$$

Note that the phase of the pulse does have an effect: It introduces a "phase factor" (in this case 1 or −1) in front of the operator, but it does not change the coherence order. Most of the time, we will be ignoring these phase factors.

The effect of delays is even simpler: the coherence order *does not change* during a delay. For example, $\mathbf{I^-}$ remains $\mathbf{I^-}$ after a delay τ:

$$\mathbf{I^-} = \mathbf{I}_x - i\mathbf{I}_y \xrightarrow{\tau} [\mathbf{I}_x\cos + \mathbf{I}_y\sin] - i[\mathbf{I}_y\cos - \mathbf{I}_x\sin]$$
$$= (\mathbf{I}_x - i\mathbf{I}_y)\cos + i(\mathbf{I}_x - i\mathbf{I}_y)\sin = (\mathbf{I}_x - i\mathbf{I}_y)e^{i\Omega_H\tau} = \mathbf{I^-}e^{i\Omega_H\tau}$$

where $\sin = \sin(\Omega_H \tau)$ and $\cos = \cos(\Omega_H \tau)$ and $e^{i\Theta} = \sin \Theta + i \cos \Theta$. The important thing is that we get back $\mathbf{I^-}$ multiplied by a phase factor. Chemical-shift evolution in the x–y plane has been reduced to the rotation of a unit vector ($e^{i\Omega\tau}$) in the complex (real, imaginary) plane.

We can also have *J*-coupling evolution from $\mathbf{I^+}$ to $\mathbf{I^+S}_z$ or from $\mathbf{I^-S}_z$ to $\mathbf{I^-}$, but the coherence order (1 and −1, respectively) does not change. Because ZQC and DQC do not undergo *J*-coupling evolution, $\mathbf{I^+S^+}$ will stay as $\mathbf{I^+S^+}$ and $\mathbf{I^+S^-}$ will stay as $\mathbf{I^+S^-}$ during a delay (times a phase factor for DQ or ZQ chemical-shift evolution) and the coherence order (5 and 3, respectively) will not change.

This allows us to diagram the coherence order pathway of an NMR experiment in a very simple way. For example, in an INEPT experiment with intermediate DQC we have

$$\underset{0}{\mathbf{I}_z} \xrightarrow{90°\,^1\mathrm{H}} \underset{4}{\mathbf{I^+}} \xrightarrow{\frac{1}{2J}} \underset{4}{\mathbf{I^+S}_z} \xrightarrow{90°\,^{13}\mathrm{C}} \underset{5}{\mathbf{I^+S^+}} \xrightarrow{90°\,^1\mathrm{H}} \underset{1=p}{\mathbf{S^+I}_z}$$

We can add gradients to this pulse sequence for selection of the coherence pathway outlined above by applying a gradient of relative strength 5 during the $\mathbf{I^+} \to \mathbf{I^+S}_z$ delay ($\mathbf{p}=4$) and another gradient of strength −4 during the $\mathbf{I^+S^+}$ period ($\mathbf{p}=5$). The amount of coherence "twist" (position-dependent phase shift) produced by the two gradients is 20 (5·4) and −20 (−4·5) for an overall twist of zero for the desired coherence pathway. Other pathways will lead to nonzero twist and will be scrambled during the acquisition of the FID, so they will not give any signal in the probe coil.

The coherence order **p** can be understood precisely if we consider the effect of a gradient on one of the spherical operators. A gradient is just like chemical-shift evolution except that the amount of evolution depends on the position in the tube (z coordinate) rather than the chemical shift.

$$\mathbf{I}^- = \mathbf{I}_x - i\mathbf{I}_y \stackrel{G_z(\tau)}{\rightarrow} [\mathbf{I}_x\cos + \mathbf{I}_y\sin] - i[\mathbf{I}_y\cos - \mathbf{I}_x\sin]$$
$$= (\mathbf{I}_x - i\mathbf{I}_y)\cos + i(\mathbf{I}_x - i\mathbf{I}_y)\sin = (\mathbf{I}_x - i\mathbf{I}_y)e^{i\gamma z G_z \tau} = \mathbf{I}^- e^{i\gamma z G_z \tau}$$

where $\sin = \sin(\gamma\, z\, G_z\, \tau)$, $\cos = (\gamma\, z\, G_z\, \tau)$, and $\mathbf{e}^{i\Theta} = \sin\Theta + \mathbf{i}\cos\Theta$. The gradient preserves the coherence order but it is now multiplied by a phase factor that depends on the position within the NMR tube. A similar analysis shows that $\mathbf{I}^+$ is twisted in the opposite direction

$$\mathbf{I}^+ = \mathbf{I}_x + i\mathbf{I}_y \stackrel{G_z(\tau)}{\rightarrow} \mathbf{I}^+ e^{-i\gamma z G_z \tau}$$

where $\mathbf{e}^{-i\Theta} = \sin\Theta - \mathbf{i}\cos\Theta$. Notice that the exponent is now negative, so that coherence is twisted in the opposite direction as a function of position (z). If we associate the imaginary term with the y' axis, we see that the helix is reversed because the imaginary term is reversed in sign. In general, if we focus on the exponent Θ in $\mathbf{e}^{i\Theta}$ as the "twist" induced by the gradient, we have

$$\text{"twist"} = \Theta = -\mathbf{p}\gamma_0 z G_z \tau$$

where γ_0 is the smallest γ in the system analyzed. For example, in the heteronuclear (^{13}C, ^{1}H) system $\mathbf{I}^-$ has p = −4 and we get $\Theta = 4\gamma_c\, z\, G_z\, \tau = \gamma_H\, z\, G_z\, \tau$. The phase factor is $e^{i\Theta} = e^{i\gamma z G_z \tau}$, exactly as derived above. For an operator product like $\mathbf{I}^+\mathbf{S}^-$, each operator undergoes twisting by the gradient and gains its own phase factor

$$\mathbf{I}^+\mathbf{S}^- \stackrel{G_z(\tau)}{\rightarrow} \mathbf{I}^+[e^{-i\gamma_H z G_z \tau}]\mathbf{S}^-[e^{i\gamma_C z G_z \tau}] = \mathbf{I}^+\mathbf{S}^- e^{-i(\gamma_H - \gamma_C)z G_z \tau}$$

The resulting "twist" is $\Theta = -(\gamma_H - \gamma_C)\, z\, G_z\, \tau = -(4-1)\gamma_C\, z\, G_z\, \tau = -3\gamma_0\, z\, G_z\, \tau$. Because we defined the twist as $\Theta = -\mathbf{p}\, \gamma_0\, z\, Gz\, \tau$, it is clear that **p** = 3 for the coherence $\mathbf{I}^+\mathbf{S}^-$. Because exponents add when we multiply the terms, the coherence orders of each term in the product simply add together. The conclusion is simple and we never have to look into the detailed math again: the sensitivity to gradient twisting is precisely defined, right down to the direction of twisting, by the coherence order **p**. The phase factors accumulate as we apply more and more gradients, and the "twists" add up according to the sum of **p** G_z (assuming that all gradients have the same duration τ). We can only observe coherence at the end if the twist is zero: $\Sigma \mathbf{p}_i G_i = 0$.

One more thing we can do with spherical operators: We can easily derive the expressions given in Chapter 8 for pure ZQC and DQC. Start with the spherical product $\mathbf{I}^+\mathbf{S}^+$ and express it in terms of the "Cartesian" ($\mathbf{I}_x$, $\mathbf{I}_y$, etc.) operators

$$\mathbf{I}^+\mathbf{S}^+ = (\mathbf{I}_x + i\mathbf{I}_y)(\mathbf{S}_x + i\mathbf{S}_y) = \mathbf{I}_x\mathbf{S}_x - \mathbf{I}_y\mathbf{S}_y + i\mathbf{I}_y\mathbf{S}_x + i\mathbf{I}_x\mathbf{S}_y$$
$$= [\tfrac{1}{2}(2\mathbf{I}_x\mathbf{S}_x - 2\mathbf{I}_y\mathbf{S}_y)] + i[\tfrac{1}{2}(2\mathbf{I}_y\mathbf{S}_x + 2\mathbf{I}_x\mathbf{S}_y)] = \{DQ\}_x + i\{DQ\}_y$$

The final step uses the analogy with $\mathbf{I}^+ = \mathbf{I}_x + i\,\mathbf{I}_y$ to define pure $\{DQ\}_x$ and $\{DQ\}_y$. From this we have: $\{DQ\}_x = 1/2(2\mathbf{I}_x\mathbf{S}_x - 2\mathbf{I}_y\mathbf{S}_y)$ and $\{DQ\}_y = 1/2(2\mathbf{I}_y\mathbf{S}_x + 2\mathbf{I}_x\mathbf{S}_y)$. Starting with $\mathbf{I}^+\mathbf{S}^-$, we get

$$\mathbf{I}^+\mathbf{S}^- = (\mathbf{I}_x + i\mathbf{I}_y)(\mathbf{S}_x - i\mathbf{S}_y) = \mathbf{I}_x\mathbf{S}_x + \mathbf{I}_y\mathbf{S}_y + i\mathbf{I}_y\mathbf{S}_x - i\mathbf{I}_x\mathbf{S}_y$$
$$= [\tfrac{1}{2}(2\mathbf{I}_x\mathbf{S}_x + 2\mathbf{I}_y\mathbf{S}_y)] - i[\tfrac{1}{2}(2\mathbf{I}_x\mathbf{S}_y - 2\mathbf{I}_y\mathbf{S}_x)] = \{ZQ\}_x - i\{ZQ\}_y$$

by analogy to $\mathbf{S}^- = \mathbf{S}_x - i\,\mathbf{S}_y$. Equating the real and imaginary parts, this gives us: $\{ZQ\}_x = 1/2(2\mathbf{I}_x\mathbf{S}_x + 2\mathbf{I}_y\mathbf{S}_y)$ and $\{ZQ\}_y = 1/2(2\mathbf{I}_x\mathbf{S}_y - 2\mathbf{I}_y\mathbf{S}_x)$. If you are not convinced yet, you can try to show that $\{ZQ\}_x$ turns into $\{ZQ\}_y$ after evolution for a time τ, for which $(\Omega_H - \Omega_C)\tau = \pi/2$ and that $\{DQ\}_x$ turns into $\{DQ\}_y$ after evolution for a time τ, for which $(\Omega_H + \Omega_C)\tau = \pi/2$. The skeptic will be rewarded after a few happy hours and many pages of paper.

10.5 DOUBLE-QUANTUM FILTERED COSY (DQF-COSY)

10.5.1 The Double-Quantum Filter

We saw in Chapter 9 that the homonuclear "front end" $90°_x–t_1–90°_x$ gives us four terms for the H_a–H_b system:

$$-\mathbf{I}_z^a\,c\,c' \quad -2\mathbf{I}_x^a\mathbf{I}_y^b\,c\,s' \quad +\mathbf{I}_x^a\,s\,c' \quad -2\mathbf{I}_y^b\mathbf{I}_z^a\,s\,s'$$

One problem with the simple COSY sequence is that the crosspeak term ($-2\mathbf{I}_y^b\mathbf{I}_z^a$) is on the y' axis whereas the diagonal term ($\mathbf{I}_x^a$) is on the x' axis of the rotating frame. This means that there is no way to phase the crosspeaks to absorption in the F_2 dimension without having the diagonal peaks in dispersion mode. The same is true in the F_1 dimension, as the crosspeak has cosine modulation in t_1 and the diagonal peak has sine modulation (Chapter 9, Section 9.5.2). These strong dispersive signals on the diagonal extend out much farther than absorptive peaks and give rise to long streaks stretching out in both F_1 and F_2, interfering with the observation of crosspeaks.

In the DQF-COSY experiment, a third 90° pulse is added immediately after the $90°$–t_1–$90°$ COSY sequence (Fig. 10.28). Instead of transferring the antiphase H_a magnetization directly into antiphase H_b magnetization with the second pulse, it is first converted into DQC ($H_a^\alpha, H_b^\alpha \leftrightarrow H_a^\beta, H_b^\beta$) as an intermediate state in coherence transfer. This is "filtered" by destroying all other coherences (ZQC, $\mathbf{I}_z$, SQC) with a phase cycle, and then the DQC is immediately converted into antiphase H_b magnetization by the third pulse and the FID is recorded. The filter assures that only magnetization that passes briefly through the double-quantum state between the second and third pulse can be observed in the FID. The "filtration" is accomplished by varying the phase of the third pulse by 90° on each successive transient (e.g., x', y', $-x'$, $-y'$, etc.) and varying the receiver phase (i.e.,

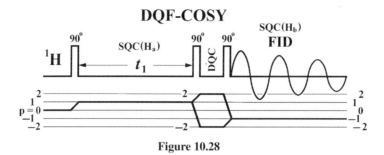

Figure 10.28

the reference axis) to select only magnetization that is converted from DQC to SQC by the final pulse.

The DQ filter destroys all terms from the homonuclear front end except for the one with both operators $\mathbf{I}^a$ and $\mathbf{I}^b$ in the x–y plane: the second term $-2\mathbf{I}^a_x\mathbf{I}^b_y$ c s'. This term will become the diagonal *and* the crosspeak in the DQF-COSY spectrum. To understand its fate we will express it in terms of pure ZQC and DQC

$$-2\mathbf{I}^a_x\mathbf{I}^b_y = -\{\mathbf{DQ}\}_y - \{\mathbf{ZQ}\}_y$$

Now we apply the final 90° pulse, but in four consecutive scans we cycle the phase of the pulse as follows: $x, y, -x, -y$. The receiver phase is cycled in the reverse sense: $x, -y, -x, y$. This is the essence of the DQ filter: We will see that only the DQ part of the $-2\mathbf{I}^a_x\mathbf{I}^b_y$ term will survive this phase cycle. The other three terms from the homonuclear front end, as well as the ZQC portion of the third term, will be canceled in the phase cycle. Let's write out the result of the phase cycle for the DQ term first

$$2\{\mathbf{DQ}\}_y = 2\mathbf{I}^a_y\mathbf{I}^b_x + 2\mathbf{I}^a_x\mathbf{I}^b_y \rightarrow$$

first scan: $90°_x$ $\rightarrow$ $2\mathbf{I}^b_x\mathbf{I}^a_z + 2\mathbf{I}^a_x\mathbf{I}^b_z$ (crosspeak and diagonal peak)

second scan: $90°_y$ $\rightarrow$ $-2\mathbf{I}^a_y\mathbf{I}^b_z - 2\mathbf{I}^b_y\mathbf{I}^a_z$ (diagonal peak and crosspeak)

third scan: $90°_{-x}$ $\rightarrow$ $-2\mathbf{I}^b_x\mathbf{I}^a_z - 2\mathbf{I}^a_x\mathbf{I}^b_z$ (crosspeak and diagonal peak)

fourth scan: $90°_{-y}$ $\rightarrow$ $2\mathbf{I}^a_y\mathbf{I}^b_z + 2\mathbf{I}^b_y\mathbf{I}^a_z$ (diagonal peak and crosspeak)

Note that both the diagonal peak and the crosspeak are antiphase and that they have the same phase (x' or y'), unlike in the simple COSY where the diagonal peak is antiphase on y' and the crosspeak is in-phase on x'. Also, note the effect of the phase cycle on the phase of the crosspeak: $x', -y', -x', y'$ (examining the left term, right term, left term, and right term above). If we cycle the receiver phase so that the reference axis is x', then $-y', -x'$, and finally y' for the four scans above, the crosspeak terms will all add together in the receiver. Now focus on the diagonal peaks: their phase is also $x', -y', -x'$ and y' for the four scans of the phase cycle. So with the receiver phase set to $x', -y', -x'$, and y' these terms will also add together in the receiver.

Another way to view the receiver phase is as a shifting of the reference axes. With receiver phase set to x' we do not alter anything in the result, but with the receiver phase (reference axis) set to y' we read an $\mathbf{I}_y$ operator as $\mathbf{I}_x$ (retarding its phase by 90°). All the other operators are retarded in phase accordingly: $-\mathbf{I}_x$ becomes $\mathbf{I}_y$, $-\mathbf{I}_y$ becomes $-\mathbf{I}_x$, and $\mathbf{I}_x$ becomes $-\mathbf{I}_y$. The effect of all four receiver phase settings are summarized below:

Receiver phase	$\mathbf{I}_x$	$\mathbf{I}_y$	$-\mathbf{I}_x$	$-\mathbf{I}_y$
x' :	$\mathbf{I}_x$	$\mathbf{I}_y$	$-\mathbf{I}_x$	$-\mathbf{I}_y$
y' :	$-\mathbf{I}_y$	$\mathbf{I}_x$	$\mathbf{I}_y$	$-\mathbf{I}_x$
$-x'$:	$-\mathbf{I}_x$	$-\mathbf{I}_y$	$\mathbf{I}_x$	$\mathbf{I}_y$
$-y'$:	$\mathbf{I}_y$	$-\mathbf{I}_x$	$-\mathbf{I}_y$	$\mathbf{I}_x$

For example, a final result of $\mathbf{I}_x$ (first column) will be received as $\mathbf{I}_y$ if the receiver phase is set to $-y'$ (fourth row). Now we can "convert" our results for the final 90° pulse, taking into account the effect of the receiver phase cycle $x', -y', -x', y'$:

$$90°_x: \quad \rightarrow \quad 2\mathbf{I}^b_x\mathbf{I}^a_z + 2\mathbf{I}^a_x\mathbf{I}^b_z \quad \rightarrow \quad 2\mathbf{I}^b_x\mathbf{I}^a_z + 2\mathbf{I}^a_x\mathbf{I}^b_z \quad \text{(receiver phase } x')$$
$$90°_y: \quad \rightarrow \quad -2\mathbf{I}^a_y\mathbf{I}^b_z - 2\mathbf{I}^b_y\mathbf{I}^a_z \quad \rightarrow \quad 2\mathbf{I}^a_x\mathbf{I}^b_z + 2\mathbf{I}^b_x\mathbf{I}^a_z \quad \text{(receiver phase } -y')$$
$$90°_{-x}: \quad \rightarrow \quad -2\mathbf{I}^b_x\mathbf{I}^a_z - 2\mathbf{I}^a_x\mathbf{I}^b_z \quad \rightarrow \quad 2\mathbf{I}^b_x\mathbf{I}^a_z + 2\mathbf{I}^a_x\mathbf{I}^b_z \quad \text{(receiver phase } -x')$$
$$90°_{-y}: \quad \rightarrow \quad 2\mathbf{I}^a_y\mathbf{I}^b_z + 2\mathbf{I}^b_y\mathbf{I}^a_z \quad \rightarrow \quad 2\mathbf{I}^a_x\mathbf{I}^b_z + 2\mathbf{I}^b_x\mathbf{I}^a_z \quad \text{(receiver phase } y')$$

Adding up all eight terms we get: $4[2\mathbf{I}^b_x\mathbf{I}^a_z] + 4[2\mathbf{I}^a_x\mathbf{I}^b_z]$. This represents the crosspeak (H_b antiphase coherence) and the diagonal peak (H_a antiphase coherence), respectively. Both have the same phase in F_2 (x') and the same phase in F_1 (cs' modulation in t_1) because they both derive from the same term: $-2\mathbf{I}^a_x\mathbf{I}^b_y \text{cs}'$.

Now let's examine the ZQC term and see how it fares in the phase cycle:

$$2\{\mathbf{ZQ}\}_y = \quad 2\mathbf{I}^a_x\mathbf{I}^b_y - 2\mathbf{I}^a_y\mathbf{I}^b_x \quad \rightarrow$$

$$90°_x: \quad \rightarrow \quad 2\mathbf{I}^a_x\mathbf{I}^b_z - 2\mathbf{I}^b_x\mathbf{I}^a_z \quad \rightarrow \quad 2\mathbf{I}^a_x\mathbf{I}^b_z - 2\mathbf{I}^b_x\mathbf{I}^a_z \quad \text{(receiver phase } x')$$
$$90°_y: \quad \rightarrow \quad -2\mathbf{I}^b_y\mathbf{I}^a_z + 2\mathbf{I}^a_y\mathbf{I}^b_z \quad \rightarrow \quad 2\mathbf{I}^b_x\mathbf{I}^a_z - 2\mathbf{I}^a_x\mathbf{I}^b_z \quad \text{(receiver phase } -y')$$
$$90°_{-x}: \quad \rightarrow \quad -2\mathbf{I}^a_x\mathbf{I}^b_z + 2\mathbf{I}^b_x\mathbf{I}^a_z \quad \rightarrow \quad 2\mathbf{I}^a_x\mathbf{I}^b_z - 2\mathbf{I}^b_x\mathbf{I}^a_z \quad \text{(receiver phase } -x')$$
$$90°_{-y}: \quad \rightarrow \quad 2\mathbf{I}^b_y\mathbf{I}^a_z - 2\mathbf{I}^a_y\mathbf{I}^b_z \quad \rightarrow \quad 2\mathbf{I}^b_x\mathbf{I}^a_z - 2\mathbf{I}^a_x\mathbf{I}^b_z \quad \text{(receiver phase } y')$$

As before, we use the table to adjust the phase according to the reference axis for each scan. Now we see that the $2\mathbf{I}^a_x\mathbf{I}^b_z$ terms alternate sign and cancel as we move from first scan, first term to second scan, second term to third scan, first term and finally to fourth scan, second term. Likewise, the $2\mathbf{I}^b_x\mathbf{I}^a_z$ terms alternate sign and cancel as we move down. So the ZQC, which exists between the second and third pulses of the DQF-COSY pulse sequence (Fig. 10.28) does not contribute anything to the observed FID after four scans. Just for completeness, we can show that all of the other terms present at the end of the $90°_x$–t_1–$90°_x$ sequence are also destroyed by the phase cycle

$$-\mathbf{I}^a_z \rightarrow 1. \mathbf{I}^a_y \rightarrow \mathbf{I}^a_y; \ 2. -\mathbf{I}^a_x \rightarrow -\mathbf{I}^a_y; \ 3. -\mathbf{I}^a_y \rightarrow \mathbf{I}^a_y; \ 4. \mathbf{I}^a_x \rightarrow -\mathbf{I}^a_y$$

$$\text{sum} = \mathbf{I}^a_y - \mathbf{I}^a_y + \mathbf{I}^a_y - \mathbf{I}^a_y = 0$$

$$+\mathbf{I}^a_x \rightarrow 1. \mathbf{I}^a_x \rightarrow \mathbf{I}^a_x; \ 2. -\mathbf{I}^a_z; \ 3. \mathbf{I}^a_x \rightarrow -\mathbf{I}^a_x; \ 4. \mathbf{I}^a_z$$

$$\text{sum} = \mathbf{I}^a_x - \mathbf{I}^a_x = 0 \ (\mathbf{I}^a_z \text{ is not observable in the FID})$$

$-2\mathbf{I}_y^b\mathbf{I}_z^a \rightarrow 1.\ 2\mathbf{I}_y^a\mathbf{I}_z^b \rightarrow 2\mathbf{I}_y^a\mathbf{I}_z^b;\ 2.\ -2\mathbf{I}_y^b\mathbf{I}_x^a;\ 3.\ 2\mathbf{I}_y^a\mathbf{I}_z^b \rightarrow\ -2\mathbf{I}_y^a\mathbf{I}_z^b;\ 4.\ 2\mathbf{I}_y^b\mathbf{I}_x^a$

sum $= 2\mathbf{I}_y^a\mathbf{I}_z^b - 2\mathbf{I}_y^a\mathbf{I}_z^b = 0$ ($-2\mathbf{I}_y^b\mathbf{I}_x^a$ and $2\mathbf{I}_y^b\mathbf{I}_x^a$ are not observable)

So the filter gets rid of everything except the two terms (one diagonal and one crosspeak) that come from the DQC portion of the second term $-2\mathbf{I}_x^a\mathbf{I}_y^b\text{cs}'$. The results carry along the $\text{c s}' = \cos(\Omega_a t_1)\sin(\pi J t_1)$ term, which makes both peaks antiphase in F_1 as well as in F_2. Singlet signals (including solvent), which cannot exist as DQC because there is no J coupling, are eliminated altogether. Furthermore, both diagonal peaks and crosspeaks have the same phase in a DQF-COSY spectrum, so that both can be phase corrected to be pure absorption mode.

10.6 COHERENCE PATHWAY SELECTION IN NMR EXPERIMENTS

An NMR experiment (pulse sequence) consists of a precisely defined series of radio frequency pulses and delays. Pulses create coherence from z magnetization or convert one coherence into another. They usually lead to multiple changes in coherence order (**p**), so that after a number of pulses there can be a bewildering array of different coherences. Delays cause coherences to undergo evolution, resulting from either chemical shift differences or J couplings. During a delay, a single coherence can lead to as many as four different components depending upon whether chemical-shift evolution, J-coupling evolution, both or neither has occurred during the delay. Delays can also lead to relaxation (T_1 and T_2) and cross relaxation (NOE). Of the myriad coherences present at the end of the pulse sequence (beginning of the FID), we are only interested in one. This coherence gives us the information we want from the experiment: specific relationships between spins that we have selected with the pulse sequence. All other coherences are artifacts: They will lead to false crosspeaks, ugly streaks that interfere with the desired information, or phase distortions in the desired peaks. The desired coherence follows a specific *pathway* throughout the pulse sequence, defined by the coherence order (**p**) at each stage of the experiment. In general, delays conserve the coherence order, whereas 180° pulses change its sign and 90° pulses cause coherences to split into several different coherence orders. To select the desired coherence pathway and to eliminate the many artifacts resulting from alternative pathways, we use coherence pathway selection.

Two methods are available for coherence pathway selection: *phase cycling* and *pulsed field gradients*. We can also use both methods working together to get even better suppression of artifacts. Phase cycling focuses on the effect of pulses on the coherence order. By changing the phase of the RF pulse, which corresponds to the orientation of the B_1 field vector in the x'–y' plane in the rotating frame of reference, we can cause the desired coherence in the FID to add together in the sum-to-memory with each successive transient or scan. The undesired signals (artifacts) can be made to cancel in the sum-to-memory with a series of scans. Gradients operate in a different way: the "twist" (position-dependent phase shift) that a gradient pulse gives to a coherence depends on its coherence order (**p**). A series of gradient pulses placed at strategic places within the pulse sequence leads to a combined "twist" that depends on the coherence order at each stage of the pulse sequence. If the gradient strength is properly adjusted for each gradient in the sequence, only the desired coherence can arrive at the beginning of the FID with no "twist." All other coherences (the artifacts) will be "twisted" and therefore not contribute to the FID.

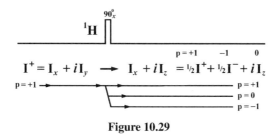

Figure 10.29

10.6.1 The Coherence Level Diagram

Below the pulse sequence we can show the desired coherence level, **p**, at each stage of the pulse sequence. This diagram defines the coherence pathway that is desired for a particular NMR experiment. Coherence order is mixed for Cartesian product operators

$$\mathbf{I}_x, \mathbf{I}_y \quad \mathbf{p} = \pm 1; \quad \{DQ\}_x, \{DQ\}_y \quad \mathbf{p} = \pm 2 \quad \text{(homonuclear)}$$

but it is pure for the spherical (raising and lowering) operators

$$\mathbf{I}^+ \quad \mathbf{p} = 1; \quad \mathbf{I}^- \quad \mathbf{p} = -1; \quad \mathbf{I}^+\mathbf{S}^+ \quad \mathbf{p} = 2 \quad \text{(homonuclear)}$$

The Cartesian operators have mixed coherence order because they are linear combinations of the spherical operators

$$\mathbf{I}_x = \tfrac{1}{2}(\mathbf{I}^+ + \mathbf{I}^-) \quad \mathbf{p} = \pm 1$$

$$\mathbf{I}_y = \tfrac{1}{2i}(\mathbf{I}^+ - \mathbf{I}^-) \quad \mathbf{p} = \pm 1$$

Using spherical operators we can see that a 90° pulse can "explode" the coherence order into several different levels. For example, a $90°_x$ pulse converts the pure coherence level $\mathbf{p} = +1$ into three different coherences, at coherence levels of $+1$, 0, and -1 (Fig. 10.29). After several pulses and delays there can be a very large number of coherences, each of which has traveled a different coherence order pathway through the experiment. The ideal coherence pathway selection will choose only one of the coherence levels after each pulse, thus defining the events of the experiment clearly and producing a clearly interpretable result in the FID. After all, the purpose of every NMR experiment is to make the spins "dance" in a particular way that reveals to us clearly their relationships and interactions: J couplings, distances, and so on. As this information can be very complex and overlapping, it is very important to select only certain pieces of information in each experiment to make the data interpretation simple. Coherence pathway selection is the editing process by which we get clean and simple results. The coherence pathway diagram summarizes the sequence of events we are interested in for the spins according to the information we want. The task of coherence pathway selection, whether by phase cycling or by gradients, is to select this desired pathway and to block all other coherence pathways.

10.6.2 Coherence Order Pathway Selection by Phase Cycling

There is a key observation that makes coherence pathway selection possible by phase cycling. Starting with a pure $\mathbf{I}^+$ ($\mathbf{p} = 1$) coherence, consider the effect of changing the phase of a 90° pulse on the phases of the resulting coherences:

$$\begin{array}{ccccc}
 & \Delta\mathbf{p}: & 0 & -2 & -1 \\
\mathbf{I}^+ \; 90°_x \rightarrow & & \tfrac{1}{2}\mathbf{I}^+ & +\tfrac{1}{2}\mathbf{I}^- & +i\mathbf{I}_z \\
\mathbf{I}^+ \; 90°_y \rightarrow & & \tfrac{1}{2}\mathbf{I}^+ & -\tfrac{1}{2}\mathbf{I}^- & -\mathbf{I}_z \\
\mathbf{I}^+ \; 90°_{-x} \rightarrow & & \tfrac{1}{2}\mathbf{I}^+ & +\tfrac{1}{2}\mathbf{I}^- & -i\mathbf{I}_z \\
\mathbf{I}^+ \; 90°_{-y} \rightarrow & & \tfrac{1}{2}\mathbf{I}^+ & -\tfrac{1}{2}\mathbf{I}^- & +\mathbf{I}_z \\
\end{array}$$

(these can all be calculated using the definitions $\mathbf{I}^+ = \mathbf{I}_x + i\,\mathbf{I}_y$, $\mathbf{I}^- = \mathbf{I}_x - i\,\mathbf{I}_y$ and the relations $\mathbf{I}_x = 1/2\,\mathbf{I}^+ + 1/2\,\mathbf{I}^-$, $i\,\mathbf{I}_y = 1/2\,\mathbf{I}^+ - 1/2\,\mathbf{I}^-$) Note that as we increment the phase of the pulse by 90° (x, y, $-x$, $-y$), the phase factor multiplying the resulting $\mathbf{I}^+$ component ($\Delta\mathbf{p} = 0$, no change in coherence order from the original $\mathbf{I}^+$ spin state) does not change, whereas the phase of the $\mathbf{I}^-$ component ($\Delta\mathbf{p} = -2$) is shifted by 180° each time. The $\mathbf{I}_z$ (sometimes written as $\mathbf{I}^0$) component ($\Delta\mathbf{p} = -1$) is shifted in phase by 90° each time if we use the complex plane (x' = real, y' = imaginary) to represent the phase (i, -1, $-i$, 1 correspond to the y', $-x'$, $-y'$ and x' axes, respectively, in the rotating frame). In general, the effect of a change in pulse phase $\Delta\Phi_p$ on the phase of the resulting coherence Φ_c depends on the change in coherence order $\Delta\mathbf{p}$ caused by the pulse

$$\Delta\Phi_c = -\Delta\mathbf{p} \times \Delta\Phi_p$$

In the above example, $\Delta\Phi_p = 90°$ and $\Delta\Phi_c = 0°, 180°,$ and $90°$ for $\Delta\mathbf{p} = 0, -2,$ and -1, respectively.

We could select the $\mathbf{I}^+$ coherence if the experiment was repeated four times, with the signal added to the sum-to-memory each time at the receiver. The $\mathbf{I}^-$ coherence (and any coherence derived from it in a longer, more complex pulse sequence) would cancel out because we would have $1/2\,\mathbf{I}^- - 1/2\,\mathbf{I}^- + 1/2\,\mathbf{I}^- - 1/2\,\mathbf{I}^-$ for the four scans. The $\mathbf{I}_z$ part would not be observable, but even if it were converted to an observable coherence later in the pulse sequence, this coherence would carry along the phase factors i, -1, $-i$, and 1, which add together to give a zero signal after four scans. If instead we alternately add and subtract the FID signals from successive scans in the sum-to-memory, the $\mathbf{I}^+$ signal would cancel ($1/2\,\mathbf{I}^+ - 1/2\,\mathbf{I}^+ + 1/2\,\mathbf{I}^+ - 1/2\,\mathbf{I}^+$), the $\mathbf{I}^-$ signal would accumulate ($+1/2\,\mathbf{I}^- - (-1/2\,\mathbf{I}^-) + 1/2\,\mathbf{I}^- - (-1/2\,\mathbf{I}^-)$), and the $\mathbf{I}_z$ component would cancel ($+i\mathbf{I}_z - (-\mathbf{I}_z) - i\,\mathbf{I}_z - (+\mathbf{I}_z)$). The net result is that we would select the $\mathbf{I}^-$ component and destroy all other components in a four-scan phase cycle. To select the $\mathbf{I}_z$ component (or some observable coherence derived from it), we would advance the receiver phase by 90° with each successive scan. This is accomplished by adding the "real" (x' axis) signal from the ADC to the "imaginary" sum in the sum-to-memory and subtracting the "imaginary" (y' axis) signal coming from the ADC from the "real" sum in the sum-to-memory. In effect, we have rotated our reference frame by 90° because the x' axis is being detected as the y' axis and the y' axis has been moved to the $-x'$ axis. For a four-scan phase cycle with $\Delta\Phi_r$ (receiver phase change) of 90°, the signal would be routed as follows:

Scan	Sum-to-memory		Reference axis
	Real	Imag.	
1	X	Y	"x"
2	Y	−X	"y"
3	−X	−Y	"−x"
4	−Y	X	"−y"

Where X and Y are the real and imaginary FIDs coming out of the ADC. Thus, we can speak of the "receiver phase" as the point of view from which we view the FID signal in the rotating frame of reference. If the receiver phase is shifted by 90° (from x' to y' axis), we mean that the "real" channel has been shifted counterclockwise by 90° from the x' axis to the y' axis, and the "imaginary" channel has been shifted counterclockwise by 90° from the y' axis to the $-x'$ axis. With this 90° shift, for example, a coherence of $\mathbf{I}_y$ would be observed as $\mathbf{I}_x$, and in a four-scan phase cycle with $\Delta\Phi_r = 90°$, we would observe a four-scan sequence of coherences $\mathbf{I}_x, \mathbf{I}_y, -\mathbf{I}_x, -\mathbf{I}_y$ as $4\mathbf{I}_x$ in the sum-to-memory. To keep track of these phase shifts, a shorthand notation is often used where 0 stands for a 0° phase shift (real part of receiver on the x' axis), 1 stands for a 90° phase shift (receiver on the y' axis), 2 stands for a 180° phase shift (receiver on the $-x'$ axis), and 3 stands for a 270° phase shift (receiver on the $-y'$ axis). In this notation, the four-scan receiver phase cycle with $\Delta\Phi_r = 90°$ would be written: 0 1 2 3.

In general, the receiver phase should follow the phase shifts that result from a shift in the phase of a pulse

$$\Delta\Phi_r = -\Delta\mathbf{p} \times \Delta\Phi_p$$

For a series of pulses (a pulse sequence), we can select the change in coherence order $\Delta\mathbf{p}$ resulting from each of the pulses if we phase cycle all of the pulses and then calculate the effect of the desired coherence pathway on the final phase. If we diagram the coherence pathway, we can note the change in coherence order $\Delta\mathbf{p}$ caused by each pulse and then calculate the receiver phase change necessary to make the desired combination of $\Delta\mathbf{p}$'s add together at the receiver while all other pathways cancel:

$$\Delta\Phi_r = \Sigma(-\Delta\mathbf{p} \times \Delta\Phi_p)$$

where $\Delta\Phi_p$ is the phase increment for each pulse and the sum is taken over all pulses that are phase cycled. The phase cycling of pulses must be done independently for this selection to work; which means that if the first pulse has a phase shift of 180° ($x, -x$ or 0 2) and the second pulse has a phase shift of 90° ($x, y, -x, -y$ or 0 1 2 3), it would be necessary to have eight scans in our phase cycle:

Pulse 1: 0 0 0 0 2 2 2 2
Pulse 2: 0 1 2 3 0 1 2 3

How selective can we be in terms of $\Delta\mathbf{p}$? If we cycle the phase of a pulse by $360°/N$, then with N scans in the phase cycle we are effectively putting up a "mask" that permits every Nth value of $\Delta\mathbf{p}$ to get through the mask. For example, if $N = 2$ (pulse phase $x, -x$ or 0 2), and we set the receiver phase to $\Delta\Phi_r = -\Delta\mathbf{p} \times \Delta\Phi_p = -(-1)(180°) = -180° = 180°$, we would permit $\Delta\mathbf{p} = -1$ as planned, but we would also allow $\Delta\mathbf{p} = -3, -5, -7$ and so on as well as $\Delta \mathbf{p} = 1, 3, 5$, and so on. That is, the "holes" in our mask are spaced every second value of $\Delta \mathbf{p}$. To block $\Delta \mathbf{p} = 1$ and permit $\Delta \mathbf{p} = -1$, we would need a four-scan phase cycle ($N = 4$, $\Delta \Phi_p = 90°$) and the "mask" would have holes every fourth value of $\Delta \mathbf{p}$, allowing $\Delta \mathbf{p} = -5, -1, 3$, and so on. This is a much more effective phase cycle, but it takes twice as many scans to complete it:

Pulse: 0 1 2 3
Receiver: 0 1 2 3

To summarize, the strategy for phase cycling is to select pulses at crucial points in the pulse sequence where the coherence order change $\Delta \mathbf{p}$ is different for the desired and undesired coherence pathways. Then decide how selective the mask must be for each pulse (N-fold mask) and construct a phase cycle so that all of the selected pulses are independently stepped through their N-fold phase progressions. Finally, the receiver phase is calculated by adding the phase shifts for the desired coherence pathway that would result from the phase cycle you have constructed for the pulses, using the rule

$$\Delta \Phi_r = \Sigma(-\Delta \mathbf{p} \times \Delta \Phi_p)$$

The effect of setting the receiver phase cycle is to *position* each N-fold mask so that it passes the desired coherence order change $\Delta \mathbf{p}$ resulting from that pulse.

As an example, consider the 2D DQF-COSY experiment. Except for the length of the delay between pulse 2 and pulse 3, the NOESY pulse sequence is identical to the DQF-COSY pulse sequence, so that in this case it is *only* the coherence pathway selection that makes it a DQF-COSY experiment! The coherence order diagram is shown in Figure 10.30, with the undesired NOESY pathway diagramed with dotted lines. The convention for 2D experiments is to always have the observed coherence in F_2 (observed in the FID) be negative and the observed coherence in F_1 (during the t_1 delay) be positive. This is called the "echo" pathway because we have opposite sign of coherence in the two dimensions. The desired pathway has $\Delta \mathbf{p} = 1$ for the first pulse, 1 or -3 for the second and -3 or 1 for the third, whereas the NOESY path has $\Delta \mathbf{p} = 1, -1$, and -1 for the same pulses. We saw in detail how phase cycling the final pulse of the DQF-COSY experiment ($N = 4$) kills everything except DQC ($\mathbf{p} = 2$ or -2) existing between pulse 2 and pulse 3. In that case we used $\Delta \Phi_p = +90°$ and $N = 4$ ($x, y, -x, -y$). As we want to select $\Delta \mathbf{p} = -3$ or $\Delta \mathbf{p} = +1$ for that pulse, we need to cycle the receiver as follows:

$$\Delta \Phi_r = -\Delta \mathbf{p} \times \Delta \Phi_p = -(-3)(90°) = 270° \quad \text{or} \quad -(+1)(90°) = -90°$$

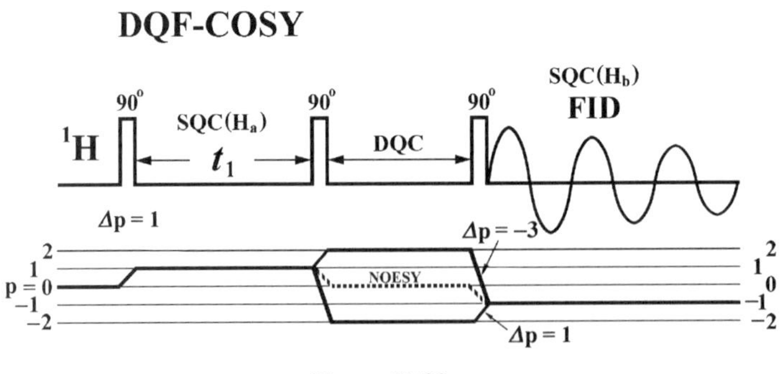

DQF-COSY

Figure 10.30

We see that these are the same, corresponding to a receiver phase increment of -1 or $+3$ in the 0, 1, 2, 3 system:

$$\text{Pulse 3:} \quad 0\ 1\ 2\ 3 = x,\ y,\ -x,\ -y$$
$$\text{Receiver:} \quad 0\ 3\ 2\ 1 = x,\ -y,\ -x,\ y$$

Moving backward from 0 you go to 3 ($270° = -90°$). Every fourth coherence change is let through: $\Delta \mathbf{p} = -7, -3, 1, 5$, and so on. The NOESY pathway ($-\mathbf{I}_z^a cc'$ term from the homonuclear "front end") goes from $\mathbf{p} = 0$ to $\mathbf{p} = -1$, with $\Delta \mathbf{p} = -1$ in the final pulse. This is blocked by the fourfold mask of the phase cycle ($\Delta \mathbf{p} = -3$ or $+1$). The ZQC part of the ZQ/DQ term is also blocked because it has the same $\Delta \mathbf{p}$ as the NOESY pathway ($\mathbf{p} = 0$ to $\mathbf{p} = -1$) in the final pulse. The COSY crosspeak and diagonal terms ($+\mathbf{I}_x^a sc' - 2\mathbf{I}_y^b \mathbf{I}_z^a ss'$) are SQC ($\mathbf{p} = 1$ or -1) and would have to undergo a change of $\Delta \mathbf{p} = 0$ or $\Delta \mathbf{p} = -2$ ($\mathbf{p} = -1$ to -1 or $\mathbf{p} = +1$ to -1) in order to be observed in the FID. These are also blocked by the phase cycle, which allows only $\Delta \mathbf{p} = -3$ and $\Delta \mathbf{p} = +1$.

To select the NOESY pathway, we use exactly the same pulse sequence but change the phase cycle of the final pulse. We want to select the pathway $\mathbf{I}_z^a \xrightarrow{90°} \mathbf{I}_a^+ \xrightarrow{90°} -\mathbf{I}_z^a \xrightarrow{\tau_m} \mathbf{I}_z^b \xrightarrow{90°} \mathbf{I}_b^-$, which has $\Delta \mathbf{p} = -1$ for the final pulse (Fig. 10.31). Using the same phase cycle $x, y, -x, -y$ ($\Delta \Phi = 90°$) for the final pulse, we can calculate the receiver phase increment:

$$\Delta \Phi_r = -(\Delta \mathbf{p})(\Delta \Phi_p) = -(-1)(90°) = 90°$$

So instead of *retarding* the receiver phase by 90° with each scan, we will *advance* it by 90°: $x, y, -x, -y$. This will allow $\Delta \mathbf{p}$ of $-5, -1, +3$, and so on, and reject the DQF-COSY pathway ($\Delta \mathbf{p} = -3$ or $+1$).

Better coherence pathway selection is achieved by cycling more than one of the pulses in the sequence. For a DQF-COSY sequence, we could set N to 2, 2, and 4 for the three 90° pulses, thus making the last pulse the most selective so that we can allow both $\Delta \mathbf{p} = -3$ and $\Delta \mathbf{p} = 1$ while blocking the NOESY $\Delta \mathbf{p} = -1$. So we need to cycle the first two pulses with 180° phase shifts (360°/2) and the final pulse through a 90° phase shift (360°/4). To do all of these phase shifts independently will require 16 scans ($2 \times 2 \times 4$) because it requires two steps to sample all possible 180° phase shifts and four steps to sample all possible 90° phase shifts. Because these must be sampled independently, the number of scans required

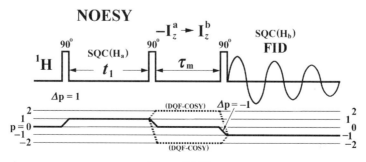

Figure 10.31

is the product of the *N*'s for each pulse. Here is one way to do it:

Pulse 1: 0 0 0 0 2 2 2 2 0 0 0 0 2 2 2 2 (*N* = 2)
Pulse 2: 0 0 0 0 0 0 0 0 2 2 2 2 2 2 2 2 (*N* = 2)
Pulse 3: 0 1 2 3 0 1 2 3 0 1 2 3 0 1 2 3 (*N* = 4)

where 0, 1, 2, and 3 correspond to phase shifts of 0°, 90°, 180°, and 270°, placing the **B**$_1$ vector of the pulse on the x', y', $-x'$, and $-y'$ axes of the rotating frame of reference, respectively. Note that the goal of quadrature image elimination is also accomplished as the final pulse is cycled through all four axes in the rotating frame. If we acquire 16 scans for each FID in the 2D experiment, we will accomplish the full selectivity of the phase cycle. If we have a high concentration of sample so that we do not need 16 scans for signal averaging, we could use eight or 4 scans per FID. If we use only eight scans, the selectivity of pulse 2 is eliminated; with only four scans per FID, we lose the selectivity of the pulse 1 and have only the "mask" created by pulse 3. Clearly, the decision to cycle pulse 3 more rapidly (0 1 2 3...) than the others means that its selection is more important than the others because it will be present even for the minimum number of scans (4).

The next step is to calculate the receiver phases. Each time we change the phase of a pulse, the receiver phase must be advanced by the appropriate amount so that the desired coherence (resulting from the selected coherence level change Δ**p**) is summed in the receiver and not canceled. For more than one pulse, we can add together the phase shifts required at the receiver according to

$$\Delta \Phi_r = \Sigma(-\Delta \mathbf{p} \times \Delta \Phi_p)$$

The easiest way to do this is to calculate the receiver phase Φ_r for each pulse independently and then add these receiver phase shifts together. For the first pulse, Δ**p** is +1 and for the second pulse it is either +1 or −3 (Fig. 10.30), whereas in both cases $\Delta\Phi_p$ is 180° (*N* = 2). We calculate $\Delta\Phi_r = -(1) \times 180° = 180°$ for Δ**p** = 1 and $\Delta\Phi_r = -(-3) \times 180° = 540°$, which is the same as 180° for Δ**p** = −3. Thus, every change of 180° in pulse phase results in a change of 180° in the desired receiver phase. For the third pulse, Δ**p** is −3 (or +1) and $\Delta\Phi_p$ is 90° (0 1 2 3), so $\Delta\Phi_r = -(-3) \times 90° = 270° = -90°$. So every time we advance the phase of the third pulse by 90°, we must also retard the phase of the receiver by 90°. Considering each pulse individually, the receiver phases are as follows:

Pulse phase	Receiver phase
Pulse 1: 0 0 0 0 2 2 2 2 0 0 0 0 2 2 2 2 →	0 0 0 0 2 2 2 2 0 0 0 0 2 2 2 2
Pulse 2: 0 0 0 0 0 0 0 0 2 2 2 2 2 2 2 2 →	0 0 0 0 0 0 0 0 2 2 2 2 2 2 2 2
Pulse 3: 0 1 2 3 0 1 2 3 0 1 2 3 0 1 2 3 →	0 3 2 1 0 3 2 1 0 3 2 1 0 3 2 1
Φ_r (total):	0 3 2 1 2 5 4 3 2 5 4 3 4 7 6 5
Φ_r (corrected):	0 3 2 1 2 1 0 3 2 1 0 3 0 3 2 1

The receiver phase is obtained simply by adding the phase shifts for the three pulses in each column. The total receiver phase can be simplified as we can subtract 360° from any phase greater than 360° to obtain a value between 0° and 360°. Thus, a phase of "4" is really 0, "5" is really 1, "6" is really 2, and "7" is really 3. If the receiver phases are calculated wrong, all or part of the desired signal will be eliminated by cancelation in the sum-to-memory. I have more than once experienced the unhappy result when a long 2D experiment results in a perfect 2D spectrum of noise! In this case, the elimination of artifacts is so efficient that even the desired signals are removed.

Now consider the 2D NOESY experiment. If we use the same selectivity for the three pulses, we could use the same pulse phases as we used for the DQF-COSY experiment, and only the receiver phases will be different:

Pulse phase	Receiver phase
Pulse 1: 0 0 0 0 2 2 2 2 0 0 0 0 2 2 2 2 →	0 0 0 0 2 2 2 2 0 0 0 0 2 2 2 2
Pulse 2: 0 0 0 0 0 0 0 0 2 2 2 2 2 2 2 2 →	0 0 0 0 0 0 0 0 2 2 2 2 2 2 2 2
Pulse 3: 0 1 2 3 0 1 2 3 0 1 2 3 0 1 2 3 →	0 1 2 3 0 1 2 3 0 1 2 3 0 1 2 3
Φ_r (total):	0 1 2 3 2 3 4 5 2 3 4 5 4 5 6 7
Φ_r (corrected):	0 1 2 3 2 3 0 1 2 3 0 1 0 1 2 3

The first two pulses give the same result: pulse phase of 0 2 and receiver phase of 0 2, selecting $\Delta \mathbf{p} = +1, -1, -3$, and so on. For the third pulse, $\Delta \mathbf{p}$ is -1 and $\Delta \Phi_p$ is $90°$, so $\Delta \Phi_r = -(-1) \times 90° = 90°$. So every time we advance the phase of the third pulse by $90°$, we must also *advance* the phase of the receiver by $90°$. Note that we have done exactly the same experiment as the DQF-COSY, but we have changed the receiver phase cycle slightly from 0 3 2 1 2 1 0 3 2 1 0 3 0 3 2 1 to 0 1 2 3 2 3 0 1 2 3 0 1 0 1 2 3, thus letting the NOESY signals accumulate in the sum-to-memory whereas the DQF-COSY signals cancel out.

For heteronuclear experiments, changing the phase of a pulse will only affect that part of the coherence that is sensitive to the pulse. For example, in the stepwise INEPT with $2\mathbf{I}_z\mathbf{S}_z$ as in intermediate step in coherence transfer

$$\mathbf{I}_z \xrightarrow{90° {}^1\mathrm{H}} \mathbf{I}^+ \xrightarrow{\frac{1}{2J}} \mathbf{I}^+\mathbf{S}_z \xrightarrow{90° {}^1\mathrm{H}} 2\mathbf{I}_z\mathbf{S}_z \xrightarrow{90° {}^{13}\mathrm{C}} \mathbf{S}^+\mathbf{I}_z$$

the second ^{1}H $90°$ pulse converts $\mathbf{I}^+\mathbf{S}_z$ into $2\mathbf{I}_z\mathbf{S}_z$, so that $\Delta p(\mathbf{I}) = -1$. If instead we used a simultaneous ^{1}H $90°$ pulse and ^{13}C $90°$ pulse to convert $\mathbf{I}^+\mathbf{S}_z$ directly into $\mathbf{S}^+\mathbf{I}_z$, we would have $\Delta p(\mathbf{I}) = -1$ for purposes of phase cycling the ^{1}H $90°$ pulse ($\mathbf{I}^+ \to \mathbf{I}_z$) and $\Delta p(\mathbf{S}) = +1$ for purposes of phase cycling the ^{13}C $90°$ pulse ($\mathbf{S}_z \to \mathbf{S}^+$). For calculating gradients, we "inflate" the ^{1}H coherence order by the ratio of magnetogyric ratios γ_H/γ_o, where γ_o is the magnetogyric ratio of the lowest frequency nucleus in the experiment, but for phase cycling we always consider the different nuclei separately because a pulse is delivered on a specific channel and affects only that nucleus. Thus, for $\mathbf{I} = {}^1$H, $\mathbf{I}^+$ has $\mathbf{p} = 1$, $\mathbf{I}_z$ has $\mathbf{p} = 0$, and $\mathbf{I}^-$ has $\mathbf{p} = -1$ regardless of the definition of $\mathbf{S}$ ($\mathbf{S} = {}^1$H, ^{13}C, ^{15}N, ^{31}P, etc.). Although a pulse can only affect one type of nucleus (e.g., ^{1}H or ^{13}C), we will see below that gradients affect all coherences, and each coherence is affected according to its type of coherence (SQC, DQC, ZQC, etc.) *and* its magnetogyric ratio.

10.6.3 Coherence Pathway Selection with Pulsed Field Gradients

A pulsed field gradient is a distortion of the normally homogeneous magnetic field in the region of the sample that leads to a linear gradient of B_0 along a single direction, usually the z axis, which is along the length of the NMR tube. This distortion is very small relative to B_0 and is typically turned on for a period of around 1 ms and then turned off, returning the field to its homogeneous state. During the gradient, the B_0 field depends on the position of a molecule within the NMR sample tube:

$$B_g = B_o + G_z z$$

where z is the position along the z axis. The effect of the gradient on an SQC is to produce a position-dependent phase shift or "twist" that scrambles the coherence so that its net value, averaged through the whole sample volume, is zero. This comes about because the precession frequency of an SQC is proportional to B_o:

$$\nu_o = \gamma B_o / 2\pi$$

During the gradient pulse

$$\nu(\text{gradient}) = \nu_g = \gamma(B_o + G_z z)/2\pi = \nu_o + \gamma G_z z/2\pi = \nu_o + \Delta\nu$$

For a gradient of duration τ, the phase of the coherence will have changed by

$$\Delta\Phi(\text{in radians}) = 2\pi\nu_o\tau + \gamma G_z z\tau$$

The first part, which is not position dependent, is just the chemical-shift evolution that would be expected for a delay of duration τ, but the second part is the result of the gradient. This is the position-dependent twist that results from the gradient pulse and is proportional to the gradient strength G_z and the gradient duration τ. Usually, the gradient duration is the same for all of the gradients applied in a pulse sequence, and the gradient strength G_z is varied.

To see the power of gradients in coherence pathway selection, consider the effect of a gradient on a homonuclear DQC $I_a^+ I_b^+$. The DQC undergoes evolution during a delay according to the sum of the two precession frequencies for the two protons H_a and H_b:

$$\nu_{DQ} = \nu_a + \nu_b$$

During a gradient, B_o is changed by the gradient field $G_z z$, leading to the following double-quantum precession rate during a gradient

$$\nu_{DQ}(\text{gradient}) = \nu_a' + \nu_b' = [\nu_a + \gamma_H G_z z/2\pi] + [\nu_b + \gamma_H G_z z/2\pi]$$
$$= \nu_{DQ} + 2\gamma_H G_z z/2\pi$$

After a gradient pulse of duration τ, the phase shift for this DQC will be

$$\Delta\Phi_{DQ} = 2\pi\nu_{DQ}\tau + 2\gamma G_z z\tau$$

The first term is the evolution that would have occurred in the absence of the gradient, and the second term is the position-dependent phase twist. Note that the twist is twice as much as that experienced by a SQC in the same gradient. Thus, the coherence order (in this case **p** = 2) is encoded in the twist, and this gives us a way to select the coherence at any point during the pulse sequence by simply applying a gradient pulse.

When the same reasoning is applied to a homonuclear ZQC, we see that the gradient has no effect:

$$\nu_{ZQ}(\text{gradient}) = \nu_a' - \nu_b' = [\nu_a + \gamma G_z z/2\pi] - [\nu_b + \gamma G_z z/2\pi]$$
$$= \nu_{ZQ}$$

The effect of gradients is especially simple to understand if we consider spherical operators. A Cartesian operator, such as $\mathbf{I}_x$ or $\mathbf{I}_y$, will precess during a gradient according to

$$\mathbf{I}_x \to \mathbf{I}_x \cos(\gamma G_z z\tau) + \mathbf{I}_y \sin(\gamma G_z z\tau)$$

$$\mathbf{I}_y \to \mathbf{I}_y \cos(\gamma G_z z\tau) - \mathbf{I}_x \sin(\gamma G_z z\tau)$$

This is the position-dependent phase shift expressed in terms of the x and y components of the magnetization. Here, we assume that the $\mathbf{I}$ spin chemical shift is on-resonance, so that ordinary chemical-shift evolution can be ignored during the gradient. Now we can look at the effect of a gradient on the spherical operators by expressing them in terms of the Cartesian operators:

$$\begin{aligned}\mathbf{I}^+ = \mathbf{I}_x + i\mathbf{I}_y &\to \mathbf{I}_x \cos(\gamma G_z z\tau) + \mathbf{I}_y \sin(\gamma G_z z\tau) + i\mathbf{I}_y \cos(\gamma G_z z\tau) - i\mathbf{I}_x \sin(\gamma G_z z\tau) \\ &= \mathbf{I}_x[\cos(\gamma G_z z\tau) - i\sin(\gamma G_z z\tau)] + i\mathbf{I}_y[\cos(\gamma G_z z\tau) - i\sin(\gamma G_z z\tau)] \\ &= \mathbf{I}^+[\cos(\gamma G_z z\tau) - i\sin(\gamma G_z z\tau)] \\ &= \mathbf{I}^+ e^{-i(\gamma G_z z\tau)} \end{aligned}$$

In general, the effect of a gradient pulse on any operator is to multiply it by a phase factor that represents the position-dependent phase twist:

$$\text{Phase factor} = e^{-i(\mathbf{p}\gamma_0 G_z z\tau)}$$

where $\mathbf{p}$ is the coherence order of the operator and γ_0 is the lowest magnetogyric ratio in the spin system. For a homonuclear (^{1}H) DQC

$$\mathbf{I}_a^+ \mathbf{I}_b^+ \to \mathbf{I}_a^+ e^{-i(\gamma_H G_z z\tau)} \mathbf{I}_b^+ e^{-i(\gamma_H G_z z\tau)} = \mathbf{I}_a^+ \mathbf{I}_b^+ e^{-i(2\gamma_H G_z z\tau)}$$

Note that the phase factor fits the general pattern $e^{-i(\mathbf{p}\gamma_0 G_z z\tau)}$, with coherence order $\mathbf{p} = +2$ and $\gamma_0 = \gamma_H$. For a homonuclear ZQC:

$$\mathbf{I}_a^+ \mathbf{I}_b^- \to \mathbf{I}_a^+ e^{-i(\gamma_H G_z z\tau)} \mathbf{I}_b^- e^{+i(\gamma_H G_z z\tau)} = \mathbf{I}_a^+ \mathbf{I}_b^-$$

As expected, homonuclear ZQC is not affected by a gradient as it has coherence order $\mathbf{p} = 0$. For heteronuclear MQC, we need to consider the difference in magnetogyric ratio γ. For example, if $\mathbf{I} = {}^1$H and $\mathbf{S} = {}^{13}$C

$$\mathbf{I}^+ \mathbf{S}^+ \to \mathbf{I}^+ e^{-i(\gamma_H G_z z\tau)} \mathbf{S}^+ e^{-i(\gamma_C G_z z\tau)} = \mathbf{I}^+ \mathbf{S}^+ e^{-i((\gamma_H + \gamma_C) G_z z\tau)}$$

For heteronuclear systems, it is convenient to redefine the coherence order $\mathbf{p}$ so that it includes the relative magnetogyric ratio γ/γ_C (or γ/γ_0 in the general case):

$$\mathbf{p}_H = \gamma_H/\gamma_C = 4 \quad \text{and} \quad \mathbf{p}_C = \gamma_C/\gamma_C = 1$$

Thus, $\mathbf{I}^+$ has $\mathbf{p} = 4$, $\mathbf{I}^-$ has $\mathbf{p} = -4$, $\mathbf{S}^+$ has $\mathbf{p} = 1$ and $\mathbf{S}^-$ has $\mathbf{p} = -1$. Using these definitions,

$$\mathbf{I}^+ \mathbf{S}^+ \to \mathbf{I}^+ \mathbf{S}^+ e^{-i((\gamma_H + \gamma_C) G_z z\tau)} = \mathbf{I}^+ \mathbf{S}^+ e^{-i((\gamma_H/\gamma_C + \gamma_C/\gamma_C)\gamma_C G_z z\tau)}$$

$$= \mathbf{I}^+ \mathbf{S}^+ e^{-i((4+1)\gamma_C G_z z\tau)} = \mathbf{I}^+ \mathbf{S}^+ e^{-i(5\gamma_C G_z z\tau)}$$

This is consistent with the rule if we define $\mathbf{p} = \mathbf{p}_H + \mathbf{p}_C = 4 + 1 = 5$ for $\mathbf{I}^+\mathbf{S}^+$. Following this rule, we have $\mathbf{p} = 3$ for $\mathbf{I}^+\mathbf{S}^-$, $\mathbf{p} = -5$ for $\mathbf{I}^-\mathbf{S}^-$, and $\mathbf{p} = -3$ for $\mathbf{I}^-\mathbf{S}^+$. For nuclei other than ^{13}C, we factor out the lowest magnetogyric ratio of the two; for example, if $\mathbf{I} = {}^1$H and $\mathbf{S} = {}^{31}$P, we can define for this case, $\mathbf{p}_H = 2.5$ and $\mathbf{p}_P = 1$ so that $\mathbf{p} = 3.5$ for $\mathbf{I}^+\mathbf{S}^+$ and 1.5 for $\mathbf{I}^+\mathbf{S}^-$. This may seem sloppy because the coherence order of proton SQC depends on the context, but it is quite convenient and easy to use. In this way, we can define the effect of a gradient on any kind of coherence, and we see that the degree of twist or the tightness of the helix formed by the position-dependent phase shift is proportional to the coherence order $\mathbf{p}$ at the time of the gradient pulse.

For several gradients applied at different points during the pulse sequence, the position-dependent phase shift accumulates:

$$\text{Phase factor} = e^{-i(\mathbf{p}_1\gamma_0 G_1 z\tau)} e^{-i(\mathbf{p}_2\gamma_0 G_2 z\tau)} e^{-i(\mathbf{p}_3\gamma_0 G_3 z\tau)} \ldots$$
$$= e^{-i[(\mathbf{p}_1 G_1 + \mathbf{p}_2 G_2 + \mathbf{p}_3 G_3 \ldots)(\gamma_0 z\tau)]}$$

where $\mathbf{p}_1, \mathbf{p}_2, \mathbf{p}_3, \ldots$ are the coherence orders at the time of the gradients of strength $G_1, G_2, G_3, \ldots$, respectively, for the desired coherence pathway. At the end of the pulse sequence (beginning of the FID), we want the phase to be the same at all levels of the NMR sample so that the signals will combine and induce a coherent signal in the probe coil. This will be the case if the exponent is zero, so that the phase factor equals 1, regardless of the position (z) in the NMR tube:

$$\mathbf{p}_1 G_1 + \mathbf{p}_2 G_2 + \mathbf{p}_3 G_3 + \cdots = \Sigma(\mathbf{p}_i G_i) = 0$$

This is the rule for selecting a coherence pathway: The gradient strengths G_i are adjusted so that the sum of coherence order times gradient strength over all of the gradients is equal to zero. In this case, the last gradient will have

$$\mathbf{p}_N G_N = -(\mathbf{p}_1 G_1 + \mathbf{p}_2 G_2 + \mathbf{p}_3 G_3 + \cdots, +\mathbf{p}_{N-1} G_{N-1})$$

and will impart a helical twist exactly equal and opposite to that accumulated so far, effectively "unwinding" the position-dependent phase twist to yield an observable coherence in the FID. The artifacts will not unwind at this point because they did not follow the same coherence pathway and will not satisfy the magic formula

$$\Sigma(\mathbf{p}_i G_i) = 0$$

These coherences will have a helical phase twist in the NMR sample tube and will add to give a net signal of zero in the probe coil during the FID.

Let's apply the gradient coherence pathway selection technique to the DQF-COSY pulse sequence. Using a phase cycle to select the coherence pathway that passes through DQC between the second and third pulses, it is required that we acquire at least four transients for each FID in the 2D acquisition. If we have sufficient sample concentration to get a good signal-to-noise ratio in a single scan, the experiment takes four times as long as is necessary based purely on sensitivity considerations. Using pulsed field gradients to select the desired coherence pathway, the experiment time could be reduced by a factor of 4!

Instead of focusing on pulses as we did for phase cycling, we will focus on the delay periods between pulses where we can insert gradient pulses. First, a gradient during the

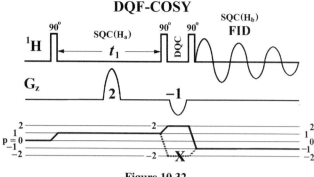

Figure 10.32

evolution (t_1) period "twists" the observable single-quantum magnetization into a helix oriented along the z axis. A second gradient is applied during the short period between the second and third pulses (Fig. 10.32) to untwist the coherence. Because the desired magnetization component during this period is DQC, the twist accumulates twice as fast as it did during the first gradient. This is due to the fact that the DQ evolution occurs at a rate that is the sum of the two resonance frequencies:

$$\Omega_{DQ} = \Omega_a + \Omega_b$$

Each of these frequencies is shifted the same amount by the gradient, so the total position-dependent change in frequency is twice as large for the DQ coherence:

$$\Delta\Omega_{DQ} = \Delta\Omega_a + \Delta\Omega_b$$

For the second gradient to undo the twisting caused by the first, it needs to be of opposite sign and half the gradient strength of the first: $G_2 = -1/2\, G_1$. For example, we could use relative gradient strengths of $G_1 = 2$ and $G_2 = -1$ (Fig. 10.32). The sensitivity to twisting by gradients is $\mathbf{p} = 1$ during t_1 (SQC) and $\mathbf{p} = 2$ during the short delay between pulses 2 and 3 (DQC). To get zero twist at the end, we use gradients G_1 and G_2 such that

$$\Sigma \mathbf{p}_i G_i = \mathbf{p}_1 G_1 + \mathbf{p}_2 G_2 = 1 \times 2 + 2 \times (-1) = 0$$

In terms of spherical operators, the DQF-COSY experiment looks like this

$$I_a^o \xrightarrow{90°} I_a^+ \xrightarrow{t_1} I_a^+ I_b^o \xrightarrow{90°} I_a^+ I_b^+ \xrightarrow{90°} I_b^- I_a^o$$

H_a magnetization starts on z (equilibrium, $\mathbf{p} = 0$) and is excited to I_a^+ (SQC, $\mathbf{p} = 1$) and I_a^- ($\mathbf{p} = -1$) by the first 90° pulse. Following just the I^+ term, during the t_1 delay we generate a mixture of I_a^+ and the antiphase term $I_a^+ I_b^o$ (still $\mathbf{p} = 1$), which is the only term that can lead to coherence transfer. The second 90° pulse generates DQC ($I_a^+ I_b^+$) with coherence order $\mathbf{p} = 2$ (along with five other undesired coherences!). The final 90° pulse "knocks down" H_a from I_a^+ ($\mathbf{p} = 1$) to I_a^o ($\mathbf{p} = 0$) and H_b from I_b^+ ($\mathbf{p} = +1$) to I_b^- ($\mathbf{p} = -1$) to generate the crosspeak $I_b^- I_a^o$. Here, we use $I^o = I_z$ for consistency with our spherical operators. In each step, we drop the phase factors (we do not bother specifying them) that accumulate because of pulse phases and evolution during delays. We are only interested in the general form of the coherence and its coherence order. This frees us from a lot of tedious bookkeeping! For example, each time an I^+ or I^- term is hit with a 90° pulse, we get three coherences: I^+, I^o, and I^- (Fig. 10.29). So a product like $I_a^+ I_b^+$ can generate nine different product operators (nine different coherences) with **p** ranging from -2 to $+2$! As long as we know the precise coherence order we want at each step, we can design the gradient strengths needed to select that "pathway" (sequence of coherences desired). We use the term "gradient selected" for any experiment where the gradients actually select the desired coherence pathway as opposed to "gradient enhanced" if they just clean up artifacts. The common feature of gradient selected experiments is the need for a precise ratio of gradient strengths (twist, untwist). In "gradient enhanced" experiments (e.g., gradient enhanced NOESY, Fig. 10.16) any old gradient strength will do because we are just "spoiling" coherences we do not want (twist, forget), rather than "selecting" the coherences we do want.

With the sequence of Figure 10.32 we would have to use magnitude mode to present the data because the gradients take up considerable time during which evolution is going on. As a typical gradient is 1–2 ms long, the t_1 period cannot start at $t_1 = 0$ but is forced to start at $t_1 = 1$ ms or $t_1 = 2$ ms, leading to large chemical-shift dependent (first-order) phase errors in F_1. Remember, it is the phase of each spin at the start of the FID that determines the phase of the peak after Fourier transformation. If we delay the start of the FID, different peaks end up with different phases because they have "fanned out" due to chemical shift differences before the FID starts. Furthermore, DQ evolution ($\Omega_a + \Omega_b$) is going on during the second gradient, leading to large phase errors in F_2. We could use magnitude mode to "band-aid" these phase errors but then we lose all the advantages of phase-sensitive 2D NMR. The solution is to construct a spin echo whenever we need some time for a gradient. The gradient fits into one of the delays of the spin echo and the other delay refocuses the evolution that goes on during the gradient. The improved sequence is shown in Figure 10.33. A spin echo is built onto the end of the evolution delay, with echo delay Δ just long enough for a gradient (plus a short delay for recovery of field homogeneity). Now the first t_1 value can truly be set to zero because no net chemical-shift evolution occurs during the spin echo. A second spin echo is built into the short delay between the second and third pulses, with the gradient "tucked in" to the second delay of the spin echo.

In terms of spherical operators, the improved DQF-COSY experiment looks like this

$$I_a^o \xrightarrow{90°} I_a^+ \xrightarrow{t_1} I_a^+ I_b^o \xrightarrow{180°} I_a^- I_b^o \xrightarrow{90°} I_a^- I_b^- \xrightarrow{180°} I_a^+ I_b^+ \xrightarrow{90°} I_b^- I_a^o$$

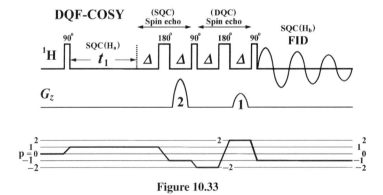

Figure 10.33

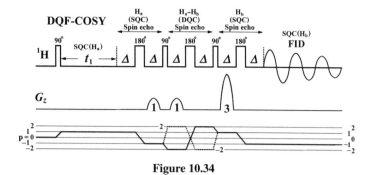

Figure 10.34

The 180° pulses in the center of the spin echoes invert the spherical operators, with I^+ becoming I^- and I^- becoming I^+. We have to take this into account when designing gradients: now $p_1 = -1$ ($I_a^- I_b^o$) and $p_2 = +2$ ($I_a^+ I_b^+$), so we need gradient strengths of $G_1 = 2$ and $G_2 = 1$ to select this coherence pathway.

We might also want to have a third gradient applied after the final 90° pulse, so that we would "cover" the SQC period (H_a), the DQC period ($H_a - H_b$), and the SQC period (H_b) just before the FID (Fig. 10.34). We will have a huge phase twist in F_2 if we simply delay starting the acquisition of the t_2 FID, so we insert a spin echo between the third 90° pulse and the FID acquisition. In this case, one solution for gradient selection is $G_1 = 1$, $G_2 = 1$, and $G_3 = 3$, selecting $p = -1$, $p = -2$, and $p = +1$, respectively. The accumulated "twist" at the end of each gradient is -1 (after G_1), -3 (after G_2), and 0 (after G_3) for the selected pathway. Note that the alternative pathway $p = -1$, $p = 2$, and $p = 1$ (Fig. 10.34, dotted line) will lead to a twisted coherence and will not be observed:

$$\Sigma(p_i G_i) = (-1 \times 1 + 2 \times 1 + 1 \times 3) = +4$$

This pathway is permitted in the phase cycled experiment, but the data from this pathway is lost in the gradient version. In this sense, gradients are sometimes *too* selective because they generally allow only one coherence level even when two or more pathways are equivalent in terms of the signal we want to see.

We have been annotating the pulse sequence to indicate what kind of evolution is going on at each stage: ν_H, $-\nu_C$, J, and so on. Another way to diagram what is happening during the pulse sequence is to show the spherical operators at each stage for the coherence we are selecting with the gradients. This is shown in Figure 10.35 for the DQF-COSY sequence with gradients and spin echoes. The best way to do this is to start at the end with the FID: We always detect negative coherence, and as we are talking about a crosspeak at $F_1 = H_a$, $F_2 = H_b$, we will be detecting H_b SQC in the FID. Because there is no refocusing in the DQF-COSY, we will start the FID with antiphase H_b coherence: $I_b^- I_a^o$ with $p = -1$. Moving backward, the 180° pulse will always convert I^- to I^+ and *vice versa*, so before the final 180° pulse we have $I_b^+ I_a^o$ with $p = +1$. Moving farther back in time, we encounter a 90° pulse. Many things happen to SQC with a 90° pulse (Fig. 10.29): The coherence order splits into three parts: it stays the same, goes to zero, or reverses sign. It is up to the gradients (or phase cycle) to select which of these three pathways is selected. A product operator with two operators such as $I_b^+ I_a^o$ can come from many different levels: I_b^+ can come from I_b^+, I_b^o, or I_b^- by a 90° pulse, and I_a^o can come from I_a^+ or I_a^-. Taken together,

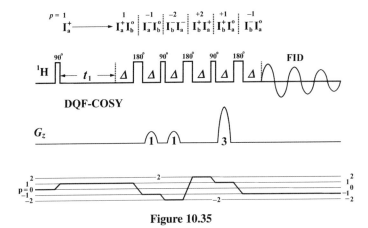

Figure 10.35

the coherence level before the 90° pulse could be $-2, -1, 0, +1,$ or $+2$! Fortunately, we know what coherence level we want to select for a DQF-COSY experiment: the coherence pathway diagram shows $\mathbf{p} = +2$ at this point, and any other level will not survive the gradients. So we know that before the 90° pulse, we have $\mathbf{I}_b^+ \mathbf{I}_a^+$ with $\mathbf{p} = 2$. From $\mathbf{I}_b^+$ the 90° pulse creates $\mathbf{I}_b^+, \mathbf{I}_b^o$, and $\mathbf{I}_b^-$, but the gradient ratio selects only $\mathbf{I}_b^+$. Likewise, from $\mathbf{I}_a^+$ we select $\mathbf{I}_a^o$ after the 90° pulse. It is important to keep in mind the big picture: DQF-COSY accomplishes coherence transfer from H_a antiphase SQC to H_b antiphase SQC by going through the obligatory intermediate state of $\{H_a, H_b\}$ DQC.

Moving back further, we come to another 180° pulse at the center of the second spin echo. This is easy: $\mathbf{I}_b^+$ comes from $\mathbf{I}_b^-$ and $\mathbf{I}_a^+$ comes from $\mathbf{I}_a^-$, so we have $\mathbf{I}_b^- \mathbf{I}_a^-$ before the 180° pulse and $\mathbf{p} = -2$. Before this we have another 90° pulse, the one that converts H_a antiphase SQC into $\{H_a, H_b\}$ DQC. $\mathbf{I}_b^-$ has to come from $\mathbf{I}_b^o$ because the starting H_a SQC is antiphase with respect to H_b, but $\mathbf{I}_a^-$ can come from either $\mathbf{I}_a^-$ or $\mathbf{I}_a^+$, both of which are H_a SQC. To decide which one, we look at the coherence pathway diagram ($\mathbf{p} = -1$) or we look further back to the evolution delay t_1, where we have to have positive coherence order for H_a SQC. This is the rule: negative coherence order during t_2 and positive coherence order during t_1 for the "echo" pathway. Because there is a 180° pulse between our state and the t_1 period, we have to have negative coherence order here: $\mathbf{I}_a^- \mathbf{I}_b^o$. We can complete the sequence by moving back before the 180° pulse: $\mathbf{I}_a^+ \mathbf{I}_b^o$ ($\mathbf{I}_z^b \rightarrow -\mathbf{I}_z^b$, but we drop the "phase factor" of -1) and then we know that during t_1 we have to get from in-phase H_a SQC to antiphase H_a SQC (i.e., selecting the $\sin(\pi J t_1)$ term) so that we can move back to $\mathbf{I}_a^+$ just after the first (preparation) pulse. The whole concept here is very different from our analysis using Cartesian ($\mathbf{I}_x, \mathbf{I}_y$, etc.) operators: We ignore the phase factors that result from evolution and we focus on the coherence order only, letting the gradients do the work of choosing the pathway we desire. As an added bonus, we can just add up the superscripts of the spherical operators at each step to obtain the coherence order: for example, $\mathbf{I}_a^- \mathbf{I}_b^o$ has $\mathbf{p} = -1 + 0 = -1$. We are definitely focusing on the big picture here, and we have come a long way from the laborious and meticulous analysis of vector rotations.

10.6.4 Quadrature Detection in F_1 Using Gradients

In Chapter 9, we saw that phase sensitive 2D NMR requires that we change the phase of the coherence observed in t_1 (Fig. 9.47) using either the States technique (analogous

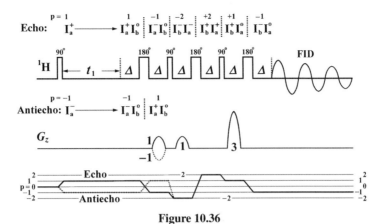

Figure 10.36

to simultaneous sampling of real and imaginary channels in t_2) or the TPPI method (analogous to alternate sampling in t_2). In either case, the phase of the t_1 FID is changed by changing the phase of the preparation pulse, the pulse just before the t_1 delay, for each new FID in the 2D acquisition. With the advent of coherence pathway selection by gradients, it was discovered that the same phase shifting can be obtained by merely changing the sign of the gradient that selects coherence order during t_1. If we select positive coherence order during t_1 (Fig. 10.36, top, and solid lines), we have the so-called "echo" pathway: positive coherence order during t_1 and negative coherence order during t_2. The term "echo" refers to the reversal of sign of coherence order that occurs in the second half of a spin echo. If we select negative coherence order during t_1 (Fig. 10.36, center, and dotted lines), we have the "antiecho" pathway: negative coherence order during both t_1 and t_2. The method ("echo–antiecho") is similar to the States method—for each t_1 value, two FIDs are acquired: one selecting the echo pathway ($G_1 = 1$) and one selecting the antiecho pathway ($G_1 = -1$). The t_1 delay is then incremented by $2\,\Delta t_1 = 1/sw(F_1)$ and the process is repeated, acquiring two more FIDs.

The data are processed in a different way, combining the echo and antiecho FIDs for each t_1 value to regenerate the real and imaginary (cosine modulated and sine modulated) FIDs. Then the data is processed just like States data. How this is done can be seen if we examine the phase factors that result from evolution during t_1

$$I_a^+ \to I_a^+ I_b^o e^{-i\Omega_a t_1} \sin(\pi J t_1) \quad \text{echo}$$
$$I_a^- \to I_a^- I_b^o e^{i\Omega_a t_1} \sin(\pi J t_1) \quad \text{antiecho}$$

The same phase factors attach themselves to the final spin state observed at the start of the FID:

$$I_b^- I_a^o e^{-i\Omega_a t_1} \sin(\pi J t_1) \quad \text{echo}$$
$$I_b^- I_a^o e^{i\Omega_a t_1} \sin(\pi J t_1) \quad \text{antiecho}$$

Now consider the signal recorded during t_2, ignoring the J-coupling evolution for simplicity:

$$E(t_1, t_2) = e^{-i\Omega_a t_1} e^{i\Omega_b t_2} \quad \text{echo FID}$$
$$A(t_1, t_2) = e^{i\Omega_a t_1} e^{i\Omega_b t_2} \quad \text{antiecho FID}$$

The chemical-shift evolution during the FID is taken care of by the exponential term in $\Omega_b\, t_2$, with a positive exponential because it is the $\mathbf{I}_b^-$ that is evolving. In this complex arithmetic, the real part corresponds to the real FID in t_2 (M_x component in the rotating frame) and the imaginary part is the imaginary FID in t_2 (M_y component). We can substitute sines and cosines for the imaginary exponentials as

$$E(t_1, t_2) = [\cos(\Omega_a t_1) - \mathbf{i}\sin(\Omega_a t_1)][\cos(\Omega_b t_2) + \mathbf{i}\sin(\Omega_b t_2)] \quad \text{echo FID}$$
$$A(t_1, t_2) = [\cos(\Omega_a t_1) + \mathbf{i}\sin(\Omega_a t_1)][\cos(\Omega_b t_2) + \mathbf{i}\sin(\Omega_b t_2)] \quad \text{antiecho FID}$$

Multiplying and substituting -1 for $\mathbf{i}^2$

	Real part		Imaginary part	
$E(t_1, t_2) =$	$[\cos\cos' + \sin\sin']$	$+$	$\mathbf{i}\,[\cos\sin' - \sin\cos']$	echo FID
$A(t_1, t_2) =$	$[\cos\cos' - \sin\sin']$	$+$	$\mathbf{i}\,[\cos\sin' + \sin\cos']$	antiecho FID

where sin and cos have the argument $\Omega_a t_1$ and sin' and cos' have the argument $\Omega_b t_2$. Now we combine the data from these two FIDs, swapping around the real and imaginary FIDs that are stored in the computer:

	Real part		Imaginary part	
$R(t_1, t_2) =$	$[-\text{Im}(A) - \text{Im}(E)]$	$+$	$\mathbf{i}\,[\text{Re}(A) + \text{Re}(E)]$	real FID
$I(t_1, t_2) =$	$[\text{Re}(A) - \text{Re}(E)]$	$+$	$\mathbf{i}\,[\text{Im}(A) - \text{Im}(E)]$	imaginary FID

where Im(A) is the imaginary part of the antiecho FID, Re(E) is the real part of the echo FID, and so on. Plugging in the real and imaginary parts from the echo and antiecho FIDs above

	Real part		Imaginary part	
$R(t_1, t_2) =$	$[-\cos(\Omega_a t_1)\sin(\Omega_b t_2)]$	$+$	$\mathbf{i}\,[\cos(\Omega_a t_1)\cos(\Omega_b t_2)]$	real FID
	$= [-\sin(\Omega_b t_2)$	$+$	$\mathbf{i}\cos(\Omega_b t_2)] \times \cos(\Omega_a t_1)$	
$I(t_1, t_2) =$	$[-\sin(\Omega_a t_1)\sin(\Omega_b t_2)]$	$+$	$\mathbf{i}\,[\sin(\Omega_a t_1)\cos(\Omega_b t_2)]$	imaginary FID
	$= [-\sin(\Omega_b t_2)$	$+$	$\mathbf{i}\cos(\Omega_b t_2)] \times \sin(\Omega_a t_1)$	

Now we have exactly the same kind of data we have with States mode acquisition: cosine modulation in t_1 for the real FID and sine modulation in t_1 for the imaginary FID. In t_2 we can equate M_x with the real part and M_y with the imaginary part, so that in each case we have $M_x = -\sin(\Omega_b t_2)$ and $M_y = \cos(\Omega_b t_2)$. This represents a vector starting on the y' axis at $t_2 = 0$ and rotating counter clockwise (positive offset Ω) at the rate of Ω_b rad s^{-1}. Once these rearrangements have been made in the computer, the data is processed just like States data, with a complex Fourier transform in t_1.

10.6.5 Disadvantages of Phase Cycling Compared to Gradient Selection

Phase cycling operates by acquiring both the desired signals and the artifact signals in the FID. Then by accumulating a number of FIDs in the sum-to-memory of the spectrometer, the desired signals add together and the artifact signals subtract and cancel. Gradients destroy the artifacts by twisting their coherences spatially in the NMR sample. These artifact coherences never reach the receiver because their vector sum is zero at the probe coil. This accounts for the fundamental advantage of gradient pathway selection: it does not depend on mathematical subtraction of signals that have already been digitized; the subtraction occurs in the NMR sample. Thus phase cycling as a method suffers from a number of drawbacks:

1. *Many Scans are Required for Each FID.* To apply an N-fold mask at each crucial pulse (P1, P2, P3, ...) in the sequence requires a phase cycle of $N_1 \times N_2 \times N_3 \times \cdots$ scans to completely cancel the blocked pathways. If the sample concentration is low, you might need 32 or 64 scans per FID in a 2D experiment just to get enough signal-to-noise ratio. In that case, a long phase cycle is not a disadvantage. But if you have a high sample concentration, as is often the case in small-molecule NMR, a single scan per FID might give you sufficient signal-to-noise in a 2D experiment. In that case, every additional scan required by the phase cycle will greatly increase the experiment time. For example, a 16 scan phase cycle will multiply the experiment time by 16 for a concentrated sample, making an 8 min experiment (with gradient selection) into a more than 2 h acquisition. This is the area where gradients give the most dramatic advantages: in time savings for 2D experiments on concentrated samples (10 mg or more of an organic molecule).

2. *The Receiver Gain Must be Reduced to Avoid ADC Overflow From Artifact Signals.* In many cases, the artifact signals are many times more intense than the desired signals. The large artifact signal would overflow the ADC if we set the receiver gain to be appropriate for the desired signal, so we have to reduce the gain by a large factor to make the FID "fit" in the input of the ADC. Since some of the noise observed in the digitized FID originates in the later stages of amplification, this means we are adding a smaller (less amplified) signal to this noise and therefore our signal-to-noise ratio is reduced as we decrease the receiver gain. Amplification is good! We do not want to turn it down, but we have to because of these large artifact signals. In a gradient-selected experiment, the artifacts never appear in the FID signal and the receiver gain can be dramatically increased, leading to higher sensitivity.

3. *Dynamic Range Problems.* Dynamic range is the complete range of signal intensities that can be observed in an NMR spectrum. The largest signal will almost fill the ADC if the receiver gain is adjusted properly, and the smallest detectable signal is then limited by the accuracy (number of bits) of the ADC, since no signal smaller than a single bit can be digitized. If the artifact signal is 100 times larger than the desired signal, then we are wasting 6.6 ($2^{6.6} = 100$) bits of our 16-bit ADC digitizing a signal that will simply be canceled out in the sum-to-memory by the phase cycle. Now the smallest signal that can be digitized is only 328 times smaller than the largest desired signal, rather than 32,768 times smaller. In a gradient-selected experiment, the entire 16 bits of the digitizer are used to measure the desired signal, since the artifact never appears in the receiver.

4. *Subtraction Artifacts.* Any time you cancel an artifact signal by subtraction, you are assuming that the signal is exactly the same with each scan so that perfect cancellation will occur. If the frequency (chemical-shift position) of the signal changes very slightly, subtraction will lead to a large "dispersive" artifact because the negative (subtracted) peak is offset slightly to the left or right of the positive peak. If the intensity is not exactly equal, there will be a residual positive or negative peak. Many things can change slightly in a spectrometer from one scan to the next: temperature

may be varying slightly; pulse frequency, intensity, or phase may not be perfectly reproducible; and vibration from air flow, building equipment, trains, and so on may reach the probe. Spinning of the sample is also a source of irreproducibility—for this reason, most experiments with a critical phase cycle are run without sample spinning. In a gradient experiment there is no subtraction of two or more separate measurements so none of these sources of irreproducibility can lead to artifacts.

10.6.6 Disadvantages of Gradient Selection

You might think that gradients are an endlessly beneficial technology, but in fact there are a few minor disadvantages of gradient coherence pathway selection:

1. *Cost of Gradient Controllers, Amplifiers and Probes.* Gradient technology is not inexpensive, and is simply not available on older NMR spectrometers. The hardware required consists of a gradient controller (digital timing control plus signal generation), a gradient amplifier (very stable source of large currents that can be accurately controlled) and a gradient probe that has gradient coils surrounding the active volume of the sample. The currents produced by the gradient amplifier run through these coils and produce the magnetic fields that add a gradient to the B_o field. In spite of this additional cost, the advantages of gradients are so powerful even for routine work that nearly all new NMR spectrometers are now purchased with gradient capability.

2. *Diffusion.* The primary assumption of the gradient approach to pathway selection is that a spin will be in the same physical location in the sample tube throughout the pulse sequence. Each spin receives a position-dependent phase shift with each gradient pulse, and if the spin changes its position it will not receive the correct "unwinding" B_o field in the last gradient. The distance between positive and negative phase in the "twisted coherence" can be on the order of microns (1 μm = 10^{-6} m, smaller than a typical eukaryotic cell) so that diffusion will occur, especially for small molecules, and will lead to a reduced signal. This can be put to use to distinguish small molecule signals from large molecule signals, but in general it means that there will be some loss of sensitivity. If you look closely at gradient-selected pulse sequences, you will see that there is an attempt to place gradients as close to each other in time as possible to minimize the distance traveled as a result of molecular diffusion in solution.

3. *Sensitivity Loss Due to T_2 Relaxation.* A typical gradient requires about 1 ms of time, usually followed by 200 μs of recovery time to allow the field homogeneity to be reestablished. Unless the gradient can be placed in a fixed delay period (e.g., $1/(4J)$ in an INEPT), a spin echo will be required to refocus any evolution that occurs during the gradient, doubling the total time required for a gradient. For small molecules an additional delay of 2.5 ms per gradient does not lead to a large signal loss since T_2 is relatively long. For large biological molecules (proteins and nucleic acids), however, or for paramagnetic molecules, the T_2 values can be quite short and the extra delays may be intolerable because of the drastic loss of signal. In these special cases it may be necessary to use the old phase-cycled experiments or to use shorter, stronger gradient pulses without refocusing delays.

4. *Loss of Sensitivity Due to Overselectivity.* Gradient selection means that only a single coherence level can be present at the time of each gradient. With phase cycling we apply a mask with "holes" at regular intervals, so that more than one coherence level

can be permitted. For example, in the DQF-COSY experiment with phase cycling we can have double-quantum coherence between the second and third pulses with coherence order +2 or −2. Both pathways are preserved and add together in the sum-to-memory since the mask ($N = 4$) used in the final pulse allows both $\Delta \mathbf{p} = -3$ and $\Delta \mathbf{p} = +1$. With gradient selection, we can allow $\mathbf{p} = 2$ during the DQ filter or we can allow $\mathbf{p} = -2$, but we cannot permit both pathways. Thus there is a loss of sensitivity in the gradient-selected experiment corresponding to a factor of $\sqrt{2}$.

10.7 THE DENSITY MATRIX REPRESENTATION OF SPIN STATES

The density matrix is based on the description of all NMR spin states as linear combinations or superpositions of the basic spin states or energy levels. For a system of identical spins, there are only two spin states: α and β. At equilibrium, there is a slight excess of population in the lower energy (α) state and a slightly depleted population in the higher energy (β) state. This can be represented as a 2×2 matrix with the diagonal elements corresponding to the populations of the α and β states:

$$\sigma_{eq} = \begin{bmatrix} (N/2)(1+\varepsilon) & 0 \\ 0 & (N/2)(1-\varepsilon) \end{bmatrix} \begin{matrix} \alpha \\ \beta \end{matrix}$$

where N is the total number of identical spins and ε ($<<1$) is a very small dimensionless number. This equilibrium state is broken down into two parts: the equal populations that play no role in NMR (identity matrix) and the population differences. The identity matrix and the factor $N\varepsilon/2$ will be ignored from now on since they will be the same in all spin states.

$$\sigma_{eq} = N/2 \left[\begin{bmatrix} 1 & 0 \\ 0 & 1 \end{bmatrix} + \varepsilon \begin{bmatrix} 1 & 0 \\ 0 & -1 \end{bmatrix} \right]$$

The effect of a radio frequency pulse is to "mix" the α and β states so that a superposition of states is obtained:

$$\Psi = c_1 \alpha + c_2 \beta \quad \text{"wave function"}$$

For a single spin, the probability of finding it in a given "pure" energy state is found by multiplying the coefficient of that state by its complex conjugate: the probability of finding the spin in the α state is thus $c_1 \cdot c_1^*$, and the probability that the spin will be in the β state is $c_2 \cdot c_2^*$. Since a spin can only be in either the α or the β state, the sum of these two probabilities ($c_1 \cdot c_1^* + c_2 \cdot c_2^*$) is 1 (a certainty that it will be in one state or the other). When these probabilities are averaged over the entire population of spins we have the populations of the α and β states. The "cross products" $c_1 \cdot c_2^*$ and $c_2 \cdot c_1^*$ represent the amount of mixing or superposition of the α and β states. When these are averaged over the entire population of spins we have the single-quantum *coherence* of the sample, which is the measurable magnetization that we represent as a vector in the x'–y' plane in the vector model. The state of the entire system can be described by the averages of these four products over the whole sample, and these are conveniently represented as a matrix:

$$\sigma = \begin{bmatrix} (c_1c_1^*)_{av} & (c_1c_2^*)_{av} \\ (c_2c_1^*)_{av} & (c_2c_2^*)_{av} \end{bmatrix} \begin{matrix} \alpha \\ \beta \end{matrix}$$

The diagonal terms will always be real numbers, and the $\alpha \rightarrow \beta$ term $((c_1c_2^*)_{av})$ will always be the complex conjugate of the $\beta \rightarrow \alpha$ term $((c_2c_1^*)_{av})$. Now we can show the density matrix representations of the three single-spin product operators:

$$\mathbf{I}_x = \tfrac{1}{2}\begin{bmatrix} 0 & 1 \\ 1 & 0 \end{bmatrix} \quad \mathbf{I}_y = \tfrac{1}{2}\begin{bmatrix} 0 & -i \\ i & 0 \end{bmatrix} \quad \mathbf{I}_z = \tfrac{1}{2}\begin{bmatrix} 1 & 0 \\ 0 & -1 \end{bmatrix}$$

In all cases, the identity matrix and the factors $N/2$ and ε have been omitted for simplicity. Note that the distinction between the x' and y' axes is made by using real and imaginary numbers for the off-diagonal terms. This is similar to the use of "real" and "imaginary" to represent the two parts of the FID signal (M_x and M_y) in the NMR spectrometer. A pulse can be represented by a matrix as well, so that the effect of any pulse can be calculated by "simple" matrix multiplication. For a general pulse of pulse angle Θ, the rotation matrices are

$$R_x(\Theta) = \begin{bmatrix} \cos(\Theta/2) & i\sin(\Theta/2) \\ i\sin(\Theta/2) & \cos(\Theta/2) \end{bmatrix} \quad R_y(\Theta) = \begin{bmatrix} \cos(\Theta/2) & \sin(\Theta/2) \\ -\sin(\Theta/2) & \cos(\Theta/2) \end{bmatrix}$$

For the 90° pulses this becomes

$$R_x(90°) = \tfrac{1}{\sqrt{2}}\begin{bmatrix} 1 & i \\ i & 1 \end{bmatrix} \quad R_y(90°) = \tfrac{1}{\sqrt{2}}\begin{bmatrix} 1 & 1 \\ -1 & 1 \end{bmatrix}$$

And for the 180° pulses

$$R_x(180°) = \begin{bmatrix} 0 & i \\ i & 0 \end{bmatrix} \quad R_y(180°) = \begin{bmatrix} 0 & 1 \\ -1 & 0 \end{bmatrix}$$

For example, starting from equilibrium ($\mathbf{I}_z$), the effect of a 90° pulse on the y axis is

$$\sigma = (R_y^{90})^{-1} \sigma_{eq} R_y^{90} = \tfrac{1}{\sqrt{2}}\begin{bmatrix} 1 & -1 \\ 1 & 1 \end{bmatrix} \times \tfrac{1}{2}\begin{bmatrix} 1 & 0 \\ 0 & -1 \end{bmatrix} \times \tfrac{1}{\sqrt{2}}\begin{bmatrix} 1 & 1 \\ -1 & 1 \end{bmatrix}$$

$$\sigma = (R_y^{90})^{-1} \sigma_{eq} R_y^{90} = \underbrace{\tfrac{1}{\sqrt{2}}\begin{bmatrix} 1 & -1 \\ 1 & 1 \end{bmatrix}}_{A} \times \underbrace{\tfrac{1}{2\sqrt{2}}\begin{bmatrix} 1 & 1 \\ 1 & -1 \end{bmatrix}}_{B} = \underbrace{\tfrac{1}{4}\begin{bmatrix} 0 & 2 \\ 2 & 0 \end{bmatrix}}_{C} = \tfrac{1}{2}\begin{bmatrix} 0 & 1 \\ 1 & 0 \end{bmatrix} = \mathbf{I}_x$$

The pulse is applied mathematically by multiplying the spin state matrix σ by the rotation matrix R and then multiplying this result by the inverse matrix R^{-1} (the product $R^{-1}R$ is the identity matrix $\mathbf{1}$). For rotation (pulse) operators, the inverse matrix is simply the rotation in the opposite direction ($\Theta' = -\Theta$). Note that the final result is the same as the representation of the product operator $\mathbf{I}_x$ given above.

Matrix multiplication involves forming a sum of products of matrix elements. For example, in the final multiplication above, the upper right-hand element of the product matrix

(C) is formed from the sum $(1 \times 1 + (-1) \times (-1)) = 2$. The first 1 and -1 came from the first row of the left-hand matrix (A) and the second 1 and -1 came from the second column of the second matrix (B). The 2 was then factored out of the entire product matrix to change the 1/4 to a 1/2. In general, the element in the ith row and the jth column of the product matrix is formed from the ith row of the first matrix and the jth column of the second matrix by forming a sum of products of elements: $a_{i1}b_{1j} + a_{i2}b_{2j} + \cdots + a_{in}b_{nj}$, where a_{i1} is the first element in the ith row of the first matrix and b_{1j} is the first element in the jth column of the second matrix. Matrix multiplication does not generally give the same result if the order of the two matrices being multiplied is reversed; for this reason the order is important. The identity matrix (**1**) is a square n × n matrix with a 1 for each diagonal element $\mathbf{1}_{ii}$ and a zero for each off-diagonal element $\mathbf{1}_{ij}$ ($i \neq j$).

What is the effect of a time delay on the density matrix? Each off-diagonal element gets multiplied by a *phase factor* that depends on the length of time of the delay and the energy difference between the two energy levels that are connected by that transition:

$$\text{phase factor} = e^{i\omega t}, \text{ where } \omega = 2\pi \nu_o$$

The exponent of an imaginary number is a shorthand for two trigonometric functions:

$$e^{i\omega t} = \cos \omega t + i \sin \omega t$$

Note that ω is just the Larmor frequency and, because real numbers are associated with the x axis and imaginary numbers with the y axis, time evolution is simply rotation in the x–y plane at the offset frequency. For double-quantum transitions, $\omega = \omega_I + \omega_S$, and for zero-quantum transitions $\omega = \omega_I - \omega_S$. For example, a 90° pulse on the y' axis followed by a delay τ would give

$$\frac{1}{2}\begin{bmatrix} 1 & 0 \\ 0 & -1 \end{bmatrix} \xrightarrow{90°_y} \frac{1}{2}\begin{bmatrix} 0 & 1 \\ 1 & 0 \end{bmatrix} \xrightarrow{\tau \text{ delay}} \frac{1}{2}\begin{bmatrix} 0 & e^{-i\omega\tau} \\ e^{i\omega\tau} & 0 \end{bmatrix}$$

Matrix elements above the diagonal get the phase factor $e^{-i\omega t}$ because their energy difference is opposite in sign (downward transition). The factor $e^{i\omega t}$ is 1, **i**, -1, $-\mathbf{i}$, and 1 for $\omega \tau = 0$, $\pi/2$, π, $3\pi/2$, and 2π. It can be represented as a vector in the complex plane of unit length rotating with angular velocity ω. The phase factor $e^{-i\omega t}$ rotates in the opposite direction with values of 1, $-\mathbf{i}$, -1, $\mathbf{i}$, and 1 for $\omega \tau = 0$, $\pi/2$, π, $3\pi/2$, and 2π. Thus the final matrix would be

$$\omega t = \quad 0 \qquad\qquad \pi/2 \qquad\qquad \pi \qquad\qquad 3\pi/2$$

$$\underset{\mathbf{I}_x}{\tfrac{1}{2}\begin{bmatrix} 0 & 1 \\ 1 & 0 \end{bmatrix}} \to \underset{\mathbf{I}_y}{\tfrac{1}{2}\begin{bmatrix} 0 & -i \\ i & 0 \end{bmatrix}} \to \underset{-\mathbf{I}_x}{\tfrac{1}{2}\begin{bmatrix} 0 & -1 \\ -1 & 0 \end{bmatrix}} \to \underset{-\mathbf{I}_y}{\tfrac{1}{2}\begin{bmatrix} 0 & i \\ -i & 0 \end{bmatrix}}$$

depending on the value of $\omega\tau$. Note that these matrices correspond to the product operators $\mathbf{I}_x, \mathbf{I}_y, -\mathbf{I}_x$, and $-\mathbf{I}_y$, which is the expected progression for chemical-shift evolution. If you focus on the numbers at the lower left of each matrix you can "read" the product operators if you associate 1 with x and **i** with y. This is a lot of work to carry out the equivalent of $\mathbf{I}_z \xrightarrow{90°_y} \mathbf{I}_x \xrightarrow{(\tau = \pi/(2\omega))} \mathbf{I}_y$, but properly programed computers just love this sort of thing and have no trouble keeping track of it all.

For a two-spin system (e.g., ^{13}C–^{1}H, ^{13}C = **S**, and ^{1}H = **I**) each pair of spins can be represented by a superposition of the four "pure" states $\alpha_I\alpha_S$, $\alpha_I\beta_S$, $\beta_I\alpha_S$, and $\beta_I\beta_S$. In the heteronuclear case the energy difference for the **S** transitions ($\alpha_I\alpha_S \to \alpha_I\beta_S$ and $\beta_I\alpha_S \to \beta_I\beta_S$) will be different from the energy difference for the **I** transitions ($\alpha_I\alpha_S \to \beta_I\alpha_S$ and $\alpha_I\beta_S \to \beta_I\beta_S$). The wave function for one spin pair is thus

$$\Psi = c_1\,\alpha_I\alpha_S + c_2\,\alpha_I\beta_S + c_3\,\beta_I\alpha_S + c_4\,\beta_I\beta_S$$

and the probabilities and coherences can be represented by a 4 × 4 matrix:

$$\Psi \otimes \Psi^* = \begin{array}{c} \\ \alpha_I\alpha_S \\ \alpha_I\beta_S \\ \beta_I\alpha_S \\ \beta_I\beta_S \end{array} \begin{array}{cccc} \alpha_I\alpha_S & \alpha_I\beta_S & \beta_I\alpha_S & \beta_I\beta_S \\ \left[\begin{matrix} c_1c_1^* & c_1c_2^* & c_1c_3^* & c_1c_4^* \\ c_2c_1^* & c_2c_2^* & c_2c_3^* & c_2c_4^* \\ c_3c_1^* & c_3c_2^* & c_3c_3^* & c_3c_4^* \\ c_4c_1^* & c_4c_2^* & c_4c_3^* & c_4c_4^* \end{matrix}\right] \end{array}$$

If the average is taken over all of the spin pairs in the sample for each term in the matrix, we have the density matrix σ that describes the state of the whole system. The superpositions (or coherences) between states can be described as follows:

$$\begin{array}{c} \\ \alpha_I\alpha_S \\ \alpha_I\beta_S \\ \beta_I\alpha_S \\ \beta_I\beta_S \end{array} \begin{array}{cccc} \alpha_I\alpha_S & \alpha_I\beta_S & \beta_I\alpha_S & \beta_I\beta_S \\ \left[\begin{matrix} N(\alpha\alpha) & SQ^*(S_1) & SQ^*(I_1) & DQ^* \\ SQ(S_1) & N(\alpha\beta) & ZQ^* & SQ^*(I_2) \\ SQ(I_1) & ZQ & N(\beta\alpha) & SQ^*(S_2) \\ DQ & SQ(I_2) & SQ(S_2) & N(\beta\beta) \end{matrix}\right] \end{array}$$

where N represents the population of an energy state, SQ represents single-quantum (observable) coherence, DQ represent double-quantum coherence, and ZQ represents zero-quantum coherence. For example, SQ*(I_2) is the complex conjugate of the coherence for the $\alpha_I\beta_S \leftrightarrow \beta_I\beta_S$ transition, with spin I changing state and spin S remaining in the β state. The spectrum (observable SQ transitions only) consists of four lines (two doublets): I_1, I_2, S_1, and S_2.

For a ^{13}C–^{1}H system at equilibrium, we have

$$\sigma_{eq} = \begin{bmatrix} (N/4)(1+5\varepsilon) & 0 & 0 & 0 \\ 0 & (N/4)(1+3\varepsilon) & 0 & 0 \\ 0 & 0 & (N/4)(1-3\varepsilon) & 0 \\ 0 & 0 & 0 & (N/4)(1-5\varepsilon) \end{bmatrix}$$

corresponding to the energy diagram shown in Figure 10.37 (units of radians/s). For $\mathbf{I} = {}^1$H and $\mathbf{S} = {}^{13}$C, the energy separation for the ^{1}H transitions is about four times as large as the energy separation for the ^{13}C transitions. As before, we factor out the $N/4$, separate out the identity matrix, and factor out the ε. Omitting the identity matrix and the $N\varepsilon/4$ factor, this gives

$$\sigma_{eq} = \frac{1}{2}\begin{bmatrix} 5 & 0 & 0 & 0 \\ 0 & 3 & 0 & 0 \\ 0 & 0 & -3 & 0 \\ 0 & 0 & 0 & -5 \end{bmatrix}$$

THE DENSITY MATRIX REPRESENTATION OF SPIN STATES   473

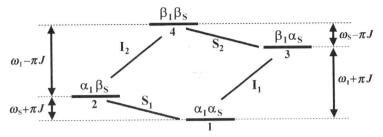

Figure 10.37

Note that the population differences across the **I** (^{1}H) transitions are 4 (0.5[5 − (−3)] and 0.5[3 − (−5)]) and the differences across the **S** (^{13}C) transitions are 1 (0.5[5 − 3] and 0.5[−3 − (−5)]). For a homonuclear system (e.g., ^{1}H$_a$, ^{1}H$_b$) the populations (diagonal elements) would be 2, 0, 0, and −2.

The density matrix representations of the 16 product operators for a heteronuclear two-spin system are given below.

$$\mathbf{I}_z = \tfrac{1}{2}\begin{bmatrix} 4 & 0 & 0 & 0 \\ 0 & 4 & 0 & 0 \\ 0 & 0 & -4 & 0 \\ 0 & 0 & 0 & -4 \end{bmatrix} \quad \mathbf{S}_z = \tfrac{1}{2}\begin{bmatrix} 1 & 0 & 0 & 0 \\ 0 & -1 & 0 & 0 \\ 0 & 0 & 1 & 0 \\ 0 & 0 & 0 & -1 \end{bmatrix}$$

The factor of 4 corresponds to the (nearly) four times greater population difference for ^{1}H compared to ^{13}C, which is a direct result of the (nearly) four times larger magnetogyric ratio of ^{1}H. Note that the equilibrium state for this two-spin system is $\sigma_{eq} = \mathbf{I}_z + \mathbf{S}_z$, so that the sum of the two matrices above equals the matrix σ_{eq}.

In-phase single-quantum coherence (SQC) is represented by nonzero values for the matrix elements that correspond to the single-quantum transitions. For example, **I** spin (^{1}H) SQC corresponds to a superposition of the $\alpha_I\alpha_S$ and $\beta_I\alpha_S$ states (row 1 and column 3), and the $\alpha_I\beta_S$ and $\beta_I\beta_S$ states (row 2 and column 4). Real numbers are used for magnetization on the x' axis, and imaginary numbers are used for magnetization on the y' axis. Notice that the "downward" transition $\beta_I\alpha_S \rightarrow \alpha_I\alpha_S$ has a matrix element that is the complex conjugate of the "upward" transition $\alpha_I\alpha_S \rightarrow \beta_I\alpha_S$.

$$\tfrac{1}{2}\begin{bmatrix} 0 & 0 & 4 & 0 \\ 0 & 0 & 0 & 4 \\ 4 & 0 & 0 & 0 \\ 0 & 4 & 0 & 0 \end{bmatrix} \quad \tfrac{1}{2}\begin{bmatrix} 0 & 0 & -4i & 0 \\ 0 & 0 & 0 & -4i \\ 4i & 0 & 0 & 0 \\ 0 & 4i & 0 & 0 \end{bmatrix} \quad \tfrac{1}{2}\begin{bmatrix} 0 & 1 & 0 & 0 \\ 1 & 0 & 0 & 0 \\ 0 & 0 & 0 & 1 \\ 0 & 0 & 1 & 0 \end{bmatrix} \quad \tfrac{1}{2}\begin{bmatrix} 0 & -i & 0 & 0 \\ i & 0 & 0 & 0 \\ 0 & 0 & 0 & -i \\ 0 & 0 & i & 0 \end{bmatrix}$$
$$\mathbf{I}_x \qquad\qquad \mathbf{I}_y \qquad\qquad \mathbf{S}_x \qquad\qquad \mathbf{S}_y$$

$$\tfrac{1}{2}\begin{bmatrix} 0 & 0 & 4 & 0 \\ 0 & 0 & 0 & -4 \\ 4 & 0 & 0 & 0 \\ 0 & -4 & 0 & 0 \end{bmatrix} \quad \tfrac{1}{2}\begin{bmatrix} 0 & 0 & -4i & 0 \\ 0 & 0 & 0 & 4i \\ 4i & 0 & 0 & 0 \\ 0 & -4i & 0 & 0 \end{bmatrix} \quad \tfrac{1}{2}\begin{bmatrix} 0 & 4 & 0 & 0 \\ 4 & 0 & 0 & 0 \\ 0 & 0 & 0 & -4 \\ 0 & 0 & -4 & 0 \end{bmatrix} \quad \tfrac{1}{2}\begin{bmatrix} 0 & -4i & 0 & 0 \\ 4i & 0 & 0 & 0 \\ 0 & 0 & 0 & 4i \\ 0 & 0 & -4i & 0 \end{bmatrix}$$
$$2\mathbf{I}_x\mathbf{S}_z \qquad\qquad 2\mathbf{I}_y\mathbf{S}_z \qquad\qquad 2\mathbf{S}_x\mathbf{I}_z \qquad\qquad 2\mathbf{S}_y\mathbf{I}_z$$

Notice that the antiphase density matrix differs from the in-phase representation only in that the sign of the two transitions (e.g., I$_1$ ($\alpha_I\alpha_S \rightarrow \beta_I\alpha_S$) and I$_2$ ($\alpha_I\beta_S \rightarrow \beta_I\beta_S$) for the ^{1}H or

I transitions) is opposite. That is the same as saying that the **I** transition $\alpha \to \beta$ ($\mathbf{S} = \alpha$) is 180° out of phase with the **I** transition $\alpha \to \beta$ ($\mathbf{S} = \beta$). You can "read" the coherences if you focus on the terms below the diagonal. These matrices are in fact the matrix multiplication products of their individual components; for example, $2\mathbf{I}_x\mathbf{S}_z$ is just twice the matrix product of the $\mathbf{I}_x$ and $\mathbf{S}_z$ matrices.

The products of the x and y operators give the following four matrices:

$$\frac{1}{2}\begin{bmatrix} 0 & 0 & 0 & 4 \\ 0 & 0 & 4 & 0 \\ 0 & 4 & 0 & 0 \\ 4 & 0 & 0 & 0 \end{bmatrix} \quad \frac{1}{2}\begin{bmatrix} 0 & 0 & 0 & -4 \\ 0 & 0 & 4 & 0 \\ 0 & 4 & 0 & 0 \\ -4 & 0 & 0 & 0 \end{bmatrix} \quad \frac{1}{2}\begin{bmatrix} 0 & 0 & 0 & -4i \\ 0 & 0 & 4i & 0 \\ 0 & -4i & 0 & 0 \\ 4i & 0 & 0 & 0 \end{bmatrix} \quad \frac{1}{2}\begin{bmatrix} 0 & 0 & 0 & -4i \\ 0 & 0 & -4i & 0 \\ 0 & 4i & 0 & 0 \\ 4i & 0 & 0 & 0 \end{bmatrix}$$

$$2\mathbf{I}_x\mathbf{S}_x \qquad\qquad 2\mathbf{I}_y\mathbf{S}_y \qquad\qquad 2\mathbf{I}_x\mathbf{S}_y \qquad\qquad 2\mathbf{I}_y\mathbf{S}_x$$

Clearly all four of these matrices represent linear combinations of zero-quantum and double-quantum coherences, since the "antidiagonal" terms near the center are ZQ terms and the ones on the outside are DQ terms:

$$\begin{array}{c} \\ \alpha_I\alpha_S \\ \alpha_I\beta_S \\ \beta_I\alpha_S \\ \beta_I\beta_S \end{array} \begin{array}{cccc} \alpha_I\alpha_S & \alpha_I\beta_S & \beta_I\alpha_S & \beta_I\beta_S \\ \begin{bmatrix} N(\alpha\alpha) & SQ^*(S_1) & SQ^*(I_1) & DQ^* \\ SQ(S_1) & N(\alpha\beta) & ZQ^* & SQ^*(I_2) \\ SQ(I_1) & ZQ & N(\beta\alpha) & SQ^*(S_2) \\ DQ & SQ(I_2) & SQ(S_2) & N(\beta\beta) \end{bmatrix} \end{array}$$

Appropriate linear combinations of the four matrices above can generate the pure ZQ and DQ matrices:

$$\frac{1}{2}\begin{bmatrix} 0 & 0 & 0 & 0 \\ 0 & 0 & 4 & 0 \\ 0 & 4 & 0 & 0 \\ 0 & 0 & 0 & 0 \end{bmatrix} \quad \frac{1}{2}\begin{bmatrix} 0 & 0 & 0 & 4 \\ 0 & 0 & 0 & 0 \\ 0 & 0 & 0 & 0 \\ 4 & 0 & 0 & 0 \end{bmatrix} \quad \frac{1}{2}\begin{bmatrix} 0 & 0 & 0 & 0 \\ 0 & 0 & -4i & 0 \\ 0 & 4i & 0 & 0 \\ 0 & 0 & 0 & 0 \end{bmatrix} \quad \frac{1}{2}\begin{bmatrix} 0 & 0 & 0 & -4i \\ 0 & 0 & 0 & 0 \\ 0 & 0 & 0 & 0 \\ 4i & 0 & 0 & 0 \end{bmatrix}$$

$$\frac{1}{2}(2\mathbf{I}_x\mathbf{S}_x + 2\mathbf{I}_y\mathbf{S}_y) \quad \frac{1}{2}(2\mathbf{I}_x\mathbf{S}_x - 2\mathbf{I}_y\mathbf{S}_y) \quad \frac{1}{2}(2\mathbf{I}_y\mathbf{S}_x - 2\mathbf{I}_x\mathbf{S}_y) \quad \frac{1}{2}(2\mathbf{I}_y\mathbf{S}_x + 2\mathbf{I}_x\mathbf{S}_y)$$
$$\{ZQ\}_x \qquad\qquad \{DQ\}_x \qquad\qquad \{ZQ\}_y \qquad\qquad \{DQ\}_y$$

The density matrix representation is actually simpler than the product operator formalism for dealing with zero and multiple quantum coherences. Note that the type of multiple quantum coherence can be "read" from the lower left elements of the "antidiagonal."

The final two product operators are the longitudinal spin order operator ($2\mathbf{I}_z\mathbf{S}_z$) and the identity operator (**1**):

$$\frac{1}{2}\begin{bmatrix} 4 & 0 & 0 & 0 \\ 0 & -4 & 0 & 0 \\ 0 & 0 & -4 & 0 \\ 0 & 0 & 0 & 4 \end{bmatrix} \qquad \begin{bmatrix} 1 & 0 & 0 & 0 \\ 0 & 1 & 0 & 0 \\ 0 & 0 & 1 & 0 \\ 0 & 0 & 0 & 1 \end{bmatrix}$$
$$2\mathbf{I}_z\mathbf{S}_z \qquad\qquad\qquad \mathbf{1}$$

By the way, all of the products commute, so the order is not important: $2\mathbf{I}_y\mathbf{S}_y = 2\mathbf{S}_y\mathbf{I}_y$. If only one of the operators in a product is observable, it is always written first: $2\mathbf{S}_x\mathbf{I}_z$.

The rotation matrices corresponding to pulses are 4 × 4 matrices of the following form:

$$R_x^I(\Theta) = \begin{bmatrix} \cos(\Theta/2) & 0 & i\sin(\Theta/2) & 0 \\ 0 & \cos(\Theta/2) & 0 & i\sin(\Theta/2) \\ i\sin(\Theta/2) & 0 & \cos(\Theta/2) & 0 \\ 0 & i\sin(\Theta/2) & 0 & \cos(\Theta/2) \end{bmatrix}$$

This matrix corresponds to pulses that only affect the **I** spins (^{1}H). Note that the **i** sin ($\Theta/2$) term appears for all of the **I** (^{1}H) transitions and the cos ($\Theta/2$) term appears for all of the diagonal elements. The rotation matrix for **S** (^{13}C) pulses has the **i** sin ($\Theta/2$) terms in matrix elements corresponding to the **S** (^{13}C) transitions:

$$R_x^S(\Theta) = \begin{bmatrix} \cos(\Theta/2) & i\sin(\Theta/2) & 0 & 0 \\ i\sin(\Theta/2) & \cos(\Theta/2) & 0 & 0 \\ 0 & 0 & \cos(\Theta/2) & i\sin(\Theta/2) \\ 0 & 0 & i\sin(\Theta/2) & \cos(\Theta/2) \end{bmatrix}$$

The corresponding rotation matrices for pulses on the y axis are formed by simply replacing the **i** sin ($\Theta/2$) terms with sin ($\Theta/2$) above the diagonal and $-$sin ($\Theta/2$) below the diagonal. A simultaneous pulse on both the *I* and *S* spins can be formed from the product of the rotation matrices for the individual pulses:

$$R_x^{I,S}(\Theta) = R_x^I(\Theta) R_x^S(\Theta)$$

$$= \begin{bmatrix} \cos(\Theta/2) & 0 & i\sin(\Theta/2) & 0 \\ 0 & \cos(\Theta/2) & 0 & i\sin(\Theta/2) \\ i\sin(\Theta/2) & 0 & \cos(\Theta/2) & 0 \\ 0 & i\sin(\Theta/2) & 0 & \cos(\Theta/2) \end{bmatrix}$$

$$\times \begin{bmatrix} \cos(\Theta/2) & i\sin(\Theta/2) & 0 & 0 \\ i\sin(\Theta/2) & \cos(\Theta/2) & 0 & 0 \\ 0 & 0 & \cos(\Theta/2) & i\sin(\Theta/2) \\ 0 & 0 & i\sin(\Theta/2) & \cos(\Theta/2) \end{bmatrix}$$

$$= \begin{bmatrix} \cos^2 & i\sin\cos & i\sin\cos & -\sin^2 \\ i\sin\cos & \cos^2 & -\sin^2 & i\sin\cos \\ i\sin\cos & -\sin^2 & \cos^2 & i\sin\cos \\ -\sin^2 & i\sin\cos & i\sin\cos & \cos^2 \end{bmatrix}$$

$$= \frac{1}{2}\begin{bmatrix} 1 & i & i & -1 \\ i & 1 & -1 & i \\ i & -1 & 1 & i \\ -1 & i & i & 1 \end{bmatrix} \begin{array}{l} \text{for} \\ \text{a} \\ 90° \\ \text{pulse} \end{array}$$

where the argument of the sin and cos functions is $\Theta/2$ in each case. Similarly for the y axis rotation matrix for a pulse that affects both the I and S spins:

$$R_y^{I,S}(\Theta) = R_y^I(\Theta)R_y^S(\Theta)$$

$$= \begin{bmatrix} \cos(\Theta/2) & 0 & \sin(\Theta/2) & 0 \\ 0 & \cos(\Theta/2) & 0 & \sin(\Theta/2) \\ -\sin(\Theta/2) & 0 & \cos(\Theta/2) & 0 \\ 0 & -\sin(\Theta/2) & 0 & \cos(\Theta/2) \end{bmatrix}$$

$$\times \begin{bmatrix} \cos(\Theta/2) & \sin(\Theta/2) & 0 & 0 \\ -\sin(\Theta/2) & \cos(\Theta/2) & 0 & 0 \\ 0 & 0 & \cos(\Theta/2) & \sin(\Theta/2) \\ 0 & 0 & -\sin(\Theta/2) & \cos(\Theta/2) \end{bmatrix}$$

$$= \begin{bmatrix} \cos^2 & \text{sincos} & \text{sincos} & \sin^2 \\ -\text{sincos} & \cos^2 & -\sin^2 & \text{sincos} \\ -\text{sincos} & -\sin^2 & \cos^2 & \text{sincos} \\ \sin^2 & -\text{sincos} & -\text{sincos} & \cos^2 \end{bmatrix}$$

$$= \tfrac{1}{2}\begin{bmatrix} 1 & 1 & 1 & 1 \\ -1 & 1 & -1 & 1 \\ -1 & -1 & 1 & 1 \\ 1 & -1 & -1 & 1 \end{bmatrix} \quad \begin{array}{l} \text{for} \\ \text{a} \\ 90° \\ \text{pulse} \end{array}$$

A short example will illustrate the usefulness (and complexity) of density matrix notation. Consider the INEPT sequence for a ^{13}C–^{1}H (**S–I**) spin system. The pulse sequence is simply

$$\sigma_{eq} \xrightarrow{90°_x(^1H)} \sigma_1 \xrightarrow{\tau=1/(2J_{CH})} \sigma_2 \xrightarrow{90°_y(^1H,^{13}C)} \sigma_3 \quad (\text{FID})$$
$$\mathbf{I}_z+\mathbf{S}_z \quad\quad -\mathbf{I}_y+\mathbf{S}_z \quad\quad 2\mathbf{I}_x\mathbf{S}_z+\mathbf{S}_z \quad\quad 4[-2\mathbf{S}_x\mathbf{I}_z]+\mathbf{S}_x$$

where σ_{eq}, σ_1, σ_2, and σ_3 represent the spin state of the system at each stage of the pulse sequence. The product operator notation is shown above for each spin state. First a 90° proton pulse on the x axis generates observable single-quantum **I** coherence:

$$\sigma_1 = \tfrac{1}{\sqrt{2}}\begin{bmatrix} 1 & 0 & -i & 0 \\ 0 & 1 & 0 & -i \\ -i & 0 & 1 & 0 \\ 0 & -i & 0 & 1 \end{bmatrix} \times \tfrac{1}{2}\begin{bmatrix} 5 & 0 & 0 & 0 \\ 0 & 3 & 0 & 0 \\ 0 & 0 & -3 & 0 \\ 0 & 0 & 0 & -5 \end{bmatrix} \times \tfrac{1}{\sqrt{2}}\begin{bmatrix} 1 & 0 & i & 0 \\ 0 & 1 & 0 & i \\ i & 0 & 1 & 0 \\ 0 & i & 0 & 1 \end{bmatrix}$$
$$\quad\quad R_{xH}(90°)^{-1} \quad\quad\quad\quad \sigma_{eq} \quad\quad\quad\quad R_{xH}(90°)$$

$$= \tfrac{1}{\sqrt{2}}\begin{bmatrix} 1 & 0 & -i & 0 \\ 0 & 1 & 0 & -i \\ -i & 0 & 1 & 0 \\ 0 & -i & 0 & 1 \end{bmatrix} \times \tfrac{1}{2\sqrt{2}}\begin{bmatrix} 5 & 0 & 5i & 0 \\ 0 & 3 & 0 & 3i \\ -3i & 0 & -3 & 0 \\ 0 & -5i & 0 & -5 \end{bmatrix}$$
$$\quad\quad R_{xH}(90°)^{-1} \quad\quad\quad\quad \sigma_{eq}R_{xH}(90°)$$

$$= \tfrac{1}{2}\begin{bmatrix} 1 & 0 & 4i & 0 \\ 0 & -1 & 0 & 4i \\ -4i & 0 & 1 & 0 \\ 0 & -4i & 0 & -1 \end{bmatrix}$$
$$\sigma_1 = -\mathbf{I}_y+\mathbf{S}_z$$

THE DENSITY MATRIX REPRESENTATION OF SPIN STATES

The intermediate product of the last two matrices is shown to illustrate the matrix multiplication. Along the diagonal of the resultant matrix σ_1 you can still see the carbon z magnetization, but the proton z magnetization is gone (compare to $\mathbf{S}_z$). The off-diagonal elements correspond to in-phase proton magnetization on the $-y$ axis (compare $\mathbf{I}_y$).

To calculate the effect of the $1/(2J)$ delay we need to know the energy differences among the four spin states $\alpha_I\alpha_S$, $\alpha_I\beta_S$, $\beta_I\alpha_S$, and $\beta_I\beta_S$ so that the phase factors $e^{i\omega t}$ and $e^{-i\omega t}$ can be applied to the off-diagonal elements of the density matrix. The frequencies for the $\alpha_I\alpha_S \rightarrow \beta_I\alpha_S(1,3)$ and $\alpha_I\beta_S \rightarrow \beta_I\beta_S$ (2,4) transitions differ only by the coupling constant $^1J_{CH}$:

$$\omega_{13} = \omega_I + \pi J \qquad \omega_{24} = \omega_I - \pi J$$

and the phase factors are $e^{i(\omega_I + \pi J)\tau}$ for the (1,3) transition and $e^{i(\omega_I - \pi J)\tau}$ for the (2,4) transition (Fig. 10.37). With a delay of $\tau = 1/(2J)$ and for an on-resonance peak ($\omega_I = 0$), these factors become $e^{i\pi/2}$ and $e^{-i\pi/2}$ or simply $\mathbf{i}$ and $-\mathbf{i}$, respectively. The elements above the diagonal have phase factors of $e^{-i(\omega_I + \pi J)\tau}$ for the (1,3) transition and $e^{-i(\omega_I - \pi J)\tau}$ for the (2,4) transition, which become $e^{-i\pi/2}$ and $e^{i\pi/2}$ or simply $-\mathbf{i}$ and $\mathbf{i}$. Thus the density matrix becomes:

$$\sigma_2 = \tfrac{1}{2}\begin{bmatrix} 1 & 0 & 4i(-i) & 0 \\ 0 & -1 & 0 & 4i(i) \\ -4i(i) & 0 & 1 & 0 \\ 0 & -4i(-i) & 0 & -1 \end{bmatrix} = \tfrac{1}{2}\begin{bmatrix} 1 & 0 & 4 & 0 \\ 0 & -1 & 0 & -4 \\ 4 & 0 & 1 & 0 \\ 0 & -4 & 0 & -1 \end{bmatrix}$$
$$2\mathbf{I}_x\mathbf{S}_z + \mathbf{S}_z$$

Note that the populations are unchanged (we are neglecting relaxation) and the proton magnetization, which was in-phase on the $-y'$ axis before the delay, is now antiphase on the x' axis.

Finally, a 90° pulse on the y' axis is applied simultaneously to both ^{1}H and ^{13}C:

$$\sigma_3 = \tfrac{1}{2}\begin{bmatrix} 1 & -1 & -1 & 1 \\ 1 & 1 & -1 & -1 \\ 1 & -1 & 1 & -1 \\ 1 & 1 & 1 & 1 \end{bmatrix} \times \tfrac{1}{2}\begin{bmatrix} 1 & 0 & 4 & 0 \\ 0 & -1 & 0 & -4 \\ 4 & 0 & 1 & 0 \\ 0 & -4 & 0 & -1 \end{bmatrix} \times \tfrac{1}{2}\begin{bmatrix} 1 & 1 & 1 & 1 \\ -1 & 1 & -1 & 1 \\ -1 & -1 & 1 & 1 \\ 1 & -1 & -1 & 1 \end{bmatrix}$$
$$R_{y\text{H,C}}(90°)^{-1} \qquad\qquad \sigma_2 \qquad\qquad R_{y\text{H,C}}(90°)$$

$$= \tfrac{1}{2}\begin{bmatrix} 1 & -1 & -1 & 1 \\ 1 & 1 & -1 & -1 \\ 1 & -1 & 1 & -1 \\ 1 & 1 & 1 & 1 \end{bmatrix} \times \tfrac{1}{4}\begin{bmatrix} -3 & -3 & 5 & 5 \\ -3 & 3 & 5 & -5 \\ 3 & 3 & 5 & 5 \\ 3 & -3 & 5 & -5 \end{bmatrix} = \tfrac{1}{2}\begin{bmatrix} 0 & -3 & 0 & 0 \\ -3 & 0 & 0 & 0 \\ 0 & 0 & 0 & 5 \\ 0 & 0 & 5 & 0 \end{bmatrix}$$
$$R_{y\text{H,C}}(90°)^{-1} \qquad \sigma_2 R_{y\text{H,C}}(90°) \qquad \sigma_3 = 4[-2\mathbf{S}_x\mathbf{I}_z] + \mathbf{S}_x$$

The resulting matrix, σ_3, has single quantum (observable) $\mathbf{S}$ (^{13}C) coherence (x axis) with relative peak heights of -3 and $+5$ for the two components of the doublet. Compare this to the result of a single ^{13}C 90° pulse on the equilibrium matrix:

$$\sigma' = \tfrac{1}{\sqrt{2}}\begin{bmatrix} 1 & -1 & 0 & 0 \\ 1 & 1 & 0 & 0 \\ 0 & 0 & 1 & -1 \\ 0 & 0 & 1 & 1 \end{bmatrix} \times \tfrac{1}{2}\begin{bmatrix} 5 & 0 & 0 & 0 \\ 0 & 3 & 0 & 0 \\ 0 & 0 & -3 & 0 \\ 0 & 0 & 0 & -5 \end{bmatrix} \times \tfrac{1}{\sqrt{2}}\begin{bmatrix} 1 & 1 & 0 & 0 \\ -1 & 1 & 0 & 0 \\ 0 & 0 & 1 & 1 \\ 0 & 0 & -1 & 1 \end{bmatrix} = \tfrac{1}{2}\begin{bmatrix} 4 & 1 & 0 & 0 \\ 1 & 4 & 0 & 0 \\ 0 & 0 & -4 & 1 \\ 0 & 0 & 1 & -4 \end{bmatrix}$$
$$R_{y\text{C}}(90°)^{-1} \qquad\qquad \sigma_{eq} \qquad\qquad R_{y\text{C}}(90°) \qquad\qquad \sigma' = \mathbf{I}_z + \mathbf{S}_x$$

In this case the result is also single-quantum (observable) $\mathbf{S}\,(^{13}\mathrm{C})$ magnetization on the x axis, but the intensity is lower because the magnetization came only from the $^{13}\mathrm{C}\ z$ magnetization of the equilibrium state. Note that the $^1\mathrm{H}\ z$ magnetization (population differences between the $\alpha_I\alpha_S$ and $\beta_I\alpha_S$ states, 1–3, and between the $\alpha_I\beta_S$ and $\beta_I\beta_S$ states, 2–4) still remains along the diagonal of the density matrix σ'. The INEPT sequence has no z magnetization left in the final density matrix σ_3, as shown by the zero values on the diagonal. The final 90° pulse of the INEPT sequence converts not only the $^1\mathrm{H}$ antiphase magnetization to $^{13}\mathrm{C}$ antiphase magnetization but also the $^{13}\mathrm{C}\ z$ magnetization to in-phase observable $^{13}\mathrm{C}$ magnetization:

$$\frac{1}{2}\begin{bmatrix} 0 & -3 & 0 & 0 \\ -3 & 0 & 0 & 0 \\ 0 & 0 & 0 & 5 \\ 0 & 0 & 5 & 0 \end{bmatrix} = \frac{1}{2}\begin{bmatrix} 0 & -4 & 0 & 0 \\ -4 & 0 & 0 & 0 \\ 0 & 0 & 0 & 4 \\ 0 & 0 & 4 & 1 \end{bmatrix} + \frac{1}{2}\begin{bmatrix} 0 & 1 & 0 & 0 \\ 1 & 0 & 0 & 0 \\ 0 & 0 & 0 & 1 \\ 0 & 0 & 0 & 0 \end{bmatrix}$$

$$\sigma_3 \qquad\qquad 4[-2\mathbf{S}_x\mathbf{I}_z] \qquad\qquad \mathbf{S}_x$$

The resulting density matrix σ_3 is just the sum of the enhanced (by a factor of 4) and antiphase $^{13}\mathrm{C}$ single quantum coherence and the normal in-phase $^{13}\mathrm{C}$ SQC.

10.8 THE HAMILTONIAN MATRIX: STRONG COUPLING AND IDEAL ISOTROPIC (TOCSY) MIXING

I have done everything I can to avoid using the "H word," but now that we have learned how to represent product operators in a matrix form it is a short step to working with the Hamiltonian. The Hamiltonian is a representation of the environment the spins find themselves in—it contains the energies of all of the interactions of spins with the B_0 field, the B_1 field, and with each other.

The symbols we have been using to represent spin states ($\mathbf{I}_x$, $\mathbf{S}_y$, $2\mathbf{I}_y\mathbf{S}_z$, etc.) of the entire ensemble of spins are actually operators: they can "operate" on a spin state (of a single spin pair in our H_a, H_b system) and spit out another spin state. We already saw this with the raising and lowering operators:

$$\mathbf{I}^+|\alpha> \to |\beta> \qquad \mathbf{I}^-|\beta> \to |\alpha>$$

The raising operator $\mathbf{I}^+$, acting as an operator, raises the α state of a single spin to the β state. All of these operators can be represented as matrices. In the case of the homonuclear two-spin system (H_a and H_b), these are 4×4 matrices. For example, the raising operator $\mathbf{I}_a^+$ can be represented by the following matrix, which acts on the "vector" that describes the $\alpha\alpha$ state to give a new vector that describes the $\beta\alpha$ state:

$$\begin{bmatrix} 0 & 0 & 0 & 0 \\ 1 & 0 & 0 & 0 \\ 0 & 0 & 0 & 0 \\ 0 & 0 & 1 & 0 \end{bmatrix} \times \begin{bmatrix} 1 \\ 0 \\ 0 \\ 0 \end{bmatrix} = \begin{bmatrix} 0 \\ 1 \\ 0 \\ 0 \end{bmatrix} \begin{matrix} \alpha\alpha \\ \beta\alpha \\ \alpha\beta \\ \beta\beta \end{matrix}$$

These vectors are just a column of numbers representing the coefficients of the four pure spin states: c_1, c_2, c_3 and c_4. They do not describe the whole ensemble of spins, just one $\mathrm{H}_a - \mathrm{H}_b$ pair.

The Hamiltonian is an operator that represents the energies of all of the interactions in the system. In quantum mechanics, the classical energy terms are replaced by the analogous operators to generate the Hamiltonian. For NMR, the classical energy can be written as

$$E/\hbar = [-\gamma(\boldsymbol{B}_o \cdot \boldsymbol{I}_a) - \gamma(\boldsymbol{B}_o \cdot \boldsymbol{I}_b)] + J[\boldsymbol{I}_a \cdot \boldsymbol{I}_b] \quad \text{(in units of Hz)}$$

Where $\boldsymbol{B}_o$ is a vector representing the external magnetic field, J is the coupling constant, $\boldsymbol{I}_a$ and $\boldsymbol{I}_b$ are the vectors representing the nuclear magnets of H_a and H_b and the products are vector dot products. The first part represents the energy of interaction of the spins with the B_o field and the second part represents the energy of interaction of the spins with each other. Since the B_o vector has only a z component, we can write the vector products as

$$E/\hbar = [-\gamma B_o I_z^a - \gamma B_o I_z^b] + J[I_x^a I_x^b + I_y^a I_y^b + I_z^a I_z^b]$$

where B_o is now just the magnitude of the B_o vector. Only the z component of the nuclear magnet contributes to its interaction with the B_o field. Now we replace the classical components I_x, I_y, and I_z of the nuclear magnet's vector with the corresponding quantum mechanical operators:

$$H = [-\nu_a \mathbf{I}_z^a - \nu_b \mathbf{I}_z^b] + J[\mathbf{I}_x^a \mathbf{I}_x^b + \mathbf{I}_y^a \mathbf{I}_y^b + \mathbf{I}_z^a \mathbf{I}_z^b] = H_z + H_J$$

Here we also account for the slightly different B_{eff} fields experienced by H_a and H_b leading to their individual Larmor frequencies ν_a and ν_b. The first part, called the Zeeman Hamiltonian, again represents the interaction of the spins with the B_o field. The second part, called the scalar coupling Hamiltonian, represents the energy of interaction of H_a with H_b, independent of the B_o field. The J-coupling interaction is isotropic: it happens in all possible directions of space depending on the relative orientation of the two nuclear magnets.

There are many other terms to the Hamiltonian but for spin-1/2 nuclei in liquids they can all be ignored. The dipole–dipole (dipolar or direct coupling) Hamiltonian is important in solids and partially oriented liquids, and the quadrupolar Hamiltonian is important for spins greater than 1/2. The dipolar interaction contains a multiplier of

$$(3\cos^2\Theta - 1)$$

where Θ is the angle between the $H_a - H_b$ vector and the B_o field (z axis). This factor averages to zero for random isotropic (equal in all directions in space) molecular tumbling, and if the motion is rapid the dipolar term can be ignored. This is what defines NMR in liquids: rapid isotopic reorientation of the molecules.

From our study of product operators and the density matrix we know what these operators $\mathbf{I}_x$, $\mathbf{I}_y$ and $\mathbf{I}_z$ do and we know how to represent them in matrix form. So we can write out the Hamiltonian matrix for the H_a, H_b system:

$$H_z = -\nu_a/2 \begin{bmatrix} 1 & 0 & 0 & 0 \\ 0 & -1 & 0 & 0 \\ 0 & 0 & 1 & 0 \\ 0 & 0 & 0 & -1 \end{bmatrix} - \nu_b/2 \begin{bmatrix} 1 & 0 & 0 & 0 \\ 0 & 1 & 0 & 0 \\ 0 & 0 & -1 & 0 \\ 0 & 0 & 0 & -1 \end{bmatrix}$$

$$= \frac{1}{2}\begin{bmatrix} -v_a-v_b & 0 & 0 & 0 \\ 0 & v_a-v_b & 0 & 0 \\ 0 & 0 & -v_a+v_b & 0 \\ 0 & 0 & 0 & v_a+v_b \end{bmatrix}$$

$$H_J = J/4 \left[\begin{bmatrix} 0 & 1 & 0 & 0 \\ 1 & 0 & 0 & 0 \\ 0 & 0 & 0 & 1 \\ 0 & 0 & 1 & 0 \end{bmatrix} \begin{bmatrix} 0 & 0 & 1 & 0 \\ 0 & 0 & 0 & 1 \\ 1 & 0 & 0 & 0 \\ 0 & 1 & 0 & 0 \end{bmatrix} + \begin{bmatrix} 0 & -i & 0 & 0 \\ i & 0 & 0 & 0 \\ 0 & 0 & 0 & -i \\ 0 & 0 & i & 0 \end{bmatrix} \begin{bmatrix} 0 & 0 & -i & 0 \\ 0 & 0 & 0 & -i \\ i & 0 & 0 & 0 \\ 0 & i & 0 & 0 \end{bmatrix} \right.$$

$$\left. + \begin{bmatrix} 1 & 0 & 0 & 0 \\ 0 & -1 & 0 & 0 \\ 0 & 0 & 1 & 0 \\ 0 & 0 & 0 & -1 \end{bmatrix} \begin{bmatrix} 1 & 0 & 0 & 0 \\ 0 & 1 & 0 & 0 \\ 0 & 0 & -1 & 0 \\ 0 & 0 & 0 & -1 \end{bmatrix} \right]$$

$$H_J = J/4 \left[\begin{bmatrix} 0 & 0 & 0 & 1 \\ 0 & 0 & 1 & 0 \\ 0 & 1 & 0 & 0 \\ 1 & 0 & 0 & 0 \end{bmatrix} + \begin{bmatrix} 0 & 0 & 0 & -1 \\ 0 & 0 & 1 & 0 \\ 0 & 1 & 0 & 0 \\ -1 & 0 & 0 & 0 \end{bmatrix} + \begin{bmatrix} 1 & 0 & 0 & 0 \\ 0 & -1 & 0 & 0 \\ 0 & 0 & -1 & 0 \\ 0 & 0 & 0 & 1 \end{bmatrix} \right]$$

$$= J/4 \begin{bmatrix} 1 & 0 & 0 & 0 \\ 0 & -1 & 2 & 0 \\ 0 & 2 & -1 & 0 \\ 0 & 0 & 0 & 1 \end{bmatrix}$$

Note that the only off-diagonal terms of H_J are the H_{32} and H_{23} terms (both equal to 2) that come from the $\mathbf{I}_x^a \mathbf{I}_x^b$ and $\mathbf{I}_y^a \mathbf{I}_y^b$ parts.

Combining the Zeeman and scalar coupling Hamiltonians,

$$H_z + H_J = 1/2 \begin{bmatrix} -v_a-v_b+J/2 & 0 & 0 & 0 \\ 0 & v_a-v_b-J/2 & J & 0 \\ 0 & J & -v_a+v_b-J/2 & 0 \\ 0 & 0 & 0 & v_a+v_b+J/2 \end{bmatrix} \begin{matrix} \alpha\alpha \\ \beta\alpha \\ \alpha\beta \\ \beta\beta \end{matrix}$$

The diagonal terms are just twice the energies of the four spin states $\alpha\alpha$ (top), $\beta\alpha$, $\alpha\beta$ and $\beta\beta$ (bottom) given in hertz. Here we are using the convention that aligned pairs of spins ($\alpha\alpha$ and $\beta\beta$) are higher in energy, meaning that the "β" components of doublets are downfield of the "α" components. Up till now in this book we have used the opposite convention, effectively assuming that the J value is negative. We will drop the 1/2 factor for simplicity, and bring it back when we need to calculate energies.

To find the stationary states ("eigenfunctions") and the energies ("eigenvalues") of the Hamiltonian we need to solve the Schrödinger equation:

$$H\Psi = E\Psi$$

This would be easy except for the off-diagonal terms. If the chemical-shift difference in hertz, $\Delta v = v_b - v_a$, is much larger than the coupling constant J, we can ignore the off-diagonal terms because they are very small compared to the diagonal terms. Then the energies are just the diagonal terms and the stationary states are just the $\alpha\alpha$, $\beta\alpha$, $\alpha\beta$ and $\beta\beta$ states. This "weak coupling" assumption is what we have been working under for most of this book:

$$\Delta\nu(\text{in Hz}) >> J \quad \text{"weak coupling"}$$

10.8.1 Strong Coupling: The AB System

We saw in Chapter 2 that when the chemical-shift difference in Hz, $\Delta\nu$, decreases and approaches the order of magnitude of J we get distortions of the peak heights ("leaning") and of the line positions. We are now in a position to derive this effect precisely for the AB system (H_a, H_b). The H_{11} and H_{44} terms of the Hamiltonian are enormous compared to the central portion of the matrix: for a 600-MHz spectrometer $\nu_a + \nu_b$ is around 1200 MHz! So we can assume that these diagonal terms are the correct energies and that the $\alpha\alpha$ and $\beta\beta$ states are true stationary states. We only need to look at the central 2×2 matrix in the Schrödinger equation:

$$\begin{bmatrix} -\Delta\nu\text{-}J/2 & J \\ J & \Delta\nu\text{-}J/2 \end{bmatrix} \begin{bmatrix} c_2 \\ c_3 \end{bmatrix} = E \begin{bmatrix} c_2 \\ c_3 \end{bmatrix}$$

where c_2 and c_3 are the coefficients of the $\beta\alpha$ and $\alpha\beta$ states, respectively, in the wave function Ψ, with $c_1 = c_4 = 0$. This can be written differently if we express E as a matrix (the identity matrix **1** times the number E):

$$\begin{bmatrix} -\Delta\nu\text{-}J/2 & J \\ J & \Delta\nu\text{-}J/2 \end{bmatrix} \begin{bmatrix} c_2 \\ c_3 \end{bmatrix} = \begin{bmatrix} E & 0 \\ 0 & E \end{bmatrix} \begin{bmatrix} c_2 \\ c_3 \end{bmatrix}$$

Then we subtract the right side of the equation from both sides to get

$$\begin{bmatrix} -\Delta\nu\text{-}J/2\text{-}E & J \\ J & \Delta\nu\text{-}J/2\text{-}E \end{bmatrix} \begin{bmatrix} c_2 \\ c_3 \end{bmatrix} = 0$$

This is just a pair of simultaneous linear equations. To solve for E we set the determinant (product of diagonal terms minus product of antidiagonal terms) to zero:

$$(-\Delta\nu - J/2 - E)(\Delta\nu - J/2 - E) - J^2 = 0$$

$$E^2 + JE + (-\Delta\nu^2 - J^2 + J^2/4) = 0; \quad E^2 + JE + (-\Delta\nu^2 - 3J^2/4) = 0$$

The solution to this quadratic equation in E is

$$E = -J/2 \pm \sqrt{\Delta\nu^2 + J^2} = -J/2 \pm \Delta\nu'$$

We can visualize $\Delta\nu'$ as the hypotenuse of a right triangle with sides equal to J and $\Delta\nu$ (Fig. 10.38, left). If $\Delta\nu > J$ (weak coupling limit), $\Delta\nu'$ becomes nearly equal to $\Delta\nu$ and the energies are just the diagonal elements H_{22} and H_{33} of the Hamiltonian. If $\Delta\nu = 0$, we have $\Delta\nu' = J$ and the two energies are $+J/2$ and $-3J/2$. The line positions are just the transition frequencies or the energy differences between the pairs of energy levels involved in the transition, so we can diagram the effect of strong coupling on the spectrum

482 ADVANCED NMR THEORY: NOESY AND DQF-COSY

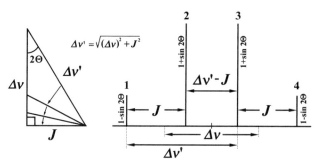

Figure 10.38

(Fig. 10.38, right). The doublets retain their separation of J Hz but they move apart from the positions we would expect based on the weak coupling assumption. $\Delta\nu'$ is the distance between lines 1 and 3 (or between lines 2 and 4). $\Delta\nu'$ can be viewed as the hypotenuse of a right triangle with sides of length $\Delta\nu$ and J, and the angle 2Θ opposite J. As $\Delta\nu$ decreases relative to J, $\Delta\nu'$ approaches J and 2Θ approaches $90°$. In the extreme case of $\Delta\nu = 0$, $\Delta\nu' = J$ and the two central lines (lines 2 and 3) coincide in the spectrum. In this case, the outer lines at $\nu + J$ and $\nu - J$ have zero intensity.

Once we have the energies we can solve for the wave functions; that is, for the coefficients c_2 and c_3. First try the solution $E = \Delta\nu' - J/2$, which in the weak coupling limit is the energy of the $\alpha\beta$ state (c_3 coefficient). We know that the probability of being in the $\beta\alpha$ state plus the probability of being in the $\alpha\beta$ state has to be 1, since c_1 and c_4 are zero. This means that $c_2^2 + c_3^2 = 1$. If we restrict c_2 and c_3 to real numbers, we can set $c_3 = \cos\Theta$ and $c_2 = \sin\Theta$ and we know that $\cos^2\Theta + \sin^2\Theta = 1$. Now we only have one variable, which along with E makes two variables to extract from our two linear equations. Setting E to the solution $\Delta\nu' - J/2$ we can solve for the "angle" Θ,

$$\begin{bmatrix} -\Delta\nu - J/2 & J \\ J & \Delta\nu - J/2 \end{bmatrix} \begin{bmatrix} c_2 \\ c_3 \end{bmatrix} = E \begin{bmatrix} c_2 \\ c_3 \end{bmatrix} \begin{matrix} \beta\alpha \\ \alpha\beta \end{matrix}$$

Multiplying the bottom row of the matrix by the column vector, we have

$$Jc_2 + (\Delta\nu - J/2)c_3 = Ec_3$$

$$J\sin\Theta + (\Delta\nu - J/2)\cos\Theta = (\Delta\nu' - J/2)\cos\Theta$$

$$J\sin\Theta = (\Delta\nu' - J/2 - \Delta\nu + J/2)\cos\Theta$$

$$\tan\Theta = \sin\Theta/\cos\Theta = (\Delta\nu' - \Delta\nu)/J$$

In the weak coupling limit, $\Delta\nu'$ becomes $\Delta\nu$, the ratio on the right side becomes zero and the angle Θ goes to zero. The wave function becomes simply $\alpha\beta$ since $c_3 = \cos\Theta = 1$ and $c_2 = \sin\Theta = 0$. As $\Delta\nu$ gets smaller and approaches J, the difference $\Delta\nu' - \Delta\nu$ gets larger relative to J and the angle Θ becomes positive. The wave function starts to be a mixture of $\alpha\beta$ and $\beta\alpha$. When $\Delta\nu$ reaches zero ($\nu_a = \nu_b$), $\Delta\nu' = J$ and we have $\tan\Theta = 1$ and $\Theta =$

45°. The wave function is now an equal mixture of $\beta\alpha$ and $\alpha\beta$:

$$\Psi_3 = \sin\Theta\,\beta\alpha + \cos\Theta\,\alpha\beta = (1/\sqrt{2})(\beta\alpha + \alpha\beta)$$

with energy $+J/2$. Like $\alpha\alpha$ and $\beta\beta$, this wave function is symmetric: if we switch the labels on H_a and H_b the wave function is the same.

A similar method shows that the other energy solution, $-\Delta\nu' - J/2$, gives a wave function $\Psi = \cos\Theta\,\beta\alpha - \sin\Theta\,\alpha\beta$ with Θ defined in the same way in terms of $\Delta\nu'$, $\Delta\nu$ and J. With $\Delta\nu > J$, the angle is close to zero and we have $\Psi \sim \beta\alpha$. As $\Delta\nu$ gets smaller we get more of the $\alpha\beta$ state mixed into the wave function, but now with a negative coefficient. When $\Delta\nu = 0$ we have the wave function

$$\Psi_2 = \cos\Theta\,\beta\alpha - \sin\Theta\,\alpha\beta = (1/\sqrt{2})(\beta\alpha - \alpha\beta)$$

with energy $-3J/2$. This wave function is antisymmetric: switching the labels on H_a and H_b makes the $\beta\alpha$ state into the $\alpha\beta$ state and *vice-versa*, and this changes the sign of the wave function. The two outer lines of the spectrum (lines 1 and 4) involve transitions from this antisymmetric state Ψ_2. The frequencies are

$$\text{line 1}: E_4 - E_2 = \tfrac{1}{2}[(2\nu + J/2) - (-3J/2)] = \nu + J$$

$$\text{line 4}: E_2 - E_1 = \tfrac{1}{2}[(-3J/2) - (-2\nu + J/2)] = \nu - J$$

The inner lines (lines 2 and 3) involve transitions from the symmetric state Ψ_3,

$$\text{line 3}: E_4 - E_3 = \tfrac{1}{2}[(2\nu + J/2) - (J/2)] = \nu$$

$$\text{line 2}: E_3 - E_1 = \tfrac{1}{2}[(J/2) - (-2\nu + J/2)] = \nu$$

A rule of quantum mechanics states that transitions between states of opposite symmetry are forbidden; this is why the intensity of the outer lines falls to zero in the limit of $\Delta\nu = 0$. In between, in the strong coupling "zone," the outer lines are diminished in intensity and this gives the "leaning" or "house shape" of the AB system.

The intensity of the lines turns out to be: $[\Psi_n(\mathbf{I}_x^a + \mathbf{I}_x^b)\Psi_m]^2$ for a transition between state m and state n, so we can calculate them using matrix math. For line 1 ($\Psi_2 \to \Psi_4$ transition) we have for $[\Psi_4(\mathbf{I}_x^a + \mathbf{I}_x^b)\Psi_2]$:

$$[0\ 0\ 0\ 1] \times \begin{bmatrix} 0 & 1 & 1 & 0 \\ 1 & 0 & 0 & 1 \\ 1 & 0 & 0 & 1 \\ 0 & 1 & 1 & 0 \end{bmatrix} \times \begin{bmatrix} 1 \\ \cos\Theta \\ -\sin\Theta \\ 0 \end{bmatrix} = [0\ 0\ 0\ 1] \times \begin{bmatrix} \cos\Theta-\sin\Theta \\ 0 \\ 0 \\ \cos\Theta-\sin\Theta \end{bmatrix}$$

$$= \cos\Theta - \sin\Theta$$

Squaring this we get $\cos^2\Theta - 2\cos\Theta\sin\Theta + \sin^2\Theta = 1 - 2\cos\Theta\sin\Theta = 1 - \sin 2\Theta$. From weak coupling to $\Delta\nu = 0$ the angle Θ goes from 0 to 45° and the intensity of the outer lines goes from 1 to 0. The same result is obtained for line 4, the other "outer" line of the AB system.

For line 2 (($\Psi_1 \to \Psi_3$ transition) we have for $[\Psi_3(\mathbf{I}_x^a + \mathbf{I}_x^b)\Psi_1]$:

$$[0 \;\; \sin\Theta \;\; \cos\Theta \;\; 0] \times \begin{bmatrix} 0 & 1 & 1 & 0 \\ 1 & 0 & 0 & 1 \\ 1 & 0 & 0 & 1 \\ 0 & 1 & 1 & 0 \end{bmatrix} \times \begin{bmatrix} 1 \\ 0 \\ 0 \\ 0 \end{bmatrix} = [0 \;\; \sin\Theta \;\; \cos\Theta \;\; 0] \times \begin{bmatrix} 0 \\ 1 \\ 1 \\ 0 \end{bmatrix}$$

$$= \cos\Theta + \sin\Theta$$

Squaring this we get $\cos^2\Theta + 2\cos\Theta\sin\Theta + \sin^2\Theta = 1 + 2\cos\Theta\sin\Theta = 1 + \sin 2\Theta$. From weak coupling to $\Delta\nu = 0$ the angle Θ goes from 0 to 45° and the intensity of the inner lines goes from 1 to 2. The same result is obtained for line 3, the other "inner" line of the AB system. These intensities are shown in Fig. 10.38.

10.8.2 The Commutator and the Evolution of Non-Stationary States

Consider the effect on H_a and H_b of the B_1 field placed on the x' axis during a pulse. The Hamiltonian can be represented as $\gamma B_1 (\mathbf{I}_x^a + \mathbf{I}_x^b)$; using the matrix operators and adding them together:

$$\gamma B_1 \left[\begin{bmatrix} 0 & 1 & 0 & 0 \\ 1 & 0 & 0 & 0 \\ 0 & 0 & 0 & 1 \\ 0 & 0 & 1 & 0 \end{bmatrix} + \begin{bmatrix} 0 & 0 & 1 & 0 \\ 0 & 0 & 0 & 1 \\ 1 & 0 & 0 & 0 \\ 0 & 1 & 0 & 0 \end{bmatrix} \right] = \gamma B_1 \begin{bmatrix} 0 & 1 & 1 & 0 \\ 1 & 0 & 0 & 1 \\ 1 & 0 & 0 & 1 \\ 0 & 1 & 1 & 0 \end{bmatrix}$$

Now what can we do with it? We put a spin state (A) into this environment and find out what happens to it. The way to find out is to calculate the *commutator*, which is the matrix:

$$\{H, A\} = HA - AH$$

Because matrix multiplication is often different depending on the order of the matrices in the product, the commutator is a measure of whether the two matrices *commute*. If they commute, the order of multiplication does not matter and the commutator is zero. This means that the spin state A is a stationary state: it is happy in the environment described by the Hamiltonian and it does not change. If the commutator is not zero, then the spin state A is not stationary and will oscillate between state **A** and a new state **B** described by the commutator:

$$\{H, A\} = HA - AH = iB$$

where i is the imaginary number, $\sqrt{-1}$. The oscillation is described by our familiar sine and cosine terms: $\mathbf{A} \to \mathbf{A}\cos + \mathbf{B}\sin$, where the argument of the cos and sin functions is the quantity in front of the Hamiltonian multiplied by τ, the time variable—in this case,

$$\sigma(\tau) = A\cos(\gamma B_1 \tau) + B\sin(\gamma B_1 \tau)$$

Let's find out what the spin state **B** is if we start with $A = I_z^a + I_z^b$, the equilibrium state:

$$\{H, A\} = (\gamma B_1/2) \left[\underbrace{\begin{bmatrix} 0 & 1 & 1 & 0 \\ 1 & 0 & 0 & 1 \\ 1 & 0 & 0 & 1 \\ 0 & 1 & 1 & 0 \end{bmatrix}}_{H} \times \underbrace{\begin{bmatrix} 1 & 0 & 0 & 0 \\ 0 & 0 & 0 & 0 \\ 0 & 0 & 0 & 0 \\ 0 & 0 & 0 & -1 \end{bmatrix}}_{A} \right.$$

$$\left. - \underbrace{\begin{bmatrix} 1 & 0 & 0 & 0 \\ 0 & 0 & 0 & 0 \\ 0 & 0 & 0 & 0 \\ 0 & 0 & 0 & -1 \end{bmatrix}}_{A} \times \underbrace{\begin{bmatrix} 0 & 1 & 1 & 0 \\ 1 & 0 & 0 & 1 \\ 1 & 0 & 0 & 1 \\ 0 & 1 & 1 & 0 \end{bmatrix}}_{H} \right]$$

$$= (\gamma B_1/2) \left[\begin{bmatrix} 0 & 0 & 0 & 0 \\ 1 & 0 & 0 & -1 \\ 1 & 0 & 0 & -1 \\ 0 & 0 & 0 & 0 \end{bmatrix} - \begin{bmatrix} 0 & 1 & 1 & 0 \\ 0 & 0 & 0 & 0 \\ 0 & 0 & 0 & 0 \\ 0 & -1 & -1 & 0 \end{bmatrix} \right]$$

$$= (\gamma B_1/2) \begin{bmatrix} 0 & -1 & -1 & 0 \\ 1 & 0 & 0 & -1 \\ 1 & 0 & 0 & -1 \\ 0 & 1 & 1 & 0 \end{bmatrix}$$

The commutator can be written as $\mathbf{i}\, (\gamma B_1)[-I_y^a - I_y^b]$.

$$-I_y^a - I_y^b = \tfrac{1}{2}\begin{bmatrix} 0 & i & 0 & 0 \\ -i & 0 & 0 & 0 \\ 1 & 0 & 0 & i \\ 0 & 0 & -i & 0 \end{bmatrix} + \tfrac{1}{2}\begin{bmatrix} 0 & 0 & i & 0 \\ 0 & 0 & 0 & i \\ -i & 0 & 0 & 0 \\ 0 & -i & 0 & 0 \end{bmatrix} = \tfrac{1}{2}\begin{bmatrix} 0 & i & i & 0 \\ -i & 0 & 0 & i \\ -i & 0 & 0 & i \\ 0 & -i & -i & 0 \end{bmatrix}$$

So the time course can be written as:

$$[I_z^a + I_z^b] \xrightarrow{\Theta_x \text{pulse}} [I_z^a + I_z^b]\cos\Theta + [-I_y^a - I_y^b]\sin\Theta$$

Since $\gamma B_1 \tau = \Theta$, the pulse rotation in radians.

That is a lot of work to find out what we already knew, but it establishes a general method to figure out what happens to a spin state when it finds itself in an environment described by a particular Hamiltonian. Here are a few Hamiltonians for common time-dependent situations we have looked at

$\gamma B_1 [I_y^a + I_y^b]$ RF Pulse on y' axis
$\Omega_a I_z^a + \Omega_b I_z^b$ Chemical shift evolution
$\pi J 2 I_z^a I_z^b$ J-Coupling evolution ($\Delta\nu \gg J$)

In each case, τ times the multiplier in front of the Hamiltonian becomes the argument (in radians) of the sine and cosine functions when we write out the time course of the spin state.

486 ADVANCED NMR THEORY: NOESY AND DQF-COSY

10.8.3 The Ideal Isotropic Mixing (TOCSY Spin-Lock) Hamiltonian

Now let's look at something we *do not* know the answer to: the ideal isotropic mixing Hamiltonian. This is the ideal TOCSY mixing sequence that leads to in-phase to in-phase coherence transfer. The ideal sequence of pulses creates this average environment expressed by the Hamiltonian. The "Zeeman" Hamiltonian that represents the chemical shifts goes away and we have only the isotropic (i.e., same in all directions) J-coupling Hamiltonian:

$$H = H_J = J[\mathbf{I}_x^a \mathbf{I}_x^b + \mathbf{I}_y^a \mathbf{I}_y^b + \mathbf{I}_z^a \mathbf{I}_z^b] \text{ (units of Hz)}$$

We already put together the matrix representation of H_J:

$$H_J = J/4 \begin{bmatrix} 1 & 0 & 0 & 0 \\ 0 & -1 & 2 & 0 \\ 0 & 2 & -1 & 0 \\ 0 & 0 & 0 & 1 \end{bmatrix}$$

Now let's try it on our spin-locked magnetization $\mathbf{I}_x^a + \mathbf{I}_x^b$:

$$\{H, A\} = (J/8) \left[\begin{bmatrix} 1 & 0 & 0 & 0 \\ 0 & -1 & 2 & 0 \\ 0 & 2 & -1 & 0 \\ 0 & 0 & 0 & 1 \end{bmatrix} \times \begin{bmatrix} 0 & 1 & 1 & 0 \\ 1 & 0 & 0 & 1 \\ 1 & 0 & 0 & 1 \\ 0 & 1 & 1 & 0 \end{bmatrix} - \begin{bmatrix} 0 & 1 & 1 & 0 \\ 1 & 0 & 0 & 1 \\ 1 & 0 & 0 & 1 \\ 0 & 1 & 1 & 0 \end{bmatrix} \times \begin{bmatrix} 1 & 0 & 0 & 0 \\ 0 & -1 & 2 & 0 \\ 0 & 2 & -1 & 0 \\ 0 & 0 & 0 & 1 \end{bmatrix} \right]$$

$$= (J/8) \left[\begin{bmatrix} 0 & 1 & 1 & 0 \\ 1 & 0 & 0 & 1 \\ 1 & 0 & 0 & 1 \\ 0 & 1 & 1 & 0 \end{bmatrix} - \begin{bmatrix} 0 & 1 & 1 & 0 \\ 1 & 0 & 0 & 1 \\ 1 & 0 & 0 & 1 \\ 0 & 1 & 1 & 0 \end{bmatrix} \right] = 0$$

The commutator is zero; in other words, the spin state commutes with the Hamiltonian so it is a stationary state. The spin-lock simply preserves this combination of H_a and H_b coherence on the x' axis. Let's try another starting state: $\mathbf{I}_x^a - \mathbf{I}_x^b$:

$$\{H, A\} = (J/8) \left[\begin{bmatrix} 1 & 0 & 0 & 0 \\ 0 & -1 & 2 & 0 \\ 0 & 2 & -1 & 0 \\ 0 & 0 & 0 & 1 \end{bmatrix} \times \begin{bmatrix} 0 & 1 & -1 & 0 \\ 1 & 0 & 0 & -1 \\ -1 & 0 & 0 & 1 \\ 0 & -1 & 1 & 0 \end{bmatrix} \right.$$

$$\left. - \begin{bmatrix} 0 & 1 & -1 & 0 \\ 1 & 0 & 0 & -1 \\ -1 & 0 & 0 & 1 \\ 0 & -1 & 1 & 0 \end{bmatrix} \times \begin{bmatrix} 1 & 0 & 0 & 0 \\ 0 & -1 & 2 & 0 \\ 0 & 2 & -1 & 0 \\ 0 & 0 & 0 & 1 \end{bmatrix} \right]$$

$$= (J/8) \left[\begin{bmatrix} 0 & 1 & -1 & 0 \\ -3 & 0 & 0 & 3 \\ 3 & 0 & 0 & -3 \\ 0 & -1 & 1 & 0 \end{bmatrix} - \begin{bmatrix} 0 & -3 & 3 & 0 \\ 1 & 0 & 0 & -1 \\ -1 & 0 & 0 & 1 \\ 0 & 3 & -3 & 0 \end{bmatrix} \right]$$

$$= J/2 \begin{bmatrix} 0 & 1 & -1 & 0 \\ -1 & 0 & 0 & 1 \\ 1 & 0 & 0 & -1 \\ 0 & -1 & 1 & 0 \end{bmatrix}$$

The commutator is not zero; it can be written as:

$$\{H, A\} = i(J/2) \begin{bmatrix} 0 & -i & i & 0 \\ i & 0 & 0 & -i \\ -i & 0 & 0 & i \\ 0 & i & -i & 0 \end{bmatrix}$$

$$= i(J) \left[\frac{1}{2}\begin{bmatrix} 0 & -i & 0 & 0 \\ i & 0 & 0 & 0 \\ 0 & 0 & 0 & i \\ 0 & 0 & -i & 0 \end{bmatrix} - \frac{1}{2}\begin{bmatrix} 0 & 0 & -i & 0 \\ 0 & 0 & 0 & i \\ i & 0 & 0 & 0 \\ 0 & -i & 0 & 0 \end{bmatrix} \right]$$

The two matrices on the bottom represent antiphase coherence on the y' axis:

$$\{H, A\} = iJ[2\mathbf{I}_y^a\mathbf{I}_z^b - 2\mathbf{I}_y^b\mathbf{I}_z^a] = iJB$$

where B is the "destination" state. Thus the spin state $\Delta_x = \mathbf{I}_x^a - \mathbf{I}_x^b$ will move towards the spin state $\Delta_{yz} = 2\mathbf{I}_y^a\mathbf{I}_z^b - 2\mathbf{I}_y^b\mathbf{I}_z^a$:

$$\Delta_x \rightarrow \Delta_x\cos(2\pi J\tau) + \Delta_{yz}\sin(2\pi J\tau)$$

In this weird environment, the individual spins states $\mathbf{I}_x^a$ and $\mathbf{I}_x^b$ are not important. The "collective spin mode" $\Sigma_x = \mathbf{I}_x^a + \mathbf{I}_x^b$ is stable and the "collective spin mode" $\Delta_x = \mathbf{I}_x^a - \mathbf{I}_x^b$ moves to Δ_{yz} and back to itself in an oscillatory manner. It is these *collective* spin modes that characterize TOCSY mixing, as opposed to the individual (independent) spin modes we normally deal with. Note that the off-diagonal terms of the Hamiltonian, which are important in strong coupling, are completely dominant in isotropic mixing. Not only is chemical shift missing from the central 2×2 region of the Hamiltonian ($\Delta v = 0$), but it is also gone from the H_{11} and H_{44} elements ($v = 0$). In this case we can look at individual spin states as a linear combination of the collective spin modes:

$$\mathbf{I}_x^a = \tfrac{1}{2}(\Sigma_x + \Delta_x) = \tfrac{1}{2}[(\mathbf{I}_x^a + \mathbf{I}_x^b) + (\mathbf{I}_x^a - \mathbf{I}_x^b)]$$

If we start with $\mathbf{I}_x^a$, for example, in a selective 1D TOCSY, it will evolve in the spin lock to give:

$$\Sigma_x + \Delta_x \rightarrow \Sigma_x + [\Delta_x\cos(2\pi J\tau) + \Delta_{yz}\sin(2\pi J\tau)]$$

since Σ_x is stationary (commutes with the Hamiltonian). Substituting the individual operators, we have:

$$\mathbf{I}_x^a \rightarrow \tfrac{1}{2}(\mathbf{I}_x^a + \mathbf{I}_x^b) + \tfrac{1}{2}(\mathbf{I}_x^a - \mathbf{I}_x^b)\cos(2\pi J\tau) + \tfrac{1}{2}(2\mathbf{I}_y^a\mathbf{I}_z^b - 2\mathbf{I}_y^b\mathbf{I}_z^a)\sin(2\pi J\tau)$$

$$= \tfrac{1}{2}\mathbf{I}_x^a(1 + \cos(2\pi J\tau)) + \tfrac{1}{2}\mathbf{I}_x^b(1 - \cos(2\pi J\tau)) + \tfrac{1}{2}(2\mathbf{I}_y^a\mathbf{I}_z^b - 2\mathbf{I}_y^b\mathbf{I}_z^a)\sin(2\pi J\tau)$$

This is the result that we showed without proof in Chapter 9. In-phase H_a coherence on x' is completely converted into in-phase H_b coherence on x' after $\tau = 1/(2J)$, with the antiphase term reaching a maximum in the middle of this period at $\tau = 1/(4J)$.

FURTHER READING

You will find that many of the sources do not use exactly the same matrix representations for some of the product operators and rotation matrices. The exact form of the density matrix depends on the numbering of the spin states and on certain conventions that are not consistent in the literature. In the above examples, the definitions are consistent with the product operator methods and with themselves.

1. Bax A. Two-Dimensional Nuclear Magnetic Resonance in Liquids. Delft University Press, D. Reidel Publishing Co; 1982 (especially pp. 12–23, introduction, pp. 129–153, multiple quantum coherence, and pp. 188–200, density matrix).
2. Subramanian C. Modern Techniques in High Resolution Fourier Transform NMR. Springer-Verlag; 1987.
3. Feynman RP, Leighton RB, Sands M. The Feynman Lectures on Physics, Volume III. Addison Wesley; 1971, Chapters 6–11.
4. Bothner-By AA, Stephens RL, Lee J-M, Warren CD, Jeanloz RW. Structure determination of a tetrasaccharide: transient nuclear Overhauser effects in the rotating frame. *J. Am. Chem. Soc.* 1984;**106**:811–813.
5. Braunschweiler L, Ernst RR. Coherence transfer by isotropic mixing: application to proton correlation spectroscopy. *J. Magn. Reson.* 1983;**53**:521–528.
6. Davis AL, Keeler J, Laue ED, Moskau D. Experiments for recording pure-absorption heteronuclear correlation spectra using pulsed field gradients. *J. Magn. Reson.* 1992;**98**:207–216.

11

INVERSE HETERONUCLEAR 2D EXPERIMENTS: HSQC, HMQC, AND HMBC

This chapter examines three two-dimensional (2D) experiments that correlate ^{13}C nuclei with ^{1}H nuclei within a molecule. Unlike the HETCOR experiment, in which ^{1}H magnetization is indirectly detected (F_1) and converted to ^{13}C magnetization that is directly detected (F_2), in these "inverse" experiments the F_1 dimension is ^{13}C and the F_2 dimension is ^{1}H. They are called "inverse" experiments because historically the ^{13}C-detected experiments were done first and therefore were considered "normal." There are numerous advantages to these experiments over the traditional HETCOR experiment, including increased sensitivity (a 0.5-mg sample is sufficient) and the ability to see long-range (two and three bond) interactions between ^{13}C and ^{1}H. The combination of HSQC and HMBC constitutes the most powerful method available for tracing out the carbon skeleton of an organic compound.

HSQC stands for heteronuclear single quantum correlation, meaning that two different types of nuclei (usually ^{1}H and ^{13}C) are correlated in a 2D experiment by the evolution and transfer of single-quantum (SQ) coherence, the simple magnetization that can be represented by vectors in the x'–y' plane. HMQC stands for heteronuclear multiple quantum correlation and does the same thing as HSQC except that it uses double-quantum (DQ) and zero-quantum (ZQ) coherence during the evolution (t_1) period. These mysterious states involve the DQ ($\alpha\alpha \leftrightarrow \beta\beta$) and ZQ ($\alpha\beta \leftrightarrow \beta\alpha$) transitions that cannot be directly observed but evolve in the x'–y' plane during the t_1 period and are then converted back to observable (SQ) coherence. HMQC and HSQC are equivalent in the appearance of the spectra and the processing of data. HMBC stands for heteronuclear multiple bond correlation, which is the same thing as HMQC except that the J value selected for coherence transfer is much smaller (10 Hz for HMBC versus 150 Hz for HMQC) so that the two- and three-bond relationships are detected ($^{2,3}J_{CH} \sim 10$ Hz) and the one-bond relationship is rejected ($^1J_{CH} \sim 150$ Hz).

NMR Spectroscopy Explained: Simplified Theory, Applications and Examples for Organic Chemistry and Structural Biology, by Neil E. Jacobsen
Copyright © 2007 John Wiley & Sons, Inc.

11.1 INVERSE EXPERIMENTS: ^{1}H OBSERVE WITH ^{13}C DECOUPLING

Because heteronuclear NMR started with the observation of ^{13}C and decoupling of ^{1}H, this is considered the "normal mode" for the two nuclei. Probes were originally designed with two concentric coils: an inner "observe" coil tuned to the ^{13}C frequency and an outer "decouple" coil tuned to the ^{1}H frequency (Chapter 4, Fig. 4.9). The inner coil was used for ^{13}C because ^{13}C is a far less sensitive nucleus to observe, so the coil needs to be as close as possible to the sample. The opposite experiment is called an "inverse" experiment for purely historical reasons: ^{1}H is observed on the inner coil whereas ^{13}C pulses and decoupling are applied to the outer (^{13}C) coil. This arrangement is called an "inverse" probe. The obvious advantage of observing ^{1}H rather than ^{13}C is sensitivity: Of the three "gammas" (Chapter 1, Section 1.4) that contribute to the amplitude of the NMR signal of a nucleus, two are involved in the observation of the FID: the strength of the nuclear magnet (γ) and the rate at which it precesses in the x–y plane (γB_o). These together determine the intensity of the FID signal induced in the probe coil. Because γ_H is four times larger than γ_C, this means that observation of ^{1}H gives a signal 16 times larger than observation of ^{13}C.

The other gamma comes from the population difference at equilibrium; we saw in Chapter 7 how ^{13}C can be observed using the population difference of ^{1}H, so this disadvantage can be overcome by coherence transfer. This is the strategy used in the 2D HETCOR experiment (Chapter 9). The low natural abundance of ^{13}C, 1.1%, is also irrelevant here—whether we observe ^{1}H or ^{13}C in a 2D correlation experiment we are still dependent upon the number of ^{13}C–^{1}H pairs in the sample, and this is limited by the natural abundance of ^{13}C. In this analysis we have ignored the intensity of the noise; to describe the sensitivity of an NMR experiment we also have to consider the noise amplitude. It turns out that noise amplitude is roughly proportional to the square root of the frequency being detected (ν_o), so this reduces S/N by a factor of $\sqrt{\gamma B_o}$. So the advantage of inverse detection in a ^{1}H-^{13}C 2D experiment is actually a factor of 8 in signal-to-noise ratio ($4 \times 4/\sqrt{4}$).

Another advantage of ^{1}H observation is that a proton can only be attached to one ^{13}C. We saw in Chapter 7 the complexities of refocusing of ^{13}C antiphase coherence: different times are optimal for CH, CH$_2$, and CH$_3$ groups. A proton coupled to ^{13}C will always be a doublet—never a triplet or quartet—and will evolve into antiphase or refocus from antiphase to in-phase in a time of exactly $1/(2J)$.

There is one disadvantage to the observation of ^{1}H. With the observation of ^{13}C, the 99% of carbon nuclei that are ^{12}C are invisible in the NMR experiment and so they do not interfere in any way. In contrast, when we observe ^{1}H we are trying to see the 1.1% of protons that are associated with ^{13}C in the presence of the 98.9% of protons that are associated with ^{12}C. Both will give a signal in the ^{1}H FID unless we use special techniques to destroy the "^{12}C artifact." In a 1D ^{1}H spectrum, the peaks we normally see are due to the ^{12}C-bound protons, which constitute the vast majority of protons ($\sim$99%, neglecting the protons bound to oxygen and nitrogen). The ^{13}C-bound protons appear as tiny "satellites," which are very wide doublets ($^1J_{CH} \sim 150$ Hz or 0.5 ppm on a 300-MHz spectrometer) centered on the ^{12}C-bound proton signal and 0.55% of its peak intensity (Fig. 11.1). In this chapter we will focus on these satellite peaks, around 150 Hz apart, and do our best to destroy the 100 times larger ^{12}C-bound ^{1}H signal in the center.

We saw in Chapter 2 (Fig. 2.18) that the intensities of these two peaks (singlet for ^{1}H–^{12}C and doublet for ^{1}H–^{13}C) are reversed for ^{13}C-labeled compounds: The wide doublet dominates, and the residual ^{12}C shows up as a tiny central singlet. Not only are the protons directly attached to ^{13}C "split" into doublets, but also those two or three bonds away (^{13}C–C$\underline{H}$ and

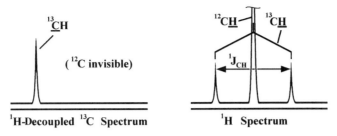

Figure 11.1

^{13}C–C–C$\underline{\text{H}}$) are split by couplings to ^{13}C similar in magnitude to the homonuclear $^2J_{HH}$ (geminal) and $^3J_{HH}$ (vicinal) couplings (Fig. 11.2). These "long range" (>1 bond) heteronuclear couplings can be transmitted through oxygen and nitrogen as well as carbon: ^{13}C–O–C$\underline{\text{H}}$ and ^{13}C–N–C$\underline{\text{H}}$. As long as the coupled proton is not exchanging too rapidly to be observed, it can even be bound to nitrogen or oxygen: ^{13}C–C–O$\underline{\text{H}}$ or ^{13}C–C–N$\underline{\text{H}}$. The three-bond couplings even show a Karplus dependence (Chapter 2, Fig. 2.11) on dihedral angle very similar to the vicinal $^3J_{HH}$ coupling: Maximum $^3J_{CH}$ coupling occurs in the *anti* configuration of ^{13}C–X–Y–^{1}H, with H and C opposite each other. All of these heteronuclear couplings are in addition to the homonuclear J couplings that determine the "multiplicity" of a ^{1}H resonance: double doublet, ddd, dt, and so on. For example, a ddt ^{1}H peak will have identical ddt peaks about 75 Hz upfield and downfield of the central ^{12}C peak, 0.55% of the intensity of the central peak.

Figure 11.3 shows the downfield portion of the ^{1}H spectrum of sucrose (g1 doublet, $J_{HH} = 3.8$ Hz) with normal vertical scaling and with the vertical scale increased by a factor of 100 to show the ^{13}C satellites. The satellites show the same doublet J_{HH} coupling observed in the ^{12}C-bound proton signal ($J = 3.8$ Hz), with an additional 169.6 Hz coupling to the ^{13}C nucleus ($^1J_{CH}$). The peak height is roughly half (actually half of 1% because the vertical scale is 100 ×) of the central peak because they are part of a doublet with concentration about 1% of the concentration of the ^{12}C species. If you look closely you will see that the center of the $^1\underline{\text{H}}$–^{13}C double doublet is not exactly the same chemical shift as the center of the ^{1}H–^{12}C doublet. There is a small isotope effect on chemical shift, so we do not expect them to be exactly the same. Figure 11.4 (top) shows the ^{1}H spectrum of

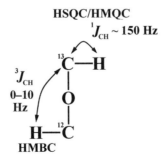

Figure 11.2

492 INVERSE HETERONUCLEAR 2D EXPERIMENTS: HSQC, HMQC, AND HMBC

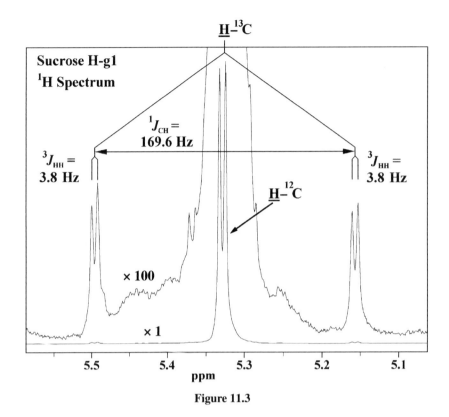

Figure 11.3

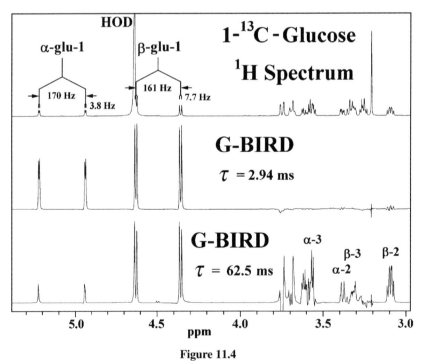

Figure 11.4

glucose labeled with ^{13}C in the C-1 (anomeric) position. Glucose exists as an equilibrium mixture of α and β anomers (Chapter 1, Fig. 1.12) in slow exchange. The α anomer (H-1 equatorial) gives a wide, 170 Hz doublet for the H-1 proton, which is directly bonded to ^{13}C. The smaller doublet coupling (3.8 Hz) is the homonuclear ($^{3}J_{HH}$) coupling to H-2, which is small because it is a *gauche* (equatorial–axial) relationship. This is just like the ^{13}C satellites observed for H-g1 of sucrose (Fig. 11.3). The β anomer (H-1 axial) gives a 161 Hz doublet for the H-1 proton, with a homonuclear coupling of 7.7 Hz. This vicinal (H–H) coupling (7.7 Hz) is relatively large because H-1 in the β form has an axial–axial (*anti*) relationship to H-2. Anomeric protons (protons on doubly oxygenated sp^3 carbons) give larger $^{1}J_{CH}$ couplings (160–170 Hz) than protons on singly oxygenated carbons (140–150 Hz). In addition, in six-membered ring sugars the α anomer tends toward the upper end of this range (~170 Hz) and the β anomer tends toward the lower end (~160 Hz). These one-bond C–H coupling constants can be useful in structure analysis in other ways as well, and we will see how they can be obtained from 2D HSQC (or HMQC) spectra.

11.1.1 Isotope Filtering: Fun With BIRDs

We saw in Chapter 9 how the BIRD building block (Fig. 9.11) can be used as a selective 180° pulse that affects only the ^{12}C-bound protons and has no effect on the ^{13}C-bound protons. We can change the selectivity by changing the phase of the central 180° ^{1}H pulse from y to $-x$ (Fig. 11.5). For a ^{12}C-bound proton, there is no net evolution during the two delays, so we can see it as three rotations about the x' axis: 90°, -180°, and 90°, adding up to zero. For a ^{13}C-bound proton, if we start with $\mathbf{I}_z$ the first 90° pulse rotates to $-\mathbf{I}_y$, and the spin-echo results in *J*-coupling evolution from $-\mathbf{I}_y$ to $2\mathbf{I}_x\mathbf{S}_z$ and on to $\mathbf{I}_y$, with the central 180° pulse on $-x$ changing this to $-\mathbf{I}_y$. The final 90° pulse rotates $-\mathbf{I}_y$ to $-\mathbf{I}_z$, so overall we have an inversion (180° pulse).

We saw in Chapter 8 how a selective 180° pulse can be placed between two gradients of the same sign and duration to give a pulsed field gradient spin echo (PFGSE) that not only selects the desired coherence but also destroys any other coherences. First, we use a hard 90° pulse to create coherence on all spins, and then the first gradient twists the coherence into a helix (Fig. 8.21). The selective 180° pulse reverses the direction of twist in

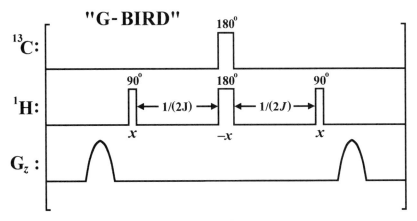

Figure 11.5

the helix for the selected spins only, and then the second gradient "untwists" the helix for the selected spins and doubles the twist for all other spins. We can put our selective 180° pulse (BIRD: ^{13}C-bound ^{1}H only) between two gradients and get the same effect: all ^{12}C-bound ^{1}H coherence is "double-twisted" and all ^{13}C-bound ^{1}H coherence is "rescued" by the second gradient (Fig. 11.5). The sequence is repeated to give a double PFGSE (DPFGSE) with a 180° ^{1}H pulse in the center to reverse J coupling evolution that happens during the gradients: ^{1}H 90°–G_1–BIRD–G_1–^{1}H 180°–G_2–BIRD–G_2–FID. This overall sequence is called "G-BIRD." After the first gradient, we have magnetization in the x'–y' plane in all directions, depending on the position along the z axis in the NMR tube. The effect of the BIRD element on the ^{12}C-bound coherence is as follows:

$$\mathbf{I}_x \xrightarrow{^1\text{H} 90°_x} \mathbf{I}_x \xrightarrow{\tau-180-\tau} \mathbf{I}_x \xrightarrow{^1\text{H} 90°_x} \mathbf{I}_x$$

$$\mathbf{I}_y \xrightarrow{^1\text{H} 90°_x} \mathbf{I}_z \xrightarrow{\tau-180-\tau} -\mathbf{I}_z \xrightarrow{^1\text{H} 90°_x} \mathbf{I}_y$$

The effect on the ^{13}C-bound proton coherence in the x–y plane is:

$$\mathbf{I}_x \xrightarrow{^1\text{H} 90°_x} \mathbf{I}_x \xrightarrow{\tau-180-\tau} -\mathbf{I}_x \xrightarrow{^1\text{H} 90°_x} -\mathbf{I}_x$$

$$\mathbf{I}_y \xrightarrow{^1\text{H} 90°_x} \mathbf{I}_z \xrightarrow{\tau-180-\tau} -\mathbf{I}_z \xrightarrow{^1\text{H} 90°_x} \mathbf{I}_y$$

In the case of $\mathbf{I}_x$, the central spin echo ($\tau-180-\tau$) leads to $^1J_{\text{CH}}$ evolution for a total time of $1/J$, which moves $\mathbf{I}_x$ to $2\mathbf{I}_y\mathbf{S}_z$ and on to $-\mathbf{I}_x$. The central 180° pulse on ^{1}H must be taken into account, but we consider it as if it happened at the beginning of the evolution, when we have $\mathbf{I}_x$, so it has no effect. Thus, the overall effect of the BIRD element is exactly the same as a 180° pulse on the y' axis for the ^{13}C-bound protons, and it has no effect on the ^{12}C-bound protons. This will reverse the sense of the coherence helix in the NMR tube for the ^{13}C-bound protons only, allowing the second gradient to "straighten them out" while it further scrambles the ^{12}C-bound protons.

Figure 11.4 (center) shows the result of the G-BIRD sequence on 1-^{13}C glucose with the spin-echo delay ($1/(2J)$) set to 2.94 ms (i.e., for $^1J_{\text{CH}} = 170$ Hz). The HOD peak is gone, as are all of the glucose peaks that are not due to the H-1 position, the position that is labeled with ^{13}C. We only see the ^{13}C-bound protons. This is an example of *isotope filtering*, a class of NMR experiments in which only those protons that are bound to ^{13}C (or ^{15}N) show up in the experiment. Filtering can work either way: We can also set it up to see only the protons not bound to ^{13}C (or ^{15}N). This is a very powerful and sophisticated tool for biological NMR. For example, an unlabeled small molecule can be bound to a ^{13}C-labeled protein and only those NOE interactions between a ^{13}C-bound proton and a ^{12}C-bound proton pass through the isotope filter. This allows observing only the interactions between the protein and the ligand, without interference from the intramolecular NOEs.

We can change the delay time of the G-BIRD to "tune" the isotope filter to different $^1J_{\text{CH}}$ values. The crucial J-coupling evolution for the time $1/J$ must give an inversion (180° rotation in the x'–y' plane from in-phase to antiphase and back to in-phase) in order for the coherence to survive the G-BIRD gradients. What happens to ^{13}C-bound ^{1}H coherence if the delay is not exactly tuned to the J_{CH} coupling? We can use product operators to predict the result:

$$\mathbf{I}_x \xrightarrow{^1\mathrm{H}\,90°_x} \mathbf{I}_x \xrightarrow{\tau} \mathbf{I}_x \cos(\pi J\tau) + 2\mathbf{I}_y\mathbf{S}_z \sin(\pi J\tau) \xrightarrow[^{13}\mathrm{C}\,180°]{^1\mathrm{H}\,180°_{-x}} \mathbf{I}_x \cos(\pi J\tau) + 2\mathbf{I}_y\mathbf{S}_z \sin(\pi J\tau)$$

$$\xrightarrow{\tau} \mathbf{I}_x \cos^2 + 2\mathbf{I}_y\mathbf{S}_z(2\sin\cos) - \mathbf{I}_x \sin^2 \xrightarrow{^1\mathrm{H}\,90°_x} \mathbf{I}_x \cos^2 + 2\mathbf{I}_z\mathbf{S}_z(2\sin\cos) - \mathbf{I}_x \sin^2$$

The first term is just like ^{12}C-bound ^{1}H coherence: It has not experienced a 180° pulse, and it will be destroyed by the second gradient. Only the last term will survive because it has experienced the equivalent of a $180°_y$ pulse. In the complete G-BIRD there are two PFGSE elements, so the term becomes $\sin^4(\pi J\tau)$. In the example of Figure 11.4 (center), τ was set to 2.94 ms, so the effect on various $^1J_{CH}$ values can be predicted, as well as on long-range ^2J and ^3J values:

J:	170	160	142	125	250	8
$\sin^4(\pi J\tau)$:	1.0	0.98	0.87	0.70	0.30	0.00003

The filter is fairly tolerant of the range of $^1J_{CH}$ values normally encountered (142 is a typical singly oxygenated carbon, 125 is a saturated hydrocarbon environment, 250 is a terminal alkyne, and 8 is the maximum for long-range $^{2,3}J_{CH}$ couplings). This is the assumption built into all of the heteronuclear coherence transfer experiments, from DEPT to HETCOR to the inverse experiments HSQC and HMQC: that the one-bond CH coupling is around 150 Hz and the range of variation is not that large.

Figure 11.4 (bottom) shows the result of the G-BIRD experiment with the τ delay set to 62.5 ms ($1/(2J)$ for $J = 8$ Hz). We see many signals now in the ^{12}C-bound ^{1}H region of the spectrum because many of these have long-range couplings to C-1. We expect a $^2J_{CH}$ coupling from C-1 to H-2 in both forms and a $^3J_{CH}$ coupling from C-1 to H-3 and H-5. The most prominent peak is the β-glucose H-2 peak at 3.1 ppm (compare the lactose β-glu-2 peak in Fig. 10.10). We still see the H-1 peaks because these are spinning around in the $x'-y'$ plane many, many times during the long 62.5 ms τ delay and where they land is more or less random. For example, H-1 of α-glucose has a J coupling of 170 Hz to C-1 and moves from $\mathbf{I}_x$ to $2\mathbf{I}_y\mathbf{S}_z$ by J coupling evolution ($1/(2J)$) in 2.94 ms, so in 62.5 ms it has made 5.3 complete cycles from $\mathbf{I}_x$ to $2\mathbf{I}_y\mathbf{S}_z$ to $-\mathbf{I}_x$ to $-2\mathbf{I}_y\mathbf{S}_z$ and back to $\mathbf{I}_x$. The term $\sin^4(\pi J\tau)$ is 0.73 for 170 Hz, but it is zero if $J\tau$ is an integer; that is, if J is any multiple of 16 Hz ($1/\tau$). A J_{CH} value of 128, 144, 160, or 176 Hz would give a null whereas J values of 136, 152, 168, or 184 would give a maximum signal. We will see how this long evolution delay leads to the same effects in the HMBC experiment.

Finally, if we observe ^{1}H we will have to consider how to remove the ^{13}C coupling during the recording of the FID; otherwise, we will have these very wide (150 Hz) doublets centered on the ^{1}H chemical shift. ^{13}C decoupling requires more power because of the much larger range of chemicals shifts (~ 200 ppm ^{13}C on a 600-MHz instrument is $200 \times 150 = 30{,}000$ Hz, whereas ~ 10 ppm ^{1}H corresponds to $10 \times 600 = 6000$ Hz) that must be "covered" by a broadband decoupling scheme. In addition, the B_1 field strength (or B_2 as it is called for decoupling) needs to be four times greater (16 times greater power) to achieve the same rate of rotation ($\gamma B_2/2\pi$) of the sample's ^{13}C magnetization because γ is only one fourth as large for ^{13}C. These problems have been solved by new, more efficient composite pulse decoupling sequences and by limiting the "duty cycle" (percent time the decoupler is on during the pulse sequence) to avoid heating the sample and overtaxing the amplifiers.

^{13}C decoupling was not a standard feature on spectrometers until the mid-1990s, so many of the inverse experiments (HSQC and HMQC) were done without ^{13}C decoupling, simply living with the wide doublets due to $^1J_{CH}$. Now this technique is routine on modern spectrometers. Computer optimization of the WALTZ-16 decoupling sequence (Chapter 4, Section 4.4) led to the GARP sequence, which uses seemingly arbitrary pulse widths: 30.5°, 55.2°, 257.8°, 268.3°, 69.3°, 62.2°, 85.0°, 91.8°, 134.5°, 256.1°, 66.4°, 45.9°, 25.5°, 72.7°, 119.5°, 138.2°, 258.4°, 64.9°, 70.9°, 77.2°, 98.2°, 133.6°, 255.9°, 65.5°, and 53.4°, with phase alternating $x, -x, x, -x$, and so on. This sequence of 25 pulses is executed four times with the phases reversed ($-x, x, -x, x$, etc.) in cycles three and four. The 100-pulse supercycle is repeated throughout the acquisition of the ^{1}H FID. Figure 11.6 (top) shows the results of a ^{13}C decoupling test on the β-anomeric proton (β-glu-1) of 1–^{13}C-glucose using GARP decoupling at a power level corresponding to $\gamma_C B_2/2\pi = 5320$ Hz $(1/(4 \times 47 \mu s))$. The decoupler offset is started at the ^{13}C resonance of C-1 of β-glucose and then moved 10 ppm farther off-resonance each time the experiment is repeated. This is exactly the reverse of the WALTZ-16 decoupler test (^{13}C observe, ^{1}H decouple) shown in Figure 4.7. Excellent decoupling is obtained for offsets up to 80 ppm on either side of the on-resonance ^{13}C frequency, representing a bandwidth of 160 ppm. This is adequate to "cover" all protonated carbons (0–150 ppm) except the carbonyl carbon of an aldehyde (~200 ppm). At 90 ppm off-resonance, we see the doublet ($^3J_{HH} = 7.7$ Hz) split into two doublets due to the $^1J_{CH}$ coupling that is not completely removed. In Chapter 8 (Section 8.10) we briefly touched on the use of a moving spin lock as an efficient broadband method for inversion ("adiabatic inversion") using a shaped pulse. The frequency of the spin lock is "swept" from far upfield to far downfield, which has the effect of starting the spin-lock axis on the $+z$ axis and slowly (i.e., on a timescale of ms) tilting it down to the x'–y' plane and on down to the $-z$ axis, maintaining the sample magnetization spin

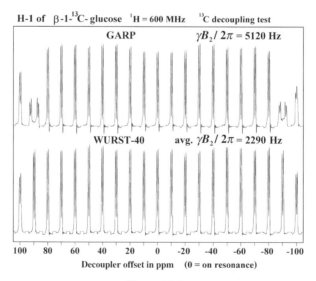

Figure 11.6

locked at all times. If this process is repeated over and over (+z to −z, −z to +z, etc.) during the acquisition of the FID, we have a decoupling sequence that can "cover" a very wide range of ^{13}C shifts with relatively low power output. After all, decoupling is just the process of "confusing" the protons by showing them an attached ^{13}C that is rapidly flipping between the α and β states. The effectiveness of adiabatic decoupling (WURST-40) is shown in Figure 11.6 (bottom). With less than half of the average B_2 amplitude used for the GARP (rectangular pulse) decoupling scheme, we see good decoupling over a bandwidth of 200 ppm. For modern instruments with the capability of generating rapid strings of shaped pulses on the second (decoupler) channel, this is clearly the method of choice.

A look at direct ($^1J_{CH}$) and long-range ($^{2,3}J_{CH}$) heteronuclear couplings of glucose in the 1D ^{1}H spectrum will prepare us for the direct (HSQC, HMQC) and long-range (HMBC) 2D experiments. Figure 11.7 shows the downfield region of the ^{1}H spectrum of 1-^{13}C-glucose in D$_2$O, without ^{13}C decoupling (top) and with GARP ^{13}C decoupling (bottom). The large (160–170 Hz) $^1J_{CH}$ couplings are collapsed in the lower spectrum and only the homonuclear ($^3J_{HH}$) couplings remain. In the upper spectrum the small amount of residual ^{12}C-bound ^{1}H is seen at the center of the wide ^{13}C-bound ^{1}H doublet. The upfield region of the ^{1}H spectrum is shown in Figure 11.8. On the right-hand side, the β-glu-2 proton shows up as a ddd ($J = 9.4, 7.9, 6.2$ Hz) without ^{13}C decoupling (top) and as a dd ($J = 9.4, 7.9$ Hz) with ^{13}C decoupling. The 6.2 Hz coupling can be assigned to the $^2J_{CH}$ coupling between H-2 and C-1 in β-glucose. This is a relatively large CH coupling that would give rise to a strong crosspeak in a 2D HMBC spectrum. In the center portion two CH couplings are evident: a 2.3 Hz coupling on the right side (C-1 to H-5 of β-glucose) added to the ddd pattern of H-5β and a barely resolved 1.3 Hz coupling in the middle (C-1 to H-3 of β-glucose) added to the triplet pattern of H-3β. In the left portion a 1.9 Hz CH coupling can be identified (C-1 to H-5 of α-glucose). All of these are $^3J_{CH}$ couplings—relatively small because the protons are in axial positions and have a *gauche* relationship to C-1. These would probably show up as weak HMBC crosspeaks or would be lost in the noise if the proton already has complex splitting like H-5α. The arrows indicate peaks that are broadened by coupling to ^{13}C with couplings too small to be resolved: C-1 to H-2 of α-glucose (middle section) and C-1 to H-3 of α-glucose (left). These couplings are probably too small to show up as crosspeaks in a 2D HMBC spectrum.

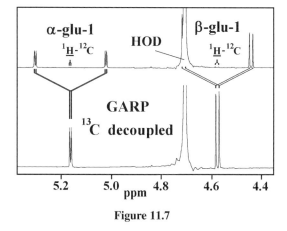

Figure 11.7

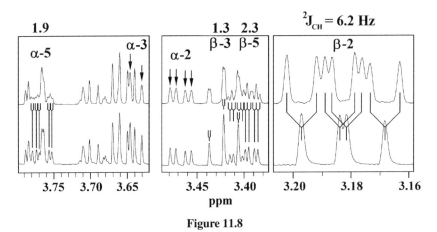

Figure 11.8

11.2 GENERAL APPEARANCE OF INVERSE 2D SPECTRA

11.2.1 2D HETCOR *versus* 2D HSQC/HMQC

We saw in Chapter 9 that the INEPT experiment can be extended to a ^{1}H–^{13}C 2D correlation experiment ("HETCOR") by simply adding an evolution (t_1) delay. The coherence flow diagram (Fig. 9.3) shows that ^{1}H coherence is created and evolves during t_1 and then is transferred to ^{13}C coherence, which is observed in t_2. To create the equivalent inverse experiment, we simply reverse the roles of ^{13}C and ^{1}H: ^{13}C coherence is created and evolves during t_1, encoding its chemical shift, and then "reverse" INEPT transfer (mixing) moves it to ^{1}H where it is observed in the FID. This would give us a factor of 16 increase in signal strength in recording the FID, but we would lose a factor of 4 because now we are starting with the population difference of ^{13}C rather than that of ^{1}H. For this reason we use a more complex preparation step: The proton is excited with the first pulse, and then this coherence is immediately transferred to ^{13}C. These two parts of the preparation step can be labeled as step 1a and 1b (Fig. 11.9).

Both experiments yield a 2D spectrum that has ^{13}C chemical shifts on one axis and ^{1}H chemical shifts on the other axis, with crosspeaks representing the one-bond relationship between ^{13}C and ^{1}H. The main difference is that the HSQC spectrum has the ^{13}C chemical shifts on the indirect (F_1) axis whereas the HETCOR spectrum has the ^{13}C chemical shifts in the directly detected (F_2) dimension. Thus, an HSQC spectrum looks like a HETCOR spectrum turned on its side (90°). We saw in the last section the consequences of detecting the ^{1}H signal directly: We have a much stronger signal (16 ×) and better resolution (complex

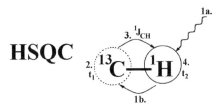

Figure 11.9

^{1}H multiplets observed in F_2), but we have to solve the technical problems of removing the ^{12}C-bound ^{1}H artifact (100 times larger than the desired signal), and we would like to be able to decouple ^{13}C. Thus, there are two important differences between HETCOR and the inverse one-bond experiments HSQC and HMQC:

1. In HMQC/HSQC, the crosspeaks appear in pairs separated in the F_2 (horizontal) dimension by the large one-bond CH coupling ($\sim$ 150 Hz) and centered on the ^{1}H chemical shift. This coupling can be eliminated by turning on a ^{13}C decoupler during the acquisition of the FID, which operates just like the ^{1}H decoupler in a ^{13}C-detected experiment. In many cases, however, this coupling gives useful information because the exact value of $^1J_{CH}$ gives us structural information.

2. Between these pairs there will be a vertical streak (parallel to the F_1 axis) that represents the ^{12}C-bound proton signal. Because the ^{12}C-bound proton signal is not modulated in t_1, the ^{13}C evolution period, it has no F_1 frequency, and so it just appears at all F_1 frequencies; that is, as a vertical streak. This problem can be solved by coherence pathway selection using phase cycling or gradients.

11.2.2 One-Bond (HSQC/HMQC) Versus Multiple-Bond (HMBC) 2D Spectra

Consider the molecular fragment H_aC_a–O–C_bH_b: In some portion of the molecules (about 1.1%), we will have $C_a = {}^{13}C$ and $C_b = {}^{12}C$, giving rise to an $F_1 = C_a$, $F_2 = H_a$ crosspeak in the HSQC (or HMQC) spectrum due to $^1J_{CH}$, and an $F_1 = C_a$, $F_2 = H_b$ crosspeak in the HMBC spectrum due to $^3J_{CH}$ (Fig. 11.10, upper dotted line). The HSQC crosspeak will be a wide doublet separated by the large $^1J_{CH}$ coupling ($\sim$150 Hz), and the HMBC peak will be a single "blob." In another portion of the molecules (again $\sim$ 1.1%), C_a will be ^{12}C and C_b will be ^{13}C, giving rise to the $F_1 = C_b$, $F_2 = H_b$ crosspeak in the HSQC (HMQC) spectrum and a crosspeak at $F_1 = C_b$, $F_2 = H_a$ in the HMBC spectrum (Fig. 11.10, lower dotted line). It is important to keep in mind that these are two different experiments, and the data are superimposed on the same 2D spectrum for comparison only. With these two 2D spectra, we can establish that H_a is three bonds or less distant from C_b, and likewise H_b is

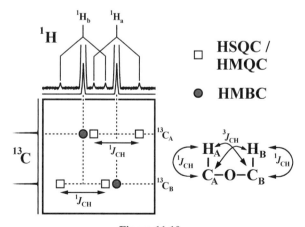

Figure 11.10

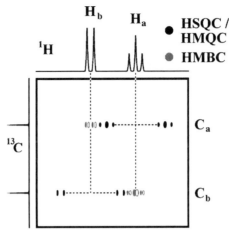

Figure 11.11

three bonds or less away from C_a. We can conclude that C_a and C_b are either directly bonded (C_a–C_b) or are separated by one intervening atom (C_a–C–C_b, C_a–O–C_b, C_a–N–C_b, etc.).

Now consider the homonuclear couplings of H_a and H_b. Let's expand the fragment to –CH_2–C_aH_a–O–C_bH_b–CH– so that H_a appears in the ^{1}H spectrum as a triplet and H_b appears as a doublet (Fig. 11.11). These multiplicities also show up in the ^{13}C satellites on either side of the main ($^1\underline{H}$–^{12}C) peaks. In the one-bond inverse correlation spectrum (HSQC or HMQC, in black), these satellite peaks appear at the F_1 position of the corresponding ^{13}C resonance: H_a satellite peaks at $F_1 = C_a$ and H_b satellite peaks at $F_1 = C_b$. In fact, the F_2 slice at $F_1 = C_a$ is a double triplet, with the doublet coupling being the large $^1J_{CH}$ coupling, and the F_2 slice at $F_1 = C_b$ is a double doublet, with one of the doublet couplings being the large coupling to C_b. The HMBC crosspeaks (in gray and white for positive and negative intensities) also contain the additional coupling to ^{13}C, but it is much smaller because it is a long-range ($^2J_{CH}$ or $^3J_{CH}$) coupling, which is of the same order of magnitude as the homonuclear (J_{HH}) couplings. If the HMBC data are processed in phase-sensitive mode, the coupling to ^{13}C will appear *antiphase*, just as the active couplings in COSY spectra appear. This is because the INEPT transfer is always antiphase to antiphase (Fig. 11.9, mixing = step 3). In the one-bond experiments (HSQC/HMQC), this antiphase coherence is refocused, so the large $^1J_{CH}$ coupling appears in-phase, but in HMBC the refocusing delay ($1/(2J)$) would be too long, so it is left antiphase.

Let's move one step farther and consider a real molecule, ethyl acetate (Fig. 11.12). The cartoon shows a superposition of the HSQC/HMQC spectrum (white) and the magnitude-mode HMBC spectrum (black). ^{13}C decoupling is not used in the HSQC/HMQC spectrum, so the one-bond crosspeaks appear as wide doublets centered on the ^{1}H chemical shift in F_2. From center-left to upper-right we see the paired one-bond crosspeaks in roughly diagonal fashion, displaying the homonuclear splitting pattern in F_2: quartet for the CH_2 group, singlet for the acetate CH_3 group, and triplet for the ethyl CH_3 group. The carbonyl carbon does not show up in the HMQC spectrum because it has no directly attached proton. In the HMBC spectrum (black), there are two crosspeaks at the ^{13}C position of the ester carbonyl carbon in F_1: A double quartet at the ^{1}H shift of the CH_2 group in F_2 and a doublet at the ^{1}H shift of the acetate CH_3 group in F_2. The additional coupling in each

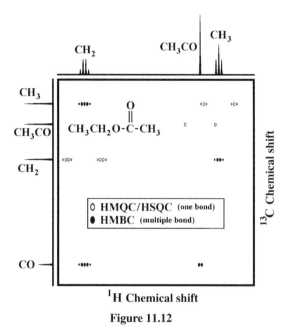

Figure 11.12

case is the long-range J_{CH}, which is comparable to the homonuclear couplings ($^3J_{HH}$) in magnitude. The crosspeak between the CH_2 protons of the ethyl group and the carbonyl carbon of the acetate portion is especially important: It connects two parts of the molecule that cannot be connected by homonuclear (COSY or TOCSY) 2D experiments. In other words, HMBC is one way to connect one spin system to another. The only other way to do this is with NOE experiments (NOESY and ROESY). Two-bond HMBC correlations are also predicted between the CH_2 carbon and the CH_3 proton of the ethyl group (double triplet, center right) and between the ethyl CH_3 carbon and the CH_2 proton (double quartet, upper left). We will see that in the real world, not all predicted HMBC correlations are observed because some of the couplings are too small to give crosspeaks above noise level. Another aspect of real HMBC spectra is that one-bond artifacts (the wide doublets observed in the HSQC/HMQC spectrum without ^{13}C decoupling) will often show up as strong peaks in the HMBC spectrum. In the worst case, the HMBC spectrum of ethyl acetate would contain all of the crosspeaks shown in Figure 11.12 (black and white).

11.3 EXAMPLES OF ONE-BOND INVERSE CORRELATION (HMQC AND HSQC) WITHOUT ^{13}C DECOUPLING

The HMQC spectrum of 3-heptanone is shown in Figure 11.13, with the ^1H spectrum shown at the top. The assignments come from the COSY analysis (Chapter 9). Note the prominent vertical streaks of noise at the F_2 frequencies of the most intense ^1H peaks: 0.78 ppm (H-7), 0.93 ppm (H-1), and 2.30 ppm (H-2 and H-4). Because this is not a gradient experiment, the ^{12}C-bound ^1H artifact is removed by subtraction using a phase cycle, and these are subtraction artifacts. The signal we are removing by subtraction is 100 times larger than the signal we are selecting in the phase cycle, so the artifacts are similar in magnitude to

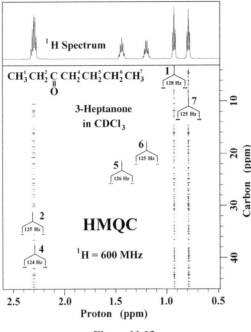

Figure 11.13

the crosspeaks. The contour threshold is set high to minimize the artifacts, so we see only the most intense peaks: triplets for H-1, H-7, and H-4, but only the central doublets of the H-6 sextet and the H-2 quartet and the central triplet of the H-5 quintet. The one-bond couplings ($^1J_{CH}$) can be readily measured from the crosspeak separations in F_2; all are in the range 125–128 Hz, typical for sp^3-hybridized carbon with no bonds to electronegative atoms. Notice the quasi-diagonal pattern of crosspeaks, extending from lower left to upper right, even for the relatively narrow range of chemical shifts represented. This is because the factors that shift protons downfield (in this case proximity to the slightly positive C-3) have a similar effect on the attached carbons.

The HMQC spectrum of sucrose in D_2O is shown in Figure 11.14. The glucose-1 crosspeak is found in the lower left because the two bonds to oxygen (anomeric position) lead to downfield shifts in both ^1H and ^{13}C relative to the more typical (for a sugar) singly oxygenated carbon. The "pack" of CH–O positions appears at the center right side, with the CH_2OH positions in a tight group at the upper right. The effect of substitution on ^{13}C shifts was mentioned in Chapter 1, Section 1.3: Every time an H is replaced by C, we get a downfield shift of about 10 ppm. Thus, for singly oxygenated sp^3-hybridized carbons, CH_3–O is roughly 50–60, CH_2–O is 60–70, CH–O is 70–80, and C_q–O is 80–90 ppm. The "upward" shift in the HMQC spectrum from CH–O to CH_2–O is thus a result of the sensitivity of ^{13}C shifts to steric crowding, which is not observed in ^1H chemical shifts. Any time there is a significant deviation from the roughly diagonal appearance of the HSQC/HMQC spectrum, it is due to different sensitivities of ^1H and ^{13}C to the chemical environment, and this can be very valuable information in structure determination.

The anomeric position (g1) shows a much larger $^1J_{CH}$ (170 Hz) than the rest of the protonated carbons, which fall in a tight range 141–149 Hz. Oxygen substitution tends to

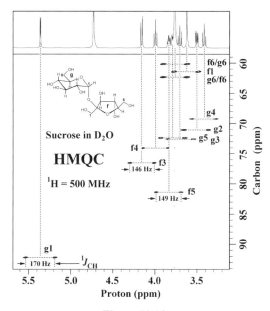

Figure 11.14

increase the one-bond coupling: We see typical values of 125–128 Hz with no bonds to oxygen (e.g., 3-heptanone), 140–150 Hz for one bond to oxygen and 160–170 Hz for two bonds to oxygen. As we saw with lactose and glucose, the anomeric $^1J_{CH}$ is 160–170 Hz with α orientation falling near 170 and β orientation near 160. The g1 position in sucrose is in the α orientation (170 Hz). Comparing to the HETCOR spectrum of sucrose (Fig. 9.7), rotating the HETCOR counterclockwise by 90° and then flipping it left-for-right gives the basic pattern seen in the HMQC (Fig. 11.14), except that in the HMQC the $^1J_{CH}$ couplings expand each crosspeak into a wide doublet in F_2 because we are not decoupling ^{13}C in this experiment.

Figure 11.15 shows the upfield (CH$_3$) region of the HMQC spectrum of cholesterol. In general, because inverse experiments are ^1H detected, methyl groups give much more intense signals than CH or CH$_2$ groups because there are three equivalent protons. It is almost always possible to increase the contour threshold to a point where only the crosspeaks due to CH$_3$ groups are observed. This is very useful, but the corollary is that CH$_3$ crosspeaks from impurities can sometimes be as strong as the CH and CH$_2$ crosspeaks from the compound being studied. Ignoring the wide $^1J_{CH}$ couplings, one can see that two of the CH$_3$ crosspeaks are singlets; that is, they have no homonuclear couplings. This means that the CH$_3$ group is attached to a quaternary carbon: CH$_3$–C$_q$ (the possibilities of CH$_3$–O or CH$_3$–N can be ruled out because of the upfield ^1H and ^{13}C chemical shifts). These singlet methyl groups are very useful in structure determination because of their unambiguous interpretation in the HMBC spectrum. The crosspeak with a ^{13}C shift of 18.7 ppm shows a clear doublet structure in F_2: It is a methyl group attached to a CH carbon. Two more "doublet" methyl groups can be seen at $F_1 = 22.58$ and 22.80 ppm, partially overlapped. The "diagonal" shape of the crosspeaks (see insets) makes it clear that the upfield ^{13}C (22.58 ppm) is connected to the upfield ^1H signal, even though these two doublets are not resolved in the ^1H spectrum (top).

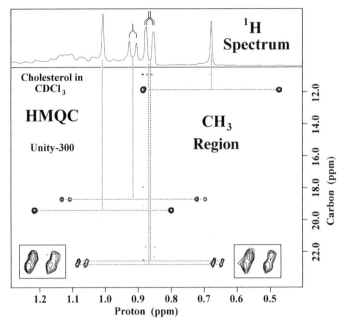

Figure 11.15

11.4 EXAMPLES OF EDITED, ^{13}C-DECOUPLED HSQC SPECTRA

We have already seen the effect of ^{13}C decoupling on 1D ^{1}H spectrum of a ^{13}C-labeled compound. In 2D HSQC or HMQC spectra, we can just add ^{13}C decoupling (e.g., GARP) during the acquisition of the ^{1}H FID. The wide pair of crosspeaks collapses into a single crosspeak at the ^{1}H chemical shift in F_2, with twice the intensity. The main practical consideration is to limit the duty cycle (percent time on) of the ^{13}C decoupling. Duty cycle is defined as AQ/(recycle delay) where AQ (acquisition time) is the length of time for acquiring the FID and the recycle delay is the total time for a single scan or transient, including the relaxation delay, the pulse sequence, and the acquisition of the FID. The duty cycle can be limited by setting a minimum relaxation delay (e.g., 1.0 s) and a maximum (e.g., 220 ms) setting for the acquisition time (Bruker: *aq*; Varian: *at*). This way the duty cycle can never exceed 0.22/1.22 = 0.18 or 18%. The acquisition time of the t_2 FID depends on the spectral width in F_2: $td(F_2) = 2 \times swh \times aq$ (Bruker) or $np = 2 \times sw \times at$ (Varian). Here *swh* is used for the Bruker parameter "spectral width in hertz" because the *sw* parameter refers to spectral width in parts per million on the modern (AMX, DRX) Bruker instruments. After setting the spectral width in F_2, the *aq/at* parameter is checked, and if it exceeds the safe limit (e.g., 220 ms) for ^{13}C decoupling, the number of points ($td(F_2)/np$) is reduced until *aq/at* is within the limits. As we saw with INEPT and DEPT, refocusing of the initially antiphase transferred coherence is required if we are to apply decoupling: Decoupling of an antiphase coherence puts the positive and negative lines on top of each other, resulting in a signal of zero.

Another very useful technique is to "edit" the HSQC data just like an APT or DEPT spectrum, so that the sign of the crosspeaks gives us information about the number of protons attached to each carbon: CH$_3$ crosspeaks are positive, CH$_2$ crosspeaks are negative, and CH crosspeaks are positive. We saw how CH$_3$ crosspeaks can usually be distinguished by their intensity, so this gives us a complete identification of the type of carbon (quaternary carbons are absent in one-bond correlations). Spectral editing is accomplished just like it is in the APT experiment: by a delay of 1/J during which we have ^{13}C SQC undergoing J-coupling evolution. Starting with in-phase coherence on the x' axis, the CH ^{13}C coherence moves from $\mathbf{S}_x$ to $-\mathbf{S}_x$; CH$_2$ coherence moves one full cycle for the outer lines of the triplet and not at all for the inner line, landing back at $\mathbf{S}_x$; and CH$_3$ coherence moves $1\frac{1}{2}$ full cycles for the outer peaks and $\frac{1}{2}$ cycle for the inner peaks, ending up in-phase on the $-x$ axis ($-\mathbf{S}_x$). The important thing is that the CH and CH$_3$ coherences are now opposite in sign to the CH$_2$ coherences. The phase is corrected in data processing to make the CH$_3$ and CH crosspeaks positive and the CH$_2$ crosspeaks negative. CH$_2$ crosspeaks are often obvious by their "pairing" (two crosspeaks with different ^{1}H shifts and the same ^{13}C shift) along horizontal lines for chiral molecules, but with spectral editing even "degenerate" CH$_2$ groups (both protons equivalent or coincidentally having the same chemical shift) are clearly identified as such by their negative sign. This places a heavier burden on the user in data processing: You have to be very light on the first-order phase correction, just like in the processing of DEPT spectra, so that you do not destroy the phase information by phase "correction." Usually some crosspeaks are known to be CH$_3$, CH$_2$, or CH from their chemical shift, intensity, or "pairing", and this knowledge helps with phase correction. Another useful trick is to watch for "waves" in the baseline when phase correcting a row or column of the matrix. Usually if the first-order phase correction is way off, the baseline will take on a slight sinusoidal "wobble," which will have more cycles the farther off the first-order phase correction becomes. Try for a flat baseline and then "touch up" the phase without making large changes to the chemical-shift dependent (first-order) correction.

Figure 11.16 shows the decoupled, edited HSQC spectrum of 3-heptanone. Positive intensity is shown in black and negative intensity in gray. With ^{13}C decoupling, the crosspeaks now line up in F_2 with the "normal" (^{12}C-bound) ^{1}H peaks in the proton spectrum. Crosspeaks are also twice the intensity relative to noise because they are no longer divided into two peaks by the wide $^1J_{CH}$ coupling. The editing feature makes all of the CH$_2$ crosspeaks negative and the CH$_3$ crosspeaks positive. The decoupled, edited HSQC spectrum of cholesterol in CDCl$_3$ is shown in Figure 11.17. Positive contours are shown in black and negative contours are shown in gray. The overall appearance of the spectrum is roughly diagonal from lower left to upper right, with the olefinic position (6 = z) in the lower left, the alcohol position (3 = y) in the center, and the "hydrocarbon" bulk of the molecule in the upper right. The carbons have been labeled according to the ^{13}C spectrum (Chapter 1, Fig. 1.26), using letters from the most upfield to the most downfield peak. In crowded regions, it is useful to compare expansions of the ^{13}C and DEPT spectra (Fig. 11.18). Although a single peak of twice the intensity is observed at 31.9 ppm in the ^{13}C spectrum of cholesterol, a slightly upfield CH peak (l) and a slightly downfield CH$_2$ peak (m) can be resolved in the DEPT-135 spectrum. In the tight group n-o-p-q in the ^{13}C spectrum, peak p can be identified as a quaternary carbon because it is missing in the DEPT spectrum. Likewise, in the very tight pair t-u (too close to be resolved in the F_1 dimension of 2D spectra), we see that the downfield peak u is quaternary. These observations will help us to correctly label the crosspeaks in the HSQC spectrum. In peak lists and assignment tables, use the precise chemical-shift values from the 1D ^{13}C spectrum if it is available, because the 1D spectrum has much higher resolution.

506 INVERSE HETERONUCLEAR 2D EXPERIMENTS: HSQC, HMQC, AND HMBC

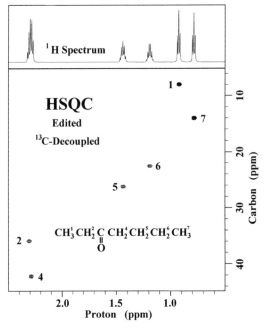

Figure 11.16

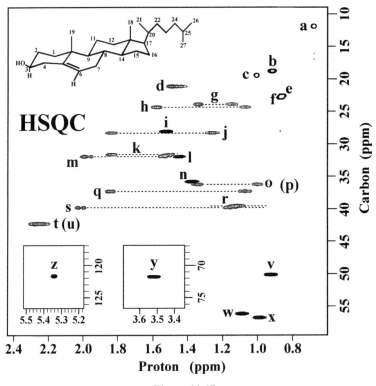

Figure 11.17

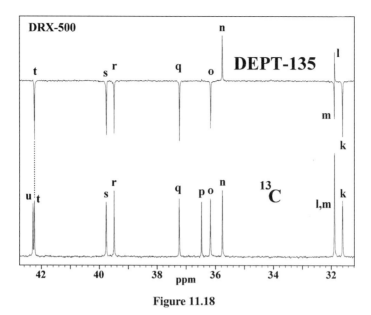

Figure 11.18

The methyl crosspeaks a, b, c, e, and f (Fig. 11.17) have such high intensity that they appear with "holes" in the center. This is because we only display a small number of contours starting with the contour threshold (in this case 10) with a constant ratio between intensity levels (in this case 1.25). So the highest contour level is 7.5 times more intense than the threshold (1.25 to the 9th power) and any intensity higher than that is not shown, leaving "holes" in the center of the most intense crosspeaks. Other peaks with positive intensity (black) include the olefinic (z) and alcohol (y) CH groups (insets) and six other CH peaks (i, l, n, v, w, and x). The general appearance is roughly diagonal from the lower left to the upper right, with the notable exception of crosspeaks v, w, and x, which are about 20 ppm below (downfield in ^{13}C shift) the other "hydrocarbon" CH groups i, l, and n. This is an example of the sensitivity of ^{13}C shifts to steric crowding: These are the CH positions next to the bridgehead quaternary carbons in the steroid framework (positions 9, 14, and 17). Their steric environment is similar to a neopentyl position (e.g., neopentyl alcohol $HO-CH_2-C(CH_3)_3$) that is known to be extremely hindered in organic chemistry (Fig. 11.19). The other three upfield CH groups in cholesterol are relatively open positions (8, 20, and 25), and their HSQC crosspeaks fall in with the rest of the "pack" (i, l, and n).

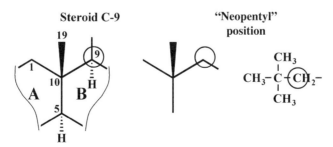

Figure 11.19

For most of the CH$_2$ groups, we see two distinct ^{1}H chemical shifts, leading to two negative crosspeaks lying on a horizontal line at the ^{13}C shift in F_1. In some cases (e.g., q and s) the difference is quite dramatic, nearly 1 ppm difference in ^{1}H chemical shift for the geminal pair. These large separations usually result from a highly anisotropic environment in a rigid molecule due to a nearby unsaturation. For a number of CH$_2$ groups (k, m, q, and s) we can see a doublet homonuclear splitting on the left side and a triplet pattern on the right side. This is a common pattern in steroids and triterpenes due to the rigid cyclohexane chair structure. In 2D spectra the F_2 dimension gives higher resolution than the F_1 dimension because we typically acquire 512–2048 complex pairs in t_2 (direct detection of the FID) and only 256–375 complex pairs (512–750 FIDs) for the best spectra in t_1 (indirect dimension). But resolution in F_2 is still poor compared to a 1D spectrum (typically 8192 or 16,384 complex pairs), so we will not see J couplings smaller than 7 Hz resolved. This has the advantage of simplifying the ^{1}H spectrum: We only see the large couplings, and these are generally the geminal (13–16 Hz) and *anti* vicinal (9–15 Hz) couplings on saturated carbons. The doublet–triplet pattern is the result of one equatorial (one large coupling—geminal) and one axial (two large couplings—geminal and *anti* vicinal) proton on the same carbon. We saw this with the H4$_{ax}$–H4$_{eq}$ pair in cholesterol in Chapter 8 (Fig. 8.35), which can be assigned to peak t in the HSQC due to its unique ^{1}H chemical shift. Even in cyclohexane itself the equatorial positions are downfield of the axial positions by about 0.5 ppm, so it is not surprising that in these six examples (j, k, m, q, s, and t) the equatorial proton is downfield of the axial proton. There are exceptions to this doublet–triplet pattern when the carbon has axial protons on both sides: for C11 we predict a "doublet" for the equatorial proton but a "quartet" (geminal plus two *anti* vicinal couplings to H12$_{ax}$ and H9) for the axial proton.

Figure 11.20 shows three F_2 slices of the HSQC spectrum, at the F_1 shifts of carbons j, q, and s. For q and s we see the classic doublet–triplet pattern, but for j there is a broad,

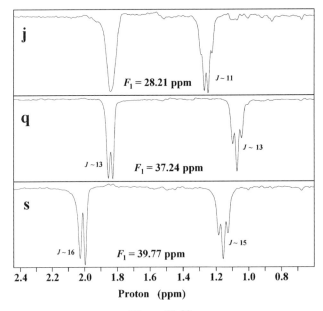

Figure 11.20

unresolved signal downfield and a quartet pattern upfield. The slices are impressive because they show good resolution in F_1 between j and i (^{13}C shift difference 0.23 ppm), and between r and s (difference of 0.27 ppm), with none of the i or r patterns "leaking" into the F_2 slices at j and s, respectively. The "quartet" pattern in slice j has been seen before in the F_2 slice of the ROESY spectrum (Chapter 10, Fig. 10.24, top, assigned to 16β) at the F_1 shift of methyl-18. All three of these slices have negative intensity because of the editing feature ($1/^1J_{CH}$ delay just before mixing) that turns all CH$_2$ crosspeaks upside down.

Some of the CH$_2$ crosspeaks are *degenerate*, meaning that the two protons have the same ^{1}H chemical shift (d and r). This can be a coincidence, but it is more likely to happen in a flexible chain, so we would suspect carbons 22–24 in cholesterol, although 22 is less likely because it is next to a chiral center.

11.5 EXAMPLES OF HMBC SPECTRA

HMBC is just an HMQC experiment with the $1/(2J)$ delay for evolution into antiphase "tuned" for a much smaller J_{CH} value: typically 8–10 Hz rather than around 150 Hz. This optimizes the experiment for observation of 2D crosspeaks between a ^{13}C and a "remote" ^{1}H nucleus two or three bonds away in the covalent structure: ^{13}C–XH or ^{13}C–X–YH where X and Y can be ^{12}C, O, S, N, and so on. Tuning to a much smaller J coupling means that the $1/(2J)$ delay will be much longer: 50–62.5 ms rather than around 3.3 ms. Loss of coherence through T_2 relaxation becomes significant with these longer delays, so the $1/(2J)$ refocusing delay at the end is left out. To mimimize T_2 losses the $1/(2J)$ delay is set as short as possible, so for routine work the optimal J value is usually set to the highest expected J_{CH}, or even a bit higher: 8–10 Hz. The FID records a ^{1}H signal that is antiphase with respect to the ^{13}C it correlates with in the 2D spectrum, and ^{13}C decoupling cannot be used because the antiphase lines would cancel each other, leading to no signal at all. This does simplify things a bit because we do not have to worry about the decoupler duty cycle, and we can use a longer acquisition time (*aq/at*) if we want. The relative span of J-coupling values is much greater for $^2J_{CH}$ and $^3J_{CH}$, ranging typically from 0 to 8 Hz, compared to the more predictable $^1J_{CH}$ (125–175 Hz). This means that for smaller J_{CH} values the crosspeak intensity is even weaker because evolution into antiphase occurs to the extent of $\sin(\pi J\tau)$. All of this contributes to HMBC being one of the less sensitive experiments, perhaps more sensitive than NOESY but much less sensitive than COSY, TOCSY, or HSQC/HMQC. With a few milligrams of a small molecule ("organic") sample on a 600-MHz instrument, a good HSQC spectrum can be obtained in 30 min or 1 h, but a good HMBC will take 3–4 h of acquisition.

The HMBC also incorporates a "low-pass" filter that tries to reject the one-bond correlations seen in HSQC/HMQC. "Low pass" means that only the low values of J_{CH} (0–10 Hz) are allowed to pass through and produce crosspeaks in the 2D spectrum. Because there is no ^{13}C decoupling, the one-bond correlations appear as wide doublets ($J \sim 150$ Hz) centered on the ^{1}H peak position in F_2 (Fig. 11.10—squares). They obscure the weak HMBC crosspeaks and can easily be misinterpreted as long-range correlations, especially if one of the two components of the doublet happens to fall at the position of another peak in the ^{1}H spectrum. The low-pass filter is set to reject a particular J value, typically 135 Hz for molecules dominated by saturated hydrocarbon (e.g., 3-heptanone, menthol, cholesterol), 142 Hz for sugars, and 170 Hz for molecules dominated by aromatic carbons. The same

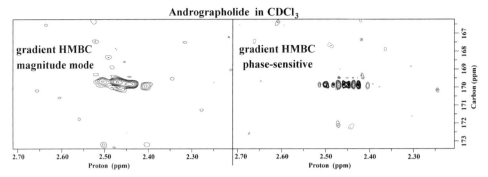

Figure 11.21

sin($\pi J\tau$) dependence applies to the rejection process, so any $^1J_{CH}$ values that do not match the filter setting will "leak" through and show up in the HMBC spectrum as wide doublets centered on the position of the ^{13}C-decoupled HSQC (HMQC) crosspeak. Because a one-bond 2D spectrum (HSQC or HMQC) is almost always acquired along with the HMBC, it is a good idea to superimpose them and find and label the one-bond artifacts right away, so you will not mistake them for true HMBC crosspeaks.

HMBC spectra have traditionally been processed and displayed in magnitude mode, but more recently HMBC sequences have been developed that permit a phase-sensitive display mode. To avoid "phase twists" the sequence is designed to refocus all but the desired chemical-shift evolution: ^{13}C in F_1. In this case, the HMBC crosspeaks appear as "candycanes": alternating positive and negative intensities along F_2 due to the relatively small J_{CH} coupling in antiphase. This makes them easy to distinguish from noise and from one-bond artifacts, which are also antiphase but with respect to the very wide $^1J_{CH}$ coupling. An HMBC crosspeak is shown in Figure 11.21 for the traditional HMBC experiment (magnitude mode) and an experiment designed to allow phase-sensitive display. Note the much sharper peaks in F_2 and the greater contrast with noise in the phase-sensitive (right side) spectrum. The crosspeak is a rare four-bond ($^4J_{CH}$) correlation observed for the natural product andrographolide (Fig. 11.22). Horizontal (F_2) slices from phase-sensitive HMBC spectra can also be analyzed by curve fitting to extract the exact coupling constant $^3J_{CH}$ from the antiphase splitting, for conformational or stereochemical studies using the Karplus relation.

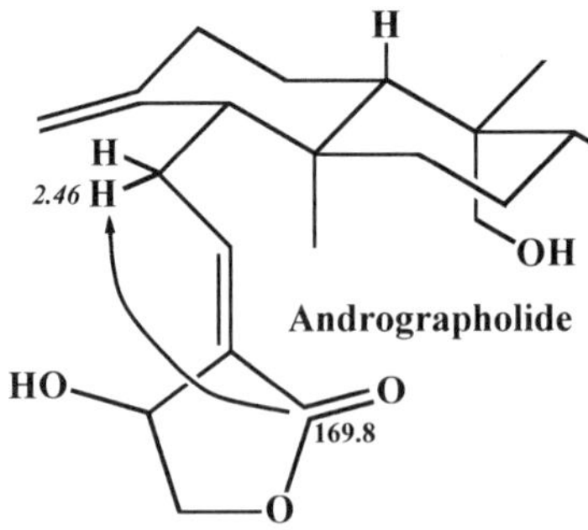

Figure 11.22

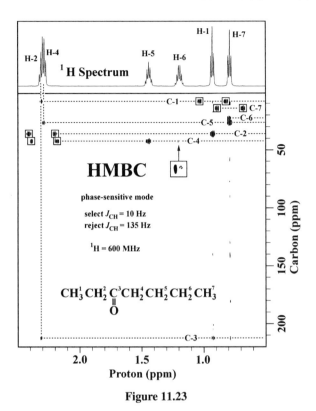

Figure 11.23

11.5.1 HMBC Spectrum of 3-Heptanone

Figure 11.23 shows the HMBC spectrum of 3-heptanone, presented in phase-sensitive mode with positive intensities shown in black and negative intensities in gray. Note that the F_1 spectral window is much wider (0–220 ppm) than in the HSQC (Fig. 11.16: 5–45 ppm) because long-range correlations can be seen to quaternary carbons, including the carbonyl carbon at 212 ppm. Strong one-bond artifacts (indicated by squares) centered on the HSQC peak positions are observed for the "strong" ^{1}H peaks: H-1 and H-7 CH_3 triplets and H-2 and H-4 CH_2 triplets. The H-5 and H-6 proton signals have more complex splitting patterns that reduce their ^{1}H peak height. This same splitting effect will reduce the peak heights of the HMBC correlations from H-5 and H-6. The one-bond artifacts could have been minimized if the low-pass filter were "tuned" to reject the narrow range of $^1J_{CH}$ values (125–128), but instead it was set to a more "generic" natural product value of 135 Hz. The ketone carbonyl carbon C-3 has a weak crosspeak to H-2 (bottom left, $^2J_{CH}$) and a strong crosspeak to H-1 (bottom right, $^3J_{CH}$). When counting the number of bonds between a ^{1}H and a ^{13}C, do not forget to count the C–H bond! There appears to be a weak crosspeak between C-3 and H-7 (bottom right), but closer examination reveals this to be part of a vertical streak at the F_2 frequency of H-7. The strong methyl triplets in the ^{1}H spectrum give these subtraction artifact streaks. In phase-sensitive spectra look for crosspeaks with alternating positive and negative intensity—even one tiny positive spot at the side of a negative spot, at the intersection of a known ^{1}H shift and a known ^{13}C shift, can be unambiguously assigned. No correlation is observed from C-3 to H-4 or H-5, even though these are two-bond and three-bond relationships, respectively. Many crosspeaks will be missing in HMBC spectra due to the lack of sensitivity and sometimes very small J values. Lack of a crosspeak should never be used to rule out a possible structure, but the presence of a crosspeak that is more than three bonds distant in the proposed structure is a serious problem.

Moving up to the C-2 frequency in F_1, we see a strong HMBC crosspeak to H-1 ($^2J_{CH}$). At the $F_1 =$ C-1 horizontal line, in addition to the one-bond artifacts, there is a weak correlation to H-2 at the far left. This is complementary to the C-2/H-1 crosspeak. Moving to the other side of the carbonyl group, start with the $F_1 =$ C-4 (42 ppm) horizontal line: There is a strong crosspeak to H-5 ($^2J_{CH}$) and a very weak crosspeak (arrow) to H-6 ($^3J_{CH}$). Close examination of the H-6 crosspeak shows a negative spot just to the right side of the positive spot (inset). On the $F_1 =$ C-5 horizontal line, there is a weak crosspeak to H-4 (left side) and a strong crosspeak to H-7 (right side). The $F_1 =$ C-6 line also has a strong crosspeak to H-7, just above the C-5/H-7 crosspeak. Often you will see these crosspeaks "stretched" in the vertical (F_1) dimension because they are spanning two nearby ^{13}C chemical-shift positions. Be careful not to interpret them as a single crosspeak at an intermediate F_1 shift value. Finally, on the $F_1 =$ C-7 line there are no true HMBC crosspeaks, only the one-bond artifacts.

A table of predicted and observed HMBC correlations confirms that many crosspeaks are missed, even in a fairly concentrated sample:

	H-1	H-2	H-4	H-5	H-6	H-7
C-1	art.	2(m)				
C-2	2(s)	art.	3			
C-3	3(m)	2(w)	2	3		
C-4		3	art.	2(s)	3(w)	
C-5			2(m)		2	3(s)
C-6			3	2		2(s)
C-7				3	2	art.

In the table, the predicted crosspeaks are indicated with 2 for $^2J_{CH}$ and 3 for $^3J_{CH}$, and the intensity is indicated in parentheses: s for strong, m for medium, and w for weak. The strong ^1H methyl signals (H-1 and H-7) virtually never fail to give crosspeaks, and usually these are strong. The worst performance is seen for the most "split" proton resonance: H-6, which is a sextet. The more splittings there are, the smaller each individual line appears, so the multiplet tends to "fall" into the noise and disappear. In rigid molecules, the $^3J_{CH}$ depends on the stereochemical relationship of ^{13}C and ^1H (Fig. 11.24) with an *anti* relationship giving a near maximum J value (~ 8 Hz) and a *gauche* relationship giving a small J value (~ 2–3 Hz). Because of the low sensitivity of HMBC, the *gauche* relationships give weak crosspeaks or no crosspeak at all. In 3-heptanone the molecule is flexible and the $^3J_{CH}$ values are probably conformationally averaged to a "medium" value (~ 4 Hz).

In this example, we had already assigned all of the carbon peaks in the ^{13}C spectrum, so there was no mystery. In the most difficult case of a complete unknown, the peaks in the ^{13}C spectrum are numbered arbitrarily from upfield to downfield, and the protons are numbered correspondingly according to the carbon they correlate to in the HSQC/HMQC spectrum. Then the HMBC correlations are tabulated (e.g., C-5–H-16a) and the puzzle solving begins.

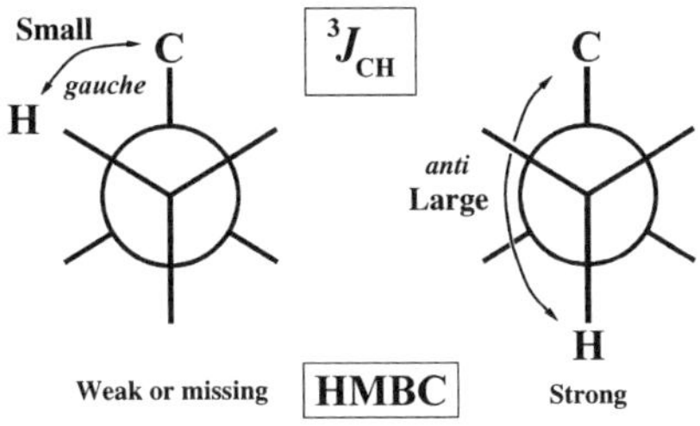

EXAMPLES OF HMBC SPECTRA   513

Figure 11.24

11.5.2 HMBC Spectrum of Cholesterol

The HMBC spectrum of cholesterol, displayed in phase-sensitive mode, is shown in Figure 11.25 with positive contours black and negative contours gray and with a portion of the HSQC spectrum at the left side for reference. The full HMBC spectrum has a very large number of crosspeaks, so for simplicity we show here only the upper right portion at a very high contour threshold. The most intense crosspeaks occur at the ^{1}H chemical shifts of the methyl groups in F_2, for the same reason that these are the most intense peaks in the HSQC spectrum: We are directly observing proton and the methyl groups have three equivalent protons. The protons of a methyl group can be correlated to no more than four carbons in an HMBC, and these are always observed due to their high intensity (Fig. 11.26). If the methyl group is attached to a carbonyl group, it can only correlate to two carbons (one if it is an ester or amide). If it is a methoxy group, it can only correlate to one carbon. If it is attached to a saturated carbon, that carbon can be a methylene (leading to a maximum of

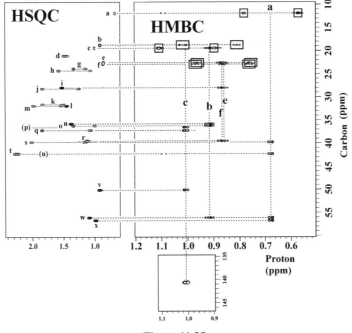

Figure 11.25

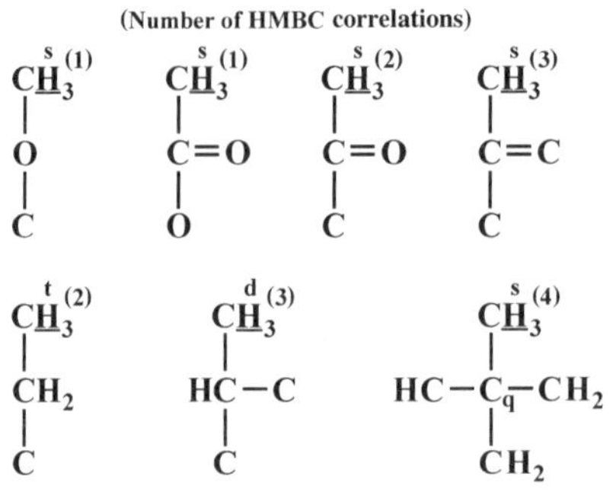

514 INVERSE HETERONUCLEAR 2D EXPERIMENTS: HSQC, HMQC, AND HMBC

(Number of HMBC correlations)

$$\underset{|}{\overset{s\ (1)}{\text{C}\underline{\text{H}}_3}} \quad \underset{|}{\overset{s\ (1)}{\text{C}\underline{\text{H}}_3}} \quad \underset{|}{\overset{s\ (2)}{\text{C}\underline{\text{H}}_3}} \quad \underset{|}{\overset{s\ (3)}{\text{C}\underline{\text{H}}_3}}$$

$$\underset{|}{\text{O}} \quad \underset{|}{\text{C}=\text{O}} \quad \underset{|}{\text{C}=\text{O}} \quad \underset{|}{\text{C}=\text{C}}$$

$$\text{C} \quad \text{O} \quad \text{C} \quad \text{C}$$

$$\underset{|}{\overset{t\ (2)}{\text{C}\underline{\text{H}}_3}} \quad \underset{|}{\overset{d\ (3)}{\text{C}\underline{\text{H}}_3}} \quad \underset{|}{\overset{s\ (4)}{\text{C}\underline{\text{H}}_3}}$$

$$\underset{|}{\text{CH}_2} \quad \underset{|}{\text{HC}-\text{C}} \quad \text{HC}-\underset{|}{\text{C}_q}-\text{CH}_2$$

$$\text{C} \quad \text{C} \quad \text{CH}_2$$

Figure 11.26

two correlations), a methine (maximum of three correlations), or a quaternary carbon (four correlations). The latter case is the most interesting, the case of a *singlet methyl group*. From the ^{1}H shift in F_2 of a singlet methyl group, we look along the vertical line in the HMBC spectrum and identify the four carbons. From the DEPT or edited HSQC, we can put them into categories: CH$_3$, CH$_2$, CH, or C$_q$. If only one of the four is quaternary, we know it has to be the one attached to our methyl group, and we know that the HMBC crosspeak is due to a $^2J_{CH}$ relationship. This means that the other three are attached to the C$_q$, giving $^3J_{CH}$ crosspeaks to our methyl protons. The beauty of this special case (singlet methyl group with only one correlation to a quaternary carbon) is that the three-bond versus two-bond ambiguity is removed: We can construct a molecular fragment consisting of five carbons from one vertical line in the HMBC spectrum (Fig. 11.26, lower right).

In Figure 11.25 the HSQC spectrum is aligned with the HMBC spectrum for comparison. In practice, this is done with NMR software in which two or more 2D spectra can be displayed simultaneously with crosshair cursors in each spectrum, controlled by the mouse and coordinated to be at the same chemical-shift position in F_2 and F_1 in all of the frames displayed. Another way to analyze the HMBC spectrum is with a printed list of chemical shifts taken from the HSQC spectrum. In this case, two lists are used: One in order of ^{1}H chemical shift and one in order of ^{13}C chemical shift (including the quaternary carbons). Both lists use the same arbitrary numbering (or lettering) system for the carbons. In this way the F_1 (^{13}C) and F_2 (^{1}H) shifts of an HMBC crosspeak can be searched in both lists, and the "nearest neighbor" shifts can be checked to see if there is any ambiguity in the assignment. If another resonance is very close, the assignment is made as a choice of possibilities, for example, H-5a/H-11b. These ambiguities can be resolved later as the structure becomes clear. The juxtaposition of the HSQC and HMBC spectra in Figure 11.25 will give you a feel for using correlated cursors in the NMR software. In this case, we use the HSQC spectrum itself as a "lookup table" for chemical shifts.

Because they are so intense, the methyl signals give strong one-bond artifacts in the HMBC spectrum. These are indicated by rectangles and have the same pattern as the methyl crosspeaks (a, b, c, e, and f) in the HSQC spectrum (left side) but with the additional $^1J_{CH}$ coupling in F_2. They are identical to the nondecoupled HMQC pattern for methyl groups (Fig. 11.15). Starting with methyl group a (upper right), the vertical line extending downward from the center of the one-bond artifact goes through four intense, antiphase doublet crosspeaks. Using the precise alignment with the HSQC spectrum, we can assign them as s (CH$_2$), t/u (CH$_2$/C$_q$), w (CH), and x (CH). Carbons t and u are too close in chemical shift to be distinguished in the F_1 dimension (0.05 ppm difference), although

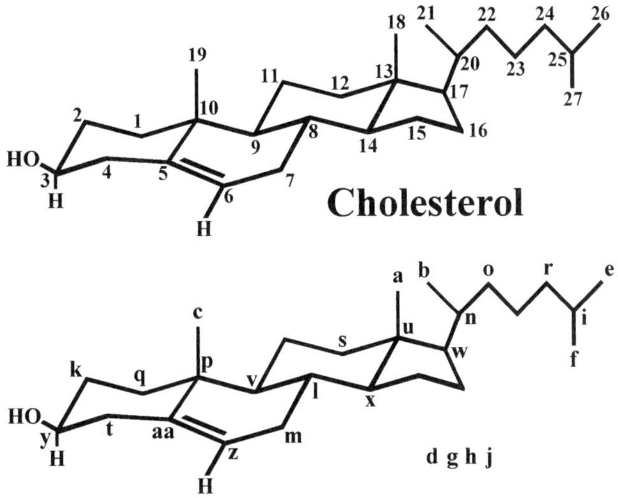

EXAMPLES OF HMBC SPECTRA 515

Figure 11.27

we could look for other HMBC crosspeaks that are very slightly higher or lower than this one to make the assignment. In this case we can use logic: Because methyl group a is a singlet in the ^{1}H spectrum, it must be bonded to a quaternary carbon. The other three HMBC crosspeaks are not quaternary, so this crosspeak has to be to carbon u, the quaternary carbon. This gives us an unambiguous five-carbon fragment: $C^aH_3-C^u{}_q-(C^sH_2, C^wH, C^xH)$. In the cholesterol structure, this has to be C-18, the angular methyl group at the C-D ring juncture (Fig. 11.27). Focus on the top of the figure and pencil in the resonances (a–aa) as we go through the assignment process. So we can assign a = 18, s = 12, and w,x = 14,17. Carbons w and x are ambiguous because both C14 and C17 are CH carbons. Note that this confirms our assessment of w and x as two of the "crowded" CH carbons 9, 14, and 17. Methyl group a cannot be C19, the other singlet methyl group, because it has no HMBC correlation to a quaternary olefinic carbon (C5).

Starting again with the other singlet methyl group, c, we have HMBC correlations to carbons q (CH$_2$), v (CH), and aa (olefinic C$_q$) (Fig. 11.25). The crosspeak at $F_1 = 36.47$ ppm (just below carbon o), however, does not line up exactly with any of the HSQC crosspeaks. Careful examination of the DEPT and ^{13}C spectra (Fig. 11.18) identifies this correlation as the quaternary carbon p. Because there are two correlations to quaternary carbons, there might be an ambiguity in how to arrange them, but one is olefinic (140.74 ppm) and cannot be directly connected to the methyl group (CH$_3$–C(=C)–C) because then there would be only three HMBC correlations. The fragment is thus $C^cH_3-C^p{}_q-(C^{aa}{}_q, C^vH, C^qH_2)$, and we can assign c = 19, p = 10, aa = 5, v = 9, and q = 1 using the structure-based numbering system (Fig. 11.27). This confirms the observation from both 1D selective NOE and 2D NOESY and ROESY spectra of a strong NOE from H4$_{ax}$ to the singlet methyl peak at 1.01 ppm (peak c in the HSQC).

Now we move to the doublet methyl groups. These have to be connected to a CH carbon because they appear as doublets in the ^{1}H spectrum, and all three are on the side chain: C-21, C-26, and C-27. Moving down from the methyl b one-bond artifact in the HMBC spectrum (Fig. 11.25), we encounter a "stretched" HMBC crosspeak at 36 ppm that is much "fatter" in the vertical (F_1) direction than the others. This is a correlation to two nearby carbons that is not resolved in F_1. These correspond to peaks n (CH) and o (CH$_2$) in the HSQC spectrum. A third and final crosspeak is observed to peak w (CH) at the bottom (a doublet methyl group can have only three correlations). Because there are correlations to two CH groups, it is not immediately clear which is directly bonded to the methyl group (of course, the COSY spectrum would tell us). But we have already assigned w and x to positions 14 and 17 (or 17 and 14) from carbon a (C-18). So it is clear that the fragment is C^{b}H$_3$–C^{n}H–(C^{o}H$_2$, C^{w}H), and we can assign b = 21, n = 20, o = 22, and w = 17 (Fig. 11.27). This also clears up the w/x ambiguity: x = 14.

The remaining two doublet methyl groups must be C-26 and C-27 of the isopropyl group. If you look closely at the center of the one-bond artifact for peaks e and f in the HMBC (Fig. 11.25), you will see a strong crosspeak that has a diagonal shape running from upper left to lower right, opposite to the direction (lower left to upper right) of the overlapping one-bond artifacts (and the overlapped HSQC crosspeaks). These are the HMBC correlations from C-26 to H-27 and from C-27 to H-26. Even when two carbons are exactly equivalent in an isopropyl group, there will be an HMBC correlation at the center of the one-bond artifact. Similarly, the pair of equivalent *ortho* or *meta* positions in a monosubstituted or *p*-disubstituted benzene ring will show three-bond HMBC crosspeaks at exactly the position of the ^{13}C-decoupled HSQC crosspeak. We know this is not a one-bond artifact because it would be widely separated by the $^1J_{CH}$ coupling. Moving down along the F_2 = H$_e$ and F_2 = H$_f$ vertical lines, there are two "double-wide" crosspeaks corresponding to carbons i (CH) and r (degenerate CH$_2$). These can be assigned to C-25 and C-24, respectively. The suspicion that degenerate CH$_2$ groups might correspond to the flexible side chain is confirmed at least in the case of C-24.

Figure 11.28 shows the HMBC correlations from the olefin and alcohol portions of the molecule: H-6 (z) in F_2 and C-3, C-6, and C-5 (y, z, and aa, respectively) in F_1, aligned with a portion of the HSQC spectrum. The olefinic proton H-6 (left side) correlates to carbons l/m, p (C$_q$), and t/u. Carbons p = C10 and u = C13 are already assigned from the methyl groups, so these three correlations can be assigned to p = C10 (C$_q$) and t = C4 (CH$_2$), with l and m remaining ambiguous (l = C8 and/or m = C7, Fig. 11.27). From the olefinic carbons (Fig. 11.28, bottom), correlations can be seen from C-3 (y), C-5 (aa), and C-6 (z) to the proton of HSQC peak t (H-4ab). C-5 = aa (but not C-6 = z) shows a correlation to a proton position that intersects peaks q (C-1), k, and j (all at the downfield or equatorial proton of the geminal pair). This must be q: H-1$_{eq}$, which bears an *anti* relationship to C-5 and is too far away (four bonds) from C-6 to give an HMBC correlation. Although four- and even five-bond correlations are common for J_{HH} when a double bond is in the path, more than three-bond correlations are very rare in HMBC spectra. Both C-5 (aa) and C-6 (z) show another correlation to a proton position that intersects the i, k' and m' crosspeaks in the HSQC (k and m at the upfield or axial positions). Peak i has already been assigned (C-25, too far away), and the alignment is best with peak m' (see enlargement, Fig. 11.28, upper left), so the correlation can be assigned to the axial proton on C$_m$: H-7$_{ax}$. Finally, the alcohol carbon y (C-3) shows correlations to proton peaks t and q (already assigned to H-4$_{eq}$, H-4$_{ax}$, and H-1$_{eq}$) and to an F_2 position a bit upfield of the one already assigned to m' (H-7$_{ax}$). This aligns perfectly with peak k', which is offset to the right side and above peak m' (see enlargement, Fig. 11.28, upper left side). Because C-1 and C-4 are already assigned, the only remaining CH$_2$ proton "within reach" of C-3 is H-2$_{ax}$ (k'). This example serves to illustrate how precise alignment and consideration of a number of overlapping possibilities are essential to the interpretation of HMBC spectra.

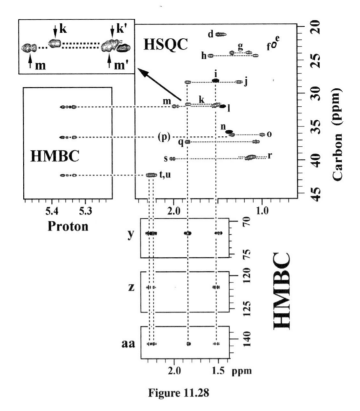

Figure 11.28

The assignments for cholesterol obtained just from these most intense HMBC crosspeaks are summarized in Figure 11.27. Seventeen of the 27 carbons of cholesterol were assigned from methyl group HMBC correlations alone. The one remaining "uncrowded" aliphatic CH group, l, can be assigned to position 8 by the process of elimination, and z = 6, y = 3, and t = 4 have already been assigned from chemical-shift arguments alone. From the olefin and alcohol functional groups, we assigned k = 2 and m = 7. With the HSQC data, including the "doublet" and "triplet" method for stereospecific assignment, we can precisely assign most of the proton resonances, even those that fall in heavily overlapped regions of the ^{1}H spectrum. This leaves only four unassigned positions: methylene groups d, g, h, and j for positions 11, 15, 16, and 23. COSY data would be the best way to finish the assignments and to confirm those already made. You might also look for protons that are "in the clear" (resolved) in the HSQC and look for correlations from these F_2 positions: j' and h have unique ^{1}H chemical shifts to search the HMBC and COSY spectra.

11.6 STRUCTURE DETERMINATION USING HSQC AND HMBC

11.6.1 Testosterone Metabolites

HMBC is a powerful tool for locating the position of a functional group within a known carbon skeleton. Oxidation of testosterone with the enzyme cytochrome P-450 (Fig. 11.29) leads to a number of hydroxylation (C–H → C–OH) and di-hydroxylation products. One

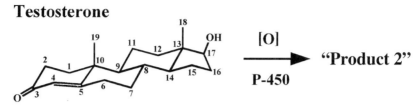

Figure 11.29

of the purified products ("Product 2") shows two ^{1}H resonances in the "alcohol" region (C$\underline{\text{H}}$–OH, 3–4.5 ppm), each with an integral of one proton. Because testosterone itself has one triplet resonance in this region (H-17), this indicates that one CH$_2$ carbon was oxidized to CH–OH to give Product 2. There are eight positions that could have been oxidized: 1, 2, 6, 7, 11, 12, 15, and 16. The "new" CH–OH resonance at 4.15 ppm is a ddd ($J = 7.5$, 5.8, 2.4 Hz), so we can test whether this coupling pattern is consistent with oxidation at each of these positions. The ddd resonance is "new" because H-17, the existing C$\underline{\text{H}}$–OH group in testosterone, cannot have more than two vicinal ^{1}H coupling partners. C-16 is ruled out because oxidation would remove one of the couplings to H-17, which appears as a triplet. To be a ddd, the C$\underline{\text{H}}$OH must be between a CH and a CH$_2$ carbon. This rules out C-1, C-2, C-6, and C-12 because they are next to quaternary carbons. Only positions 7, 11, and 15 are possible: Each is between a CH and a CH$_2$ group.

Figure 11.30 shows a portion of the HMBC spectrum of Product 2, including the "alcohol" region in both dimensions (3.2–4.4 in $F_2 = \,^1$H and 67–85 ppm in $F_1 = \,^{13}$C). The HSQC peaks are superimposed and shown in parentheses, and the ^{1}H resonances are shown above or below these peaks with an expansion of the ddd at 4.15 ppm. One-bond artifacts are clearly visible in the HMBC spectrum as wide doublets centered on the position of the ^{13}C-decoupled HSQC crosspeaks (compare to the diagram in Fig. 11.11). In addition a "fat" HMBC crosspeak is observed between the "new" ^{1}H resonance at 4.15 ppm in F_2 and the C-17 ^{13}C resonance at 82 ppm in F_1 (Fig. 11.30, lower left side). This means that the C$\underline{\text{H}}$–OH proton at the position of oxidation is two or three bonds away from C-17: Of the three possibilities (7, 11, and 15), it must be H-15. H-11 would be four bonds away and

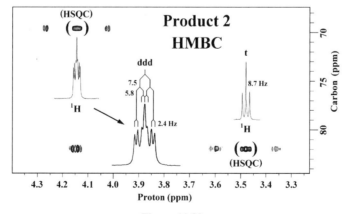

Figure 11.30

H-7 would be five bonds away, so an HMBC crosspeak would be impossible. Note that the corresponding crosspeak between H-17 and C-15 ($F_2 = 3.48$ ppm, $F_1 = 70$ ppm, upper right side) is not observed; this is because the H–C_{17}–C_{16}–C_{15} dihedral angle is different from the C_{17}–C_{16}–C_{15}–H dihedral angle, leading to a small $^3J_{CH}$ for the former and a large $^3J_{CH}$ for the latter. HMBC does not exhibit the same symmetry as the COSY spectrum because the relationships are not equivalent.

Having established the position of hydroxylation (the regiochemistry), the stereochemistry is the next important question. The orientation of the activated oxygen species in the enzyme with respect to the plane of the steroid is expected to lead to hydroxylation on one side only, and determining which side could lead to an understanding of the geometry of substrate binding at the enzyme active site. Although we normally think of NOE experiments to determine stereochemistry, it is actually more useful to look at dihedral angles and J couplings. Two energy-minimized models of this relatively rigid molecule, one for oxidation at C-15 on the α face and one for oxidation on the β face, give the following dihedral angles:

	15α-OH	15β-OH	Result
H–C_{15}–C_{16}–C_{17}:	110° (0.8 Hz)	124° (2.5 Hz)	HMBC Observed
H–C_{17}–C_{16}–C_{15}:	−96° (−0.4 Hz)	−94° (−0.5Hz)	No HMBC
H–C_{14}–C_{15}–H_{15}:	−162° (8.3 Hz)	−37° (5.1 Hz)	$J_{HH} = 5.8$
H–C_{15}–C_{14}–C_{13}:	84° (−0.7 Hz)	−150° (6.3 Hz)	HMBC Observed
H–C_{15}–C_{14}–C_8:	−44° (2.5 Hz)	79° (−0.6 Hz)	No HMBC

J values in parentheses are predicted from the dihedral angles using the Karplus relation, modified in the case of C–H couplings for the smaller $^3J_{CH}$ range. The "fat" HMBC crosspeak observed in Figure 11.30 between H-15 and C-17 is more consistent with β hydroxylation (predicted $^3J_{CH} = 2.5$ Hz) than with α hydroxylation (0.8 Hz), while in neither case would we expect to see the corresponding H-17 to C-15 crosspeak ($^3J_{CH} = -0.4$ or -0.5 Hz). The observation of an H-15 to C-13 crosspeak and the absence of an H-15 to C-8 crosspeak both support the assignment of β stereochemistry for the hydroxyl group. Even the observed $^3J_{HH}$ coupling of 5.8 Hz is consistent with the predicted coupling (5.1 Hz) from the β isomer and too small for the predicted coupling (8.3 Hz) for the α isomer. NOE intensities would not be useful in this case: The H-17 to H-15 distance in the models is 3.91 Å for the α isomer (*trans* relationship) and 3.48 Å for the β isomer (*cis* relationship). Both are within the range (< 5 Å) to give a measurable NOE—the simple $1/r^6$ relationship predicts an NOE twice as large for the *cis* relationship. The simple notion that *cis*-related protons will have large NOEs and *trans*-related protons will not show NOEs at all is clearly false. Only by isolating both isomers and carefully comparing the NOE intensities could we attempt to assign the stereochemistry based on NOE data alone.

11.6.2 Oxidation of β-Carotene

To mimic the metabolism of β-carotene (Fig. 11.31), a sample was oxidized with *m*-chloroperbenzoic acid (mcpba) and a product with biological activity was partially purified. Due to higher molecular weight impurities, the 1D ^{1}H spectrum was overlapped and difficult to interpret, but a small molecule component was easily identified in the HMQC spectrum due to its narrow linewidths. Figure 11.32 shows four crosspeaks from the downfield (lower left) region of the 500 MHz HMQC spectrum of the oxidation product

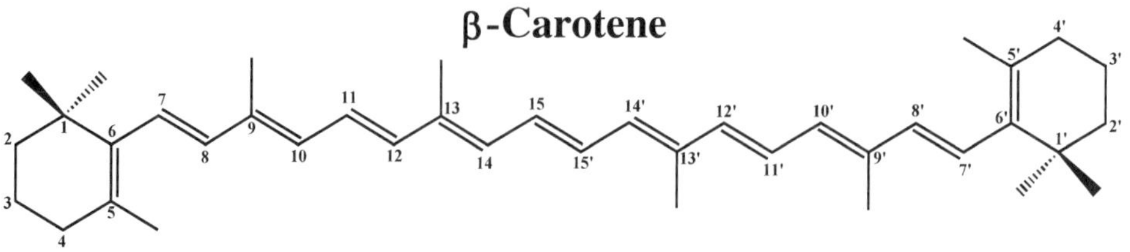

β-Carotene

Figure 11.31

(insets) with an F_2 slice through each crosspeak at the indicated ^{13}C chemical shift in F_1. The HMQC was acquired without ^{13}C decoupling over a 24-h period, and the component that gives rise to these crosspeaks was estimated to be about 0.7 mg. In addition to the large (~ 154–160 Hz) one-bond J_{CH} couplings, the larger homonuclear (J_{HH}) couplings can also be measured from the F_2 slices. From these alone, a spin system can be identified in which one proton is coupled to two others with coupling constants of 14 Hz and 18 Hz: CH–CH–CH. Because the ^1H and ^{13}C shifts (5.5–7.4 ppm and 115–139 ppm, respectively) are in the olefinic region these large couplings suggest a *trans* olefin or an *anti* relationship between protons on two sp²-hybridized carbons connected by a single bond. Because the spin system is isolated (no other homonuclear J couplings), we can "cap" the spin system on either end with a quaternary carbon: C_q=CH–CH=CH–C_q. Another spin system can be deduced from the olefinic CH with a 4 Hz coupling to another proton (Fig. 11.32, top). An HMQC crosspeak at F_2 = 4.67 ppm, F_1 = 82.1 ppm also shows a single 4 Hz J_{HH} coupling (not shown). These chemical shifts are in the region of an alcohol (CH-OH) group, shifted downfield by its proximity to the olefinic CH. Again we can "cap" the spin system at both ends: C_q=CH–CH(OH)–C_q. These fragments are shown in Figure 11.33, with the corresponding parts of the β-carotene structure from which they could be derived indicated by circles (C6–C9 and C9–C13). This would indicate that oxygen was introduced at C8, with the double bond moving from C5–C6 to C6–C7.

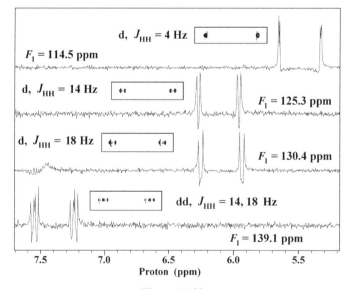

Figure 11.32

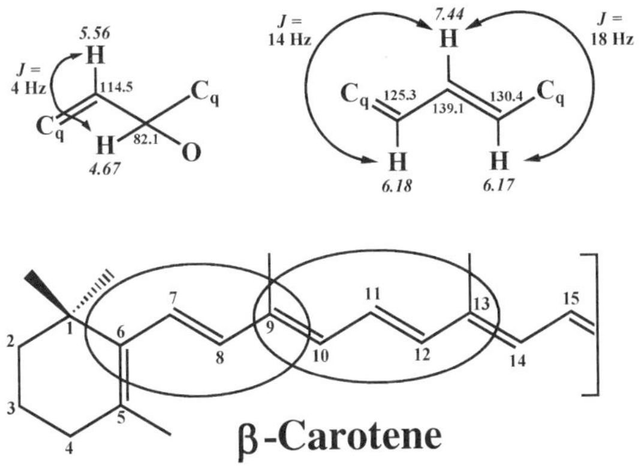

STRUCTURE DETERMINATION USING HSQC AND HMBC   521

β-Carotene

Figure 11.33

An HMBC spectrum with phase-sensitive data presentation was acquired on a purer sample over a 60-h period to tie these fragments together and complete the structure. At an F_1 chemical shift of 198.4 ppm three crosspeaks were observed (Fig. 11.34: insets and F_2 slice). The F_1 chemical shift indicates a ketone carbonyl carbon, and one of the crosspeaks, an antiphase doublet ($J = 5$ Hz) at $F_2 = 2.27$ ppm, is at the right ^{1}H shift for a methyl ketone: CH_3–CO. The lack of any homonuclear couplings in addition to the 5 Hz active ($^2J_{CH}$) coupling is consistent because the methyl ketone would be a singlet in the ^{1}H spectrum. The crosspeak at $F_2 = 6.17$ ppm (Fig. 11.34, left of center) has an active (antiphase) coupling of 3.4 Hz and a passive (in-phase) coupling of 18 Hz. This is the doublet CH proton found in the HMQC spectrum at $F_2 = 6.17$, $F_1 = 130.4$ (Fig. 11.32) corresponding to the right-hand CH of the five-carbon fragment (Fig. 11.33). The crosspeak at $F_2 = 7.44$ (Fig. 11.34, left) has a more complex coupling indicating more than one passive (J_{HH}) coupling. This is the CH proton found in the HMQC spectrum at $F_2 = 7.44$, $F_1 = 139.1$, corresponding to the middle CH of the five-carbon fragment. So the ketone carbonyl of the CH_3–CO fragment must be the quaternary carbon at the right side of the five-carbon fragment (Fig. 11.33), with two-bond correlations from the CO carbon to the CH_3 proton and the right side CH proton (6.17 ppm) and a three-bond correlation to the middle CH proton (7.44 ppm). The downfield shift of this center CH (139.1 ppm ^{13}C, 7.44 ppm ^{1}H) relative to typical olefin values (120–130 ppm ^{13}C, 5–6 ppm ^{1}H) is explained by its β position in an α,β-unsaturated ketone ($-C_\beta^+-C_\alpha=C-O^-$ resonance structure).

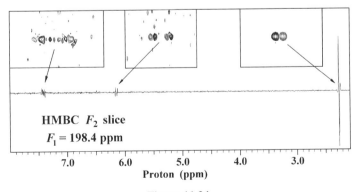

Figure 11.34

522 INVERSE HETERONUCLEAR 2D EXPERIMENTS: HSQC, HMQC, AND HMBC

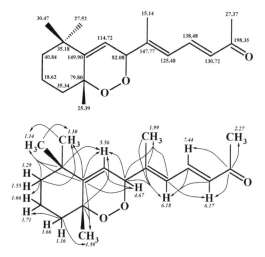

Figure 11.35

Thus it appears that in addition to oxidation at C8 and shifting of the C6–C7 double bond, we have cleaved the β-carotene structure at C13–C14, with oxidation of C13 to a ketone. Another quaternary carbon was identified in the HMBC spectrum at $F_1 = 79.8$ ppm, with correlations to the proton at 5.56 ppm next to the CHOH (Fig. 11.33) and to a methyl group at 1.59 ppm. This corresponds to C5 of β-carotene, which must be oxygenated and sp³-hybridized because its ^{13}C chemical shift (near 80 ppm) is characteristic of C_q–OH rather than an olefinic carbon. Analysis of many more crosspeaks in the HMBC spectrum, as well as a mass spectrum with molecular ion (M+) of nominal mass 290 ($C_{18}H_{26}O_3$, 6 unsaturations) led to the cyclic peroxide structure shown in Figure 11.35. The mass spectral data was inconsistent with a diol structure (predicted m/z 292) so the cyclic peroxide was proposed. In all, 30 HMBC correlations were identified (arrows), making the structural assignment quite solid, and all ^{1}H and ^{13}C positions were assigned chemical shifts. All of this was accomplished on an impure sample without enough material for a simple ^{13}C spectrum.

11.7 UNDERSTANDING THE HSQC PULSE SEQUENCE

Now that we know most of the basic building blocks of NMR pulse sequences, we should be able to use the coherence flow diagram (Fig. 11.9) to design an HSQC pulse sequence. It needs to accomplish the following steps:

1a. Create ^{1}H magnetization in the x'–y' plane (preparation).
1b. Transfer this magnetization from ^{1}H to ^{13}C via the one-bond J_{CH} (preparation).
2. Let the ^{13}C magnetization rotate in the x'–y' plane for a period t_1, allowing us to indirectly measure the ^{13}C chemical shift (evolution).
3. Transfer the ^{13}C magnetization back to ^{1}H magnetization (mixing).
4. Observe the ^{1}H magnetization directly (t_2) so that the ^{1}H chemical shift can be determined (detection).

Note that we do not start directly with ^{13}C magnetization because we want to take advantage of the larger (by a factor of 4) equilibrium population difference of ^1H compared to ^{13}C, as well as the shorter T_1 (faster relaxation) of ^1H, which will permit shorter relaxation delays. We now know a lot of tricks, and the main one we need here is the heteronuclear INEPT transfer:

1a. 90° pulse on the ^1H channel (e.g., 500 MHz).

1b. INEPT transfer: A delay of $1/(2J)$ to generate an antiphase ^1H doublet, followed by simultaneous 90° pulses on both the ^1H and ^{13}C channels (e.g., 500 and 125 MHz, respectively).

2. Evolution period: simply insert a delay of t_1 s and repeat the experiment many times with increasingly larger t_1 delays. Increment t_1 each time by $\Delta t_1 = 1/(2 \times swl)$, where swl is the spectral width in the ^{13}C dimension in hertz.

3. Mixing: INEPT transfer ("back" transfer). Assuming that ^{13}C magnetization is still antiphase with respect to the directly bound proton, simultaneous 90° pulses on both the ^1H and ^{13}C channels will convert the antiphase ^{13}C coherence back into antiphase ^1H coherence. This signal differs from that at the end of the first $1/(2J)$ delay (step **1b**) in that its intensity has been modulated by the chemical-shift evolution that occurred during step **2**. In other words, the ^{13}C chemical shift has been encoded within the ^1H signal.

4. Acquisition: simply turn on the analog-to-digital converter and record a ^1H FID. Fourier transformation of this signal will give an antiphase doublet whose amplitude is modulated by a factor $\cos(\Omega_c t_1)$.

This pulse sequence is diagrammed in Figure 11.36. Below the pulse sequence are shown the spectra that *would be obtained* if an FID were acquired at each stage of the pulse sequence, with ^1H spectra above and ^{13}C spectra below. The antiphase ^{13}C signal is shown with a phase shift of 180°, for example, resulting from evolution of the ^{13}C chemical shift during t_1.

Now that we have the basic concept of the pulse program written down, we can start to customize and enhance it, and to consider details of making it work correctly. First of

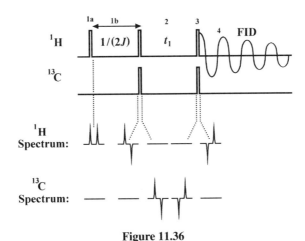

Figure 11.36

all, we have to consider the fact that the evolution periods $1/(2J)$ and t_1 serve for very different purposes: during the $1/(2J)$ delay we want the ^{1}H doublet to evolve from in-phase to antiphase under the influence of the one-bond $\{^1\text{H}-^{13}\text{C}\}$ J coupling. During the t_1 period, however, we only want to see rotation of the ^{13}C magnetization vector in the $x'-y'$ plane under the influence of the ^{13}C chemical shift. In other words, we want J-coupling evolution to occur during the first delay and chemical-shift evolution during the second delay. As the pulse program is designed so far, both kinds of evolution will occur during both delays, leading to lots of complications. The way to prevent certain kinds of evolution while allowing others to occur is to make each delay period into a spin echo by dividing the delay time into two equal parts separated by a 180° pulse. How we apply the 180° pulse (i.e., on which nucleus or nuclei) will determine which type of evolution is allowed and which type is refocused.

We spent a lot of time in Chapter 6 (Section 6.10) using the vector diagrams to understand the effect of 180° pulses in the center of a spin echo. This is easy to understand now that we have the product operator tools. In general, consider the effect of a 180° pulse on the in-phase and antiphase ^{1}H and ^{13}C operators:

$\mathbf{I}_x$	$\rightarrow$	$-\mathbf{I}_x$	^{1}H 180°$_y$: refocus ^{1}H chemical-shift evolution.
$\mathbf{S}_x$	$\rightarrow$	$-\mathbf{S}_x$	^{13}C 180°$_y$: refocus ^{13}C chemical-shift evolution.
$2\mathbf{I}_x\mathbf{S}_z$	$\rightarrow$	$-2\mathbf{I}_x\mathbf{S}_z$	^{1}H 180°$_y$ or ^{13}C 180°: refocus J-coupling evolution.
$2\mathbf{I}_x\mathbf{S}_z$	$\rightarrow$	$2\mathbf{I}_x\mathbf{S}_z$	^{1}H 180°$_y$ and ^{13}C 180°: no effect on J-coupling evolution.

Reversing the sign of these operators in the center of a spin echo leads to refocusing of the evolution. This is an easy way to remember how to design a heteronuclear spin echo: use a ^{1}H 180° pulse alone to refocus all but ^{13}C chemical shift evolution; use a ^{13}C 180° pulse alone to refocus all but ^{1}H chemical-shift evolution; use simultaneous ^{1}H and ^{13}C 180° pulses to refocus all but J_{CH} evolution. You can go through the full product operator analysis of each kind of spin echo, and you will find that the sign changes shown above are the crucial differences that control what refocuses and what continues to evolve in the second half of the spin echo.

For the first delay of $1/(2J)$ in Figure 11.36, we insert simultaneous 180° pulses in the center on both ^{1}H and ^{13}C (Fig. 11.37). The ^{1}H chemical-shift evolution "sees" the ^{1}H 180° pulse in the middle and reverses direction (v_H in the first half, $-v_H$ in the second half). The J coupling evolution "sees" the simultaneous 180° pulses on both channels and their effects cancel out $(2\mathbf{I}_y\mathbf{S}_z \rightarrow 2[-\mathbf{I}_y][-\mathbf{S}_z])$ so the J-coupling evolution continues unabated in the second half (J in the first half, J in the second half). We used the same strategy in the INEPT experiment (Chapter 7). At the end of the $1/2J$ period we have antiphase ^{1}H coherence, which is transferred to antiphase ^{13}C coherence by the simultaneous 90° pulses on ^{1}H and ^{13}C $(2\mathbf{I}_y\mathbf{S}_z \rightarrow -2\mathbf{S}_y\mathbf{I}_z)$.

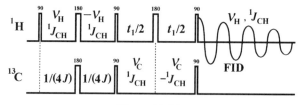

Figure 11.37

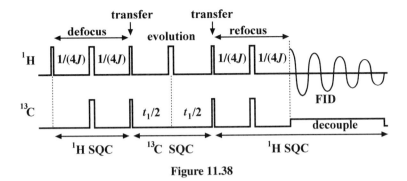

Figure 11.38

You can probably guess how we can control the evolution during the t_1 delay. We start this period with antiphase ^{13}C magnetization, and we want to have ^{13}C chemical-shift evolution during t_1 without any J-coupling evolution. The solution is to convert the t_1 delay into a spin-echo. We divide the t_1 delay period into two delays of $t_1/2$ each, with a 180° ^{1}H pulse in the middle (Fig. 11.37). The J-coupling evolution "sees" the 180° ^{1}H pulse in the middle and turns around ($2S_yI_z \rightarrow 2[S_y][-I_z]$), but the ^{13}C chemical shift continues because the S_x and S_y terms are unaffected by the ^{1}H pulse. Now this sequence will do what our original design was supposed to do, even in the real world where peaks are not on-resonance.

One final refinement we might want to add: a refocusing delay of $1/(2J)$ to allow the antiphase ^{1}H signals to come back together into in-phase signals. This makes all of our crosspeaks positive in the 2D spectrum. This is essential if we plan to use ^{13}C decoupling, because the doublet collapses into a singlet that will have no intensity at all if the doublet is antiphase. Of course, we will need to use the spin echo $1/(2J)$ delay as we did at the beginning of the sequence (Fig. 11.38). We can call the first $1/(2J)$ period "defocusing" because the ^{1}H SQC goes from in-phase to antiphase. Then we transfer from ^{1}H SQC to ^{13}C SQC and let the ^{13}C SQC evolve under the influence of the ^{13}C chemical shift. After transferring back to ^{1}H SQC, we "refocus" for a period of $1/(2J)$ to bring the ^{1}H SQC from antiphase back to in-phase. The arrows at the bottom of the diagram indicate the type of coherence (magnetization in the $x'-y'$ plane) that we have at each stage of the pulse sequence. The pulse sequence uses single-quantum coherence (SQC) throughout, which is why it is called HSQC (heteronuclear single quantum correlation).

11.7.1 Product Operator Analysis of the HSQC Experiment

This is quite simple if we take into account only the type of evolution that is not refocused during each stage of the pulse sequence. Starting with ^{1}H z magnetization:

Preparation: $I_z \xrightarrow{90_x{}^1H} -I_y \xrightarrow{1/(2J)} 2I_xS_z \xrightarrow{90_y{}^1H,\,90_x{}^{13}C} 2[-I_z][-S_y] = 2S_yI_z$

Evolution: $2S_yI_z \xrightarrow{t_1} 2S_yI_z\cos(\Omega_c t_1) - 2S_xI_z\sin(\Omega_c t_1)$

Mixing: $2S_yI_z\cos(\Omega_c t_1) \xrightarrow{90_y{}^1H,\,90_x{}^{13}C} 2I_xS_z\cos(\Omega_c t_1)$

$-2S_xI_z\sin(\Omega_c t_1) \xrightarrow{90_y{}^1H,\,90_x{}^{13}C} -2S_xI_x\sin(\Omega_c t_1)$

Detection: $2I_xS_z\cos(\Omega_c t_1) \xrightarrow{1/(2J)} I_y\cos(\Omega_c t_1)$ (observed in FID)

In the mixing step the cosine term is transferred back from ^{13}C SQC to ^{1}H SQC, but the sine term is converted to a mixture of DQC and ZQC, which is not observable and can be ignored from here on. Refocusing of the cosine term gives in-phase ^{1}H SQC modulated as a function of t_1 and the ^{13}C shift because of the $\cos(\Omega_c t_1)$ multiplier. During the recording of the FID we use ^{13}C decoupling so the ^{1}H SQC remains in-phase and simply rotates in the x–y plane at the rate Ω_H in the rotating frame.

11.7.2 Cancelation of the ^{12}C–H Signal

As already mentioned, the ^{1}H signal from the protons bound to carbon-12 is about 200 times as intense as the ^{13}C-bound proton signals (satellites), so we need a way of removing this artifact. As always we have a choice of phase cycling or gradients (or both!) to remove the undesired signals. Phase cycling is a subtraction method, so the whole mess is recorded in the FID (^{12}C-bound ^{1}H and ^{13}C-bound ^{1}H signals) and by recording multiple FIDs (scans or transients) and subtracting them we remove the ^{12}C-bound ^{1}H signal. Gradients kill the undesired signal by "twisting" its coherence and leaving it twisted during the FID acquisition. The ^{12}C-bound ^{1}H signal never reaches the receiver so it is removed in a single scan.

The phase cycling method works like this: the phase of the second 90° ^{13}C pulse is alternated between x (B_1 field aligned along the x' axis) and $-x$ (B_1 field aligned along the $-x'$ axis) with each signal-averaged acquisition. Because this pulse is essential for the transfer of magnetization (mixing) from ^{13}C back to ^{1}H, inverting its phase will have the effect of inverting the detected FID signal. In terms of product operators ($\mathbf{I} = {}^1$H, $\mathbf{S} = {}^{13}$C):

$$2\mathbf{S}_y\mathbf{I}_z \xrightarrow{90_x{}^{13}\text{C},90_y{}^1\text{H}} 2[\mathbf{S}_z][\mathbf{I}_x] = 2\mathbf{I}_x\mathbf{S}_z \xrightarrow{1/(2J)} \mathbf{I}_y$$

$$2\mathbf{S}_y\mathbf{I}_z \xrightarrow{90_{-x}{}^{13}\text{C},90_y{}^1\text{H}} 2[-\mathbf{S}_z][\mathbf{I}_x] = -2\mathbf{I}_x\mathbf{S}_z \xrightarrow{1/(2J)} -\mathbf{I}_y$$

If we alternately add and subtract FID signals as the signal averaging progresses, these signals will reinforce and build up as we acquire a number of scans for each t_1 value. The ^{12}C-bound proton signal, however, is not affected by the ^{13}C pulses, and it gives rise to observable signals that do not alternate in sign:

$$\mathbf{I}_z \xrightarrow{90_x{}^{13}\text{C},90_y{}^1\text{H}} \mathbf{I}_x \xrightarrow{1/(2J)} \mathbf{I}_x \quad \text{first scan(add)}$$

$$\mathbf{I}_z \xrightarrow{90_{-x}{}^{13}\text{C},90_y{}^1\text{H}} \mathbf{I}_x \xrightarrow{1/(2J)} \mathbf{I}_x \quad \text{second scan(subtract)}$$

$$\mathbf{I}_y \xrightarrow{90_x{}^{13}\text{C},90_y{}^1\text{H}} \mathbf{I}_y \xrightarrow{1/(2J)} \mathbf{I}_y \quad \text{first scan(add)}$$

$$\mathbf{I}_y \xrightarrow{90_{-x}{}^{13}\text{C},90_y{}^1\text{H}} \mathbf{I}_y \xrightarrow{1/(2J)} \mathbf{I}_y \quad \text{second scan(subtract)}$$

As these signals are alternately added and subtracted into the summed FID, they cancel as long as we are careful to acquire an even number of scans for each t_1 increment. This method depends for its success on precise subtraction of a very large signal, so it is sensitive to any instability (temperature change, vibration, variation in pulse widths, etc.) that occurs between one scan and the next.

To use the more formal analysis of phase cycling developed in Section 10.6, we first need to describe the coherence pathway in terms of spherical operators ($\mathbf{I}^+$, $\mathbf{S}^-$, etc). Starting at the end and working backward and using the convention of positive coherence order during

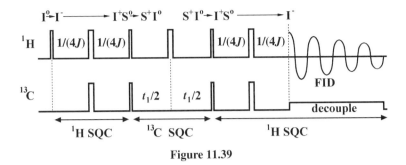

Figure 11.39

t_1 and negative coherence order during t_2 (Fig. 11.39), we have $\mathbf{I}^-$ at the start of the FID (in-phase ^{1}H SQC with $\mathbf{p} = -4$) and $\mathbf{I}^+\mathbf{S}^\circ$ ($= \mathbf{I}^+\mathbf{S}_z$) before the refocusing period (antiphase ^{1}H SQC with $\mathbf{p} = +4$). The change in sign of coherence order is a result of the ^{1}H 180° pulse in the center of the refocusing delay. The mixing step (simultaneous ^{1}H and ^{13}C 90° pulses) converts $\mathbf{S}^+\mathbf{I}^\circ$ to $\mathbf{I}^+\mathbf{S}^\circ$, so we have $\mathbf{S}^+\mathbf{I}^\circ$ throughout the evolution (t_1) delay. The first coherence transfer step converts $\mathbf{I}^+\mathbf{S}^\circ$ to $\mathbf{S}^+\mathbf{I}^\circ$, and so we start with $\mathbf{I}^-$ just after the initial 90° pulse.

In the mixing step where we apply the phase cycle (^{13}C SQC → ^{1}H SQC), the desired coherence pathway is $\mathbf{S}^+\mathbf{I}^\circ \to \mathbf{I}^+\mathbf{S}^\circ$. Considering the 90° ^{13}C pulse, the effect it has on coherence order is $\Delta p = -1$ because the ^{13}C operator goes from $\mathbf{S}^+$ ($\mathbf{p} = 1$) to $\mathbf{S}^\circ$ ($\mathbf{p} = 0$) as a result of the pulse. So if we alternate the phase of this pulse ($\Delta \Phi_p = 180°$) we will have to alternate the phase of the receiver:

$$\Delta \Phi_r = -\Delta p \Delta \Phi_p = -(-1)(180°) = 180°$$

The ^{12}C-bound ^{1}H signal cannot be affected by the ^{13}C 90° pulse, so $\Delta \mathbf{p} = 0$ regardless of where it is ($\mathbf{I}_z$, $\mathbf{I}_y$, or $\mathbf{I}_x$) when the pulse is executed. $\Delta \Phi_r = 0$ for this signal, and thus it will be canceled if we alternate the receiver phase.

The same phase cycle can be used for the first coherence transfer step (^{1}H → ^{13}C) by alternating the phase of the first 90° ^{13}C pulse. For example, we could use $x, x, -x, -x$ (0 0 2 2) for the first 90° ^{13}C pulse and $x, -x, x, -x$ (0 2 0 2) for the second 90° ^{13}C pulse in a four-step phase cycle. The receiver phase must follow the sum of the phase changes of the signal:

$$\Phi_r(\text{1st pulse alone}) : \quad x, \quad x, \quad -x, \quad -x \quad (0022) \quad \Delta \mathbf{p} = +1$$
$$\Phi_r(\text{2nd pulse alone}) : \quad x, \quad -x, \quad x, \quad -x \quad (0202) \quad \Delta \mathbf{p} = -1$$
$$\Phi_r(\text{both pulses}) : \quad \quad x, \quad -x, \quad -x, \quad x \quad (0220)$$

Of course, the experiment will take four times as long to acquire the same number of FIDs (the same number of t_1 values) and this time is wasted if there is enough sample to get the desired signal-to-noise ratio with one scan per FID.

11.7.3 Gradient Coherence Pathway Selection

The disadvantage of using a phase cycle to cancel the ^{12}C-bound ^{1}H signal is that you have to do a minimum of two or four transients for each FID collected in the 2D experiment.

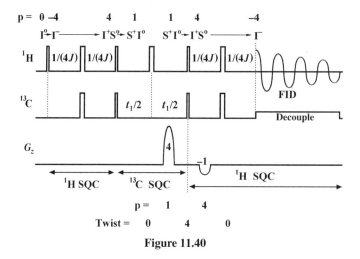

Figure 11.40

Furthermore, since the FID observed in each scan contains the large ^{12}C-bound ^{1}H signal as well as the weak ^{13}C-bound ^{1}H signal, the receiver gain must be turned down to prevent this signal from overloading the digitizer, even though 99% of the signal will be canceled when the second FID is added to the sum in memory. Less amplification of the FID can result in lower signal-to-noise in the spectrum. Finally, subtraction of two FIDs is not always perfect, and subtraction artifacts will give vertical streaks in the spectrum (see Fig. 11.13). The solution to all of these problems is to use gradients in place of a phase cycle to eliminate the ^{12}C-bound ^{1}H signal. Because the "twist" imparted by a gradient is proportional to the magnetogyric ratio, a ^{1}H SQC signal is twisted four times as much by a given gradient pulse as a ^{13}C SQC signal. Thus if a gradient of relative strength 4 is applied during the evolution time (t_1), the ^{13}C SQC signal ($\mathbf{p} = 1$) will acquire a twist down the gradient axis (usually z) of 4 units (Fig. 11.40). A second gradient of relative strength -1 is applied after this ^{13}C magnetization is transferred back to ^{1}H SQC ($\mathbf{p} = 4$) by the INEPT sequence, imparting an additional twist of -4 units, leading to a total of 0 units of twist:

$$\Sigma \mathbf{p}_i G_i = 1(4) + 4(-1) = 0$$

Thus the signal will be perfectly "untwisted" and will be observed in the FID. The ^{12}C-bound ^{1}H signal cannot transfer from ^{13}C SQC to ^{1}H SQC during the INEPT sequence, so it will end up with a net twist at the beginning of the FID and will not be observed. With gradient selection of the coherence pathway, no ^{12}C-bound ^{1}H "streaks" are observed even with a single scan per FID. There is no harm in combining methods, so the phase cycle is always included even in the gradient experiment if you want to use more than one scan per FID.

11.7.4 Gradient-Selected HSQC with Phase-Sensitive Data Presentation

We saw with the gradient DQF-COSY experiment that the relatively long gradients (~ 1 ms) allow chemical shift evolution that will produce large chemical-shift dependent phase errors in the final spectrum. In the sequence of Figure 11.40, the gradient placed in the second half of the t_1 period will set a minimum value for t_1 of twice the gradient time (and

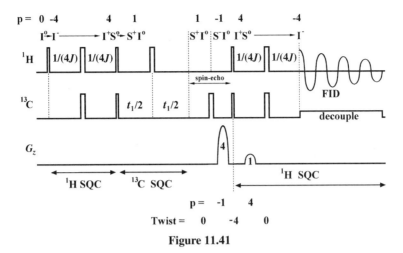

Figure 11.41

recovery delay). Any time an FID is "started late" there will be chemical-shift evolution before it starts, leading to huge chemical-shift dependent phase errors ("phase twists"). In this case the phase errors will show up in F_1, since we are starting the t_1 FID late. The solution, as always, is to build a spin-echo with the gradient in one of the spin echo delays (Fig. 11.41). Because we have ^{13}C SQC during the t_1 delay, we use a ^{13}C $180°$ pulse in the center of the spin echo to refocus both ^{13}C shift evolution and J-coupling evolution. The second gradient is already contained in a spin echo, so there is a big enough "gap" ($1/(4J) = 1/(4 \times 150$ Hz$) = 1.67$ ms) to fit a typical gradient pulse (1 ms) and its recovery delay (200 µs). Using the spherical operators to describe the coherence pathway, it becomes clear that the gradients must now be of the same sign to select the desired (echo) pathway. If echo-antiecho phase encoding is used in t_1, the first gradient (relative strength +4) would alternate sign to select $\mathbf{S}^+\mathbf{I}°$ during t_1 for the first FID ($G_1 = +4$, echo signal) and to select $\mathbf{S}^-\mathbf{I}°$ during t_1 for the second FID ($G_1 = -4$, antiecho signal) acquired for each value of t_1.

A different approach to destroying the ^{12}C-bound ^{1}H artifact is to use the gradient in a simpler way—as a "spoiler" that just kills all of the magnetization in the x'-y' plane while our desired signal is "stored" briefly on the z axis. To do this we go back to our discussion of intermediate states in INEPT coherence transfer (Section 7.10) and recall that instead of using simultaneous $90°$ pulses on ^{1}H and ^{13}C to effect coherence transfer, we can start with the ^{1}H $90°$ pulse and then, after a short delay, complete the INEPT transfer with the ^{13}C $90°$ pulse:

$$2\mathbf{I}_x\mathbf{S}_z \xrightarrow{90°_y \, ^1\text{H}} -2\mathbf{I}_z\mathbf{S}_z \xrightarrow{90°_x \, ^{13}\text{C}} 2\mathbf{S}_y\mathbf{I}_z$$

(again we omit the factor of 4 reflecting our change to ^{13}C M_o as a standard of comparison because coherence will be transferred back to ^{1}H in the end). The $-2\mathbf{I}_z\mathbf{S}_z$ product operator will not be affected by a gradient, nor will it undergo any evolution during the time of the gradient. ^{12}C-bound ^{1}H coherence cannot achieve this spin state, so any coherence it has at this point ($\mathbf{I}_x$ or $\mathbf{I}_y$) will be destroyed. We call this a "spoiler" (or homospoil) gradient because the twist it produces is never untwisted in the pulse sequence—we are just getting rid of stuff. The same strategy can be used in the back transfer (Fig. 11.42), usually with a

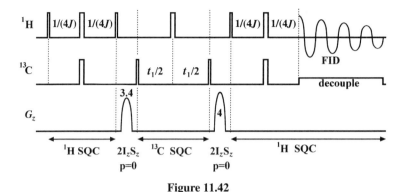

Figure 11.42

different gradient strength to prevent any kind of untwisting from occuring. The advantage of this approach is that no spin echoes are required to refocus evolution that happens during the gradients. The disadvantage is that echo-antiecho phase encoding cannot be incorporated because there is no selection of coherence order during the t_1 period.

11.7.5 Edited HSQC

Editing (modification of the sign of crosspeaks based on the type of carbon—CH_3, CH_2, or CH) is easily accomplished by allowing a period of $^1J_{CH}$-coupling evolution of time $1/J$ while we have ^{13}C SQC. We saw in Chapter 6 (Section 6.8) how the in-phase ^{13}C coherence of a CH or CH_3 group reverses sign after a $1/J$ delay while a CH_2 group does not. This is the strategy of the APT experiment. Adding a spin echo that allows only J-coupling evolution at the end of the t_1 period will accomplish this without interfering in any other way with the pulse sequence (Fig. 11.43). The fact that we are starting with antiphase rather than in-phase coherence does not change the effect of the $1/J$ delay. For example, for a CH carbon:

$$2\mathbf{S}_y\mathbf{I}_z\cos(\Omega_c t_1) \stackrel{1/(2J)}{\rightarrow} -\mathbf{S}_x\cos(\Omega_c t_1) \stackrel{1/(2J)}{\rightarrow} -2\mathbf{S}_y\mathbf{I}_z\cos(\Omega_c t_1)$$

For a CH_2 group we have to consider coupling to both protons–H_1 and H_2:

$$2\mathbf{S}_y\mathbf{I}_z^1 \stackrel{1/(2J)}{\rightarrow} -2\mathbf{S}_y\mathbf{I}_z^2 \stackrel{1/(2J)}{\rightarrow} 2\mathbf{S}_y\mathbf{I}_z^1$$

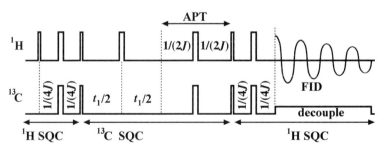

Figure 11.43

omitting the cosine multiplier and considering only coherence that transferred from H_1 in the first coherence transfer step. For the first delay, the antiphase relationship to H_1 refocuses from antiphase to in-phase ($2\mathbf{S}_y\mathbf{I}_z^1 \xrightarrow{J_1} -\mathbf{S}_x$) and the in-phase relationship to H_2 defocuses from in-phase to antiphase ($-\mathbf{S}_x \xrightarrow{J_2} -2\mathbf{S}_y\mathbf{I}_z^2$). During the second delay they reverse roles again: $-2\mathbf{S}_y\mathbf{I}_z^2 \xrightarrow{J_2} \mathbf{S}_x \xrightarrow{J_1} 2\mathbf{S}_y\mathbf{I}_z^1$. As always, each time we change from antiphase to in-phase or *vice versa* as the phase of the observable operator (^{13}C in this case) advances by 90° in the counterclockwise direction. Finally, for a CH_3 group we consider coupling to H_1, H_2, and H_3:

$$2\mathbf{S}_y\mathbf{I}_z^1 \xrightarrow{1/(2J)} 4\mathbf{S}_x\mathbf{I}_z^2\mathbf{I}_z^3 \xrightarrow{1/(2J)} -2\mathbf{S}_y\mathbf{I}_z^1$$

For each delay there is a phase advance of 270° because there are three J couplings that are active: J_1, J_2, and J_3 (all equal). In the first step the ^{13}C coherence refocuses with respect to H_1 and defocuses with respect to H_2 and H_3 as the $\mathbf{S}$ operator rotates by 90° three times: $\mathbf{S}_y \rightarrow -\mathbf{S}_x \rightarrow -\mathbf{S}_y \rightarrow \mathbf{S}_x$. The normalization factor of 4 is used because we are multiplying by two $\mathbf{I}_z$ operators, each of which carries a factor of 1/2 in the density matrix representation. Overall we see the editing effect:

$$\begin{aligned} \text{CH}: \quad & 2\mathbf{S}_y\mathbf{I}_z \xrightarrow{1/(J)} -2\mathbf{S}_y\mathbf{I}_z & \text{(reversed)} \\ \text{CH}_2: \quad & 2\mathbf{S}_y\mathbf{I}_z^1 \xrightarrow{1/(J)} 2\mathbf{S}_y\mathbf{I}_z^1 & \text{(unchanged)} \\ \text{CH}_3: \quad & 2\mathbf{S}_y\mathbf{I}_z^1 \xrightarrow{1/(J)} -2\mathbf{S}_y\mathbf{I}_z^1 & \text{(reversed)} \end{aligned}$$

This editing is carried through the rest of the sequence and shows up in the sign of the observed in-phase ^{1}H coherence. In the 2D data processing we phase the CH and CH_3 crosspeaks positive, which makes the CH_2 crosspeaks negative.

11.7.6 Sensitivity Enchancement by Preservation of Equivalent Pathways

In every 2D experiment we have looked at, the chemical-shift evolution during the t_1 delay produces two terms—sine and cosine—and in each case only one of them survives the mixing step to reach the FID as observable magnetization. The HSQC experiment is no exception:

$$\text{Evolution}: \quad 2\mathbf{S}_y\mathbf{I}_z \xrightarrow{t_1} 2\mathbf{S}_y\mathbf{I}_z\cos(\Omega_c t_1) \; -2\mathbf{S}_x\mathbf{I}_z\sin(\Omega_c t_1)$$

$$\text{Mixing}: \quad 2\mathbf{S}_y\mathbf{I}_z\cos(\Omega_c t_1) \xrightarrow{90_y{}^1\text{H},90_x{}^{13}\text{C}} 2\mathbf{I}_x\mathbf{S}_z\cos(\Omega_c t_1) \quad \text{(observable}\,^1\text{H SQC)}$$

$$-2\mathbf{S}_x\mathbf{I}_z\sin(\Omega_c t_1) \xrightarrow{90_y{}^1\text{H},90_x{}^{13}\text{C}} -2\mathbf{I}_x\mathbf{S}_x\sin(\Omega_c t_1) \quad \text{(nonobservable ZQC/DQC)}$$

The second (sine) term produced by the evolution delay has all the information we need—it is antiphase ^{13}C coherence labeled with the ^{13}C chemical shift in t_1—but it is lost because its phase ($\mathbf{S}_x$) causes it to be unaffected by the $90°_x$ ^{13}C pulse at the end of t_1. Effectively we are throwing away half of our signal at this point. A new method was developed to "save" this wasted signal and boost the sensitivity of HSQC and many other experiments. These modified pulse sequences are called "sensitivity enhanced" or "sensitivity improved" (Bruker adds "si" to the pulse sequence name) and the strategy is called "preservation of equivalent pathways" (PEP) because the two terms are equivalent except for their phase.

The strategy is quite simple: after the mixing step, the cosine term is refocused as before with a spin echo of duration $1/(2J)$ whereas the sine term "sits out" the refocusing delay on the sidelines because ZQC and DQC do not undergo J-coupling evolution with respect to the active coupling ($^1J_{CH}$):

$$2\mathbf{I}_x\mathbf{S}_z\cos(\Omega_c t_1) \xrightarrow{1/(2J)} \mathbf{I}_y\cos(\Omega_c t_1)$$
$$-2\mathbf{I}_x\mathbf{S}_x\sin(\Omega_c t_1) \xrightarrow{1/(2J)} -2\mathbf{I}_x\mathbf{S}_x\sin(\Omega_c t_1)$$

Then the cosine term is flipped to the z axis where it "sits out" another $1/(2J)$ refocusing delay while the sine term, converted by the same pulse to antiphase ^{1}H coherence (completing the coherence transfer), is refocused to in-phase ^{1}H coherence:

$$\mathbf{I}_y\cos(\Omega_c t_1) \xrightarrow{90_x{}^1H, 90_y{}^{13}C} \mathbf{I}_z\cos(\Omega_c t_1) \xrightarrow{1/(2J)} \mathbf{I}_z\cos(\Omega_c t_1)$$
$$-2\mathbf{I}_x\mathbf{S}_x\sin(\Omega_c t_1) \xrightarrow{90_x{}^1H, 90_y{}^{13}C} 2\mathbf{I}_x\mathbf{S}_z\sin(\Omega_c t_1) \xrightarrow{1/(2J)} \mathbf{I}_y\sin(\Omega_c t_1)$$

Finally, a 90° ^{1}H pulse flips the cosine term back to the x'–y' plane:

$$\mathbf{I}_z\cos(\Omega_c t_1) \xrightarrow{90_y{}^1H} \mathbf{I}_x\cos(\Omega_c t_1)$$
$$\mathbf{I}_y\sin(\Omega_c t_1) \xrightarrow{90_y{}^1H} \mathbf{I}_y\sin(\Omega_c t_1)$$

Now we have observable in-phase ^{1}H coherence coming from both terms. This is some pretty fancy footwork in the world of spin choreography: taking one partner for a spin while the other sits out, then reversing the roles. The complete pulse sequence is shown in Figure 11.44. Note that each of the $1/(2J)$ delays is a spin echo with simultaneous ^{1}H and ^{13}C 180° pulses in the center to prevent any chemical-shift evolution and permit only J-coupling evolution. Some fancy phase-cycling tricks are required to get the desired pure cosine and sine terms with the same phase in t_2. This is equivalent to acquiring two FIDs in which one has the opposite sign for the second (sine) term. Adding the two FIDs gives the pure cosine term: $2\mathbf{I}_x\cos(\Omega_c t_1)$, and subtracting with a 90° shift in receiver phase gives the pure sine term: $2\mathbf{I}_x\sin(\Omega_c t_1)$. Thus we have the same two FIDs we would get with a States mode experiment acquiring the "real" and "imaginary" FIDs for each t_1 value. Because we also acquire noise in the two FIDs we have increased the noise by $\sqrt{2}$ (because it is random), and the increase in sensitivity (S/N) is $2/\sqrt{2} = \sqrt{2}$ or 1.414. An increase in sensitivity of 41% is worth a little bit of trouble!

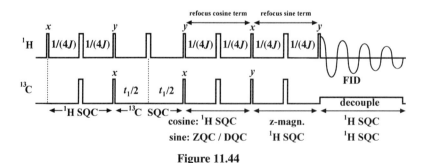

Figure 11.44

11.8 UNDERSTANDING THE HMQC PULSE SEQUENCE

The HMQC experiment gives exactly the same result as the HSQC, and the data is processed in the same way. There are some differences in sensitivity and peak shape that depend on the size and complexity of the molecule, and the pros and cons of the two experiments are the subject of some debate in the literature. Because it relies on double-quantum and zero-quantum coherences (DQC and ZQC) during the evolution (t_1) period, the HMQC is more difficult to explain and understand than HSQC, which uses only the familiar single-quantum transitions that can be diagramed and analyzed using vectors. We discuss it here because it forms the basis of the HMBC (multiple-bond) experiment.

The sequence is similar to the HSQC sequence, but much simpler—there are only four pulses (Fig. 11.45). The preparation period is the same, except that we do not bother to refocus chemical-shift evolution—this ends up being corrected by the final (refocusing) delay. Neglecting chemical-shift evolution, we have ^{1}H magnetization at the end of the first $1/(2J)$ delay which is antiphase with respect to the directly bound ^{13}C. Instead of subjecting this to simultaneous 90° pulses on both ^{1}H and ^{13}C channels, which would cause INEPT transfer of magnetization to antiphase ^{13}C single-quantum coherence, we have a single 90° pulse on ^{13}C only. We saw before that this leads to an intermediate state in coherence transfer, a combination ZQC and DQC:

$$2\mathbf{I}_x\mathbf{S}_z \xrightarrow{90°_x {}^{13}\text{C}} -2\mathbf{I}_x\mathbf{S}_y = -0.5(\{\mathbf{DQ}\}_y + \{\mathbf{ZQ}\}_y)$$

Now we have both the ^{13}C and the ^{1}H magnetization in the x'–y' plane, not as independent magnetization vectors but tied up together in a *product* of operators. This is a combination of ZQC and DQC called collectively multiple-quantum coherence (MQC). In the HMQC experiment, the ^{1}H–^{13}C DQC and ZQC precess in the x'–y' plane at rates of $\nu_H + \nu_C$ and $\nu_H - \nu_C$, respectively, during the evolution (t_1) delay (Fig. 11.46). Note that this coherence cannot be called "^{1}H" or "^{13}C" coherence—it involves the entanglement of both nuclei in a mutual dance. The effect of the 180° ^{1}H pulse in the center is to convert DQC into ZQC and *vice versa*:

$$\{\mathbf{ZQ}\}_x = 2\mathbf{I}_x\mathbf{S}_x + 2\mathbf{I}_y\mathbf{S}_y \xrightarrow{180°_x {}^1\text{H}} 2\mathbf{I}_x\mathbf{S}_x - 2\mathbf{I}_y\mathbf{S}_y = \{\mathbf{DQ}\}_x$$

$$\{\mathbf{ZQ}\}_y = 2\mathbf{I}_x\mathbf{S}_y - 2\mathbf{I}_y\mathbf{S}_x \xrightarrow{180°_x {}^1\text{H}} 2\mathbf{I}_x\mathbf{S}_y + 2\mathbf{I}_y\mathbf{S}_x = \{\mathbf{DQ}\}_y$$

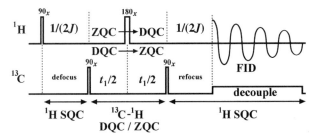

Figure 11.45

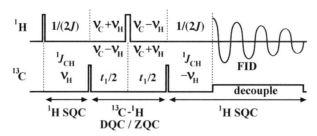

Figure 11.46

because only the $\mathbf{I}_y$ operator is affected by the $180°_x$ ^{1}H pulse. Thus, for the second half of the t_1 period, the ZQC that was evolving at a rate of $\nu_C - \nu_H$ is now evolving at a rate of $\nu_C + \nu_H$, for a total evolution of $(\nu_C - \nu_H) \times (t_1/2) + (\nu_C + \nu_H) \times (t_1/2) = \nu_C \times t_1$. The net evolution depends only on t_1 and ν_C, so the t_1 delay serves the same purpose as the t_1 delay in HSQC: to indirectly measure the ^{13}C chemical shift and encode this information as a modulation of intensity of the ^{1}H SQC that is detected in the FID. Even though we only go "halfway" in the process of coherence transfer from ^{1}H SQC to ^{13}C SQC, we still accomplish the goal of labeling the final ^{1}H SQC with the chemical shift of the ^{13}C.

The second 90° ^{13}C pulse converts the ZQC and DQC back into antiphase ^{1}H SQC, which is refocused by the final $1/(2J)$ delay just as it is in the HSQC experiment. This is an important theme that occurs in many of the more sophisticated NMR experiments: multiple-quantum coherences (DQC and ZQC) cannot be directly observed, but they can be created from SQC, allowed to precess, and converted back to measurable SQC. The multiple-quantum coherences *can* be detected then, but only indirectly. Multiple quantum coherence is essential to the DQF-COSY and DEPT experiments as well, so even though it is difficult to understand it is very important in modern NMR and cannot be ignored.

The spherical operator description of the HMQC experiment is shown in Figure 11.47, along with a diagram of the coherence order. As always, we observe negative coherence order during t_2 and positive coherence order during t_1 (echo pathway). Because only the ^{13}C chemical-shift evolution is observed during t_1, we choose $\mathbf{S}^+$ in the DQC/ZQC product operators. Working backward from $\mathbf{I}^-$, we have $\mathbf{I}^-\mathbf{S}°$ (antiphase ^{1}H coherence) at the start

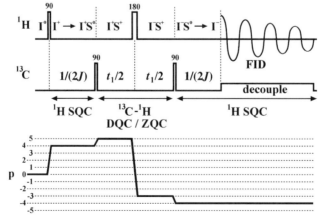

Figure 11.47

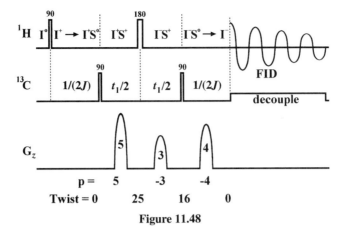

Figure 11.48

of the refocusing period and $\mathbf{I}^-\mathbf{S}^+$ just before the second 90° ^{13}C pulse. Because there is only a pulse on ^{13}C, the ^{1}H coherence order cannot change. In the center of the evolution period, the ^{1}H 180° pulse converts $\mathbf{I}^+\mathbf{S}^+$ (DQC, $\mathbf{p} = 5$) to $\mathbf{I}^-\mathbf{S}^+$ (ZQC, $\mathbf{p} = -3$). The first 90° ^{13}C pulse converts $\mathbf{I}^+\mathbf{S}^\circ$ (antiphase ^{1}H coherence) to $\mathbf{I}^+\mathbf{S}^+$ (DQC). Again, because there is no ^{1}H pulse at this point the $\mathbf{I}^+$ operator does not change. During the first $1/(2J)$ delay, $\mathbf{I}^+$ undergoes J-coupling evolution to the antiphase product $\mathbf{I}^+\mathbf{S}^\circ$. The coherence order at each stage can be easily determined by adding the coherence order contribution of each operator in the product, with $+4$ for $\mathbf{I}^+$, -4 for $\mathbf{I}^-$, $+1$ for $\mathbf{S}^+$, -1 for $\mathbf{S}^-$, and so on. For example, $\mathbf{I}^-\mathbf{S}^+$ has coherence order -3: -4 ($\mathbf{I}^-$) $+1$ ($\mathbf{S}^+$). The chemical-shift evolution can also be easily predicted from the operators: $\mathbf{I}^+\mathbf{S}^+$ will evolve at the rate $\nu_H + \nu_C$ and $\mathbf{I}^-\mathbf{S}^+$ will evolve at the rate $-\nu_H + \nu_C$ (Fig. 11.46, top).

Once we have diagramed the desired coherence order pathway, it is easy to add gradients to select that pathway. One simple solution is to use the 3, 4, 5 relationship of a right triangle: $3 \times 3 + 4 \times 4 = 5 \times 5$. Put a gradient in the first half of t_1 of relative amplitude $G_1 = 5$, another in the second half with amplitude $G_2 = 3$, and a third in the refocusing delay with amplitude $G_3 = 4$. As the coherence order $\mathbf{p}$ is 5, -3, and -4, respectively, during these three periods, we have a total twist of:

$$\Sigma \, \mathbf{p}_i G_i = 5 \times 5 + (-3) \times 3 + (-4) \times 4 = 25 - 9 - 16 = 0$$

The pulse sequence is shown in Figure 11.48. The net twist will be zero only for the desired pathway: DQC → ZQC → ^{1}H SQC. Of course, we have created a new problem: the minimum t_1 delay is now twice the time required for a gradient and its recovery. This will lead to a very large phase twist in F_1, so we can either present the data in magnitude mode, where phase is not an issue, or insert the appropriate spin echoes to refocus the chemical-shift evolution that occurs during the gradients.

11.9 UNDERSTANDING THE HETERONUCLEAR MULTIPLE-BOND CORRELATION (HMBC) PULSE SEQUENCE

The HMBC experiment is just an HMQC experiment with the $1/(2J)$ delay set for a J value of about 10 Hz (typical for two- and three-bond J_{CH}) rather than 150 Hz (typical for one-bond J_{CH}). This corresponds to a much longer $1/(2J)$ delay of 50 ms for remote (multiple-bond) couplings compared to the 3.33 ms used for HMQC and HSQC (direct or one-bond couplings). The protons we are interested in observing are two or three bonds away from a ^{13}C, and the carbon they are directly bonded to is very likely (99%) to be a ^{12}C. The pulse sequence differs from the HMQC sequence in two ways:

1. Because of the signal loss due to T_2 relaxation during the long $1/(2J)$ defocusing delay, the $1/(2J)$ refocusing delay is omitted and we observe antiphase coherence in the FID just as we do in the COSY and DQF-COSY experiments. One consequence of this is that we cannot use ^{13}C decoupling during the acquisition of the FID because the antiphase lines in the F_2 spectrum would cancel and there would be no signal.

2. To suppress the one-bond (HMQC) cross peaks, which would lead to wide doublets in F_2 centered on the position of the crosspeaks in the (^{13}C-decoupled) HMQC or HSQC spectrum, a simple trick is applied. After the first ^1H 90° pulse a delay of $1/(2\,^1J_{CH})$ (or ~3.33 ms) is followed by a 90° pulse on ^{13}C. This is identical to the start of the HMQC sequence, right down to the length of the $1/(2J)$ delay. This converts the ^1H magnetization from the proton directly attached to the ^{13}C into ZQC and DQC, but the 90° ^{13}C pulse is phase-alternated ($+x, +x, -x, -x$) whereas the receiver is not (individual FIDs from the first two scans are added to, not subtracted from, the FIDs from the second two scans). This means that any observable ^1H magnetization that shows up in the FID due to this pathway will be canceled out by the phase cycle. The ^1H magnetization from protons that are two or three bonds away from the ^{13}C, however, is separating into antiphase much more slowly ($1/(2\,^{2,3}J_{CH}) \sim 50$ ms) so that after 3.33 ms, it is essentially still in-phase ^1H magnetization and is not affected by the 90° ^{13}C pulse ($\cos(\pi J\tau) = 0.995$, so only 0.5% of the signal is lost). What follows is a much longer (50 ms) delay to allow these protons that are two or three bonds away from the ^{13}C to evolve into antiphase with respect to the ^{13}C. Then a 90° ^{13}C pulse converts this ^1H magnetization into ZQC and DQC that continues through the pulse sequence as it does for the HMQC. This second 90° ^{13}C pulse is phase cycled *with* the receiver so that the coherences it generates are ultimately added together and appear in the final FID. The pulse sequence is diagramed in Figure 11.49.

The refocusing delay just before acquisition has been eliminated, so the peaks in F_2 will be antiphase doublets separated by the long-range J_{CH} (2–15 Hz). This splitting is in addition to any $^1H-^1H$ splitting pattern already present in the 1D proton spectrum (Fig. 11.11). The second ^{13}C 90° pulse is phase cycled as before to make sure that only ^1H magnetization

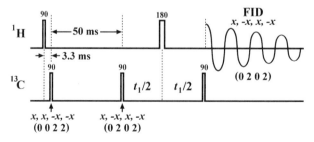

Figure 11.49

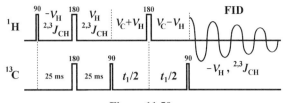

Figure 11.50

that is converted to ZQC/DQC by that pulse is detected. Each FID coming into the receiver is alternately added to and subtracted from the sum in memory. Another way to say this is that the receiver phase is x, $-x$, x, $-x$ in this scheme, and the desired signals, which are also modulated in this way, will combine and add in memory. The first ^{13}C 90° pulse, which is a trap put there to "spoil" the signal of protons directly bound to ^{13}C (the one-bond signals) is phase cycled x, x, $-x$, $-x$ for each set of four scans. Because the receiver phase cycle is x, $-x$, x, $-x$, which does not match this pattern, the signals that arise from ^{1}H magnetization following this pathway are canceled by subtraction and eliminated from the sum FID in memory. We can call this dual-purpose phase cycle a "four step" phase cycle; the number of transients (Bruker: ns, Varian: nt) will have to be a multiple of four in order for the cancelation of undesired signals to work.

Because the ^{1}H chemical-shift evolution during the long (∼50 ms) $1/(2J)$ defocusing delay is not refocused as it is in HMQC, HMBC data is normally presented in magnitude mode. To make the data suitable for phase-sensitive presentation, it is necessary to insert spin echoes to allow for refocusing of all unwanted shift evolution. For example, one can insert simultaneous 180° pulses on ^{1}H and ^{13}C in the center of the long $1/(2J)$ delay, refocusing the ^{1}H chemical-shift evolution (Fig. 11.50). To eliminate the ^{13}C-bound ^{1}H signal a different strategy can be used: a variant of the BIRD sequence called TANGO delivers a 90° pulse to the ^{13}C-bound protons without exciting the ^{12}C-bound protons (Fig. 11.51). The resulting ^{13}C-bound proton coherence is then destroyed with a "spoiler" gradient, which has no effect on the ^{12}C-bound proton magnetization which is on the z axis. As long as we are adding gradients, we can do coherence pathway selection by adding gradients during the two halves of the evolution delay. Now we have a sequence that blocks the one-bond artifacts and eliminates the "streaks" due to ^{1}H coherence that is not correlated to any ^{13}C, all in a single scan. But the gradients during t_1 create a new problem: again we can not start with $t_1 = 0$ in

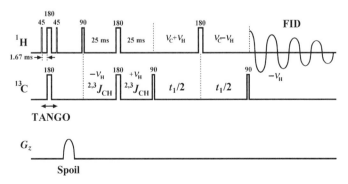

Figure 11.51

538 INVERSE HETERONUCLEAR 2D EXPERIMENTS: HSQC, HMQC, AND HMBC

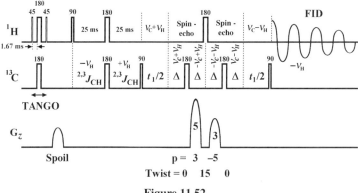

Figure 11.52

our experiment; the minimum t_1 value is twice the gradient time. This can be remedied by introducing two ^{13}C spin echoes in the center of the evolution delay to "make room" for the two gradients (Fig. 11.52). Note that all chemical-shift evolution during the four Δ delays cancels, for ^{1}H and for ^{13}C. This is just one example of how you can get creative with pulse sequences: there are many ways to design an HMBC sequence and this is a very active area of research (e.g., "CIGAR," "ACCORDION," "ADEQUATE," "HSQMBC," etc.). Some sequences are designed for accurate measurement of long-range ($^{2,3}J_{CH}$) heteronuclear coupling constants, and there are many different strategies for getting rid of the one-bond artifacts ("low-pass filters"). Two big problems remain to be solved: removing the ambiguity of two-bond and three-bond correlations, and eliminating homonuclear (J_{HH}) J-coupling evolution during the long ($\sim$50 ms) 1/(2J) delay. The complex logic of puzzle solving in interpreting HMBC data would be vastly simplified if we knew for each crosspeak whether it is a two-bond or a three-bond relationship. The homonuclear J evolution distorts the peak shapes by introducing antiphase terms, and this makes it difficult to accurately measure the long-range J_{CH} coupling.

11.10 STRUCTURE DETERMINATION BY NMR—AN EXAMPLE

A case study of covalent structure deterimation by 2D NMR will help to illustrate the central importance of HSQC and HMBC, used in conjunction with the homonuclear COSY and ROESY experiments. Oxidation of the natural product Pristimerin with DDQ in dioxane gave four products which were separated and purified (Fig. 11.53). The third fraction (LGJC3, 4.1 mg) was dissolved in 0.5 mL of deuterated chloroform (CDCl$_3$) and analyzed by one-dimensional (^{1}H and ^{13}C) and two-dimensional (HSQC, HMBC, TOCSY, COSY and ROESY) NMR using a Bruker DRX-600. In this case, we have a great deal of information about the sample since we know its origin and we can imagine that most of the carbon skeleton of Pristimerin is conserved in the product. In the following discussion, however, the problem will be approached initially as a "white powder" unknown to show how much we can conclude from NMR alone.

Carbon resonances in the ^{13}C spectrum were numbered 1–32 in order of chemical shift from the farthest upfield (c1) to the farthest downfield (c32). Using the HSQC spectrum, the protons were named according to the carbon to which they are correlated, using "a" and

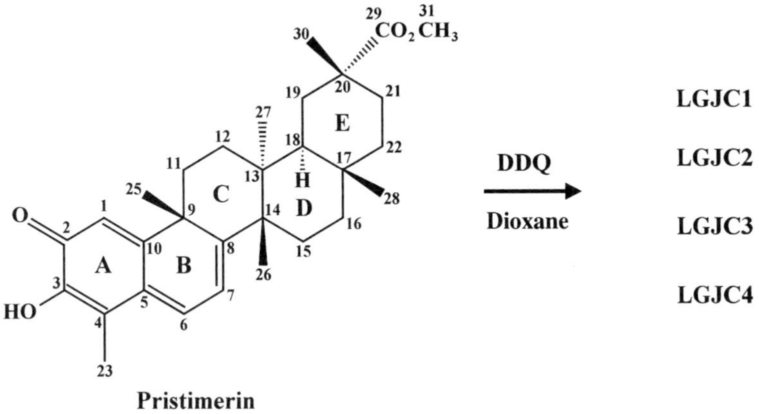

STRUCTURE DETERMINATION BY NMR—AN EXAMPLE 539

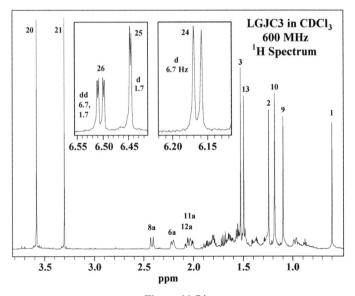

Figure 11.53

"b" to indicate the downfield and upfield resonances of a geminal pair (CH_2 group): h1, h6a, h6b, and so on. To avoid confusion between this arbitrary numbering system and the structure-based numbering system of Pristimerin (Fig. 11.53), lower case "c" and "h" will be used for the arbitrary, chemical-shift based numbering (e.g., h23, c16, etc.) and upper case "C" and "H" will be used for structure-based references (C-5, H-8, etc.). The advantage of arbitrary numbering is that it does not bias us toward any particular structure and it does not have to be revised as our structural model evolves.

11.10.1 1D ^{1}H spectrum (Fig. 11.54)

There are clearly three one-proton olefinic peaks (6.1–6.6 ppm), two coupled to each other with a 6.7 Hz coupling and a third coupled to one of the first two with a long-range coupling

Figure 11.54

of 1.7 Hz. These might correspond to the original H-1, H-6, and H-7 of Pristimerin, with the *cis* vicinal H-6 to H-7 coupling being the large coupling (6.7 Hz) and the long-range coupling (1.7 Hz) being between H-1 and H-6, a 5-bond coupling through the extended π system of the transoid diene (C1–C10–C5–C6). There are two OCH$_3$ singlets, one at 3.30 ppm, typical of a methyl ether, and one at 3.58 ppm, closer to the typical chemical shift for a methyl ester. A methyl ether is unexpected, and in fact for the first ^{1}H spectrum, run on a less pure sample at 200 MHz, this peak was ignored. In the cleaner sample at 600 MHz it is clear that both peaks integrate to around three protons. In the upfield region (0.6–1.6 ppm) there are six methyl singlets, corresponding to the six methyl groups in the triterpene skeleton of Pristimerin, attached at carbons 4, 9, 13, 14, 17, and 20. Most of the other peaks are complex and overlapped.

11.10.2 1D ^{13}C spectrum (Fig. 11.55)

Ignoring the TMS peak and the three CDCl$_3$ peaks, there are 32 peaks in the ^{1}H-decoupled ^{13}C spectrum. One of these peaks, c22 at 77.23 ppm, was eventually assigned to the CHCl$_3$ resonance (residual CHCl$_3$ in the deuterated solvent and possible CHCl$_3$ solvent left over from extraction and purification), leaving 31 carbons in the compound of interest. Of these 31 peaks, nine are in the downfield region typical for olefinic, aromatic or carbonyl carbons. Four of these nine peaks are in the typical olefinic region (116–131 ppm). Six are "short" peaks, indicating that they are quaternary carbons, including one of those in the olefinic region. Pristimerin would be expected to have ten downfield ^{13}C peaks (sp^2 carbons), nine from the A-B ring system (C-1–C-8 and C-10) and one from the methyl ester carbonyl

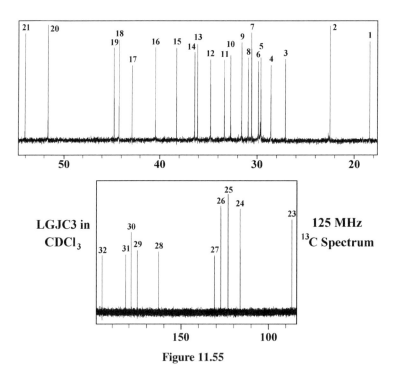

Figure 11.55

(C29), so it appears that one is missing. Perhaps one of these olefinic/aromatic/carbonyl carbons was converted by DDQ oxidation to an sp^3-hybridized carbon that appears in the upfield ($<$ 100 ppm) region (Fig. 11.55). LGJC3 (C$_{31}$) has one more carbon than Pristimerin (C$_{30}$H$_{40}$O$_4$), consistent with the observation of an extra methyl group (methyl ether) in the proton spectrum. Combined with the observation of one fewer sp^2-hybridized carbon in LGJC3, this suggests the addition of methanol to one of the sp^2-hybridized carbons in the extended π system (C-1–C-8 and C-10) of Pristimerin. A carbon peak observed at 86.6 ppm in LGJC3 is typical for an oxygenated sp^3-hybridized carbon, and this could be the carbon in the triterpene backbone where the methoxy group is attached. If so, the chemical shift puts it in the range of quaternary, singly oxygenated carbons as opposed to CH–O (70–80 ppm) or CH$_2$–O (60–70 ppm).

11.10.3 2D HSQC Spectrum (Upfield Region: Fig. 11.56; Center and Downfield Regions: Fig. 11.57)

This is an edited, ^{13}C-decoupled HSQC spectrum with negative intensities (CH$_2$/methylene groups) shown in gray and positive intensities (CH/methine and CH$_3$/methyl groups) in black. The carbons with no hydrogens attached (quaternary carbons) do not show up in this spectrum. A broad CH$_2$ crosspeak was observed at $\delta_H = 1.26$, $\delta_C = 29.7$ due to long-chain hydrocarbon grease ((CH$_2$)$_n$), a common contaminant in solvents used for extraction and chromatography. The methyl carbon crosspeaks (c1, 2, 3, 9, 10, and 13) can be easily distinguished from the methine carbon crosspeaks because they are much more intense. Thus 8 CH$_3$ peaks (6 upfield and 2 methoxy), 7 CH$_2$ peaks (one degenerate and 6 in "pairs" at the same ^{13}C shift), and 4 CH peaks (three olefinic and one aliphatic) were observed in the HSQC spectrum, leaving 31 -8 -7 -4 = 12 quaternary carbons. By comparison with the 1D ^{13}C spectrum, the 12 quaternary carbons can be identified and fall into three categories: five sp^3-hybridized carbons without oxygen (30–45 ppm); one sp^3-hybridized carbon with one oxygen (86.6 ppm); and six sp^2-hybridized carbons (130–196 ppm: the "short" ones in the downfield region of the ^{13}C spectrum). Primisterin has 7 CH$_3$, 7 CH$_2$, 4

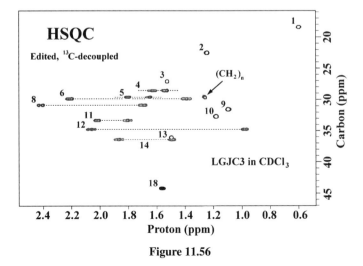

Figure 11.56

CH, and 12 quaternary carbons, of which five are sp^3-hybridized carbons without oxygen and seven are sp^2-hybridized carbons.

	Pristimerin	LGJC3
CH$_3$	7	8
OCH$_3$	(1)	(2)
CH$_2$	7	7
CH	4	4
C$_q$:	12	12
sp^3 (C)	(5)	(5)
sp^3 (O)	(0)	(1)
sp^2	(7)	(6)
Total	30	31

Thus the HSQC data confirm that, compared to Pristimerin, LGJC3 has one additional CH$_3$ group (a methoxy) and one oxygenated quaternary sp^3-hybridized carbon that corresponds to a sp^2-hybridized carbon in Pristimerin.

The only upfield (sp^3-hybridized without oxygen) CH group in the spectrum (Fig. 11.56) is the positive peak c18/h18. This would correspond to the aliphatic CH group C-18/H-18 at the D/E ring juncture in the Pristimerin structure. In the mid- and downfield regions (Fig. 11.57), we see the methoxy CH$_3$ groups 20 and 21 and the olefinic CH groups 24, 25, and 26, all as positive crosspeaks.

The primary value of the HSQC spectrum is that it provides us with an accurate list of all ^{1}H and ^{13}C chemical shifts. This would not be possible from 1D data alone. This list (Table 11.1, columns 1, 2, and 4) is then our reference for assigning correlations from other 2D spectra: HMBC (2-3 bond C to H), DQF-COSY (2-3 bond, occasionally longer, H to H), and ROESY (< 5 Å in space, H to H). We can use this list by sorting it into two lists: one in order of ^{1}H chemical shift and one in order of ^{13}C chemical shift. If, for example, we have an HMBC correlation with $F_1 = 30.9$ ppm, we would look down the list of ^{13}C shifts until we find something close to this value: c8. This is easy if the list is sorted by ^{13}C shift, and we can then look to the "neighbors" in the list to see if there is any ambiguity in the assignment. If there are other carbons very near to 30.9 ppm (e.g., c7 at 30.6 ppm), we will have to consider them as possible assignments as well; for example, we might assign the crosspeak "c8/c7." Keep in mind that the resolution of 2D spectra is considerably lower than that of

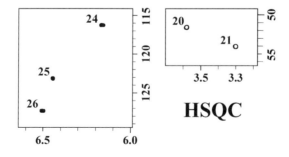

Figure 11.57

1D spectra—especially in the F_1 dimension. Two peaks may be clearly resolved in the ^{13}C spectrum but the corresponding crosspeaks in the HMBC spectrum may be too close to be distinguished. Another way to use the HSQC spectrum is as a graphical "list" of chemical shifts: many modern 2D data processing software packages allow the user to display more than one 2D spectrum and align them so that a crosshair can be used to precisely compare peak positions in the horizontal (F_2) and vertical (F_1) dimensions. To get a feel for this capability a number of the figures below show a portion of the HSQC spectrum above or to the side of the 2D spectrum (HMBC, COSY, ROESY, etc.) of interest so that the peaks can be visually aligned. This makes it clear when there is overlap or near overlap so we can determine if an assignment is unique (only one resonance at that chemical shift) or if we need to put down multiple possible assignments. Accurate referencing is essential and all spectra need to be acquired with the same temperature, solvent and concentration.

The assignment list (Table 11.1) includes lots of other information that will be discussed as we come to the other correlation spectra. Carbon shifts come initially from the HSQC but are then replaced by the more accurate values from the 1D ^{13}C spectrum. Multiplicity, where indicated other than "m" (multiplet) or "s" (singlet), comes from 1D slices of the HSQC peaks along F_2, so it only picks up the large (>4 Hz) couplings. The only exceptions are the olefinic protons and h8a and h6a, which are resolved in the 1D ^{1}H spectrum (Fig. 11.54).

11.10.4 2D HMBC spectrum (Fig. 11.58)

The figure is an expansion of the lower right corner of the full spectrum. Note that quaternary carbons *will* show up in the HMBC spectrum, as long as there is a hydrogen within two or three bonds of the carbon. Using our two chemical-shift lists, the HMBC crosspeak at $\delta_H = 3.30$, $\delta_C = 86.6$ (Fig. 11.58 center left) can be assigned to h21 and c23, indicating that the CH$_3$O group (ether methoxy) is connected to the oxygenated quaternary carbon (three bonds from the CH$_3$O hydrogen to the quaternary carbon):

$$H_3C - O - C_q$$

This quaternary carbon (c23) also correlates to the singlet methyl group h3 (1.53 ppm), which correlates to two other downfield quaternary carbons: c27 (Fig. 11.58, center right) and c32 (Table 1). One of the singlet methyl groups of Pristimerin (C-23) is unique in that it is bound to an sp^2-hybridized carbon and should give a ^{1}H peak around 2.1 ppm (compare to acetone, toluene, etc.). None of the carbon-bound singlet methyl groups of LGJC3 gives a ^{1}H peak downfield of h3 (1.53 ppm), so it appears that the "missing" sp^2-hybridized carbon in LGJC3 is derived from C-4 of Pristimerin (Fig. 11.53), which was changed into the sp^3-hybridized c23. This is consistent with the downfield quaternary carbons expected in the A ring (c27, c32 = C-3, C-5) that are correlated to the singlet methyl group C-23/c3.

For a more systematic approach to solving the puzzle, a large number of these correlations was tabulated and are included in the chemical-shift table from the HSQC spectrum (Table 11.1, column 6). Using the HMBC data we can now begin to construct structural pieces of the molecule. Since we have a very good idea of the general type of structure, this is an easy process, but for the sake of illustration we will proceed at first without considering the Primisterin skeleton, as if we were dealing with a true unknown. The 1D ^{1}H spectrum (Fig. 11.54) shows eight singlet methyl groups, six of them in the upfield region. These are excellent "handles" to start our analysis because they give very strong HMBC

Table 11.1. Chemical shift assignments for LGJC3 in CDCl$_3$

No.	^{1}H (ppm)	Mult.	^{13}C (ppm)	Type	HMBC Carbons	COSY Protons
1	0.60	s, 3H	18.4	CH$_3$	5, 15, 18, 19	
2	1.24	s, 3H	22.5	CH$_3$	4, 15, 19, 28	
3	1.53	s, 3H	27.1	CH$_3$	23, 27, 32	
4a	1.63	m	28.6	CH$_2$		
4b	1.54	m	"	"		
5a	1.80	m	29.6	CH$_2$	19	
5b	1.65	m	"	"		11a
6a	2.21	br. d, 13.5	29.9	CH$_2$		6b, 12a
6b	1.39	m	"	"	30	12b, 6a, 12a
7	—	—	30.6	C$_q$		
8a	2.42	d, 15.7	30.9	CH$_2$		8b
8b	1.70	dd, 16, 8	"	"	15, 30	18, 8a
9	1.10	s, 3H	31.6	CH$_3$	7, 12, 14, 18	
10	1.18	s, 3H	32.7	CH$_3$	6, 8, 16, 30	
11a	2.02	dd, 14, 5	33.4	CH$_2$		5b, 11b/5a
11b	1.81	m	"	"	17	11a
12a	2.06	dt, 4, 14	34.8	CH$_2$		12b, 6b, 6a
12b	0.98	br. dt, 14, 3	"	"		6b, 12a
13	1.50	s, 3H	36.1	CH$_3$	11, 17, 28, 29	
14a	1.87	dt, 6, 14	36.4	CH$_2$		
14b	1.49	m	"	"		4b
15	—	—	38.3	C$_q$		
16	—	—	40.5	C$_q$		
17	—	—	42.9	C$_q$		
18	1.56	br. d, 8	44.2	CH	7, 15, 16	8b
19	—	—	44.8	C$_q$		
20	3.58	s, 3H	51.6	CH$_3$	30	
21	3.30	s, 3H	54.1	CH$_3$	23	
23	—	—	86.6	C$_q$		
24	6.17	d, 6.7	116.2	CH	17, 19, 27	26
25	6.45	d, 1.7	123.1	CH	17, 27	(26)
26	6.50	dd, 6.7, 1.7	127.3	CH	23, 28, 29	24, (25)
27	—	—	131.0	C$_q$		
28	—	—	163.1	C$_q$		
29	—	—	175.2	C$_q$		
30	—	—	178.8	C$_q$		
31	—	—	182.0	C$_q$		
32	—	—	196.0	C$_q$		

crosspeaks (three equivalent protons in a singlet) and must be attached to a quaternary carbon (singlet pattern in ^{1}H spectrum). For example, h9 (CH$_3$) shows HMBC crosspeaks to c7 (C$_q$), c12 (CH$_2$), c14 (CH$_2$), and c18 (CH) (Table 11.1). This CH$_3$ group (^{1}H singlet) must be connected to c7 (quaternary carbon) and this carbon must be attached to c12, c14, and c18 (Fig. 11.59). Note that the correlation to c7 (C$_q$) is a two-bond relationship and the correlations to c12, c14, and c18 are three-bond relationships. Because we know the methyl group has to be attached to a quaternary carbon, the ambiguity (2 versus 3 bonds)

STRUCTURE DETERMINATION BY NMR—AN EXAMPLE   545

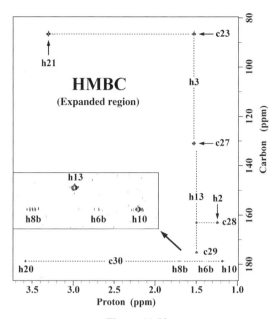

Figure 11.58

does not cause any confusion. In a similar way, the bonding networks can be constructed for h1, h2, h3, h10, and h13 (Fig. 11.59). In the case of h1 and h2 there is clearly a shared carbon since both CH$_3$ proton signals are correlated to c15 and c19. Since both c15 and c19 are quaternary, however, it is not possible to determine which one is connected to CH$_3$ (h1) and which is connected to CH$_3$ (h2). Note that c18 is shared by the h1 network and the h9 network, and c28 is shared by the h2 network and the h13 network. In the case of h3, there are only three HMBC correlations: CH$_3$ (h3) is attached to c23 (C$_q$), which is attached to c27 and c32 *and an oxygen atom*. This is the same quaternary carbon that is connected to the OCH$_3$ (methyl ether) group (see above). The h3 methyl group could also have been connected to either c27 or c32, since these are also quaternary, but other HMBC evidence shows that the structural fragment in Figure 11.59 is correct. Other ambiguities

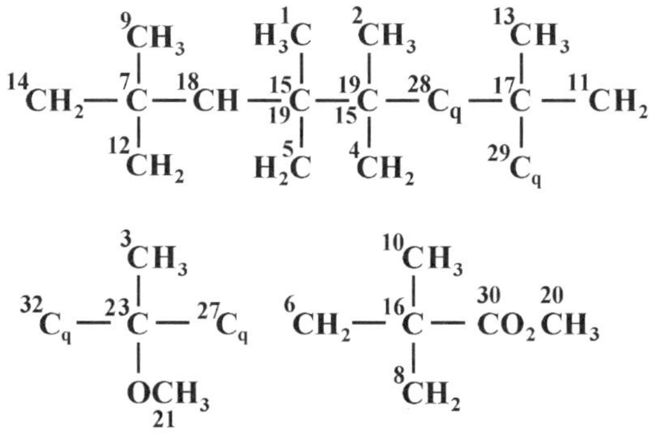

Figure 11.59

546 INVERSE HETERONUCLEAR 2D EXPERIMENTS: HSQC, HMQC, AND HMBC

can be easily resolved: CH_3 (h13) must be connected to c17 (C_q) since this is the only sp^3-hybridized quaternary carbon correlated to h13. If CH_3 (h13) were connected to c28 or c29 it could only show three HMBC correlations since c28 and c29 are sp^2-hybridized and can only be bonded to two carbons in addition to CH_3 (h13). Likewise, CH_3 (h10) must be bonded to c16 for the same reasons (Fig. 11.59). In this fragment, we can add a methyl ester since h20 (CH_3) shows an HMBC correlation to c30, which has a chemical shift (178.8 ppm) consistent with an ester carbonyl group.

11.10.5 2D COSY Spectrum

Figure 11.60 shows a portion of the DQF-COSY spectrum of LGJC3 with part of the HSQC spectrum aligned above it and, turned sideways, to the left side. There are correlations from h18 (1.56 ppm) to h8b (1.70 ppm, strong) and h8a (2.42 ppm, weak). Since these cannot be geminal couplings, we can assume that c8 is attached to c18 and the relationship is vicinal (three bonds) between h18 and h8b and between h18 and h8a. We can now connect the h1/h2 fragment to the h10 fragment (Fig. 11.61). Likewise, the h6a COSY crosspeak to h12a (Fig. 11.60) allows us to connect the h9 fragment to the h10 fragment. Because of the shared HMBC correlations from h1 (CH_3) and h9 (CH_3) to

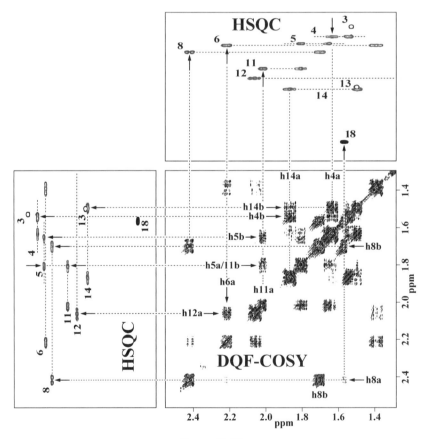

Figure 11.60

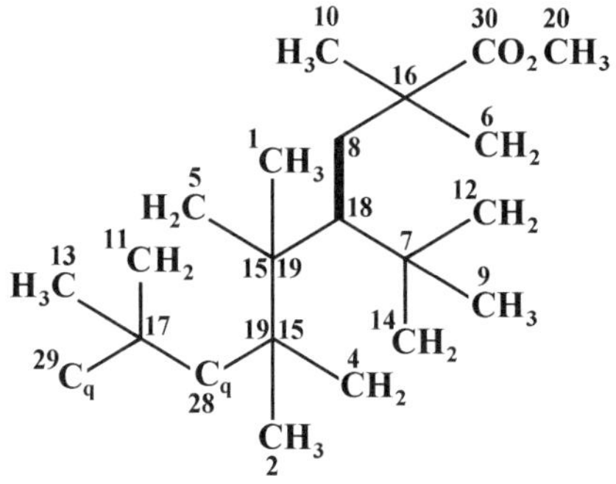

STRUCTURE DETERMINATION BY NMR—AN EXAMPLE 547

Figure 11.61

c18, the connection of c6 to c12 completes a six-membered ring: ring E of the triterpene (Fig. 11.62).

Some crosspeaks are ambiguous due to overlap of chemical shifts. For example, a COSY crosspeak is observed at 2.02 ppm (h11a) and 1.80 ppm (Fig. 11.60). Since h5a is in the list at 1.80 ppm, we might assume that c5 is connected to c11. But h11b has a chemical shift of 1.81 ppm, so the crosspeak could also be a geminal (two-bond) correlation between h11a and h11b. Sometimes these ambiguities can be resolved by careful alignment and comparison of the COSY and HSQC spectra, but in this case the ambiguity can be easily resolved since another crosspeak is observed at 2.02 ppm (h11a) and 1.65 ppm (h5b). Since h5b is a unique chemical shift, not overlapped with any other proton, we can safely conclude that c5 is connected to c11. This completes another six-membered ring due to the shared HMBC correlation from h2 (CH$_3$) and h13 (CH$_3$) to c28 (Fig. 11.63). COSY crosspeaks between h14a and h4b, and between h14b and h4a (Fig. 11.60) close the D ring of the triterpene (Fig. 11.64).

Finally, let's examine the olefinic/aromatic fragment starting with the three olefinic protons in the 1D 1H spectrum, identified from the HSQC spectrum as h24, h25 and h26. There is a 6.7 Hz coupling between h24 and h26, confirmed by a COSY crosspeak and consistent with a vicinal coupling in a *cis* olefin. The long-range coupling between h26 and h25 (1.7 Hz) is confirmed by a weak COSY cross peak and indicates that h25 is located farther away than three bonds from h26 on the h26 side of the olefin. In six-membered ring aromatic systems, we usually see strong three-bond HMBC correlations and weak or missing

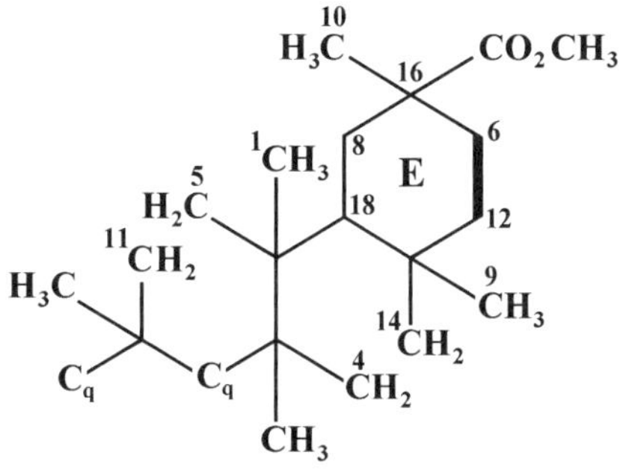

Figure 11.62

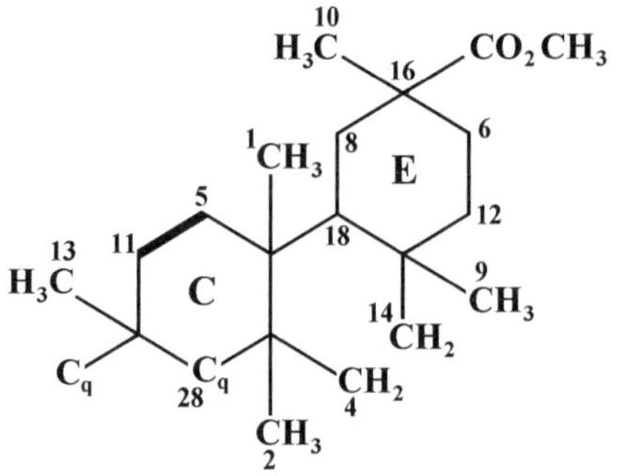

Figure 11.63

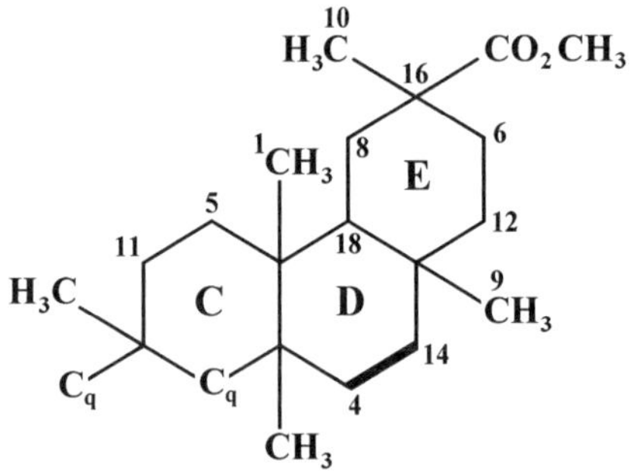

Figure 11.64

two-bond correlations. Thus from h24, h25, and h26 the fragment shown in Figure 11.65 can be deduced from the HMBC correlations, which are shown as arrows.

Many of the aliphatic fragments are now connected to this olefinic/aromatic portion, including the h13 (CH$_3$) fragment and the quaternary carbon c19 shared by h1 (CH$_3$) and h2 (CH$_3$). The important methyl ether fragment identified from h3 (CH$_3$) is connected, and the correlation from h3 to c32 is particularly interesting since the chemical shift of c32 (196.0 ppm) is typical for a ketone carbonyl group. Comparison with Pristimerin, using

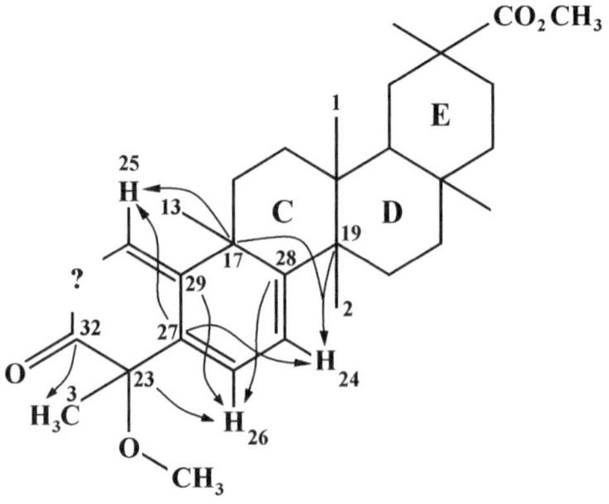

Figure 11.65

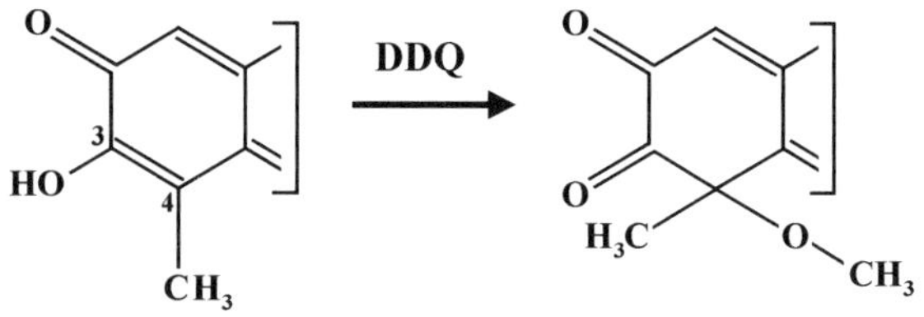

STRUCTURE DETERMINATION BY NMR—AN EXAMPLE

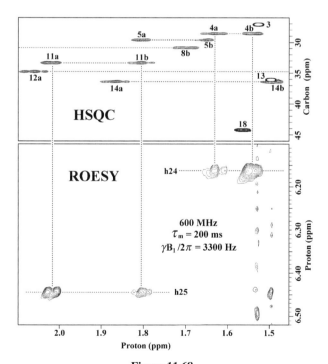

Figure 11.66

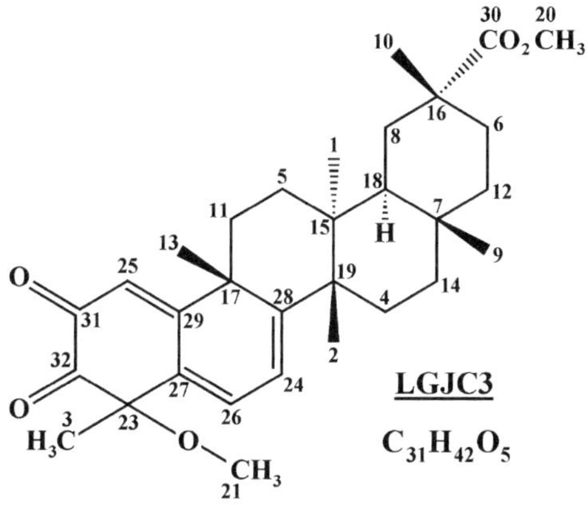

Figure 11.67

Figure 11.68

the h24, h25, h26 system for alignment, shows that this same carbon (C-3 of Pristimerin numbering) is bonded to a phenolic OH group. Thus, it appears that the C-3–C-4 double bond of Pristimerin has undergone addition of methanol and oxidation to the C-3 ketone (Fig. 11.66). If this were the only change between Pristimerin and LGJC3, the mass would increase by 30 mass units (methanol = 32, oxidation removes H_2). In fact, the mass spectrum of LGJC3 does show *m/z* 494 as the parent ion, 30 mass units heavier than the nominal mass of Pristimerin ($C_{30}H_{40}O_4$ = 464 Daltons). Thus, we can complete the structure of LGJC3 by inserting c31 (the only fragment not accounted for so far) as an α,β unsaturated ketone carbonyl (182.0 ppm) between c25 and c32 (Fig. 11.67). The ROESY spectrum (Fig. 11.68) confirms the location of h25 close to h11a and h11b and h24 close to h4a and h4b (Fig. 11.67). The crosspeaks are negative relative to the positive diagonal (negative NOE). The presence of correlations to both members of a geminal pair is a good confirmation of an NOE assignment.

12

BIOLOGICAL NMR SPECTROSCOPY

12.1 APPLICATIONS OF NMR IN BIOLOGY

NMR initially grew from a curiosity for physicists into an essential tool for organic chemists, but the more recent explosion in NMR technology and applications has been driven by the use of NMR for structural biology. The drive for higher field strength and greater sensitivity, as well as new technologies (such as gradients and shaped pulses) has made possible the study of ever larger and more complex molecules, breaking the barrier from "small" organic molecules (natural products, peptides, oligosaccharides, etc.) into the realm of proteins, DNA, and RNA: the "informational" biomolecules. The precise three-dimensional (3D) structure of these molecules and the geometry of their noncovalent interactions with small molecules and with other biomolecules are the keys to understanding how biology functions on the molecular level. Formerly the exclusive realm of X-ray crystallography, this structural information is increasingly derived from NMR experiments without the need for crystals. This field is extremely demanding and has not only pushed the development of NMR hardware but also created an explosion of new NMR pulse sequences and techniques.

Biological NMR is a very broad field and we will attempt here only to list some of the applications and then focus on just one: protein NMR spectroscopy in solution. Solid-state NMR has been a very active area for biological molecules—either using fixed samples such as multiple aligned sheets of lipid bilayers to study membrane-bound biomolecules, or using very rapid spinning of powder samples tilted at the "magic angle" (54.74°) to the B_o field. This magic angle spinning (MAS) technique mimics the rapid, isotropic motion of small molecules in solution to give narrow lines similar to solution-state NMR spectra. Isotopic labeling of specific positions in a biological molecule, for example, with one ^{13}C nucleus in one position and one ^{15}N nucleus in another, allows very precise direct through-space distance measurements. NMR imaging is a very active area of research in the

---

NMR Spectroscopy Explained: Simplified Theory, Applications and Examples for Organic Chemistry and Structural Biology, by Neil E. Jacobsen
Copyright © 2007 John Wiley & Sons, Inc.

medical field as well as in animal, plant, and materials studies. Applying gradients during the acquisition of the FID replaces the chemical-shift scale with a distance (position within the sample) scale, so that a physical map of H_2O concentration in the human body can be constructed. New methods give contrast in these images from not only water concentration but also flow rate and direction, diffusion, relaxation (T_1 or T_2), and targeted injectable contrast agents that can "light up" the locations of specific biomolecules in the body. By synchronizing a fast imaging experiment to the patient's heartbeat, a real time sequence of images can be constructed for each stage of the heart's cycle. Microimaging is made possible by very strong field gradients, so that an ordinary wide-bore vertical NMR spectrometer can be used for microscopy of plant or animal tissues. Imaging can also be combined with spectroscopy, so that a separate NMR spectrum (e.g., ^{31}P spectrum) can be obtained for each volume unit ("voxel") of the brain. Dispensing entirely with the imaging component, *in vivo* spectroscopy can be used to study metabolism in cell cultures. Special NMR tubes and infusion systems have been developed to grow cell cultures directly in the NMR tube, using microimaging to monitor cell growth. Specifically labeled compounds can be introduced and their conversion to metabolites monitored by ^{13}C or ^{31}P NMR spectroscopy.

In the solution state, NMR allows us to study molecular motion in detail. Pulsed-field gradients allow the measurement of diffusion along any of the three axes (x, y, or z) as well as the direction and speed of flow. Rotation of the entire biomolecule can be observed by its effect on the relaxation parameters (T_1, T_2, and NOE) of any of the nuclei within the molecule. By focusing on specific pairs of nuclei that are oriented in a fixed relationship to the entire molecule (e.g., $^{15}N-^1H$), we can measure the rotational diffusion rate about any of the molecular axes. These random rotational motions are on the timescale of μs to ns, but we can also look at internal motions that are faster than the overall molecular motion (ns and faster). Because we can measure this for every pair ($^{15}N-^1H$ or $^{13}C-^1H$) in the molecule, this "flexibility" can be mapped onto the three-dimensional structure of a molecule to locate disordered regions or regions that become ordered upon binding of other molecules. Slower motions (ms timescale) can also be detected and mapped to specific regions, and these usually correspond to conformational changes that are essential to the function of a biomolecule. In this sense, biological NMR is about biochemical function as much as it is about three-dimensional structure.

In the 1980s the foundations were laid for a series of solution NMR techniques that allow the determination of three-dimensional structure (conformation) of biomolecules in solution. These methods were pioneered by Kurt Wüthrich at ETH in Zürich, using small proteins such as bovine pancreatic trypsin inhibitor (BPTI). This approach has been applied to increasingly more complex proteins and nucleic acids (DNA and RNA) and more recently to the structure of molecular complexes (protein–protein, protein–DNA, drug–DNA, etc.). Because NMR relies on local "reporters" (specific ^1H, ^{13}C, and ^{15}N nuclei within a molecule) it can be used to "map" the binding of small molecules or other biomolecules onto the surface of a protein by observing perturbations of chemical shifts. This technique, called "SAR by NMR" (structure-activity relationships by NMR), is now used for screening of large libraries of small molecules in the drug discovery process. In this book, we will be limited to outlining the general process of structure determination of proteins, starting with small proteins at natural abundance and moving to newer techniques involving uniformly ^{15}N- and/or ^{13}C-labeled proteins.

There are two kinds of solution-state NMR problems and up to now we have been concerned entirely with the first: determining the covalent structure or bonding network of a molecule. We are confronted with an unknown or incompletely known covalent structure of a "small" molecule, and we need to find out which atoms are connected to which and in what stereochemical orientation (*trans*, *cis*, E, Z, equatorial, axial, α, β, etc.). These problems come up in organic synthesis (confirmation of an expected product structure), synthetic methods (ratios of products with different regiochemistry and stereochemistry), in the oxidation or metabolism of natural products or drugs (known starting materials), and in the "blue sky" or "white powder" pursuit of new natural products isolated from plants and microorganisms. The tools for this pursuit are the experiments we have discussed in the previous chapters: HSQC, HMBC, COSY, TOCSY, NOESY/ROESY, and the 1D selective NOE and TOCSY experiments. From these experiments we get basic information about chemical shifts, 1, 2, and 3 bond relationships between ^{1}H and ^{13}C, 2-3 bond relationships between protons, ^{1}H–^{1}H distances, and ^{1}H–^{1}H and ^{1}H–^{13}C dihedral angles.

The second type of structural problem in NMR is the subject of this chapter: determining the conformation or specific three-dimensional fold of a molecule with a known covalent structure. The conformation must be relatively rigid, held together by a large number of noncovalent interactions and hydrophobic forces, and specific with little or no heterogeneity of structure. While conformation is sometimes of interest with small molecules, this type of problem is found mostly in the area of biopolymers—large molecules composed of a specific covalent sequence of unlike monomer building blocks and "folded" into a specific three-dimensional shape. These systems include carbohydrates, peptides, glycopeptides, proteins, double-stranded DNA, RNA, and the noncovalent complexes of any pair of these molecules. The techniques include all of the 2D NMR tools used for small molecules (except HMBC) and a number of new methods we will describe in this chapter: ^{1}H/^{2}H exchange, uniform ^{15}N, ^{13}C and/or ^{2}H labeling, 3D NMR expanding our homonuclear 2D experiments (TOCSY and NOESY) into a third dimension using the ^{15}N or ^{13}C chemical shift, and 3D and higher-dimensional "triple-resonance" experiments that rely on doubly-labeled (^{15}N and ^{13}C) protein samples. We will also briefly mention two new techniques that extend the size limit of molecules we can study (TROSY) and cross the boundary with solid-state NMR by measuring the direct through-space dipolar (dipole-dipole) interaction, which is normally averaged to zero by rapid isotropic tumbling in the solution state. The information we obtain from these experiments includes sequence-specific chemical-shift assignments for all spins (^{1}H, ^{15}N, and ^{13}C) in the molecule, chemical shift deviations as indicators of secondary structure (α-helix or β-sheet), degree of protection of amide N–H groups from solvent, thousands of proton–proton dihedral angles and distances, sequence-specific dynamics (order vs. flexibility), and in the case of the newest experiments the orientation of N–H and C–H vectors relative to the rest of the molecule. This vast and heterogeneous store of information is used in the process of structure calculation, which attempts to arrive at a single three-dimensional structure (conformation) of the molecule that is most consistent with all of these measurements.

12.2 SIZE LIMITATIONS IN SOLUTION-STATE NMR

12.2.1 Crystallography

X-ray diffraction (crystallography) has been around for much longer than NMR and has been used to determine the precise three-dimensional structure of biomolecules as large as viruses (molecular weight in the tens of millions!). This technique requires a high-quality crystal and calculates a three-dimensional map of electron density for the molecule. The molecular structure (atoms and bonds) is then fit to this electron density map to generate the precise coordinates of all atoms, which is what we call a "structure." Because hydrogen does not contribute much to the electron density, the X-ray technique effectively only sees the "heavy" atoms (N, C, O, P, S, etc.). In contrast, NMR determines the molecular structure (conformation) directly in the solution state in the native environment for biomolecules (water) by detecting primarily the hydrogens and measuring the short-range distances (<5 Å) and dihedral angles (3-bond relationships) between them. Until recently, solution NMR did not offer any way to determine global relationships (geometry of the molecule relative to the outside world).

12.2.2 NMR

In terms of molecular size, NMR is the poor relation of X-ray crystallography. For natural-abundance proteins we can use only homonuclear techniques (^{1}H–^{1}H J values and NOEs), and we are limited to proteins of less than about 125 residues or a molecular weight of around 12,000 amu (12 kiloDaltons or kD). These are very small proteins, and in the early days of protein NMR there was quite a bit of competition for the relatively small number of biologically significant proteins that were within the reach of the technique, especially for those that did not already have an X-ray structure. A suitable NMR sample must also be soluble and monomeric at a concentration very high compared to physiological concentrations: a 1 mM sample with a volume of 0.5 mL is desirable, requiring a large amount of pure protein.

Why are we so limited in the size of molecules we can study? The enemy of NMR spectroscopy is overlap: we have a finite amount of "turf" to put our NMR peaks in—about 10 ppm for ^{1}H—and we have to fit all the resonances of the molecule into that space. For a specific type of proton with a particular relationship to electronegative atoms the space available is even less: for example, an H$_\alpha$ proton in an amino acid residue will fall between 4 and 5 ppm, with an extreme variation of 3–6 ppm. The "footprint" of the resonance is determined by a combination of the linewidth (width of each line in the multiplet) and the coupling constants (separation of those lines) and might typically be about 30 Hz for an H$_\alpha$ in a small protein. With a 600 MHz instrument, that corresponds to 0.05 ppm (30/600), so we might be able to "fit" 20 of these in each ppm of real estate for a total of 60 in the extreme "window" of 3–6 ppm for an H$_\alpha$ proton if we are extremely lucky and their chemical shifts are spread out evenly. More likely the majority will be near the center of this window and fewer will be near the edges. For a 100 residue protein, with 100 different H$_\alpha$ protons, there will be overlap. Using 2D NMR techniques is helpful but eventually even in two dimensions we will have overlap of crosspeaks, and unambiguous assignment of a resolved crosspeak still requires that the 1D chemical shift be unique.

What happens to these difficulties as we increase the molecular weight of the protein (Table 12.1)? Clearly, the number of resonances within the same chemical shift range will increase and inevitably there will be more overlap and ambiguity of assignments. For example, a NOESY crosspeak at $F_1 = 3.56$ ppm and $F_2 = 9.28$ ppm can be interpreted in nine different ways if there are three H$_N$ protons with chemical shift 9.28 ppm and three H$_\alpha$ protons with chemical shift 3.56 ppm. But this is not the only problem. As molecular weight increases, T_2 gets shorter and linewidth increases. The FID decays faster and the Fourier transform of a fast-decaying signal is a broad Lorenztian peak. So the footprint of each resonance is now larger, and we can fit even fewer unique chemical-shift positions into the fixed range of chemical shifts for each type of proton (H$_N$, H$_\alpha$, aromatic, etc.). It's like

Table 12.1. Typical proteins and size limits for NMR

Protein	MW	Residues	Chains	Mass (1 mM/0.5 mL)
Insulin	5733	51	2	2.9 mg
Heregulin-EGF	7200	63	1	3.6
Approximate Limit with ^{1}H only				
RNase	12,640	124	1	6.3
Lysozyme	13,930	129	1	7.0
Myoglobin	16,890	153	1	8.3
Chymotrypsin	22,600	241	1	11
Approximate Limit with ^{1}H, ^{15}N and ^{13}C Labeling				
Hemoglobin	64,500	574	4	32
Hexokinase	96,000	~800	4	48
Glycogen Phosphorylase	495,000	4100	4	248 mg
Fatty Acid Synthetase	2,300,000	20,000	21	1.15 g
Tobacco Mosaic Virus	40×10^6	336,500	2,130	20 g

we have a fixed area of land, and we not only have more houses to put in it but the houses are getting larger as well. In addition to the overlap problem, the peak height is reduced because the intensity is spread out over a larger range of frequencies. Because the "peak height" of the noise remains the same, the signal-to-noise ratio decreases and eventually the peak "disappears" in a sea of noise.

12.2.3 Chemical-Shift Anisotropy

Why does linewidth increase with molecular weight? One reason has to do with chemical-shift anisotropy (CSA, Chapter 2, Section 2.6), which is the dependence of chemical shift on orientation of the molecule with respect to the B_o field. We saw that chemical shifts in the benzene ring are particularly sensitive to orientation, but the same applies to the ^{1}H, ^{15}N, and ^{13}C chemical shifts in an amide linkage and to a lesser extent at all positions in a protein. In the extreme case of a solid sample, the molecules are all fixed in their orientation and each resonance would be very broad because of the variation of chemical shift with orientation over the whole ensemble of molecules in the sample. In the solution state, small molecules are rapidly reorienting ("tumbling") all the time with no orientational preference ("isotropic" motion), so we see a single sharp line at the average chemical shift, just as we saw in fast exchange (Chapter 10, Section 10.2) when different environments are switched back and forth rapidly. As molecules get larger, though, they tumble more slowly and the lines begin to broaden due to incomplete averaging of the different orientations, just like exchange broadening. Because CSA (the range of variation of chemical shift with orientation) is measured in ppm, the actual variation in hertz gets larger as we go to larger B_o field. It is the variation in hertz that determines the tumbling rate necessary to average it to a sharp line: $1/(2.22 \, \Delta v)$ is the NMR timescale for exchange coalescence, where Δv is expressed in Hz. So if we try to get around the overlap problem by spending millions for a very high field (e.g., 900 MHz) instrument, we are also broadening our lines due to this CSA effect!

12.2.4 Dipole-Dipole Relaxation

Even if we consider only the dipole–dipole interactions, T_2 is determined entirely by the tumbling rate of the molecule (Chapter 5, Fig. 5.14) and always decreases as the molecular size increases because the tumbling rate decreases. T_2 relaxation is dominated by the ZQ pathway ($\alpha\beta \leftrightarrow \beta\alpha$), and the population of molecules tumbling at this very slow rate gets larger and larger as the molecular size increases (Chapter 5, Fig. 5.13). Linewidth is inversely proportional to T_2, so that even with perfect shimming the linewidth will increase steadily with molecular weight:

Molecular weight	Typical T_2	Minimum $\Delta\nu_{1/2}$ (Hz)
100	1.3 s	0.13
1,000	200 ms	0.80
10,000	50 ms	3.2
100,000	5 ms	32

Thus we cannot escape the depressing reality that T_2 will get shorter and linewidth will get bigger as we increase the size of the protein studied. The reduced T_2 is not only a problem for linewidth, but also causes loss of sensitivity as coherence decays during the defocusing and refocusing delays ($1/(2J)$) required for INEPT transfer in our 2D experiments. The only ray of hope comes in the form of a new technique called "TROSY" (transverse relaxation optimized spectroscopy), which takes advantage of the cancellation of dipole–dipole relaxation by CSA relaxation to get an effectively much longer T_2 value; we will briefly discuss TROSY at the end of this chapter.

12.2.5 Sample Size

Another consideration is the amount of sample (in mg) required for NMR. Because it is the concentration (mM) of molecules that determines the signal strength in NMR, as the molecular size increases the desired concentration of around 1 mM corresponds to larger and larger amounts of protein (Table 12.1). For a small protein (e.g., RNase at 12.6 kD) the sample size is around 6 mg. Talk to a biochemist or molecular biologist if you think this is a small amount of pure protein! Now move to chymotrypsin (22.6 kD), which would require 11 mg of pure protein to be soluble and monomeric in 0.5 mL of water. For fatty acid synthase, with 20,000 amino acid residues and 21 polypeptide chains, we would need to dissolve 1.15 g of protein in our 0.5 mL sample volume, clearly an impossibility.

The mention of multimeric proteins brings up another issue: symmetry. As we already said, the size problem in NMR is due to two factors: the complexity problem (too many resonances to fit in a fixed range of chemical shifts without overlap) and the linewidth problem (decreasing T_2 with increasing size). If we have a protein consisting of 10 identical subunits arranged in a symmetrical fashion, there will only be 1/10 as many unique positions within the molecule and the complexity problem is reduced by a factor of 10. But the linewidth problem is still there because the molecule tumbles slowly due to its large physical size. This is also true if you want to study a peptide or protein bound to a phospholipid micelle, even if the micelle is fully deuterated and therefore "invisible" to NMR: the linewidth is determined by the tumbling rate of the entire molecular assembly, in this case the peptide plus the micelle.

12.2.6 Uniform Labeling

There are still other tricks available to extend the size limit of protein NMR. Uniform labeling with ^{15}N allows us to bring in another chemical-shift scale—the ^{15}N chemical shift—and to the extent that different nitrogens in the molecule have a variation or spread ("dispersion") of chemical shifts, we can "pull apart" the overlap by introducing a third dimension in our experiments: 3D NMR. A 3D experiment has the same beneficial effect on overlap that we saw in going from 1D to 2D NMR, and we can now "cram" more resonances into our spectra without ambiguity. Labeling both ^{15}N and ^{13}C allows us to transfer coherence across peptide bonds and into the amino acid side chains with nothing but one-bond INEPT transfers. Because INEPT transfer requires a delay of $1/(2J)$, the T_2 loss associated with the delay is greatly reduced if we rely only on the very large (30–150 Hz) one-bond couplings. This is the basis of the "triple resonance" (^{1}H, ^{15}N, and ^{13}C) experiments. With all of these improvements we can extend the limit up to around 25–30 kD. Keep in mind that determining the structure of a 30 kD protein is no picnic even with ^{15}N and ^{13}C labeling. Research groups that specialize in this sort of thing have one subgroup focused on sample preparation, one subgroup running NMR experiments, one analyzing the NMR data, and one doing structure calculations. Even so, the whole process can take well over a year to complete.

12.2.7 Deuterated Proteins

One of the things that shortens T_2 in larger molecules is the dipole–dipole interaction, and the biggest and most abundant dipole around is ^{1}H. One way to reduce the dipole–dipole relaxation is to replace the ^{1}H with ^{2}H (i.e., with deuterium, D). The magnet strength (γ) of deuterium is about one seventh of that of proton, so the dipole–dipole relaxation is much less. Even partial deuteration (e.g., 50% randomly distributed) will give a significant improvement. In the extreme case of 100% deuteration we would have no ^{1}H signals, but even then we can exchange back the "labile" NH protons with H$_2$O and have at least one proton per residue. This not only radically reduces the complexity of the spectrum, even within the NH region because aromatic protons are removed, but it does so at the cost of a great reduction in information content of the NMR data: only NH–NH interactions are observed in the NOESY.

In 3D experiments where ^{13}C SQC of a C$_\alpha$ carbon is evolving (equivalent to t_1 in a 2D HSQC) we normally "decouple" the attached protons by inserting a ^{1}H 180° pulse in the center of the evolution period to reverse any J-coupling evolution. We could accomplish the same thing by turning on ^{1}H decoupling (e.g., waltz-16) during the ^{13}C evolution period. In deuterated proteins, the one-bond ^{2}H–^{13}C coupling, though only about one seventh (~20 Hz) of the $^1J_{CH}$, can lead to significant broadening of the crosspeaks in the ^{13}C dimension, reducing their signal-to-noise ratio. The situation is even worse for a β-CH$_2$ carbon. The solution is to apply broadband ^{2}H decoupling during the ^{13}C evolution period: linewidths of C$_\alpha$ and C$_\beta$ resonances in the ^{13}C dimension are significantly reduced, bringing the peak heights up and out of the noise.

12.2.8 Why Bother With NMR?

With so many disadvantages to large-molecule NMR, you might wonder why we all don't trade in our magnets and spectrometers for area detectors and start doing X-ray crystallography. There are many unique advantages of NMR, the most important being the lack of crystal packing forces. NMR structures are obtained in the native environment of biomolecules without the artifacts of neighboring molecules packed together in a crystal array. In some cases, the NMR structure has been shown to be quite different from the crystal structure, in a way that provides more relevant insights into biological function. Another advantage of NMR is that the feasibility of a project is known fairly early: all you need to do is put it in an NMR tube and record a ^{1}H spectrum. You can see right away if the linewidth and the spread (dispersion) of chemical shifts is good enough to proceed with 2D experiments, isotopic labeling, and so on. Crystallography requires getting a crystal, a difficult and time-consuming process, before any data can be obtained. There are also, of course, "uncrystallizable" proteins that can never be studied by X-ray diffraction. NMR also is a much more flexible technique, allowing for simple modifications of the medium such as pH adjustment, temperature changes, and addition of small molecules. For example, an active site histidine can be titrated to determine its pK_a without interference of any of the other histidine residues in the protein. To study the binding of proteins to other proteins or to DNA we can just add the other biomolecule to the solution; the crystallographer needs to start all over and cocrystallize the molecules, usually leading to a completely new and different problem to solve. Finally, although the size limitations are severe for NMR, more and more large proteins are being discovered that are covalent combinations of a number of small, specifically folded polypeptide segments ("domains") connected by short, flexible linkers. The linkers can be cut by mild and selective protease digestion, releasing the domains as biologically active small proteins amenable to solution structure determination by NMR. More recently new techniques, such as residual dipolar couplings (RDCs), are being applied to determine the relative orientations of these domains in the whole protein.

NMR and crystallography should be viewed as complementary rather than competing methods. Different kinds of structural information can be obtained, and information gleaned from one technique can aid in the process for the other. Still, NMR is the "younger sister" in the family and must meet a higher standard of proof to justify big spending and job security in corporate research and development.

12.3 HARDWARE REQUIREMENTS FOR BIOLOGICAL NMR

Because of the demands of complexity and linewidth, the highest possible field strength is required. Currently a serious biological NMR group will have an 800 MHz spectrometer and possibly a group of 600s and 800s to accommodate the long experiment times (up to 3.5 days) and large array of experiments required for each sample. Organic chemists are no longer driving the research and development of NMR spectrometers—it is the more demanding experiments and deeper pockets of biological NMR research that is pushing the envelope. Now a few research institutions are investing in 900 MHz spectrometers, which cost many millions of dollars and require construction of an entire building to contain them, all for a 0.5 mL solution of a biological molecule!

The simplest hardware requirement is a probe capable of doing water suppression experiments in 90% H_2O samples. "Water suppression" probes have carefully designed shielding on the wires leading up to the proton coil, eliminating the possibility that these wires can "pick up" an NMR signal from the solvent. Any solvent signal needs to be sharp so that it can be effectively suppressed, so this "lead pickup" can be a big problem because it leads to intense, broad solvent signals. Modern probes for biological NMR are "triple-resonance" inverse probes: the ^{1}H coil (also used for ^{2}H lock) is on the inside and the outer coil is double-tuned to ^{13}C and ^{15}N frequencies (1/4 and 1/10 of the ^{1}H frequency, respectively). Of course, sensitivity of the probe is critical, and biological researchers will pay a lot of money to get even a 10 or 20% increase in signal-to-noise. Gradients are essential, and in some cases "triple axis" gradients (three gradient channels that produce gradients along the x, y, and z axes) are desirable for optimal water suppression. The engineering limitations of putting three gradient coils in a probe limit the sensitivity a bit, so a single (z) axis gradient probe is usually preferred.

12.3.1 Cryogenic Probes

The latest fad in biological NMR is the cryogenic probe, which has a transmit/receive coil cooled to 25 K. This reduces thermal noise in the coil and leads to an increase of up to 3–4 times in signal-to-noise ratio for ^{1}H detection. The ^{1}H preamplifier is also cooled to 25 K so that thermal noise is minimized in the first stage of amplification of the FID. The sample is still around room temperature, so the technical challenge of a distance of a millimeter or so between the room temperature solution sample ($\sim$25°C) and the receiver coil at 25 K ($-$248°C) is considerable. This is accomplished by insulation with a high vacuum ($\sim 10^{-8}$ torr) between the outside of the probe and the cold inner workings, maintained by a turbo vacuum pump that runs continuously. A helium gas refrigerator (two stages of helium compression and expansion) sits away from the magnet and sends cold He gas ($\sim$10 K) into the probe, returning warmer He gas. A heater block in the probe maintains the desired 25 K temperature and is in thermal contact with the probe coil and the preamplifier. One of the problems with the cryogenic probe is that the advantages are considerably reduced in polar solvents (such as water!) and particularly if salts are present in the solution. For biological NMR the increase in signal-to-noise ratio is typically more like 2–2.5 times rather than 3–4 times, but this is still a major improvement. The main advantage is that we can go considerably lower than the recommended 1 mM concentration of protein. Sometimes a few hundreds of μM is all the protein you can obtain, or all that will dissolve and remain in the monomeric state.

The spectrometer console has to have at least three separate channels to accommodate triple-resonance experiments in which we detect ^{1}H but use pulses on ^{1}H, ^{15}N, and ^{13}C. This leads to a problem in terminology because the older two-channel instruments have only two boxes that produce RF energy: the "transmitter" and the "decoupler." Varian uses the term decoupler 2 (dec2) for the additional channel, whereas Bruker sticks to F1, F2, and F3 for naming the three channels (not to be confused with the frequency axes of a 3D experiment: F_1, F_2, and F_3). Shaped pulse capability on all three channels is desirable. Many spectrometers have a fourth channel for ^{2}H decoupling (dec3 or F4). ^{2}H decoupling is a hardware challenge because the deuterium channel of the spectrometer is busy transmitting and receiving the lock signal in order to stabilize the field strength over time with the lock feedback loop (Chapter 3, Section 3.3). Spectrometers used for biological research often have a "lock switch" that allows rapid switching between transmitting ^{2}H pulses and decoupling sequences and the continuous transmit/receive of the ^{2}H lock feedback loop.

12.3.2 Gradient Shimming

The availability of pulsed-field gradients makes it possible to automatically shim using NMR imaging techniques. In MRI we rely on the dominance of water in the human body to obtain

560 BIOLOGICAL NMR SPECTROSCOPY

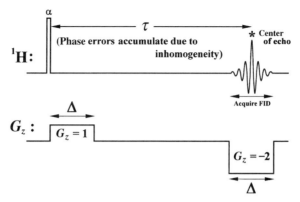

Figure 12.1

a single, very strong ^{1}H NMR signal. By applying a gradient during the acquisition of the FID, the chemical-shift scale is transformed into a scale of physical position because there is only one peak in the normal ^{1}H spectrum. Biological NMR samples are similar in that they have one enormous and dominant peak: the H_2O peak. In the absence of water suppression techniques, this signal can be used for NMR imaging to "map" the inhomogeneity of the magnetic field along the gradient axis. The software then calculates precisely how much each shim value will have to be changed and applies these changes to remove the inhomogeneity. In principle, this would be the end and you would have perfect homogeneity, but in reality, it takes several rounds of an iterative process: map the inhomogeneity, calculate and apply the shim changes, and repeat. While gradient shimming is not limited to biological samples, it is especially useful because the traditional manual shimming method is especially difficult in D_2O or 90% H_2O samples. The D_2O line in the lock system is broad and the lock level (height of the ^{2}H peak of D_2O) does not respond much when shims are changed. Water suppression techniques are sensitive to errors in higher order shims (e.g., Z^4, Z^5), and these are nearly impossible to shim by hand.

The pulse sequence for gradient shimming is shown in Figure 12.1. This is an imaging experiment, so the gradient is on during the acquisition of the FID. Consider a single-axis (z-axis) gradient and a water signal that is precisely on-resonance. A small-angle pulse creates H_2O coherence that is then "twisted" into a coherence helix by the first gradient. During the delay τ, water coherence will remain stationary in a perfectly homogeneous field and the phase twist will be preserved exactly as it was at the end of the first gradient. The second gradient, of opposite sign and twice the amplitude as the first, is applied during the acquisition of the FID to "untwist" the helix (Fig. 12.2, left). The water molecules at the top of the tube ($z = +8$ mm) are experiencing a magnetic field reduced by the gradient so their magnetization vector is moving clockwise in the rotating frame during the recording of the FID. The water molecules at the bottom of the tube ($z = -8$ mm) are in a region of increased field due to the gradient so their magnetization vector is moving counterclockwise. So the actual frequencies of these vectors are detected in the FID as distinct chemical-shift values, each giving a peak at a different part of the ^{1}H spectrum (Fig. 12.3, top). Water molecules at the center of the tube experience an unaltered B_o field so they give rise to a stationary magnetization vector during the FID, resulting in a peak at the center of the spectral window. The helical phase twist caused by the first gradient unravels during the second gradient and exactly halfway through the second gradient all of the vectors are

HARDWARE REQUIREMENTS FOR BIOLOGICAL NMR   561

Phase relationships at center of echo

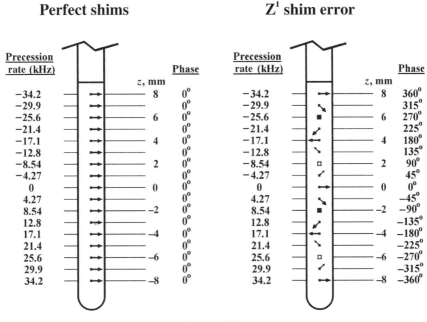

Figure 12.2

exactly aligned again (Fig. 12.2, left). At this moment there is an "echo" because the FID signal reaches a maximum when all the vectors align and add together from all levels of the tube. The phase of each peak in the spectrum is determined by the position of the vector at each level in the tube *at this moment*, halfway through the FID. Because they are all aligned on the y' axis (choosing y' as the phase reference), all of the peaks in the spectrum

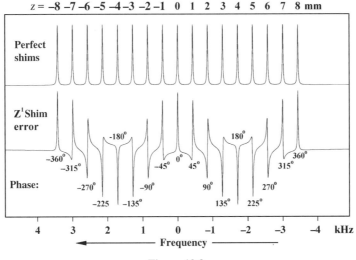

Figure 12.3

are positive absorptive (Fig. 12.3, top). This tells us that during the delay τ the B_o field was exactly the same at all levels of the tube and we have perfect field homogeneity.

Consider now what happens if the B_o field is not homogeneous. Let's assume that there is a Z^1 shim error, which means a linear gradient in B_o along the z axis. During the τ delay the vectors at each level will precess slightly away from the perfect helix created by the first gradient because they are not perfectly on-resonance at each level. Suppose that the linear Z^1 gradient leads to an additional 45° rotation for each level, relative to the level above it. This means that at the crucial moment at the center of the FID the H_2O magnetization vectors will not be aligned, but rather will have a helical twist of 45° phase change for each 1 mm of vertical distance (Fig. 12.2, right). This twist will lead to phase differences in the peaks in the spectrum. Because each peak represents one of the levels in the NMR tube, we see a progression of phase errors from left to right in the spectrum (Fig. 12.3, bottom): 0° (absorptive positive), 45°, 90° (dispersive), 135°, 180° (absorptive negative), and so on. These phases can be directly "read off" the spectrum as a map of the B_o field along the z axis. We know how long the τ delay is, so we can calculate back from the amount of rotation (the phase difference) to obtain the deviation in rotation rate (in hertz) at each level. Assuming that the Z^1 shim has been "calibrated" so we know how much field change we get for a given change in the Z^1 setting, we could calculate exactly how much and in which direction we have to change the Z^1 setting to "erase" the difference in B_o field along the z axis. This is automatic gradient shimming.

If the strength of the gradient used during the acquisition of the FID is 10 gauss/cm (0.001 T/cm), the B_o field is changed by 1 gauss (0.0001 T) for each mm of distance along the z axis. We know that a 500 MHz (^{1}H) NMR instrument has a B_o field of 11.7 T, so 0.0001 T corresponds to (0.0001/11.7) × 500 MHz or 4.27 kHz for protons. The entire height of the NMR receiver coil (16 mm) corresponds to a range of 16 × 4.27 = 68.4 kHz. This is the width of our NMR signal (Fig. 12.3). A typical proton spectral window is 12.5 ppm wide, corresponding to 6.25 kHz on a 500 MHz instrument, so for gradient shimming, we are using a spectral window more than 10 times wider. Note also that it is the length of the receiver coil, not the depth of the water in the sample tube, that determines the width of the NMR signal in frequency domain. The water above and below the receiver coil is not detected so it does not contribute to the spectrum. The amount of "inhomogeneity" at each level can be calculated from the phase difference: if the delay τ is set to 3 ms (0.003 s), a 45° phase error corresponds to a difference in Larmor frequency of

$$\text{cycles of rotation} = 0.125 = \Delta\nu\,(\text{Hz}) \times \tau(s) = \Delta\nu\,(0.003); \Delta\nu = 0.125/0.003 = 41.7\,\text{Hz}$$

Thus we have a linear B_o field difference of 41.7 Hz per mm or 417 Hz/cm along the z axis. We can describe field differences in hertz because we are talking about ^{1}H frequencies, just the same way we refer to an 11.7 T magnet as a "500 MHz" magnet. The field differences along the z axis due to field inhomogeneity (bad shims) create phase differences in the signals at each level in the NMR tube as a result of the 3 ms delay time, and the imaging experiment (gradient on during the FID) separates these levels into different frequencies in the spectrum. The phase differences at these different levels can be converted into a precise map of the field strength difference (inhomogeneity) as a function of the z coordinate in the NMR sample.

By now you may have realized that there is a continuum of water molecules in the sample at all levels, not just at the 17 discreet levels we considered above. We can simulate the spectrum we expect by adding together a very large number of NMR peaks, starting at the

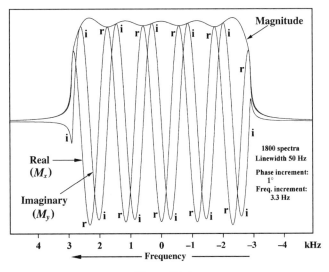

Figure 12.4

bottom of the receiver coil ($z = -8$ mm) and moving in small steps to the top ($z = +8$ mm), while incrementing the phase as we move up to model the Z^1 shim error. The result is shown in Figure 12.4, with five complete cycles of phase error from the bottom of the coil to the top. The real spectrum and the imaginary spectrum, as well as the magnitude ($\sqrt{(\text{real}^2 + \text{imag.}^2)}$) spectrum are shown. The phase at any position in the spectrum can be determined by setting the intensity of the real spectrum (relative to the magnitude value) equal to sin Φ and the intensity of the imaginary spectrum (relative to magnitude) equal to cos Φ. The angle Φ is the phase.

$$\text{I(real)}/\text{I(imag.)} = \sin \Phi / \cos \Phi = \tan \Phi; \quad \Phi = \arctan[\text{I(real)}/\text{I(imag.)}]$$

If the arctan (inverse tangent) function is defined to give values between $+90°$ and $-90°$, we would have Φ moving linearly from $-90°$ to $+90°$ and then suddenly jumping back to $-90°$, but the software can unravel these discontinuities to generate a smooth field map. A "control" experiment is done with $\tau = 0$, and the control field map is subtracted from the field map generated with the delay τ.

To calculate the changes in shim settings, we need an accurate "map" of the effect of each shim on the B_0 field as a function of the z coordinate. We need to know how many Hz difference in field is created at each level of the sample by changing, say, the Z^1 shim by $+1$ DAC unit (shim units are called "DAC units" because they are integers that drive the digital-to-analog converter to produce an analog current in the shim coil). The shims can be mapped by changing each shim by a significant amount (e.g., 100 DAC units) and remapping the field. By subtracting this field map from the map obtained before changing the shim value, we know exactly what effect is produced for each 100 units of change in that shim setting. Figure 12.5 shows an ideal shim map for shims Z^1–Z^6; from it we could calculate the change in field at each level for any change in one of these six shim settings. The mathematical problem is then to determine, for any arbitrary map of field inhomogeneity, how much we need to change each of the six shims to generate a function exactly opposite to this inhomogeneity to cancel it out and give us a perfectly homogeneous field. This is a relatively simple fitting problem and the computer can solve it very quickly, automatically

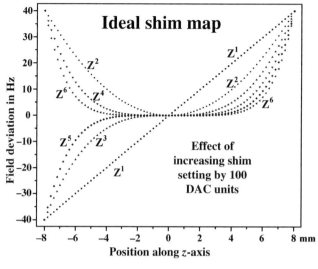

Figure 12.5

applying the shim changes to the six shim DACs and changing the currents in the shim coils to correct for the inhomogeneity.

Some gradient probes have three gradient coils, oriented in the x, y, and z directions. With a "triple axis" gradient probe and three pulsed-field-gradient amplifiers, one can generate a 3D map of the field inhomogeneity, and using 3D maps of all of the shims, including those involving the x and y axes (e.g., X, Y, XZ, YZ, XZ^2, YZ^2, XY, X^2-Y^2, X^3, Y^3, etc.) corrections can be calculated for all of the shims, not just the Z shims. In a matter of minutes, a spectrometer with 40 shim settings can be shimmed from zero shim current in all 40 coils to nearly perfect homogeneity without using any human skill or judgement. The only thing that is required is a sample with a single, very strong, and dominant peak. For biological samples in 90% H_2O, this single peak is the water peak, but the technique has now been extended to samples in deuterated solvents (D_2O, $CDCl_3$, etc.) by using the deuterium spectrum. In $CDCl_3$, for example, the deuterium spectrum consists of a single, very stong, and dominant peak: the ^2H peak due to the solvent $CDCl_3$. A bit of switching hardware is needed to shut off the lock circuit and apply the same pulse sequence (Fig. 12.1) at the deuterium frequency. Now automatic gradient shimming is routine even for samples in fully deuterated organic solvents.

12.4 SAMPLE PREPARATION AND WATER SUPPRESSION

Water is the most relevant biological solvent. The amide protons (one for each peptide linkage -HN–CO-) exchange with solvent water so it is not desirable to use D_2O because the HN becomes DN and can no longer participate in NMR experiments. Usually the solvent is a mixture of 90% H_2O and 10% D_2O, with the D_2O used for locking and shimming in the NMR magnet. This requires some very fancy methods of water signal suppression to remove the enormous H_2O peak at 4.8 ppm (about 100 M in H_2O protons vs. about 1 mM in protein).

12.4.1 Buffers and pH

Although the most relevant pH for most biological molecules is near 7, the exchange of amide protons ("H_N") with solvent H_2O is very rapid at neutral pH, leading to exchange broadening and weak or nonexistent NMR peaks for these protons. If we have fast enough exchange, the H_N protons spend the vast majority of their time in the much larger pool of H_2O protons and the average chemical shift is identical to the water chemical shift. The minimum exchange rate occurs at pH 2–3, and early protein NMR studies were done at this pH, but most of the work is now done between pH 4.5 and pH 7. Even at pH 7.5 life is getting difficult and 8.0 is a real challenge. Buffers for NMR include deuterated acetate (CD_3CO_2Na + HCl titrated to pH 4.5), sodium phosphate (NaH_2PO_4/Na_2HPO_4), and deuterated tris (($HOCD_2)_3CNH_2$ + HCl), typically at concentrations around 50 mM. Salt (NaCl) sometimes has to be added up to 100 mM in order to provide the proper ionic strength to solubilize the protein or prevent it from aggregating. Excessive salt is to be avoided if at all possible because it leads to poor matching of the probe circuit and lengthens the 90° pulse. With cryogenic probes salt is death to the sensitivity advantage you paid so much for, and special care should be taken to use the best buffer for cryogenic probe work. Usually sodium azide (NaN_3, 1 mM) is added to prevent bacterial growth in the sample. Samples are stored in a refrigerator (5°C) and not frozen because freezing can break the NMR tube and denature the protein. Special tubes (Shigemi tubes) are often used to reduce the sample volume from 0.5 mL to around 0.3 mL (300 μL) just within the probe coil, filling the remaining volume above and below the sample with a special glass whose magnetic susceptibility is matched to that of water.

12.4.2 Referencing

TMS cannot be used in aqueous solution because it is not water soluble. For a chemical-shift reference, a water-soluble equivalent (such as sodium d_4-3-trimethylsilylpropanoate (($CH_3)_3SiCD_2CD_2CO_2^-Na^+$, "TSP")) can be added; the single ^1H peak is defined as zero ppm in water. The water peak itself can also be used as a chemical-shift reference, but care must be taken to correct for the temperature dependence of its chemical shift. Referencing of ^{13}C and ^{15}N chemical shifts can be done by using an accurate ^1H reference. If the exact chemical shift is known at the center of the ^1H spectral window (usually the water resonance), the precise radio frequency can be calculated for the zero point of the ^1H chemical-shift (ppm) scale. For example, on a 600 MHz spectrometer with a reference frequency of 600.13231564 MHz and a water chemical shift of 4.755 ppm:

$$\nu(\delta = 0, ^1H) = \nu_r - 600.13 \text{ Hz/ppm} \times 4.755 \text{ ppm}$$
$$= 600.13231564 \text{ MHz} - 2853.62 \text{ Hz} = 600.12946202 \text{ MHz}$$

Using the exact values for the magnetogyric ratios of ^1H, ^{13}C, and ^{15}N, we can calculate the frequencies corresponding to 0 ppm ^{13}C and 0 ppm ^{15}N:

$$\nu(\delta = 0, ^{13}C) = 600.12946202 \text{ MHz} \times 0.25144953 \; (\gamma_C/\gamma_H)$$

$$\nu(\delta = 0, ^{15}N) = 600.12946202 \text{ MHz} \times 0.101329118 \; (\gamma_N/\gamma_H)$$

It is important to use the most accurate values available for these ratios and to do the calculation on a computer spreadsheet to avoid truncation errors. Once you have the frequency corresponding to 0 ppm, the chemical shift at the center of the spectral window ("the carrier") can be calculated from the reference frequency of the ^{13}C or ^{15}N channel as we did above for the ^{1}H channel.

12.4.3 Radiation Damping

The water signal from a 90% H_2O sample is unlike any other NMR peak in that it is incredibly intense, so intense that the FID signal in the probe coil is strong enough to turn around and act on the sample as a pulse! This "pulse" then rotates the net magnetization vector of water back toward the $+z$ axis, effectively accelerating its transverse relaxation and broadening the water peak. The worst thing you can do to water is to put it on the $-z$ axis: after a 180° pulse the water net magnetization is very close to $-z$, but never exactly on it. It begins to precess around the $-z$ axis, and the tiny component in the x–y plane induces a strong FID in the probe coil, which in turn starts to rotate the net magnetization away from the $-z$ axis (Fig. 12.6). As it rotates away, the component in the x–y plane increases, the FID signal increases, and the rate of rotation increases as a result. The process accelerates until the water magnetization reaches the x–y plane, where the FID signal in the probe is at a maximum and the rotation rate is the greatest. It continues precessing and rotating until it reaches the $+z$ axis. If we view it in the rotating frame of reference with the water peak on-resonance, we see only the rotation part and no precession: the vector starts near the $-z$ axis and rotates around the x' axis at a rate that accelerates until it reaches the y' axis and then slows down as it approaches the $+z$ axis. Keep in mind that relaxation by the normal T_1 and T_2 processes cannot generate coherence: after a 180° pulse the z magnetization recovers from $-M_o$ to $+M_o$ in an exponential fashion, without any rotation of the net magnetization vector. Dilute solutions of water (e.g., in D_2O) behave normally with quite long T_1 and T_2 times due to the small size of the H_2O molecule (see Fig. 5.17), but pure water (or 90% water) relaxes very rapidly due to radiation damping, leading to a very broad water peak in the spectrum.

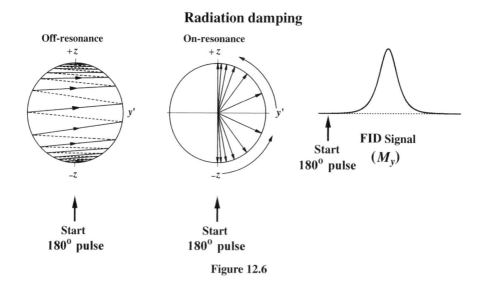

Figure 12.6

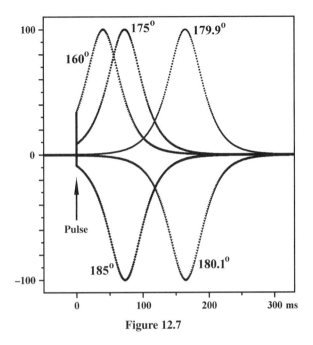

Figure 12.7

Radiation damping is strongest when the probe is matched perfectly and has a very high "Q" factor (a very sharp tuning "dip"), so one way to minimize it is to detune the probe a bit.

You can use radiation damping to your advantage as a very fast method for ^{1}H pulse calibration. With the water peak on resonance and the receiver gain set to a minimum, give a single pulse near 180° and watch the FID (separate real and imaginary displays). At least one of these displays (real or imaginary) will show a bell-shaped curve indicating the slow rise of M_{xy}, rapid passage through a maximum and the slowing decay to zero. Figure 12.7 shows a simulation of the FID after a pulse on the $-x'$ axis, based on the assumption that the radiation damping rotation rate (equivalent to a pulse on the x' axis) is proportional to the FID signal at each point, ignoring the normal T_1 and T_2 relaxation processes that are much slower. A pulse less than 180° will leave the sample magnetization short of the $-z$ axis and M_y will start with a positive value (e.g., a 160° pulse). From there it immediately starts rotating back toward $+z$, generating an FID that abruptly jumps up at the beginning. A pulse closer to 180° (e.g., a 175° pulse) will leave the magnetization very close to $-z$, so that the FID will start with a smaller value of M_y and then increase to the maximum and fall to zero as the net magnetization rotates to $+y$ and on to $+z$. A pulse greater than 180° (e.g., a 185° pulse) rotates the water net magnetization past the $-z$ axis and starts the FID abruptly with a small negative M_y value. This signal is of opposite sign, and so it now rotates the water net magnetization in the opposite direction—toward the $-y$ axis and on to the $+z$ axis. The FID signal has flipped upside down, with the same abrupt jump at the beginning as it had for the 175° pulse. If the pulse is exactly 180°, in a perfect world, there would be no radiation damping and the water signal would very slowly recover from $-z$ to $+z$ by normal T_1 relaxation. But it is not a perfect world because the B_1 field is notoriously inhomogeneous and different parts of the sample experience a bit more than 180° and other parts a bit less. Even in this perfect world, however, a pulse very slightly less than 180° (e.g. 179.9°) leaves the magnetization slightly short of $-z$ and thus starts with a tiny component in the x–y plane that generates a tiny FID. This FID acts on the water magnetization and rotates it slightly, leading to a bigger FID. It takes a while for this to build up, but eventually the FID becomes large enough to rotate all the way past $+y$ and on to $+z$. The same lag period occurs if we rotate just a hair beyond 180° (e.g., 180.1°) except that now the rotation builds up in the opposite direction and eventually passes $-y$ on its way to $+z$. For pulse calibration, we only need to adjust the pulse width to minimize the abrupt jump at the start of the FID, and when we pass through 180° we will see the FID "flip" to the opposite sign.

Pulse calibration and careful probe tuning are important for water samples because the ionic strength varies from sample to sample and can greatly affect the probe matching and the pulse widths. You may need to calibrate the ^{1}H pulse at more than one power level: for example, high power for hard pulses, medium power for TOCSY mixing, low power for waltz decoupling, and so on. In biological NMR, we invest a great deal of time in each sample, sometimes acquiring many 2D or 3D datasets for a total experiment time of many weeks. It is definitely worth the time to tune, match, and calibrate carefully. For X-nucleus pulse calibration (^{13}C, ^{15}N), you will need to calibrate a hard 90° pulse as well as a decoupling (GARP, WURST, etc.) 90° pulse at a lower power level. Usually this is done on a standard sample, such as ^{13}C-methyl iodide or ^{15}N-benzamide, rather than on the biological sample itself.

12.4.4 Water Suppression Techniques

The H$_2$O signal is an enormous problem in biological NMR—water protons are about 100 M in concentration whereas the protein sample is about 1 mM, a difference of five orders of magnitude! Even with good water suppression the water signal usually dominates the FID, with the protein signal a "fuzzy growth" on the smooth curve of the water signal. The receiver gain is a good measure of the success of water suppression: the smaller the water signal in the FID, the more we can amplify the FID without exceeding the digitizer limits. For simple presaturation (Chapter 5, Section 5.11), a receiver gain of 64 (*rg*, Bruker) or 18 (*gain*, Varian) is about as high as you can get. If the receiver gain is set very low, the noise that accumulates after the original FID received in the probe coil (including digital noise in the ADC) dominates and the protein signal-to-noise ratio is drastically reduced. Most of the water suppression achieved with presaturation is actually achieved in the phase cycle, in the cancellation of the water signal after a number of scans.

The jump-return or "1, $\bar{1}$" method is a very simple and elegant solution because rather than destroying the water signal it simply does not excite water in the first place. We saw in Chapter 8, Figure 8.19 that a null in excitation occurs at the center of the spectral window, and this can be adjusted to put the water peak exactly on-resonance. A jump-return NOESY spectrum of a small protein will be shown later in this chapter. Jump-return and some more complicated variations ("1, $\bar{1}$" - echo" and "binomial") are not applicable to all experiments, however, and require some careful tuning and adjustment to work well. They also distort the peak intensities throughout the spectrum and greatly reduce the intensities near the water resonance.

The Watergate method (Chapter 8, Section 8.6) uses gradients to "crush" the water signal very effectively in a single scan. The advantage of Watergate is that the water net magnetization is not destroyed until the end of the pulse sequence, so there is not much time for saturation transfer (by exchange) to "bleach" the H$_N$ signals in the spectrum. If we fight water as an enemy, we tend to destroy other signals that exchange with water or have an NOE with water in the process of destroying the water signal. Watergate can be viewed as a "water-friendly" sequence because it leaves water alone until the very end. The disadvantage of Watergate is that it cuts a fairly wide swath around the water signal (see Chapter 8, Figs. 8.22 and 8.23), greatly reducing the intensity of H_α signals near the water peak. This may not be a problem in many experiments where we are only interested in the H_N resonances in the F_2 dimension.

Another commonly used technique is the water "flip-back" pulse, a shaped pulse designed to selectively rotate only the water magnetization by 90°, putting it back on the $+z$ axis after a "hard" (nonselective) pulse has rotated all of the sample magnetization into the x–y plane. Water can be viewed as a wild and powerful bucking bronco—it must be tamed and never allowed to get out of its pen. The best place for water is on the $+z$ axis where it will not do any harm. This is the rationale behind the flip-back pulse: every time water is moved from the $+z$ axis, use a selective pulse to put it back there.

Water suppression is not a routine technique or a technique for beginners! Everything has to be perfect, and it is worth the trouble to make some adjustments and fine tuning. If you do not do it right, you will get an enormous signal and the most common result is receiver overflow (exceeding the limits of the ADC). This leads to "clipping" of the FID and terrible distortions of the baseline of the spectrum. If the receiver gain is turned down to correct for this, the signal-to-noise ratio can suffer so much that you do not even see the protein signals.

In a 2D experiment, be sure to start the experiment and check that the first scan of the first FID does not exceed the digitizer limits. Then let the experiment continue until you see the first scan of the *second* FID. This may be different because the F_1 phase encoding (TPPI or States) requires that the preparation pulse phase be changed by 90° for the second FID (x' for FID 1, y' for FID 2). Especially in spin-lock experiments (TOCSY or ROESY), the spin-lock axis might be destroying the water signal in the first FID (water magnetization perpendicular to the spin-lock axis) and preserving it in the second (water magnetization co-linear with the spin-lock axis). If you optimize the receiver gain only for the first FID (Bruker *rga* command; Varian *gain* = 'n'), you might end up with a huge ADC overflow in the second FID (and all subsequent even numbered FIDs) of the 2D experiment. This can be the cause of complete failure of many experiments.

12.4.5 Other Solvents

Peptides (short polypeptides) often function as important biological ligands, binding to receptors at a membrane surface. Many are too hydrophobic to be soluble in water at millimolar concentrations. The argument has been made that molecules that bind to membrane receptors should be studied in a medium that mimics the hydrophobic membrane environment in order to obtain the relevant conformation that binds to the receptor. One option is to use d_6-DMSO (CD_3SOCD_3), an excellent solvent that practically freezes exchange and makes NH and even OH protons give sharp resonances with J coupling to neighboring protons. Sometimes trifluoroethanol (TFE) is added to water to increase the strength of intramolecular hydrogen bonds and increase helicity of peptides for NMR studies of conformation. Another approach is to add deuterated micelles to 90% H_2O in order to provide a "membrane" environment for a peptide solute. Fully deuterated lipids, such as deuterated dodecylphosphocholine (DPC-d_{38}), can be added up to a concentration above the critical micellar concentration, solubilizing the most hydrophobic peptides. Because the molecular weight of a micelle is quite large, specific tight binding of a peptide at the micelle surface will drastically broaden the NMR lines. Often the binding is nonspecific, however, simply providing a "drop of grease" for the peptide, and the NMR lines remain sharp. The biological relevance is limited because the exact nature of binding to the micelle is seldom determined; in some cases,

12.5 ¹H CHEMICAL SHIFTS OF PEPTIDES AND PROTEINS

Proteins are linear polymers of amino acids, with each amino acid unit ("residue") chosen from the 20 natural amino acids, which differ only in the side chains (Fig. 12.8). From the point of view of NMR, we can describe each amino acid residue spin system as starting with the proton on the backbone nitrogen ("H_N") and moving to the proton on the α-carbon (H_α) and out to the side chain (H_β, H_γ, H_δ, H_ε, etc.). Typical regions of proton chemical shifts are backbone amide H_N (peptide linkages, 7–11 ppm), side-chain amide H_N (Asn and Gln, 6–7.5 ppm), aromatic protons (6.5–8 ppm), alpha protons ($C_\alpha H$, 3.5–5.5 ppm), side-chain protons (−0.5 to 3.3 ppm unless close to oxygen), and methyl groups (−0.3 to 1.3 ppm unless connected to S). Surveying the types of amino acid spin systems, we start with glycine, which has two nonequivalent H_α protons and no side chain, followed by the "hydrocarbon" side chains (alanine, valine, leucine, and isoleucine). These all have methyl groups that give prominent strong upfield peaks in the NMR spectrum. The pair of methyl groups in valine and leucine is nonequivalent and gives rise to two nearby resonances. Proline is

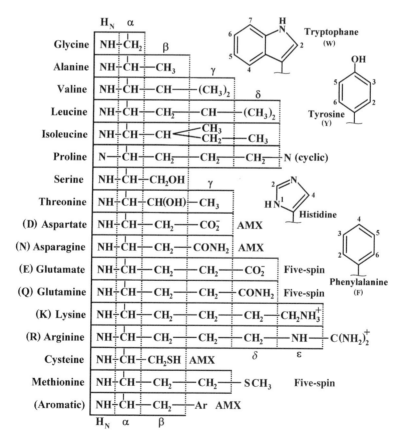

Figure 12.8

unique in that it lacks a backbone H_N proton—the δ-carbon of the five-membered ring takes its place. Serine and threonine have alcohols in the β-position, pulling the H_β resonance downfield to a position near the H_α resonance. Many of the amino acids have the spin system CH–CH$_2$ that can be called a three-spin or AMX system (ignoring the H_N): serine, aspartate, asparagine, cysteine, and all of the aromatic amino acids. The two β-protons nearly always have different chemical shifts (H_β and H'_β) due to the chiral environment. Another group of amino acids have the spin system CH–CH$_2$–CH$_2$ known as five-spin (again ignoring H_N): glutamate, glutamine, and methionine. The basic side chains of arginine and lysine lead to long spin systems: CH–CH$_2$–CH$_2$–CH$_2$–N and CH–CH$_2$–CH$_2$–CH$_2$–CH$_2$–N. These are complex but can usually be identified by the CH$_2$ next to the side-chain nitrogen, which is shifted downfield. As the side chains get longer, one generally sees separate resonances only for the β-protons—the γ and δ CH$_2$ groups often give a single "degenerate" chemical shift. The aromatic rings of phenylalanine, tyrosine, tryptophane, and histidine form their own spin systems, separate from the H_N–CH–CH$_2$ spin system, as do the side-chain NH$_2$ groups of asparagine and glutamine and the methyl group of methionine.

Figure 12.9 displays graphically the proton chemical shifts of all 20 amino acids in an unstructured peptide context. These are called "random coil" chemical shifts because they are not influenced by the through-space effects observed in specifically folded proteins. In this environment, there is not much chemical-shift dispersion: H_N falls between 8 and 9 ppm, H_α between 4 ppm and the water resonance (~4.8 ppm), and the side-chain H_N resonances

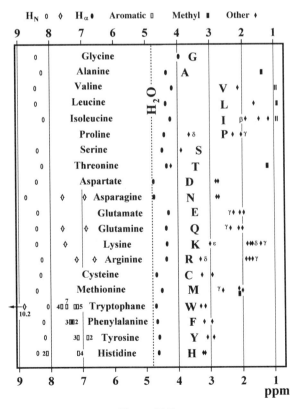

Figure 12.9

are between 6.5 and 7.6 ppm. Note that the $CONH_2$ groups at the end of the Asn (N) and Gln (Q) side chains have distinct chemical shifts for the two H_N protons due to hindered rotation of the amide linkage (circles with a cross). As we move outward in the "hydrocarbon" side chains, the chemical shifts move upfield because the distance to the nearest functional group (α-amino and carbonyl of the backbone) increases. The methyl groups appear at the "classic" hydrocarbon positions of about 0.8 ppm. Long side chains of repeated methylene units show the same behavior, but the trend then reverses when there is a nitrogen at the end: $CH-(CH_2)_n-N$. We see this for Lys (K), Arg (R), and Pro (P), which have shifts of 3–3.6 ppm for the last CH_2 before the nitrogen. The most upfield methylene group is the γ-CH_2 because it is far from both the backbone functional groups and the nitrogen at the other end of the chain (backbone N in the case of Pro). Proline lacks an H_N but the position of the δ-CH_2 is similar to where the H_N would be located, so the same kinds of NOE interactions can be observed.

We see this effect of the last CH_2 group shifting downfield to a lesser extent in the five-spin systems Glu, Gln, and Met. The carbonyl group (Glu, Gln) or sulfur (Met) "pulls" the γ-CH_2 group downfield to the 2–3 ppm range. Even farther downfield than the final CH_2 of the chains ending in nitrogen are the β-protons of Ser and Thr, which are on an oxygenated carbon. These H_β resonances are near the H_α at around 4 ppm, standing out clearly from the other amino acids.

The first goal of an NMR study is to identify the spin systems present and assign each one to one of the 20 amino acids or at least to a group of amino acids. In a 2D TOCSY spectrum in 90% H_2O we can observe the entire spin system at the F_2 position of the H_N proton (see Chapter 9, Fig. 9.45). Many of these patterns of chemical shifts are unique to one of the 20 amino acids. For example, valine has a β-proton at 2.1 ppm and two diastereotopic methyl resonances at around 0.8 ppm. This pattern is clearly recognizable in a TOCSY spectrum, and so we know it is a valine residue. A protein will typically have a number of valines in the sequence, so at this point we only know that it represents one of these residues. Other spin systems fall into a group of amino acids. For example, a number of amino acids have the spin system CH_α–CH_2–R where R is a "dead end" for J coupling: a quaternary carbon or a heteroatom (such as oxygen or sulfur). These are called AMX or three-spin systems, and they include Asp, Asn, Cys, Phe, Tyr, Trp, and His. All have, in addition to the backbone H_N and H_α, two H_β resonances in the vicinity of 3 ppm (2.6–3.4 ppm). If we observe this pattern in a TOCSY spectrum along an H_N line, we can conclude that it belongs to this "AMX" group of amino acids. Serine is technically an AMX spin system (CH_α–CH_2OH), but the β-protons are shifted farther downfield by the oxygen, closer to 4 ppm and just upfield of the H_α resonance. This makes Ser a recognizable "unique" spin system rather than part of the AMX group. Another group can be recognized from chemical shifts as "five-spin" systems (Fig. 12.9): CH_α–CH_2–CH_2–R, where again R is a "dead end": Glu, Gln, and Met. The two β-proton resonances appear around 2 ppm and the γ-protons (which may or may not be degenerate) are farther downfield (2.3–2.6 ppm). This pattern can usually be distinguished from the AMX pattern.

12.5.1 Sequence-Specific Chemical Shifts in Structured Proteins

The random-coil chemical shifts shown in Figure 12.9 have very little variation in H_N or H_α chemical shifts among the 20 amino acids. Furthermore, if we have more than one of a particular amino acid in an unstructured protein, they will be indistinguishable by chemical shift. Sometimes proteins are observed in an "unfolded" state, and this can be clearly seen

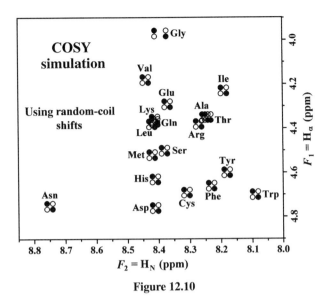

Figure 12.10

from the COSY or TOCSY spectrum, looking at the "fingerprint region" of $F_2 = H_N$ and $F_1 = H_\alpha$. If all of the amino acid residues are showing the random-coil shift values, we would see all of the H_N–H_α crosspeaks in a small region less than one ppm wide in each dimension (H_N 8–9 ppm, H_α 4–4.8 ppm) and there could be no more than 20 crosspeaks regardless of the size of the protein (Fig. 12.10). Even if a protein has folded only to the extent of "hydrophobic collapse" (aggregation of the hydrophobic side chains to protect them from water) we will see this very poor dispersion of chemical shifts. This state is called the "molten globule" state because these hydrophobic clusters are like drops of oil in a liquid state—they do not have the very specific packing arrangements of side-chains characteristic of folded (native) proteins. An unfolded or molten globule form of a protein can easily be identified by NMR, even from a 1D proton spectrum, because of this very poor dispersion. If part of a protein is disordered, those residues will fall in the narrow "random coil" chemical-shift range and will also give much sharper and stronger crosspeaks because their greater flexibility gives them the long T_2 values of a smaller molecule.

In a folded protein, the random-coil chemical shifts are changed slightly by the immediate environment of the spin system in a protein: the precise orientations of nearby aromatic rings and peptide bonds lead to specific changes in these chemical shifts due to through-space effects of unsaturated "ring currents" (anisotropic effects). Thus in a protein there may be many serine residues but each one will have slightly different chemical shifts for the H_N, H_α, and two H_β protons. This is illustrated by some of the ^1H chemical shifts for a small (63 residue) globular protein, the Heregulin-α EGF domain (Fig. 12.11). Heregulin-α is a protein ligand for a membrane receptor associated with breast cancer, and the EGF domain is a part of this protein "cut out" for structure determination by NMR. The "generic" chemical shifts for each residue type (e.g., glycine) are shown above the specific residue chemical shifts for each occurrence of that amino acid in the protein. There are seven lysine residues, for example, and each one has a unique pattern of chemical shifts similar to the random coil values (in the same general region of the spectrum) but not identical. Because the Lys side chain is charged and likely to be exposed to solvent, we do not see a great deal of variation in the side-chain chemical shifts, but the backbone (H_α and H_N)

shifts are widely separated, ranging from 7.7–9.6 ppm for the H_N resonances (random coil H_N is 8.4). Similar variations are seen for each of the amino acids found in the protein (e.g., 4 Gly, 3 Ala, 4 Leu, 5 Val, 3 Pro, 6 Cys, 3 Phe, and 2 Tyr). This sequence-specific variation of chemical shifts is what makes protein NMR possible: each residue in the sequence has its own "address" (set of precise chemical shifts) that allows us to measure distances and dihedral angles from a vast number of precisely defined positions within the protein. Before we can use these addresses, however, we need a "phone book" that pairs up the chemical shifts with the sequence-specific locations within the protein. For example, we need to know that the resonance in the ^1H spectrum at 9.80 ppm (farthest left in Fig. 12.11) is not just an H_N resonance and not just the H_N of a phenylalanine residue, but that this resonance is the H_N of Phe 21: a single specific proton in the entire molecule. It's like looking in the phone book and finding several pages of "Jones": we need the address of one particular Jones, Samuel P. Jones, in order to find his house and make a map of his neighbors. We have already encountered this process of *assignment* with natural products, but it is more complicated in a

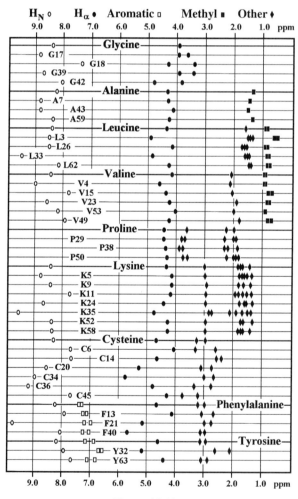

Figure 12.11

protein because of the regular repeating units and the multiple copies of identical monomers in a biopolymer. The chemical-shift dispersion due to specific environments in a protein makes sequence-specific assignment possible. Some proteins are more difficult than others due to the lack of chemical-shift dispersion even in a fully folded structure. A sequence with few or no aromatic amino acids will be a challenge from an NMR perspective because the aromatic rings are a major source of chemical-shift dispersion. A protein with primarily α-helix secondary structure is more difficult than a β-sheet protein because the β-sheet produces larger deviations in chemical shift.

12.5.2 Secondary Structure

The variation in chemical shifts for the multiple copies of a particular amino acid is the result of two factors: one is the essentially random effect of aromatic rings oriented in a precise relationship to the proton in question (Chapter 2, Figure 2.15), shifting it upfield (shielding region above and below the ring) or downfield (deshielding region in the plane of the ring). This is a result of the tertiary structure of the protein: the precise three-dimensional folding of the polypeptide backbone and all of the side chains. The second factor is the relative position of the backbone carbonyl groups of the peptide bonds. This is also a through-space (anisotropic) effect due to unsaturation, but it correlates with certain common medium-range folding motifs in proteins: the α-helix (Fig. 12.12) and the (antiparallel) β-sheet (Fig. 12.13). In the figures, hydrogen bonds are shown as dotted lines. These motifs are called *secondary structure*, and they constitute the structural building blocks of proteins. Note that in the β-sheet the carbonyls and the NH groups alternate direction along the strand,

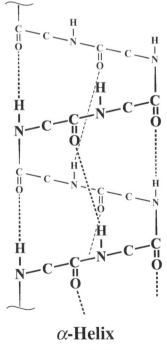

α-**Helix**

Figure 12.12

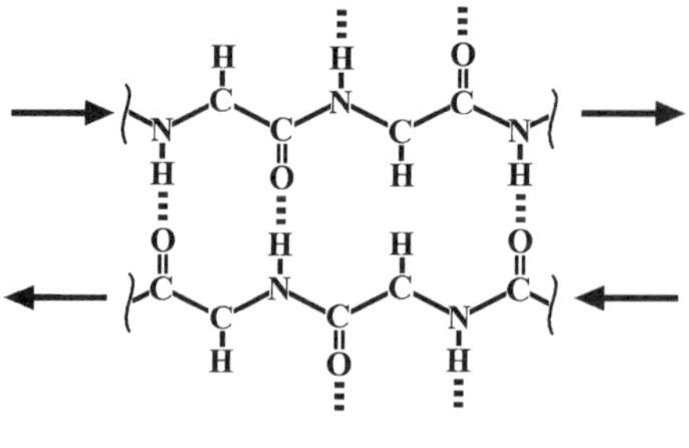

576 BIOLOGICAL NMR SPECTROSCOPY

β-Sheet

Figure 12.13

whereas in the α-helix all carbonyls point in one direction and all NH groups point in the opposite direction. These repeating geometric arrangements lead to a correlation between secondary structure and chemical shift. Both the H_N and the H_α chemical shifts move downfield from their random coil values when the residue is in a β-sheet structure, and upfield when it is in an α-helix. For example, of the seven Lys residues in Heregulin-α EGF domain (Fig. 12.11), the H_N shift of K35 (9.6) is shifted far downfield from the lysine random-coil value (8.4) and the H_N shift of K11 is shifted far upfield (7.6 ppm). In fact, K35 is located in a β-sheet and K11 is part of an α-helical portion of the protein structure. The same patterns are seen for the H_α protons: H_α of F40 (5.6) is shifted downfield from the phenylalanine random-coil H_α value (4.7), and H_α of F13 (4.1) is shifted upfield. In the 3D structure, F40 is in a β-sheet and F13 is in the single α-helix. Note that we have to compare sequence-specific chemical shifts to the random-coil shifts for the *same* amino acid because there is some dependence on the amino acid type even when there is no specific conformation (Fig. 12.9). This predictive tool can be formalized by subtracting the random-coil shift from the actual shift for each residue to obtain the "chemical-shift deviation" (CSD) for that residue. For example, the H_α CSD is -0.6 for F14 in Heregulin-α EGF domain (4.1–4.7 ppm). To simplify the prediction even further, one can define a "chemical-shift index," which is $+1$ if the CSD is greater than or equal to 0.1 ppm, -1 if CSD is less than or equal to -0.1, and zero if it is between $+0.1$ and -0.1. A bar graph of these CSI indicators *versus* the residue number will show a string of $+1$ values for a strand of a β-sheet and a string of -1 values for an α-helix. Thus with nothing more than sequence-specific assignments and a list of chemical shifts, we can begin to identify the secondary structure building blocks of the 3D protein structure. Figure 12.14 shows the CSI values for Heregulin-α EGF domain with the β-strands and α-helix identified from the final 3D structure. A fairly long "run" of negative CSI values (residues 5–13) even with a few gaps, identifies an α-helix. Long runs of positive CSI values (residues 19–24 and

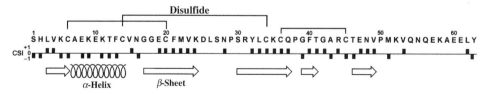

Figure 12.14

32–37) indicate extended strands of a β-sheet. Blank regions (25–31 and 54–61) may be unstructured or at least lacking in regular secondary structure. Because we are ignoring the "random" factors—the orientation of nearby aromatic rings—which can significantly affect chemical shifts and which have no relationship to secondary structure, the CSI is only a broad indicator and will not correlate perfectly with the final structure. Still, it is extremely useful and requires nothing more than sequence-specific assignments.

Ideally, every proton in the protein structure would have a unique chemical shift, different from every other proton. In NMR, we can only identify a proton by its chemical shift, so if two protons have the same chemical shift all of the information associated with them—NOE distances and dihedral angles—becomes ambiguous. One way to overcome this "overlap" problem is to measure the chemical shifts of heavy atoms: nitrogen and carbon. If we replace every N and C (normally ^{14}N and ^{12}C) with spin-½ ^{15}N and ^{13}C, we can now distinguish two protons with identical chemical shifts by the (usually) different chemical shifts of the carbon or nitrogen they are connected to. This has led to the possibility of studying larger proteins with a complex array of new experiments in which magnetization is "tossed around" between ^{1}H, ^{15}N, and ^{13}C nuclei. These experiments, mostly 3D, are called "triple-resonance" experiments because they include pulses on all three of these nuclei.

12.6 NOE INTERACTIONS BETWEEN ONE RESIDUE AND THE NEXT RESIDUE IN THE SEQUENCE

Using DQF-COSY and TOCSY we can link all of the protons within a single spin system, which corresponds to a single amino acid residue. We can classify each spin system as a pattern of chemical shifts unique to one amino acid or as a member of a class: AMX or five spin. In order to get sequence-specific assignments, however, we have to have some way to correlate protons in one residue to protons in the *next residue* in the sequence. For unlabeled proteins this is done by NOE interactions: certain protons in one residue are constrained by the peptide bond to be close in space to certain protons in the next residue. These NOE correlations are called sequential or "$i, i + 1$" because they correlate a proton in residue "i" with a proton in the next residue in the sequence, residue "$i + 1$." Specifically, we expect to see NOE correlations between H$_\alpha$ of residue i and H$_N$ of residue $i + 1$ (Fig. 12.15) and sometimes between the H$_\beta$ protons of residue i and the H$_N$ of the next residue. Because the DQF-COSY and TOCSY spectra correlate protons within a residue, we can move from

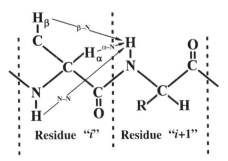

Figure 12.15

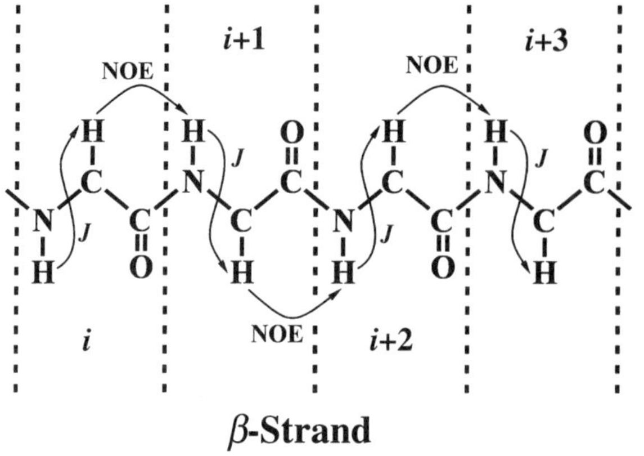

[Figure 12.16: β-Strand diagram showing NOE connectivities between residues i, $i+1$, $i+2$, $i+3$ with J couplings]

β-Strand

Figure 12.16

H_N of residue $i + 1$ to H_α of residue $i + 1$ via the J coupling. This sets us up for the next sequential NOE "jump" from the H_α of residue $i + 1$ to the H_N of residue $i + 2$ (Fig. 12.16). This is similar to the "walk" along a carbon chain using COSY data, except that now we alternate between NOE interactions (2D-NOESY) crossing the peptide bond (a dead end for J couplings) and J couplings (2D-COSY) to move to the next residue's H_α. Using this strategy we can "walk" along the polypeptide backbone and directly "read off" the chemical shifts of each H_α and H_N, assigning them to specific residues in the protein. These sequential NOEs, known as "α,N" and "β,N" NOEs, are directional and will <u>not</u> be seen in the other direction: $H_\alpha (i) \leftrightarrow H_N (i - 1)$. The geometry of the peptide bond only brings the H_α into proximity with the next residue's H_N ($i \rightarrow i + 1$).

Sequential α,N and β,N NOEs are commonly very strong for extended conformations such as the extended strand of a β sheet (Fig. 12.16). The H_α of residue i is clearly very close to the H_N of residue $i + 1$, but the H_N of residue i is farther away from H_N of residue $i + 1$ (the N–H vector points in the opposite direction), so the sequential "N,N" NOE is weak or missing in this conformation (Fig. 12.15). In contrast, the α-helix conformation (Fig. 12.17) orients all of the N–H vectors in the same direction so that the H_N of residue i is now close to the H_N of residue $i + 1$ in space. Within an α-helix we can "walk" along the peptide backbone directly using only the NOESY spectrum, using J-coupling information (COSY and TOCSY) to link each H_N chemical shift to a specific spin system associated with an amino acid (unique pattern) or amino acid category (AMX or five spin). In the α-helical conformation the H_α of residue i is farther away from the H_N of residue $i + 1$ (Fig. 12.17), so the α,N (and β,N) sequential NOEs are weak or missing. This is a nice consequence of the secondary structure: β-sheet regions will give strong sequential α,N crosspeaks in the NOESY spectrum and weak or missing N, N($H_N^i \rightarrow H_N^{i+1}$) crosspeaks, whereas α-helical regions will give strong sequential N,N crosspeaks and weak or missing α,N crosspeaks. It is always an added bonus if the primary sequential NOE observed (e.g., α,N in a β-sheet) can be confirmed with a weak sequential NOE of the other type (e.g., N,N).

Figure 12.18 shows a portion of the $F_2 = H_N$, $F_1 = H_\alpha$ region of the NOESY spectrum of *B. subtilis* HPr, a phosphotransferase that uses an active-site histidine side chain to transfer a high-energy phosphate group. A large number of sequential $H_\alpha (i) \rightarrow H_N (i + 1)$ NOE crosspeaks are shown. The H_N chemical shifts are indicated by vertical dotted

NOE INTERACTIONS BETWEEN ONE RESIDUE AND THE NEXT RESIDUE   579

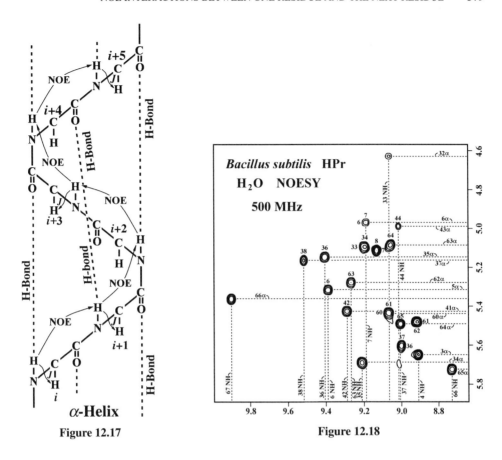

Figure 12.17

Figure 12.18

lines and the H_α shifts are shown as horizontal dotted lines, and in crowded regions the H_N assignments are shown above or below the crosspeaks with the H_α assignments to the right or left. The protein is mostly β-sheet, as shown in Figure 12.19, with four antiparallel β-strands forming the "bread" and two α-helices the "sausages" of an open-faced sandwich. The NOE crosspeaks shown in Figure 12.18 are from the A, D, and B strands (1–10, 60–70 and 30–40, respectively) of the β-sheet. Figure 12.20 shows an expansion of this region of the NOESY spectrum, with a portion of the β-sheet shown at the right. The three sequential

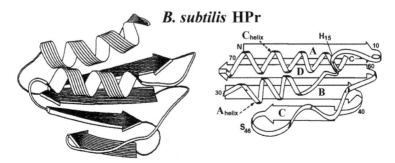

Figure 12.19. Reprinted with permission (see reference 2). Copyright 1990, American Chemical Society.

580 BIOLOGICAL NMR SPECTROSCOPY

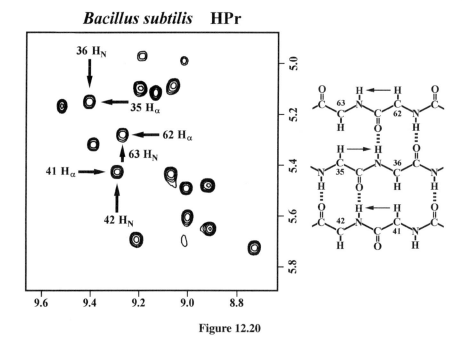

Figure 12.20

α,N connectivities indicated on the structure (arrows) correspond to the NOE crosspeaks identified in the spectrum. Of course, when you first obtain a NOESY spectrum, the crosspeaks do not have names, so before we can interpret the data we must first assign all of the protein resonances to specific protons in the covalent structure.

12.7 SEQUENCE-SPECIFIC ASSIGNMENT USING HOMONUCLEAR 2D SPECTRA

Figure 12.21 shows the basic strategy for sequential assignment. Consider a sequential pair of amino acid residues: serine followed by valine (S–V). In the TOCSY spectrum, we see the characteristic patterns of crosspeaks on the $F_2 = H_N$ vertical lines for the two residues. Because both patterns are unique, we know that the one on the left side is a valine residue and the one on the right side is a serine. In the DQF-COSY spectrum, we see only the H_α crosspeaks on the vertical lines corresponding to the H_N chemical shifts in F_2 because COSY mixing involves only a single "jump" of INEPT transfer via the H_N–H_α J coupling. This region ($F_2 = H_N = 7$–10 ppm, $F_1 = H_\alpha = 3$–5 ppm) is called the "fingerprint region" of the COSY spectrum and can be used to count up the number of crosspeaks, which should be equal to one fewer than the number of amino acid residues (not counting prolines). The first residue (N-terminal) will not show up because the amino-terminus is a protonated amine (H_3N^+) rather than an amide (HN–CO) and is exchanging with water far too rapidly to be observed. The COSY crosspeaks have a fine structure that is an antiphase doublet in the F_2 dimension because the H_N proton is only coupled to the single H_α proton (except for glycine residues where there are two H_α protons). Each crosspeak can be analyzed by curve fitting to extract the H_N–H_α coupling constant from the antiphase F_2 slice (Fig. 12.21, lower left). Usually the raw 2D data is reprocessed by zero filling and cutting out the relevant regions

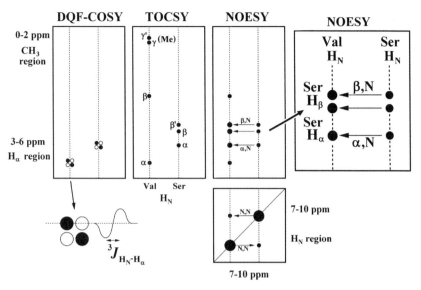

Figure 12.21

of the spectrum (H_N region in F_2 and H_α region in F_1) to generate a 2D matrix including only the fingerprint region, with much greater digital resolution. Accurate values for the H_N–H_α ^3J couplings can be used to determine the dihedral angle (related to the Φ angle in biochemistry) defined by the path H–N–C_α–H_α. This angle, along with the N–C_α–CO–N or Ψ dihedral angle, defines the conformation of the polypeptide backbone and is a crucial input for NMR structure calculations.

The NOESY spectrum (Fig. 12.21, right) gives the sequential connectivity, the proof that the Ser residue on the right side is followed in the primary sequence by the Val residue on the left side. On the vertical H_N = Val line we see the intraresidue crosspeaks to the H_α and H_β of valine (which also appear in the TOCSY spectrum), but there is a new crosspeak that connects the H_α of the Ser spin system (in F_1) with the H_N of the Val spin system (in F_2). This crosspeak lines up with the intraresidue crosspeak on the serine H_N line (H_N = Ser in F_2 and H_α = Ser in F_1) that is in the exact position of a crosspeak in the DQF-COSY spectrum. These intraresidue H_N–H_α crosspeaks may be weak or missing in the NOESY spectrum, so it may be necessary to mark the position of the COSY crosspeaks in the NOESY spectrum. With modern NMR software this is usually done by using correlated cursors (crosshairs) displayed in the separate COSY and NOESY spectra. In Figure 12.21 (right side), we also see sequential crosspeaks on the vertical H_N = Val line corresponding to the β and β' protons of the preceding (Ser) residue. These β,N NOE crosspeaks are very useful in confirming the sequential connection, especially if the H_α chemical shift in F_1 falls in a crowded region, overlapped with the H_α shifts of other residues. Finally, the N,N region of the NOESY spectrum, near the diagonal, can be searched for sequential connectivities (Fig. 12.21, lower right). Unlike the α,N and β,N correlations, the $H_N \leftrightarrow H_N$ connections work in both directions, i to $i + 1$ and i to $i - 1$. These will be the primary connections for α-helical regions of the protein, but will be weak for the β-sheet regions. Because the crosspeaks in a NOESY spectrum are weak relative to the diagonal peaks (inefficient magnetization transfer), it may be difficult to see the N,N crosspeaks if they are close to the diagonal.

In the real world, we do not know beforehand which spin system preceeds the valine spin system shown in the figure. There may be a number of Val residues in the protein, so we do not even know which kind of spin system to look for. What we have to do is to identify the NOE crosspeaks on the valine H_N vertical line in the H_α region that are not found in the TOCSY spectrum. These have to be interresidue NOE crosspeaks. Then we search the COSY spectrum to find a crosspeak with the exact F_1 chemical shift of the interresidue NOE peak on the valine H_N line. There may be more than one candidate due to overlap, and we can rule some of them out because we know the amino acid sequence. For example, if none of the valine residues are preceded by an AMX residue in the amino acid sequence, we can rule out any AMX residue even if it has the right H_α chemical shift. To find the spin system *following* the valine, locate the intraresidue H_α crosspeak on the valine H_N line and search horizontally for an NOE crosspeak at the same level (same F_1 shift) that is not found in the TOCSY spectrum. This sequential crosspeak will occur at the H_N chemical shift (in F_2) of the next residue in the sequence. The process will inevitably reach a dead end at some point—especially when you get to a proline residue in the sequence (no H_N). But even before that you may get stuck because of overlap and ambiguity. There are many starting and stopping points in the sequence and eventually when enough spin systems are assigned you can assign others by process of elimination.

To illustrate the process with real data, we will assign three segments of Heregulin-α EGF domain using sequential α,N correlations. Figure 12.22 shows a portion of the 70-ms TOCSY spectrum of the protein in 90% H_2O. The most downfield-shifted H_N resonances (probably in a β-sheet) can be easily classified: AMX, Lys, five-spin, Leu, Thr, AMX, Val, AMX, and so on, moving from left to right. Figure 12.23 shows the fingerprint region of the 150-ms jump-return NOESY spectrum. The positions of selected crosspeaks in the

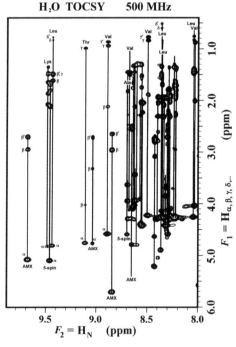

Figure 12.22

SEQUENCE-SPECIFIC ASSIGNMENT USING HOMONUCLEAR 2D SPECTRA 583

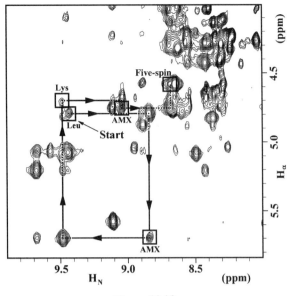

Figure 12.23

DQF-COSY spectrum are marked as squares. Starting at the position of the COSY crosspeak for the leucine residue identified in the TOCSY (Fig. 12.22, $H_N = 9.44$ ppm) we move to the right to a "fat" sequential crosspeak at $H_N = 8.84$ ppm. This H_N shift corresponds to an AMX spin system in the TOCSY spectrum. Moving down to the position of the COSY peak for this AMX system (square: $H_\alpha = 5.69$), we start the search again horizontally for another sequential NOE crosspeak. This is found on the left side at $H_N = 9.48$ ppm, corresponding to a lysine spin system in the TOCSY spectrum. Moving up to the H_α position of this system (square: $H_\alpha = 4.69$ ppm), we set off again horizontally looking for another nice, fat, well-resolved sequential crosspeak. Here we run into some regions of overlap and the going gets rough. So far we have the sequence: Leu–AMX–Lys. Searching the amino acid sequence (Fig. 12.14), there are four sequences that start with leucine: LVK, LSN, LCK, and LY. Only LCK (Leu-Cys-Lys) fits the pattern Leu-AMX-Lys, so we can assign this to the region 33-34-35 in the sequence, which is followed by Cys-36 and Gln-37. An AMX system at $H_N = 9.04$ has a "stretched" crosspeak corresponding to the sequential peak at $H_\alpha = 4.69$ and the intraresidue peak (square) below it at $H_\alpha = 4.76$ ("AMX"). Moving to the right from this square we come to an overlapped but strong sequential crosspeak at $H_N = 8.68$, corresponding to a five-spin system overlapped with an Ala. Because we are expecting Gln-37, we choose the five-spin system. This completes the assignments for the stretch LCKCQ from residue 33 to residue 37.

Figure 12.24 shows the same region of the NOESY spectrum with these assignments written in. Now we begin another "walk," starting with the well-resolved valine residue in the TOCSY at $H_N = 8.49$ ppm (Fig. 12.22, top right) and working backward in the sequence. On this vertical line in the NOESY spectrum (Fig. 12.24) there is a lovely sequential peak all alone at $H_\alpha = 5.08$ ppm, corresponding to the H_α chemical shift of a five-spin system with $H_N = 9.45$ ppm. At the position corresponding to the COSY crosspeak of this five-spin system (square, left side) there is a "stretched" crosspeak with the interresidue peak at $H_\alpha = 5.08$ and the sequential peak at $H_\alpha = 5.07$ ppm, corresponding to an AMX spin system with $H_N = 9.68$ ppm. Checking the amino acid sequence, there are five Val residues contained

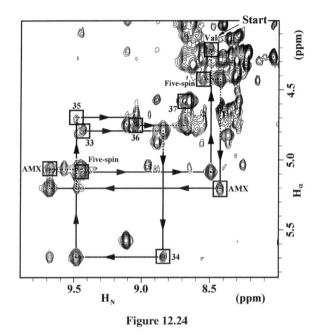

Figure 12.24

in the tripeptides HLV, FCV, FMV, ENV, and MKV. Only the tripeptide FMV corresponds to the sequence AMX–five-spin–Val found in the NOESY, so we can assign this to Phe-21, Met-22, and Val-23. From the AMX (F21) at $H_N = 9.68$, $H_\alpha = 5.07$ we can move backward (down) along the $H_N = 9.68$ line to a big sequential crosspeak at $H_\alpha = 5.20$ ppm, another AMX system with $H_N = 8.43$ ppm (moving to the square on the right side). The sequence tells us that this corresponds to Cys-20. Moving up along the $H_N = 8.43$ line leads to a sequential crosspeak at $H_\alpha = 4.43$, corresponding to a five-spin system ($H_N = 8.55$) that we can assign to Glu-19.

A short, three-residue walk through the NOESY spectrum is shown in Figure 12.25. Although there is a significant overlap in this region, two prominent sequential NOE peaks connect a Leu to a Val (H_α (Leu) = 4.89 to H_N (Val) = 8.88) and the same Val to a Lys

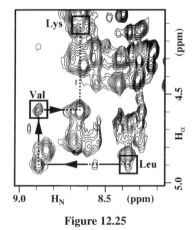

Figure 12.25

system (H_α (Val) = 4.58 to H_N (Lys) = 8.64). The dipeptide LV occurs only once in the sequence, followed by lysine: LVK at position L3-V4-K5.

Figure 12.26 shows the TOCSY spectrum with all of the sequence-specific assignments indicated. There are some horrendously overlapped regions in the H_N and H_α regions, and sorting all of this out requires some special tricks. One of these is to vary the temperature. The main effect of changing the sample temperature is to move the water peak because the H_2O chemical shift is temperature-dependent due to the change in the extent of hydrogen bonding. The chemical shift of water changes with temperature according to the formula

$$\delta(H_2O) = 5.013 - T(°C)/96.9$$

This means that the water chemical shift moves downfield by 1/96.9 or 0.0103 ppm, which is 10.3 parts per billion (ppb) with every decrease in temperature of 1 °C. The H_N chemical shift "contains" a certain contribution due to the H_2O chemical shift because of exchange with water: fast exchange leads to a chemical shift that is the weighted average of the chemical shifts experienced by the spin over time. Some H_N protons in the protein exchange rapidly with water because they are "exposed" to solvent on the surface of the protein; others exchange slowly because they are "buried" in hydrophobic regions and tied up in stable hydrogen bonds to protein groups such as backbone carbonyls. If an H_N proton is solvent exposed, it will exchange with water rapidly and its chemical shift will show a large downfield shift (6–8 ppb/°C) as temperature is decreased. A "protected" H_N proton, however, will experience much lower temperature shifts, less than 4 ppb/°C. If an overlapped region has a mixture of exposed and protected H_N resonances, changing the temperature will change the

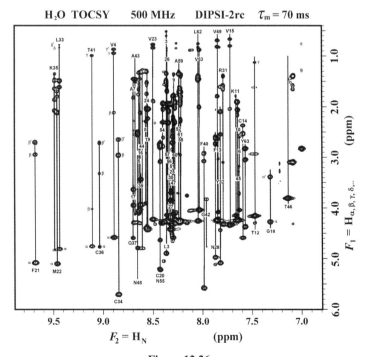

Figure 12.26

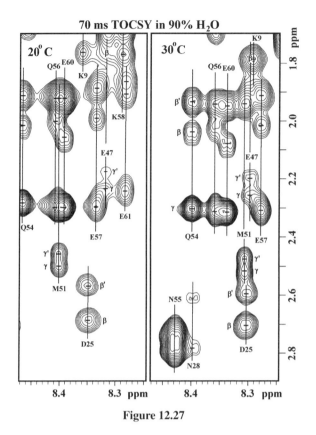

Figure 12.27

relative H_N chemical shifts in a way that may be very helpful in sorting out the assignments. Figure 12.27 shows a small portion of the TOCSY spectrum of Heregulin-α EGF domain at 20 °C (left) and 30 °C (right) with the assignments indicated. The H_N of E47, for example, moves only slightly (it is part of a short β-sheet) but the H_N of M51 moves almost 0.10 ppm (it is in the disordered C-terminal region). With NOESY spectra at both temperatures many ambiguities in the sequential connectivities can be sorted out. The three H_N resonances Q56, M51, and E60 are almost completely overlapped at 20 °C, but they are well separated at 30 °C. Keep in mind that a great deal of human effort and judgement is involved in data interpretation in protein NMR. A week of data acquisition can be followed by a year or two of tedious, eyesight-destroying analysis in front of a computer screen in a darkened room. After a while even raindrops on the car windshield begin to look like NOE crosspeaks!

12.8 MEDIUM AND LONG-RANGE NOE CORRELATIONS

So far we have not learned anything from the NOE data except to assign all of the protons. This is an essential first step, but the real goal is to extract distances between specific pairs of protons from the NOE data. Sequential NOEs are not very useful because we already know that an amino acid residue is close in space to its nearest neighbor in the sequence. The real "mother lode" of the NOESY data consists of the medium-range ($i \to i+2$, $i \to i+3$ and $i \to i+4$) and long-range connectivities. These define the secondary structure

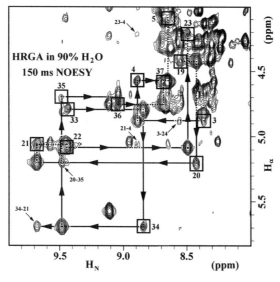

Figure 12.28

(α-helix and β-sheet) elements and the precise spatial relationships between them, which constitute the tertiary structure. Figure 12.28 shows the NOESY spectrum of Heregulin-α EGF domain with the assignments indicated for the three fragments assigned earlier. The CSI data (Fig. 12.14) predicts these three sections (3–5, 19–23, and 33–37) to be in an extended or β-strand conformation. Weak long-range crosspeaks are also indicated, connecting one strand with another (34–21, 20–35, 3–24, 21–4, and 23–4). These long-range correlations indicate that the strands are antiparallel because increasing the residue number by one (34 to 35) decreases the residue number on the other strand by one (21 to 20). The precise alignment of the β-strands can be determined from the H$_\alpha$ to H$_\alpha$ long-range NOE correlations. Figure 12.29 shows the $F_2 = $ H$_\alpha$, $F_1 = $ H$_\alpha$ region of the D$_2$O NOESY spectrum. To see the H$_\alpha$ protons in the F_2 dimension the experiment is performed in D$_2$O because the H$_\alpha$ chemical shifts are close to the water resonance. The H$_\alpha$ to H$_\alpha$ NOE interactions are directly across the interface between strands in the β-sheet, so the alignment of strands in unambiguous: Leu-3 aligns with Val-23 and Cys-20 aligns with Cys-34 (Fig. 12.30). Now we have a complete 2D picture of the major β-sheet part of the protein. Figure 12.30 summarizes a large body of information that supports the β-sheet structure. A large number of cross-strand H$_\alpha$-H$_\alpha$ and H$_\alpha$-H$_N$ NOEs are observed, and hydrogen bonds are implied by a slow exchange of H$_N$ with D$_N$ when the protein is dissolved in D$_2$O. Finally, the temperature coefficient (change in H$_N$ chemical shift with a temperature change) measures the degree of exposure of the H$_N$ to solvent. Amide protons with a small coefficient are probably buried in the protein interior with little access to solvent. In proteins with a well-defined single conformation this can be used as a criterion for hydrogen bonding. In peptides, it is not as useful because the equilibrium between a "folded" peptide and a random coil form is shifted as the temperature is raised, and this also contributes to the temperature coefficient.

Deuterium exchange is measured by recording a series of fast TOCSY experiments (30 min each) immediately after dissolving the protein in D$_2$O. As the amide protons

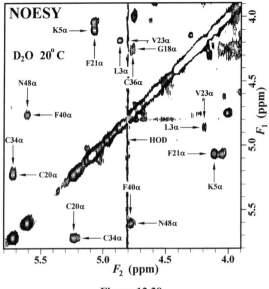

Figure 12.29

exchange with solvent they are replaced with ^{2}H and disappear from the spectrum. Figure 12.31 shows the TOCSY spectrum in 90% H$_2$O (left) and the first TOCSY spectrum acquired in D$_2$O (right). Already the majority of H$_N$–H$_\alpha$ crosspeaks has disappeared and those that remain are the "buried" H$_N$ protons. The rate of loss of the H$_N$–H$_\alpha$ crosspeaks can be quantified by measuring the crosspeak volume at each time point and plotting against time (Fig. 12.32). The loss of signal is exponential, and curve-fitting yields the rate constant (or half-life) for the exchange. In the plots shown, the half-lives range from just slow enough to observe (Val-4, 19 min) to very long-lived (F21, 7.4 h). These can be compared to the inherent (random coil) exchange rates for each of the 20 amino acids to obtain a rate ratio or "protection factor" that indicates the degree of "burial" of the H$_N$ proton in the protein core.

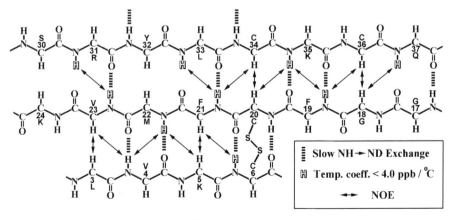

Figure 12.30

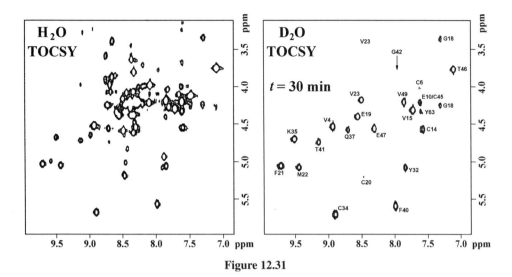

Figure 12.31

Figure 12.33 summarizes all of the secondary structure evidence for Heregulin-α EGF domain. Exchange protection is indicated by black circles (very slow H/D exchange) and gray circles (slow H/D exchange). ^3J coupling between H_α and H_N is indicated by an arrow pointing up ($J > 8$ Hz, β-strand Φ angle) or pointing down ($J < 6$ Hz, α-helical Φ angle). Sequential and medium-range NOEs are indicated by bars connecting the two residues horizontally (short bar: weak NOE, tall bar: strong NOE, gray bar: NOE ambiguous due to overlap). The chemical-shift index (for H_α protons) is shown at the bottom along with the secondary structure elements found in the final calculated structure. Strong αN sequential connectivity is found in the β-sheet regions, with NN connections limited to the helix and various turns. The rare $i \rightarrow i+2$ NOEs are indicative of turns, which occur between elements of regular secondary structure. Only two $i \rightarrow i+3$ NOEs are found (K5-E8 and E8-K11), both within the α-helix. These medium-range (i to $i+3$ and i to $i+4$) NOEs

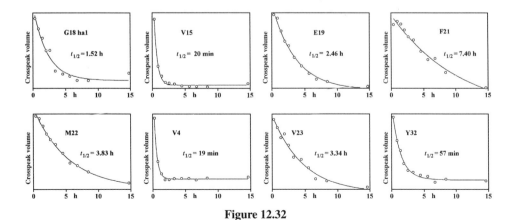

Figure 12.32

590 BIOLOGICAL NMR SPECTROSCOPY

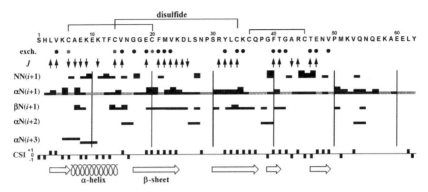

Figure 12.33

result from the proximity of one rung of the helix to the next rung in space (Fig. 12.12) because the helix repeat is about 3.6 residues per turn.

12.9 CALCULATION OF 3D STRUCTURE USING NMR RESTRAINTS

The ultimate goal of most biological NMR studies is to obtain an accurate 3D structure (conformation) of the molecule. This cannot be done by human judgement and analysis of individual pieces of evidence; there is far too much data and we need an unbiased method for finding the best 3D structure that is most consistent with the NMR data. Structure calculations have been done since computers became available on a variety of organic compounds and biological molecules. The various forces exerted by covalent bonds (bond lengths, bond angles, planarity of double bonds, Van der Waals attraction, hard sphere repulsion, etc.) are summarized in a *force field* (e.g., Amber or cvff). The goal is to search the "conformational space" (the total of all possible conformations) to find the minimum of energy as defined by the force field. This is a big challenge for a linear polymer like a protein because the number of possible conformations is astronomical, as defined by the various dihedral angles (Φ, Ψ, X_1, X_2, etc.) for each residue resulting from free rotation around single bonds. Even the most sophisticated structure calculations cannot define a protein's conformation without restraints defined by experimental observations. The "folding problem" in proteins is far from solved: no one can predict the 3D shape of a protein simply by knowing its amino acid sequence.

12.9.1 NMR Experimental Restraints

The NMR data enter into the structure calculation in the form of "restraints": limitations placed on H–H distances and H–N–C–H$_\alpha$ dihedral angles based on the observation of NOE crosspeaks or J couplings. The specific form of the restraints is a "penalty function" that adds to the total energy of a conformation if the distance or dihedral angle is outside the limits defined by the NMR data. For example, a strong NOE between two protons indicates that the distance should be less than 2.8 Å, so any time that distance is exceeded we "penalize" the structure by adding to its total energy. Exceeding the restraint distance is called a "violation," and the larger the violation the more energy is added. The energy gradient behaves like a force, pulling the two protons together. It's like tying a rubber band between the two protons: if they are within the restraint distance (in this case 2.8 Å), the rubber band is slack and there is no force. If the restraint is violated (distance greater than 2.8 Å) the rubber band is taut and exerts an attractive force between the two protons. As the violation increases the force also increases. A similar torsional force is introduced for dihedral angles by adding an energy penalty any time the angle gets outside the range of angles indicated by the measured J coupling. In this way the process of energy minimization is simultaneously maintaining the bond distances and angles defined by covalent geometry and trying to satisfy all of the NOE distances and dihedral angles defined by the NMR data.

NOE distance restraints are determined from the intensities of the NOESY crosspeaks. There is a theoretical relationship between the initial rate of NOE buildup (as mixing time is increased) and the inverse sixth power of the distance between two protons. In practice, it is very difficult to measure accurate distances in protein NMR, so NOESY crosspeaks are sorted into "bins" representing, for example, very strong (<2.9 Å), strong (<3.3 Å), medium (<4.0 Å), weak (<5.0 Å), and very weak (<6.5 Å). The dividing points for these intensity categories are determined by measuring NOE intensities corresponding to well-known distances such as cross-strand H_α-H_α in a β-sheet or $H_\alpha(i)$ to $H_N(i+3)$ in an α-helix, a process known as calibration of the NOE intensities. Once a few accurate distances are associated with specific NOE crosspeak volumes, we can use the $1/r^6$ rule to calculate the volume cutoffs for sorting NOEs into the distance bins. Notice that these categories are merely upper limits of distance between two protons; lower limits are not used. In the case of overlapped NOE crosspeaks it is best to reduce the restraint to the next bin (e.g., from strong to medium) or to the least restrictive bin (weak) for severe overlap. The temptation to define the distance restraints very tightly must be resisted; the molecule will tie up itself like a tangled ball of yarn if we force the NOE restraints too hard. Keep in mind that it is not the precision of any one NOE restraint that gives us an accurate structure, but rather the combination of a very large number (often in the thousands) of relatively imprecise and loosely enforced measurements that, taken together, can lead to a very well defined 3D structure. Figure 12.34 shows a typical penalty function for NOE distance restraints. The penalty is a quadratic function defined by the square of the violation distance:

$$E = k(r - r_0)^2 \quad \text{if } r > r_0; \qquad E = 0 \quad \text{if } r < r_0$$

where r is the distance between the two protons and r_0 is the maximum distance set by the NOE restraint. The force constant k is the same for all distance restraints and determines the tightness of the "spring" or "rubber band" connecting the two protons. It is important to realize that the NOE is not an attractive force! It is an interaction observed by NMR that allows us to apply an artificial force in the structure calculation.

The nondegenerate geminal pairs are usually named according to their chemical shifts (e.g., β downfield of β') rather than their stereochemical relationships (pro-R and pro-S). In structure calculations, this usually is dealt with by creating a "pseudo-atom" right between the pro-R and pro-S positions in 3D space. The NOE restraints are applied to the pseudo-atom and not to the real atoms, and the distance limit is increased a bit to account for the ambiguity (we do not really know which restraint applies to which of the two positions in space). Similarly, a pseudoatom is created at the center of the three equivalent protons of a CH_3 group, and the distance restraint is applied to the pseudoatom.

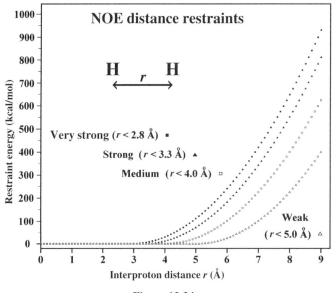

Figure 12.34

Similar restraints are generated for dihedral angles based on measured J coupling constants using the Karplus relation. The H_N–N–C_α–H_α dihedral angle is determined by measuring the H_N–H_α J value from the DQF-COSY crosspeak fine structure of each residue. The H_N–H_α dihedral angle is related to the Φ angle of protein conformational analysis: for an ideal α-helix the J coupling should be small (3.9 Hz) and for an extended strand of a β-sheet the J coupling should be large (8.9 Hz). Typically, if the coupling constant $^3J_{HN–H\alpha}$ is greater than 8 Hz, we assume an extended conformation and the Φ dihedral angle is restricted to the range $-150°$ to $-90°$ (Fig. 12.35, left ↔). If the coupling constant is less than 6 Hz we assume a helical conformation and the Φ angle is restricted to the range $-90°$ to $-40°$ (Fig. 12.35, right ↔). If the coupling constant is between 6 and 8 Hz the conformation is probably changing rapidly between these two extremes so that the J value is averaged over all conformations to an intermediate value. In this case no restrictions are placed on the Φ angle. Figure 12.35 shows the energy penalty function for a "helical" dihedral restraint (□) and for an "extended" or β-sheet dihedral restraint (■).

The mathematics of computational chemistry is very complicated and we will only attempt to describe the steps in the process in general terms. The first step is to generate a group of structures that satisfy the NMR restraints while providing the largest possible diversity of starting conformations. This process is called "embedding" and uses a technique called *distance geometry*. We might generate 50 different structures to ensure a wide variety of starting points. At this point, we have some very unhappy molecules with bonds stretched and twisted in bizarre ways. The next step is *energy minimization*, where we allow each structure to "relax" under the influence of the total energy function (covalent force field plus NMR restraint energy) and move down the energy hill (gradient) toward a minimum of energy. There are many pitfalls in this journey because the molecule might find a "local minimum" of energy, like a dry lake basin in the desert, such that all directions leading

CALCULATION OF 3D STRUCTURE USING NMR RESTRAINTS   593

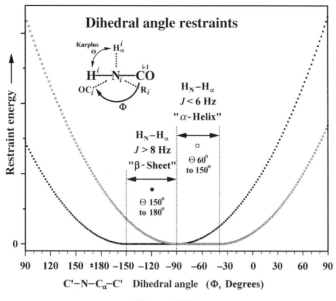

Figure 12.35

away from it are uphill. This is not the global miminum (ocean) we are searching for, but it is easy to get "stuck" in these local minima and move no further. To avoid this, a process of "simulated annealing" is used, which involves heating up the molecule and cooling it down again in a number of cycles. The heating gives the molecule more random energy so it can "bounce out of" these local traps and continue the journey downhill. Gradual cooling then allows the many conformational degrees of freedom to relax in a smooth fashion down to the energy minimum. The final stage of NMR structure calculation usually involves *restrained molecular dynamics* (rMD), where we actually try to model the motion of the molecule in real time (picoseconds) by calculating the motion of all atoms moving around under the limits of the force field and the NMR restraints. This is extremely demanding of computer capabilities, especially if we try to include the solvent in the calculation.

In a perfect world, all 50 of our calculated structures would be identical conformations differing only in their orientation in 3D space. To compare them and see how close we have come to this ideal, the structures are moved around and rotated in space to align them as much as possible to the same location and orientation in space. Figure 12.36 shows the 10 structures of Heregulin-α EGF domain with the lowest total energy, superimposed by aligning residues 3–6, 17–24, and 30–37 (the major β-sheet diagramed in Fig. 12.30). Only the trace of the polypeptide backbone (N–C$_\alpha$–C'–N–C$_\alpha$–C' $\cdots$) is shown for each structure and the last 13 residues (disordered region, 51–63) are not shown. The three strands of the major β-sheet are well defined, with the 10 polypeptide chains nearly on top of each other (lower center), and the α-helix is fairly well defined (lower left side). The upper portion (a minor β-sheet) is not well defined relative to the rest of the molecule.

To quantify how well the group of structures fit each other, we calculate the "RMSD." For any atom in one of the structures of the "ensemble" of ten structures (e.g., H$_\alpha$ of Phe-21), the distance to the same atom (H$_\alpha$ of Phe-21) in each of the nine other structures is

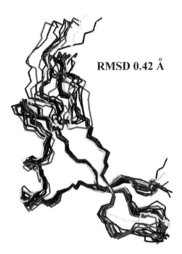

Figure 12.36

measured and the average of the squares of these distances ("deviations", d_i) is calculated. The square root of this average is called the "root mean square" (RMS) deviation or RMSD for that particular atom in the structure, measured in Angstroms.

$$\text{RMSD of } H_\alpha(\text{Phe-21}) = \left[\left(\sum d_i^2\right)/9\right]^{1/2}$$

where d_i is the distance in Å from structure 1 to structure i, with i running from 2 to 10. If the RMSD is calculated in the same way for *all* atoms, or for all atoms of a certain type (all backbone atoms, or all "heavy" or non-hydrogen atoms, or all atoms in the β-sheet) we can calculate an average RMSD that describes how well the whole group of structures ("ensemble" of structures) agree with each other.

$$\text{average RMSD} = \left(\sum \left[\left(\sum d_i^2\right)/9\right]^{1/2}\right)/N$$

where the first sum is taken over all N atoms under consideration. The goal of structure calculation is to explore as much of the conformational space as possible, minimizing the energy function and maximizing the RMSD (the diversity of structures that have low energy). The latter may seem counter-productive because the best calculated ensemble of structures will have a very low RMSD, but it is necessary to prove that within the experimental restraints the RMSD can be no larger than this value—any other approach would be dishonest.

In Figure 12.36 the average RMSD for the backbone (N, C_α, and C') atoms of the major β-sheet (residues 3–6, 17–24, and 30–37) is 0.42 Å, an excellent fit. If agreement is bad, we have not done a very good job of determining the 3D structure of the protein. If agreement is very good, we can say that any one of the structures or the average of all structures is a good description of the 3D structure (conformation) of the protein in solution. Usually if the aligned ensemble of structures is viewed together we can see that some regions align very well (the structures pack closely into a tight multistranded wire) and other regions have poor alignment (the structures fan out like spaghetti on a plate). The "disordered" regions may be truly flexible in solution (usually corresponding to solvent-exposed loops between areas of defined secondary structure) or we may simply lack the number of restraints needed to define the conformation because of overlap or other limitations of the NMR method. The only way to differentiate between these two possibilities is to study the motion (dynamics) of the protein on different timescales by a detailed study of its relaxation (T_1, T_2, and NOE) properties.

Structure calculation is an iterative process—it is not over after a single calculation on an ensemble of structures. After selecting the "best" structures of an ensemble (e.g., the 10 lowest energy structures of the 50 calculated), a careful analysis is made of the violations of NMR restraints. Let's say that a certain NOE distance restraint, K5 H_α to T12 H_N, is violated in all 10 of the "best" structures. You have a big problem: the assignment is probably wrong in the NOESY spectrum. In the calculation this "rubber band" is pulling two parts of the protein close together that really are not close in the actual 3D structure. This pulls the "bad" NOE restraint out of its range and into violation, but it also distorts a lot of other "good" restraints, pulling them up their parabolic energy curves into violation. Even the covalent connections (bond lengths, planarity of peptide linkages, etc.) will "bend" a little to try to accommodate the incorrect restraint. The result is that the energy penalty is distributed among a number of restraints, both covalent and NMR, and it may even be difficult to see which one is the "bad" restraint. You need to go back to the NMR data and see if the restraint is ambiguous. For example, if H_N of T12 is overlapped with H_N of E8, we see that the assignment was ambiguous and should not have been used for an NOE restraint. Keep in mind that even though the NOESY is a 2D experiment, the assignment in each dimension is based only on one dimension: its chemical shift compared to the shifts of all other protons in the molecule. If it is not ambiguous, continue the process of examining any restraint that is violated, with special attention to the "habitual" violators that are in violation in a large proportion of the 10 structures, or which have large violations. It is helpful to consider NOE distance violations with a cutoff: for example, all violations greater than 0.2 Å above the restraint limit, then all violations greater than 0.1 Å, and so on. The same applies to dihedral restraints: those that violate have to be checked by going back to the DQF-COSY

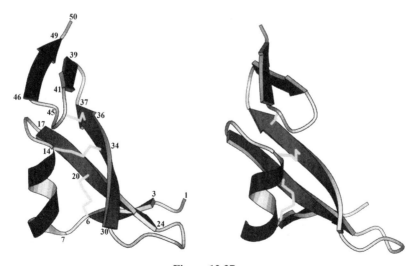

Figure 12.37

spectrum to see if the *J*-coupling measurement is correct. Don't just "throw out" restraints that violate—you need to find out what is wrong.

Once all the violations have been checked and some corrections have been made, the process of structure calculation starts all over again with a new, improved restraint list. It may take more than 20 cycles of calculation and reexamination of NMR data before the restraint violations are reduced to a minimum. The results of this process after 27 cycles of calculation are shown in Figure 12.37 for Heregulin-α EGF domain. The average of the 20 lowest energy structures is shown on the left side, and on the right side is the same structure rotated a bit about the vertical axis. The major β-sheet (3–6, 17–24, 30–37) is clearly visible but is twisted in the 3D structure. A short α-helix is inserted between two strands of the β-sheet, and a minor two-strand β-sheet (39–41, 46–49) is seen at the top. Residues 51–63 are disordered and are not shown in the figure.

12.10 ^{15}N-LABELING AND 3D NMR

Beyond about 10 kD the complexity (number of resonances) and the linewidth increase to a point where the homonuclear ^{1}H methods described up to this point are no longer effective. If the protein can be labeled with ^{15}N by expression (bacterial growth) in a medium that has ^{15}NH$_4$Cl as the only nitrogen source, the problem can be greatly simplified because overlapped H$_N$ protons can be distinguished by the chemical shift of their attached ^{15}N. If every nitrogen in the protein is replaced by ^{15}N ("uniformly labeled"), each residue can now be uniquely identified by its H$_N$ proton chemical shift *and* its backbone nitrogen ^{15}N chemical shift. It is much less likely that *both* of these chemical shifts would be the same for any two residues, so together they provide a unique "address" for each residue.

Protein expression requires that you have the gene that codes for the protein with a suitable promotor, inserted into the genome of a bacterial cell (e.g., *E. coli*) that is easy to grow in cell culture. The cells have to grow well in "minimal media"—a bare-bones food source that has no nitrogen in it at all until you add the ^{15}NH$_4$Cl. "Good" protein expression means two entirely different things to an NMR spectroscopist and a biochemist/molecular biologist. We need milligrams of pure protein, not micrograms! The "bugs" are grown to a certain density and then the promoter is "switched on" by adding something to the medium that activates it. Protein is cranked out in huge quantities, and then you "harvest it" from the cells and purify it. If you are lucky you find it in the soluble fraction, but you may find it precipitated in "inclusion bodies" of the cell. Then you have to resolubilize it, and if you have disulfide bonds you hope that they re-form in the correct way. Labeling with ^{13}C or with both ^{15}N and ^{13}C is similar except that the medium is much more expensive: the cost of isotopes can exceed $100,000 for a purified NMR sample. ^{13}C is usually supplied to the bugs from U–^{13}C glucose or U–^{13}C acetate. Again, you cannot use rich media that have natural abundance carbon—just the bare bones plus your ^{13}C-labeled food source. Although we boasted earlier about the ease of getting started in biological NMR compared to biological crystallography, it must be admitted here that people have spent two or three frustrating years preparing an NMR sample!

Peptides are much harder to prepare with uniform ^{15}N or ^{13}C labeling because they are normally prepared by solid-phase synthesis rather than by expression in cells. Solid-phase synthesis uses a very large excess of a protected amino acid at each stage, and the cost of these reagents in labeled form is prohibitive.

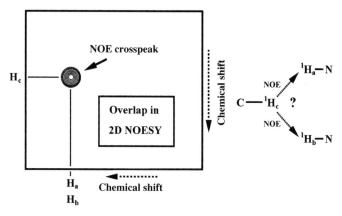

Figure 12.38

12.10.1 Why Label?

The big advantage of isotopic labeling of proteins is in overcoming the ambiguity of resonance overlap. Consider the case of two H_N protons with exactly the same ^1H chemical shift: H_a and H_b. In the 2D NOESY spectrum we have a crosspeak at the F_1 chemical shift of H_c, a carbon-bound proton with a unique chemical shift, and the F_2 chemical-shift position shared by H_a and H_b (Fig. 12.38). We cannot use this information in structure calculations because we do not know whether the distance information applies to H_a or to H_b. Now consider a ^{15}N-labeled protein sample so that we have H_a–$^{15}N_a$ and H_b–$^{15}N_b$ with different ^{15}N chemical shifts for $^{15}N_a$ and $^{15}N_b$. We can expand the 2D-NOESY spectrum into a third dimension, the ^{15}N chemical-shift scale. Consider the 2D-NOESY spectrum to be the "floor" of a cube, and each 2D crosspeak is "lifted" above the floor by a distance corresponding to the ^{15}N chemical shift of the nitrogen of the NH group (Fig. 12.39). Now the crosspeak is a "chunk" in 3D space, centered vertically at the ^{15}N chemical shift of

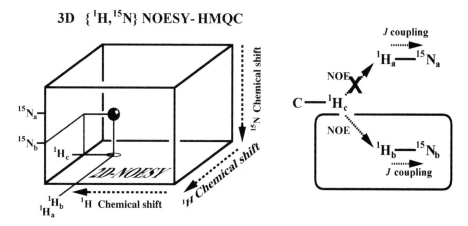

Figure 12.39

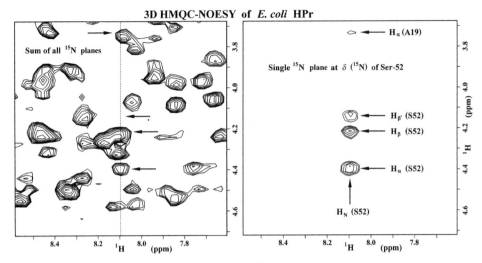

Figure 12.40

$^{15}N_b$, so we can be sure that the NOE correlation is to H_b rather than to H_a. What we have done is to "explode" the heavily overlapped 2D-NOESY spectrum into a 3D version where crosspeaks are now indexed by the ^{15}N chemical shift of the nitrogen attached to the H_N proton.

Looking at the front wall of the cube, we have a 2D correlation between the H_N chemical shift (^1H shift, horizontal scale) and the ^{15}N shift of its attached nitrogen (^{15}N shift, vertical scale). This is a one-bond correlation based on the J coupling between ^1H and ^{15}N: in other words, it is a 2D HSQC spectrum. We have seen lots of 2D HSQC spectra correlating ^1H with directly attached ^{13}C (Chapter 11), and the same kind of correlation can be made between ^1H and directly attached ^{15}N by using the large ($\sim$90 Hz) $^1J_{NH}$ coupling. We are not familiar with the ^{15}N chemical-shift scale, but right now it is sufficient to know that there is chemical-shift dispersion between the various ^{15}N positions in a protein. We cannot get away with natural abundance ^{15}N the way we used natural abundance ^{13}C because the abundance is about five times lower, and the protein sample concentration is fairly low (1 mM). Isotopic enrichment (labeling) is the only way to do these ^{15}N experiments.

The 3D HMQC-NOESY spectrum of *E. Coli* HPr in 90% H_2O is shown in Figure 12.40. On the left side is the sum of all levels in the cube: this is the view in Figure 12.39 looking straight down on the cube, equivalent to a 2D NOESY spectrum. The intraresidue H_N to H_α crosspeak of serine 52 is resolved (lowest arrow), but severe overlap at the H_N chemical shift of Ser-52 makes it impossible to assign any other crosspeaks from this residue in the region displayed. On the right side is just one plane of the 3D data matrix, corresponding to the ^{15}N chemical shift of the backbone ^{15}N of Ser-52, shown at a lower contour threshold. This is one floor of the multistory cube in Figure 12.39. Now the intraresidue NOEs are clearly resolved for the entire Ser-52 spin system—H_α, H_β, and $H_{\beta'}$—as well as a long-range NOE from H_N of Ser-52 to the H_α proton of Ala-19. In fact, no other NOE crosspeaks are observed in the displayed 2D region. The "contact" between Ser-52 and Ala-19 is very useful, connecting the loop between β-strands C and D with the loop between β-strand A and the A-helix (Fig. 12.19).

12.10.2 2D Heteronuclear Correlation Using Nitrogen-15

The 2D {^{1}H–^{15}N} HSQC spectrum of U–^{15}N Heregulin-α EGF domain is shown in Figure 12.41. This is the view of the 3D spectrum cube (Fig. 12.39) from the front, with the ^{1}H chemical shift on the horizontal (F_2) axis and the ^{15}N chemical shift on the vertical (F_1) axis. The pulse sequence is the same as the {^{1}H–^{13}C} HSQC (Chapter 11, Section 11.7) except that the second (decoupler) channel is set to the ^{15}N frequency (about 1/10 of the ^{1}H frequency). For gradient selection, the coherence order is calculated using 1 for ^{15}N and 10 (instead of 4) for ^{1}H. ^{15}N decoupling (GARP) is used during acquisition of the FID and a 180° ^{1}H pulse in the center of the evolution (t_1) period provides decoupling in the F_1 dimension. Thus each residue gives a single crosspeak for the backbone ^{1}H–^{15}N pair, and in addition each residue with a side-chain amide group (Asn and Gln) gives a pair of crosspeaks at the same ^{15}N shift due to the NH$_2$ group in the side chain. These pairs are seen on the right side of the spectrum (*N16, N28, N55, Q37*). Other side-chain NH groups also show up: N$_\varepsilon$H of arginine (R31ε and R44ε, lower right side, aliased from above the spectral window). Four of the residues give no crosspeaks in the {^{1}H–^{15}N} HSQC spectrum: S1 because it has no amide (the N-terminal H$_3$N$^+$ group exchanges much more rapidly with water and is not observed at all); and P29, P38, and P50 because the backbone ^{15}N of proline has no ^{1}H attached. Assignments in parentheses indicate that the crosspeak is aliased in F_1—it is common to cut the spectral window narrower in F_1 and let the "outliers" alias. With the proper phase-encoding method in F_1 (States-TPPI, a variant of the States method), these crosspeaks will be opposite in sign (negative intensity) to the rest of the peaks, so it will be clear that they are aliased.

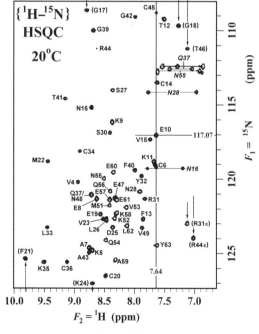

Figure 12.41

With >90% abundance ("enrichment") of ^{15}N, this 2D spectrum can be acquired in about 4 min, so it is the ideal spectrum for measuring H/D exchange. It can also be used as a rapid measurement of chemical-shift perturbations due to change in pH (sequence-specific pK_a determinations of side-chain acidic or basic groups) or binding of ligands or other proteins. If a crosspeak "moves" in the 2D HSQC when these conditions are changed, the residue it repesents must be close to the site of the perturbation (titratable group, ligand binding site, protein–protein interface, etc.). If the 3D structure of the protein is known, these shift perturbations can be used to physically map binding sites of other molecules on the surface of the protein. All that is required is that the ^{15}N and ^{1}H chemical shifts of the backbone amide NH groups be assigned; the 3D structure can come from an X-ray crystal structure or by analogy to a homologous protein of known 3D structure. This method has been called "SAR by NMR" (structure-activity relationships by NMR) and is used extensively for screening small molecules in drug development.

In the 2D $\{^{15}N, ^{1}H\}$ HSQC spectrum of Heregulin-α EGF domain, there are a number of H_N protons at or very near 7.64 ppm (Fig. 12.41, vertical line): C45, C14, E10, C6, and Y63. But of these only the H_N of E10 is bonded to a ^{15}N with a chemical shift of 117.07 ppm (Fig. 12.41, horizontal line). This dual "address" (^{1}H = 7.64 ppm, ^{15}N = 117.07 ppm) is unique, with only V15 being anywhere near E10 in the 2D HSQC. In the

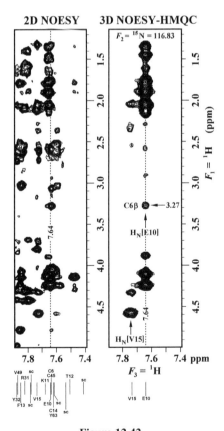

Figure 12.42

2D NOESY spectrum in H$_2$O (Fig. 12.42, left) there are a very large number of overlapped NOE crosspeaks at or very near the $F_2 = 7.64$ line, and it would be impossible to assign any of them to a unique H$_N$ proton. In the 3D NOESY-HMQC spectrum (Fig. 12.42, right), we can select the plane corresponding to a ^{15}N chemical shift of 116.83 ppm (close to E10's backbone ^{15}N shift of 117.07) and see only the NOE crosspeaks from the H$_N$ of E10 on the $F_2 = 7.64$ vertical line. Because there is no ambiguity in the 3D spectrum, the crosspeak at $F_1 = 3.27$ ppm can be confidently assigned to one of the β protons of Cys-6, yielding an important medium-range $(i, i + 4)$ restraint for the α-helix (Fig. 12.37). In the 2D NOESY this crosspeak is buried by an intraresidue NOE from the H$_N$ of C6 to the β proton of C6. But in the 3D spectrum this interfering crosspeak is in a distinctly different (lower) horizontal plane of the "cube" (Fig. 12.39) with a ^{15}N shift of 119.3 ppm (Fig. 12.41). Below the horizontal (^{1}H) chemical-shift scale of both spectra in Figure 12.42, the ^{1}H chemical-shift positions of all resonances in the range 7.4–7.9 ppm are shown. In this narrow range there are 16 closely spaced resonances, all contributing to NOE crosspeaks in the 2D NOESY spectrum, but only two of these (H$_N$ of Val-15 and H$_N$ of Glu-10) contribute to crosspeaks in the single plane of the 3D spectrum shown at the right side. Val-15 shows up because it is close to Glu-10 in *both* ^{1}H shift and ^{15}N shift (Fig. 12.41), but its NOE crosspeaks can be clearly distinguished in the 3D spectrum because of the difference in ^{1}H shift (7.73 ppm for V15 vs. 7.64 ppm for E10).

12.11 THREE-DIMENSIONAL NMR PULSE SEQUENCES: 3D HSQC–TOCSY AND 3D TOCSY–HSQC

2D NMR experiments consist of four steps: preparation, evolution, mixing, and detection. To create a third dimension we need to add another evolution step and another mixing step: preparation, evolution(1), mixing(1), evolution(2), mixing(2), and detection. The journey of the coherence pathway gets a little longer and more complicated, and we need to sample the chemical shift indirectly (evolution) at one more point in the journey. First, let's consider the extra mixing step. Any two 2D NMR experiments can be combined by just adding the mixing sequence of one experiment between the mixing and detection steps of another. Consider, for example, the 2D TOCSY and the 2D HSQC experiments (Fig. 12.43). The TOCSY mixing sequence is a series of low-power pulses (pulsed spin lock), which transfers

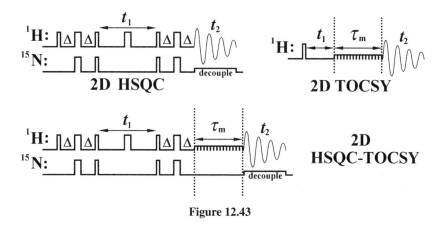

Figure 12.43

2D HSQC-TOCSY

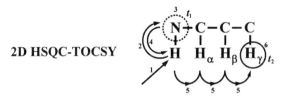

Figure 12.44

coherence within a spin system. The HSQC sequence uses an "out and back" pathway: ^{1}H SQ coherence is transferred to ^{15}N SQC, the ^{15}N chemical shift is recorded indirectly during the t_1 delay, and then ^{15}N SQC is transferred back to ^{1}H SQC, which is refocused and recorded in the t_2 FID. We can add a TOCSY transfer step at the end of the HSQC experiment to get a "relayed" transfer from ^{15}N to ^{1}H (H$_N$) and then from H$_N$ to H$_\alpha$, H$_\alpha$ to H$_\beta$, H$_\beta$ to H$_\gamma$, and so on, within an amino acid residue via the TOCSY mixing step. Because the H$_N$ magnetization is in the x–y plane at this point, it is in the correct place to be "locked" by the TOCSY spin lock and start the TOCSY mixing process. This 2D "HSQC–TOCSY" experiment (Fig. 12.43, bottom) will have the backbone ^{15}N chemical shift in the F_1 dimension of the 2D spectrum, and in the F_2 dimension on a single horizontal line we will see the H$_N$ proton (failed TOCSY transfer) and the rest of the amino acid spin system (H$_\alpha$, H$_\beta$, H$_\gamma$, etc.) due to TOCSY transfer from the H$_N$. The coherence flow diagram is shown in Figure 12.44, for observation of a γ-proton in F_2 and the backbone ^{15}N in F_1. Although H$_\gamma$ is shown as the final destination, any of the protons in the spin system can be observed in F_2 – H$_N$, H$_\alpha$, H$_\beta$, or H$_\gamma$ – because TOCSY transfer is not 100% efficient. This is a useful experiment because the spin systems are separated by ^{15}N chemical shift (F_1) rather than by the ^{1}H shift of the H$_N$ as they are in the 2D TOCSY (Fig. 12.26). While it is still just a 2D experiment, there may be an overlap of certain H$_N$ protons that can be resolved because the ^{15}N shifts are not overlapped. The right side (^{1}H signals upfield of water) is "wasted space" in the 2D HSQC because only H$_N$ protons will appear in the F_2 dimension, so the 2D HSQC-TOCSY simply uses this space to "fill out" the 2D matrix with additional useful information. The only cost is a loss of sensitivity because the intensity of the HSQC crosspeaks is now divided among a number of TOCSY crosspeaks.

Figure 12.45 shows a portion of the 2D HSQC–TOCSY spectrum of U–^{15}N Heregulin-α EGF domain in 90% H$_2$O at 30°C. The left side is the H$_N$ region, downfield of the water resonance, and the right side is the upfield region (H$_\alpha$ and side-chain resonances). The left side is just like the lower part of the 2D HSQC spectrum (Fig. 12.41), and on the right side we see the spin system of each amino acid residue spelled out along the horizontal line corresponding to the ^{15}N shift of the backbone amide nitrogen of that residue.

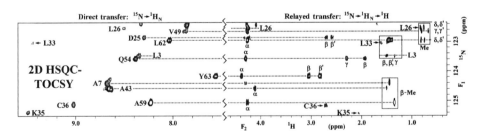

Figure 12.45

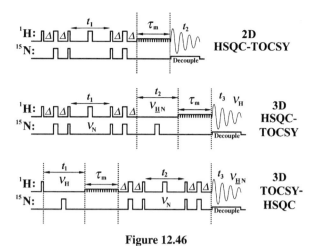

Figure 12.46

To make this into a 3D experiment we need to create a third time domain, in this case a time domain that encodes the chemical shift of the H_N proton. We simply stop for a moment on our journey from ^{15}N SQC to H_N SQC to H_α and side-chain H coherence, at the point where we have an H_N coherence, and insert an evolution delay to indirectly record the chemical shift of the H_N. The pulse sequence is shown in Figure 12.46 (center) and the coherence pathway is diagramed in Figure 12.47. The new evolution delay is called "t_2" because it is the second independent time domain, forcing us to rename the direct time domain of the FID as "t_3." In the center of the t_2 evolution delay there is a ^{15}N 180° pulse to reverse the $^1J_{NH}$ coupling evolution so that the H_N will not be split by ^{15}N in the F_2 dimension, just as the t_1 evolution delay includes a ^1H 180° pulse in the center to "decouple" the ^{15}N peak in F_1. All three time domains must be independent, meaning that for every value of t_1 we use in the first time domain we have to run the experiment and acquire an FID for *each value* of t_2 in the incremented series. If we choose to have 64 values of t_1 (for indirectly recording the FID of ^{15}N) and 128 values of t_2 (for indirectly recording the FID of ^1H_N), we will need to repeat the experiment $64 \times 128 = 8192$ times, generating 8192 separate FIDs. If each FID requires 4 scans (transients), we have in all 32768 scans that will take 18.2 h to acquire if each scan takes 2 s. You can see that we have to sacrifice a lot of digital resolution to keep the experiment this short. Many 3D experiments are run for as long as 3.5 days to get better resolution in the two indirect dimensions.

The experiment can also be run in the other sense, with the TOCSY part first and the HSQC second (Fig. 12.46, bottom). This 3D experiment is called a TOCSY-HSQC, and the coherence flow is diagramed in Figure 12.48. Coherence is created on all protons (H_N,

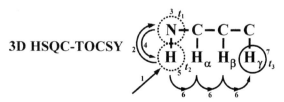

Figure 12.47

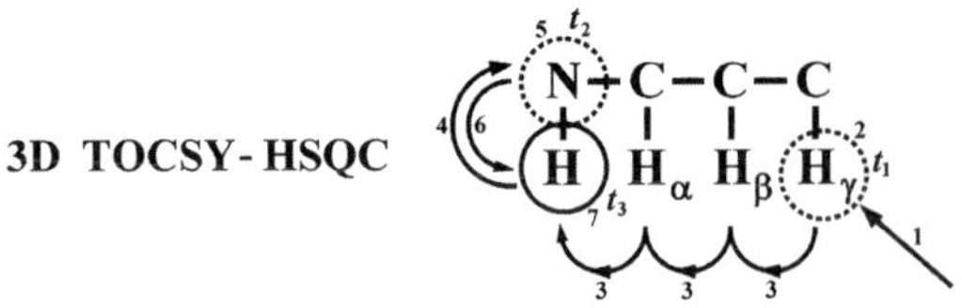

604 BIOLOGICAL NMR SPECTROSCOPY

Figure 12.48

H_α, H_β, etc.) and the ^1H chemical shift is encoded during the first evolution (t_1) delay. The TOCSY spin lock then transfers coherence to the H_N proton. H_N magnetization is in the x–y plane at the end of the spin lock, just right for beginning the HSQC sequence. Essentially, the first 90° ^1H pulse of the 2D HSQC sequence has been replaced by the 2D TOCSY sequence. In either case, TOCSY-HSQC or HSQC-TOCSY, we end up with a 3D data matrix with ^1H chemical shift on two of the axes and ^{15}N chemical shift on the third (Fig. 12.49). The floor of this data "cube" can be compared to the 2D HSQC spectrum, which correlates the ^1H shift of each H_N proton with the ^{15}N chemical shift of the nitrogen it is connected to. At the position of each HSQC crosspeak on the "floor," we extend a vertical line upwards and find the entire spin system of that amino acid residue along that "vector," including the H_N resonance itself. Each of these vertical lines is part of a strip, a narrow lane of a 2D plane parallel to the front face of the cube. These strips can be cut out of the 3D cube and we can put each one in a category (unique, three-spin, or five-spin) based on the TOCSY pattern. With sequence-specific assignments we can arrange all of these strips (one for each residue in the protein) side-by-side in order of residue number to make a strip plot of the 3D HSQC-TOCSY (or 3D TOCSY-HSQC) data (Fig. 12.49, right).

12.11.1 3D Data Processing

The mechanics of addressing the three dimensions of the cube are shown in Figure 12.50. Because of the vast amount of data in a 3D data matrix, a lower digital resolution is used compared to a 2D matrix. Whereas a typical 2D data matrix might have 2048 columns and 1024 rows (about 2 million data values), a 3D matrix might have 256 columns, 256 rows

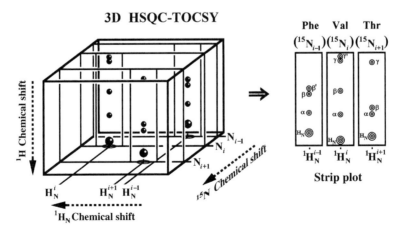

Figure 12.49

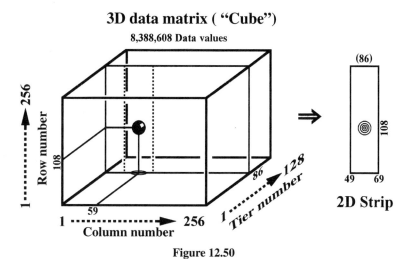

Figure 12.50

and 128 "tiers" (i.e., 128 2D planes) for a total of over 8 million data values or "intensities." Some software can actually display the entire 3D cube and manipulate it interactively, turning it in space, but in fact this is not a useful display. We always look at 2D planes and display them as contour plots, just as we do for 2D NMR data. These planes can be orientied in any of the three directions by fixing the column number (to display a vertical plane parallel to the right face of the cube), the row number (for a horizontal plane), or the tier number (for a vertical plane parallel to the front of the cube). It is also possible to run a line through the matrix in any of the three directions by specifying the chemical shifts in two of the dimensions and plotting the intensities along the third dimension as a 1D spectrum or "vector." A more useful way to display a small part of the data, however, is to plot a 2D plane while limiting one of the two dimensions to a narrow range around the crosspeak of interest. For example, the crosspeak shown in Figure 12.50 is centered at column 59, row 108, and tier 86. If we select the vertical plane defined by tier 86, we can display the full height of this 2D plane (row 1–256) but only a narrow width (column 49–69) centered on column 59. This "strip," a narrow lane of the 2D plane, can be displayed side by side in a "strip plot" with other strips cut out of the matrix in order of residue number.

The 3D Fourier transform is performed in three steps, starting with the directly detected time domain t_3. For example, for a 3D TOCSY-HSQC we might have 100 t_1 values (200 FIDs: 100 real and 100 imaginary) and 32 t_2 values (64 FIDs) for a total of 12,800 FIDs. The t_1 dimension is ^{1}H and the t_2 dimension is ^{15}N. The "acquisition order" is the order in which the evolution delays are incremented; in this example the first delay (t_1, ^{1}H evolution) is incremented first and constitutes the "inner loop." That means that we go through all 100 t_1 values first, keeping the t_2 delay fixed at the first value (usually zero). Then we repeat the whole process with the t_2 delay set to the second value, and so forth.

We also have to think about the phase sensitive detection in both indirect dimensions. For example, if the phase is encoded in States mode in t_1 and in echo–antiecho mode (using gradients) in t_2, we have to acquire four FIDs for each combination of t_1 and t_2 delay values. States mode real or imaginary is selected by using an x or y phase for the pulse just before the t_1 delay (^{1}H SQC), and echo or antiecho mode is selected by using a negative gradient (selecting **p** = +1, echo mode) or a positive gradient (selecting **p** = −1, antiecho mode) during the t_2 evolution period (^{15}N SQC). For each combination of t_1 and t_2 delays, we could acquire four FIDs in the order real-echo, imaginary-echo, real-antiecho, imaginary-antiecho. For data processing it is essential to know the acquisition order and the manner and order in which phase is encoded in the two indirect dimensions.

The first step in data processing is to do a Fourier transform on each of the raw FIDs, loading them into the appropriate rows of the 3D matrix (Fig. 12.50). For example, we might tranform all 200 FIDs in order of t_1 for the first t_2 value, loading the resulting ^{1}H spectra into rows 1–200 of the first tier of the 2D matrix. Then for the second t_2 value we would transform all 200 FIDs and load them into rows 1–200 of the second tier of the matrix. When all of the of the 64 sets of 200 are finished (12,800 Fourier transforms), we will have filled the matrix up to row 200 and back to tier 64. Then we pull out each column and do the Fourier transform in the t_1 time domain, zero filling each FID (100 complex pairs) to the size of the matrix (256 rows in this example) and discarding the imaginary spectrum after the FT. The resulting real spectrum is put back into the same column. This requires 256 columns × 64 tiers = 16,384 Fourier transforms. Finally, when all 256 columns in each of the 64 tiers containing data is transformed, we proceed to the t_2 time domain Fourier transform. Each vector (1D FID) corresponding to a specific row and column is pulled out of the matrix, zero filled from 32 complex points to the number of tiers (128), Fourier transformed and replaced in the same position. This final step requires 256 × 256 = 65,536 Fourier transforms.

Figure 12.51 shows the 2D {^{1}H, ^{15}N} HSQC spectrum of U–^{15}N *Rhodobacter capsulatus* Ferrocytochrome c_2 in 90% H$_2$O. The F_1 planes of the corresponding 3D TOCSY-HSQC spectrum are indicated on the left side of the 2D spectrum. Plane 15 (lower dotted line) corresponds to a ^{15}N chemical shift of 131.59 ppm, very close to the ^{15}N shift of the backbone nitrogen of Ala-66 (131.66). Plane 16 (upper dotted line) corresponds to

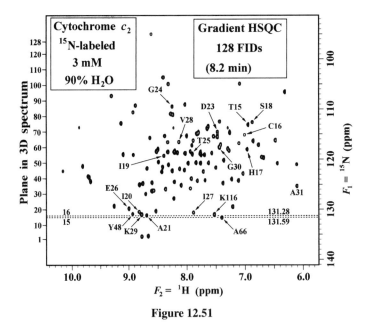

Figure 12.51

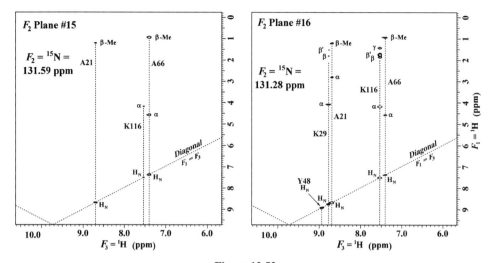

Figure 12.52

$F_1 = 131.28$ ppm and goes through the crosspeak for Ala-21 (131.21), grazing the bottoms of the Lys-116 (131.03), Lys-29 (131.09), and Tyr-48 (130.96) crosspeaks. Figure 12.52 shows F_1–F_3 planes number 15 (left, $\delta_N = 131.59$ ppm) and 16 (right, $\delta_N = 131.28$ ppm) of the 3D TOCSY-HSQC spectrum. The 3D data matrix has 512 data points (rows) along F_1 (^{1}H), 128 points (tiers) in F_2 (^{15}N) and 256 points (columns) in F_3 (the directly detected dimension, H_N). The diagonal ($F_1 = F_3$) represents coherence that failed to transfer in the TOCSY mixing step, causing the H_N resonance of each residue to appear in both F_1 and F_3. A close look at these two planes will show what kind of resolution can be achieved in the 3D spectrum. These two planes would be adjacent vertical planes in Figure 12.49, with plane 15 in front of plane 16. The Ala-66 spin system is very strong in plane 15 on the $F_3 = H_N = 7.40$ ppm vertical line, corresponding to the rightmost vertical line in Figure 12.52 (left). Weak crosspeaks are seen for the two resonances that are nearby in the ^{15}N (F_2) dimension: H_N and H_α for Lys-116 ($F_3 = 7.54$) and H_N and H_β for Ala-21 ($F_3 = 8.69$). In plane 16 (Fig. 12.52, right) the Ala-66 spin system is very weak and the Lys-116 system appears very strong, along with H_N, H_α, and H_β crosspeaks for Ala-21 and Lys-29. Only the H_N resonance is visible for Tyr-48. The really amazing thing about the 3D planes shown in Figure 12.52 is how sparse the data is—only one residue shows up in plane 15 with very faint contributions from two others. As long as the backbone amide H–N pairs are resolved (single, separated crosspeaks) in the 2D HSQC spectrum, we can expect to find the TOCSY spin system without overlap in the 3D TOCSY-HSQC along the F_1 dimension.

One way to look at these TOCSY spin systems is to draw a vertical line through the 3D data matrix at the specific column (^1H_N shift) and tier (^{15}N shift) of a particular residue and look at the intensities along the rows (^{1}H shift) as a 1D proton spectrum, corresponding to a vertical line drawn through one string of crosspeaks in Figure 12.49. This is shown for the cytochrome c_2 TOCSY-HSQC in Figure 12.53 for five of the residues. For example, at the $F_3 = H_N$ (6.88 ppm) and $F_2 = {}^{15}$N (112.61 ppm) shifts of Serine 18 we see along the F_1 dimension a 1D proton spectrum corresponding to the Ser-18 spin system: H_N at 6.88 ppm (on the $F_1 = F_3$ diagonal), H_α at 4.36 ppm, H_β at 3.75 ppm, and $H_{\beta'}$ at 3.28 ppm.

A section of the strip plot (residues 15–31) taken from the 3D TOCSY-HSQC spectrum is shown in Figure 12.54. Each strip is taken from the nearest F_2 (^{15}N) plane to the backbone

608 BIOLOGICAL NMR SPECTROSCOPY

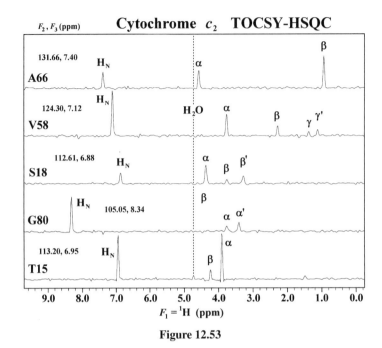

Figure 12.53

amide nitrogen of that residue (Fig. 12.49), and includes a very small range of chemical shifts near the H_N resonance in F_3 and the whole spectral width in F_1 (^1H). Any crosspeaks that are not centered on the ^1H chemical shift of the H_N in F_3 and the ^{15}N chemical shift of the backbone N in F_2 can be ignored in the strip plot, and for clarity these peaks have been removed (this would not be done in a research publication, of course!). Note that Pro-22 has a blank strip because it has no backbone H_N. The F_2 and F_3 chemical shifts used to obtain each strip are indicated; for example, T15 is found at $H_N = 6.95$ ppm and

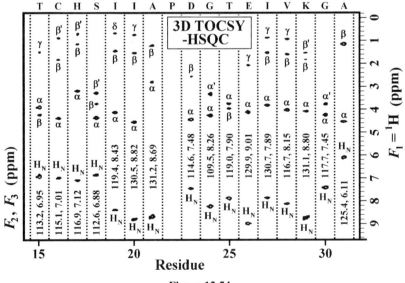

Figure 12.54

^{15}N = 113.2 ppm in the 2D HSQC (Fig. 12.51). In practice, the 3D matrix is projected onto the floor of the 3D cube (Fig. 12.49) and the point values (column and tier numbers) are read off of this low-resolution 2D HSQC spectrum for each crosspeak: T15 is centered at point numbers 189 (column = F_3 = H_N) and 74 (tier = F_2 = ^{15}N). From this position on the "floor," the spin system rises up in a vertical line along the F_1 (rows) dimension.

Figure 12.54 shows fairly complete systems for a variety of residues. The H_N shifts range from 6 to 9 ppm (bottom), and coherence is spread to the H_γ protons of valine, isoleucine, and threonine residues and even to H_δ (CH$_3$) of one isoleucine. Some unusual chemical shifts are observed for residues 16 and 17, with β-protons of Cys and His appearing in the 0–2 ppm range instead of the normal 3.0–3.5 range. This is undoubtedly due to a location above or below the heme aromatic system, which has a large ring current (see Chapter 2, Fig. 2.15). Large differences in crosspeak intensity are seen from residue to residue, with the side chains that are "flopping in the wind" (exposed to solvent on the outside of the protein) having longer T_2's and thus holding onto their coherence during the TOCSY spin lock. This is seen in Figure 12.52, where K116 shows a very strong spin system and Y48 barely shows its H_N resonance. Flexible regions in large proteins always give strong crosspeaks and the rigid portions (shorter T_2) lose a great deal of signal during the long and complicated 3D pulse sequences.

12.11.2 ^{13}C Labeling

Proteins can be expressed in a medium containing U–^{13}C glucose or U–^{13}C acetate as the sole carbon source. This is quite a bit more expensive than ^{15}N labeling because of the need for a labeled organic food source. With ^{13}C at every carbon position in the protein, the same strategies can be used to minimize overlap in NOESY spectra, and the number of protons attached to the labeled heavy atom is much greater than with ^{15}N, offering a larger number and variety of unambiguous NOEs.

The complete side-chain ^{13}C and ^{1}H assignments can be obtained by another 3D experiment that uses TOCSY mixing of ^{13}C coherence: the HCCH-TOCSY. Because ^{13}C is no longer a "dilute" nucleus, there is a continuous chain of ^{13}C nuclei in each residue: C_α, C_β, C_γ, and so on, so that an appropriate ^{13}C isotropic mixing spin lock will move ^{13}C coherence from C_α out to all the carbons of the spin system. The coherence flow is diagramed in Figure 12.55. Starting with any ^{13}C-bound proton (e.g. H_α), an evolution period (t_1) records its chemical shift, coherence transfer occurs via INEPT to its ^{13}C, followed by a ^{13}C evolution period (t_2) that measures the ^{13}C chemical shift of its directly bound carbon. The TOCSY mixing period spreads this ^{13}C coherence to another carbon in the spin system, and a final INEPT transfer moves the coherence to the ^{1}H of this last carbon (hence the name, H→C→C→H). A ^{1}H FID is recorded (t_3) to measure the ^{1}H chemical shift of

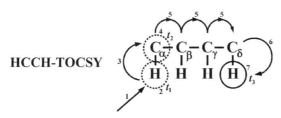

Figure 12.55

the last proton. The large $^1J_{CC}$ couplings (~33 Hz) make this a much more efficient means of coherence transfer than the H–H TOCSY transfer ($^3J_{HH}$ ~ 7 Hz).

12.11.3 Typical ^{13}C Chemical Shifts in Proteins

The α-carbon (CH), with one bond to nitrogen and a carbonyl group next door, is typically shifted downfield to the region of 50–60 ppm. Glycine's α-carbon (CH$_2$) is less substituted (less crowded) so it appears upfield of this region at around 45 ppm. Amino acids with a branched β-carbon (Thr, Val, and Ile) or a ring connecting back to the backbone nitrogen (Pro) have a more crowded environment near the α-carbon and resonate a bit downfield of 60 ppm. The β-carbon is typically found at 30–40 ppm but is more sensitive to the type of side chain. Alanine (C$_\beta$ = CH$_3$) is found around 20 ppm, whereas Threonine (C$_\beta$ = CHOH) and Serine (C$_\beta$ = CH$_2$OH) have β-carbon resonances downfield of the α-carbon, at about 65 ppm (Ser) and 70 ppm (Thr). These are in the range of typical values for CH$_2$OH (60–70) and for CHOH (70–80). Cystine (Cys with a disulfide linkage) and Leucine have β-carbon resonances a bit downfield of the typical values at 40–45 ppm. Carbonyl carbons in proteins are only found in amides (backbone and side chain) and resonate in the range 165–180 ppm, around the typical value for carboxylic acid derivatives (170–175 ppm).

12.11.4 Assignment of ^{15}N- or ^{13}C-Labeled Proteins

Strip plots can only be constructed when the crosspeaks have already been assigned in the 2D HSQC spectrum. In a ^{15}N-labeled protein, sequence-specific assignments come from sequential NOE (α,N, β,N and N,N) crosspeaks located in the 3D HSQC-NOESY spectrum. The "walk" through the protein backbone is done in the same way as with unlabeled proteins, except that overlap in NOESY spectra is greatly reduced by spreading the crosspeaks out in the ^{15}N dimension of a 3D spectrum.

Assignment by NOE interactions relies on very inefficient (a few percent at most) magnetization transfer, and even in a 3D HSQC-NOESY the assignment of each pair of protons involved in the NOE requires one unique chemical shift. For this reason as molecules get larger we are inclined to rely more on magnetization transfer based on J coupling (INEPT and TOCSY transfer) rather than NOE. Of these, the one-bond (large J coupling, short 1/(2J) delays) jumps are preferred to long-range ^2J and ^3J (smaller J coupling, longer 1/(2J) delays) relationships. This reasoning leads us to the idea that if ^{15}N labeling is good, why not replace all (or nearly all) atoms in the protein with NMR-active spin-½ nuclei? Then we could "walk" along the polypeptide backbone using only one-bond J-coupling relationships to assign the protein.

12.12 TRIPLE-RESONANCE NMR ON DOUBLY-LABELED (^{15}N, ^{13}C) PROTEINS

If a protein is produced (expressed) by bacterial growth in a medium with ^{15}NH$_4$Cl as the sole source of nitrogen *and* uniformly labeled ^{13}C-glucose or ^{13}C-acetate as the only source of carbon, it will be uniformly labeled with both ^{15}N and ^{13}C ("double-labeled"). A vast alphabet soup of NMR experiments has been developed to exploit the possibilities provided by a continuous path of high-abundance spin-½ nuclei along the peptide backbone: ^{15}N–^{13}C$_\alpha$–^{13}CO–^{15}N–^{13}C$_\alpha$–^{13}CO–and so on. These are called "triple-resonance" experiments because they use pulses on three channels: ^{1}H (e.g., 600 MHz), ^{13}C (150 MHz), and ^{15}N (60 MHz) with appropriate delays to bring about INEPT transfer along the chain. For 3D triple-resonance experiments, the three dimensions are normally the ^{1}H$_N$, ^{15}N, and ^{13}C chemical shifts, and as before we "index" the 3D data cube using the "address" of each backbone amide H–N pair: its ^{1}H$_N$ and ^{15}N chemical shifts. The third dimension then gives us a ^{13}C spectrum of carbons associated in a specific way with the H–N pair by a series of INEPT transfers.

The general strategy for coherence transfer from A to B to C is as follows: for the first coherence transfer step (A to B), we start with A coherence and have a defocusing delay of $1/(2J_{AB})$ (usually with simultaneous 180° pulses on both A and B in the center to prevent chemical-shift evolution) and then simultaneous 90° pulses on A and B to bring about the antiphase-to-antiphase INEPT transfer. Then a period of refocusing ($1/(2J_{AB})$) is allowed for getting the B coherence back into phase with respect to A, and another defocusing delay ($1/(2J_{BC})$) to get the B coherence antiphase with respect to C. Simultaneous 90° pulses on spins B and C bring about the next coherence transfer, resulting in antiphase C coherence. A, B, and C can be any combination of ^{15}N, ^{13}C, and ^{1}H. Thus the basic HSQC strategy can be extended to a number of transfers. Each evolution period is just a delay (t_1 or t_2) with 180° pulses in the middle for all nuclei that are coupled to the one whose chemical shift is being encoded during that period. Sometimes the 180° pulse in the middle is replaced by continuous decoupling during the whole evolution period. The 180° pulses (or decoupling) prevent the J coupling from showing up in the frequency domain (F_1 or F_2) associated with that evolution delay (t_1 or t_2).

Because of the large gaps in ^{13}C chemical shifts between the aliphatic (10–70 ppm) and carbonyl (160–180 ppm) chemical shifts, one can treat these categories as separate "channels" by using selective ^{13}C pulses (shaped pulses or low-power rectangular pulses) that excite only one range (or "band") of ^{13}C shifts. Thus most triple-resonance experiments are diagramed with four channels: ^{1}H, ^{15}N, ^{13}C$_{\alpha,\beta}$, and ^{13}C' (carbonyl), even though there are only three hardware channels.

12.12.1 3D HNCO

To illustrate how a complex triple-resonance pulse sequence can be understood relatively easily using the concepts developed in this book, consider the HNCO experiment. The H$_N$ and ^{15}N chemical shifts of one residue are correlated with the chemical shift of the carbonyl group of the previous residue (Fig. 12.56). This is accomplished by a simple out-and-back scheme of coherence transfer (Fig. 12.57): ^{1}H$_N$ → ^{15}N → ^{13}C' (t_1) → ^{15}N (t_2) → ^{1}H$_N$ (t_3) (here we use C' for the backbone carbonyl carbon). The pulse sequence is shown in Figure 12.58, with the dotted line showing the path of coherence transfer from H$_N$ to N to C' and back again and the letters a–j referring to specific times within the sequence. Separate channels are shown for C$_\alpha$ and C' even though there is only one ^{13}C hardware channel. Rectangular C' 90° pulses can be applied with frequency at the center of the carbonyl region (175 ppm) and with low power (e.g., $\gamma_C B_1/2\pi = 4$ kHz, 62.5 μs 90° pulse) so that the sinc-shaped excitation profile has a null in the C$_\alpha$ region. The single 180° pulse on the C$_\alpha$ "channel" is applied with frequency at the center of the C$_\alpha$ region (54 ppm) and with low power (e.g., $\gamma_C B_1/2\pi = 9$ kHz, 55.6 μs 180° pulse) to minimize excitation in the C' region. In the following discussion, **N** and **C** will be used instead of **S** to indicate ^{15}N and ^{13}C' product operators, respectively.

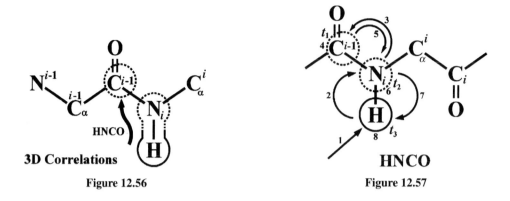

Figure 12.56

Figure 12.57

a to b. First, a proton 90° pulse creates ^{1}H coherence ($\mathbf{I^+}$). This is step 1 of Figure 12.57. During the 2Δ delay ($1/(2J_{NH}) = 1/(2 \times 90 \text{ Hz}) = 5.6$ ms) we have evolution of the J_{NH} coupling until the proton coherence is antiphase ($\mathbf{I^-N_z}$). The simultaneous 180° pulses on ^{1}H and ^{15}N in the center of this delay prevent ^{1}H chemical-shift evolution.

b to c. Simultaneous 90° pulses on ^{1}H and ^{15}N bring about coherence transfer (INEPT transfer) from antiphase ^{1}H to antiphase ^{15}N coherence ($\mathbf{I^-N_z} \to \mathbf{N^-I_z}$). This is step 2 of Figure 12.57.

c to d. During the 2Δ ($1/(2J_{NH}) = 5.6$ ms) delay, refocusing occurs with respect to ^{1}H coupling ($\mathbf{N^-I_z} \to \mathbf{N^-}$), and during the 2τ ($1/(2J_{NC'}) = 1/(2 \times 15 \text{ Hz}) = 33.3$ ms) delay defocusing occurs with respect to C' coupling ($\mathbf{N^-} \to \mathbf{N^+C_z}$). The simultaneous ^{15}N and C' 180° pulses at the center of the 2τ delay prevent ^{15}N chemical-shift evolution while still allowing J-coupling evolution with respect to C'. In order to save time (proteins have short T_2 so time is critical), the refocusing with respect to ^{1}H occurs simultaneously

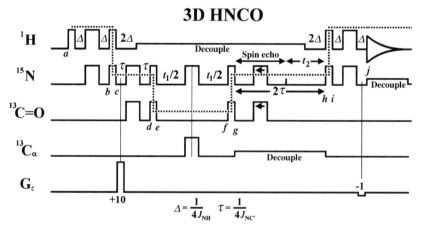

Figure 12.58

with the defocusing with respect to C′. Because $2\Delta < \tau$ (the drawing is not to scale), the ^{1}H refocusing is complete before the 180° pulses. After the 2Δ delay proton decoupling is started, so that no ^{1}H coupling will be active until we reach a point 2Δ before time h.

d to e. Simultaneous 90° pulses on ^{15}N and ^{13}C′ bring about coherence transfer (INEPT transfer) from antiphase ^{15}N to antiphase ^{13}C′ ($\mathbf{N^+C_z \rightarrow C^+N_z}$). This is step 3 of Figure 12.57.

e to f. C′ evolution: During the t_1 delay, C′ single quantum coherence rotates in the x'–y' plane at a rate determined by its rotating frame chemical shift ($\Omega_{C'}$), introducing a factor of $\exp(-i\Omega_{C'}t_1)$ into the coherence ($\mathbf{C^+N_z \rightarrow C^+N_z} \exp(-i\Omega_{C'}t_1)$). This encoded chemical shift will be decoded during data processing (3D Fourier transform). The simultaneous ^{15}N and C_α 180° pulses in the center of the evolution period refocus J coupling evolution with respect to ^{15}N ($^1J_{NC'}$) or C_α ($^1J_{C\alpha C'}$) so that only chemical-shift evolution occurs and these couplings do not appear in the F_1 dimension. This first evolution step corresponds to step 4 in Figure 12.57.

f to g. Simultaneous 90° pulses on ^{15}N and ^{13}C′ bring about coherence transfer (INEPT transfer) back from antiphase ^{13}C′ to antiphase ^{15}N ($\mathbf{C^+N_z \rightarrow N^+C_z}$, leaving out the $\exp(-i\Omega_{C'}t_1)$ term). This is step 5 in Figure 12.57.

g to h. ^{15}N SQC evolves during the t_2 period, encoding the backbone ^{15}N chemical shift in the second indirect time dimension. This is carried out in a different way, including a spin echo with a "moving" pair of 180° pulses. This "constant time" evolution will be discussed further below; for now we just need to know that ^{15}N is allowed to refocus with respect to C′, its chemical shift is encoded, and it is allowed to defocus (evolve into antiphase) with respect to its ^1H_N partner. Continuous $^{13}C_\alpha$ decoupling during the entire constant time period (2τ) prevents any J coupling with respect to C_α ($^1J_{NC\alpha}$) from appearing in the F_2 dimension, and a 180° pulse on ^{13}C′ in the center of the active t_2 evolution period "decouples" the $^1J_{NC'}$ coupling in F_2. This ^{15}N evolution part corresponds to step 6 in Figure 12.57.

h to i. Simultaneous 90° pulses on ^{15}N and ^{1}H bring about coherence transfer (INEPT transfer) back from antiphase ^{15}N to antiphase ^{1}H ($\mathbf{N^-I_z \rightarrow I^+N_z}$, leaving out the $\exp(-i\Omega_{C'}t_1 + i\Omega_N t_2)$ term). This is step 7 in Figure 12.57.

i to j. A 2Δ ($1/(2J_{NH})$) delay refocuses ^{1}H coherence with respect to coupling with ^{15}N, with simultaneous 180° pulses on ^{1}H and ^{15}N to prevent ^{1}H chemical-shift evolution ($\mathbf{I^+N_z \rightarrow I^-}$).

Detection. The ^{1}H FID of H_N (starting with $\mathbf{I^-}$) is recorded with decoupling of ^{15}N ($^1J_{NH}$ = 90 Hz). The C′ chemical shift ($\exp(-i\Omega_{C'}t_1)$) and ^{15}N chemical shift ($\exp(i\Omega_N t_2)$) terms are "decoded" in the 3D Fourier transform to display a crosspeak at $F_1 = \delta(C'_{i-1})$, $F_2 = \delta(N_i)$ and $F_3 = \delta(H_{N_i})$. This is step 8 in Figure 12.57. Gradient selection of the coherence pathway is provided by two gradient pulses, the first while we have ^{15}N SQC and the second while we have ^{1}H SQC. Using a definition of coherence order that includes the γ, we would have $\mathbf{p} = 1$ for ^{15}N and $\mathbf{p} = 10$ for ^{1}H, so that $\mathbf{p} = -1$ for $\mathbf{N^-I_z}$ (immediately after point c) and

p = −10 for $\mathbf{I^-N_z/I^-}$ just before point j. Thus the total "twist" is:

$$\sum \mathbf{p}_i G_i = -1(10) + (-10)(-1) = 0$$

for the desired coherence pathway.

12.12.2 Constant-Time (CT) Evolution

This is a common trick in modern triple-resonance pulse sequences. In order to save time (and minimize T_2 relaxation), evolution occurs within a constant time period corresponding to the $1/(2J)$ delay required for refocusing after the previous coherence transfer step (in this case C′ → N). During this constant delay period of $2\tau = 1/(2J_{NC'}) = 33.3$ ms, J-coupling evolution with repect to C′ refocuses antiphase ^{15}N coherence to in-phase ($\mathbf{N^+C_z \to N^+}$) while simultaneous 180° pulses on ^{15}N and C′ create a spin echo for part of this time. The spin-echo blocks ^{15}N chemical-shift evolution while allowing $^1J_{NC'}$ evolution, so that only in the remaining time t_2 of the constant 2τ delay does ^{15}N chemical-shift evolution occur. The pair of 180° pulses are in the center of the 2τ delay at the start of the experiment ($t_2 = 0$) and for each increment of t_2 these pulses are "moved" to the left within the constant-time period, shortening the spin echo and "exposing" at the right side a larger t_2 period for ^{15}N evolution. Since $^1J_{NC'}$ is active during the whole constant-time period, we have full refocusing of antiphase ^{15}N coherence regardless of the t_2 value ($\mathbf{N^+C_z \to N^-} \exp(i\Omega_N t_2)$). The ^{1}H decoupling is turned off 2Δ ($1/(2J_{NH}) = 5.6$ ms) before the end of the constant-time period, allowing J coupling with repect to ^{1}H so that the ^{15}N coherence ends up antiphase with respect to ^{1}H (overall $\mathbf{N^+C_z \to N^-I_z} \exp(i\Omega_N t_2)$). Again, because $2\Delta < \tau$, the two 180° pulses do not interfere with this defocusing. This is amazingly efficient: within a single constant time period of 2τ we refocus with respect to C′, encode the ^{15}N shift in t_2 and defocus with respect to H_N in preparation for the next coherence transfer step.

The 3D HNCO data matrix is very sparsely populated with crosspeaks because there is only one correlation per residue in the protein (Fig. 12.59). The main purpose of this experiment is to count residues and make sure all of the peaks can be found and identified. Once we have the assignments for each H–N pair, the data can be arranged in a strip plot in order of residue number.

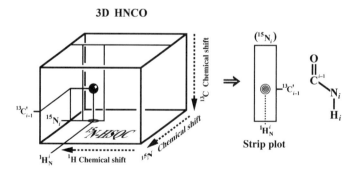

Figure 12.59

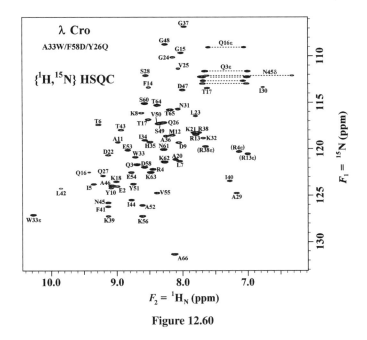

Figure 12.60

We will use a doubly labeled (^{15}N, ^{13}C) sample of a triple mutant of the 66 residue Cro protein from bacteriophage λ to illustrate the triple-resonance experiments and the strategy for sequence-specific assignment without NOE experiments. The 2D {^{1}H, ^{15}N} HSQC spectrum, the road map for analyzing all triple-resonance experiments, is shown in Figure 12.60. One crosspeak is observed for each backbone amide (H$_N$–N) pair (Met-1, Pro-57, and Pro-59 lack a backbone amide H–N). In addition, pairs of crosspeaks are observed for each of the Gln and Asn side-chain NH$_2$ groups (e.g., N45, Q3, and Q16 in the upper right side), and crosspeaks are also observed for the H–N pairs of Arg, Trp, and His side chains (e.g., R4ε, R13ε, and R38ε, which are aliased from the region of F_1 = ~85 ppm). To locate the crosspeaks in the 3D HNCO data matrix (Fig. 12.59), the data in all of the horizontal planes is summed and projected onto the "floor" to give a low-resolution {^{1}H, ^{15}N} 2D HSQC spectrum similar in appearance to the 2D HSQC shown in Figure 12.60. The column (H$_N$) and tier (^{15}N) number is recorded for each resolved crosspeak and this "address" is used to cut out a strip from the 3D data matrix (Fig. 12.59). Once the assignments are obtained, these strips can be lined up in order of residue number, but for now they are just counted to make sure we have one for each backbone H–N pair in the protein.

Figure 12.61 shows a strip plot of the HNCO spectrum of the λ-Cro protein for residues 34–48. The point values in the 3D matrix used to extract each strip are shown at the bottom of the strip. For example, the Ile-40 H–N pair is found in the matrix at column 321 (F_3 = H$_N$ = 7.28 ppm) and tier 111 (F_2 = ^{15}N = 123.50 ppm) and shows a crosspeak for the C′ (backbone carbonyl) carbon of Lys-39 at F_1 = ^{13}C′ = 171.30 ppm.

12.12.3 3D HNCA and HN(CO)CA

One of the earliest triple-resonance experiments is the HNCA, which correlates the chemical shifts of H$_N$(i), ^{15}N(i), and either ^{13}C$_\alpha$(i) (strong crosspeak, Fig. 12.62 major path) or

616 BIOLOGICAL NMR SPECTROSCOPY

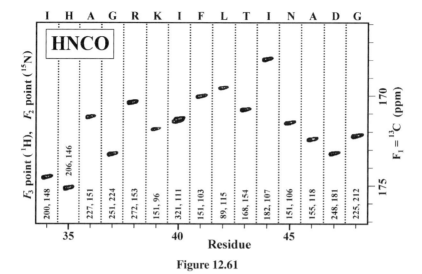

Figure 12.61

$^{13}C_\alpha(i-1)$ (weak crosspeak, Figure 12.62 minor path). The experiment uses the same out-and-back strategy (Fig. 12.63) starting with ^1H_N coherence, with INEPT transfer to ^{15}N and then either to $^{13}C_\alpha(i)$ via the one-bond coupling $^1J_{NC\alpha}$ (major path, $J = 7$–11 Hz) or to $^{13}C_\alpha(i-1)$ via the two-bond coupling $^2J_{NC\alpha}$ (minor path, $J = 4$–9 Hz). A t_1 evolution period records the ^{13}C chemical shift and then INEPT transfer moves the coherence back to ^{15}N, where the ^{15}N shift is recorded in the t_2 evolution delay. Finally, coherence is transferred back to ^1H_N for recording of the FID (t_3). If one extends a 1D vector (line) through the cube at the "address" of the unique $F_3 = H_N^i$, $F_2 = {}^{15}N_i$ location, extending along the F_1 (^{13}C) dimension, there will be a strong crosspeak at the chemical shift of $C_\alpha(i)$ and a weak crosspeak at the chemical shift of $C_\alpha(i-1)$ due to the minor pathway (steps 3, 4, and 5 using

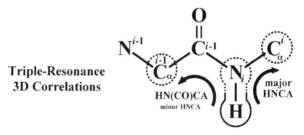

Figure 12.62

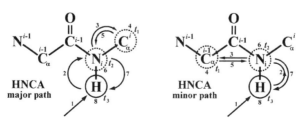

Figure 12.63

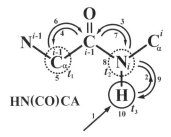

Figure 12.64

the smaller J value). Knowing the chemical shift of $C_\alpha(i)$, we can look for a weak crosspeak at this chemical shift at a different $F_3 = H_N$, $F_2 = {}^{15}N$ vector, and we will have the ${}^1H_N$, ${}^{15}N$ pair from residue $(i + 1)$. In this way we can "walk" through the backbone (until we encounter a proline) using only coherence transfer (${}^1J_{NH}$, ${}^1J_{NC\alpha}$, and ${}^2J_{NC\alpha}$) steps.

If the hardware is available to distinguish the C_α region (45–75 ppm) from the carbonyl region (165–180 ppm) of the ${}^{13}C$ spectrum, we can treat these as separate radio frequency "channels" using band-selective pulses that excite only one region or the other. This allows us to take advantage of the much larger coupling constants ${}^1J_{C_\alpha C'}$ (~55 Hz) and ${}^1J_{NC'}$ (~15 Hz), so that the defocusing and refocusing delays are shorter (here we use C' for the backbone carbonyl carbon). This experiment goes out and back using only one-bond jumps: $H_N(i)$ to $N(i)$ to $CO(i - 1)$ to $C_\alpha(i - 1)$ and back again (Fig. 12.64). The chemical shift of the carbonyl carbon is not recorded in an evolution period, so it is shown in parentheses in the experiment name: HN(CO)CA. This experiment correlates H_N to the C_α of the previous residue in an unambiguous way (Fig. 12.62), so it complements the HNCA experiment, which primarily correlates H_N with the C_α of its own residue. Using these two experiments, the assignments (sequence-specific chemical shifts of all of the backbone H_N, N, and C_α) can be obtained without using NOEs for a protein of up to 25 kD.

12.12.4 Adding the β-Carbon: HNCACB and CBCA(CO)NH

In a large protein it is dangerous to rely on a single chemical shift (e.g., C_α) shared by two H–N pairs to make a sequential assignment, especially since C_α shift does not vary that much among the 20 amino acid types. If we could record the ${}^{13}C$ chemical shifts of *both* the C_α and C_β carbons for each residue, the "fingerprint" would be much more likely to be unique. The range of C_β shifts (20–70 ppm) is larger than the range of C_α shifts (45–75 ppm), and having two shifts makes it much less likely to have overlap of both. This

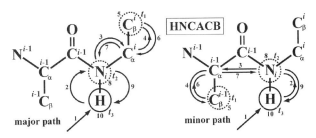

Figure 12.65

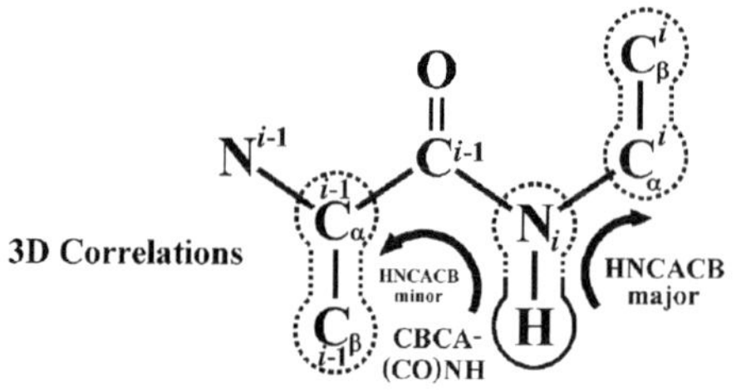

Figure 12.66

is accomplished by adding a simple homonuclear INEPT (i.e., COSY) transfer between the C_α and C_β carbons before and after recording the ^{13}C shift (t_1) and before transferring coherence from C_α to N. The experiment exactly analogous to HNCA is called HNCACB (Fig. 12.65). From a given H–N pair, strong correlations are observed to the C_α and C_β of the same residue, and weak correlations are observed to the C_α and C_β of the previous residue. Figure 12.65 shows the pathways for observing C_β in the F_1 dimension, but it could just as well be C_α since the homonuclear INEPT transfer steps (4 and 6) are not directional; they simply mix the C_α and C_β coherences. The correlations are shown in Figure 12.66: at the address of a particular H–N pair we see strong crosspeaks to the C_α and C_β resonances of the same residue (intraresidue correlations) and weak crosspeaks to the C_α and C_β resonances of the previous residue (interresidue or sequential correlations). If we collect all of the strips from the 3D matrix (Fig. 12.67), we can match up the weak correlations in each strip with the strong correlations of another strip to establish sequential connectivity. We have the additional advantage that C_β shifts will stand out for Ala residues (~20 ppm) and for Ser and Thr residues (~65 and ~70 ppm), allowing us to quickly locate a unique series in the amino acid sequence. The Glycine residues will also stand out due to the lack of a C_β resonance.

Figure 12.68 (top spectrum, lower trace) shows the 1D vector taken from the H_N and ^{15}N "address" of Ala-66 in the 3D HNCACB data matrix for λ-Cro protein. In this HNCACB experiment the crosspeaks are edited (*cf.* Chapter 11, edited HSQC) so that the C_β resonances (which had two COSY transfers, steps 4 and 6 in Fig. 12.65) have negative intensity and the C_α resonances (for which these two transfers "failed") have positive intensity. This 1D vector corresponds to the vertical dotted line passing through the crosspeaks in Figure 12.67. In the 1D vector, strong peaks are seen for the C_α of A66 (positive peak at 54.1 ppm) and the C_β of A66 (negative peak at 20.2 ppm) and weaker peaks are seen for the C_α of T65 (positive peak at 61.7 ppm) and the C_β of T65 (negative peak at 70.2 ppm). If we were in

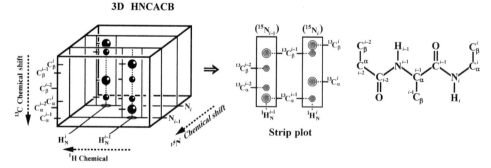

Figure 12.67

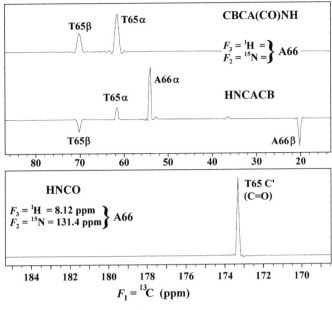

Figure 12.68

the process of assigning the protein, we would know that this vector represents an Alanine residue with a preceding Serine or Threonine (C_β downfield of C_α). In the sequence this could be T19-A20, S28-A29, or T65-A66. To resolve the ambiguity we could search out a vector with major peak chemical shifts of $C_\alpha = 61.7$ (positive) and $C_\beta = 70.2$ (negative), and look for its minor C_α and C_β peaks. These could represent either K18, Q27, or T64. From the chemical shifts we would easily identify T64.

Once we have sequence-specific assignments, the strips from the HNCACB matrix can be arranged in order of residue number. Figure 12.69 (right) shows the strip plot for λ-Cro

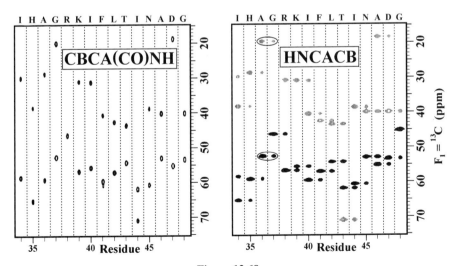

Figure 12.69

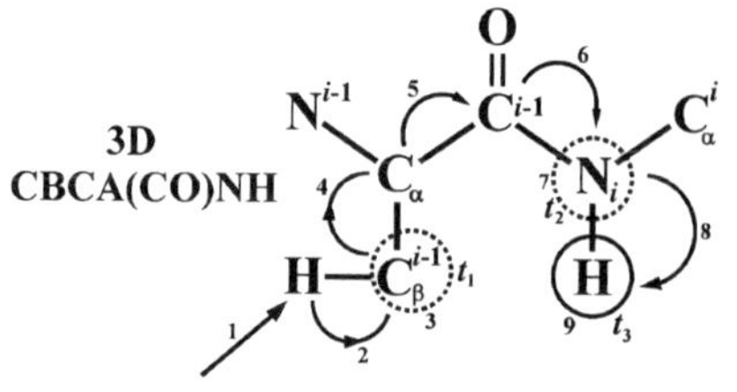

Figure 12.70

protein residues 34–48. Negative peaks are shown in gray and positive peaks in black. Note that the crosspeaks appear in pairs, with the more intense peak on the left side of each pair. In each strip, the more intense peaks represent the C_α (positive) and C_β (negative) resonances for that residue, and the weaker peaks represent the C_α and C_β resonances for the previous residue. From these weak crosspeaks we can move to the left along a horizontal line to the previous strip and see the major (strong) crosspeak for that strip. For example, the weak positive peak in the Gly-37 strip (C_α-36, 52.6 ppm) matches the strong positive peak in the Ala-36 strip (Fig. 12.69, right side, lower ellipse). Likewise the weak negative peak in the Gly-37 strip (C_β-36, 19.7 ppm) is exactly aligned with the strong negative peak in the Ala-36 strip (upper ellipse). The Glycine strips lack a strong negative peak (intraresidue C_β) and the strips following Glycines (e.g., R38) lack a weak negative peak (interresidue C_β) because Glycine has no β-carbon. The Alanines stand out (C_β's at 19.7 and 18.5 ppm) as does the Threonine (C_β at 71 ppm). Both Gly-37 and Gly-48 have only one negative (C_β) peak, the weak one "carried over" from the previous residue. The only α-carbon resonances below 60 ppm are I34, T43, and I44, all with branched β-carbons.

In some cases the distinction between "strong" and "weak" peaks may be ambiguous, or the "weak" peaks may fall into the noise. We need an experiment analogous to the HN(CO)CA that uses an unambigous path through the C' carbon to the C_α and C_β carbons of the previous residue. There are two such experiments: the out-and-back experiment directly analogous to HN(CO)CA is called HN(CO)CACB, and another experiment, called CBCA(CO)NH, starts with the H_α and H_β protons and moves in one direction to the H_N of the next residue. The simple out-and-back extension of the HN(CO)CA would involve 12 steps, 8 of them INEPT transfers! The alternative one-way ticket (H_α and $H_\beta \rightarrow C_\alpha$ and $C_\beta \rightarrow C_\alpha \rightarrow C' \rightarrow N \rightarrow H_N$) is only 9 steps (CBCA(CO)NH, Fig. 12.70). The path is unambiguous because it involves only one-bond INEPT transfers (^1J) and passes through the carbonyl (CO) carbon on its way to the previous residue. The experiment correlates each H–N pair to the C_α and C_β of the *previous residue only* (Fig. 12.66). Each vector (vertical line) in the 3D data matrix corresponding to a backbone H–N pair passes through only two crosspeaks (one if the previous residue is Glycine) representing the ^{13}C chemical shifts of the previous residue's C_α and C_β carbons (Fig. 12.71).

Figure 12.68 (top spectrum, upper trace) shows the 1D vector from the CBCA(CO)NH 3D data matrix corresponding to the Ala-66 H-N pair. There are only two crosspeaks, corresponding to the C_α and C_β carbon resonances of the previous residue, Thr-65. In this version there is no editing, so both peaks appear with positive intensity. Also in Figure 12.68 (bottom spectrum) is the corresponding vector from the HNCO 3D matrix, which has only one crosspeak corresponding to the C' (carbonyl) carbon resonance of Thr-65. The CBCA(CO)NH strip plot for λ-Cro protein residues 34–48 is shown in Figure 12.69 (left). Comparison to the HNCACB strip plot (Fig. 12.69, right) clearly shows that the crosspeaks found in each

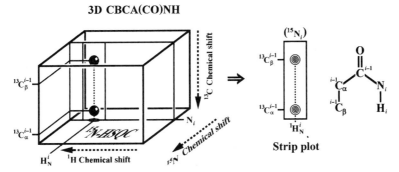

Figure 12.71

strip of the CBCA(CO)NH correspond to the minor crosspeaks (coming from the previous residue) of the HNCACB strip. These crosspeaks, weak in the HNCACB, are now strong and can be unambiguously identified as coming from the previous residue's C_α and C_β resonances. Working with these two datasets, the HNCACB and the CBCA(CO)NH, it is relatively straightforward to assign a double-labeled protein up to about 25–30 kD. The HNCO dataset completes the backbone assignments by adding the carbonyl (C') chemical shifts.

12.13 NEW TECHNIQUES FOR PROTEIN NMR: RESIDUAL DIPOLAR COUPLINGS AND TRANSVERSE RELAXATION OPTIMIZED SPECTROSCOPY (TROSY)

12.13.1 Residual Dipolar Couplings

Methods for biological NMR are constantly extending the limits of size and complexity of proteins that can be studied, as well as simplifying the process of structure determination. Until recently, determination of the 3D structure of proteins was dependant exclusively on measurements of NOE distances and dihedral angles, relationships that are very short-range (5 Å and three bonds, respectively). As we move from one part of a protein to another, the errors in these local relationships begin to accumulate so that there may be serious overall errors in the relationship of one end of a molecule to the other, especially if it is elongated. To overcome this, we would like to be able to orient each part of the molecule relative to an absolute frame of reference such as the B_o field. Consider, for example, an ^{15}N–^{1}H pair in a ^{15}N-labeled protein (Fig. 12.72). We saw in Chapter 5, Section 5.7, that the nuclear magnet

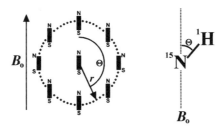

Figure 12.72

of one spin modifies the B_o field at the position of a nearby spin in a through-space effect called the dipole–dipole interaction. The sign of the effect on B_{eff} (increased or decreased) depends on the orientation of the spin: α or β. The direct or dipolar coupling between a ^{15}N and its directly bound ^{1}H is very large, on the order of tens of kilohertz, and it depends on the orientation of the N–H vector with respect to the B_o field:

$$D = D_o(1 - 3\cos^2\Theta)$$

where Θ is the angle between the N–H vector and the B_o field and D_o depends on the distance (fixed by the single bond length) and the product $\gamma_H\gamma_N$. Of course, in solution the protein is rapidly tumbling (reorienting) and samples all possible orientations in 3D space equally ("isotropic" tumbling) so that the dipolar coupling is the average of D over all these orientations. It turns out that the average value of $\cos^2\Theta$ over the surface of an evenly populated sphere of vector orientations is exactly 1/3, so that the dipolar coupling averages to zero and the only coupling left is the indirect or scalar coupling, J, which is about 90 Hz.

This is good because all of those enormous (~10 kHz) couplings would really mess up the spectrum, but it would be nice if just a tiny fraction of the dipolar coupling could be "brought back" so that we could measure the angle Θ and orient each of the N–H vectors in the protein with respect to the absolute laboratory frame. This would give us a third and fundamentally different kind of NMR restraint based on reference to the fixed B_o direction rather than the relative relationships of NOE distances and dihedral angles. What we would like to do is to skew the molecule's tumbling so that there is a very slight preference for a particular orientation with respect to the B_o field direction, while keeping the tumbling rapid so that T_2 remains relatively long. A number of methods have been developed to do this. A liquid crystal solution can be used, made up of "coin-shaped" lipid bilayers called "bicelles" (Fig. 12.73, left). These orient in the magnetic field below a critical transition temperature near room temperature, and are randomly oriented above that temperature. The protein, unless it is perfectly spherical, tends to line up with the oriented bicelles simply because of the restricted space it has available in solution. This is easier to visualize if the protein is elongated, so that the long axis will tend to align with the long axis of the bicelles, which is the direction of the B_o field (Fig. 12.74). Another method is to put the protein

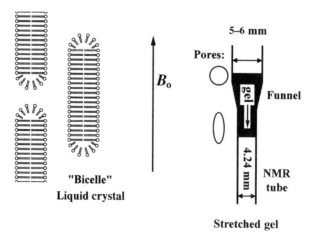

Figure 12.73

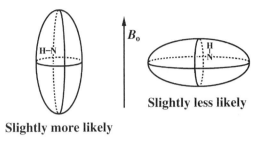

Figure 12.74

solution in a polymerized gel, such as SDS-polyacrylamide, which has tiny spherical voids containing the protein solution. If the gel is stretched along the direction of the NMR tube and inserted into the tube (Fig. 12.73, right), the spherical voids are stretched and become elongated along the z axis. The protein, again assuming it is elongated along one axis, has a slight tendency to put its longer axis parallel to the long axis of the voids. In either case, the tendency to align may be only a small fraction of 1%, but this is enough to make the average of $(1-3\cos^2\Theta)$ deviate from zero so that the *residual* dipolar coupling (RDC) is on the order of a few hertz. The exact value of the RDC gives us an idea of the orientation of the N–H vector relative to the alignment axis (the long axis in Fig. 12.74) of the protein. The RDCs are measured by recording a 2D $\{^1H,^{15}N\}$ HSQC spectrum with the central ^1H 180° pulse (in the center of the t_1 evolution delay) removed so that the $^1J_{NH}$ coupling appears in the F_1 dimension as a doublet (Fig. 12.75). Each $^1J_{NH}$ value can be read by the vertical separation in the pair of crosspeaks that appears for each residue in the protein. If the experiment is performed twice, once without orientation (Fig. 12.75, center: above the liquid crystal melting point or in the absence of the stretched gel) and then again with the slight orientation (Fig. 12.75, right), the difference in the measured $^1J_{NH}$ value is the residual dipolar coupling, which can be either negative (decreased J value) or positive (increased J value). The challenge is to determine the orientation axis (basically, the long axis) of the protein, a complex calculation involving all of the measured RDCs. Then limits can be placed on the individual Θ angles for each N–H vector relative to this orientation axis. This technique is still very new and the calculation methods are not standardized, but the power to "refine" 3D protein structures using these absolute orientational restraints has been clearly demonstrated. There have even been some structures determined with RDCs alone, without any local dihedral or NOE restraints. This technique borrowed from the field of solid-state NMR is blurring the boundaries between the solution-state and solid-state NMR fields, just as pulsed field gradients have borrowed from NMR imaging (MRI) technology.

12.13.2 TROSY

Two problems limit the size of proteins that can be studied: the complexity problem (more and more nuclei that must be identified uniquely by a fixed range or dispersion of chemical shifts) and the linewidth problem (decreasing T_2 due to slower molecular tumbling in solution leads to broader NMR peaks). The complexity problem is dealt with by adding NMR-active nuclei and using 3D and 4D experiments, and occasionally by the good fortune of having molecular symmetry (symmetrical dimers, trimers, etc.). The linewidth problem can be dealt with by replacing all but the N–H protons with deuterium (^2H), which has one seventh of the γ of ^{1}H and therefore is less effective at relaxing (reducing the T_2) of nearby nuclei, especially directly bonded ^{13}C nuclei. This can be accomplished by replacing all of the hydrogen in the growth medium with deuterium. It drastically reduces the number of ^{1}H nuclei that can be studied, resulting in far fewer NOE restraints, and it requires a hardware capability of ^{2}H decoupling in a way that will not interfere with the ^{2}H lock channel. Another even more promising approach to the linewidth problem is a new NMR experiment called TROSY (transverse relaxation optimized spectroscopy). The core experiment is a modified $\{^1\text{H},^{15}\text{N}\}$ HSQC in which no decoupling is done in *either* the F_1 or the F_2 dimensions. This is accomplished in F_1 by omitting the ^{1}H 180° pulse in the center of the t_1 evolution delay (as we did for RDC measurement) and in F_2 by turning off the ^{15}N decoupling during acquisition of the ^{1}H FID. This turns each crosspeak, which represents a single amino acid residue, into a "square" of four crosspeaks.

If we look at the 1D ^{1}H and ^{15}N spectra, we see that one component of the doublet is broader and one is sharper. This is the result of the interaction of two separate mechanisms of T_2 relaxation: dipole–dipole relaxation, in which the proton's magnetic field modulates the field experienced by the nitrogen as the molecule tumbles, and chemical-shift anisotropy (CSA) relaxation, in which the field experienced by the nitrogen is slightly modified by the circulation of electrons around the nucleus. Because the ^{15}N chemical shift at any instant in time is dependent on the orientation of the molecule ("chemical-shift anisotropy," Chapter 2, Section 2.6) with respect to the B_0 field direction, the chemical-shift perturbation of the field experienced by the nitrogen oscillates as the molecule tumbles. Thus we have two perturbations of the magnetic field experienced by the ^{15}N nucleus in an N–H pair: dipole–dipole from the ^{1}H, which has a fixed magnitude determined by the bond distance and the product $\gamma_H\gamma_N$, but a sign that depends on the spin state of the ^{1}H nucleus (α or β state); and CSA from the electrons around the ^{15}N nucleus, which has a magnitude proportional to B_0 and a sign that is independent of the proton spin state (Fig. 12.76). Both of these perturbations are the direct result of the molecular tumbling, so for an individual N–H pair in one molecule they have a frequency equal to the tumbling rate. If the molecule tumbles at the zero-quantum frequency ($\nu_H-\nu_N$) we have effectively a radio frequency oscillation of the magnetic field that can induce spin exchange ($\alpha\beta \leftrightarrow \beta\alpha$) and bring about T_2 relaxation. If we look at the contributions of dipole–dipole and CSA to the effective field experienced by the ^{15}N nucleus in a particular tumbling molecule as a function of time, we see that for one of the components of the ^{15}N doublet (the ^{1}H = β component in this example), the two effects are opposite in sign and tend to cancel each other (Fig. 12.76). This leads to a reduction in ZQ transitions and a longer T_2 (sharper peak). For the other component of the doublet, the

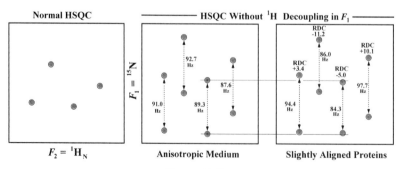

Figure 12.75

NEW TECHNIQUES FOR PROTEIN NMR   625

Effect of molecular tumbling on field at ^{15}N

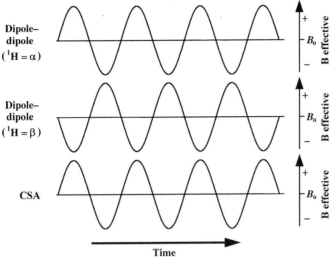

Figure 12.76

two effects add together to give an increased rate of ZQ transitions and a shorter T_2 (broader peak). The same thing happens to the ^{1}H nucleus when we examine the effect of the ^{15}N dipole–dipole relaxation and the CSA relaxation of the ^{1}H, so that one component of the doublet is broad and one is narrow. The amplitude of the CSA perturbation is proportional to B_o, so there is, in principle, a particular field strength where the CSA exactly balances the dipole–dipole perturbation and we have essentially no T_2 relaxation for one component of the doublet ("immortal coherence"). Of course the balance is never quite perfect, and there are other relaxation mechanisms besides dipole–dipole and CSA, but the result would be a very sharp peak. This optimal field is predicted to be around 1000 MHz ^{1}H frequency (23.5 T), a field that has not yet been reached in commercial spectrometers, although 900 MHz is available. Thus the "TROSY principle" gives an impetus for pushing the magnet technology to the 1 GHz level and beyond.

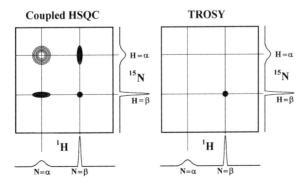

Figure 12.77

Even at 600 or 800 MHz (14.1 or 18.8 T), for very large proteins, if we "turn off" decoupling in both dimensions of an HSQC experiment one of the four components of each crosspeak will be a very sharp peak, with good signal-to-noise ratio (Fig. 12.77). The TROSY experiment selects the pathway leading to this one component only and suppresses the other three. Thus for each amino acid residue we still see only one crosspeak in the ^{1}H,^{15}N HSQC, and we have to some extent overcome the linewidth problem in protein NMR. Since the HSQC is the basis of all of the 3D experiments (the final step is an HSQC-type transfer from ^{15}N to ^{1}H, or from ^{13}C to ^{1}H), all of these experiments can be converted to TROSY versions with greatly improved linewidth for large proteins. One example of how large a molecule can now be studied is the Gro-EL complex, an 800 kD symmetrical complex of 14 identical subunits. This is cheating a bit, of course, since the complexity level is the same as a 60 kD protein, but it shows how far TROSY can extend into the world of large biological molecules. A 0.13 mM sample was prepared with uniform ^{15}N and ^{2}H labeling and studied using a cryogenic probe on a 900 MHz spectrometer. A separate sample was prepared with ^{15}N labeling only in the leucine residues to help with the assignment process. These large proteins are enormously ambitious projects, and just completing the assignments is a huge achievement, but it shows how far NMR has come as a tool in structural biology.

FURTHER READING

1. Cavanagh J, Fairbrother WJ, Palmer AG, Skelton NJ. Protein NMR Spectroscopy, Principles and Practice. Academic Press; 1996.
2. Wittekind M, Reizer J, Klevit RE. Sequence-specific ^{1}H NMR resonance assignments of *Bacillus subtilis* HPr: use of spectra obtained from mutants to resolve spectral overlap. *Biochemistry* 1990;**29**:7191–7200.
3. Gooley PR, Caffrey MS, Cusanovich MA, MacKenzie NE. Assignment of the ^{1}H and ^{15}N NMR Spectra of *Rhodobacter capsulatus* ferrocytochrome c_2. *Biochemistry* 1990;**29**: 2278–2290.
4. Kay LE, Muhandiram DR. Gradient-enhanced triple-resonance three-dimensional NMR experiments with improved sensitivity. *J. Magn. Reson. B* 1994;**103**:203–216.
5. Jacobsen NE, Abadi N, Sliwkowski MX, Reilly D, Skelton NJ, Fairbrother WJ. High resolution solution structure of the EGF-like domain of heregulin-α. *Biochemistry* 1996;**35**: 3402–3417.
6. Bax A, Tjandra N. Direct measurement of distances and angles in biomolecules by NMR in a dilute liquid crystalline medium. *Science* 1997;**278**:1111–1113.
7. Newlove T, Atkinson KR, Van Dorn LO, Cordes MHJ. A trade between similar but nonequivalent intrasubunit and intersubunit contacts in Cro dimer evolution. *Biochemistry* 2006;**45**:6379–6391.
8. Brunner E. Residual dipolar couplings in protein NMR. *Concepts Magn. Reson.* 2001;**13**: 238–259.
9. Pervushin K, Riek R, Wider G, Wüthrich K. Attenuated T_2 relaxation by mutual cancellation of dipole–dipole coupling and chemical shift anisotropy indicates an avenue to NMR structures of very large biological macromolecules in solution. *Proc. Natl. Acad. Sci. USA* 1997;**94**:12366–12371.

APPENDIX A

A PICTORIAL KEY TO NMR SPIN STATES

The following pages show the 15 Cartesian product operators for a spin system consisting of two *J*-coupled protons **I** (H_a) and **S** (H_b) (Fig. A.1). Each operator is represented in six ways: the product operator symbol, an energy diagram with transitions, a vector diagram, a spectrum, a density matrix, and the coherence order.

1. *Product Operator.* For multiple-quantum coherences, the pure zero-quantum and double-quantum states are shown with their Cartesian product operator equivalents.

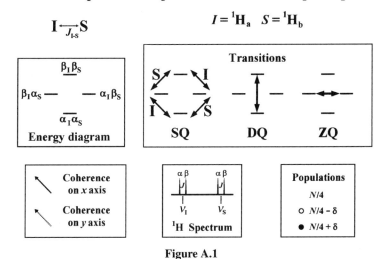

Figure A.1

NMR Spectroscopy Explained: Simplified Theory, Applications and Examples for Organic Chemistry and Structural Biology, by Neil E. Jacobsen
Copyright © 2007 John Wiley & Sons, Inc.

628   APPENDIX A: A PICTORIAL KEY TO NMR SPIN STATES

The sixteenth operator is the identity operator (zero net magnetization); it is not shown.

2. *Energy Diagram.* The four energy levels are labeled according to the spin states of the **I** spin (α or β) and the **S** spin (α or β). The S spin transitions are on the lower right side and upper left side of the diagram; the I transitions are on the upper right side and lower left side (Fig. A.1). Solid arrows indicate coherences between energy levels with phase aligned along the x' (real) axis; dotted arrows indicate coherences with phase aligned along the y' (imaginary) axis. Population differences are indicated using open circles to indicate a slight deficit in population and closed circles to indicate a slight excess, relative to an even distribution of $N/4$ spins in each spin state.

3. *Vector Diagram.* Two vectors are shown, one representing approximately 50% of spins whose coupling partner is in the α spin state and the other representing the other half of the spins whose partner is in the β spin state. For multiple-quantum coherences, there is no way to represent the spin state using vector diagrams.

4. *Spectrum.* If collection of the FID data begins with the system in the given spin state and the FID is Fourier transformed, the diagram shows how the spectrum would look. The normal spectrum (starting from $\mathbf{I}_x + \mathbf{S}_x$, which would result from a $90°_y$ pulse on the equilibrium state $\mathbf{I}_z + \mathbf{S}_z$) would consist of a doublet at v_I and doublet at v_S, both with the same coupling constant J. The reference axis is x': Absorptive phase is shown for signals that are aligned along the x'-axis at the start of the FID, and dispersive phase is shown for signals that start along the y'-axis.

5. *Density Matrix.* Each row and column is labeled with the spin state corresponding to the energy diagram at the top. A coherence between two different energy levels appears as a complex number in the row and column corresponding to the two levels. Coherences on the x'-axis are real and those on the y'-axis are imaginary. Population differences (excess or deficit) show up in the diagonal elements.

6. *Coherence Order.* The coherence order, **p**, is zero for z magnetization and zero-quantum coherence, 1 or -1 for single-quantum coherence, and 2 or -2 for double-quantum coherence. The coherence order is useful for diagraming the coherence pathway in a pulse sequence and for predicting the effect of gradient pulses on the sample magnetization.

The pure **I** ($\mathrm{H_a}$) spin states (Fig. A.2) are the equilibrium state, $\mathbf{I}_z$, which is not observable, the in-phase single-quantum coherence $\mathbf{I}_x$, which gives rise to an absorptive doublet at v_I, and the in-phase single-quantum coherence $\mathbf{I}_y$, which gives rise to a dispersive doublet at v_I. The vector diagrams are labeled "**I**" to make clear that they represent I-spin net magnetization only. The pure **S** spin states (Fig. A.3) are similar, except that they give rise to peaks in the spectrum at v_S instead of v_I. For $\mathbf{S}_z$, the population differences exist only along the two **S** transitions (equilibrium difference $= 1 - (-1) = 2$). The antiphase spin states for spin **I** (Fig. A.4) include the absorptive antiphase state, with vectors aligned along the x' and $-x'$ axes for the **I** spins with their coupling partners in the α and β spin states, respectively, and the dispersive antiphase state, with vectors aligned along y' and $-y'$. Note that the density matrix elements in the lower left are 1 and -1 (x' and $-x'$) for the absorptive antiphase state, indicating that the two transitions (S $= \alpha$ and S $= \beta$) have opposite sign. The dispersive antiphase state has matrix elements i and $-i$ (y' and $-y'$

APPENDIX A: A PICTORIAL KEY TO NMR SPIN STATES 629

I: In-Phase SQC and z magnetization

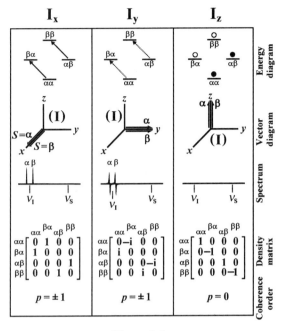

Figure A.2

S: In-Phase SQC and z magnetization

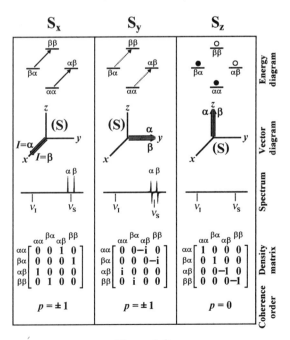

Figure A.3

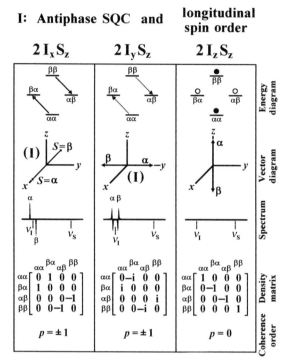

Figure A.4

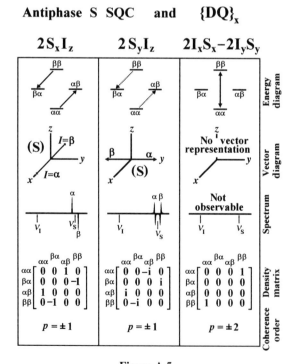

Figure A.5

axes). The third state (Fig. A.4, right) is called two-spin order, which can be considered a halfway-point in INEPT-type coherence transfer. Note that the population differences for the **I** transitions are $+2$ (**S** $= \alpha$, lower left) and -2 (**S** $= \beta$, upper right). The first two spin states in Figure A.5 are the antiphase **S** spin states, with magnetization vectors pointing in opposite directions depending upon the spin state (α or β) of the coupling partner (**I** spin). The third state (Fig. A.5, right) is double-quantum coherence aligned along the x' axis. This is a coherence between the $\alpha\alpha$ and the $\beta\beta$ energy levels that cannot be represented in the vector model. It is not directly observable, so there is no spectrum shown. Note that the density matrix elements connect the $\alpha\alpha$ state to the $\beta\beta$ state, with matrix element 1 (x'-axis). The remaining three multiple-quantum states are shown in Figure A.6. Zero-quantum coherences represent a superposition of the $\alpha\beta$ and $\beta\alpha$ energy levels. Coherences along the x' axis are given real numbers for density matrix elements, and coherences along the y' axis are given imaginary numbers. The zero-quantum density matrix has nonzero elements in row 2/column 3 and row 3/column 2, indicating a superposition of the $\alpha\beta$ and $\beta\alpha$ states. Note that the pure ZQ and DQ states must be represented as a sum or difference of Cartesian product operators. For the homonuclear system, ZQ coherence has coherence order 0 and DQ coherence has coherence order ± 2.

Figure A.7 shows the time evolution of product operators as a result of chemical shift only. Each "wheel" represents one full cycle of evolution (rotation in the x'–y' plane), starting on the right side and moving to the top (90° or $\pi/2$ rotation counterclockwise), to the left side, and to the bottom. The center of the wheel is the rotation angle in radians: Ω_I is the I-spin offset (distance from the center of the spectral window in radians s^{-1}) and Ω_S is the S-spin offset. I-spin SQC rotates at a rate Ω_I; S-spin SQC rotates at rate Ω_S; DQC

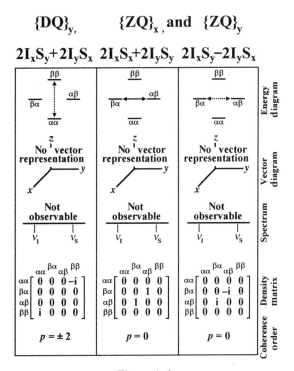

Figure A.6

Chemical shift evolution only: $\Omega = 2\pi(\nu_0 - \nu_r)$

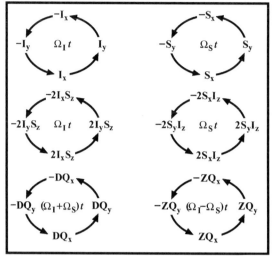

Example: $DQ_y \rightarrow DQ_y \cos(\Omega_I + \Omega_S)t - DQ_x \sin(\Omega_I + \Omega_S)t$

Figure A.7

rotates at the sum of these frequencies; and ZQC rotates at the difference. To write out the general time course of chemical-shift evolution, simply multiply the starting spin state by the cosine of the angle (value in the center of the wheel) and add the next spin state going around the wheel times the sine of the angle. With chemical-shift evolution alone, in-phase spins states remain in-phase and antiphase states remain antiphase.

Figure A.8 shows the time evolution of product operators as a result of J-coupling evolution alone. In-phase coherence continuously progresses into antiphase and back to in-phase as the two component vectors (α and β) counterrotate in the $x'-y'$ plane. The angle of rotation in all cases is πJt in radians (where J is in hertz). This corresponds to $\pi/2$ or one fourth of a full rotation for time equal to $1/(2J)$. To write out the general time

J-coupling evolution only: J_{IS} in Hz

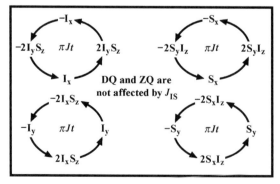

Example: $2I_xS_z \rightarrow 2I_xS_z \cos(\pi Jt) + I_y \sin(\pi Jt)$

Figure A.8

course of J-coupling evolution, multiply the starting spin state by $\cos(\pi Jt)$ and add the next spin state going around the wheel, multiplied by $\sin(\pi Jt)$. Because in DQC and ZQC both spins undergo a transition ("spin-flip"), the energy difference between $\alpha\alpha$ and $\beta\beta$ levels (or between $\alpha\beta$ and $\beta\alpha$) is not affected by J, so there is no J-coupling evolution due to J_{IS}. Coupling to a third spin (passive coupling), however, *will* lead to J-coupling evolution (e.g., $2\mathbf{I}_x\mathbf{S}_x \rightarrow 4\mathbf{I}_x\mathbf{S}_x\mathbf{I}'_z$).

APPENDIX B

A SURVEY OF TWO-DIMENSIONAL NMR EXPERIMENTS

All two-dimensional experiments have the same general strategy: correlate one nucleus to another nearby nucleus in the same molecule by a process of magnetization transfer. Consider the transfer of magnetization from nucleus **I** to nucleus **S**:

1. *Preparation:* Create coherence on spin **I**.
2. *Evolution:* Let spin **I** coherence precess in the x–y plane for an incremented time delay t_1. This indirectly measures the frequency of precession, which can be converted into the chemical shift of spin **I**.
3. *Mixing:* Transfer magnetization from spin **I** to spin **S**. This is the only part that is different for different 2D experiments. Transfer may be by transfer of z magnetization (NOE) or by transfer of coherence (J coupling).
4. *Detection:* The **S** spin coherence is measured by recording an FID in the normal way. Fourier transformation of this FID gives a spectrum, and the **S** spin gives a peak in this spectrum at its chemical-shift position. This peak varies in intensity in a sinusoidal fashion as we step through the different FIDs recorded with gradually increasing values of the t_1 delay. The frequency of this oscillation in intensity is the frequency of the **I** spin, which can be converted into its chemical shift.

The variety of 2D NMR experiments, each with its own acronym, can easily be understood by knowing the nature of the **I** and **S** spins (e.g., ^{1}H and ^{13}C) and the type of mixing used (e.g., coherence transfer via one-bond J coupling). Two general distinctions can be made:

  A. *Homonuclear* ($\mathbf{I} = {}^1\text{H}, \mathbf{S} = {}^1\text{H}$) and *Heteronuclear* ($\mathbf{I} = {}^1\text{H}, \mathbf{S} = \text{X}$), where X is any nucleus other than ^{1}H. Homonuclear 2D spectra have a diagonal that is a trace of the

---

NMR Spectroscopy Explained: Simplified Theory, Applications and Examples for Organic Chemistry and Structural Biology, by Neil E. Jacobsen
Copyright © 2007 John Wiley & Sons, Inc.

1D ^{1}H spectrum, and crosspeaks are arranged symmetrically around the diagonal. There is only one radio frequency channel in a homonuclear experiment, the ^{1}H channel, so the center of the spectral window (set by the exact frequency of pulses and of the reference frequency in the receiver) is the same in F_2 and F_1 (Varian *tof*, Bruker *o1*). The spectral widths should be set to the same value in both dimensions, leading to a square data matrix. Heteronuclear experiments have no diagonal, and two separate radio frequency channels are used (transmitter for F_2, decoupler for F_1) with two independently set spectral windows (Varian *tof* and *dof*, *sw*, and *sw1*, Bruker *o1* and *o2*, $sw(F_2)$, and $sw(F_1)$). Heteronuclear experiments can be further subdivided into direct (HETCOR) and inverse (HSQC, HMQC, HMBC) experiments. Direct experiments detect the X nucleus (e.g., ^{13}C) in the directly detected dimension (F_2) using a direct probe (^{13}C coil on the inside, closest to the sample, ^{1}H coil on the outside), and inverse experiments detect ^{1}H in the F_2 dimension using an inverse probe (^{1}H coil on the inside, ^{13}C coil outside).

B. *NOE* (z magnetization transfer) and *J coupling* (coherence transfer) experiments. If the mixing involves transfer of z magnetization through cross relaxation (NOE), the correlation is based on a through-space distance between protons of less than 5 Å (2D NOESY). If the mixing involves transfer of magnetization in the x–y plane (coherence) through a J coupling, the correlation is based on a through-bond separation of 1–3 bonds. Coherence transfer may occur by an INEPT sequence (antiphase to antiphase) or by the TOCSY mixing sequence (in-phase to in-phase). Coherence transfer can involve an intermediate state that is double-quantum or zero-quantum coherence (DQC or ZQC) as in the DQF-COSY and HMQC/HMBC experiments, but it is still essentially an INEPT transfer. One experiment that is difficult to classify in this scheme is the 2D ROESY, which involves NOE transfer but not on the z-axis. ROESY mixing occurs on the spin-lock axis, which is in the x–y plane, but the mechanism of magnetization transfer is by cross relaxation dominated by DQ transitions. Essentially, ROESY mixing is just like NOESY mixing with the z-axis (M_z and B_0 parallel) of the laboratory frame moved to the y' axis (M_y and B_1 parallel) of the rotating frame. The effect is to reduce the B_0 field by a factor of around 10^5 to the strength of the B_1 field, making DQ relaxation the dominant pathway for molecules of all sizes.

In the following survey, the pulse sequence is shown for each experiment, along with a diagram of the expected 2D spectrum and a structure fragment showing a typical interaction that will lead to a crosspeak in the 2D spectrum. Positive crosspeaks are shown as filled circles and negative crosspeaks are shown as open circles. In some cases the coherence pathway is diagramed on a structure fragment, showing the sequence of events in the pulse sequence.

B.1 HOMONUCLEAR (^{1}H–^{1}H) 2D EXPERIMENTS

B.1.1 Coherence-Transfer Experiments

COSY (correlation spectroscopy) is the first and the simplest 2D experiment. It correlates ^{1}H to ^{1}H via a single J coupling that may be two-bond (geminal), three-bond (vicinal), or in rare cases four-bond (long-range). Crosspeaks are antiphase in both F_2 and F_1 with respect to

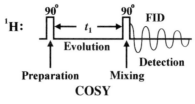

Figure B.1

the active coupling (the one that gives rise to the crosspeak). The pulse sequence (Fig. B.1) consists of two ^{1}H pulses separated by the t_1 delay. Mixing is achieved by the second 90° pulse, which can be viewed as a simultaneous 90° pulse with respect to spins **I** (H$_a$) and **S** (H$_b$). The antiphase state is achieved during the t_1 delay to a varying extent depending on J and t_1. Thus, COSY uses an INEPT transfer without a specific $1/2J$ delay. One can "walk" through a spin system by moving from diagonal to crosspeak vertically, back to the diagonal horizontally, and repeating this process (Fig. B.2). The intense crosspeaks result from geminal or vicinal couplings, and weak crosspeaks may be observed for long-range (allylic, "W," bis-allylic, *meta*, etc.) couplings.

A simple variant of the COSY experiment is *COSY-35* (sometimes called COSY-45), in which the second 90° pulse is reduced from a 90° pulse to a 35° or 45° pulse (Fig. B.3). The result is that the fine structure of crosspeaks is simplified, with half the number of peaks within the crosspeak. This makes it much easier to sort out the coupling patterns in both dimensions and to measure couplings (active and passive) from the crosspeak fine structure. A more important variant of the COSY experiment is the *DQF* (double-quantum filtered)—*COSY* (Fig. B.4), which adds a short delay and a third 90° pulse. The INEPT transfer is divided into two steps: antiphase **I** spin SQC to **I,S** DQC, and **I,S** DQC to antiphase **S** spin SQC. The "filter" enforces the DQC state during the short delay between the second and third pulses either by phase cycling or with gradients. DQF-COSY spectra have better phase characteristics and weaker diagonal peaks than a simple COSY, so this has become the standard COSY experiment.

TOCSY (total correlation spectroscopy) is an extension of the COSY experiment, in which the coherence transfer is not limited to a single "jump" from one proton to another via a J coupling. Instead, coherence is spread out over an entire "spin system" of coupled protons via multiple J-coupled jumps. For example, in a string of carbons CH$_a$–CH$_b$–CH$_c$–CH$_d$, coherence can be transferred by the TOCSY mixing sequence from H$_a$ to H$_c$ or from H$_a$ to H$_d$. Thus, crosspeaks will be observed at $F_2 = \nu_a$ and $F_1 = \nu_b$, ν_c or ν_d (Fig. B.5).

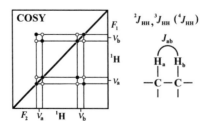

Figure B.2

APPENDIX B: A SURVEY OF TWO-DIMENSIONAL NMR EXPERIMENTS 637

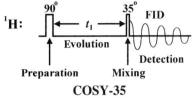

Figure B.3

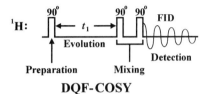

Figure B.4

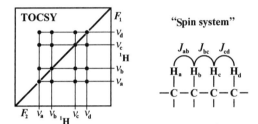

Figure B.5

The TOCSY mixing sequence (Fig. B.6) is a spin lock that consists of a long sequence of pulses of varying pulse width and phase without any delays (MLEV-17, DIPSI-2, etc.). The B_1 amplitude for the spin lock is about three times higher than that for ROESY and is not continuous as it is with the ROESY spin lock. The goal of the TOCSY mixing sequence is to reduce the chemical-shift differences to zero while retaining the J-coupling interactions. In the spin-lock in the absence of chemical-shift differences ($\gamma B_1 \ll \gamma B_0$), the protons within a spin system behave as if they are all coupled together ("virtual coupling"), and the coherence spreads throughout the spin system during the spin lock. The important parameters are the duration of the spin lock (mixing time τ_m), which gives simple COSY-type transfer ("one jump") with a mixing time of about 35 ms, and maximum mixing throughout a spin system with a mixing time of about 70 ms ("many jumps"), and the amplitude of the spin-lock pulse (typically $\gamma B_1 / 2\pi \sim 8$ kHz).

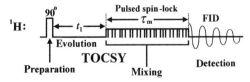

Figure B.6

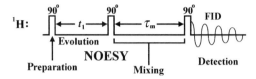

Figure B.7

B.1.2 NOE (z-Magnetization Transfer) Experiments

NOESY (nuclear Overhauser and exchange spectroscopy) is the simplest 2D NOE experiment, consisting of three 90° pulses (Fig. B.7). The first pulse creates H_a coherence, which then precesses during the t_1 period to indirectly record the H_a chemical shift. The second 90° pulse flips the H_a magnetization back to the z axis, where the z magnetization can be anywhere from $+M_0$ to $-M_0$ depending on the precession that occurred during t_1. This perturbed z magnetization (nonequilibrium population difference) propagates during the mixing time τ_m to the nearby (<5 Å) proton H_b, so that its z magnetization is slightly enhanced (a bit more than M_0). The third 90° pulse flips the H_b z magnetization into the x–y plane and the H_b FID is recorded. Normally, the diagonal peaks are phased to positive absorptive lineshape so that the crosspeaks appear as negative peaks for small molecules (relaxation is primarily DQ) and as positive peaks for large molecules (relaxation primarily ZQ) or for chemical exchange (Fig. B.8). Chemical exchange (change of the chemical shift of a proton due to conformational change or bond breaking and bond making) also leads to an effective "transfer" of z magnetization during τ_m and thus gives a crosspeak in the NOESY spectrum. Molecules of intermediate size ($\sim$ 2000 Da.) can have very small or zero NOE as the effect passes from negative to positive, and in these cases a ROESY spectrum is preferred. The NOESY mixing time should be adjusted for molecular size, because the timescale for both self-relaxation (T_1) and cross relaxation (NOE) is dependent on molecular size (shorter time for bigger molecules). The intensity of the transient NOE increases linearly with τ_m (slope is proportional to $1/r^6$, where r is the direct through-space distance between the two protons) and then levels off and falls to zero. A typical optimal mixing time is 350 ms for small molecules (e.g., sucrose in D_2O).

ROESY (rotating frame Overhauser effect spectroscopy) is a variant of NOESY, in which the transfer of magnetization occurs on the spin-lock axis in the x–y plane rather than on the z axis (Fig. B.9). A continuous low-power radio frequency pulse provides the mixing by effectively reducing the field strength (B_0 in the laboratory frame on the z axis to B_1 in

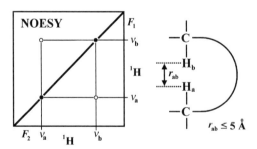

Figure B.8

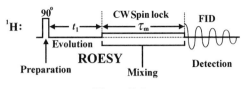

Figure B.9

the rotating frame on the spin-lock axis) by a factor of about 10^5 (e.g., from 600 MHz to 3300 Hz). At this field strength the Larmor frequency ν_1 is so small that even very large molecules have a significant population tumbling at $2\nu_1$ (the DQ frequency), and relaxation is dominated by DQ relaxation. Thus, the NOE is negative (negative crosspeaks when the diagonal is positive) regardless of molecular size, and there is no null in the NOE at intermediate molecular size (MW $\sim$ 2000). The optimal mixing time for a ROESY is about half of the optimal mixing time for a NOESY, so a typical value of τ_m is 200 ms for small organic molecules. TOCSY crosspeaks often show up in ROESY spectra, but because they are positive they are easily distinguished from the negative ROESY crosspeaks.

B.2 HETERONUCLEAR (USUALLY ^{1}H–^{13}C) EXPERIMENTS

B.2.1 Coherence-Transfer Experiments

HETCOR (heteronuclear correlation) is the simplest heteronuclear 2D experiment. It correlates a proton with the carbon it is bonded to, using the very large (125–180 Hz) one-bond *J* coupling between ^{1}H and ^{13}C. The pulse sequence (Fig. B.10) is based on the simple INEPT transfer from ^{1}H to ^{13}C. Coherence is created on ^{1}H and then allowed to precess during the t_1 period, recording the ^{1}H chemical shift indirectly. Then a delay of $1/(2J)$ is followed by simultaneous ^{1}H and ^{13}C 90° pulses, which leads to INEPT transfer of antiphase ^{1}H SQC to antiphase ^{13}C SQC. A refocusing delay allows the antiphase ^{13}C SQC to evolve into in-phase SQC, which is then detected during the FID with ^{1}H decoupling. In the HETCOR spectrum (Fig. B.11), ^{1}H chemical shift appears on the vertical (F_1) axis and ^{13}C chemical shift appears on the horizontal (F_2) axis. For CH$_2$ groups in chiral molecules, there are usually two crosspeaks on the vertical line corresponding to the ^{13}C chemical shift because the two protons usually have different chemical shifts. For CH and CH$_3$ groups, there is only one crosspeak at each ^{13}C chemical shift.

HETCOR has been largely replaced by the far more sensitive "inverse" experiments, HSQC and HMQC. Because HETCOR is a ^{13}C-detected experiment, it is called "direct"

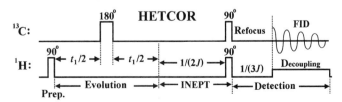

Figure B.10

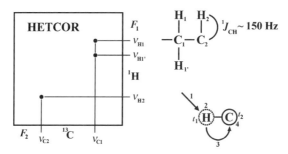

Figure B.11

because this older way of doing ^{1}H–^{13}C correlation is by direct detection of ^{13}C in the t_2 FID. *HSQC* (heteronuclear single quantum correlation) is similar to HETCOR, except that the ^{1}H chemical shift axis is F_2 (horizontal) and the ^{13}C chemical shift axis is F_1 (vertical). Thus, it is called an "inverse" experiment because ^{1}H is directly detected and ^{13}C is indirectly detected, the inverse of the old way. Correlation is based on the one-bond coupling ($^1J_{CH}$) between a ^{13}C and the ^{1}H directly bonded to it. The experiment uses an "out and back" coherence transfer (Fig. B.12): ^{1}H SQC is created, allowed to evolve into antiphase with respect to its directly bonded ^{13}C ($1/(2J) \sim 3.3$ ms), and simultaneous ^{1}H and ^{13}C 90° pulses lead to coherence transfer from ^{1}H to ^{13}C (INEPT transfer). The ^{13}C SQC, antiphase with respect to its directly bonded ^{1}H, precesses at the frequency of the ^{13}C chemical shift during the evolution (t_1) period, and a ^{1}H 180° pulse in the center refocuses any J-coupling evolution. Because the coherence is ^{13}C SQC during t_1, the experiment is called *HSQC*. At the end of the evolution period, simultaneous ^{13}C and ^{1}H 90° pulses transfer the coherence back to ^{1}H SQC (INEPT transfer), antiphase with respect to ^{13}C, and a refocusing period ($1/(2J)$) allows J-coupling evolution from antiphase back to in-phase ^{1}H SQC. The FID is recorded with ^{13}C decoupling to collapse the very wide ^{1}H doublets (split by ^{13}C) into normal proton signals. The simultaneous 180° pulses on ^{1}H and ^{13}C in the middle of the $1/(2J)$ delays prevent any ^{1}H chemical-shift evolution, limiting evolution to J-coupling evolution during these "dephasing" and "refocusing" periods. The HSQC signal 16 times stronger than HETCOR because it is proton detected, and proton is a much more sensitive nucleus than ^{13}C due to its four times higher "magnet strength" (γ). Gradients can be used to select the coherence pathway (^{1}H to ^{13}C to ^{1}H, all SQC), and an APT-like delay of $1/J$ can be inserted during the ^{13}C SQC period to "edit" the crosspeaks, making the CH$_3$ and CH crosspeaks positive and the CH$_2$ crosspeaks negative. The only significant parameter that needs to be adjusted is the $^1J_{CH}$ value, which ranges from 125 to 180 Hz depending on

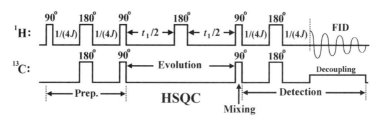

Figure B.12

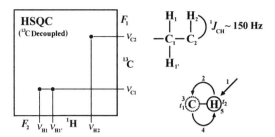

Figure B.13

the hybridization (sp^3, sp^2, sp) of carbon and the presence of bonds to electronegative atoms (oxygen). The $^1J_{CH}$ value sets the J-coupling evolution delay $1/(2J)$, and miscalibration can lead to difficulties in phasing the HSQC spectrum. The HSQC spectrum (Fig. B.13) is like the HETCOR, only turned on its side with ^{13}C chemical shift on the vertical axis and ^{1}H shift on the horizontal axis. The best resolution is obtained in the ^{1}H dimension (F_2), so that horizontal slices contain ^{1}H–^{1}H coupling information for the stronger couplings.

HMQC (heteronuclear multiple quantum correlation) is a variant of the HSQC spectrum that gives essentially the same results with a slightly different strategy (Fig. B.14). Instead of converting antiphase ^{1}H SQC into antiphase ^{13}C SQC, a single 90° pulse on ^{13}C alone converts it into ^{1}H–^{13}C multiple quantum coherence (DQC and ZQC). DQC ($\mathbf{I}^+\mathbf{S}^+$) is selected during the first half of t_1, and it evolves at the rate of $\nu_H + \nu_C$ in the x–y plane. A 180° pulse on ^{1}H then converts the DQC into ZQC ($\mathbf{I}^-\mathbf{S}^+$), which evolves during the second half of the t_1 period at the rate of $-\nu_H + \nu_C$, so that the net evolution during t_1 is only due to ν_C, just as in the HSQC experiment. The ZQC is then converted back into antiphase ^{1}H SQC, which is refocused by a $1/(2J)$ delay just as in the HSQC experiment. The ^{1}H FID is then recorded with ^{13}C decoupling. As with HSQC, the only significant parameter to set is the $^1J_{CH}$ value.

HMBC (heteronuclear multiple bond correlation) is a variant of the HMQC experiment, designed to focus on the long-range (multiple bond) J couplings between ^{1}H and ^{13}C: $^2J_{CH}$ and $^3J_{CH}$. These couplings are much smaller than $^1J_{CH}$, in the range of 0–8 Hz typically (Fig. B.15). The strategy is to maximize coherence transfer from ^{1}H to the "remote" ^{13}C (two or three bonds away) and attempt to destroy any coherence transfer from ^{1}H to the "direct" (one bond away) ^{13}C. ^{1}H SQC is created as in the HMQC, but the $1/(2J)$ period is lengthened from the 3.3 ms value ($^1J_{CH} = J_d = 150$ Hz) to a value of 50 ms ($^{2,3}J_{CH} = J_{lr} = 10$ Hz) to maximize the remote transfer (J_d is the "direct" coupling and J_{lr} is the "long range" coupling). To minimize losses due to T_2 relaxation (decay of coherence) during the much longer $1/(2J)$ delays, the refocusing delay is left out and the FID records the antiphase ^{1}H SQC directly. This means that ^{13}C decoupling cannot be used while recording the FID.

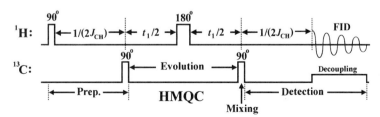

Figure B.14

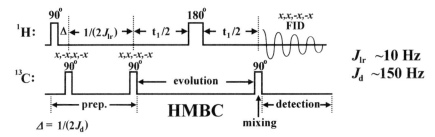

Figure B.15

The direct coherence transfer is destroyed by placing a ^{13}C 90° pulse after a J-coupling evolution time of only 3.3 ms (1/(2 × 150 Hz)) and phase cycling this pulse (x, $-x$) so as to alternate the sign of any coherence created by the pulse in consecutive scans (Fig. B.15). The FIDs are added together in the sum to memory, and this signal cancels because it is alternating in sign. The remote coherence transfer is carried out after the full 50-ms delay and is phase cycled in concert with the receiver so that the signals add together in the sum to memory. An alternative method for destroying the ^{13}C-bound ^{1}H signal is to use a TANGO sequence to selectively excite only the ^{13}C-bound protons, and then kill this coherence with a "spoiler" gradient. The relevant parameters to set in an HMBC experiment are the $^1J_{CH}$ value (to block the direct correlations) and the $^{2,3}J_{CH}$ value (to optimize the remote correlations). Inevitably, there will be some "bleed through" of the one-bond correlations, and these artifacts will appear as wide antiphase doublets separated by the $^1J_{CH}$ coupling constant (because there is no ^{13}C decoupling during acquisition), and centered on the positions of the HSQC crosspeaks (Fig. B.16).

HMBC data are normally processed in magnitude mode (magn = (real^2 + imag^2)$^{1/2}$) so that all the data values are positive. This can make it difficult to distinguish weak signals from noise, so some newer sequences are designed to show HMBC data in phase-sensitive mode (real part only displayed after phase correction). The long-range crosspeaks are narrow, single absorptive peaks in the F_1 (^{13}C) dimension and wider peaks with alternating positive and negative intensities in the F_2 (^{1}H) dimension. It is this regular alternation of sign, due to the small antiphase $^{2,3}J_{CH}$ couplings, which helps to distinguish signals from the noise. The three-bond couplings are sensitive to the H–X–Y–^{13}C dihedral angle, with a Karplus relation similar to the one for vicinal ^{1}H–^{1}H couplings. When the $^3J_{CH}$ is small (*gauche* relationship of C and H), the HMBC crosspeak is weak or missing because the J-coupling evolution into antiphase takes much longer than 50 ms. When the $^3J_{CH}$ is large (*anti* relationship of C and H), the crosspeak is strong. In this way, HMBC data can be used to determine stereochemistry in rigid systems.

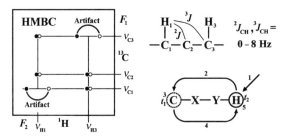

Figure B.16

INDEX

Note: Italicized letters *f*, and *t* following page numbers indicate figures and tables, respectively.

1,1̄ sequence, 312
1,1̄-echo, 568
^{12}C-H artifact in HSQC, cancellation in phase cycle, 526–527
^{13}C decoupling, 338, 536
   limits on duty cycle, 504
^{13}C inversion-recovery, 178
^{13}C solvent peak, 131
^{13}C-acetate, 137
^{13}C-glucose, 137
15-β-hydroxytestosterone, 376–377, 377*f*
   DQF-COSY spectrum, 378*f*
^{15}N-labeling and 3D NMR, 596–601
180° pulse, 166, 170, 207
180° pulse width, 208
180° pulse, poor performance with off-resonance, 294
1D selective NOE (DPFGSE), 515
   on lactose, 421
1D TOCSY, 185, 289
^{1}H decoupling, 132
^{1}H inversion-recovery, 178
^{1}H preamplifier, 559
$^{1}J_{CH}$, 137, 138, 139

2,2′,3,3′-d$_4$-3-trimethylsilylpropionate (TSP), 76
270° pulse, 208
2D (two dimensional) NMR, 44, 99, 108, 119, 126, 192, 347
2D COSY, 143, 185
2D COSY spectrum of pristimerin oxidation product, 546–547
2D experiments, 238, 338
   overview, 364
2D Fourier transform, 368
2D HETCOR, 138, 143
2D HETCOR spectrum, cartoon, 358, 359*f*
2D HMBC, 497
2D HSQC, 148
2D NOESY and ROESY, assignment of H19 of cholesterol, 515
2D NOESY spectrum, sign of crosspeaks relative to diagonal, 414
2D ROESY,
   sign of crosspeaks relative to diagonal, 414
   TOCSY artifacts in, 342
2D spectral window, 398
2D TOCSY (total correlation spectroscopy), 393–398

644 INDEX

2D TOCSY,
  pulse sequence, 394f
  window functions for, 404
$2I_zS_z$ (longitudinal two-spin order), 257, 260, 266, 317
$^2J_{CH}$, 138
$2S_zI_z$ (longitudinal two-spin order), 260
3–9–19 sequence, 312
  vector analysis, 313, 313f, 314
3D (3-dimensional) NMR, 370, 553, 557
3D (3-dimensional) structures, 151
3D data cube, 611
3D HNCO, 611–615
3-heptanone, 374, 375f, 396, 503,
  decoupled edited HSQC, 505, 506f
  HMBC spectrum, 501–512
  HMQC spectrum, 501, 502f
$^3J_{CH}$, 138
4-hydroxy-3-methyl-2-butanone, 220, 277, 279
4-isopropylacetophenone, 10
$4S_xI_z^1I_z^2$, 271
^{57}Fe, 299
6 dB rule, 301, 349
90° pulse, 92, 162, 170, 206, 207, 241
90° pulse width, 182, 301
90° read pulse, 199
90°-shifted sine-bell, 403

A_2B_2 system, 73
AA′BB′ system, 72, 73
AB pattern, 68, 72, 136
AB system, 64, 65, 328, 483
abs command (Bruker), 133
Absolute frame of reference for structure restraints in NMR, 621
Absorptive lineshape, 127, 129, 212, 215, 405
  mathematical formula, 390
Abundance, 3, 38
ABX system, 68
ABX_3 system, 69
Accelerating frame of reference, 202
ACCORDIAN, 538
Accumulation of phase twist, with multiple gradients, 460
Acetate, 136
Acetone, 374
Acorn-NMR, 99
acqi window, 86
acqu file (Bruker), 119
Acquisition order, in 3D experiments, 605

Acquisition time, 103, 104, 126, 139, 146, 504
  in t_1 time domain, 401, 404
Active coupling, 284, 380, 500
ad180 adiabatic inversion pulse, 338
ADC, 94, 98, 102, 110, 344, 398
Address,
  ^{1}H and ^{15}N shifts in 3D dataset, 596, 611
  precise chemical shift viewed as, 574
ADEQUATE, 538
Adiabatic condition, 336
Adiabatic decoupling, 338
Adiabatic inversion, 296, 337, 337f, 496
Advancing the receiver phase, 455
Aggregation, 76
  of proteins, and NaCl concentration, 565
Alachlor herbicide, 63
Alanine, 570
Alcohol OH proton, exchange of, 423
Aldopyranose, 15
Aldrich, 77
Aliasing, 101, 102, 110, 111, 117, 129
  in F_1 dimension of 2D, 398, 400, 599
Alignment, of multiple calculated structures, 593
Alkyne, terminal, one-bond CH coupling constant, 495
Allylic coupling, 25, 346
Alpha (α) state, 33, 34, 156
Alpha-helix, 575, 575f
Alternating positive and negative intensity, in phase-sensitive HMBC crosspeaks, 511
Alternating sampling, 98, 99, 102, 119, 399
AM Spectrometer, Bruker, 134
Amber force field, 590
Ambiguity, two-bond vs. three-bond correlations, in HMBC, 514, 538
Amide NH proton, 315, 316
Amino acids, identification from TOCSY, 397, 398
Ammonium chloride, ^{15}N labeled, 596
Amplifier and sample heating, 182
Amplifier compression, 350
Amplitude modulation, in 2D NMR, 357
Amplitude setting unit (ASU), Bruker, 320
Amplitude, of RF pulse, 348, 349
AMX (3-spin) spin system, 68, 383, 398, 571, 572
Analog audio filter, 111, 117, 297
Analog mixing, 399
Analog noise, 108
Analog-to-digital converter (ADC), 8, 94, 353, 367, 523

Andrographolide, HMBC, 510
Angular methyl groups, 25
Angular momentum, 31, 156
Angular velocity, 31, 156
Annihilation, with gradients, 303, 311
Anomeric, 12, 13, 14
Anomeric equilibrium, in ^{1}H spectrum of glucose, 493, 493*f*
Anomeric proton, 184, 197, 340
  stereochemistry, 493
anti relationship, 24, 493
Antidiagonal, 394
Antiecho mode, 606
Antiecho pathway, 465
Antiparallel β sheet, 587
Antiphase coherence, 214, 247, 254, 317
Antiphase doublet, 215, 380, 386
Antiphase peaks,
  in COSY, 389
  in HMBC, 500
Antiphase relationship, of vectors, 217, 218, 234, 242, 243, 244*f*, 250
Antiphase signals, 270
Antiphase terms, in TOCSY mixing, 487
Antiphase to antiphase, coherence transfer, 342, 500
Antisymmetric state, 483
aph command (Varian), 129
Appearance, of inverse 2D spectra, 498–501
Applications of NMR in biology, 551–553
Aprotic solvents, 423
APT (attached proton test) experiment, 18, 138, 255, 276, 278, 363, 504, 530
AQ parameter, Bruker, 105, 126, 144, 504, 509
Arbitrary numbering, in structure determination by NMR, 539
Archiving, of NMR data, 134
Arcing, in probe coil, 299
Area, under pulse shape, 297, 320
Arginine, 571
Aromatic amino acids, role in chemical shift dispersion, 575
Aromatic protons, 9, 316
Array, in Varian software, 119, 178, 184, 279, 366
Artifacts, removal, 264, 450
Asparagine, 571
Aspartate, 571
Assignment, 39, 278, 360, 370, 575
Assignment list, 514
Assignment, of β-carotene oxidation product, 520*f*, 522

AT parameter (Varian), 105, 126, 144, 404, 504, 504, 509
Attached proton test (APT), 218, 220
Attached proton, of CH group, 12
Attenuation, in Bruker decibel scale, 349
Audio filter, 94, 98, 111
Audio frequency signal, in receiver, 42, 90, 96, 102, 103, 104, 105, 108, 110, 201, 296
Automatic baseline straightening (abs) command, Bruker, 133
Automatic gradient shimming, 562
Averaged conformations, effect on *J* coupling, 68, 385
Avogadro's number, 33
Axial (Z axis) shims, 87
Axial position, in cyclohexane chair, 12

B. subtilis HPr, 578
B_1 field inhomogeneity, 290, 334, 334*f*, 393
B_1 field strength, 208, 297, 430
B_1 field, vector, 240, 450
Back transfer, in HSQC pulse sequence, 523
Backbone carbonyl groups, 575
Back-exchange of H_N protons, 557
Band selective shaped pulse, 320
Bandpass, of audio filter, 111
Bandwidth test, for ^{13}C decoupling, 496, 496*f*
Bandwidth,
  of decoupling, 150
  of RF pulse, 140, 296, 298, 338, 341
Baseline correction, 108, 118, 132, 406, 437
Baseline separation, of lines in multiplet, 49
bc(1) command, Varian, 133
Benzene, INEPT spectrum, 264, 265*f*, 271
Benzoyl, slow bond rotation, 422
Benzyloxycarbonyl, slow bond rotation, 422
Beta position of α,β-unsaturated ketone, downfield shift, 521
Beta sheet, 575, 576*f*, 578
Beta (β) state, 33, 34, 156
Beta-carotene, 519, 520*f*
  oxidation product, 519–522, 521*f*, 522*f*
bf2 parameter, Bruker, 183
Bicelles, 622
Bicyclo[2.2.1] system, 328
Bilinear rotation decoupling (BIRD), 363, 363*f*
Binomial water suppression techniques, 568
Bins, of NOE crosspeak volumes, 591
Biological NMR spectroscopy, 551–626
Biomolecules, 551
Biopolymers, 135, 553
Biosynthetic studies by NMR, 137

BIRD (bilinear rotation decoupling), 363, 363f, 493, 537
Blanking, of gradients, 319
Bleaching, of H_N peaks, 186, 311, 315, 568
B_0 field strength,
  effect on ^1H spectrum, 40–45
  effect on CSA relaxation rate, 625
  effect on exchange, 418
Boltzmann condition, 409
Boltzmann population distribution, 34–36, 91, 159, 164, 165, 176, 187, 208, 240, 258, 324
Bore, of superconducting NMR magnet, 77
Bottleneck, in TOCSY transfer of magnetization, 345, 347
Bovine pancreatic trypsin inhibitor (BPTI), 552
B_{res}, residual field in rotating frame, 296
Brick wall digital filter, 117
Broadband decoupling of ^{13}C, 143, 495
Broadband shaped pulses, 294, 296, 320
Bruker "smile" baseline, 117
Bruker AM spectrometer, 74, 75, 102, 119, 280
Bruker AMX spectrometer, 74, 89, 107, 119
Bruker DRX spectrometer, 42, 74, 89, 102, 119, 216, 281
Bruker Fourier transform, 120
Bruker WM spectrometer, 74
Buffers and pH, 568
Building blocks, of pulse sequences, 234
Buildup study, of lactose anomeric exchange, 421
Buildup, NOE, 322
Bulk magnetization, 158, 239

Calculated shift reference, based on magnetogyric ratios, 565
Calibration,
  of crosspeak volumes, in 2D NOESY, 436, 591
  of RF pulse, 208, 301, 334, 351, 351f, 567
CAMELSPIN (2D ROESY), 430
Cancelation,
  in DQF-COSY phase cycle, 449–450
  of dipole-dipole and CSA relaxation, 624
Candycanes, in phase-sensitive HMBC, 510
Carbohydrates, 13, 372
Carbon-14, in biosynthetic studies, 136
Carrier frequency, 291, 566
Carr-Purcell-Meiboom-Gill (CPMG), 232
Cartesian product operators, 443, 451

Categories, of amino acid spin systems, 572
Caveats, for distance measurement using NOESY, 436
cawurst adiabatic inversion pulse, 338
CBCA(CO)NH 1D vector, 619f, 620
CBCA(CO)NH experiment, 620
CBCA(CO)NH,
  coherence flow diagram, 620f
  data cube, 620, 621f
Chemical environment, 3, 4, 33, 40, 56, 69
Chemical equivalence, 54, 55, 67, 71
Chemical exchange, 408, 414–425
Chemical kinetics, 164
Chemical reaction, 414
Chemical shift, 3, 32, 33, 39, 104, 105, 118, 127, 130, 156, 200, 297, 342
Chemical shift anisotropy (CSA), 60–61, 176, 358, 555, 624, 625
Chemical shift correlation, 39, 356
Chemical shift dependent phase correction, in 2D HSQC, 505
Chemical shift dependent phase errors, 462, 529
Chemical shift deviation (CSD), 553, 576
Chemical shift evolution, 212, 216, 226, 228, 232, 233, 246, 247, 250, 270, 304–307, 312, 313, 354, 361, 458
Chemical shift index (CSI), 576, 576f, 587
Chemical shift labeling, in t_1 of 2D experiment, 368
Chemical shift reference, 76, 565
Chemical shift refocusing, 234
Chemical shifts,
  of ^{13}C in proteins, 610
  of ^1H in peptides and proteins, 570–577
Chirality, effect on NMR spectrum, 54–56
Cholesterol, 20, 220, 280, 282, 331, 331f
  ^1H spectrum, 24–25, 24f
  ^{13}C spectrum, 25–27, 26f
  2D TOCSY spectrum, 396–397, 396f
  600 MHz NOESY spectrum, 431f
  biosynthesis 20, 21f
  decoupled edited HSQC, 505–509, 506f
  HMBC spectrum, 513–517, 513f, 517f
  HMQC spectrum, upfield region, 503f
  ROESY spectrum, 432–435
  structure, 23, 25f
Chymotrypsin, 556
CIGAR experiment, 538
Classical model, 1, 3, 31, 33, 155
Clipping of FID, 94, 109, 569
Coalescence spectrum, 416f

Coalescence time, 415
Coefficients, of pure energy states in wave function, 441
Coherence, 157, 161, 166, 207, 237, 239–241, 252, 439
Coherence annihilator (gradient), 308
Coherence flow diagram,
  2D HSQC, 498, 498f, 522
  2D HSQC-TOCSY, 602f
  3D HSQC-TOCSY, 603f
Coherence helix, 307, 311, 319, 343
Coherence level diagram, 451, 451f
Coherence order, 316–319, 408, 428, 443, 444, 450, 599
Coherence order change $\Delta \mathbf{p}$, in phase cycling, 452, 454
Coherence order pathway, 445
Coherence order, heteronuclear, 460, 613
Coherence pathway, 263
Coherence pathway selection, 316–319
  using gradients, 446, 450–469, 465
Coherence transfer, 238, 241, 253, 254, 276, 319, 389
  in TOCSY spin-lock, Hamiltonian description, 487
Coherence, quantum mechanical definition, 441–443
Coherences, in density matrix, 472
Collective spin mode, in TOCSY mixing, 487
Column number, of 3D data matrix, 605
Combined twist, due to gradients, 450
Commutation of matrices, 474
Commutator, 484
Compass needle, 155
Complex conjugate, 441, 469, 472
Complex Fourier transform, 120, 210, 400
Complex pairs, 99
Complexity problem in NMR, 556, 623
Composite (sandwich) pulses, 294, 295
Composite pulse decoupling, 495
Computational chemistry, 592
Computer simulation, of TOCSY mixing sequences, 341
Conformation of biomolecules, 197
Conformational averaging, 54
Conformational change, 175, 414, 552
Conformational space, in structure calculation, 590
Connectivity of monosaccharide units, 197
Connectivity, sequential, 581
Composite pulse decoupling, 144
Constant time evolution, 613–614

Context-dependent definition of coherence order, 460
Continuous wave (CW) irradiation, 140, 193, 200, 322, 334, 341–342
Continuous wave (CW) spectrometer, 6, 36, 37, 40, 74
Contour interval or multiplier, 365
Contour map, 353
Contour plot, 365
Contour threshold, 365, 384, 406, 507
Contrast, in MRI, 552
Convolution, 114
Convolution theorem, 113, 115, 402f, 403
Coriolis forces, 202
Correlated crosshairs, 366
Correlation, chemical shift, 356
Cosine-bell window function, 403
Cost, xii, 8, 41, 468
COSY, 181, 242, 353, 354, 369, 370–393
COSY crosspeak, cancellation in DQF-COSY phase cycle, 455
COSY spectrum, cartoon, 371f
COSY transfer, between $^{13}C_\alpha$ and $^{13}C_\beta$ in HNCACB, 618
COSY vs. TOCSY, order of resonances in spin systems, 374
COSY, coherence flow diagram, 371f
COSY-35, 143, 369
  crosspeak fine structure, 383, 384f
  product operator analysis, 391–393, 392f
COSY-45, 393
COSY-type 2D spectra, for 2D TOCSY, 345
Coupling constants, 4, 14, 15, 52, 125
  measurement, 510, 538
  measurement in 2D COSY, 379
Covalent force field, 592
cpdprg2 parameter (Bruker), 149
CPMG (Carr-Purcell-Meiboom-Gill), 229
Critical micellar concentration (CMC), 569
Cro protein from bacteriophage λ,
  2D (^{15}N, ^{1}H) HSQC, 615, 615f
  3D HNCO strip plot, 615, 615f
  CBCA(CO)NH, 619f, 620–621
  HNCACB, 618, 619f
Crosspeak fine structure,
  in 2D NOESY, 429
  of peptide H_α-H_β in COSY, 383
Crosspeak volumes,
  accuracy and baseline correction, 437
  in 2D NOESY, 429
Crosspeaks, 19, 358
  phase in DQF-COSY, 450

Cross-relaxation, 168, 187–188, 194, 198, 253, 323–324, 408, 426
Cross-relaxation pathways, 411
Cross-relaxation rate, 327, 436
Crowded CH carbons, in steroids, 515
Cryogenic probes, 93, 95, 559
Crystal packing forces, 558
Crystallography, 553
Crystals, 551
CSA (chemical shift anisotropy), 60–61, 176, 357, 555, 624, 625
CSI (chemical shift index), 576, 576f, 587
Curvature, of NMR integral, 133
Cutoff tumbling frequency, 192
cvff force field, 590
CW (continuous wave) irradiation, 140, 193, 200, 322, 334, 341–342
CW (continuous wave) spectrometer, 6, 36–37, 40, 74
Cyclic octapeptide, 186
   COSY-35 spectrum, 383–384, 385f
   DQF-COSY, 382, 383f
Cycloheptanone, ^{13}C spectrum, 4, 4f
Cyclohexane chair, 12, 508
Cyclohexane, axial vs. equatorial proton chemical shifts, 508
Cyclopentadiene, 328
Cyrogenic probe, 626
Cysteine, 571

$d0$ delay, Bruker, 366
$d1$ parameter, 91, 175
$d16$ parameter (Bruker), 320
$d2$ parameter (Varian), 366, 401
D$_2$O sample, of protein or peptide, 315
d$_6$-benzene, 59
d$_6$-DMSO, as solvent for hydrophobic peptides, 569
Data cube,
   of 3D CBCA(CO)NH, 620f
   of 3D HNCACB, 618f
   of 3D HNCO, 614f
   of 3D HSQC-TOCSY, 604, 604f
   of 3D spectrum, 605f
Data matrix, of 2D spectrum, 358, 364, 366
Data processing,
   of 1D NMR spectra, 118–134
   of 3D NMR spectra, 604–606
Data sampling in t_1, 398–401
$dconi$ command (Varian), 365
$dcpd2$ parameter (Bruker), 149, 150
DDQ oxidation of pristimerin, 538

$dec2$ parameter (Varian), 559
Decible scale (dB), of pulse power, 144, 301, 349
Decimation, 111–113, 116
Decimation factor, 111, 117
Decoupled refocused INEPT, 271
Decoupler, 11, 74–75, 142, 146, 148–150, 185, 559
Decoupler duty cycle, 509
Decoupler field strength, 140
Decoupler mode (dm) parameter (Varian), 149, 154
Decoupler modulation (dmm) parameter (Varian), 149
Decoupler modulation frequency (dmf) parameter (Varian), 149
Decoupler nucleus (dn) parameter (Varian), 149
Decoupler offset (dof) parameter (Varian), 149
Decoupler power ($dpwr$) parameter (Varian), 149
Decoupler pulses, 150
Decoupler,
   90° ^{1}H pulse, 279
   a and b (Varian), 149
Decoupling, 11, 138, 140, 149, 152–153
   during evolution period, 611
   of ^{13}C, 496
   of antiphase signals, 270
   of deuterium (^{2}H), 557
   viewed as exchange process, 423
Defocusing, from in-phase to antiphase, 270, 271, 525
Degenerate CH$_2$ groups, 17, 55, 505, 571
Degree of overlap, of pure energy states, 442
Delay, 200–201, 212, 306
   effect on spherical operators, 445
Delta (δ) scale of chemical shift, 3
Delta t_1 (Δt_1), 398
Density matrix, 168, 243, 269, 408, 443, 469–478
   for heteronuclear two-spin system, 472
DEPT, 18, 27, 119, 138, 150, 238, 242, 276, 504
DEPT spectrum, of cholesterol, 280, 280f, 281, 281f, 507f
DEPT, product operator analysis, 283–287
Deshielding, 57–59
Destination state, 273
Destroying coherence, 290
Detection, 343, 354, 357, 361, 522
Detector, 94, 95, 201, 399

Deuterated proteins, 557
Deuterium decoupling, during ^{13}C evolution periods of deuterated proteins, 557
Deuterium exchange, in proteins, 553, 587
Deuterium gradient shimming, 564
Deuterium labeling of proteins, 623–624
Dewar, of superconducting NMR magnet, 8
dfrq parameter (Varian), 184
dg command (Varian), 149
Diagonal, 369
Diagonal appearance of HSQC/HMQC, 502
Diagonal peaks, 371
   phase in DQF-COSY, 450
Diagonal symmetry, 369
Diastereotopic methyl resonances, of valine, 572
Diastereotopic protons, 56
Diaxial (1,3) relationship, NOE, 428
Diels-Alder reaction, 328
Difference experiment, 1D NOE, 333
Difference in population, 241
Difference spectrum, NOE, 194
Diffusion in gradient experiments, 302, 304, 468, 552
Diffusion-like process, of TOCSY mixing, 345
Digital filtering, 75, 110–114, 116
Digital resolution, 118, 122, 401, 581, 603
Digital-to-analog converter (DAC), 563
Digitizer (ADC), 98, 185, 201
Digitizer noise, 109
Dihedral angle, xi, 6, 12, 15, 52, 370, 491
Dilute nucleus, 135, 609
Dimethyl fumarate, 328
Dimethyl methylphosphonate, 61
Dioxane, 145
Dipolar (direct coupling) Hamiltonian, 479
Dipolar (direct) coupling, 138, 622
Dipole-dipole interaction, 171, 553, 556, 622
Dipole-dipole relaxation, 171, 176, 556, 624
DIPSI, 341
DIPSI-2, 394, 396
Direct (dipolar) coupling, 138
Direct (one-bond) C-H couplings, 497, 536
Direct probe, 89, 90, 147
Disaccharide, 419
Disadvantages,
   of gradient selection, 468–469
   of phase cycling, 466–468
Disequilibrium, 165, 327, 410
Disordered regions, of protein, 552, 595
Dispersion, of chemical shifts, 59, 557, 573

Dispersive lineshape, 79, 126, 127, 208, 212, 215, 402, 447
   mathematical formula, 390
Display of 2D NMR data, 364–366, 365*f*
Distance geometry, 592
Distance measurement, from 2D NOESY, 370, 436
Distances, H to H in cholesterol, 331, 331*f*
Distortionless enhancement by polarization transfer (DEPT), 276
Distribution of tumbling rates, 173, 173*f*, 192, 409
Ditches, on sides of NMR peak, 126, 404
Divergence of multiplets, by J coupling evolution, 228
Diversity of starting conformations, 592
dm parameter (Varian), 149, 154
dmf parameter (Varian), 144, 149, 149, 150
dmm parameter (Varian), 149
dn parameter (Varian), 149
Dodecylphosphocholine, fully deuterated, 569
dof parameter (Varian), 149, 150, 183, 195
Double doublet, 6, 44, 45, 50
Double pulsed field gradient spin-echo (DPFGSE), 309
Double quantum coherence (DQC), quantum definition, 442
Double quantum filter, 447–450
Double triplet, 47
Double-double triplet, 49
Double-double-double-doublet, 51
Double-quantum (DQ) relaxation, 187, 189, 191, 198, 266, 429, 430
Double-quantum coherence (DQC), 266, 439
   heteronuclear, in HMQC, 533
Double-quantum coherence, sensitivity to gradients, 318
Double-quantum filtered COSY (DQF-COSY), 267, 447–450, 448*f*
Doublet, 10, 14, 44
   J coupling evolution, 223
Doublet-triplet, 508
Doubly-labeled (^{15}N, ^{13}C) proteins, 610
Downfield, 8, 37
Downfield handles for TOCSY transfer, 340, 345, 340*f*
DP parameter (Bruker), 144
dp parameter (Varian), 107, 134
DPFGSE, 309, 321, 343, 393
DPFGSE 1D NOE, 328, 333, 425
dps (display pulse sequence) command (Varian), 154

dpwr parameter (Varian), 144, 149, 150, 350
DQ artifacts, 375–376, 377*f*
DQ filter, in DQF-COSY, 448
DQ relaxation, 199, 323, 323, 324, 325, 335, 409, 324*f*
DQC, 357
   effect of gradient on homonuclear, 458
DQC/ZQC, 266, 285
   definitions, derived from spherical operators, 446
DQF-COSY, 143, 242, 320, 369, 370, 376, 389, 404, 408
   described using spherical operators, 461
   gradient pathway selection, 460
   peak phase *vs.* COSY, 391
DQF-COSY spectrum of sucrose, 600 MHz, 381, 382*f*
   upfield region (300 MHz), 380, 381*f*
DQF-COSY vs NOESY, selection in phase cycle, 454
Drawbacks, of phase cycling, 466
Drift,
   of B_0 field strength over time, 75, 78
   of NMR integrals, 133
Droop, of excitation profile, 299
Drug discovery, 553
DRX spectrometer (Bruker), 134
Drying NMR tubes, 77
DS parameter (Bruker), 107
Dual address ($^{15}N/^{1}H$) of chemical shifts, 600
Dual broadband RF sources, 147
Dummy scans, 107
Duty cycle, of decoupler, 495, 504, 509
DW parameter (Bruker), 99–101, 107, 110
Dwell time, 99, 103, 104, 110, 398
Dynamic processes in NMR, 414–425
Dynamic range, 76, 94, 108–109, 111, 117
   in phase cycling, 466
Dynamics, 137, 595
Dynamics simulation *vs.* B_0, DMF, 417*f*
Dynamics simulation *vs.* temperature, DMF, 416*f*

E. Coli HPr, 598
Echo mode, 606
Echo pathway, in 2D experiments, 454, 464–465, 529, 534
Echo, in gradient imaging experiment, 561
Echo-antiecho phase encoding, 465, 529, 605
eda command (Bruker), 149
Edited HNCACB, 618

Edited HSQC, 618, 504–509, 530–531, 530*f*
Edited spectra, 220, 283
EF command (Bruker), 126
Effective field (B_{eff}), 156, 292, 292, 341
Efficiency of transfer of magnetization, 366, 368
   TOCSY, 344
Eigenfunctions, 480
Eigenvalues, 480
Electron density map, from X-ray crystallography, 553
Electronegative atoms, 8, 57
EM command (Bruker), 126
Embedding, in structure calculations, 592
Encoding, of chemical shift in t_1 time domain, 356, 523
Energies, of two-spin system, 268
Energy diagram, 33, 189, 408
Energy gap (energy difference), 167, 169, 190, 269, 342, 477
Energy minimization, 592
Energy-minimized structures, 53, 54, 199, 332
Enhancement by polarization transfer, 255, 262
Enhancement, NOE, 198
Ensemble of calculated structures, 593
Ensemble of spins, 157, 158
Entanglement of two nuclei, in MQC, 533
EPT (enhancement by polarization transfer), 261
Equatorial position, in cyclohexane chair, 12
Equilibrium constant, for anomeric exchange, 419
Equilibrium population difference, 162, 180, 199, 261, 523
Equilibrium, lack of coherence at, 159
Equivalence, 9, 10, 52, 67
Ethanol, 201
Ethernet, 74, 108
Ethyl acetate, cartoon of 2D HSQC/HMQC and HMBC spectra, 500–501, 501*f*
Ethyl group, 375
Ethylbenzene, 108
Ethylene glycol, temperature standard for VT work, 425
Eukaryotic cell, 468
Evolution, 201, 203, 215, 219, 243, 245, 450
Evolution delay (t_1), 354–355, 355*f*, 357, 361, 366, 387, 522
Evolution delay, in triple-resonance experiments, 601, 611
Evolution of DQC and ZQC, 267

Evolution of non-stationary states (density matrix), 471, 484–485
Evolution,
  during gradients, 462
  J coupling, into antiphase state, 249
Exchange broadening, 417
  compared to CSA broadening, 555
Exchange peak, in 2D NOESY of lactose, 439
Exchange rate, calculated from NOE buildup, 421
Exchange spectroscopy (EXSY), 437–439
Excitation bandwidth, 349
Excitation profile, 297, 299, 308, 349
  Gaussian pulse, 309, 310*f*
Excitation pulse, shaped, 320
Excitation sculpting, 333, 308–309
Exercise,
  RF power level settings, 350
  SPT experiment, 262
  transient NOE populations for large molecules, 325
  ZQ relaxation and NOE, 191
Exotic nuclei, 90
Experiment time for 2D, reduction using gradients, 460
Exponential decay, 164
Exponential multiplier (window function), 126
Expression of proteins, in cell culture, 137, 596
EXSY (exchange spectroscopy), 437–439
Extended conformation (β sheet), 578
Extinction profile of Watergate, 314, 316, 316*f*

F_1, 118, 357
F1, F2, F3 channels (Bruker), 149, 559
F_2, 118, 357
F_2 spectra, vs. t_1, 356*f*
Failed magnetization transfer, 371
  in 2D HSQC-TOCSY, 602
Fanning out, of individual spin vectors, 163
Fast exchange, 415, 422–424
Fast exchange spectrum, 416*f*
Fast Fourier transform (FFT), 7, 122
Fast TOCSY, for H-D exchange rate measurement, 587
Feasibility, in biomolecular NMR studies, 558
Felix software, 99, 117, 120, 134, 406
Ferrocytochrome c_2, *Rhodobacter capsulatus*,
  2D {^{1}H, ^{15}N} HSQC, 606, 606*f*
  3D TOCSY-HSQC, 607, 607*f*
  3D TOCSY-HSQC vector plot, 607, 608*f*
Fictitious field (pseudofield), 291, 292, 334

Fictitious forces, 202
FID (free induction decay), 7, 8, 41–42, 75, 77–78, 88, 90, 92–94, 98, 103, 106, 108–110, 116, 118, 125–127, 134, 158, 185, 201, 207, 239, 241, 296, 353, 370, 439
  interactive shimming on, 85
FID button (Varian acqi window), 86
Field gradient, 290, 301
Field setting (Bruker), 79, 80
Field strength (B_0), in biological NMR, 558
File transfer protocol (ftp), 134
Filter function, of digital filter, 115
Fine structure, of COSY crosspeaks, 378–386
Fingerprint region, of COSY or TOCSY spectrum of proteins, 573, 580
First-order (chemical shift dependent) phase correction, 128–129, 130, 405, 505
First-order (exponential) decay, 164, 167, 411
First-order (weak) coupling, 45, 63, 67
First-order phase errors in F_1, due to gradient in t_1 period, 462
Five-spin system, in amino acids, 571, 572
FLATT, program for baseline correction, 407
Flexibility,
  effect on T_1 and T_2, 175, 180
  in disordered regions of proteins, 552
Flip-back pulse, 569
Floor of the 3D cube, of 3D TOCSY-HSQC, 609
Flow of magnetization, diagram for HETCOR, 357*f*
Flow, studied by pulsed-field gradients, 552
Fluorobenzene, J_{CF} couplings, 61
FM radio, 156
fn parameter (Varian), 123, 402
fn1 parameter (Varian), 402
Folded (native) proteins, 573
Folding (aliasing), 101, 102, 111, 129
  in F_1 dimension of 2D, 400
Folding motifs in proteins, 575
Folding problem, in proteins, 590
Footprint on a crosspeak, for volume measurement, 437
Footprint, of a ^{1}H resonance, 41, 135, 381, 554
Forbidden transitions, between states of opposite symmetry, 483
Force constant, for NOE restraints, 591
Force field, in structure calculations, 590
Four channels, in triple-resonance experiments, 611
Four step phase cycle, 537

Four-bond W couplings, 330
Fourier pair, 86
Fourier transform, 37, 42, 108, 119, 126, 133, 201, 212, 299, 353
  2D, 366, 367f
  of the RF pulse envelope, 297
Fourth channel, for ^{2}H decoupling, 559
Fragment analysis, of pristimerin oxidation product, 545–548
Free induction decay (FID), 6, 74, 169
Frequency domain, 7, 79, 114, 116, 118, 119, 126, 298, 299
Frequency list (Bruker text file), 184
Frequency response curve, of analog filter, 111
Frequency-shifted laminar pulse, 309–311, 310f
Front end, of homonuclear 2D experiments, 426, 428
Fructose, 16
FT-80 spectrometer, 40
ftp (file transfer protocol), 134
Fumarate-cyclopentadiene adduct, 1D transient NOE, 328–330, 329f
Furanose, 16

gain parameter (Varian), 94, 109, 568, 569
Galactose, in lactose structure, 419
Gamma (magnetogyric ratio γ), 155, 317
GARP sequence, for ^{13}C decoupling, 496, 504, 568, 599
Gated decoupling, 153
gauche relationship, three bond, 22, 24
Gaussian function, 125, 300, 300f
Gaussian pulse, 308
  calibration, 351f
Gaussian window function, 125
G-BIRD (pulse sequence building block), 494
Geminal relationship, of two protons, 17, 25, 28, 508
gf parameter (Varian), 126
Gigahertz (GHz), NMR magnets, 625
Gimbles, 157
Global minimum, in structure calculation, 593
Global relationships, in protein structure, 554
Glucose, 12, 496
  ^{13}C-labeled, ^{1}H spectrum, 493, 493f
Glutamate, 571
Glutamine, 571
Glycopeptide YTGFLS(Lactose),
  2D ROESY spectrum, 435–436, 435f
  2D TOCSY spectrum, 397–398, 397f

Glycosidic linkage, 13, 14, 19, 196
  NOE across, 428
go command (Varian), 148
gpz1 parameter (Bruker), 320
Grab and drag, by adiabatic shaped pulse, 334
Gradient amplifiers, 319
Gradient coherence pathway selection, 460, 527
Gradient coils, in probe, 319
Gradient echo, 304–305
Gradient-enhanced experiments, 462
Gradient parameters, 320
Gradient pulse, 305
Gradient-selected experiments, 317, 319, 462
Gradient shimming, 87, 88, 559–564
  pulse sequence, 560f
Gradient twisting, total for a pulse sequence, 446
Gradient-enhanced INEPT, 317, 317f
Gradients *vs.* phase cycling, definition of coherence order, 457
Gradients, pulsed-field, 186, 316
  during the acquisition of the FID, 552
  effect on spherical operators, 444
  practical aspects, 319–321
Gradient-selected HSQC, 528–530
Gradient-selected refocused INEPT, 318, 318f
Graphical list of chemical shifts, HSQC spectrum as, 543
Gro-EL complex, 626
Group delay of digital filter, 113, 117
gs command (Bruker), 85
gstab parameter (Varian), 320
gt1 parameter (Varian), 320
gzlvl1 parameter (Varian), 320

H_α proton region, 315
H/D exchange experiments, 589, 589f, 600
Habitual violators of NOE restraints, 595
Half-life, of T_1 relaxation, 177
Hamiltonian matrix, 409, 478–488
Handles,
  for NOE difference, 197
  for TOCSY, 340, 340f
Hard pulse, 145, 150, 182, 185, 192, 252, 291, 301, 321, 348–350
Hardware, 320
Hardware requirements, for biological NMR, 558–564
Hardware, for pulsed-field gradients, 319
Hartmann-Hahn match, 341, 394
HCCH-TOCSY experiment, 609

Heartbeat, of the gradient experiment, 320
Heat exchanger, liquid nitrogen, 424
Heat flow analogy, 164, 167, 194, 194f, 325–326, 325f
Heater block, in cryogenic probe, 559
Heavy atoms, in structure calculation, 554
Helical coherence, after gradient pulse, 302–303, 306, 307
Helium gas refrigerator, in cryogenic probe, 559
Heme aromatic system of cytochrome c_2, 609
Heregulin-α EGF domain,
 ^{1}H chemical shifts, 573–574, 574f
 2D ^{1}H,^{15}N HSQC, 599–601, 599f
 2D HSQC-TOCSY, 602, 602f
 2D NOESY, 600f
 3D NOESY-HMQC, 600f, 601
 assignment of major β-sheet, 582–586
 D$_2$O NOESY spectrum, 587, 588f
secondary structure, 576, 576f
 secondary structure evidence, 589, 590, 590f
HETCOR experiment (2D), 242, 353, 369, 370, 390, 489, 354–364
 coherence flow diagram, 357f
 compared to HSQC/HMQC, 498
HETCOR pulse sequence, 361f, 362f, 364f
 design of, 361–364
HETCOR spectrum of sucrose, 360, 360f
 compared to HMQC spectrum, 503
HETCOR spectrum, diagram, 359f
Heteronuclear coherence order, 318
Heteronuclear decoupling, 139
Heteronuclear energy diagram, 258, 259f
Heteronuclear multiple bond correlation (HMBC), 489
 examples, 509–517
 pulse sequence, 535–538
Heteronuclear multiple quantum correlation (HMQC), 489
 examples, 501–504
 pulse sequence, 533–535
Heteronuclear NOE, 151, 153, 198, 228
Heteronuclear population diagram, 258, 259f
Heteronuclear single quantum correlation (HSQC), 524
 examples, 504–509
 pulse sequence, 522–532
Heteronuclear spin echo, 232–237
Heteronuclear spin-echo, product operator analysis, 524
High vacuum, in cryogenic probe, 559
Higher order shims, 560

Hindered rotation of the amide linkage, 572
Histidine, 571
 pK_a measurement by titration, 558
Histogram of tumbling rates, 172
HMBC, 19, 25, 27, 138, 143, 242, 354, 369, 489, 497, 533
 phase sensitive, 537–538, 537f, 538f
 sensitivity, compared to other 2D experiments, 509
HMBC pulse sequence, 535–538
HMBC spectra, 509–517
HMBC spectrum of β-carotene oxidation product, 521–522, 521f
HMQC, 138, 143, 369, 489, 497
HMQC pulse sequence, 533–535
HMQC spectrum of β-carotene oxidation product, 519–520, 520f
HMQC vs. HMBC, 536
HMQC, spherical operator analysis, 534
HN(CO)CA experiment, 617
HN(CO)CACB experiment, 620
HNCA experiment, 615
HNCACB experiment, 617–620
 1D vectors, 618, 619f
 data cube, 618, 618f
 strip plot, 618f, 619, 619f
HNCO data matrix (cube), 614, 614f
HNCO,
 3D, 611–615
 coherence flow diagram, 612f
 correlations, 612f
 pulse sequence and product operator analysis, 611–614, 612f
HOHAHA (homonuclear Hartmann-Hahn), 341
Holes,
 in center of HSQC methyl crosspeaks, 507
 in coherence order mask, 454
homo parameter (Varian), 149
Homogeneity,
 of B$_1$ field, 290, 334, 334f, 393
 of B$_0$ magnetic field, 34, 41, 77, 79, 80, 82
Homonuclear (J_{HH}) J-coupling evolution in HMBC, eliminating, 538
Homonuclear decoupling, 142, 143, 148, 181–182, 184
Homonuclear front end, 386, 448, 455
Homonuclear Hartmann-Hahn (HOHAHA), 341
Homonuclear NOE, 198
Homonuclear product operators, 252

Homonuclear splitting pattern in HSQC/HMBC, 500
Homonuclear two-spin system, coherence order, 445
Homospoil gradient, 529
Horizontal stacked plot, 178
Hot stone, in heat flow analogy, 325
House shape, in second-order splitting patterns, 483
HPr,
  B. subtilis, 3D structure, 579*f*
  B. subtilis, NOESY spectrum, 578–580, 579*f*, 580*f*
  E. Coli, 3D HMQC-NOESY, 598, 598*f*
HSQC, 18, 23, 25, 27, 138, 143, 242, 354, 364, 369, 489, 497
HSQC and HMQC, compared to HETCOR, 498–499
HSQC spectrum, 2D {^{1}H-^{15}N}, 598, 599, 599*f*, 606*f*, 615*f*
HSQC,
  ^{1}H-^{13}C pulse sequence, 522–532, 523–525*f*, 527–530*f*, 532*f*
  ^{1}H-^{15}N, and digital filtering, 117
  ^{1}H-^{15}N, pulse sequence, 601, 601*f*
  coherence flow diagram, 498
  product operator analysis, 525–532
HSQC-NOESY spectrum, 3D for sequence specific assignment, 610
HSQC-TOCSY,
  2D, pulse sequence, 601–602, 601*f*
  3D, coherence flow diagram, 603*f*
  3D, pulse sequence, 603, 603*f*
HSQMBC, 538
Hydrocarbon positions of chemical shift, 27, 572
Hydrophobic collapse, 573

i to i+1 NOEs (sequential), 435, 577
Ideal isotropic mixing, 394
Identity operator, 269, 474
IF (intermediate frequency), in receiver, 96, 108
ii command (Bruker), 148
Imaginary FID, 93, 96, 103
Imaginary part, of coherence, 442
Imaginary spectrum, 122, 405
Imidazole ring of a histidine residue, NMR titration, 424
Immortal coherence, in TROSY experiment, 625
Impedance, 89

Improved APT, 234, 235*f*
in0 parameter (Bruker), 366, 401
Inclusion bodies, in protein expression, 596
Incremented delay (t_1), 353
Independent time domains, in 3D NMR, 603
Indirect or scalar coupling (J), 622
Indirect second frequency domain, 353
Indirect time domain t_1, 353, 356, 367
INEPT, 138, 238, 241, 253, 339, 342, 361, 610
  advantages, 256
  compared to COSY, 387
  compared to SPT, 262
  density matrix description, 476–478
  refocused, 270–276
  refocused, in 2D HETCOR, 362
  with trim pulse, 335
INEPT coherence transfer, 242, 263, 354, 356, 357, 429, 523
INEPT-45, INEPT-90 and INEPT-135, 276
Informational biomolecules, 551
Inhomogeneity map, in gradient shimming, 560, 562
Inhomogeneity,
  of B_0 field, 230
  of B_1 field, 567
Initial slope, in NOE buildup, 322
Injectable contrast agents, 181
Inner coil, of probe, 89, 147, 148
Inner loop, of 3D experiment, 605
In-phase component, 250
In-phase COSY diagonal peak in F_1, 390
In-phase doublet, 386
In-phase peaks, 375
In-phase relationship of vectors, 215, 236, 243, 244*f*
In-phase to in-phase transfer, 339, 342
Integrated area, 10
  in exchange, 419
Integration, 9, 112, 118, 133
Intensity plot, of 2D data, 353, 365, 365*f*
Interferogram, 357, 358*f*
Interleaved data acquisition, 195
Intermediate frequency (IF), in receiver, 96, 108
Intermediate state, in coherence transfer, 257, 265–267, 283, 317, 389, 429, 529, 533
Intraresidue crosspeaks, in ROESY of peptide, 436
Intrinsic decay (T_2 relaxation), 230
Inverse experiment, 147, 363, 489, 490, 489–550

Inverse gated decoupling, 153
Inverse matrix, 470
Inverse mode, hardware setup, 148, 148f
Inverse probe, 89, 90, 148, 490
Inverse sixth-power rule, for NOE intensity, 198, 241, 591
Inversion of populations, analogy, 260
Inversion of spherical operators, with 180° pulse, 445, 463
Inversion profile, 294
Inversion pulse, 144, 207, 237, 320
   vs. refocusing pulse, 228, 251
Inversion-recovery experiment, 176–181, 177f, 231, 266
   sucrose ^{13}C, 178, 178f
   sucrose ^{1}H, 179–180, 179f
Ionic strength and probe matching, 568
ipso position, in aromatic ring, 29
Isochromats, in spin-echo, 230
Isolated spin-pair hypothesis, 436
Isoleucine, 570
Isoprene, 20
Isopropyl group, mutual HMBC correlations, 516
Isotope companies, 131
Isotope effect on chemical shift, 491
Isotope filtering, 493
   applied to NOE of protein-ligand complex, 494
   using G-BIRD, 494
Isotopic labeling (enrichment), 32, 38, 62, 136, 551, 557
Isotopomer, 62, 136–137
Isotropic chemical shift, 60
Isotropic *J* coupling interaction, 479
Isotropic mixing (TOCSY), 341, 342, 396, 409, 486–488
Isotropic mixing Hamiltonian, 486–487
Isotropic tumbling, 622

J coupling, 135, 138, 156, 200, 208, 216, 218, 228, 342
J coupling evolution,
   density matrix representation, 477
   product operator representation, 248
J coupling vs. NOE for stereochemical assignment, 519
J coupling, effect on ZQC & DQC, 268
J couplings,
   systematic errors in measurement, 385–386, 386f

J_{CH} value, dependence on chemical environment, 495
J-coupling (ZQC) artifacts, in transient NOE, 328
J-coupling evolution, 232–233, 247
   refocusing, 362
J-coupling patterns in F_1 of 2D spectrum, product operator analysis, 389–390
JPEG graphics format, 134
Jump, of coherence, 242
Jump-return (1,1), 312, 568
Jump-return NOESY spectrum, 582

Karplus relation, 53–54, 53f, 241, 330, 491, 510, 519

Laboratory frame of reference, 203, 213, 219, 241, 296, 335
Lactose, 419, 421, 438, 495
Lambda-Cro protein, see Cro protein from bacteriophage λ
Laminar pulse, frequency shifted, 309–311, 310f
Large *J* couplings, in steroids, 331, 432
Larmor frequency, 1, 3, 31, 33, 35–37, 41, 56, 156–158, 161–162, 169–173, 190, 201–202, 204, 213, 216, 241, 291, 306, 337, 471
   in exchange process, 420
Lattice, 167
Lattice temperature, 194
LB parameter, 123, 126
LC (tuned) circuit, 204
Leaning (second-order) multiplets, 64–65, 70, 481, 483
Least-significant digit (bit), 185
Lenz's law, 57, 58
Leucine, 570
LGJC3 (pristimerin oxidation product), 538–550
Line broadening, by exchange in fast exchange regime, 422
Linear gradient, 83
Line-broadening (LB) parameter, 123, 124f, 125f, 126f
Linewidth, 84, 142, 229
   as indicator of disordered regions of a protein, 573
   vs. molecular size, 554
Linewidth problem, for large molecules, 556, 623
Linked differential equations, 338

Lipid bilayers, 551
Liquid crystal solution, 622
Liquid nitrogen, 109
Local minima, in structure calculation, 593
Local oscillator (LO), 96
Local reporters, 552
Lock gain, 81
Lock level, 82, 85
   effect of gradient, 320
Lock parameters, 80
Lock phase, 79, 83, 86
Lock power, 81
Lock shift (Bruker), 102
Lock switch, for ^{2}H decoupling, 559
Lock, ^{2}H feedback loop, 75, 78, 79, 80, 101, 185, 195, 320, 559
Logarithmic scale (dB), of pulse power, 301
Longitudinal (T_1) relaxation, 166–167, 170, 171
Longitudinal spin order, 269, 474
Long-range couplings (C-H), 137, 489, 495, 497
Long-range couplings (H-H), 61
Long-range NOE correlations, 586–588
   β-sheet cross-strand, 587, 588f
Lookup table for chemical shifts, 514
Lorentzian lineshape, 51, 229, 365, 386, 390, 554
Loss of sensitivity, in gradient experiments, due to diffusion, 468
Lower cone, in vector model, 160
Lowering operators, definition, 444
Low-γ nuclei, 299
Low-pass filter,
   analog audio, 112
   in HMBC, 509, 511, 538
Low-power irradiation, 138, 181, 353
lp parameter (Varian), 130, 406
Lysine, 571

Macroscopic z magnetization, 269
Magic angle spinning, 551
Magnet leg (Varian), 95
Magnetic equivalence, 54, 71
Magnetic resonance imaging (MRI), xi, 301
Magnetic susceptibility, 565
Magnetic vector, 157
Magnetization transfer, 192, 214, 238, 241, 369, 427
Magnetogyric ratio (γ), 1, 30–31, 38, 46, 56, 141–142, 155, 318, 473, 528, 565

Magnitude mode, 362, 390, 402, 462, 500, 535, 537
Magnitude mode *vs.* phase sensitive, for HMBC, 510, 510f
Map,
   of all NMR interactions, 367
   of binding sites on protein surface, 600
Mask, in selection of coherence order changes, 453, 468
Matching of the probe circuit, 88
   and ionic strength (salt), 565
Matrix multiplication, 471
Matrix representation (density matrix), 243
MC2 parameter (Bruker), 401
m-chloroperbenzoic acid (mcpba), oxidation of β-carotene, 519
Mechanical relays, 185
Medical NMR imaging (MRI), 181
Medium, category of NOE crosspeak volume, 437
Medium-range NOE correlations, 586–588
Membrane-bound biomolecules, 551
Memory overflow, in sum-to-memory, 107
Menthol, 20
   ^{1}H spectrum, 21–23, 22f, 42f, 43f, 44f
   ^{13}C spectrum, 23, 23f
   for calibration of 90° ^{1}H pulse in DEPT, 279, 282
   structure, 20f
MestRec software, 99
meta positions in benzene ring, mutual HMBC correlations, 516
Metabolic studies by NMR, 137
   in cell cultures, 552
Metabolism of natural products, 553
Metabolites of testosterone, 517–519
Methanol, temperature standard for VT work, 425
Methionine, 571
Methyl group,
   ^{13}C product operators and vectors, 272f, 273
   in HMQC/HSQC, 503
Methyl iodide, ^{13}C-enriched, 62, 217
Methylene group,
   ^{13}C product operators and vectors, 271
   J coupling evolution, 273
MI parameter (Bruker), 132
Micelles, 569
Microimaging, 552
Microscopic gas constant, 34
Microscopic spin state, 243
Microscopic z magnetization, 269

Microscopy of plant or animal tissues, 552
Minimal media, in protein expression, 596
Minimum intensity, 132
Miscalibration of ^{1}H pulse in DEPT, 279, 281f, 282f
Miscalibration of 90° pulse, 295
Mixed coherence order, for Cartesian operators, 451
Mixer (phase sensitive detector, ring modulator), in receiver, 95
Mixing, 343, 354, 357, 361, 369, 522, 601
Mixing time (τ_m), 193, 195, 198, 242, 322, 323, 330, 339
    and spin diffusion in 2D NOESY, 436
    random variation, in 2D NOESY, 429
Mixing,
    for analog subtraction of frequencies, 95
    in homonuclear 2D, product operator analysis, 388
    of pure quantum states, 441, 469
MLEV-17 TOCSY mixing sequence, 341, 345, 394, 396
M_0 (equilibrium net magnetization), 160–162, 164–166, 169–170, 175, 177, 179, 195, 241
Modulator (phase-sensitive detector), 95
Molecular dynamics (motion), 170, 175
Molten globule state, of proteins, 573
Monosaccharide, 12
Monoterpene, 20
Moving pair, of 180° pulses, in constant-time evolution, 613
Moving spin lock for adiabatic inversion, 496
MQC, see multiple quantum coherence
MRI (magnetic resonance imaging), xi, 83, 86, 87, 319
mult parameter (Varian), 279
Multiple jumps of magnetization, 339, 340, 369
Multiple-pulse experiment, 176
Multiple-quantum coherence, 251, 266, 267, 533, 534
    from antiphase SQC, 440
    heteronuclear, in HMQC, 533
Multiplicity, ^{1}H, from F_2 slices of HSQC, 543
Multiplier (window) function, 401, 403
multizg command (Bruker), 279

N,N NOE, 578
N,N-dimethylformamide (DMF), 415f, 422
Natural abundance, 11, 135, 136
Natural products, xi, 12, 55, 135, 137

n-butyl group, 375
nd0 parameter (Bruker), 401
Nearest neighbor shifts, in sorted assignment list, 514
Negative absorptive, 208
Negative frequency, in rotating frame, 297
Negative NOE, 195, 198, 199, 325, 335, 409, 414, 429–430
Neighbors, in sorted chemical shift list, 542
Neopentyl alcohol, 507
Net chemical shift evolution, 234
Net magnetization, 34–35, 151, 157–158, 162–164, 168–169, 181, 238, 289, 439
Net magnetization vector, 159, 201, 206–207, 222, 240
Net magnetization, at equilibrium, 160
Net transfer of magnetization, in NOESY and TOCSY, 429
Net z-magnetization, 151
Newman projections, 384
N-fold mask, positioning, in phase cycle, 454
ni parameter (Varian), 401–402, 404
Nicolet, 40
Nitrogen-15 natural abundance, 598
NMR imaging (MRI), 289, 551, 559
NMR restraint energy, 592
NMR time scale, 15, 60, 139, 176, 183, 338, 414, 415, 555
NMR-pipe software, 99
NOE (nuclear Overhauser effect), 6, 138, 150–152, 152f, 153–154, 162, 168, 171, 173, 175, 180, 219, 241–242, 408, 577
NOE buildup, 152, 193, 195, 198–199, 326
    sucrose, 330–331, 330f
NOE difference, 148, 181, 192–198, 193f, 289, 321
    heat-flow analogy, 194f
    time course of z-magnetization, 195f
NOE enhancement, 152–153, 194, 335
NOE experiment, for detecting exchange, 420
NOE mixing time, *vs.* molecular size, 331
NOE transfer of magnetization, in spin-lock, 430
NOE transfer, efficiency compared to TOCSY, 347
NOE,
    homonuclear, 187
    of large molecules, 188
    sign for small molecules *vs.* exchange, 421
    vs. J coupling for stereochemical assignment, 519

NOESY, 181, 196, 242, 354, 369, 389, 406, 408
NOESY buildup study, 436
NOESY spectrum of heregulin-α EGF domain, in D_2O, 587, 588f
NOESY spectrum,
 coherence flow diagram, 425
 sign of crosspeaks relative to diagonal, 426–427
NOESY vs. DQF-COSY, selection in phase cycle, 454
NOESY, gradient-enhanced, 426, 428f
Noise figure, 108
Noise-modulated decoupling, 144
Noncovalent interactions, 551
Nonlinear least-squares fit, 132
Nonobservable states, 252
Normal mode (vs. inverse mode), 490
Normalization factor, for product operators, 243, 247
Normalizing the peak areas, 133
NP parameter (Varian), 99, 104–105, 123, 134, 402, 504
n-propanol, 136, 137, 138
n-propyl benzoate, 339
NS parameter (Bruker), 537
NT parameter (Varian), 537
Nu-1 (ν_1, precession rate around B_1 vector), 206, 208, 292–293, 297, 334, 341
nuc2 parameter (Bruker), 149
Nuclear Overhauser effect, see NOE
Nuclear quadrupole moment, 31
Null in excitation, for soft rectangular pulse, 568
Number of attached protons, in 2D HSQC, 505
Number of contours, 365
Number of scans (transients) in selectivity of phase cycle, 456, 526
Nyquist theorem, 398

O1 parameter (Bruker), 101–102, 150
O2 parameter (Bruker), 149–150, 183, 195
Observable operator, 252
Off-axis shims, 87–88
Off-diagonal terms of the Hamiltonian, in strong coupling and TOCSY mixing, 487
Off-resonance, 219, 222, 349
Off-resonance effects, in spin-lock, 336
Off-resonance pulses, 291–292
Offset (rotating-frame frequency Ω), 201, 320
Offset parameter (Bruker O1 or Varian TO/tof), 101

Offset, decoupler parameter (Bruker O2 or Varian DO/dof), 150, 183
Ohm's law, 88
Olefinic proton, olefinic carbon, 9
Oligosaccharides, 12
On resonance, 201–202, 206, 215, 219, 222, 225, 309, 349, 350
One-bond artifacts, in HMBC spectrum, 509, 514, 536
One-bond couplings, in labeled proteins, 557
One-bond INEPT transfers, in protein assignment, 557, 610
On-resonance pulses, 291
Operators, 242, 478
Order of pulses, in INEPT transfer, 267
Order, of matrix multiplication, 471
Organic synthesis, 553
O-ring vacuum seals, 425
ortho positions in benzene ring, mutual HMBC correlations, 516
Oscillatory behavior, in TOCSY transfer, 344
Out-and-back scheme of coherence transfer, 602, 611, 616
Outer coil, of probe, 89, 90, 147–148
Overexpressed protein, 137
Overlap, 554, 598
 of pure quantum states, 441
Overpopulated state, 410
Oversampling, 74, 109–110
Overselectivity of gradients, 463, 468
Oxidation of β-carotene, 519–522
Oxygen substitution, effect on one-bond C-H coupling constant, 502

p (coherence order), 317
p16 parameter (Bruker), 320
p3 parameter (Bruker), 150
Pairing of CH_2 peaks, in 2D HSQC and HMQC, 505
Pancake analogy, for 180° pulse, 226
Pandora's Box (Varian program Pbox), 351
Paramagnetic molecules, 180
Parameters for 2D NMR, 407t
Parts per million (ppm), 3, 33, 201
Pascal's triangle, 5, 132
Passive couplings,
 in 2D COSY, 380
 in multiple-quantum coherences, 285
pcpd2 parameter (Bruker), 144
Peak lists, 132
Peak phase, 209
Pedestals, in lineshape, 83–84, 108

Penalty function, in structure calculation, 590–591
Peppermint oil, 20
Peptide bond,
 H_N proton exchange rate *vs.* pH, 423
 NOE across, 428
Perturbations of chemical shifts, by unsaturations, 552
PFG (pulsed-field gradient), 301
PFGSE (pulsed-field gradient spin echo), 307–308, 311
PFGSE, for calibration of 180° shaped pulse, 351, 351f
Phase coherence, 158, 161–164, 169–170, 439
 quantum definition, 442
Phase correction, 102, 108, 126–130, 128, 218, 405
 in two dimensions, 405–406
Phase cycling, 107, 209, 211, 263, 304, 427, 450, 501, 536
Phase cycling *vs.* gradients, definition of coherence order, 457
Phase cycling,
 combining for multiple pulse sequence, 456
 in DPFGSE-NOE, 333
 in DQF-COSY, 447–450
Phase encoding, in t_1 time domain of 2D experiment, 400f
Phase error, 117–118, 126
Phase error in F_1, due to gradient in t_1 period, 462
Phase factor, 445–446, 464, 471, 477
Phase labeling, in edited ^{13}C spectra, 277
phase parameter (Varian), 401
Phase ramp, of frequency-shifted laminar pulse, 310–311, 320, 337
Phase reference, 248
Phase rotation angle, in phase correction, 128
Phase-sensitive 2D NMR, 399–401, 462, 464
Phase-sensitive data presentation in HSQC, 528–530
Phase-sensitive detection (quadrature detection), 211
Phase-sensitive detector, 95
Phase-sensitive HMBC, 500, 537–538, 537f, 538f
Phase twist,
 due to chemical shift evolution, 307
 due to gradient pulse, 304
 due to improper phase correction, 129

Phase,
 in time domain and frequency domain, 126
 quantum definition, 442
phc0 parameter (Bruker), 406
phc1 parameter (Bruker), 130, 406
Phenetole, 140, 142, 153
Phenylalanine, 571
Phi (Φ) angle, *vs.* H_N-H_α dihedral angle, 53, 580
Phospholipid micelle, effect on linewidth of bound molecules, 556
Phosphotransferase, 578
pl17 parameter (Bruker), 149–150
Piano analogy, 7
PIN diodes, for RF power switching, 149
Pivot peak, in phase correction, 128–129, 405
pK_a value, measurements by NMR, 424
pl1, *pl2* parameters (Bruker), 150, 350
Planck's constant, 3, 33–34
Polarization, 256, 261, 263
Polynomial baseline correction, 133
Population, 33, 91, 150, 239, 323
Population diagram, 190
 4-state, 189
 after selective inversion, 323, 323f
Population difference, 12, 34, 138, 151, 159, 170, 181, 191, 198, 207–208, 256, 269, 408, 410
Population, in α state, 159
Porches, in lineshape, 83, 84, 108
Position-dependent phase shift, 450, 458
 mathematical form, for spherial operators, 459
Positive absorptive peak, 208, 210
Positive crosspeaks, in NOESY, due to exchange, 437
Positive NOE, 195, 198, 199, 409, 414, 429
PostScript graphics format, 134
Power attenuation, 301
Power level, 348, 349
Power level for shaped pulse, calculation of, 351
pp parameter (Varian), 150, 279
pplvl parameter (Varian), 150
ppm (parts per million), 3
Preamplifier, 89, 94–95
Precession, 1, 31, 156–157, 159, 168, 201, 204, 212, 222, 245, 296
Precession frequency, 3
Precession rate, 156
Preparation, 343, 354, 357, 361, 370, 498, 522
Preparation pulse, 399

Presaturation, 181, 186, 302, 311, 315, 383, 568
Preservation of equivalent pathways (PEP), 531–532
Pristimerin, 538, 538f
Pristimerin oxidation product,
 ^{13}C spectrum, 540–541, 540f
 ^{1}H spectrum, 539–540, 539f
 assignment table, 544t
 COSY spectrum, 546f
 DQF-COSY spectrum, 546–547, 546f
 fragment analysis, 545–548
 HMBC spectrum, 545f
 HSQC spectrum, 541–543, 541f, 542f
 ROESY spectrum, 549f
Probability, in quantum mechanics, 441, 469, 472
Probe, 8, 75, 77–78, 92
Probe coil, 8, 77, 83, 89–90, 93, 95, 142, 169–170, 240–241, 299, 368, 370, 439, 446, 460
Probe matching *vs.* ionic strength, 568
Probehead, 77
procpar file (Varian), 119
Product operator, 168
 analysis of COSY, 386–393
 analysis of DEPT, 283–287
Product operator formalism, 242–253
Product operator representations of pure ZQC and DQC, 440
Proline, 570
 δ-CH$_2$ analogy to H$_N$, 572
Promotor, in protein expression, 596
Pro-R and pro-S protons in CH$_2$ group, 385
Protease digestion, 558
Protection factor, for H/D exchange, 587
Protection of amide N-H groups from solvent, 553
Protein expression, in labeled media, 596
Protein NMR spectroscopy, 551, 570–626
Proteins and peptides, 315
Proton decoupling, 228, 362
Pseudo-atom, in protein structure calculation, 591
Pseudofield (fictitious field), 202, 293
Pseudorotation, of pentavalent phosphorus, 422
Psi (Ψ) angle, 53, 581
pulprog parameter (Bruker), 154
Pulse (radiofrequency), 8, 37, 90, 105, 200, 203–206, 204f
Pulse calibration, shaped pulse, 351, 351f

Pulse phase, 92, 203, 205, 291
Pulse power, 299
Pulse sequence, 93, 204, 450
 1D TOCSY, 343f
 1D transient NOE, 321
 2D NOESY, 427–428f, 455f
 2D TOCSY, 394f
 3D NMR, 601–610, 601f, 603f, 612f
 APT, 228f
 COSY, 370, 370f
 DEPT, 276f
 DQF-COSY, 448f, 454f, 461–465f
 HETCOR, 361–364
 HMQC, 533–535, 533–535f
 HSQC, 522–532, 523–525f, 527–530f, 532f
 improved APT, 235f
 INEPT, 256f, 317–318f
 inversion-recovery, 177f
 NOE difference, 193f
 ROESY, 430f
 triple-resonance, 610–621
 Watergate, 314f
Pulse sequence building blocks, 200
Pulse width, 88, 92, 104–105, 119, 203, 297
Pulse width calibration, 92, 92f, 208, 334
 based on radiation damping, 567, 567f
Pulse width ratio, conversion to dB difference, 350
Pulse, (B$_1$) amplitude, 205, 293, 297, 299
Pulsed field gradient spin echo (PFGSE), 307, 493
Pulsed field gradients, 74, 181, 196, 200, 263, 265, 289–290, 301, 308, 319, 450
 for coherence pathway selection, 457–464
Pulsed Fourier transform, 6, 7, 37, 135
Pulse-FID sequence, 91
PulseTool program (Bruker), 351
Pure INEPT spectrum, 264
Pure subspectra, calculated from DEPT spectra, 280
Pure-phase excitation, using PFGs, 301
Purge pulse, 334–335
Puzzle solving, with HMBC data, 512
PW parameter, 91
Pyranose sugar, 15, 340

Q factor of NMR probe, 567
Quadrature artifacts, 211–212
Quadrature detection, 93, 96, 100–102, 104, 111–112, 209–210, 398
Quadrature detection in F_1, 399–401
 using gradients, 464–465

Quadrature image, 210
Quadrature phase cycling, 211
Quadrupolar Hamiltonian, 479
Quantum mechanical, 64
Quantum mechanics, xii, 157, 251, 439–443
Quantum model, 2, 155
Quantum view of coherence, 441–443
Quartet,
   for methyl group ^{13}C resonance, 12
   J coupling evolution, 224
Quaternary carbon,
   in HETCOR, 360
   in HMBC, 511, 543
Quench, 425
Quintet, 47

R_1 (longitudinal relaxation rate), 170, 176
R_2 (transverse relaxation rate), 170
Radiation damping, 566–568
Radio frequency (RF) coils, of probe, 78
Radio frequency (RF) power level, 182, 208
Radio frequency (RF) pulse, 90–93, 133, 157, 170, 203, 206, 240
   effect calculated using Hamiltonian, 485
   effect on density matrix, 469
   effect on product operators, 251
Radioactive decay, 164
Radioactivity, 137
Raging boil, analogy for decoupling, 152
Raising and lowering operators, 443–447
Raising operators, definition, 444
Random coil chemical shifts, 571–572
Random coil exchange rates for H/D exchange, 588
Random isotropic molecular tumbling, 479
Rapid isotropic tumbling, 59, 138, 551, 553
Rate constant for NMR relaxation, 410
Rate equation for self-relaxation, 411
Rates of relaxation, ratios for small and large molecules, 411
RD parameter (Bruker), 91, 175
Read pulse, 185, 192
Real and imaginary FIDs, 532
Real FID, 93, 96, 103
Real Fourier transform, 400
Real part, of coherence, 442
Real spectrum, 122, 405
Receiver, 8, 94–107, 201, 296, 304, 399
Receiver gain, 76, 94, 109, 185–186, 304
   in HSQC, 528
   in phase cycling, 466

Receiver gain as a measure of water suppression, 568
Receiver gain imbalance, 211
Receiver gain setting, in 2D experiments in 90% H_2O, 569
Receiver phase, 208–209, 211, 248, 333
Receiver phase cycle, 454
Receiver phase,
   in DQF-COSY phase cycle, 448
   routing of FID signals, 453
Recovery delay for gradient, 462, 529
Rectangular excitation profile, 300
Rectangular pulse, 92, 203, 297
Recycle delay, 504
Reduced coupling constant (J_R), with decoupling, 139–140, 183
Reducing sugar, 15, 419
Reference axis, 208, 210, 213, 217–218, 225
   in DQF-COSY phase cycle, 448
Reference frequency (v_r), 95, 100–102, 104–105, 185, 213, 296, 308
Reference peak, 118, 130
Referencing, chemical shift scale, in biological NMR, 565–566
Reflected power, 89
Refocused INEPT, 270, 283,
   product operator analysis (CH_2), 273
   product operator analysis (CH_3), 275
   vector analysis (CH_2), 274, 274f
   vector analysis (CH_3), 275, 275f
Refocusing, 227, 248, 251, 271, 284, 304, 504
Refocusing delay, 270, 283
   in HSQC pulse sequence, 525, 536
Refocusing pulse, *vs.* inversion pulse, 228
Refocusing,
   of CH coherence, 270
   of chemical shift evolution, 234
   of J-coupling evolution, 235, 250
Regiochemistry by NOE, 434
Regiochemistry of hydroxylation of testerone, 519
Relaxation, 11–12, 26, 36, 60, 91, 107, 133, 138, 151, 162, 168, 170–171, 187, 241, 246
Relaxation after a 180° pulse, 166
Relaxation after a 90° pulse, 162
Relaxation compensated DIPSI (DIPSI-2rc), 396
Relaxation delay, 11, 90–91, 104–105, 107, 133, 138, 151, 175, 180–181
Relaxation rates, two-spin system, 410f
Relayed transfer, 602

Relays, electrical, 277
Remote (multiple-bond) couplings in HMBC, 536
Remote proton, in HMBC, 509
Remote status unit (Varian), 320
Residual ^{1}H peak, from solvent, 61, 76, 131
Residual dipolar couplings (RDCs), 558, 621–623
Residual field (B_{res}), 292, 296
Residue, of biopolymer, in TOCSY, 340, 346
Resolution, 41
Resolution enhancement, 22, 124, 132, 404
Resolution in 2D spectra, F_1 vs. F_2, 375, 508
Resolving power of 2D NMR, 377
Resonance offset, 201–203, 213, 226, 245, 293
Resonance structure of α,β-unsaturated ketone, 521
Restrained molecular dynamics (rMD) calculations, 593
Restraints, in structure calculations, 590
Retarding the receiver phase, 455
Reverse INEPT transfer, 498
RF, see radio frequency
RG parameter (Bruker), 94, 109, 568
RGA command (Bruker), 95, 569
Ribonuclease A, 411
Rich media, in protein expression, 596
Ring current, 328, 573, 609
RMSD (root mean square deviation), 593
Rnase, 556
ROESY, 199, 242, 290, 336, 369, 414
   for detecting exchange in large molecules, 438
   spin lock in, 338
   TOCSY artifacts in, 394
ROESY mixing, 338
ROESY mixing times vs. NOESY mixing times, 430
ROESY spectrum of cholesterol, 432–435, 433f, 434f
   F_2 slices, 432
   compared to HSQC, 509
ROESY spectrum of glycopeptide YTGFLS(Lactose), 435–436, 435f
ROESY transfer of magnetization, 354
Rolling boil analogy for decoupling, 150
Room temperature shim coils, 77, 81
Root mean square deviation (RMSD), 593
Rotating frame of reference, 92–93, 96, 142, 201–202, 206, 222, 240, 293, 297, 450
Rotating-frame angular velocity, 297
Rotating-frame NOE, 335

Rotation matrices, 470
Rotational correlation time (τ_c), 172, 180
Rotational diffusion rate, 552
Row number, of 3D data matrix, 605
rp parameter (Varian), 406
Rubber band analogy, for NOE restraints, 591

s2pul (simple 2-pulse) sequence (Varian), 154
Salts, effect on cryogenic probe sensitivity, 559
Sample concentration, in comparison of phase cycling and gradients, 466
Sample heating, from TOCSY spin-lock, 341
Sample magnetization, 92
Sample preparation, in biological NMR, 564–565
Sample size, in biological NMR, 556
Sample spinning, 77
Sample-and-hold, lock, 320
Sampling limitations, due to Nyquist theorem, 398
Sampling of FID, by ADC, 91, 98
Sampling rate, 101, 110
Sampling rate in t_1, 398
Sandwich (composite) 180° pulse, 294–295, 338
SAR by NMR (structure-activity relationships by NMR), 552, 600
Satellites, ^{13}C, in ^{1}H spectrum, 62, 137, 490
Saturation, 36, 81, 91, 150, 152, 181, 185, 187–188, 194, 198, 241, 289, 322–323, 383
Saturation transfer from H_2O to H_N, in protein samples, 568
sb parameter (Varian), 404
sb1 parameter (Varian), 405
sbs parameter (Varian), 404
sbs1 parameter (Varian), 405
Scalar (J) coupling evolution, 213–216
Scalar (J) coupling Hamiltonian, 479
Scans, 90, 91, 104–106
Schrödinger equation, 480
Screening, for protein ligands, 552
SCSI interface, 74, 108
SDS polyacrylamide, 623
Second order (strong) coupling in ^{1}H spectra, 45, 52, 54, 63
Secondary structure, in proteins, 575, 587
Selection rule, for quantum transitions, 266
Selective 1D TOCSY, 181, 343
Selective 1D transient NOE, 181, 321, 321f
Selective annihilation, 311

Selective excitation, 298, 308, 321, 340, 353
   using rectangular pulses, 611
Selective heteronuclear decoupling, 142–143, 143*f*
Selective homonuclear (^{1}H) decoupling, 148, 148*f*
Selective ID NOE, 328–333
Selective inversion pulse, 323
Selective one-dimensional (1D) experiments, 238
Selective population transfer (SPT), 257
Selective pulses, 186, 200, 299–301
Selectivity of CW irradiation, 182
Self-relaxation, 326
selfrq parameter (Varian), 321
selpw parameter (Varian), 320
selpwr parameter (Varian), 320
selshape parameter (Varian), 320
Send-receive switch, in preamplifier, 95
Sensitivity, 3, 7, 34, 37–38, 76, 93, 112, 466
Sensitivity enhancement, by preservation of equivalent pathways (PEP), 531–532
Sensitivity improved experiments, 531
Sensitivity loss, due to short T_2 in large molecules, 556
Sensitivity of HMBC, 509
Sensitivity to twisting, by gradient, 317
Sequence specific assignments, of peptide using ROESY, 436
Sequence-specific assignments, 553, 572, 577, 580–586
   for ^{15}N or ^{13}C-labeled proteins, 610, 617
Sequential (alternating) data sampling, 120
Sequential assignment, 580
Sequential NOEs, 435, 577–580
Serial file (Bruker), 119, 366
Serine, 571
Shaka, A. J., 308
Shaped (selective) RF pulses, 257
Shaped gradient pulses, 320
Shaped pulse, 74, 181, 196, 289, 296, 308, 319, 337, 348, 353, 299–301
   calculation of power level, 351
   for decoupling, 497
   practical aspects, 319–321
Shaped pulse capability, 559
Shielding, 57, 59, 328
Shielding constant, 33
Shigemi tubes, 565
SHIM button (Varian acqi window), 86
Shim coils, room temperature, 77, 81
Shim file, 87

Shim gradient, 81–82
Shim map, for gradient shimming, 88, 563
Shimming, 41, 44, 80–82, 84, 108, 131, 229, 302, 305, 334
   and presaturation, 186
Short peaks in ^{13}C spectrum (quaternary carbons), 540
Shoulder,
   due to Z^2 shim error, 84
   in ^{1}H multiplet, 49
Shutter speed, 139, 414
Shutter time, 415
si in pulse sequence name (Bruker), 531
SI parameter (Bruker), 122
$si(F_1)$ and $si(F_2)$ parameters (Bruker), 402
Side-chain ^{13}C and ^{1}H assignments, 609
Side-chain H_N resonances, 571
Sigmatropic rearrangements, 422
Sigmoidal shape of NOE buildup, in spin diffusion, 436
Sign of crosspeaks in 2D NOESY, 426–427
Signal-to-noise ratio (S/N), 91, 93, 98, 106–109, 111, 123, 129, 136, 185, 222, 241, 278, 466, 528, 559
Signal-to-noise ratio (S/N) *vs.* molecular size, 555
Silicon Graphics (SGI), 74, 108
Simmering analogy, for saturation, 152
Simplex autoshimming, 87
Simulated annealing, in structure calculations, 593
Simulation of the transient NOE experiment, 327, 327*f*
Simultaneous 180° pulses, 255
Simultaneous 90° pulses, 242, 254, 523
Simultaneous sampling method for ADC, 98–99, 102, 119–120, 399
Sinc artifacts, 402
Sinc function, 86, 112, 114, 116–117, 297, 299, 349, 402
Sinc-shaped excitation profile, in 3D HNCO, 611
Sine-bell window function, 50, 125, 132, 403
Sine-shaped gradient pulse, 320
Single-quantum (SQ) transitions, 189, 252, 266
Single-quantum coherence (SQC), 266, 469
   in HSQC, 525
Singlet, 10, 11, 223
Singlet methyl group, in HMBC, 503, 514, 543, 545*f*

Size limitations in solution-state NMR, 553–558
Slow exchange, 415, 417, 418–422
Slow exchange of glucose anomers, 493
Slow exchange spectrum, 416f
Sodium azide, 565
Sodium d_4-3-trimethylsilylpropanoate, 565
Soft pulse, 298, 301, 348–349
Solenoid, in superconducting magnet, 8
Solid sample, CSA, 555
Solid-phase synthesis, 596
Solid-state NMR, 60, 138, 167, 341, 551, 623
Soluble fraction, in protein expression, 596
Solvent ^{13}C peak, 131
Solvent exposed H_N, temperature dependence of chemical shift, 585
Solvent peak,
　^{13}C, 131
　residual ^{1}H, 61
Solvent signal suppression, 185
Solvent-exposed side-chains, increased T_2 values, 609
SOLVNT parameter (Varian), 102
Sorted chemical shift lists, in structure determination, 542
sp1 parameter (Bruker), 320
Specific pulse excitation, using shaped pulses, 289
Spectral density function J(ν), 172, 173f, 175
Spectral editing, 270, 275–276
　in 2D HSQC, 505
Spectral width (SW), 101–102, 104, 110, 122, 133, 298
Spectral window, 101, 111, 299
　center, 201
Spherical product operator analysis of HMQC, 535
Spherical product operator analysis of HSQC, 526–527
Spherical product operators, 269, 408, 443–447
　definition, 444
　effect of 180° pulse, 445
　effect of gradients, 444
Spin, 30
Spin-1, 31, 131
Spin-1/2, 2, 30, 33
Spin choreography, 532
Spin diffusion, 199, 433, 436
Spin echo, 200–237, 222, 226, 229, 276, 305, 342, 462, 468, 524, 529–530, 532
　in evolution period of HSQC, 525

Spin kinetics, of relaxation, 409
Spin lock, 200, 289–290, 296, 333–338, 334f, 339, 341, 344, 348, 393, 602
Spin lock axis, 336
　tilt in ROESY, 431
Spin lock field, analogy to B_0 field, 335
Spin lock power levels, 334
Spin lock, continuous-wave in ROESY, 430
Spin system, 134, 184, 336, 339, 341, 375
Stacked plot, 134, 365
States-Haberkorn mode, 99, 400, 532, 605
States-TPPI mode, 599
Stationary state, 442, 480, 486
　definition in terms of commutator, 484
Statistical mechanics, 157
Steady state, 91, 107, 180, 192–195, 322
Steady state NOE, 152, 192, 194, 198, 325
Steady state NOE difference, 196
Stereochemistry,
　by HMBC crosspeak intensity, 519
　by NOE experiments, 434
　by transient NOE, 329
Steric crowding, effect on ^{13}C chemical shifts, 9, 11, 27, 29, 221, 360, 502, 507, 610
Steroids, 20, 51, 281
　doublet/triplet pattern in CH_2 groups, 508
Stimulated emission, 36, 172, 181
Stochastic switching of chemical shift in exchange, 418
Storing maganetization on the z axis, 395
Straight-chain hydrocarbon, 342
Strategy for coherence transfer, in triple-resonance experiments, 611
Streaks, in COSY, due to dispersive lineshape, 390
Stretched crosspeaks in HMBC, due to nearby ^{13}C resonances, 512
Stretched gel, for residual dipolar coupling measurement, 623
Strip plot, 605, 605f
　3D CBCA(CO)NH, 619f, 620–621, 621f
　3D HNCACB, 618f, 619–620, 619 f
　3D HNCO, 615, 616f
　3D HSQC-TOCSY, 604
　3D TOCSY-HSQC, 608f
Strong (second-order) coupling, 45, 52, 54, 67, 69, 331, 336, 341–342, 409, 481–484
Strong, category of NOE crosspeak volume, 437
Structural biology, 353, 551
Structure calculation, 553
　using NMR restraints, 590–596

Structure determination by NMR, an example, 538–550
Structure determination using HSQC and HMBC, 517–522
su command (Varian), 148
Subtraction artifacts,
  in NOE difference, 195
  in phase cycling, 304, 466, 501
Subtraction of artifacts, by phase cycle, 263, 466
Sucrose, 16, 153, 178, 183, 196, 200, 220, 256, 303–304, 308–309, 311, 316, 330, 345
  ^{13}C satellites, 491, 491*f*
  ^{1}H assignments, 184
  ^{1}H assignments, from COSY, 377–382, 379*f*, 381*f*, 382*f*
  DQF-COSY spectrum, 377–382, 379*f*, 381*f*, 382*f*
  HETCOR spectrum, 360, 360*f*
  HMQC spectrum, 502, 503*f*
  NOE difference, 196*f*, 197*f*
Sudden perturbation, in transient NOE, 322
Sugar-phosphate backbone, 14
Sum-to-memory, 86, 91, 94, 106, 108, 211, 450, 466
  addition and cancellation in phase cycle, 453
Sun Microsystems, 74, 108
Superconducting magnet, 8, 77, 156
Superconductivity, xii, 40
Superposition, of pure quantum states, 269, 441, 469, 472
Supression of one-bond correlations in HMBC, 536
SW parameter, 98, 110, 111, 122, 298, 398
sw1 parameter (Varian), 398, 404, 523
Swapping the labels, in heteronuclear spin-echo, 233
swh parameter (Bruker), 504, 504
swh(F_1) parameter, Bruker, 401
Switching of ^{1}H power level, in DEPT, 277
Symmetric state, 483
Symmetrically disposed crosspeak, 370
Symmetry of biomolecules, 556
Symmetry, molecular, and NMR, 4, 54–56
Synthetic methods, 553

t_1 (evolution delay), 354, 357
T_1 (longitudinal relaxation time), 91, 133, 152–153, 164–165, 167, 170–171, 173, 175–177, 179
  compared to T_2, 167
  ^{13}C, effect of bound ^{1}H, 178
  effect of molecular size on, 173, 174*f*
  effect on experiment time, 175
  of quaternary carbons, 180
  practical significance, 180
t_1 FID, 357
t_1 noise, 375, 429, 432
T_1 relaxation, 194, 229, 326
t_2 (FID time domain), 357
T_2 (transverse relaxation time), 41, 44, 163–165, 167, 170, 173–175, 229
  compared to T_1, 167
  effect of molecular size on, 174*f*, 554, 556*t*
  effect on peak width, 175, 229–231, 230*f*
  in solids, 174
T_2 relaxation, 222, 230, 235
  in gradient experiments, 468
  in HMBC, 509
TANGO pulse sequence building block, 364, 537
Tau (τ) scale, for chemical shifts, 37
Taxonomy of 2D experiments, 369–370
TD parameter (Bruker), 99, 105, 123, 134, 402, 504
td(F_1) parameter (Bruker), 401
Temperature dependence of H_2O chemical shift, 565
Temperature dependence of H_N chemical shifts, 585
Temperature standard, for VT work, 425
Temporary bad shimming (pulsed-field gradient), 266
Terpenoids, 12, 20
Tertiary structure, of proteins, 575, 587
Test function, in Fourier transform, 120
Testosterone, 50, 281–282
Testosterone metabolite, structure determination, 517–519
Tetramethylsilane (TMS), 76, 104–105, 130
th parameter (Varian), 132
Theory of NMR, Advanced, 408–488
Thermal equilibrium, 2, 34, 36, 90, 164, 166, 169, 190
Thermal noise, in probe coil and preamplifier, 95, 559
Third party software, for NMR data processing, 366
Three gammas, contributing to FID intensity, 37

Three-bond HMBC correlations in aromatic rings, 547
Three-bond vs. two-bond ambiguity, in HMBC, 514
Three-channel spectrometer console, 559
Three-dimensional fold, of biomolecule, 553
Three-dimensional NMR pulse sequences, 601–610
Three-dimensional structure, of biomolecules, 137, 552
Three-spin (AMX) system, in amino acids, 383, 398, 571–572
Threonine, 571
Threshold intensity, for peak picking, 132
Through-bond transfer of magnetization, 341
Tier number, of 3D data matrix, 605
TIFF graphics format, 134
Tight AB system, 67
Tilt of B_{eff} vector for off-resonance excitation, 292
Time domain, 7, 79, 114, 118–119, 298–299
Time evolution, 471
Time for refocusing of antiphase coherence, ^1H vs. ^{13}C, 490
Time reversal, in spin-echo, 227
Timescale of the NMR experiment, 54
Titration curve, of chemical shifts, 423
TMS, see tetramethylsilane
TO parameter (Varian), 101–102
TOCSY (total correlation spectroscopy), 290, 336, 353, 369, 430, 610
  2D, pulse sequence, 601
  coherence flow diagram, 394f
  distortion of peak shape, 395, 395f
  for spin system identification in proteins, 572
  mixing, 409
  mixing efficiency, 344
  spin lock in, 338
TOCSY mixing,
  described using Hamiltonian, 486–487
  product operator analysis, 394
TOCSY spectrum of 3-heptanone, 375, 376f
TOCSY spectrum of cholesterol, 369f
TOCSY spectrum of glycopeptide YTGFLS(Lactose), 397f
TOCSY spectrum of heregulin-α EGF domain, 582f, 585–586f
TOCSY transfer, 354
  efficiency compared to NOE, 348
TOCSY vs. COSY, for resolving ambiguous assignments, 373, 374f

TOCSY-HSQC spectrum, 3D, of ferrocytochrome c_2, 607–609, 607–609f
TOCSY-HSQC,
  3D, coherence flow diagram, 604f
  3D, pulse sequence, 603–604, 604f
tof parameter (Varian), 150
Total energy function, in structure calculations, 592
Total z-magnetization in transient NOE, 327
t_p (pulse duration), 292, 293, 297
TPPI (time proportional phase incrementation), 99, 103
tpwr parameter (Varian), 350
trace parameter (Varian), 406
Train of pulses, 334
trans-diaxial relationship, in cyclohexane chair, 15
Transfer of magnetization, 323–324, 354, 370
  in spin-lock, 334
Transient nuclear Overhauser effect (NOE), 196, 198, 321–322, 354, 388, 425
  time course of z magnetization, 326, 326f
Transients, 90, 91, 104, 106
Transmitter, 8, 36, 74–75, 142, 146, 148–149, 185, 279, 559
Transverse relaxation, 167, 170–171
Transverse relaxation optimized spectroscopy (TROSY), 556, 623–626
Transverse relaxation rate (R_2), 173
Transverse relaxation time (T_2), 164
Trap, to spoil one-bond correlations, in HMBC, 537
Trifluoroacetic acid, 61
Trifluoroethanol, 569
Trim pulse, 335
Triphenylphosphine oxide, 61
Triple axis (x,y,z) gradients, 319, 559
Triple axis gradient probe, 88, 564
Triple-resonance (^1H, ^{15}N and ^{13}C) experiments, 553, 557, 577, 610–621
Triple-resonance inverse probe, 90, 559
Triplet, 5, 44–45, 47
  J coupling evolution, 223
Triple-triplet, 48
Triterpene, 20
Triterpenes, doublet/triplet pattern in CH_2 groups, 508
TROSY (transverse relaxation optimized spectroscopy), 553, 623–626
TROSY principle, 625
Trunction of t_1 FID, 402f
Tryptophane, 571

TSP (2,2′,3,3′-d$_4$-3-trimethylsilylpropionate), 131
Tumbling, 138, 157, 163, 170
    rates, 172
Tune interface (Varian), 89
Tuning knob, 90
Tuning rods, 89
Tuning the probe, 88, 204
Turbine, spinner, 77
Turbo vacuum pump, for cryogenic probe, 559
Turning off the B$_0$ field, in TOCSY mixing, 342
Twist (position-dependent phase shift), 302, 446, 450, 458, 528
    of artifact coherences, 466
Two-dimensional (2D) NMR, 18, 24, 221, 242, 353
Tyrosine, 571

Uncrystallizable proteins, 558
Underpopulated state, 410
Unfolded proteins, 572
Uniform labeling of biomolecules, xi, 32, 38, 137, 553, 557, 596
UNIX operating system, 74–75, 134
Unresolved splitting, 22
Unsaturated groups, effect on chemical shift, 8
Unscrambling of coherence helix, 304
Unshifted sine-bell, 403
Untwisted magnetization, 304
    in gradient HSQC, 528
Unwinding the antiphase relationship, 275
Unwinding the coherence helix, by final gradient, 460, 468
Upfield, 8, 37
Upper cone of individual vectors, 160
Upper limits of distance, in structure calculations, 591

ν_1, see nu-1
Valine, 570
Variable capacitance, in probe circuit, 204
Variable temperature (VT) operation, 424–425
Variable temperature study, 418
Varian A-60 instrument, 40
Varian EM-390 instrument, 40
Varian Gemini instrument, 3, 74–75, 95, 103
Varian Inova instrument, 74, 95
Varian T-60 instrument, 37, 40
Varian Unity instrument, 95, 107, 145, 147–178
Varian Unity-Plus instrument, 74

Varian VXR instrument, 74
Varian XL-100 instrument, 40
Vector diagrams, of antiphase states with 3 or more spins, 272
Vector dot product, 479
Vector model, 91, 155, 222, 242–243, 247, 269, 408
    summary, 168–170
Vector,
    from 3D data matrix, 605
    in density matrix representation, 478
    net magnetization, 168
Vertical streak in 2D HSQC/HMQC, 499
Vicinal (3-bond) relationship, 12, 14, 28, 52, 242, 372
Violation, of NMR restraints, 590, 595
Virtual coupling, 70, 336, 341–342
VNMR software (Varian), 401
Voxel, 552

W coupling, 330, 377, 377f
W_0, W_1, W_2 relaxation rates, 410
Walk along backbone of peptide, 578
    using 2D ROESY, 436
    using coherence transfer only, 617
Walk, through each spin system, 372, 372f
Waltz-16 heteronuclear decoupling, 144–145, 149, 182, 200, 338, 496
Water suppression, 302, 320, 566–569
Water suppression probe, 90, 558
Water suppression, by jump-return, 312
Water-friendly water suppression methods, 568
Watergate, 181, 186, 311–316, 568
    simulation, 315, 315f
Wave function, 441, 469
Waveform generator (Varian), 320
Waves, in baseline of spectrum, 505
wdw parameter (Bruker), 405
Weak (first-order) coupling, 45, 63, 67, 71
Weak, category of NOE crosspeak volume, 437
Weighting (window or multiplier) functions, 108, 398
wft command (Varian), 126
White powder unknown, 538, 553
Whitewash, in stacked plots, 365
Wiggles, sinc, 299, 402, 402f, 403f
Wilmad, 77
Winding up the antiphase state, 275

Window (weighting or multiplier) functions, 118, 123
  for 2D, 404, 404f
  Varian parameters, 404
Wobble tuning (Bruker), 89
WURST (adiabatic shaped pulse), 337, 568
WURST-40, compared to GARP for ^{13}C decoupling, 497, 497f

X nuclei, 149
X-ray crystallography, xi, xii, 151, 332, 551

Z filter, 395
Z magnetization, 239, 241
Z0 (field) coil, in room temperature shims, 79
Z0 setting (Varian), 79, 80
Z-axis pulse, 296, 313
Zeeman energy, 342
Zeeman Hamiltonian, 479
Zero audio frequency, 398
Zero filling, 118, 122, 401
Zero-order phase correction, 128, 130, 218, 405
Zero-quantum (ZQ) artifact,
  in 1D transient NOE, 328, 330
  in 2D NOESY, 428
  removal with variable mixing time, 395
Zero-quantum (ZQ) relaxation, 188–189, 192, 198, 199, 266, 409, 429
Zero-quantum coherence (ZQC), 266, 357, 395, 439, 440
  effect of gradient on homonuclear, 458
  heteronuclear, in HMQC, 533
  sensitivity to gradients, 318
zg command (Bruker), 148
Z-magnetization transfer, 338
ZQC and DQC, 283, 389
  product operator definition, 268